Chemical Reaction Engineering

Chemical Reaction Engineering

The First International Symposium on Chemical Reaction Engineering co-sponsored by the American Institute of Chemical Engineers, the European Federation of Chemical Engineering, and the American Chemical Society held at the Carnegie Institution, Washington, D.C., June 8-10, 1970.

Kenneth B. Bischoff,

Symposium Chairman

ADVANCES IN CHEMISTRY SERIES **109**

AMERICAN CHEMICAL SOCIETY

WASHINGTON, D. C. 1972

ADCSAj 109 1-685 (1972)

ISBN 8412-0149-8

Library of Congress Catalog Card 72-84256

PRINTED IN THE UNITED STATES OF AMERICA

Advances in Chemistry Series

FOREWORD

ADVANCES IN CHEMISTRY SERIES was founded in 1949 by the American Chemical Society as an outlet for symposia and collections of data in special areas of topical interest that could not be accommodated in the Society's journals. It provides a medium for symposia that would otherwise be fragmented, their papers distributed among several journals or not published at all. Papers are refereed critically according to ACS editorial standards and receive the careful attention and processing characteristic of ACS publications. Papers published in ADVANCES IN CHEMISTRY SERIES are original contributions not published elsewhere in whole or major part and include reports of research as well as reviews since symposia may embrace both types of presentation.

CONTENTS

FLUIDIZED BED REACTORS

P. N. Rowe, *Chairman*

OPTIMIZATION OF REACTOR PERFORMANCE

J. M. Douglas, *Chairman*

PHYSICAL PHENOMENA AND CATALYSIS IN GAS–SOLID SURFACE REACTIONS

M. Boudart, *Chairman*

TWO-PHASE AND SLURRY REACTORS

J. G. van de Vusse, *Chairman*

CATALYST DEACTIVATION

V. W. Weekman, *Chairman*

INDUSTRIAL PROCESS KINETICS
AND PARAMETER ESTIMATION

H. M. Hulburt, *Chairman*

PREFACE

The papers in this volume represent the proceedings of the First International Symposium on Chemical Reaction Engineering. This was an outgrowth of the previous four European symposia and will be continued on an international basis every two years. The symposium consisted of 12 sessions whose chairmen and authors were balanced between Americans and Europeans, and industrial and academic workers as much as possible. Each subject area was covered by an authoritative review of the state of the art by an international expert, followed by research-paper presentations. This structure gave the audience both an up-to-date survey of each area and details of the latest research.

Session topics were chosen with both fundamental and applied orientations of chemical reaction engineering, and the ideas which emerged underscore this point. For example, using the basic information available for standard reactor types, computers are now doing the complicated simulations routinely. The basic chemical kinetics required to design polymerization reactors are reasonably well understood, but practical computational methods and further study of interactions with physical processes (mixing and transport) are still required. Much has been learned in the past several years about the operating behavior and transport in fluidized beds, and applications with reactions now need to be considered; this could lead to more extensive use for more different processes than at present. Although great effort has been expended to develop optimization theory and principles, a present need is to concentrate on practical applications in order to realize the full potential of the most important methods. The session on catalytic reactions emphasized the physical interactions which can cause problems in studying and applying catalytic kinetics, and they need to be evaluated carefully.

For two-phase reactors the main problem resides in the physical contacting of the phases, which can cause widely varying results, especially with complex reactions; this fact leads to the use of unique reactor types which are often difficult to analyze and design. Catalyst deactivation has only recently been extensively studied rigorously, and there is still a need for information on basic kinetics, the effects on reactor operation, and optimization studies to assess the importance and handling of the problem. Industrial process kinetics contains many of the above problems since the work must often be done in complicated reactors. Parameter estimation is a complex problem in its own right, in terms of both statistical measurement variations and calculation methods. The basic idea of reaction/reactor stability is not difficult, but application in detail to real systems in a simple and workable fashion often is not easily realized. Biochemical reaction engineering is a relatively new field and was considered here for the first time in this type of symposium. We are beginning to grasp some of the essential difficulties of the quantitative kinetics and are using them for more rational reactor design.

Following in a small way the outstanding review of van Krevelen at the Third European Symposium, the table shows the distribution of papers from this symposium. The various types of chemical reaction engineering studies for the papers are given first, and although there was

DISTRIBUTION OF PAPERS AS

	Session																									
	1						*2*				*3*					*4*					*5*					
Paper	*a*	*b*	*c*	*d*	*e*	*f*	*a*	*b*	*c*	*d*	*a*	*b*	*c*	*d*	*e*	*a*	*b*	*c*	*d*	*e*	*a*	*b*	*c*	*d*	*e*	*f*
Chemical Reaction Engineering Study Type																										
Process analysis and design	●	●				●	●		●				●	●		●	●						●	●		
Method/theory			●		●			●					●		●				●	●	●	●		●	●	●
Kinetic study						●		●								●		●								
Physical study				●		●				●	●	●								●						
Reaction Type																										
Single				●	●	●								●	●									●	●	●
Parallel	●								●												●					
Consecutive		●					●	●	●	●			●			●	●	●	●							
Complex		●														●						●	●		●	
Type of Chemistry or Industry																										
General			●		●			●		●	●	●			●			●		●	●				●	●
Oil Refining													●									●	●			
Inorganic				●		●								●										●		
Organic	●	●					●		●							●	●		●							
Biochemical																										
Metallurgical															●											

a rather good distribution of coverage, more actual applications to useful processes would have been welcome. Reaction type is considered next, showing that simple, single reactions are studied more than the complex cases which are often found in practice. This is especially true of the theoretical and "method" papers although some progress is being made in moving toward realistic cases. The last part of the table shows the type of chemistry or industry involved. The "general" row signifies that abstract cases were considered and is not meant to imply that many of the other papers do not also have general application.

Ithaca, N. Y.
January 1972

KENNETH B. BISCHOFF

PRESENTED AT THE SYMPOSIUM

	Session																																	
	6				*7*					*8*					*9*				*10*					*11*						*12*				
	a	*b*	*c*	*d*	*a*	*b*	*c*	*d*	*e*	*a*	*b*	*c*	*d*	*e*	*a*	*b*	*c*	*d*	*a*	*b*	*c*	*d*	*e*	*a*	*b*	*c*	*d*	*e*	*f*	*a*	*b*	*c*	*d*	*e*
Chemical Reaction Engineering Study Type																																		
			●								●			●							●													
					●				●	●	●		●	●	●			●	●		●		●				●							●
	●	●		●				●				●				●	●			●		●		●	●	●		●	●	●	●	●	●	●
	●	●				●	●	●											●	●		●												
Reaction Type																																		
	●		●		●			●		●	●	●	●	●		●	●		●	●	●	●		●		●	●	●			●	●		●
									●	●		●													●								●	
		●							●																									
	●			●					●						●			●											●	●				
Type of Chemistry or Industry																																		
					●	●	●		●	●			●		●					●			●				●							
												●		●											●			●	●					
			●														●					●		●										
	●	●		●				●			●					●			●															
																		●												●	●	●	●	●
																										●								

Organizing Committee

for the

First International Symposium on

Chemical Reaction Engineering

K. B. Bischoff, *Chairman*

Registration: D. E. Gushee

Program: J. M. Douglas
V. W. Weekman

Members: R. Aris
J. J. Carberry
G. F. Froment
R. L. Gorring
C. van Heerden
J. G. van de Vusse
J. Wei

1

Analysis and Design of Fixed Bed Catalytic Reactors

G. F. FROMENT

Laboratorium voor Petrochemische Techniek, Rijksuniversiteit, Ghent, Belgium

The models used to describe fixed bed catalytic reactors are classified in two broad categories: pseudo-homogeneous and heterogeneous models. In the former the conditions on the catalyst are considered to equal those in the fluid phase; in the latter this restriction is removed. The pseudo-homogeneous category contains the ideal one-dimensional model, the one-dimensional model with effective axial transport, and the two-dimensional models with axial and radial gradients. Particular emphasis is placed on such problems as parametric sensitivity, runaway, and instabilities induced by axial mixing. In the heterogeneous category attention is given to the effect of transport phenomena around and inside the catalyst particle on the behavior of the reactor. Finally, a new, general two-dimensional heterogeneous model is set up and compared with previously discussed models.

This brief review of the analysis and design of fixed bed catalytic reactors does not allow us to concentrate on specific cases and processes. Instead, an attempt is made to discuss general models and the principles involved in the design of any type of reactor, no matter what the process is.

In Table I the models are grouped in two broad categories: pseudo-homogeneous and heterogeneous. Pseudo-homogeneous models do not account explicitly for the presence of the catalyst, in contrast to heterogeneous models, which lead to separate conservation equations for fluid and catalyst. Within each category the models are classified according to increasing complexity. The basic model, used in most of the studies until now, is the pseudo-homogeneous one-dimensional model, which only considers transport by plug flow in the axial direction (A.I). Some type of mixing in the axial direction may be superposed on the plug flow

Table I. Classification of Fixed Bed Reactor Models

	A. *Pseudo-Homogeneous* ($t = t_s$; $c = c_s$)	B. *Heterogeneous* ($t \neq t_s$; $c \neq c_s$)
One-dimensional	A.I Basic, ideal	B.I + interfacial gradients
	A.II + axial mixing	B.II + intraparticle gradients
Two-dimensional	A.III + radial mixing	B.III + radial mixing

to account for nonideal flow conditions (A.II). If radial gradients must be accounted for, the model becomes two-dimensional (A.III). The basic heterogeneous model considers only transport by plug flow, but it distinguishes between conditions in the fluid and on the solid (B.I). The next step towards complexity is to take the gradients inside the catalyst into account (B.II). Finally, the most general models used today—*viz*, the two-dimensional heterogeneous models—are discussed in B. III.

Even within this framework the paper does not give a complete bibliographic survey. It focuses on what the author believes are the essential points or on some aspects which have received extensive coverage in recent years and upon which our viewpoints need clarification or correction.

Pseudo-Homogeneous Models

The Basic One-Dimensional Model. The basic or ideal model assumes that concentration and temperature gradients occur only in the axial direction. The only transport mechanism operating in this direction is the over-all flow itself, and this is considered to be of the plug flow type. The conservation equations may be written, for the steady state and a single reaction carried out in a cylindrical tube.

$$u_s \frac{dc}{dz} = \rho_B r_A \tag{1}$$

$$u_s \rho_f c_p \frac{dT}{dz} = (-\Delta H)\, \rho_B r_A - 4\, \frac{U}{d_t}\, (T - T_w) \tag{2}$$

with initial condition: at $z = 0$, $c = c_o$
$T = T_o$

In most cases the pressure drop in the reactor is relatively small so that a mean value for the total pressure is used in the calculations. Pressure drop correlations for packed beds were set up by Leva (*1*) and Brownell (*2*). They lead to predictions which are in excellent agreement. The correlations for the heat transfer coefficient, U, show considerable spread (*3, 4, 5*). Recently, De Wasch and Froment set up correlations for U which are linear with respect to the Reynolds number and which have also a static term (*6*). Integration of Reactions 1 and 2 is straightforward, either on a digital or an analog computer. Questions which can be answered by such simulation and which are important in catalytic reactor design are: what is the tube length required to reach a given conversion; what will the tube diameter have to be; or the wall temperature? An important problem encountered with exothermic reactions is how to limit the hot spot in the reactor and how to avoid excessive sensitivity to variations in the parameters? This problem was treated analytically by Bilous and Amundson (*9*) and more empirically, but more directed towards practical appplication, by Barkelew (*7*). Barkelew's results are represented in Figure 1, which is based on many numerical integrations and which has general validity for single reactions. N/S is the ratio of the rate of heat transfer per unit volume at $\tau = 1$, where $\tau =$

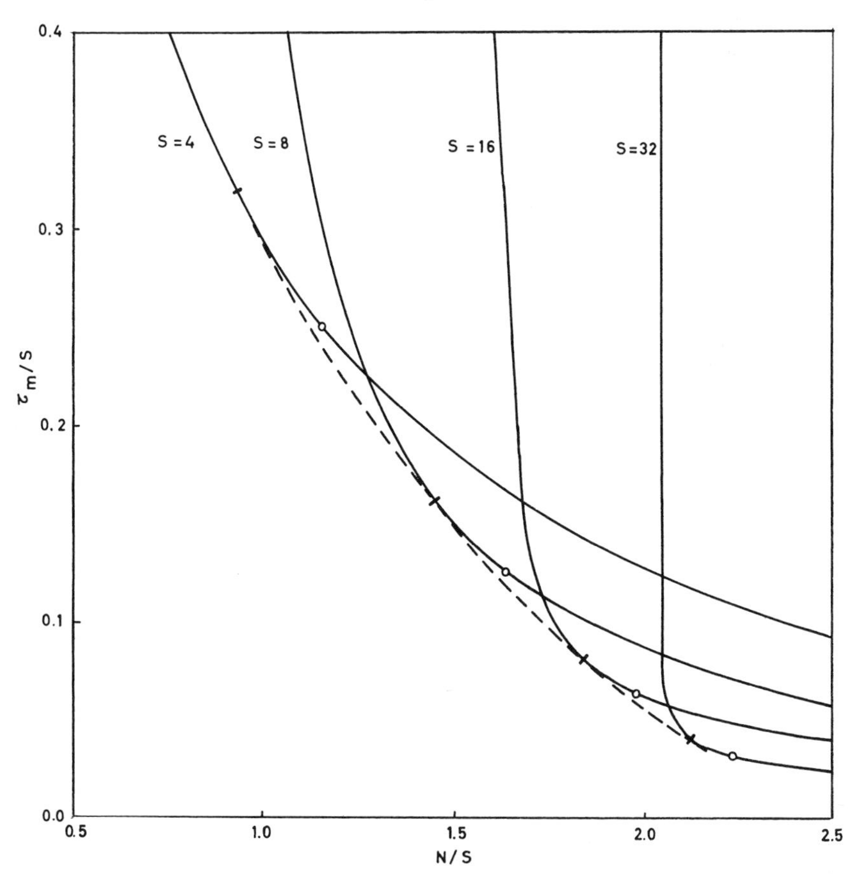

Figure 1. Barkelew plot for the parametric sensitivity of an ideal tubular reactor

$(E/RT_w^2)(T - T_w)$, to the rate of heat generation per unit volume at $\tau = 0$ and zero conversion—*i.e.*, at the reactor inlet. The ratio τ_{max}/S is that of the dimensionless maximum temperature to the adiabatic temperature rise above the coolant temperature. A set of curves is obtained with S as a parameter. They have an envelope, occurring very close to the knee of an individual curve. Above the contact point with the envelope, τ_{max} changes rapidly with N/S but not below. Therefore, Barkelew proposed a criterion according to which the reactor is stable to small fluctuations if its maximum temperature is below the value at the contact point to the envelope. Recently, Van Welsenaere and Froment analyzed the problem in a different way (8). By inspecting temperature and partial pressure profiles in a fixed bed reactor they concluded that extreme parametric sensitivity and runaway may be possible (1) when the hot spot exceeds a certain value and (2) when the temperature profile develops inflection points before the maximum. They transposed the peak temperature and the inflection points into the p–T phase plane. The locus of the maximum temperatures, called the "maxima curve" and the locus of the inflection points before the hot spot are shown as p_m and $(p_i)_1$ respectively. The symbol $(p_i)_2$ represents the locus of the inflection point beyond the hot spot, which is of no further interest in this analysis.

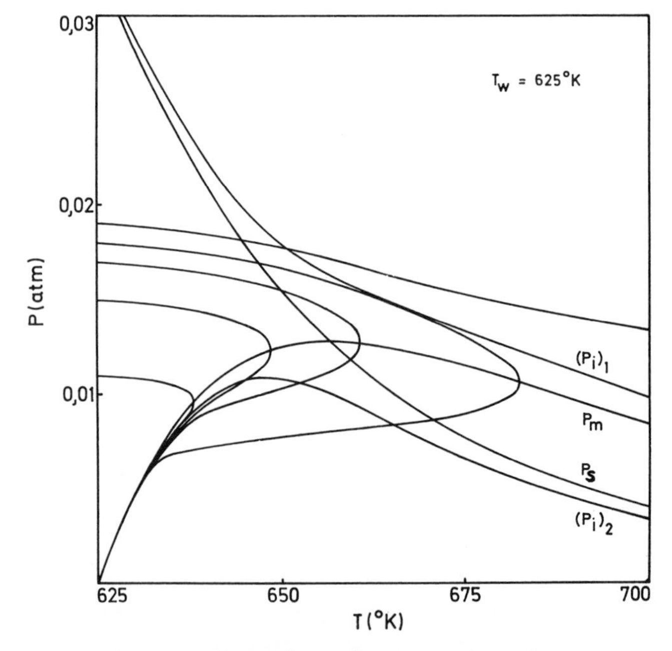

Figure 2. p-T *phase plane showing trajectories, maxima curve, and loci of inflection points according to Van Welsenaere and Froment*

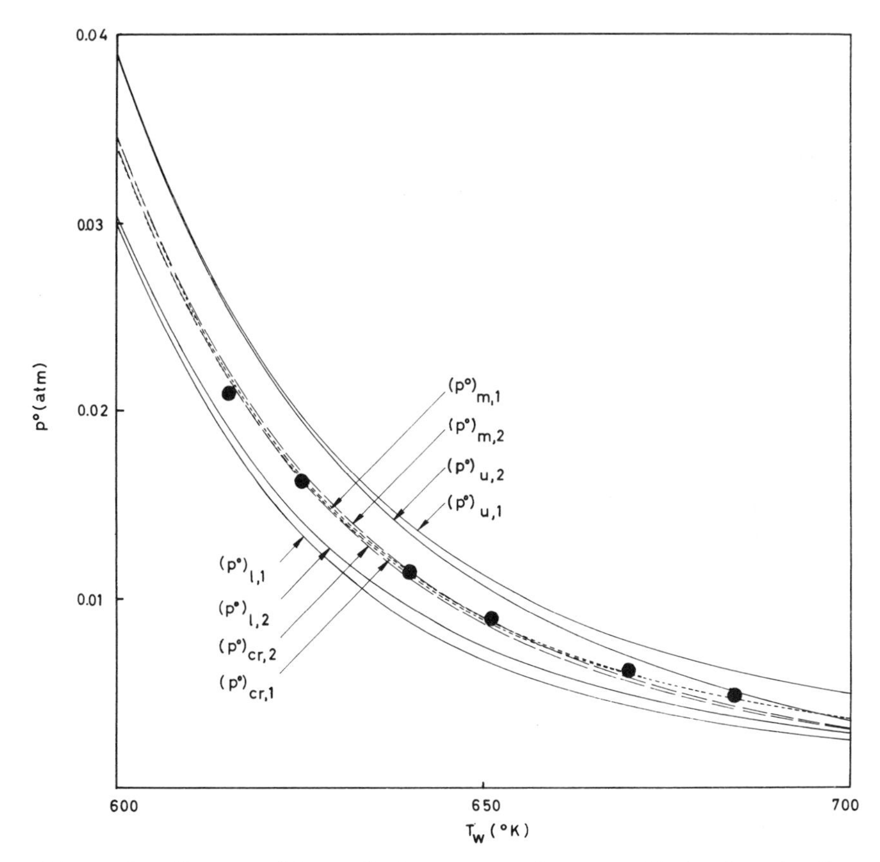

Figure 3. Upper and lower limits, mean and exact critical values for the inlet partial pressure predicted by the Van Welsenaere and Froment criterion. Filled points are values derived from Barkelew's criterion.

Two criteria were derived from this. The first is based on the observation that extreme sensitivity is found for trajectories—the p–T relations in the reactor—intersecting the maxima curve beyond its maximum. Therefore, the trajectory going through the maximum of the maxima curve is considered as critical and as the locus of the critical inlet conditions for p and T corresponding to a given wall temperature. This is a criterion for runaway based on an intrinsic property of the system—not on an arbitrarily limited temperature increase. The second criterion states that runaway will occur when a trajectory interesects $(p_i)_1$, the locus of inflection points arising before the maximum. Therefore, the critical trajectory is tangent to the (p_i) curve. A more convenient version is based on an approximation for this locus represented by p_s in Figure 2.

Representation of the trajectories in the p-T plane requires numerical integration, but the critical points involved in the criteria—the maximum of the maxima curve and the point where the critical trajectory is tangent to p_s are located easily by elementary formulas. Two simple extrapolations from these points to the reactor inlet lead to upper and lower limits for inlet partial pressures and temperatures above which safe operation is guaranteed. Figure 3 shows some results for a specific reaction. They are compared with those obtained from Barkelew's criterion (filled dots), which is more complicated to use, however.

In Figure 3 $p^\circ_{l,1}$; $p^\circ_{u,1}$ and $p^\circ_{l,2}$; $p^\circ_{u,2}$ are lower and upper limits, based upon the first and second criteria respectively; $p^\circ_{m,1}$ and $p^\circ_{m,2}$ are their mean values; $p^\circ_{cr,1}$ and $p^\circ_{cr,2}$ are the exact values obtained by numerical back integration from the critical points defined by the first and second criteria. These latter values are given here to show the error introduced by simple extrapolation methods. The mean values $p^\circ_{m,1}$ and $p^\circ_{m,2}$ agree remarkably well with $p^\circ_{cr,1}$ and $p^\circ_{cr,2}$. The reaction considered here has pseudo-first-order kinetics and a heat effect suggested by gas phase hydrocarbon oxidation. For a specific set of conditions the first criterion limits the hot spot to 31.6°C, with $p_{l,1} = 0.0135$ atm and $p_{u,1} = 0.0197$ atm; the second criterion limits ΔT to 29.6°C, while $p_{l,2} = 0.0142$ atm and $p_{u,2} = 0.0195$ atm. By numerical integration of the system of differential equations, what could be called "complete" runaway is obtained with $p_o = 0.0183$ atm.

The above criteria are believed to be of great help in first stages of design since they permit a rapid and accurate selection of operating conditions before any computer calculations are done. Their application is limited, however, to single reactions. Nothing like this is available for complex reactions which have many parameters. Complex cases will probably always be handled individually.

Objections may be raised against this model. First, it can be argued that the flow in a packed bed reactor deviates from the ideal plug flow pattern because of radial variations in flow velocity and mixing effects. Second, it is an oversimplification to assume that temperature is uniform in a cross section. The first objection led to a development which is discussed in the next section, the second to models discussed later.

One-Dimensional Model with Axial Mixing. Accounting for the velocity profile is practically never done since it immediately complicates the computation seriously. In addition very few data are available to date, and no general correlation could be set up for the velocity profile (*10, 11, 12, 13*). The mixing in an axial direction which is caused by turbulence and the presence of packing is accounted for by superposing an "effective" mechanism upon the over-all transport by plug flow. The flux arising from this mechanism is described by a formula analogous to Fick's law for mass transfer or Fourrier's law for heat transfer. The proportionality constants are "effective" diffusivities and conductivities. Because of the assumptions involved in their derivation they contain implicitly the effect of the velocity profile. This field has been reviewed and organized by Levenspiel and Bischoff (*14*). The principal experimental results concerning the effective diffusivity in axial direction are shown in Figure 4 (*12, 15, 16, 17, 18, 19*).

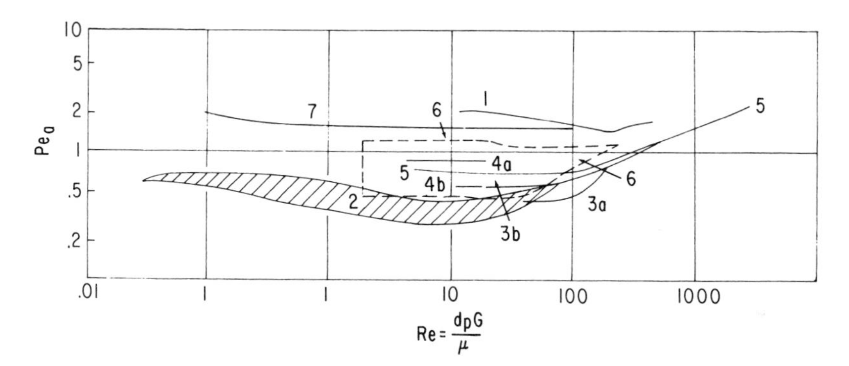

Figure 4. Axial mixing in packed beds—Peclet vs. *Reynolds numbers diagram*

Curve 1: McHenry and Wilhelm (15)
Curve 2: Ebach and White (16)
Curve 3: Carberry and Bretton (17)
Curve 4: Strong and Geankoplis (18)
Curve 5: Cairns and Prausnitz (12)
Curve 6: Hiby (19)
Curve 7: Hiby, without wall effect (19)

For design purposes Pe_a based on d_p may be considered to lie between 1 and 2. Little information is available on λ_{ea}. Yagi, Kunii, and Wakao (*20*) determined λ_{ea} experimentally, while Bischoff derived it from the analogy between heat and mass transfer in packed beds (*21*).

The continuity equation for a component A may be written in the steady state:

$$\varepsilon D_{ea} \frac{d^2c}{dz^2} - u_s \frac{dc}{dz} - r_A \rho_B = 0 \tag{3}$$

and the energy equation:

$$\lambda_{ea} \frac{d^2T}{dz^2} - \rho_f u_s c_p \frac{dT}{dz} + (-\Delta H)\, r_A \rho_B - \frac{4U}{d_t}(T - T_w) = 0 \tag{4}$$

The boundary conditions have given rise to extensive discussion (*29, 30, 31, 32*). Those generally used are

$$u_s(c_0 - c) = -\varepsilon D_{ea} \frac{dc}{dz} \qquad \text{for } z = 0$$

$$\rho_f u_s c_p (T_0 - T) = -\lambda_{ea} \frac{dT}{dz} \tag{5}$$

$$\frac{dc}{dz} = \frac{dT}{dz} = 0 \qquad \text{for } z = L$$

This leads to a two-point boundary value problem requiring trial and error in the integration. For the flow velocities used in industrial practice the effect of axial dispersion of heat and mass upon conversion is negligible when the bed depth exceeds about 100 particle diameters (*27*). Despite this the model has received great attention recently, more particularly the adiabatic version. The reason is that the introduction of axial mixing terms into the basic equations leads to an entirely new feature —namely, the possibility of more than one steady-state profile through the reactor (*28*).

Indeed, for a certain range of operating conditions three steady-state profiles are possible with the same feed conditions, as shown in Figure 5. The outer two steady-state profiles are stable, at least to small pertubations while the middle one is unstable. Which steady state profile will be predicted by steady-state computations depends on the initial estimates of c and T involved in the integration of this two-point boundary value problem. Physically this means that the steady state actually experienced depends on the initial profile in the reactor. For all situations where the initial values are different from the feed conditions transient equations must be considered to make sure the correct steady-state profile is predicted. To avoid those transient computations when they are unnecessary, it is useful to know *a priori* if more than one steady-state profile is possible. Figure 5 shows that a necessary and sufficient condition for uniqueness of the steady-state profile in an adiabatic reactor is that the curve

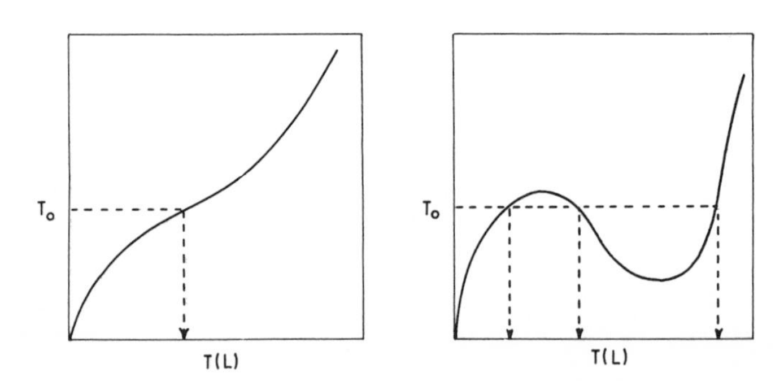

Figure 5. One-dimensional tubular reactor with axial mixing. Outlet vs. *inlet temperature.*

$t_{in} = f[t(L)]$ has no hump. Mathematically this means that Equation 6

$$\frac{d^2T}{dz'^2} - \mathrm{Pe}'_a \frac{dT}{dz'} + f(T) = 0 \qquad (6)$$

$$\text{where } z' = \frac{z}{L} \qquad \mathrm{Pe}'_a = \frac{U_i L}{D_{ea}}$$

$$\text{and } f(T) = \frac{k_o L^2}{\varepsilon D_{ea}} \rho_B (T_{ad} - T) \exp\left[\frac{E}{RT_o}\left(1 - \frac{T_o}{T}\right)\right]$$

has no bifurcation point, whatever the length of the reactor. This led Luss and Amundson (29) to the following conditions:

$$S_{up}[f'(T) - Pe'^2_a/4] \leq 0 \qquad (7)$$
$$T_o \leq T \leq T_{ad}$$

which can be satisfied by diluting the reaction mixture.

Another way of realizing a unique profile is to limit the length of the adiabatic reactor so that

$$S_{up}[f'(T) - Pe'^2_a/4] \leq \mu_1 \qquad (8)$$
$$T_o \leq T \leq T_{ad}$$

where μ_1 is the smallest positive eigenvalue of

$$\Delta v + \mu v = 0 \qquad (9)$$

and where $v(z)$ is the difference between two solutions $T_1(z)$ and $T_2(z)$. Uniqueness is guaranteed only if the only solution to Equation 9 is $v(z) = 0$.

When applied to a first-order irreversible reaction carried out in an adiabatic reactor these conditions lead to Equations 10 and 11, respectively.

$$\frac{k_o L^2}{\varepsilon D_{ea}} \rho_B \left\{\frac{E}{R}\left(\frac{T_{ad} - T_o}{T_p^2}\right) - 1\right\} \exp\left[\frac{E}{RT_o}\left(1 - \frac{T_o}{T_p}\right)\right] - \frac{u_i^2 L^2}{4D_{ea}^2} \leq 0 \qquad (10)$$

$$\text{or } < \mu_1 \qquad (11)$$

where $T_{ad} - T_o = \frac{(-\Delta H)}{\rho_f c_p} C_o$ is the adiabatic temperature rise and T_p is the value for T for which $f'(T)$ becomes a maximum.

A sufficient but not necessary condition for Equation 10 is that

$$\frac{E}{RT_o} \cdot \frac{T_{ad} - T_o}{T_o} \leq 1 \tag{12}$$

$$\text{or } \gamma\beta \leq 1$$

Luss later refined these conditions (*30*) and arrived at:

$$(T - T_o)\,\frac{d \ln f(T)}{dT} \leq 1 \tag{13}$$

The magnitude of the axial effective diffusivity determines which of the two conditions, Equations 7 or 13, is stronger. For a first-order irreversible reaction carried out in an adiabatic reactor Equation 13 leads to

$$\frac{E}{RT_o} \cdot \frac{T_{ad} - T_o}{T_o} \leq 4\,\frac{T_{ad}}{T_o} \quad \text{or} \quad \gamma\beta \leq 4\,\frac{T_{ad}}{T_o}$$

which is far less conservative than Equation 12, based on Equation 7. For adiabatic operation Hlavacek and Hoffman (*31*) found:

$$\frac{E}{RT_o} \cdot \frac{T_{ad} - T_o}{T_o} < \frac{4}{1 - 4/(E/RT_o)} \tag{15}$$

They also defined necessary and sufficient conditions for multiplicity, for a simplified rate law of the type Barkelew used and equality of the Peclet numbers for heat and mass transfer.

The necessary and sufficient conditions for multpilicity, which must be fulfilled simultaneously are:

(1) The group $\frac{E}{RT_o} \cdot \frac{(-\Delta H)c_o}{\rho_f c_p T_o} = \gamma\beta$ has to exceed a certain value.

(2) The group $\frac{Lk_o\rho_s}{u_s} = Da$ has to lie within a given interval.

(3) The Peclet number based on reactor length $\frac{u_i L}{D_{ea}}$ has to be lower than a certain value.

From a numerical study Hlavacek and Hofmann derived the results represented in Figure 6. This figure illustrates clearly that the range within which multiple steady states can occur is very narrow. It is true that, as Hlavack and Hofmann calculated, the adiabatic temperature rise is sufficiently high in ammonia-, methanol-, and oxo synthesis and in ethylene-, naphthalene-, and *o*-xylene oxidation. None of these reactions is carried out in adiabatic reactors, however, much less in multibed adiabatic reactors. According to Beskov (mentioned by Hlavacek and Hofmann) in methanol synthesis the effect of axial mixing would have to be

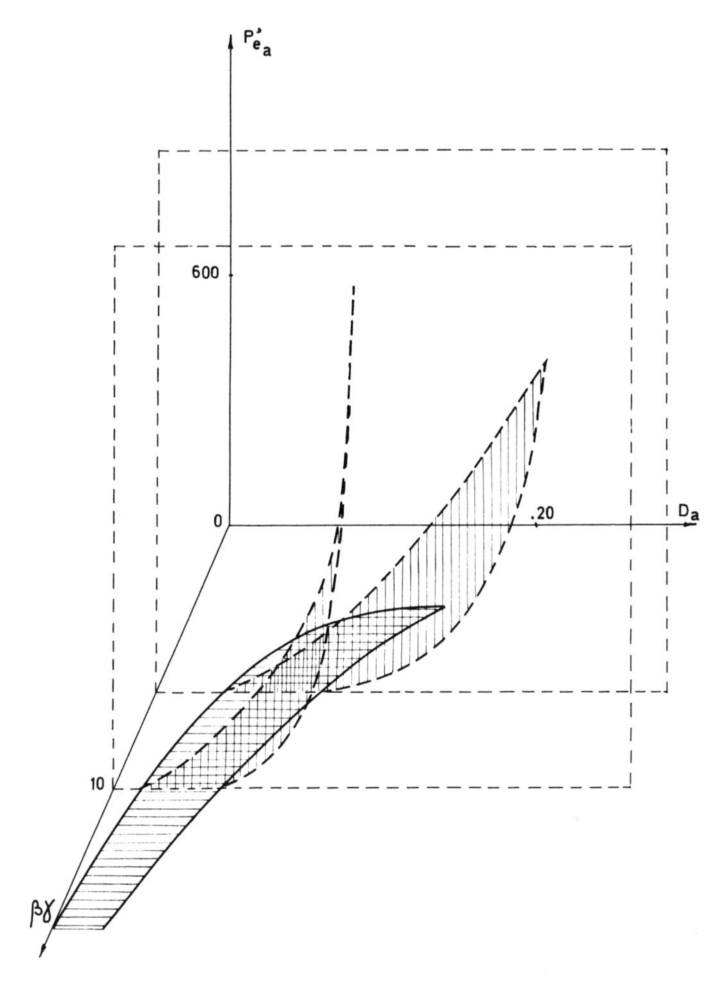

Figure 6. One-dimensional tubular reactor with axial mixing. Multiplicity of steady states. Relation between Peclet, Damkohler, and $\beta\gamma$ group according to Hlavacek and Hofmann (31).

taken into account when $Pe'_a < 30$. In industrial methanol synthesis reactors Pe'_a is of the order of 600 or more. In ethylene oxidation Pe'_a would have to be smaller than 200 for axial effective transport to be of some importance, but in industrial practice Pe'_a exceeds 2500. Therefore, the length of industrial fixed bed reactors removes the need for reactor models including axial diffusion and the risks involved with multiple steady states, except perhaps for very shallow beds. In practice, shallow catalytic beds are only encountered in the first stage of multibed adiabatic reactors. One may question if very shallow beds can be described by effective transport models in any event. The question remains as to whether shallow beds really exhibit multiple steady states. The answer probably requires a completely different approach, based on a better knowledge of the hydrodynamics of shallow beds. In our opinion there is no real need for further detailed study of the axial transport model; there are several other effects, more important than axial mixing, which must be accounted for.

Two Dimensional Pseudo-Homogeneous Models. The one-dimensional models discussed so far neglect the resistance to heat and mass transfer in the radial direction and therefore predict uniform temperatures and conversions in a cross section. This is obviously a serious simplification when reactions with a pronounced heat effect are involved. For such cases there is a need for a model that predicts the detailed temperature and conversion pattern in the reactor so that the design can be directed towards avoiding eventual detrimental over-temperatures on the axis. This then leads to two-dimensional models. The model discussed here uses the effective transport concept to formulate the flux of heat or mass in the radial direction. This flux is superposed upon the transport by over-all convection, which is of the plug flow type.

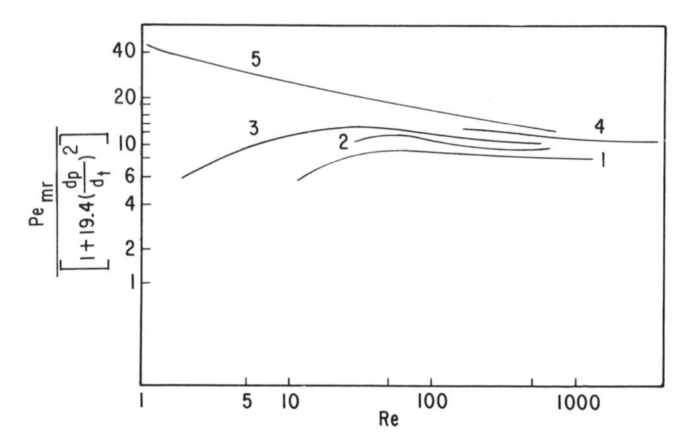

Figure 7. Radial mixing in packed beds—Peclet vs. *Reynolds numbers diagram*

Curve 1: Fahien and Smith (34)
Curve 2: Bernard and Wilhelm (32)
Curve 3: Dorrweiler and Fahier (33)
Curve 4: Plautz and Johnstone (40)
Curve 5: Hiby (19)

Since the effective diffusivity is determined mainly by the flow characteristics, packed beds are not isotropic for effective diffusion so that the radial component is different from the axial mentioned above. Experimental results concerning D_{er} are shown in Figure 7 (*19, 32, 33, 34,*

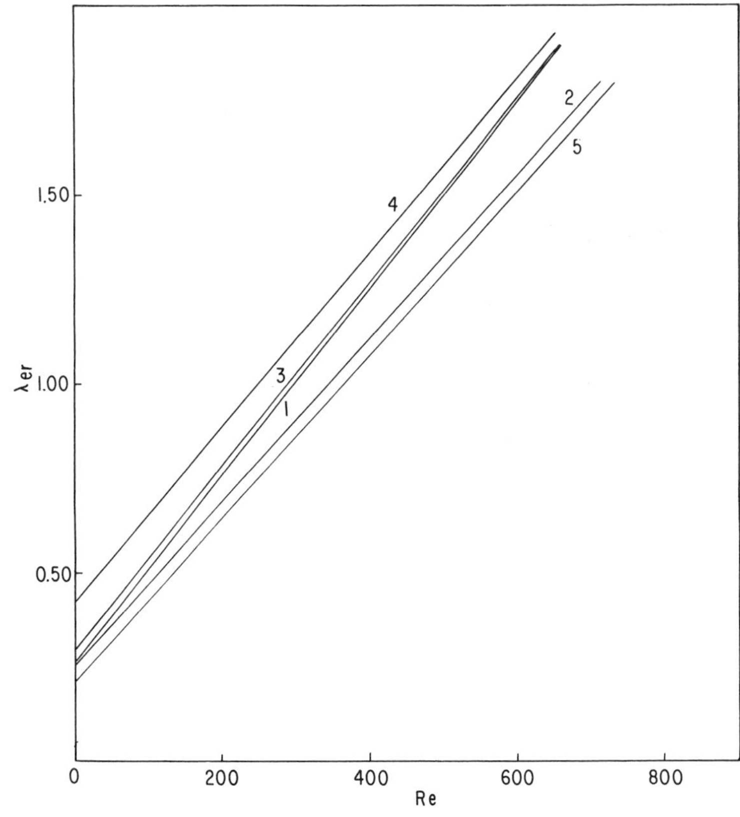

Figure 8. Heat transfer in packed beds. Effective thermal conductivity as a function of Reynolds number.

Curve 1: Coberly and Marshall (37)
Curve 2: Campbell and Huntington (36)
Curve 3: Calderbank and Pogorsky (35)
Curve 4: Kwong and Smith (38)
Curve 5: Kunii and Smith (39)

40). For practical purposes Pe_r may be considered to lie between 8 and 10. When the effective conductivity, λ_{er}, is determined from heat transfer experiments in packed beds it is observed that λ_{er} decreases strongly near the wall. It is as if a supplementary resistance is experienced near the wall, which is probably caused by variations in the packing density and flow velocity. Two alternatives are possible: either to use a mean λ_{er} or to consider λ_{er} constant in the central core and introduce a new coefficient accounting for the heat transfer near the wall, α_w, defined by

$$\alpha_w(T_R - T_w) = -\lambda_{er}(\partial T/\partial r)$$

When it is important to predict point values of the temperature with the greatest possible accuracy, the second approach is preferred, so that two parameters are involved to account for heat transfer in the radial direction.

Figures 8 and 9 show some experimental results for λ_{er} and α_w. The data for α_w are very scattered. Recently De Wasch and Froment (*6*) obtained data which are believed to have the high degree of precision needed to predict accurately severe situations in reactors (investigated Re range: from 30 to 1000). The correlations for air are of the form:

$$\lambda_{er} = \lambda^{\circ}_{er} + \frac{0.0025}{1 + 46(d_p/d_t)^2} \mathrm{Re}$$

$$\alpha_w = \alpha^{\circ}_w + \frac{0.0115 d_t}{d_p} \mathrm{Re}$$

where λ°_{er} and α°_w are static contributions, dependent on the type and

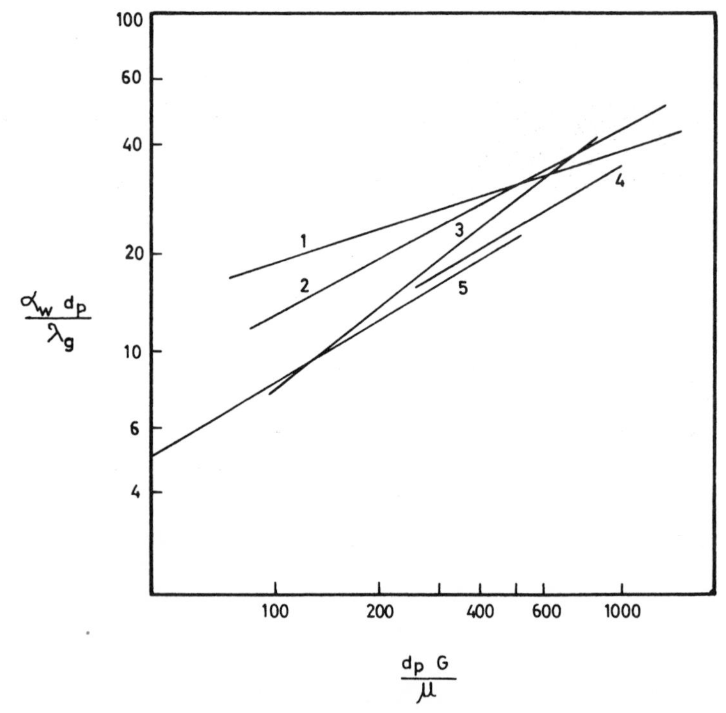

Figure 9. Heat transfer in packed beds. Wall heat transfer coefficient vs. *Reynolds number.*

Curve 1: Coberly and Marshall (37)
Curve 2: Hanratty (cylinders) (42)
Curve 3: Hanratty (spheres) (42)
Curve 4: Yagi and Wakao (44)
Curve 5: Yagi and Kunii (43)

size of the catalyst. The correlation for α_w is of an entirely different form from those published until now, but it confirms Yagi and Kunii's theoretical predictions (*43*). Since both solid and fluid are involved in heat transfer, λ_{er} is usually based on the total cross section and therefore upon the superficial velocity, in contrast with D_{er}. This is reflected in Equation 14 (below).

Yagi and Kunii (*41, 43*) and Kunii and Smith (*39*) have set up heat transfer models which allow one to predict λ_{er} and α_w from basic data. They distinguish between a static contribution which contains all the mechanisms not involving flow, such as conduction through solids, bulk fluid and stagnant fluid, radiation and a dynamic contribution, depending only upon the flow conditions.

The continuity equation for the key reacting component and the energy equation can now be written for a single reaction and steady state

$$\left\{\begin{aligned} &\varepsilon D_{er}\left(\frac{\partial^2 c}{\partial r^2} + \frac{1}{r}\frac{\partial c}{\partial r}\right) - u_s \frac{\partial c}{\partial z} - \rho_B r_A = 0 \\ &\lambda_{er}\left(\frac{\partial^2 T}{\partial r^2} + \frac{1}{r}\frac{\partial T}{\partial r}\right) - u_s \rho_f c_p \frac{\partial T}{\partial z} - \rho_B(-\Delta H)\, r_A = 0 \end{aligned}\right. \tag{14}$$

with boundary conditions:

$$c = c_o, \quad T = T_o \qquad \text{at } z = 0 \qquad 0 \leq r \leq R_t$$

$$\partial c/\partial r = 0 \qquad \text{at } r = 0 \text{ and } r = R_t \qquad \text{all} \rightarrow z$$

$$\partial T/\partial r = 0 \qquad \text{at } r = 0$$

$$\partial T/\partial r = -\frac{\alpha_w}{\lambda_{er}}(T_R - T_w) \quad \text{at } r = R_t$$

The term accounting for the effective transport in the axial direction has been neglected in this model for the reasons given above. This system of nonlinear second-order partial differential equations was integrated by Froment using a Crank-Nicolson procedure (*45, 46*). A computational scheme claimed to be faster was presented recently by Liu (*47*).

Froment used this model to simulate a multitubular fixed bed reactor for a reaction which involved yield problems and was fairly representative of hydrocarbon oxidation. Figure 10 illustrates the importance of radial gradients, even for a mild situation and a tube diameter of only 2.54 cm. These calculations revealed great sensitivity of the reactor performance with respect to D_{er}. These conclusions were confirmed by Carberry and

White (*48*). Therefore, it does not seem worthwhile refining our data on D_{er} any further.

Figure 11 compares predictions based upon the present model with those based on the basic one-dimensional model discussed previously. For drastic situations the difference is quite significant. For such a comparison to be valid and reflect only the effect of the model itself the over-all heat transfer coefficient U of the one-dimensional model has to be derived from λ_{er} and α_w as indicated by Froment (*45, 46*). Beek (*49*) and Kjaer (*50*) have also discussed features of this model.

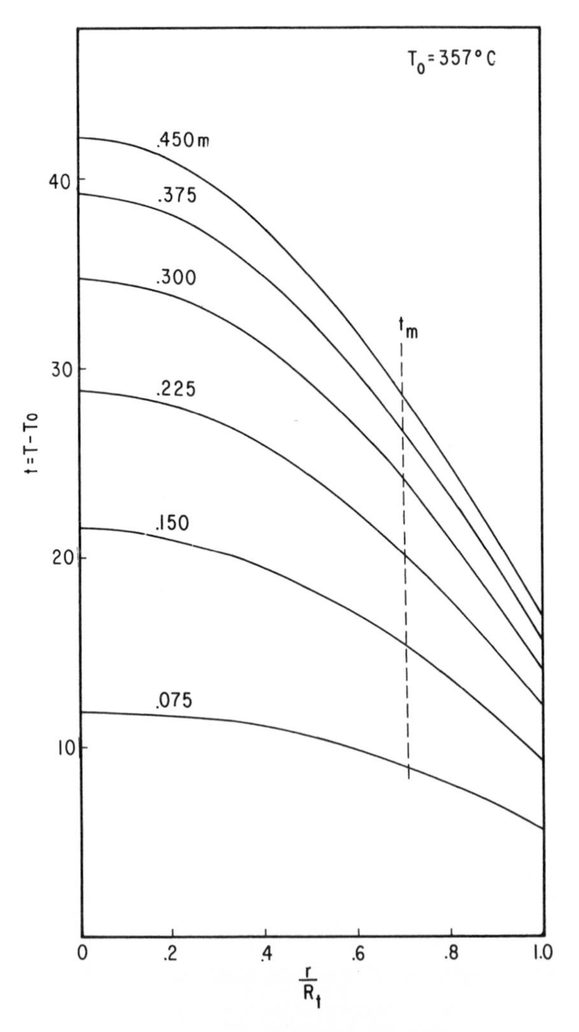

Figure 10. Two-dimensional pseudo-homogeneous model. Radial temperature profiles.

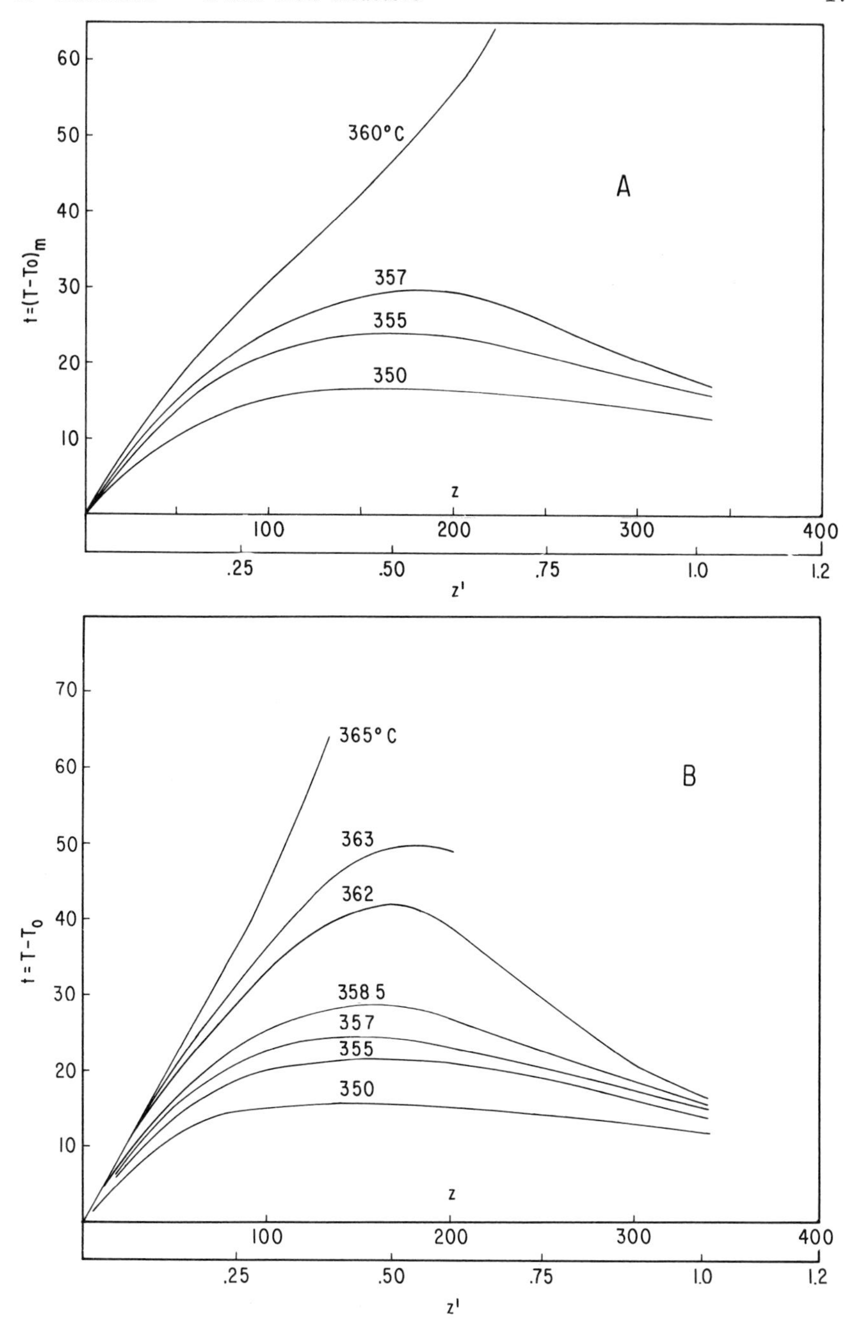

Figure 11. Sensitivity with respect to inlet temperature

A: Two-dimensional pseudo-homogeneous model. Radial mean temperature vs. *bed length.*
B: Basic one-dimensional model

The present model could be refined by introducing a velocity profile. This was done by Valstar (*51*) who used the velocity profiles of Schwartz and Smith (*10*) which exhibit a maximum at 1.5 d_p of the wall. Valstar's exploratory calculations indicate that the influence of the velocity profile is worth considering. Progress in this field will require more extensive basic knowledge of the packing pattern and hydrodynamics of fixed beds.

This discussion of the tubular reactor with radial mixing is based on a continuum model leading to a system of differential equations with mixing effects expressed in terms of effective diffusion or conduction. A different approach considers the bed to consist of a two-dimensional network of perfectly mixed cells with two outlets to the subsequent row of cells. Alternate rows are offset half a stage to allow for radial mixing. In the steady state a pair of algebraic equations must be solved for each cell. This model was proposed by Deans and Lapidus (*52*) and applied by McGuire and Lapidus (*53*) to non-steady-state cases. Agnew and Potter (*54*) used it to set up runaway diagrams of the Barkelew type. In fact, the model is not completely analogous to the one discussed above since it considers heat to be transferred only through the fluid. It is clear already from the correlations for λ_{er} given above that this is a serious simplification (as illustrated below). More elaborate cell models with a coupling between the particles to account for conduction or radiation, are possible, but the computational problems become overwhelming. The effective transport concept keeps the problem within tractable limits.

The possibilities of present day computers are such that there is no longer any reason for not using two-dimensional models for steady-state calculations, provided the available reaction rate data are accurate enough. The one-dimensional model will continue to be used for on-line computing and process control studies.

Heterogeneous Models

For rapid reactions with an important heat effect it may be necessary to distinguish between conditions in the fluid and on the catalyst surface or even inside the catalyst. As before the reactor models may be either one- or two-dimensional.

One Dimensional Model Accounting for Interfacial Gradients. For a singlet reaction carried out in a cylindrical tube and with the restrictions already mentioned for the basic case, the steady-state equations are:

Fluid
$$-u_s \frac{dc}{dz} = k_g a_v (c - c_s^s) \tag{16}$$

$$u_s \rho_f c_p \frac{dT}{dz} = h_f a_v (T_s^s - T) - 4 \frac{U}{d_t} (T - T_w) \tag{17}$$

Solid
$$\rho_B r_A = k_g a_v (c - c_s^s) \tag{18}$$

$$(-\Delta H)\, \rho_B r_A = h_f a_v (T_s^s - T) \tag{19}$$

with boundary conditions: $c = c_o$, $T = T_o$ at $z = 0$

This model does not provide any axial coupling between the particles. Consequently, heat is transferred axially only through the fluid.

In industrial fixed bed reactors the flow velocity is generally so high that the temperature drop and concentration drop over the film surrounding the catalyst is small in steady-state operation. This may not be true with very exothermic and fast reactions involving a component of the catalyst or deposited on the catalyst such as encountered in catalyst regeneration by burning off coke or in catalyst reoxidation as sometimes required in ammonia synthesis or steam-reforming plants.

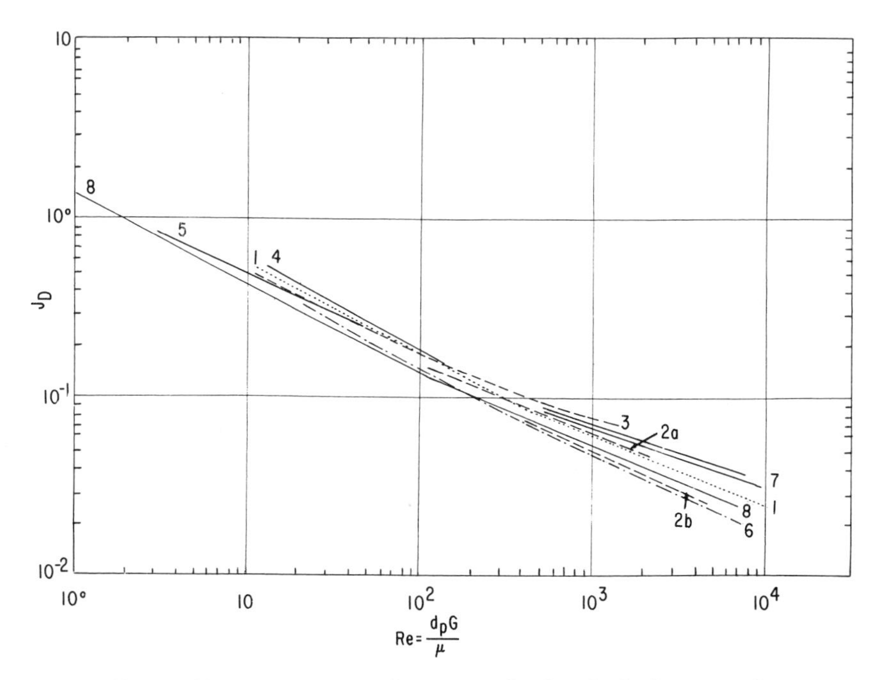

Figure 12. Mass transfer between a fluid and a bed of particles

Curve 1: Gamson et al. (56), *Wilke and Hougen* (57)
Curve 2: Taecker and Hougen (a) Berl saddles (b) Rashig rings
Curve 3: McCune and Wilhelm (59)
Curve 4: Ishino and Otake (60)
Curve 5: Bar Ilan and Resnick (61)
Curve 6: deAcetis and Thodos (62)
Curve 7: Bradshaw and Bennett (63)
Curve 8: Hougen (64), *Yoshida, Ramaswami, and Hougen* (65)

Figures 12 and 13 show most of the correlations available to date for k_g and h_f (*55*). Except perhaps for the most stringent conditions these parameters are now defined with sufficient precision. The distinction between conditions in the fluid and on the solid leads to an essential difference with respect to the basic, one-dimensional model—namely, the problem of stability, which is associated with multiple steady states.

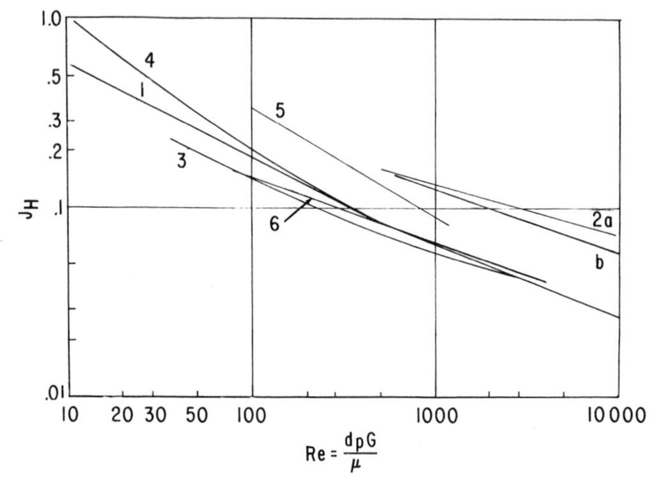

Figure 13. Heat transfer between a fluid and a bed of particles

Curve 1: Gamson et al. (56), *Wilke and Hougen* (57)
Curve 2: Baumeister and Bennett (a) for $d_t/d_p > 20$ *(b) mean correlation* (66)
Curve 3: Glaser and Thodos (67)
Curve 4: de Acetis and Thodos (62)
Curve 5: Sen Gupta and Thodos (68)
Curve 6: Handley and Heggs (69)

This aspect was studied first independently by Wicke (*73*) and by Shean-Lin-Liu and Amundson (*70–72*). They compared the heat produced in the catalyst, which is a sigmoid curve when plotted as a function of the particle temperature, with the heat removed by the fluid through the film surrounding the particle, which leads to a straight line. The steady state for the particle is given by the intersection of both lines. For a certain range of gas and particle temperatures three intersections and therefore three steady states are possible. From a comparison of the slopes of the sigmoid curve and the straight line in these three points it follows that the middle steady state is unstable to any perturbation, while the upper and lower are stable to small perturbations but not necessarily to large ones. Hence, when multiple steady states are possible, the steady state the particle actually operates in also depends on its initial temperature. When this is extended from a particle to adiabatic reactor, the concentration and temperature profiles are determined not only by the feed conditions but also by the initial solid temperature profile from which the reactor was started up. If this is not equal to the fluid feed

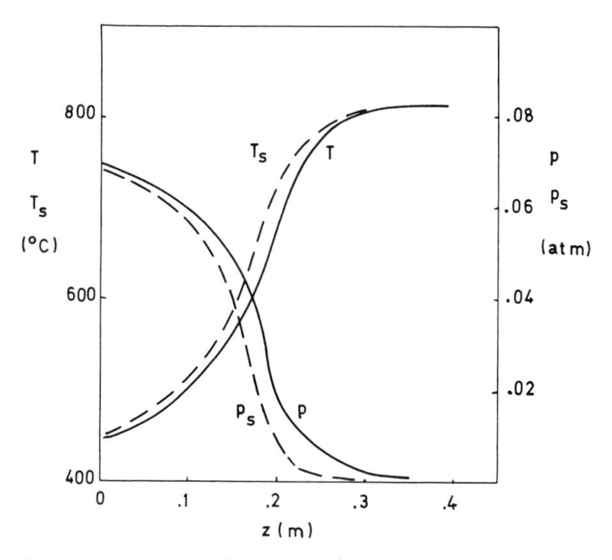

Figure 14. One-dimensional heterogeneous model with interfacial gradients. Unique steady-state case; $p_0 = 0.007$ *atm;* $T_0 = 449°C$.

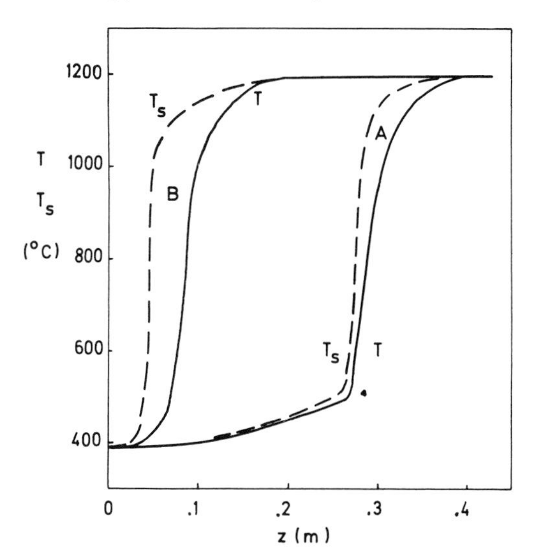

Figure 15. One-dimensional heterogeneous model with interfacial gradients. Non-unique steady-state case; $p_0 = 0.15$ *atm;* $T_0 = 393°C$; T_s = *initial: A* $\leqslant$ *393°C. B: 560°C.*

temperature, transients are involved. The design calculations would then have to be based upon the Reactions 16–19 complete with non-steady-state terms. Figures 14 and 15 illustrate this for an adiabatic reactor (*70, 71, 72*). Figure 14 shows a situation with a unique steady-state profile. In Figure 15 the gas is first heated along the lower steady state and then jumps to the upper steady state as soon as its temperature exceeds 480°C. The higher the initial temperature profile, the earlier the profile jumps from the lower to the higher steady state. From a comparison with the unique steady-state case of Figure 14 it follows that the shift from one steady state to another leads to temperature profiles which are much steeper. The reactor of Figure 15 may be unstable while the reactor of Figure 14 is stable, which does not exclude parametric sensitivity and runaway, as discussed earlier, however.

Are these multiple steady states possible in practical situations? Inspection of Figures 14 and 15 shows that the conditions chosen for the reaction are rather drastic. It would be interesting to correlate the limits on the operating conditions and reaction parameters within which multiple steady states could be experienced. These limits will probably be extremely narrow so that the phenomena discussed here would be limited to special reactions or to localized situations in a reactor, which would probably have little effect on its over-all behavior.

One-Dimensional Model Accounting for Interfacial and Intraparticle Gradients. When the resistance to mass and heat transfer inside the catalyst particle is important, the rate of reaction is not uniform throughout the particle. Equations 16–19 then no longer adequately describe the system. It must be completed with equations describing the concentration and temperature gradients inside the particle; the complete set may be written as:

Fluid:

$$-u_s \frac{dc}{dz} = k_g a_v (c - c_s^s) \tag{16}$$

$$u_s \rho_f c_p \frac{dT}{dz} = h_f a_v (T_s^s - T) - 4 \frac{U}{d_t} (T - T_w) \tag{17}$$

Solid:

$$\frac{D_s}{\xi^2} \frac{d}{d\xi}\left(\xi^2 \frac{dc_s}{d\xi}\right) + \rho_s r_A(c_s, T_s) = 0 \tag{21}$$

$$\frac{\lambda_s}{\xi^2} \frac{d}{d\xi}\left(\xi^2 \frac{dT_s}{d\xi}\right) - \rho_s(-\Delta H)\, r_A(c_s, T_o) = 0 \tag{22}$$

with boundary conditions

$$c = c_o \qquad \text{at } z = 0 \tag{23}$$
$$T = T_o$$

$$\frac{dc_s}{d\xi} = \frac{dT_s}{d\xi} = 0 \qquad \text{at } \xi = 0 \tag{24}$$

$$k_g(c_s^s - c) = -D_s \left.\frac{dc_s}{d\xi}\right|_{\xi = b} \tag{25}$$

$$h_f(T_s^s - T) = -\lambda_s \left.\frac{dT_s}{d\xi}\right|_{\xi = b} \tag{26}$$

D_s and λ_s are the effective diffusivity and conductivity inside the particle. Numerical values for D_s are given in Satterfield and Sherwood's book on diffusion in catalysis (*74, 75*). Equations 21 and 22 form a pair of second-order nonlinear differential equations which must be integrated at each node of the computational grid used in integrating the fluid field equations (16 and 17). This is feasible on present day computers but is still very lengthy. The computations may be simplified with the help of the effectiveness factor concept. In its classical sense η is a factor which multiplies the rate at the particle surface conditions to give the rate actually experienced when not all the reactant is converted at those surface conditions. The effectiveness factor η is expressed as a function of a modulus which is related to the ratio of the reaction rate at surface conditions to the mass transfer rate towards the inside of the particle. The use of this concept in essence permits the separate integration of Equations 21 and 22. The system is now reduced to Equations 16 and 17 with boundary condition (Equation 23) while Equations 21–22 and 25–26 are reduced to the algebraic equations:

$$k_g a_v(c - c_s^s) = \eta \rho_B r_A(c_s^s, T_s^s) \tag{27}$$

$$h_f a_v(T_s^s - T) = \eta \rho_B(-\Delta H)\, r_A(c_s^s, T_s^s) \tag{28}$$

$$\text{with } \eta = f(c_s^s, T_s^s)$$

Equations 27 and 28 differ only by the factor η from Equations 18 and 19.

The effectiveness factor depends on the local conditions which are introduced through the modulus; therefore, it has to be computed at each point of the grid. The saving in computational effort now depends on the relation between the effectiveness factor η and the modulus, ϕ. An analytical expression is only possible for isothermal particles and simple power law rate equations. For a first-order irreversible reaction and an isothermal particle the relation between η and ϕ is:

$$\eta = 3/\phi^2 (\phi \coth \phi - 1)$$

$$\text{where } \phi = \frac{V_p}{A_p}\sqrt{\frac{k(T_s^s)\,\rho_s}{D_s}} \tag{29}$$

For other cases there is little or no gain in using η. Fortunately, even with strongly exothermic reactions the particle is nearly isothermal; the main resistance inside the pellet is to mass transfer, and the main resistance in the film surrounding the particle is to heat transfer. Until recently, η had always been expressed as a function of conditions at the particle surface—c_s^s and T_s^s. When, however, the concept is used with models that account for a difference between bulk fluid and particle surface conditions, it may be preferable to base the effectiveness factor on bulk fluid conditions. The equivalent of Equation 29 then is (for isothermal conditions in both film and particle) (*74*):

$$\eta^* = \frac{3\text{Sh}}{\phi^2} \cdot \frac{\phi \cosh \phi - \sinh \phi}{\phi \cosh \phi + (\text{Sh}/2 - 1) \sinh \phi} \tag{30}$$

With this version of the concept η in Equations 27 and 28 has to be replaced by η^*, and r_A has to be expressed as a function of fluid conditions (c, T).

In the general case the relation between the effectiveness factor and the modulus can be obtained only by numerical integration of Equations 16–26. The result is shown in Figure 16 (*76*). With isothermal situations η tends to a limit of 1 as ϕ decreases. With non-isothermal conditions η or η^* may exceed 1. Curve 1 corresponds to the classical concept with $T_s^s = T$ and $c_s^s = c$ while Curves 2, 3, and 4 include gradients over the film into η, thus leading to η^*. The dotted portion of Curve 4 corresponds to a region of conditions within which multiple steady states inside the catalyst are possible. The range of parameters leading to this possibility is very narrow although it is widened somewhat when the film is included into the concept. Consequently, non-unique profiles are possible in the reactor. The steady state actually experienced depends on the initial conditions, so that transient calculations have to be performed. To avoid these when they are unnecessary, recent, considerable effort has gone into defining criteria for the uniqueness of the steady state of a particle. The reasoning and treatment is entirely analogous to that explained already for the tubular reactor with axial mixing. By restricting the treatment to the particle itself—the classical approach—Weiss and Hicks (*78*) arrived at the following condition for uniqueness for a first-order irreversible

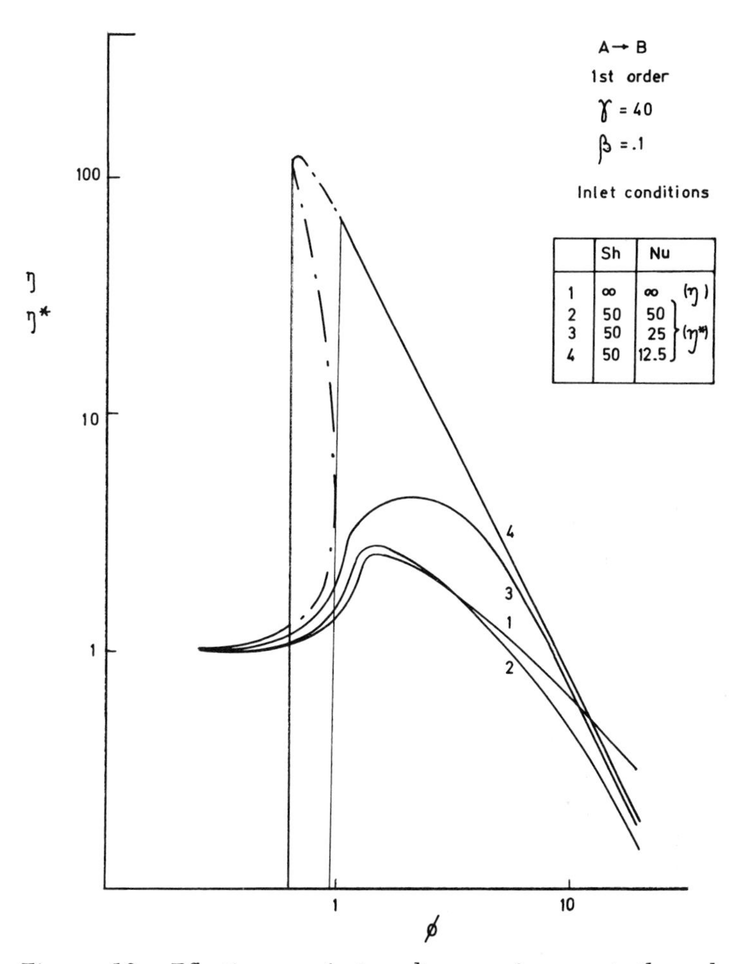

Figure 16. Effectiveness factor diagram for non-isothermal situations

reaction:

$$\frac{T_s^{ad} - T_s^s}{T_s^s} \cdot \frac{E}{RT_s^s} \leq 5 \tag{31}$$

$$\text{or } \beta \cdot \gamma \leq 5$$

Luss (*30*) found that uniqueness is guaranteed for an adiabatic reaction in a porous catalyst particle when

$$(T_s - T_s^s) \frac{d \ln f(T_s)}{dT_s} \leq 1 \tag{32}$$

$$\text{for all } T_s^s \leq T_s \leq T_s^{ad}$$

where $f(T_s)$ is the expression of the reaction rate as a function of temperature—*e.g.*, for a first-order irreversible rate equation:

$$f(T_s) = \frac{k(T_s^s)\,\rho_s}{D_s}\,(T_s^{ad} - T_s)\exp\left[\frac{E}{RT_s^s}\left(1 - \frac{T_s^s}{T_s}\right)\right]$$

This analysis does not include the external film. When applied to a first-order irreversible reaction, Equation 32 leads to

$$\beta\gamma \leq 4(T_s^{ad}/T_s^s) \tag{33}$$

Even with isothermal conditions in the particle multiple steady states may occur when the rate increases with conversion, as may happen with some Langmuir-Hinshelwood rate equations. In such cases the condition for uniqueness of the concentration profile in the particle is, according to Luss

$$(c_s - c_s^s)\,\frac{d\ln f(c_s)}{dc_s} \leq 1 \tag{34}$$

$$\text{for all } c_s^s \geq c_s \geq c_s^{ad}$$

Recently Cresswell (*79*) and McGreavy and Thornton (*80*) included the film in their analysis of the multiplicity of solution. Clearly, when the film is included, multiple steady states are possible even with isothermal situations, no matter what the kinetics are. Cresswell came to the following criterion for uniqueness:

$$\frac{E}{RT}\cdot\frac{c(-\Delta H)\,D_s}{\lambda_s T} < 8\left(\frac{\text{Nu}}{\text{Sh}} + \frac{c(-\Delta H)\,D_s}{\lambda_o T}\right)$$

$$\text{or } \gamma\beta < 8(\text{Nu/Sh} + \beta) \tag{35}$$

where T and c are the temperature and concentration in the bulk fluid surrounding the particle considered. With the large ratio of Sh/Nu found in practice it is apparent that $\text{Nu}/\beta\text{Sh} \leqslant 1$, even for small values of β, so that Equation 35 can be simplified into: $\gamma < 8$. Cresswell's work shows that even when the film is included, the region in which multiple solutions can occur is very narrow. The range of parameters investigated seems realistic (γ: 10–40; β: 0–0.1; Nu: 0.1–10; Sh: 100–500). Luss and Lee (*81*) developed a method for obtaining stability regions for the various steady states, based on the knowledge of the steady-state profiles. Thus, it is possible to predict which steady state a particle will tend toward, starting from given initial conditions. This whole field of uniqueness and stability has been reviewed recently by Aris (*82*).

As mentioned previously the possibility of multiple steady states complicates the design of the reactor seriously. Transient computations have to be performed to ensure the correct steady-state profile throughout the reactor is predicted. Another way would be to check the possibility of multiple steady states on the effectiveness factor chart for every point in the reactor. This would, in principle, require an infinite set of such charts, because β and γ vary throughout the reactor. McGreavy and Thornton (*80*) reformulated the problem to enable a single graph to be used for the whole reactor and to reduce the effectiveness factor curve to a single point in the new chart. For this purpose they introduced a new parameter

$$\theta = (d_p/2)\sqrt{(A\rho_s/D_s)}$$

which replaces the Thiele modulus and is based on the preexponential factor A rather than on the rate coefficient itself. Another convenient group is

$$\frac{\beta}{\gamma \mathrm{Nu}} = \frac{(-\Delta H)\ cD_sR}{d_p h_f E}.$$

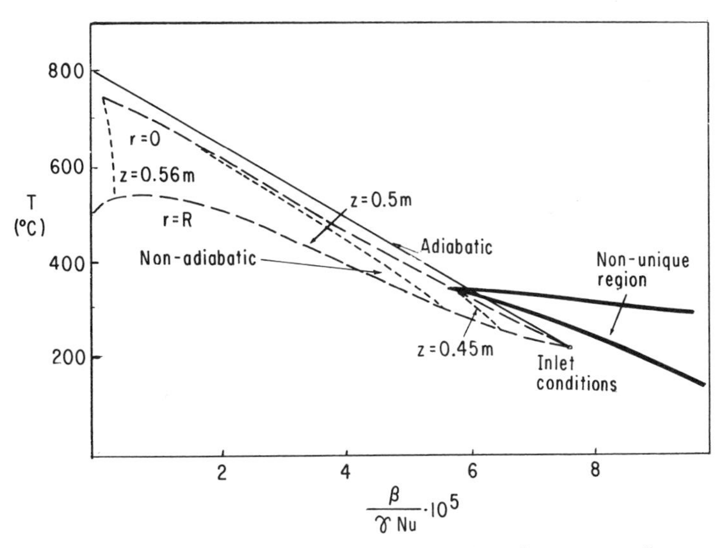

Figure 17. Tubular reactor with interfacial and intraparticle gradients. c-T *phase plane and region of multiple steady states.*

For a given system this group depends only on the reactant concentration in the fluid. Therefore, it is an implicit function of axial position. Taking advantage of the fact that the particle is generally isothermal, a relatively simple formula for the bounds on the fluid temperature within which multiple steady states may occur can be derived. It is represented graphically in Figure 17. With a reactant concentration corresponding to a β/γNu value of 8×10^{-5}, for example, multiple steady states are possible when the gas temperature is between 265° and 320°C. Figure 17 is essentially a c–T phase plane and allows a trajectory through the reactor to be plotted. When a trajectory intersects the non-unique region, multiple profiles are possible. Figure 17 shows a trajectory for adiabatic conditions which just intersects the zone. In the corresponding reactor zone the temperature may jump to the higher steady state, but since the region is very narrow, it is possible that instabilities will be damped. As soon as conditions are no longer adiabatic, only unique profiles are possible in the example considered. Since McGreavy and Thornton really used a two-dimensional model for the reactor in the non-adiabatic case, Figure 17 shows longitudinal profiles in the axis and at the wall. The parameter values used appear to be realistic although rather drastic.

Two-Dimensional Heterogeneous Models. In the last couple of years attempts have been made to develop two-dimension heterogeneous models. McGreavy and Cresswell proceeded by adding to the one-dimensional model accounting for interfacial and intraparticle gradients (discussed above) the terms accounting for radial heat and mass transfer in the bed (*76*). From an inspection of the equations it is clear, however, that it is assumed that heat transfer in the radial direction occurs only through the fluid phase. Figure 8 shows that even for typical industrial flow rates the solid and stagnant films contribute at least 25% in the radial heat flux. With regard to the extreme sensitivity of the profiles to λ_{er} (*46, 48*) the model used by McGreavy and Cresswell can be considered only as a rough approximation. The model by Carberry and White (*48*) is hybrid in the sense that it distinguishes between conditions in the gas and on the solid, but nevertheless makes use of the λ_{er} and α_w concept of Yagi and Kunii and Kunii and Smith, which lumps gas and solid (explained before).

To account correctly for heat transfer through the solid the equations concerning the solid should not be limited to a single particle as is generally done; they should be extended to the complete cross section occupied by the catalyst. This was done for the one-dimensional model (Equations 18 and 19) but without accounting for eventual radial temperature gradients. In addition, one must distinguish between the effective thermal conductivity for the fluid phase, λ_{er}^{f} and that for the solid phase λ_{er}^{s} (*84*). Strangely enough, this concept of λ_{er}^{f} and λ_{er}^{s} was introduced as early as 1953 by Singer and Wilhelm (*83*). All subsequent work in this field made use of the global λ_{er} concept, however. The preceding considerations led DeWasch and Froment (*84*) to the following mathematical model:

$$
\left\{
\begin{aligned}
& u_s \frac{\partial c}{\partial z} = \varepsilon D_{er}\left(\frac{\partial^2 c}{\partial r^2} + \frac{1}{r}\frac{\partial c}{\partial r}\right) - k_g a_v (c - c_s^s) \\
& u_s \rho_f c_p \frac{\partial T}{\partial z} = \lambda_{er}^f\left(\frac{\partial^2 T}{\partial r^2} + \frac{1}{r}\frac{\partial T}{\partial r}\right) + h_f a_v (T_s^s - T) \\
& k_g a_v (c - c_s^s) = \eta \rho_B r_A \\
& h_f a_v (T_s^s - T) = \eta \rho_B (-\Delta H)\, r_A + \lambda_{er}^s\left(\frac{\partial^2 T_s}{\partial r^2} + \frac{1}{r}\frac{\partial T_s}{\partial r}\right)
\end{aligned}
\right. \tag{36}
$$

with boundary conditions

$$
\begin{aligned}
& c = c_o && \text{at } z = 0 \\
& T = T_o && \\
& \partial c/\partial r = 0 && \text{at } r = 0 \quad \text{all } z \\
& \partial T/\partial r = \partial T_s/\partial r = 0 && \\
& \partial c/\partial r = 0 && \text{at } r = R_t \quad \text{all } z \\
& \alpha_w^f (T_w - T) = \lambda_{er}^f (\partial T/\partial r) && \\
& \alpha_w^s (T_w - T_s) = \lambda_{er}^s (\partial T_s/\partial r) &&
\end{aligned}
$$

The distinction between solid and fluid also appears in the boundary conditions for heat transfer at the wall. There are several possibilities for the boundary condition for the "solid" phase at the wall. The simplest is to set the temperature of the solid equal to that of the wall itself. A better approximation is to consider the temperature profile in the "solid" phase to be linear near the wall ($\partial^2 T_s/\partial r^2 = 0$). Still another possibility is to use a boundary condition for the "solid" analogous to that for the fluid.

These different possibilities and the numerical values to be given to the parameters are discussed by DeWasch and Froment. Figure 18 shows radial mean temperature profiles through a reactor for the three boundary conditions discussed above and no intraparticle resistance ($\eta = 1$). The influence of the boundary condition is obviously quite important. The boundary conditions for the "solid phase" $\partial^2 T_s/\partial r^2 = 0$ and $\alpha_w^s(T_w - T_s) = \lambda_{er}^s \frac{\partial T_s}{\partial r}$ lead to results which are in excellent agreement. Assuming no heat transfer through the solid predicts far too important hot spots. Such a model is no improvement at all with respect to the two-dimensional

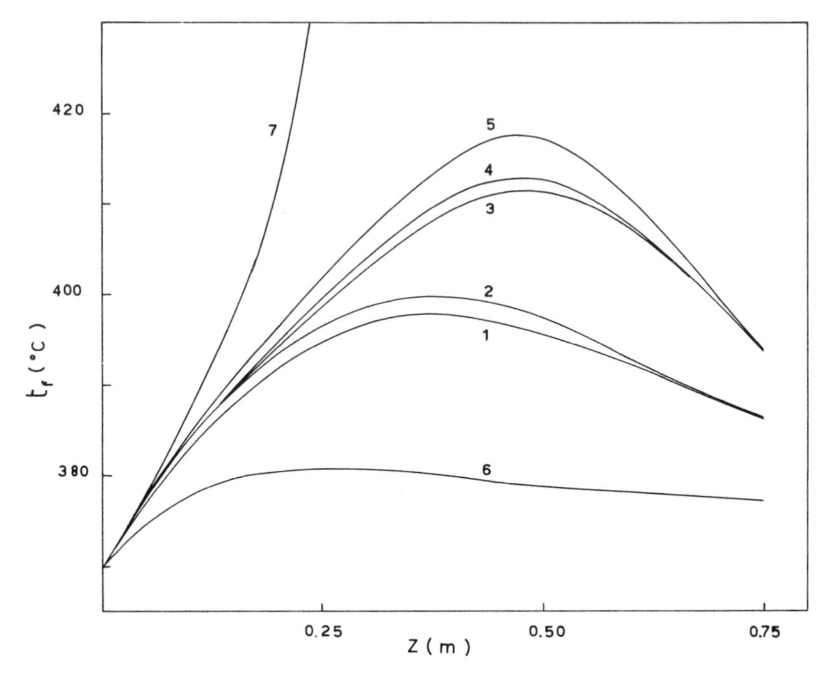

Figure 18. Two-dimensional heterogeneous model. Radial mean temperature as a function of bed length. Comparison with other models.

Curve 1: Basic pseudo-homogeneous one-dimensional model
Curve 2: One-dimensional heterogeneous model with interfacial gradients
Curve 3: Two-dimensional pseudo-homogeneous model
Curve 4: Two-dimensional heterogeneous model. Boundary conditions as in Ref. 36.
Curve 5: Two-dimensional heterogeneous model. Boundary condition at the wall: $\frac{\partial^2 t_s}{\partial r^2} = 0$.
Curve 6: Two-dimensional heterogeneous model. Boundary condition at the wall: $t_s = t_w$.
Curve 7: Two-dimensional heterogeneous model. Radial heat transfer only through the fluid [McGreavy and Cresswell (76)].

pseudo-homogeneous model discussed earlier. For the conditions used in these calculations the solid temperature exceeds the gas temperature only by 1° or 2°C. This is generally the case in industrial reactors.

Conclusion

Before 1960 little more than the basic pseudo-homogeneous model had been investigated. Owing to the increasing possibilities of computers the modelling of fixed bed catalytic reactors has been a rapidly developing field in the last decade. Models with more and more complexity have been set up and compared with more elementary ones so that it became possible to assess the effect of simplifying assumptions on the predicted results. This has now paved the way to judicious model reduction which would lead to models sufficiently accurate and simple to be used in transient simulation and process control—a field in which the application of the recent complex models is still out of the question. Finally, it is hoped that model studies and analysis of phenomena in reactors will be based on real data and real reactor configurations so that a bridge can be laid between researchers in the field and those involved in design and operation of reactors.

Nomenclature

A	preexponential factor in Arrhenius law
A_p	external surface area of a particle (m_s^2)
a_v	particle surface area per unit bed volume (m_s^2/m^3)
b	particle radius (m_s)
c	concentration of reference component A in the fluid (kmoles/ m^3 fl)
c_o	inlet concentration in fluid (id)
c_s	concentration on the surface of the particle (id)
D_A	molecular diffusivity (m^3fl/m hr)
D_{ea}, D_{er}	effective diffusivities in axial and radial direction (m^3 fl/m hr)
D_s	effective diffusivity inside the particle (m^3 fl/m_s hr)
d_p	particle diameter (m_s)
d_t	tube diameter (m)
E	activation energy (kcal/kmole)
G	mass flow velocity (superficial) (kg/m^2 hr)
h_f	film heat transfer coefficient (kcal/m_s^2 hr °C)
$(-\Delta H)$	heat of reaction (kcal/kmole)
j_D	$= \frac{k_g}{u_s} Sc^{2/3}$
j_H	$= \frac{h_f}{c_p G} Pr^{2/3}$
k_o	first order reaction velocity coefficient at inlet temperature (m^3 fl/kg catalyst hr)
$k(T_s^s)$	*idem* at catalyst surface temperature (id)
k_g	film mass transfer coefficient (m^3 fl/m_s^2 hr)
L	reactor length (m)
Nu	Nusselt number for film heat transfer $= h_f d_p/\lambda_g$
p	partial pressure (atm)
Pe_a	Peclet number for effective mass transfer in axial direction based on particle diameter, $d_p u_i/D_{ea}$
Pe'_a	*idem*, but based on reactor length $u_i L/D_{ea}$
Pe_r	Peclet number for effective mass transfer in radial direction, $u_i d_p/D_{er}$
Pr	Prandtl number $c_p \mu/\lambda_g$
r_A	rate of reaction based on component A (kmoles/kg catalyst hr)
r	radial coordinated (m)
R	gas constant (kcal/kmole, °K)
R_t	tube radius (m)

Sc	Schmidt number, $\mu/\rho_f D_A$
Re	Reynolds number, $d_p G/\mu$
Sh	Sherwood number for film mass transfer, $k_g d_p/D_A$
T_R	bed temperature at radius R (°C or °K)
T	fluid temperature (*idem*)
T_s	solid temperature (*idem*)
T_s^s	temperature at surface of particle (*idem*)
T_o	inlet fluid temperature (*idem*)
T_w	wall temperature (*idem*)
T_{ad}	adiabatic temperature (*idem*)
U	global heat transfer coefficient (kcal/m² hr °C)
u_s	superficial velocity (m³ fl/m² hr)
u_i	interstitial velocity (m/hr)
V_p	volume of a particle (m³)
z	axial coordinate (m)
z'	dimensionless axial coordinate $= z/L$
α_w	wall heat transfer coefficient (kcal/m² hr °C)
α_w^s	wall heat transfer coefficient for solid phase (*idem*)
α_w^f	wall heat transfer coefficient for fluid phase (*idem*)
β	group: $\dfrac{(-\Delta H)c_o}{\rho_f c_p T_o} = \dfrac{T_{ad} - T_o}{T_o}$ in Ref. *12;* $\dfrac{c_s^s(-\Delta H)D_s}{\lambda_s T_s^s}$ in Ref. *31;* $\dfrac{c(-\Delta H)D_s}{\lambda_s T}$ in Ref. *35*
ϵ	void fraction of the bed (m³/m³)
γ	group: E/RT_o in Ref. *12;* E/RT_s^s in Ref. *31;* E/RT in Ref. *35*
$\lambda_{ea}, \lambda_{er}$	effective thermal conductivity in axial and radial direction (kcal/m hr °C)
λ_s	solid thermal conductivity (*idem*)
λ_{er}^f	effective thermal conductivity for the fluid phase only (*idem*)
λ_{er}^s	effective thermal conductivity for the solid phase only (*idem*)
μ	dynamic viscosity (kg/m hr)
ρ_f	fluid density (kg/m³ fl)
ρ_B	catalyst bulk density (kg catalyst/m³)
ρ_s	catalyst density (kg catalyst/m_s^3)
ζ	coordinate inside the particle (m_s)
η	effectiveness factor based on particle only
η^*	effectiveness factor including the film
ϕ	Thiele modulus $= (V_p/A_p)\sqrt{\dfrac{k(T_s^s)\rho_s}{D_s}}$ for a first-order reaction

Literature Cited

(1) Leva, M., *Chem. Eng.* (1959) **56,** 115.
(2) Brownell, L. E., Dombrowski, H. S., Dickey, C. A., *Chem. Eng. Progr.* (1950) **46,** 415.
(3) Leva, M., *Ind. Eng. Chem.* (1948) **40,** 747.
(4) Maeda, S., *Techn. Dep. Tohoku Univ.* (1952) **16,** 1.
(5) Verschoor, H., Schuit, G., *Appl. Sci. Res.* (1950) **A2,** 97.
(6) DeWasch, A. P., Froment, G. F., *Chem. Eng. Sci.* (1971) **26,** 629.
(7) Barkelew, C. R., *Chem. Eng. Progr., Symp. Ser.* (1959) **55**(25), 38.

(8) Van Welsenaere, R. J., Froment, G. F., *Chem. Eng. Sci.* (1970) **25**, 1503.
(9) Bilous, O., Amundson, N. R., *AIChE J.* (1956) **2**, 117.
(10) Schwartz, C. S., Smith, J. M., *Ind. Eng. Chem.* (1953) **45**, 1209.
(11) Schertz, W. W., Bischoff, K. B., *AIChE J.* (1969) **15**, 597.
(12) Cairns, E. J., Prausnitz, J. M., *Ind. Eng. Chem.* (1959) **51**, 1441.
(13) Mickley, H. S., Smith, K. A., Korchak, E. I., *Chem. Eng. Sci.* (1965) **20**, 237.
(14) Levenspiel, O., Bischoff, K. B., *Advan. Chem. Eng.* (1963) **4.**
(15) McHenry, K. W., Wilhelm, R. H., *AIChE J.* (1957) **3**, 83.
(16) Ebach, E. A., White, R. R., *AIChE J.* (1958) **4**, 161.
(17) Carberry, J. J., Bretton, R. H., (1958) **4**, 367.
(18) Strang, D. A., Geankoplis, C. I., *Ind. Eng. Chem.* (1958) **50**, 1305.
(19) Hiby, J. W., "Interaction between Fluids and Particles," Institution of Chemical Engineers, London, 1962.
(20) Yagi, S., Kunii, D., Wakao, N., *AIChE J.* (1960) **6**, 543.
(21) Bischoff, K. B., *Can. J. Chem. Eng.* (1963) **40**, 161.
(22) Danckwerts, P. V., *Chem. Eng. Sci.* (1953) **2**, 1.
(23) Wehner, J. F., Wilhem, R. H., *Chem. Eng. Sci.* (1956) **6**, 89.
(24) Pearson, J. R. A., *Chem. Eng. Sci.* (1959) **10**, 28.
(25) Bischoff, K. B., *Chem. Eng. Sci.* (1961) **16**, 731.
(26) Van Cauwenberghe, A. R., *Chem. Eng. Sci.* (1966) **21**, 203.
(27) Carberry, J. J., Wendel, M., *AIChE J.* (1963) **9**, 132.
(28) Raymond, L. R., Amundson, N. R., *Can. J. Chem. Eng.* (1964) **42**, 173.
(29) Luss, D., Amundson, N. R., *Chem. Eng. Sci.* (1967) **22**, 253.
(30) Luss, D., *Chem. Eng. Sci.* (1968) **23**, 1249.
(31) Hlavacek, V., Hofmann, H., *Chem. Eng. Sci.* (1970) **25**, 173, 187.
(32) Bernard, R. A., Wilhelm, R. H., *Chem. Eng. Progr.* (1950) **46**, 233.
(33) Dorrweiler, U. P., Fahien, R. W., *AIChE J.* (1959) **5**, 139.
(34) Fahien, R. W., Smith, J. M., *AIChE J.* (1955) **1**, 25.
(35) Calderbank, P. H., Pogorsky, L. A., *Trans. Inst. Chem. Eng. (London)* (1957) **35**, 195.
(36) Campbell, T. M., Huntington, R. L., *Hydrocarbon Processing Petrol. Refiner* (1952) **31**, 123.
(37) Coberly, C. A., Marshall, W. R., *Chem. Eng. Progr.* (1951) **47**, 141.
(38) Kwong, S. S., Smith, J. M., *Ind. Eng. Chem.* (1957) **49**, 894.
(39) Kunii, D., Smith, J. M., *AIChE J.* (1960) **6**, 71.
(40) Plautz, D. A., Johnstone, H. F., *AIChE J.* (1955) **1**, 193.
(41) Yagi, S., Kunii, D., *AIChE J.* (1957) **3**, 373.
(42) Hanratty, T. J., *Chem. Eng. Sci.* (1954) **3**, 209.
(43) Yagi, S., Kunii, D., *AIChE J.* (1960) **6**, 97.
(44) Yagi, S., Wakao, N., *Chem. Eng. Sci.* (1959) **5**, 79.
(45) Froment, G. F., *Chem. Eng. Sci.* (1961) **7**, 29.
(46) Froment, G. F., *Ind. Eng. Chem.* (1967) **59**, 18.
(47) Shean Lin Liu, Preprints New Orleans A.I.Ch.E. meeting, March 1969.
(48) Carberry, J. J., White, D., *Ind. Eng. Chem.* (1969) **61**, 27.
(49) Beek, J., *Advan. Chem. Eng.* (1962) **3.**
(50) Kjaer, J., "Measurement and Calculation of Temperature and Conversion in Fixed Bed Catalytic Reactors," Gjellerup, Copenhagen, 1958.
(51) Valstar, J., Ph.D. Thesis, Delft University (1969).
(52) Deans, H. A., Lapidus, L., *AIChE J.* (1960) **6**, 656.
(53) McGuire, M., Lapidus, L., *AIChE J.* (1965) **11**, 85.
(54) Agnew, J. B., Potter, O. E., *Trans. Inst. Chem. Eng. (London)* (1966) **44**, T216.
(55) Froment, G. F., *Gen. Chim.* (1966) **95**, 41.
(56) Gamson, B. W., Thodos, G., Hougen, O. A., *Trans. AIChE* (1943) **39**, 1.
(57) Wilke, C. R., Hougen, O. A., *Trans. AIChE* (1945) **41**, 445.
(58) Taecker, R. G., Hougen, O. A., *Chem. Eng. Progr.* (1949) **45**, 188.

(59) McCune, L. K., Wilhelm, R. H., *Ind. Eng. Chem.* (1949) **41**, 1124.
(60) Ishino, T., Ohtake, T., *Chem. Eng. (Tokyo)* (1951) **15**, 258.
(61) Bar Ilan, M., Resnick, W., *Ind. Eng. Chem.* (1957) **49**, 313.
(62) de Acetis, J., Thodos, G., *Ind. Eng. Chem.* (1960) **52**, 1003.
(63) Bradshaw, R. D., Bennett, C. O., *AIChE J.* (1961) **7**, 50.
(64) Hougen, O. A., *Ind. Eng. Chem.* (1961) **53**, 509.
(65) Yoshida, F., Ramaswani, D., Hougen, O. A., *AIChE J.* (1962) **8**, 5.
(66) Baumeister, E. B., Bennett, C. O., *AIChE J.* (1958) **4**, 70.
(67) Glaser, M. B., Thodos, G., *AIChE J.* (1958) **4**, 63.
(68) Sen Gupta, A., Thodos, G., *AIChE J.* (1963) **9**, 751.
(69) Handley, D., Heggs, P. J., *Trans. Inst. Chem. Eng.* (1968) **46**, 251.
(70) Liu, Shean Lin, Amundson, N. R., *Ind. Eng. Chem., Fundamentals* (1962) **1**, 200.
(71) Liu, Shean Lin, Aris, R., Amundson, N. R., *Ind. Eng. Chem., Fundamentals* (1963) **2**, 12.
(72) Amundson, N. R., *Ber. Bunsen Gesellschaft* (1970) **4**, 90.
(73) Wicke, E., *Acta Technol. Chim. Acad. Nazionale Lincei Varese* (1960) (Sept. 26-Oct. 8, 1960).
(74) Satterfield, C. N., Sherwood, T. K., "The Role of Diffusion in Catalysis," Addison Wesley, 1963.
(75) Satterfield, C. N., "Mass Transfer in Heterogeneous Catalysis," M.I.T. Press, Cambridge, Mass., 1970.
(76) McGreavy, C., Cresswell, D. L., *Can. J. Chem. Eng.* (1969) **47**, 583.
(77) Butt, J. B., Kehoe, P., *Chem. Eng. Sci.* (1970) **25**, 345.
(78) Weiss, P. B., Hicks, J. S., *Chem. Eng. Sci.* (1962) **17**, 265.
(79) Cresswell, D. L., *Chem. Eng. Sci.* (1970) **25**, 267.
(80) McGreavy, C., Thornton, J. M., *Can. J. Chem. Eng.* (1970) **48**, 187.
(81) Luss, D., Lee, J. C., *Chem. Eng. Sci.* (1968) **23**, 1237.
(82) Aris, R., *Chem. Eng. Sci.* (1969) **24**, 149.
(83) Singer, E., Wilhelm, R. H., *Chem. Eng. Progr.* (1950) **46**, 343.
(84) DeWasch, A. P., Froment, G. F., *Chem. Eng. Sci.*, in press.

RECEIVED November 2, 1970.

Contributed Papers

A Mathematical Model for Simulating the Behavior of Fauser-Montecatini Industrial Reactors for Methanol Synthesis

A. CAPPELLI, A. COLLINA, and M. DENTE,[1] Montecatini Edison, Direzione Centrale delle Ricerche, Milano, Italy

This paper describes a mathematical model for simulating industrial reactors of Fauser-Montecatini type for methanol synthesis. These reactors consist of quasi-adiabatic catalytic layers, with indirect cooling between one layer and the next, and of a final non-adiabatic layer with tubes immersed in the catalyst. The model has the main features discussed below.

The chemical reactions considered are:

$$CO + 2H_2 \rightleftarrows CH_3OH$$

$$CO_2 + H_2 \rightleftarrows CO + H_2O$$

The kinetic equation of the first reaction, studied by Natta *et al.* (*1, 2, 3, 4, 5*) has been modified on the basis of new experimental data to take into account the presence of CO_2 in the reacting gas and is of the type:

$$R_1 = \frac{f_{CO}P_{CO}f^2_{H_2}P^2_{H_2} - f_{CH_3OH}P_{CH_3OH}/K_{eq}}{A^3(1 + Bf_{CO}P_{CO} + Cf_{H_2}P_{H_2} + Df_{CH_3OH}P_{CH_3OH} + Ef_{CO_2}P_{CO_2})^3}$$

Under synthesis conditions the rate of the second reaction is relatively high; hence, it has been assumed that for this reaction the conditions of thermodynamic equilibrium are reached at the catalyst surface. This reaction is significant, particularly in cases where the gas flowing into the reactor has a relatively high CO_2 content according to the present tendency for industrial plants.

Owing to the presence of this reaction, the phenomena of mass and heat transfer between the gas and the surface of the catalyst, which would be negligible on the basis of the main reaction only, are important and thus taken into account. Deviations from ideal behavior are estimated

[1] Politecnico di Milano, Ist. di Chimica Industriale, Milano, Italy.

by introducing fugacity coefficients and using suitable corrections in calculating the enthalpies of the mixtures (*6, 7, 8, 9, 10*). The effects of the diffusion of reagents and products within the porous catalyst are calculated without applying the simplified estimates generally used for complex kinetic reactions (*11, 12, 13, 14*). In this case, in view of the presence of a reaction influenced by the equilibrium and since low efficiency factor values are expected, it was considered preferable to evaluate this coefficient by solving numerically the differential equation which describes the diffusion phenomena. This equation, in integral form, has been solved by successive approximations, introducing a suitable normalizing factor to hasten convergence (*15, 16*). On the basis of the mathematical model described a calculation program in Fortran V has been prepared for a Univac 1108 computer. The program, set down according to modular technique, can be modified easily to simulate reactors in which cooling between one layer and the next is achieved with cold reacting gas. The calculation time for checking a reactor with four adiabatic plus one non-adiabatic layer is about 1 minute. This calculation program is currently used to check the operation of synthesis reactors in various Montecatini-Edison plants. Moreover, it may be used conveniently for design calculations.

We report experimental and calculated data of a test carried out in a small capacity industrial reactor with the following operating conditions:

Total feed flow-rate:	26245 Nm^3/hr	
Pressure:	254 atm	
Composition of the gas at the reactor inlet:	CO	11.20%
	CH_3OH	0.11
	H_2	65.46
	H_2O	0.15
	CH_4	13.75
	N_2	7.72
	CO_2	1.60

Inlet temperature to the	1st layer:	335 °C
	2nd layer	369
	3rd layer	368
	4th layer	364
	last layer	368

The compositions of the gas going out from each layer, compared with the experimental values, are given in Table I.

The production measured experimentally is 25.6 tons/day, and the calculated production is 25.7 tons/day. The model appears capable of supplying results which agree well with the experimental data, and in particular the hypothesis of considering the conversion reaction at equi-

Table I. Calculated and Experimental Gas Composition Values

		Layer				
		1st	2nd	3rd	4th	5th
CO	exp.	10.85	10.46	10.05	9.64	9.15
	calc.	10.99	10.70	10.22	9.65	9.11
CH_3OH	exp.	1.11	1.61	2.14	2.55	n.d.
	calc.	0.93	1.48	2.15	2.90	3.15
CO_2	exp.	1.08	1.09	1.11	1.17	1.33
	calc.	1.20	1.09	1.07	1.07	1.43
CH_4	exp.	13.98	14.10	14.25	14.50	14.65
	calc.	13.98	14.13	14.31	14.52	14.59

librium and the modification made to Natta's kinetic equation appear justifiable.

This model can be improved along the following lines, which are now being pursued:

(a) Introduction of empirical kinetic equations which describe the formation of methane and dimethyl ether.

(b) Modifications to perform reactor simulations with cooling by direct injection of fresh gas.

Finally, the same model can be used to perform optimization calculations in the industrial reactor design stage.

(1) Natta, G., Pino, P., Mazzanti, G., Pasquon, I., *Chim. Ind.* (1953) **35**, 705.
(2) Natta, G., "Catalysis," P. H. Emmet, Ed., Vol. 3, p. 345, Reinhold, New York, 1955.
(3) Natta, G., Mazzanti, G., Pasquon, I., *Chim. Ind.* (1955) **37**, 1015.
(4) Cappelli, A., Dente, M., *Chim. Ind.* (1965) **47**, 1068.
(5) Pasquon, I., Dente, M., *J. Catalysis* (1962) **1**, 508.
(6) Perry, J., "Chemical Engineers Handbook," McGraw-Hill, New York, 1963.
(7) Hougen, O., Watson, K., "Chemical Process Principles," Wiley, New York, 1959.
(8) Hamming, R. W., "Numerical Methods for Scientists and Engineers," McGraw-Hill, New York, 1962.
(9) Newton, R., *Ind. Eng. Chem.* (1935) **27**, 302.
(10) Clayton, J., Gianque, W., *Ind. Eng. Chem.* (1932) **54**, 2610.
(11) Kramers, H., Westerterp, K. R., "Elements of Chemical Reactor Design and Operation," Chapman and Hall, London, 1963.
(12) Weeler, A., *Advan. Catalysis* (1957) **3**, 249.
(13) Weiss, P., Prater, C., *Advan. Catalysis* (1954) **6**, 143; (1957) **9**, 957.
(14) Aris, R., *Chem. Eng. Sci.* (1957) **6**, 265.
(15) Ralston, A., Wilf, H., "Mathematical Methods for Digital Computers," Wiley, New York, 1960.
(16) Dente, M., Biardi, G., Ranzi, E., *Ing. Chim. It.*, in press.

Phthalic Anhydride Production by the Catalytic Oxidation of *o*-Xylene

P. H. CALDERBANK and A. D. CALDWELL, Department of Chemical Engineering, University of Edinburgh, Mayfield Rd., Edinburgh, Scotland

Commercial catalyst was obtained with the following specification: promoted V_2O_5 on SiC support; mean particle diameter, 0.6 cm; surface area, *ca.* 0.2 m^2/gram. This was sometimes diluted with inert alundum spheres of the same size before being loaded into the fixed bed reactor described below. Kinetic measurements were made using the spinning catalyst basket (CSTR) reactor. The catalyst was evaluated (with and without dilution by inert support material) in a stainless steel tubular reactor of 1-inch internal diameter and length 6 ft; this was submerged in an isothermal air-fluidized sand bath, provided with controlled electrical heating. The reactor contained many thermocouples along its length at the axis. *o*-Xylene was pumped and metered into a preheated air stream entering the reactor, and the product stream leaving the reactor was cooled to allow the phthalic anhydride to be collected, weighed, and analyzed. The range of variables studied was as follows,

air flow rates	0.84–1.26 ft^3/minute at room temperature
o–xylene, vol %	0.59–1.35
bath temperatures,	370–418°C.

Results and Discussion

The fixed bed reactor showed the phenomena of (a) ignition to a high temperature steady state where phthalic anhydride was abundantly produced and (b) hysteresis when the bath temperature was reduced since extinction occurred below the ignition point when a slow reaction unprolific of phthalic anhydride was established. This behavior suggests that mass and heat transfer are influential as discussed in detail later.

The temperature profiles in the fixed bed reactor showed that reaction was nearly complete at a short distance from the inlet. Thus, the reactor was substantially infinite in length and gave no kinetic information other than the selectivity, which remained remarkably constant at 58–60%, being independent of both temperature or flow rate. This behavior was also found by Bhattacharyya and Gulati (*1*), who record a yield of 55–60%, and is close to the value previously predicted by us (*2*) from Froment's kinetics.

Yields of phthalic anhydride obtained with the isothermal spinning catalyst basket reactor were much lower than those with the fixed bed, suggesting that an increasing temperature sequence might be optimal in contrast to our conclusions (2) arrived at on the basis of Froment's kinetics which favor nearly isothermal operation. The reaction temperature sequence in the fixed bed can be varied by diluting the catalyst with inert particles and by changing the air–hydrocarbon ratio. When these operations were carried out, the yield of phthalic anhydride remained unaltered, showing that the energies of activation for the production of phthalic anhydride and combustion of *o*-xylene are the same while the combustion of phthalic anhydride occurs to a negligible extent.

Theory

The steady-state "redox" model of Mars and van Krevelen (3) arose principally because the rate of oxidation of aromatic hydrocarbons over vanadium pentoxide catalysts of several formulations is directly proportional to the oxygen partial pressure and almost independent of the type of the type of aromatic hydrocarbon or its partial pressure (*4, 5, 6*).

The "redox" model is simply stated,

$$r = kP_S(1 - \theta) = \frac{k^*}{\beta} P_{O_2}\theta \tag{1}$$

Hence, the rate of hydrocarbon oxidation depends on the fraction of catalyst surface in the oxidized state which is determined by the steady-state rates of reduction and oxidation. From the above,

$$\theta = \frac{kP_S}{kP_S + \dfrac{k^*}{\beta} P_{O_2}} \tag{2}$$

giving,

$$r = \frac{1}{\dfrac{\beta}{k^* P_{O_2}} + \dfrac{1}{kP_S}} \tag{3}$$

showing that $r \to \dfrac{k^*}{\beta} P_{O_2}$ when $k \gg \dfrac{k^*}{\beta}$ (4)

the result which has been observed experimentally. The energy of activation associated with k^* has been reported as about 40 kcal/mole (*7, 8*), while that associated with k is much less and probably of the order of 27 kcal/mole. The condition $k >> k^*/\beta$ apparently arises at the relatively low temperatures used experimentally and could conceivably be reversed at the local high temperatures encountered in nonisothermal industrial fixed bed reactors. However, at such high reaction rates, the rate limitation imposed by the finite rate of mass transfer of hydrocarbon to the catalyst surface is likely to become important as also, by analogy, is the rate of heat transfer from the catalyst. These rates are given by,

$$r = k_g a(P_B - P_S) = \frac{ha}{\Delta H}(T_S - T_B)$$

where

$$T_S - T_B = \frac{(k_g \Delta H)}{(h)}(P_B - P_S) \tag{5}$$

We have an almost impervious catalyst particle and hence no diffusion or reaction in pores.

Substituting Equation 5 into Equation 3, and setting $R_1 = \frac{k}{k_g a}$; $R_2 = \frac{k^*}{k_g a}$,

$$\frac{1}{\dfrac{1}{R_1\left[P_B - \dfrac{T_S - T_B}{\left(\dfrac{k_g \Delta H}{h}\right)}\right]} + \dfrac{\beta}{R_2 P_{O_2}}} = \frac{T_S - T_B}{\left(\dfrac{k_g \Delta H}{h}\right)} = \phi \tag{6}$$

When Equation 6 is solved numerically, it is found that multiple steady-state solutions are sometimes possible. Thus when the L.H.S. is calculated from typical data for a series of fixed values of T_B as a function of $T_S - T_B$ and plotted, a corresponding series of curves is obtained which intersect the plot of the R.H.S. against $(T_S - T_B)$ at either one or sometimes three points (*see* Figure 1). In the former case, at high values of T_B, "ignition" is obtained at a high reaction rate which is determined almost completely by the hydrocarbon mass transfer rate when,

$$T_S - T_B \rightarrow \frac{k_g \Delta H}{h} P_B \text{ as } P_S \rightarrow 0$$

and the catalyst particle is some 200°C above its environmental temperature, T_B.

As T_B is reduced, three solutions are possible: a relatively slow reaction rate-controlled steady state where $T_S \rightarrow T_B$, an unstable intermediate state, and the high temperature physical rate controlled state previously mentioned.

When T_B is further reduced, only the relatively slow reaction rate-controlled steady state is possible corresponding to relative "extinction."

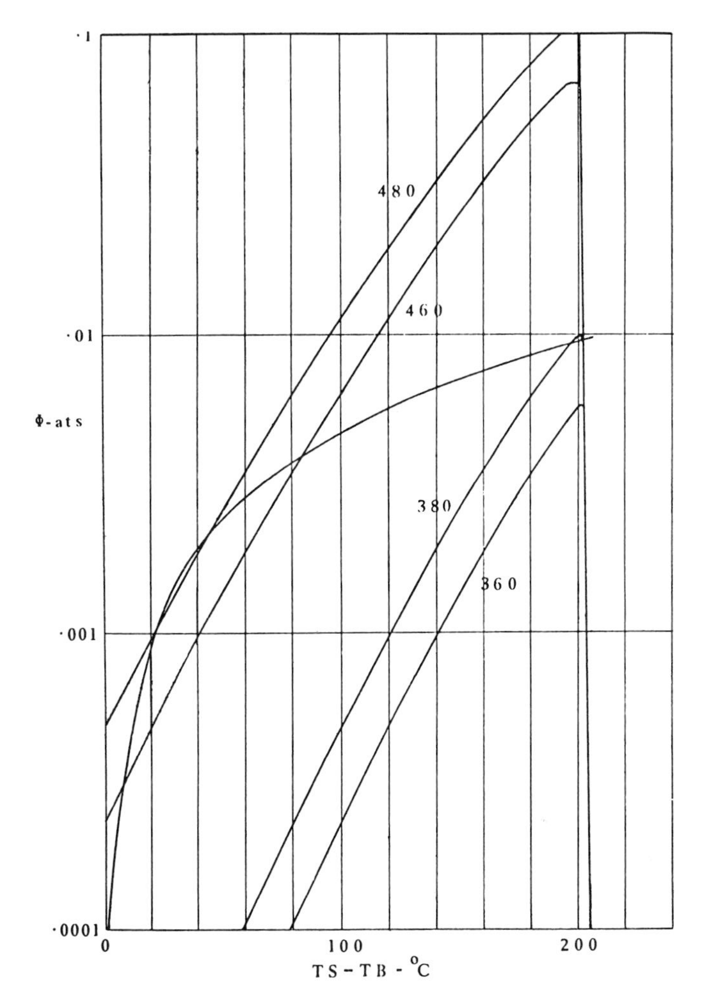

Figure 1. Multiple states for a single catalyst pellet

P_{O_2} = *0.210 atm;* P_B = *0.0097 atm;* ΔH = *359,300 cal/gram mole*
S = *5.475 × 10*[6] *gram moles/gram-sec-atm;* A = *27,450*
S* = *4.143 × 10*[6] *gram moles/gram-sec-atm;* A* = *4000*
$k_g a$ = *0.00143 gram mole/gram-sec-atm;* h/k_g = *17.0 cal atm/ gram mole °C*

Thus the phenomena of ignition, extinction, and hysteresis as observed here with the fixed bed reactor are predicted by the "redox" model with an imposed mass transfer limitation.

The application of this theory to the results with the spinning basket reactor follow since with this apparatus, physical resistances are removed and the rate of a chemical step is observed. The latter rate would be that previously observed using isothermal, differential plug-flow reactors and believed to be the rate of reoxidation of the catalyst (Equation 4). When this rate-limiting step obtains, $\theta \rightarrow 1$, and there is evidence that this reduced form of vanadium oxide is ineffectual for producing phthalic anhydride, favoring instead complete combustion of *o*-xylene (9). This is observed to be the case with air as oxidant, but when the rate of catalyst reoxidation is substantially increased by using pure oxygen in the spinning basket reactor at high temperatures, the rate-limiting step presumably changes to the rate of surface oxidation of *o*-xylene, $\theta \rightarrow 0$, and high yields of phthalic anhydride are obtained. In theory this desirable state is obtained in any reactor system when a rate-limiting step, other than surface reoxidation, prevails.

Nomenclature

Any consistent system of units may be used.

a — area of a catalyst particle
ΔH — heat of reaction per mole of *o*-xylene reacted
h — heat transfer coefficient from catalyst surface to bulk gas phase
k — surface reaction velocity constant for *o*-xylene oxidation
k_g — mass transfer coefficient of *o*-xylene at the catalyst surface
k^* — reaction velocity constant for surface reoxidation
P_B — partial pressure of *o*-xylene in the bulk fluid
P_{O_2} — partial pressure of oxygen
P_S — partial pressure of *o*-xylene at the catalyst surface
r — rate of reaction of *o*-xylene per unit mass of catalyst
T_B — temperature in the bulk fluid
T_S — temperature at the catalyst surface
A — energy of activation } $k = se^{-A/RT}$
S — pre-exponential factor }
β — moles O_2 consumed per mole of *o*-xylene reacted
θ — fraction of catalyst surface in the reduced state
ϕ — *see* Equation 6

(1) Bhattacharyya, S. K., Gulati, I. B., *Ind. Eng. Chem.* (1958) **50,** 1719.
(2) Caldwell, A., Calderbank, P. H., *Brit. Chem. Eng.* (1969) **14,** 1199.
(3) Mars, J., van Krevelen, D. W., *Chem. Eng. Sci.* (1954) **3,** 41.
(4) Calderbank, P. H., *Ind. Chemist* (1952) 291.
(5) Shelstad, K. A., Downie, J., Graydon, W. F., *Can. J. Chem. Eng.* (1960) **38,** 102.
(6) Mann, R. F., Downie, J., *Can. J. Chem. Eng.* (1968) **46,** 71.
(7) Simard, G. L., Steger, I. F., Arnott, J., Siegel, L. A., *Ind. Eng. Chem.* (1955) **47,** 1424.
(8) Ross, G. L., Calderbank, P. H., in press.
(9) Hughes, M. F., Adams, R. T., *J. Phys. Chem.* (1960) **64,** 781.

Radiant Analysis in Packed Bed Reactors

D. VORTMEYER, Technische Universität München, 8 München 13 Tengstrasse 38, München, West Germany

While it seems acceptable to neglect radiative transfer in homogeneous gaseous reactors, the situation is different in packed beds. A large amount of heat is emitted by the surface of the solid particles, particularly at higher temperatures, since the emitted energy is proportional to T^4. Therefore, the temperature gradients in the reaction zones of exothermic reactions lead to radiative fluxes as they produce conductive fluxes. For radiation the magnitude of the fluxes depends mainly on temperature level and surface emissivity. In the field of heat transfer several authors have been concerned with radiative transfer in packed beds. Since their work has been appreciated in earlier papers by the author, the important contributions are mentioned without discussion (*1, 2, 3, 4, 5, 6, 7*).

Equations for Radiative Transfer and Energy Conservation

A one-dimensional treatment of radiative transfer must include absorption, emission, and reflection or scattering. The previously developed theory (*8*) includes these effects. In this theory the radiative flux Q_r is calculated from the partial energy fluxes I and K

$$Q_R = I - K \tag{1}$$

$$\frac{dI}{dx} = -aI + b\sigma T^4 + gK \tag{2}$$

$$\frac{dK}{dx} = aK - b\sigma T^4 - gI \tag{3}$$

These fluxes are assumed to transverse the catalyst bed from the left to the right (I) and vice-versa (K). The difference between I and K at any point of the reactor leads to the net flux Q_R (*9, 10*).

One of the main achievements of the author's analysis is the fact that expressions for a, b, and g were derived so that those coefficients can be calculated from fixed bed data.

$$a = \frac{(1+B)^2 + (1+B)(1-B)(1-\varepsilon)^2}{(1+B)^2 - (1-B)^2(1-\varepsilon)^2} \frac{(1-B)2}{(1+B)d} \tag{4}$$

$$b = \frac{(1+B)^2 - (1+B)(1-B)(1-\varepsilon)}{(1+B)^2 - (1-B)^2(1-\varepsilon)^2} \varepsilon \frac{(1-B)2}{(1+B)d} \tag{5}$$

$$g = \frac{(1+B)^2 + (1+B)(1-B)}{(1+B)^2 - (1-B)(1-\varepsilon)^2} (1-\varepsilon) \frac{(1-B)2}{(1+B)d} \tag{6}$$

Coefficients a, b, and g depend on the surface emissivity ϵ; for example, the reflectivity, g, disappears if $\epsilon = 1$. Further, the coefficients depend on B, a dimensionless number which is called radiation transmission number. This number had to be introduced because radiation penetrates the bed by the void volume. The numerical value of B may be taken as *ca.* 0.1 for average conditions (*11*).

From photon theory $1/a$ can be considered the average mean beam length Δ for photons. For an emissivity $\epsilon = 1$ and a particle diameter of $d = 0.5$ cm a numerical value of $\Delta = 0.3$ cm is obtained. The radiative transport Equations 2 and 3 are derived for the homogeneous model of a packed bed. The equations for a discrete model may be taken from Ref. *8* or *12*. For the simple model of an exothermic fixed bed reactor the steady-state energy balance may be written as

$$\lambda_{ax} \frac{d^2T}{dx^2} - u\rho c_p \frac{dT}{dx} - \frac{d(I - K)}{dx} + |\Delta H\dot{r}| = 0 \tag{7}$$

where I and K are given by Equations 2 and 3 and $r = r\ (c_i, T)$. The conservation equations of the chemical species are not included since they are well known and radiation does not affect them directly.

Steady-State Boundary Conditions

The combined solution to Equations 2, 3, and 7 requires four boundary conditions. Following the normal procedure for evaluating these conditions, Equations 8 to 11 are obtained.

Entrance section: $x = O_+$

Energy balance:

$$u\rho\bar{c}_p(T_{o+} - T_\infty) + (I_{o+} - K_{o+}) - \lambda_{ax}\left(\frac{dT}{dx}\right)_{o+} = 0$$

or

$$\left(\frac{dT}{dx}\right)_{o+} = \frac{Pe}{d}\,(T_{o+} - T_\infty) + \frac{I_{o+} - K_{o+}}{u\rho\bar{C}_p\lambda_{ax}} \tag{8}$$

$$I_{o+} \text{ known} \tag{9}$$

Exit section: $x = L$

Energy balance:

$$\int_{O_+}^{L_-} |\Delta H\dot{r}|\, dx = u\rho\bar{c}_p\,(T_{L-} - T_\infty) + (I_{L-} - K_{L-}) - \lambda_{ax}\left(\frac{dT}{dx}\right)_{L-} \tag{10}$$

$$K_{L-} \text{ known} \tag{11}$$

Without radiation, Equations 8 and 10 lead to the well-known expressions

$$\left(\frac{dT}{dx}\right)_{o+} = \frac{Pe}{d}(T_{o+} - T_{\infty}) \text{ and } \left(\frac{dT}{dx}\right)_{L-} = 0$$

These conditions depend only on local gradients and properties. The inclusion of the radiative fluxes I_{o+} and K_{L-} introduce some changes since these terms depend not only on local properties but also on longer range transmission effects. Thus, I_{o+} and K_{L-} depend on the statement of the problem and must be evaluated for each particular case—*e.g.*, if the catalyst particles are imbedded between inactive particles with the emissivity ϵ,

$$I_{o+} = \varepsilon\sigma T_{o+}^4 + (1 - \varepsilon)K_{o+}$$

and

$$K_{L-} = \varepsilon\sigma T_{L-}^4 + (1 - \varepsilon)I_{L-}$$

are not the precise but quite good assumptions. Further, Equation 10 usually can be simplified for an optically thick medium as a packed bed. Since then long range effects of radiation are absent, T_{L-} may be regarded as T_{ad}, and Equation 10 simplifies to:

$$(I_{L-} - K_{L-}) \equiv \lambda_{ax}\left(\frac{dT}{dx}\right)_{L-} = 0 \quad (12)$$

Solutions of Equations 2, 3, and 8 together with the boundary conditions are presented in Figure 1, where all three possible steady-state solutions are presented with and without radiation for an Arrhenius type reaction function.

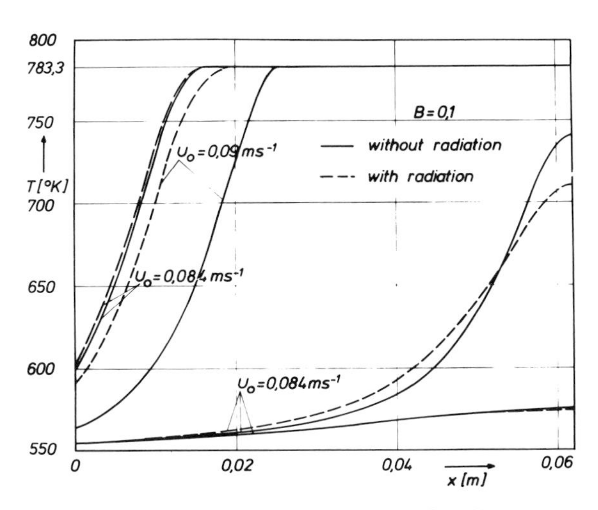

Figure 1. Boundary conditions and solutions to Equations 2, 3, and 8

Pronounced effects are observed in two cases:

(1) The increase of heat diffusion by radiation decreases the end temperature of the unstable middle solution.

(2) At a velocity of 0.09 $msec^{-1}$ the upper solution is fixed at the reactor entrance. Without radiation it is just on the verge of being blown to the reactor end.

Fourier-Type Transport Equation for Radiation

A rearrangement (*8*) of Equations 2 and 3 results in the following expression

$$Q_R = \frac{8}{a+g}\sigma T^3 \frac{dT}{dx} + \frac{1}{a^2 - g^2}\frac{d^2Q_R}{dx^2} \tag{13}$$

If the second term on the right is zero or much smaller than the first term on that side, Q_R may be calculated for the steady state from a Fourier-type equation

$$Q_R = -\lambda_R \, dT/dx \tag{14}$$

with

$$\lambda_R = \frac{2B + \varepsilon(1-B)}{2(1-B) - \varepsilon(1-B)} 4\sigma T^3 d = \psi 4\sigma T^3 d \tag{15}$$

If there are no chemical reactions, calculations (*12*) show that there are no dramatic changes of d^2Q_R/dx^2 and $|Q_R''/(a^2 - g^2)| << |\lambda_R dT/dx|$. However, for the upper solutions in Figure 1 with steep temperature gradients, Figure 2 shows that there might be an error of up to 20% if the radiation flux is calculated from Equation 14. However, the radiation

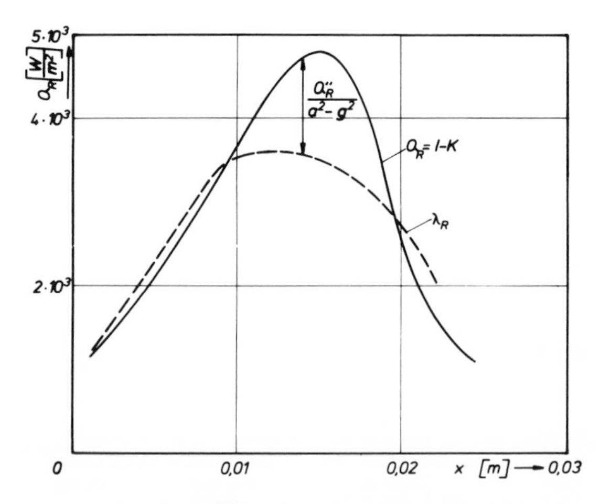

Figure 2. Possible error in using Equation 14 to calculate radiation flux (see text)

conductivity approximation reduces the problem of *bc*'s considerably since then the conventional conditions are valid with $(\lambda_{ax} + \lambda_R)$ instead of λ_{ax}.

Figure 3 shows plots of the ψ factor as a function of ϵ and B and compares them with results of other authors. The author's analysis covers most of the previous results as special cases.

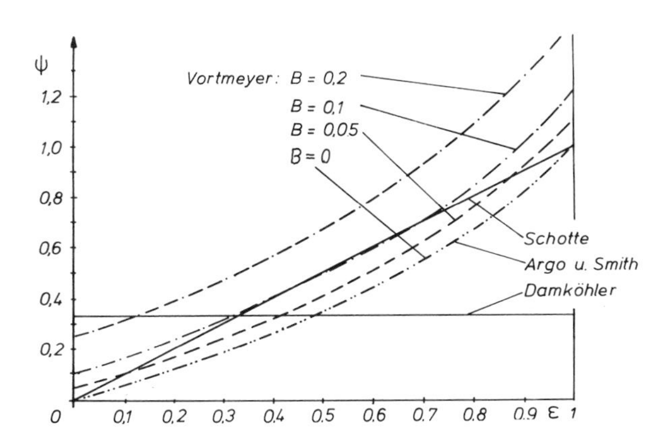

Figure 3. ψ as a function of ε and B *for this work and that of others*

Moving Reaction Zones

From the solution of the unsteady-state energy equation the moving velocity of reaction zones could be calculated with and without the influence of radiation. Results obtained so far agree well with experimental work on this subject.

(1) Damköhler, G., "Der Chemie-Ingenieur," Bd. III, 1. Teil, pp. 445–6, Akad. Verlagsgesellschaft mbH, Leipzig, 1937.
(2) Argo, W. B., Smith, J. M., *Chem. Eng. Progr.* (1953) **49,** 443.
(3) Yagi, S., Kunii, D., *A.I.Ch.E. J.* (1957) **3,** 373.
(4) Schotte, W., *A.I.Ch.E. J.* (1960) **6,** 63.
(5) Hill, F. B., Wilhelm, R. H., *A.I.Ch.E. J.* (1959) **5,** 486.
(6) van der Held, E. F. M., *Appl. Sci. Res.* (1952) **A3,** 237.
(7) Chen, J. C., Churchill, S. W., *A.I.Ch.E. J.* (1963) **9,** 35.
(8) Vortmeyer, D., "Fortschritt-Berichte VDI-Zeitung," Reihe 3, Nr. 9 (1966); Habilitation vom 28.7.1965; *Chem. Ing. Tech.* (1966) **38,** 404.
(9) Schuster, A., *Astrophys. J.* (1905) **21,** 1.
(10) Hamaker, H. C., *Phillips Res. Rept.* (1947) **2,** 55, 103.
(11) Vortmeyer, D., Börner, C. J., *Chem. Ing. Tech.* (1966) **38,** 1077.
(12) Vortmeyer, D., *Ber. Bunsenges. Phys. Chem.* (1970) **74,** 127.
(13) Wicke, E., Vortmeyer, D., *Z. Elektrochem., Ber. Bunsenges. Phys. Chem.* (1959) **63,** 145.

Experimental Evaluation of Dynamic Models for a Fixed-Bed Catalytic Reactor

J. A. HOIBERG, B. C. LYCHE, and A. S. FOSS, Department of Chemical Engineering, University of California, Berkeley, Calif.

Several dynamic experiments were performed on a laboratory fixed bed reactor to evaluate the importance of the various physical and physicochemical phenomena influencing reactor behavior. The reactor's observed frequency response was compared with that predicted by three different mathematical models. The models comprised both one- and two-dimensional continuum representations, one of which accounted for resistance to intraparticle transport of reactant. By this approach it was possible to investigate the importance of such effects as heat transport processes, nonlinear phenomena, and intraparticle transport resistance.

The highly exothermic reaction between hydrogen and oxygen was carried out in a 1-inch diameter glass tube reactor 20 inches long. The catalyst was platinum and was deposited in small amounts on silica gel granules about 0.5 mm in diameter. Typically, the reactor was operated with a feed of 1 mole % oxygen and 99% hydrogen at 100°C. Dynamic experiments were performed by perturbing the feed concentration or feed temperature in a sinusoidal manner. Temperature and oxygen concentration were measured every two inches along the reactor centerline and occasionally across the radius.

In experiments in which the peak-to-peak temperature excursion reached 114°C within the reactor, we observed definite distortions of the sinusoids, but these were very simple in character. The peaks of the sinusoids were either flattened or sharpened, and the waves were slightly skewed, with the back side the steepest. The distortion of the peaks can be traced to the convexity of the reaction rate-temperature function; the origin of the skewness was not investigated.

No noticeable distortion in the sinusoids was observed when peak-to-peak temperature excursions everywhere in the bed were less than 30°C. Some typical results of experiments under those conditions are shown in Figure 1. Amplitude and phase of both the temperature and concentration waves resulting from temperature forcing are displayed as a function of position in the bed at three forcing frequencies. Similar data were obtained for the forcing of feed concentration but are not presented here. These results show at a glance the complex behavior of disturbances in this type of reactor. A detailed interpretation of such behavior has been given (*1*). Figure 1 also shows that the two-dimensional model investigated is remarkedly descriptive of the physical process. A one-dimensional model was found to do almost as well.

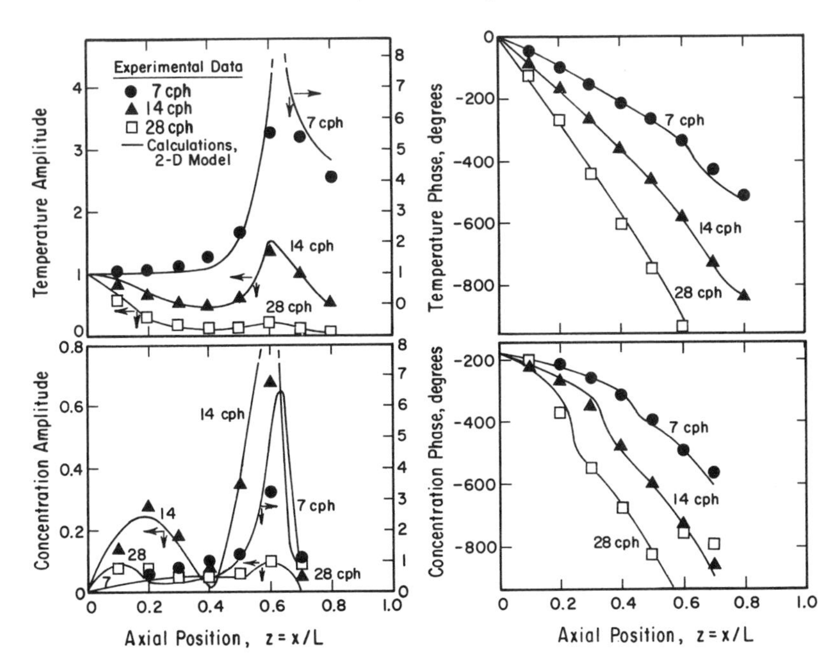

Figure 1. Amplitude and phase behavior of temperature and concentration waves at the centerline: temperature forcing

All three mathematical models investigated were of the plug-flow, locally linear, continuum type. The two-dimensional model, whose predictions of reactor behavior are shown in Figure 1, accounted for the radial transport of heat and mass by diffusive processes, radial velocity profile, variable gas density, pressure variations, thermal capacitance of catalyst and reactor wall, and finite rates of heat exchange with these capacitances. In this model, the concentration and temperature within the catalyst particles were considered uniform. The influence of intraparticle concentration gradients was investigated, however, by studying a second model. This model was one dimensional, but otherwise retained the features of the two-dimensional model. The simplest model considered was one-dimensional with no account taken of variable gas density or gradients within the catalyst particles.

A study of these models and comparison of calculations with experiments have led us to the following conclusions.

(1) The only dynamic element of consequence in these reactors is the transport of heat at finite rates to and from the stationary thermal capacitances. One must therefore take care in modeling the thermal dynamics of the bed. The other physical and physicochemical processes

interact with and are influenced by the heat transport process in a way that may be considered nondynamic.

(2) Locally linear, plug-flow models are remarkably descriptive of the principle dynamic characteristics of these reactors, even with temperature excursions that cause a doubling of the chemical reaction rate (30°C in these experiments). Nonlinear effects that appear for larger excursions are simple in nature. These models require accurate modeling of the steady state, which implies care in representing heat exchange with surroundings and heat generation by reaction.

(3) Evidence of intraparticle diffusion limitations was found in the most reactive locations of the bed. However, these effects under dynamic conditions can be treated in a purely quasi-static manner.

(4) A major concern in choosing between a one- and a two-dimensional model is accuracy in representing the steady-state profiles. A two-dimensional representation seems necessary when radial temperature differences exceed 20% of the absolute tempreature level. The dynamic behavior in these experiments was represented well by a one-dimensional model because radial temperature gradients were small.

(5) The influence of a nonuniform fluid velocity across the bed radius was found to be negligible. Under dynamic conditions, coupling between temperature and fluid velocity owing to density changes was also found to be negligible.

(1) Sinai, J., Foss, A. S., *A.I.Ch.E. J.* (1970) **16**, 658.

Fixed Bed Reactor Analysis by Orthogonal Collocation

BRUCE A. FINLAYSON, Department of Chemical Engineering, University of Washington, Seattle, Wash. 98105

Consider the equations governing a fixed bed catalytic reactor under the assumptions of constant physical properties and plug flow.

$$\frac{\partial c}{\partial z} = \alpha \frac{1}{r} \frac{\partial}{\partial r}\left(r \frac{\partial c}{\partial r}\right) + \beta\, R(c, T) \tag{1}$$

$$\frac{\partial T}{\partial z} = \alpha' \frac{1}{r} \frac{\partial}{\partial r}\left(r \frac{\partial T}{\partial r}\right) + \beta'\, R(c, T)$$

$$\frac{\partial c}{\partial r} = 0, \; -\frac{\partial T}{\partial r} = Bi(T - T_w) \text{ at } r = 1$$

where $\alpha = Ld_p/R^2Pe_M$, $\alpha' = Ld_p/R^2Pe_H$, L and R are the length and radius of the reactor, d_p is the particle diameter, $Pe_M = Gd_p/D_e\rho$ and $Pe_H = Gd_pC_p/k_e$ are the Peclet numbers for mass and heat, G is the mass flow rate, D_e and k_e are the effective diffusivity and conductivity. The Biot number h_wR/k_e is defined in terms of the effective thermal conductivity and tube radius, whereas the Nusselt number, h_wd_p/k_f is defined in terms of the particle diameter and fluid thermal conductivity.

The orthogonal collocation method is applied as outlined by Villadsen and Stewart (*1*) and Ferguson and Finlayson (*2*). It reduces the system of partial differential equations (*1*) to a set of ordinary differential equations in z, which are integrated numerically.

$$\frac{dc_j}{dz} = \alpha \sum_{i=1}^{N+1} B_{ji}c_i + \beta R_j \qquad j = 1,2, \ldots, N$$

$$\frac{dT_j}{dz} = \alpha' \sum_{i=1}^{N+1} B_{ji}c_i + \beta' R_j \qquad (2)$$

$$-\sum_{i=1}^{N+1} A_{N+1,i}T_i = Bi(T_{N+1} - T_w)$$

$$\sum_{i=1}^{N+1} A_{N+1,i}c_i = 0$$

where T_j and c_j are the temperature and conversion at the collocation points and the matrices A and B are easily calculated (*1*). In the first approximation the equations are identical to a lumped parameter model (no radial variations in c and T) with the equivalent heat transfer coefficient, U.

$$\frac{1}{Nu'} = \frac{1}{2\alpha'}\left(\frac{1}{Bi} + \frac{1}{3}\right) \qquad (3)$$

$$\frac{1}{U} = \frac{1}{h_w} + \frac{R}{3k_e} \text{ (Jacobi)} \qquad (4)$$

$$\frac{1}{U} = \frac{1}{h_w} + \frac{R}{4k_e} \text{ (Legendre)} \qquad (5)$$

where $Nu' = 2UL/GC_pR$. If Legendre polynomials (*2*) are used, in the first approximation the concentration is constant in r, the temperature is parabolic in r, and the reaction rate is evaluated at the average temperature. The equivalence (5) has been shown before for a parabolic temperature profile (4, 5, 6, 7, 8, 9).

The ratio of $1/Bi$ compared with $1/Bi + 1/3$ gives the percentage of the thermal resistance occurring at the boundary: $Bi = 1$, 75%; $Bi = 20$, 13%. Thus for small Biot numbers most of the resistance occurs at the wall, and only a one-dimensional treatment is necessary. Since the one-dimensional model corresponds to the first approximation in the collocation method, as Bi increases, the number of terms in the approximate solution can be increased to account for the two-dimensional nature of Equation 1.

Numerical computations were made for $Bi = 1$ or 20, $\alpha = \alpha' = 1$, $\beta = 0.3$, $\beta' = 0.2$, $R = (1 - c) \exp(\gamma - \gamma/T)$, $\gamma = 20$, $T_w = 0.92$ or 1. When $Bi = 1$, the first approximation predicted the hot spot within 2%. The Jacobi polynomials gave better results than Legendre polynomials in the first approximation, whereas the Legendre results converged faster (in higher approximations). A two-term collocation solution was as accurate and four times as fast as a six-term finite difference solution. For $Bi = 20$ the first approximation was less suitable, but a four-term collocation solution was as accurate and three times as fast as an eleven-term finite difference solution. When using 200 steps in the z direction, four collocation points, and the Runge-Kutta method of integrating the ordinary differential equations, the computation time on the CDC-6400 computer was 2.1 seconds.

Computations were also done using the experimental velocity profiles of Schwartz and Smith (10) for a 4-inch diameter tube, 80 inches long, packed with 5/32-inch diameter spheres. The effective diffusivity and conductivity were calculated using

$$G(r)d_p/D(r)\,\rho = 10 \tag{6}$$

$$k_e/k_f = 8.1 + 0.09\,Re\,Pr, \quad Re = G(r)d_p/\mu \tag{7}$$

$$k_e/k_f = 6.9 + 0.01\,Re\,Pr, \text{ within } \tfrac{1}{2}\,d_p \text{ of wall} \tag{8}$$

Equation 6 and the form of Equation 7 have been verified experimentally by Schertz and Bischoff (11), while Equations 7 and 8 are predicted using the methods of Baddour and Yoon (3). Experimental comparisons are given by these authors and Yagi and Kunii (12).

When using Equation 8, the heat transfer resistance at the wall is accounted for by the lower value of k_e rather than a heat transfer coefficient. During the calculations, a Biot number was calculated using

$$Bi = \frac{-\,[\partial T/\partial r]_{r=1}}{\langle T\rangle - T_w} \tag{9}$$

where <T> is the average temperature within one-half particle diameter of the wall. This distance was chosen because the void fraction rises rapidly from 0.4 to 1.0 there. This *Bi* varied only a few percent down the bed, and typical values are given in Table I. This *Bi* was then used in

Table I. Typical *Bi* Values

Re	$k_e(1)/k_{e,avg}$	Bi(predicted)
375	0.23	15
189	0.35	20
85	0.52	29

the model with a flat velocity profile and heat transfer resistance (Equation 1), and good agreement was obtained between the two models. This suggests that it may be possible to predict the heat transfer coefficient which causes the two mathematical models to agree.

(1) Villadsen, J. V., Stewart, W. E., *Chem. Eng. Sci.* (1967) **22**, 1483.
(2) Ferguson, N. B., Finlayson, B. A., *Chem. Eng. J.* (1970) **1**, 327.
(3) Baddour, R. F., Yoon, C. Y., *Chem. Eng. Prog. Symp. Ser.*, No. 32 (1961) **57**, 35.
(4) Crider, J. E., Foss, A. S., *A.I.Ch.E. J.* (1965) **11**, 1012.
(5) Barkelew, C. H., *Chem. Eng. Prog. Symp. Ser.*, No. 25 (1959) **55**, 37.
(6) Beek, J., Jr., Singer, E., *Chem. Eng. Sci.* (1951) **47**, 534.
(7) Froment, G. P., *Chem. Eng. Sci.* (1962) **17**, 849.
(8) Hoelscher, H. E., *Chem. Eng. Sci.* (1957) **6**, 183.
(9) Quinton, J. H., Storrow, J. A., *Chem. Eng. Sci.* (1956) **5**, 245.
(10) Schwartz, G. S., Smith, J. M., *Ind. Eng. Chem.* (1953) **45**, 1209.
(11) Schertz, W. W., Bischoff, K. B., *A.I.Ch.E. J.* (1969) **15**, 597.
(12) Yagi, S., Kunii, D., *A.I.Ch.E. J.* (1957) **3**, 373.

Oxidation of SO_2 in a Trickle-Bed Reactor Packed with Carbon

M. HARTMAN, J. R. POLEK, and ROBERT W. COUGHLIN, Lehigh University, Bethlehem, Pa.

Both the Lurgi "sulfacid" and the Hitachi processes for removing SO_2 from flue gas involve the reaction of SO_2 and oxygen at the surface of a carbon catalyst to form SO_3, which is washed away with water as sulfuric acid. The steady-state, continuous, countercurrent contacting of

SO_2-bearing flue gas with water in a column packed with carbon can be modeled by combining the classical treatment for gas absorption with expressions for the rate of transport through the liquid and for the rate of catalytic reaction within the porous carbon.

The rate of chemical reaction was determined by carrying out experiments in a well-stirred batch reactor saturated with sulfur dioxide and oxygen into which catalyst is introduced and the rate of sulfate production measured. By doing this with both powdered and pelletized or granular carbon, effectiveness factors can also be estimated. Some representative results are given in Table I.

Table I. Typical Experimental Results

Type of Carbon	*Oxidation Rate, mg SO_4^{2-}/min gram*	*Effectiveness Factor*
North American carbon extruded pellets, 1/16 by 3/16 inch	190	0.55
Westvaco WV–G, 12 × 40 mesh	240	0.27
Pittsburgh BPL 12 × 30 mesh	490	0.78

The model of steady-state, countercurrent contacting of gas and water in a carbon-packed column is based on the rate of transfer from gas to liquid and to packing in a differential section of the column, *dz:*

$$Gdy_1 = fK_La_1(C_{1i} - C_1)Adz + (1 - f)K_r\eta C_{1i}C_{2i}^{0.5}Adz$$

and

$$Gdy_2 = fK_La_2(C_{2i} - C_2)Adz + \tfrac{1}{2}(1 - f)K_r\eta C_{1i}C_{2i}^{05.}Adz$$

G is the molar gas flow rate, y is the mole fraction of SO_2 (subscript 1) or O_2 (subscript 2) in the gas, K_La is the composite mass transfer coefficient, f is the fraction of the carbon packing wet by the liquid, C_{1i} and C_{2i} are the liquid compositions at the gas–liquid interface, C_1 and C_2 are the bulk liquid compositions, A is the cross sectional area of the column, K_r is the reaction rate constant per unit volume of packed column, and η is the effectiveness factor. The rate of transport into the liquid equals the rate of increase of bulk liquid composition plus the rate of chemical reaction:

$$fK_La_1(C_{1i} - C_1)Adz = LdC_1 + fK_r\eta C_{s1}^{0.5}C_{s2}Adz$$

and

$$fK_La_2(C_{2i} - C_2)Adz = LdC_2 + \tfrac{1}{2}fK_r\eta C_{s1}C_{s2}^{0.5}Adz$$

where L is the liquid flow rate and C_{s1} and C_{s2} are the liquid compositions at the liquid–solid interface at the exterior surface of the carbon pellets.

The rate of production of H_2SO_4 (subscript 3) may be written:

$$LdC_3 = fK_r\eta C_{s1}C_{s2}^{0.5}Adz + (1 - f)K_r\eta C_{1i}C_{2i}^{0.5}Adz$$

The variables C_{s1} and C_{s2} may be eliminated using the expressions:

$$K_r\eta C_{s1}C_{s2}^{0.5} = K_{Ls}a_1(C_1 - C_{s1})$$

and

$$K_r\eta C_{s1}C_{s2}^{0.5} = 2K_{Ls}a_2(C_2 - C_{s2})$$

where $K_{Ls}a$ is the composite mass transfer coefficient for transport from bulk liquid to the interface between liquid and the external area of the carbon catalyst. Aside from the usual kinds of assumptions, the equations above imply that transport takes place only in directions perpendicular to the external surface of the carbon packing, that the non-wet portions of the carbon are coated with an infinitestimally thin layer of liquid in equilibrium with the gas, and that the pores in the carbon are filled uniformly with liquid.

When the above equations are solved numerically using vapor–liquid equilibrium data, experimental reaction rate data, and correlations for mass transfer coefficients it is found that if f is assumed equal to unity, the model and experiment do not agree at the lower liquid rates. The experimental data indicate more reaction than does the model when liquid rates are low. At larger liquid rates, however, the predictions of the model tend to converge with the experimental results. This implies that the carbon packing is not completely wet ($f < 1$) at the lower liquid flow rates. Choosing values of $f < 1$ can force the model to agree with experiment at lower liquid flow rates.

2

Polymerization Kinetics and Reactor Design

REUEL SHINNAR and STANLEY KATZ

Department of Chemical Engineering, The City College of the City University of New York, New York, N. Y. 10031

An informal review is given of the author's view of the present state of engineering knowledge in polymerization kinetics and the present state of the art in the design of polymerization reactors. The discussion focuses on key areas where information for reactor design is needed and relates them to the design and scale-up of the reactor. Special emphasis is given to the effect of reactor configuration. Optimization methods and the importance of proper control strategy are also discussed.

The recent advances in polymerization kinetics and design of polymerization reactors have been aptly summarized in recent reviews (*1–4*) which include most of the published articles. We therefore decided that instead of repeating these efforts we would devote this review to taking stock of our opinion of where we stand today and what problems must be solved to bring the design of polymerization reactors to the same level as conventional reactor design.

Basically, polymerization is a complex chemical reaction, and therefore most of the concepts developed with respect to the design of reactors for such reactions apply also to the design of polymerization reactors. Thus, what are the significant differences of polymerization as compared with other complex chemical reactions that merit special attention? Let us briefly consider what is the information generally needed for the successful design and control of a chemical reactor.

Assume we already have a process giving a desired product. What we need for design is:

(1) Specifications and tolerances for the product quality, the yields needed, etc.

(2) An engineering kinetic model allowing us to predict the effect of changes in the process parameters (temperature, inlet concentration, transport processes) on the yield and desired products.

(3) A description of the transport processes occurring in the reactor; in other words, a flow model and an evaluation of the effect of transport processes on the kinetics.

With this information in hand, we can then undertake to consider:

(4) A suitable design configuration allowing safe scale-up and easy control.

(5) Optimization of design and operating conditions.

(6) Control strategy to ensure conformance to specifications.

The body of this review takes up these points in order.

Product Specifications

Perhaps the most characteristic feature of polymerization reactor engineering is the fact that in a majority of cases the specifications are not given in terms which are suitable for reactor design. Let us assume for a moment that we have a simple polymerization process, where only one type of polymer is possible and that we know the kinetics sufficiently to predict the effect of all variables. What we are going to predict are molecular weight distributions (MWDs). However, product specifications are given in completely different terms: mechanical strength, electrical properties, corrosion resistance, performance in molding and extrusion equipment, etc. While for this simple case all these properties should be completely specified by the MWD, what these relations are is in no way obvious. Nor is it clear that anybody is ready to undertake the research and testing necessary to collect the data allowing one clearly to translate the product specifications into limits on the allowable MWD.

What is often done is to set arbitrary specifications on the leading moments of the MWD, usually on the weight average molecular weight μ_2 which is accessible by intrinsic viscosity measurements, and sometimes on the ratio μ_2/μ_1 (weight average to number average molecular weight) which is related to the variance of the distribution. For quality control, where perturbations are very small, this is satisfactory. As a general specification this is more questionable since some properties depend more strongly on the tail of the MWD (*5, 6, 7*).

It has been shown that non-Newtonian properties of solutions depend strongly on the higher moments of the MWD. In general one would expect that viscoelastic properties in the liquid state (melt or solution) which affect the processing properties of the polymer are strongly affected by the high molecular weight tail of the distribution. Stability and chemical reactivity, on the other hand, might often depend on the low molecular weight tail. Some work has been published on the effect of molecular weight distributions on melt fracture (*8*).

Sometimes even a qualitative knowledge as to the effect of the tails of the MWD could be valuable. People sometimes tend to say that a more uniform molecular weight is desirable, but there is often no evidence that this is true.

In no way do we intend to present here a comprehensive review of this important subject. What we do want to point out is that this is an important area which must be much further advanced than it is today before we can really think about optimization strategies in both design and control since any such optimization procedure will affect the MWD as well as the degree of branching or copolymerization. Without a more precise knowledge of what is desirable or permissible such optimization procedures will often remain mathematical games.

The problem of proper specifications becomes much more difficult if we are dealing with more complex systems involving branching, copolymerization, etc., in which the MWD alone does not specify the polymer.

There are many chemical processes in which similar problems exist, such as for example in the oil and drug industry. However, in the oil industry the final properties are easier to measure and less critical, and in the pharmaceutical industry purification processes are often possible. This brings us to our last point—namely, that polymerization normally leads to a product which cannot be further corrected by purification, though other correction methods are often possible.

Kinetic Models

Kinetic modelling for purposes of reactor design and operation includes a wide range of approaches. Seldom do we need (or can we obtain) an accurate kinetic description of the elementary steps involved in the process. The level of complexity needed often depends on how the model is to be applied. For example, consider a complex reaction from another field—*i.e.*, fluidized bed cracking. An extremely simple model of this process

$$\text{fuel oil} \rightarrow \text{gasoline} \rightarrow \text{gaseous products}$$

contains most of the essential features for safe scale-up. It tells us that we are dealing with a consecutive reaction and that deviations from plug flow might have undesirable effects. Evaluating the temperature dependence of the two pseudo-reaction rates might give us a good idea as to the allowable tolerance on the temperature. On the other hand, the model might be much too crude to answer questions about optimum process conditions. Thus, if we want a kinetic model to predict the performance of different fuel stocks in a reactor, we need a much more elaborate model. In no case do we even attempt to describe the behavior of the several hundred identifiable chemical compounds involved, and as in any simplified model one important question is to understand the errors introduced by simplifications. There have been considerable advances in this area with respect to complex chemical reactions, and similar work might be useful in polymerization kinetics. However, the problems arising in polymerization kinetics are somewhat different and merit discussion.

To illustrate these problems, we consider in turn several areas of polymerization study, first, from homogeneous polymerization, and second, from heterogeneous polymerization.

Homogeneous Polymerization. In most homogeneous polymerization processes, there has been tremendous progress in recent years in understanding and modelling the kinetics both in batch processes and in continuous reactors. These include linear free radical polymerizations (*9, 14*), polycondensation (*15*), as well as polymerization processes involving branching (*16*).

These modelling methods have some problems in common. All of them deal with the development of a distribution in time, a problem which is not normally encountered in the study of kinetics of complex chemical processes mentioned before. These problems are similar to those encountered in the study of particulate processes such as crystallization, fog formation, etc. (*17*). One of the problems in a study involving distributions is the dependence of rate parameters on particle size (in our case radical size) and on the state of the system.

Consider a simple free radical polymerization with mutual termination, described by the following set of equations:

$$\text{Initiation:} \quad \mathrm{M} + \mathrm{I} \xrightarrow{k_{\mathrm{I}}} \mathrm{R}_1$$

$$\text{Propagation:} \quad \mathrm{R}_i + M \xrightarrow{k_{\mathrm{P}_i}} \mathrm{R}_{i+1}$$

$$\text{Termination:} \quad \mathrm{R}_i + \mathrm{R}_j \xrightarrow{k_{\mathrm{T}_{ij}}} \mathrm{P}_{i+j}$$

M denotes monomer, I, initiator (catalyst), and R_i and P_i, respectively, growing radical and terminated polymer of the indicated number of monomer units. The simplest thing is to assume that all the rate constants k are functions of temperature alone, independent of i, j, and the state of the system, but it is hard to put up convincing prior argument for these assumptions. First, one would expect from purely kinetic considerations that rates of elementary steps would depend on the complexity of the molecules and their steric configuration. Thus the chance of two large complex molecules terminating on collision should be considerably lower than that of a large molecule with a smaller one, or of two smaller ones.

Termination at least should be size dependent. For some complex molecules one could also make a similar argument with respect to propagation. On the other hand as the polymerization progresses, the presence of dead polymer dissolved in either the monomer or the solvent will affect the diffusivities and the viscosity of the system and in some cases this will affect the rate constant.

Mathematically one can find ways to handle these effects. Some mathematical approximation for the dependence of k_T and k_P on i and j lead to still manageable expressions. One can also make k_P and k_T depend on the conversion, as given. Some of these problems are discussed in more detail in this volume (*14*).

Similar arguments can be made with respect to more complex polymerization models. The amount of complexity that is desirable is limited not just by mathematical considerations but much more by what is experimentally accessible and further by what are the goals of our modelling.

If only a weight average molecular weight based on intrinsic viscosity measurements and perhaps a number average molecular weight based on osmotic pressure is available, then obviously there is little sense in making very refined assumptions with respect to the size dependence of the kinetic constants. On the other hand, if those averages change significantly with conversion, one can still make some reasonable deductions as to the dependence of the rate coefficients on the state of the system.

With the availability of more accurate methods of measuring a complete MWD, by gel permeation chromatography (GPC) or by other methods of fractionation, we are coming to the state where we can get data to test the validity of our basic kinetic models and the simplifying assumptions involved.

Most of the work until now has dealt with summary properties of the MWD and trends predicted by the kinetic models, and there has been little work to confirm the results by accurate measurements of the MWD. There has also been little attempt to measure and estimate the individual rate constants for the more complex models, and here one faces a formidable task for which there is a need not only for more accurate experimental method and data but also for more basic theoretical work. One of the summaries (*18*) in this volume deals with an interesting attempt to estimate rate constants together with their activation energies for the homogeneous polymerization of polyethylene. Since the full results are not published, it is premature to discuss these results, but we can discuss the inherent difficulties of such an undertaking.

The basic difficulty is to determine confidence limits on our estimates of the rate constants, which is always a problem when we deal with the simultaneous estimation of several rate constants from integral data. We ask over what domain the set of rate constants, k_1, k_2 etc. can vary and still be statistically consistent (at an assigned probability level) with our experimental results.

To determine these limits we need first an idea of how accurate our measurements are and how many independent measurements we have. Assume at first that we have only intrinsic viscosity and osmotic pressure measurements. These give us data related to the first two moments of the distribution. We know something about the accuracy of these methods, but not enough about the dependence of this accuracy on the form of the distribution. These are not direct measurements of the moments, and their accuracy does indeed depend strongly on the form of the distribution. A low molecular weight tail will not affect the osmotic pressure since this part of the distribution often passes the membrane or is washed away in sample preparation. In the same way, while the intrinsic viscosity is related to weight average molecular weight, the averaging process is not especially straightforward, and the accuracy depends strongly on the form of the distribution.

The same applies to methods of measuring distributions, like GPC, which normally, unless special efforts are made, are adjusted to be more accurate in the middle of the distribution than at the tails.

When fitting a measured distribution, how many independent observations do such measurements really contain? If the distribution were really measured accurately and independently over the whole size range, the number of (statistical) degrees of freedom would be infinite. In GPC, with its limited experimental accuracy, the number might however be no more than three or four. In some ways the problem is similar to the estimation of the spectral distribution of a random signal, and the analysis of Tukey (*19*) and others might be suggestive in this context.

It is just in extrapolation to conditions far from those studied experimentally that it is important to have tight confidence limits on the kinetic parameters. In the range of the experimental data, the fit will usually be quite good.

Heterogeneous Polymerization. We turn now to a consideration of some problems from heterogeneous polymerization, noting that many of the industrially important polymerization processes—emulsion precipitation, and suspension polymerization—are carried out in heterogeneous systems.

Two idealized models of heterogeneous processes have been studied well. One is the case where the second phase is treated as a uniformly mixed single continuum with mass transfer between the phases (*20*). The second is the case of emulsion polymerization, in which it is assumed that the termination rate is high and the number of particles very large so that each particle contains either one or zero growing radicals (*21*, *22*). Experimentally, it has been shown that in the non-ideal state of emulsion polymerization the number of radicals is larger than one-half but still small. The first model involves the assumption that the number of free growing radicals in a single droplet or particle is very large. The necessity of this assumption was checked in a recent paper (*23*) in which it was shown that it is sufficient that the number of free radicals per particle exceeds 2 for the resulting distribution to be approximated reasonably by the case of an infinite number of radicals. The intermediate case is also treated.

The main difficulty with either assumption is that the second phase is assumed to be uniformly mixed, with the rates of growth and termination (or probabilities, in stochastic models) independent of the size of the radical or its position inside the particle. This is not a very reasonable assumption in suspension polymerization in those cases where the free radicals are generated in the water phase. There have been very few data as to how justified this assumption is in emulsion polymerization or precipitation polymerization, and more work is needed to show if these models lead to sensible over-all predictions for these processes.

There has been considerable work on the molecular weight of emulsion polymer as a function of processes variables, though not too much has been published in the way of experimental results on the actual distribution. In many cases there seems to be an initial time span during which the ideal assumption that the mean number of radicals per particle $\bar{n} = 1/2$ seems to hold. One common check of this assumption is the sudden introduction of additional initiator into the system. If $\bar{n} = 1/2$, the rate of polymerization should remain constant. This effect is often used to test the applicability of the basic theory.

The basic Smith-Ewart theory not only predicts the molecular weight distribution for this case but also the number of particles as a function of initiator and soap concentration and seems to lead to correct estimates of the magnitude of these parameters. However, recent, more detailed measurements throw some doubts on the validity of the theory.

Williams and others (*24*) have followed the developing molecular weight as a function of time and found that in some cases the molecular weight of the polymer formed increases strongly as the particles grow while initiator perturbation experiments indicate that $\bar{n}$ is still 1/2. One possible explanation would be that the transfer of a small growing radical to a polymer particle depends on the surface conditions and perhaps also on soap concentration at the surface. As these change, the effective number of free radicals generated changes since the rest terminate with a very small size in the water phase. However there are several other hypotheses that could explain the same phenomenon, and more experiments are needed to clarify the problem. Thus, this case illustrates a point mentioned previously—namely, that a model can fit some over-all data fairly well and break down when some more detailed predictions are required.

Another test of a kinetic model is to predict, for example, the behavior of a continuous stirred tank reactor from batch data. Here the data presented by Poehlein (*25*) are very interesting. Most of the data were obtained with low conversion in a system where one would expect the Smith-Ewart theory to hold. With respect to some important parameters like average molecular weight the theory does not even predict a correct trend. The molecular weight seems to be completely independent of residence time, whereas the theory predicts a strong dependence (*25*). The size distribution of the particles themselves is also completely different in shape from what one would expect and is much flatter, again indicating some basic difficulties. On the other hand, the dependence of the average size of the particles on the system parameters is predicted with remarkable accuracy. We note therefore that even in the simplest case of emulsion polymerization we need much more work before we can apply the results with confidence to the prediction of the steady state and dynamic behavior of a reactor.

Another problem in which one needs considerable additional information is the mechanism of nucleation in both precipitation and emulsion polymerization (*21, 26, 27*). Such reactions involve continuous simultaneous nucleation of new particles as well as growth and in that sense are similar to crystallizers. It has been shown for continuous crystallizers that prediction of their behavior requires a good understanding of the dependence of the rate of new particle formation on the process parameters.

Most theoretical studies of free radical polymerization deal with cases in which the reaction does not depend on the degree of polymerization or in the state of the product. In many real cases these are dominating effects on the MWD. Thus in homogeneous polymerization the gel effect becomes important at higher viscosities (*14, 28*). In emulsion polymerization there are several effects. The effectiveness of capture depends on particle size and soap concentration (*24*), the termination rate depends on particle size (*29*), and furthermore the propagation and termination of captured radicals will depend on their mobility inside the particle, which changes with particle size and degree of polymerization (*30*), all effects which have had little theoretical or experimental treatment till now. Similar arguments apply to precipitation polymerization (*31*).

The fact that most of our theoretical models are only imperfect idealizations is often not mentioned when dealing with practical problems (*32*). In experimental research we often concentrate on those conditions where one would expect that our assumptions have the best chance to be correct, such as low viscosity systems in homogeneous polymerization and the initial phase of emulsion polymerization. This is justified in research, but one should be very careful when applying the results to real systems in which these non-idealities may have a dominating effect on the MWD.

Transport Processes

The critical transport process for reactor performance is commonly material mixing, and we accordingly present here a brief discussion of how mixing and mixer design affect the performance of polymerization reactors. These effects may be classified under two headings. First, in heterogeneous systems, the mixing affects the physical state of the system, and thus the particle size, etc. Second, mixing affects the apparent kinetics and the resulting MWD in homogeneous as well as heterogeneous systems.

We discuss first the specific effects in heterogeneous systems. In suspension polymerizations the monomer is dispersed in water and kept dispersed by the combined action of turbulent agitation and a protective colloid. To prevent occlusion of water and impurities, it is important that the individual droplets do not coalesce and redisperse throughout the polymerization process. The prevention of coalescence of such large droplets by agitation was studied by Shinnar and Church (*33*), but the authors are not aware of any industrial data published since then. It is, however, well known that correct design of the agitator is important in scale-up of such reactors, and that scale-up can lead to changes in the size and stability of the droplets. In continuous reactors the droplet stability problem becomes somewhat more complex since fresh monomer is immediately exposed to partially polymerized droplets of higher viscosity. No experimental data have been published with respect to this problem. In some applications uniform size is important, but there are no published data on the effect of colloid and agitation on the size distribution, though the authors are aware that there is considerable industrial know-how in existence and that the variance of the droplet sizes can be strongly reduced by proper choice of the above parameters.

There have been several attempts to study the effect of agitation on emulsion polymerization (*34*). However there is little evidence that agitation has any effect as long as the intensity is above a minimum threshold value that prevents separation of the monomer or any large scale sedimentation. There is no evidence for a maximum allowable intensity, nor is the process as sensitive to agitation as suspension polymerization. This is not surprising since normal agitation has little effect on emulsion particles in the submicron size, which are much smaller than the smallest eddy. However, agitation affects the monomer droplets, but it seems that their size is not critical. There seems to be little indication that further work would lead to a big payoff. In continuous emulsion polymerization there is, however, one aspect of agitation that might need attention, and that is the effect of locally higher concentration of emulsifier on nucleation. Nucleation phenomena are often sensitive to agitation, because nucleation is a highly nonlinear phenomenon and might be controlled by higher concentrations near the inlet.

The considerations noted above about nucleation and about the effect of agitation on particle size apply equally to precipitation polymerization.

We turn now to the question of how mixing affects the apparent kinetics and the resulting MWD in polymerization reactors. These effects arise in both homogeneous and heterogeneous polymerization, although in systems where nucleation must be considered they are complicated by the effect of mixing on the nucleation rate.

The strong effect of mixing on the MWD has been documented by Denbigh (*35*) and others (*34, 36, 37*) but in these studies, the mixing properties and the flow regime interrelate so strongly with the choice of reactor configuration that this discussion is carried to the next section.

Reactor Configuration

Questions of reactor design and scale-up in quite general contexts can be formulated as follows:

(a) Given a complex chemical reaction, what is the best reactor configuration?

(b) Given pilot plant experiments in a given reactor configuration, how can we with confidence predict the behavior of the large scale plant? Sensible answers to these questions depend most strongly on knowledge of flow patterns and mixing processes.

Let us start with the second question. Reading the literature one might somehow get the impression that we can take any complex reactor —*e.g.*, a fluidized bed—and predict its scale-up for any complex reaction by admittedly complex but reliable computations based on measurement of the residence time distribution. In fact what we can really do is much less ambitious though often satisfactory. The only reactors that we can really scale up with good confidence are plug flow reactors (or near plug flow reactors) and stirred tanks which are close to being ideally stirred tanks. First, we seldom know the reaction mechanism for complex reactions that well, and secondly, even if we did, it would be hard to predict the exact flow regime of a complex reactor for a new system unless we built one. Let us not confuse our ability to explain the behavior of existing reactors with confident scale-up. As long as we deal with small deviations from either plug flow or a stirred tank, it really does not matter too much what the exact nature of such deviations is since most analyses show that for small perturbations the nature of the perturbation has only minor effect. Indeed, we can often put a bound on the possible magnitude of such deviations and estimate their effect. We can base such estimates on an approximate knowledge of the kinetics, or we can simulate such deviations experimentally to see if they have an effect. The possibility of simulating small deviations from complete mixing in a pilot plant is often overlooked. In most cases such simulation cannot be done by reducing agitator speed. Large vessels have considerable mixing times. To simulate the same ratio of mixing time to residence time in a small reactor of the same design will often change the flow regime or call for impractical mixer design, a fact that limits many of the experimental investigations of imperfect mixing. However, mixing patterns involving small deviations from ideal mixing can be simulated under perfectly controlled conditions using highly stirred tanks such as in Figure 1.

These difficulties do not mean that we cannot scale up a simple fluidized bed or any other complex flow. Many reactions are not sensitive to mixing patterns or to contact time distribution. This insensitivity can be deduced for simple reactions by theoretical considerations or by experimental perturbations of the system. We can often decide quite confidently on the basis of previous experience what the risks of a scale-up are.

It is the combination of complex mixing and a flow-sensitive reaction which we cannot confidently handle. Luckily, these cases are not that

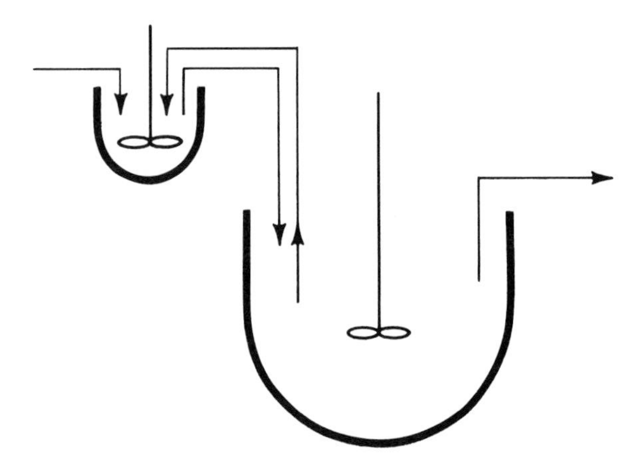

Figure 1. Controlled simulation of finite initial mixing times in a pilot plant reactor

frequent, and it will probably always be cheaper to redesign the reactor to avoid such problems than to perform the research necessary for confident scale-up. Thus, we can if necessary modify a fluidized bed so that it approaches a plug flow reactor. We can do this by baffling or by using a multistage reactor. Admittedly, we lose some of the advantages of a fluidized bed, but this can be minimized by intelligent design.

Plug flow can always be approached by suitable staging. The residence time distribution of a system with 10 stages will be close to plug flow even if each stage deviates considerably from any clearly defined flow model. The deviations can again often be estimated by suitable tracer experiments. The reasoning behind this is similar to that leading to the law of large numbers in probability theory, and need not be discussed here.

It is much more difficult to achieve a completely stirred tank, although luckily in most chemical reactors the viscosity is not too high, and even large reactors will be fairly well mixed, with the mixing time small as compared with the residence time. There are few cases of chemical reactions in which simultaneously a single ideal stirred tank is preferable to plug flow, the reaction is sensitive to micromixing, and the mixture is so viscous that mixing is slow compared with through-put. If we were to look for such an example, the first thought that would cross one's mind would probably be to look for a polymerization reaction.

A second problem in scale-up is nonuniform temperature. With a knowledge of the reaction rates, we can again predict the maximum temperature deviations with reasonable confidence and investigate experimentally the effect of such temperature dependencies and modify our design accordingly.

The foregoing is a general discussion of difficulties in scale-up. If we ask which aspects apply especially to polymerization reactors, two points immediately emerge. One is that in polymerization reactors, we much more frequently deal with very viscous and often non-Newtonian mixtures. Heat transfer is slow, and it is harder to achieve uniform temperatures. Complete and rapid mixing is often unobtainable. Luckily, however, many polymerizations are not too sensitive to micromixing effects (*34, 36, 37*). Those that are remain a challenging problem. It might be that when mixing is slow, the exact pattern is also not important, and that we could apply perturbations around segregated flow, just as we can do it with perfect mixing. The plug flow case is much simpler as we can always modify our design by suitable staging. Thus, for example it has been shown that flow in a tube at high viscosity deviates considerably from plug flow. This can easily be remedied by having several small remixing sections, arranged by means of baffles or flow distributors. The effectiveness of such devices can be checked by tracer experiments and from fluid dynamic considerations. What is more difficult is to judge if such added expenses are justified.

The other difficulty that arises especially in polymerization reactors is that it is much harder to evaluate the results of any change by chemical analysis. That is, perturbation experiments in polymer pilot plants are often hindered by the fact (mentioned earlier) that the specifications are not defined in terms suitable for the reactor engineer. Indeed, it is often very hard to pin anybody down to a confident answer that two samples of polymers are exactly equivalent. This will sometimes involve time consuming testing programs, which increase the complexity of the problem considerably.

Another approach to scale-up is the prediction of the behavior of a large scale continuous reactor from batch data. Thus for many reactions we can with confidence predict the behavior of a continuous stirred tank from batch data. There are very few industrially important polymerizations for which this is possible. It has been done in homogeneous polymerizations but for very constrained conditions. We know far too little about emulsion polymerization or precipitation polymerization to do this with any confidence. The paper by Poehlein in this volume (*25*) demonstrates this quite clearly. Surprisingly, in his results, the observed molecular weight is independent of residence time, whereas our present kinetic models would all predict a strong dependence. The paper by Graessely on branched polymers (*16*) also shows that such extrapolations are very difficult and might lead to considerable errors.

We turn now to the first question formulated at the beginning of this section, the search for the best reactor configuration for given reaction kinetics.

As long as we deal with the simpler cases such as an isothermal plug flow reactor *vs.* a stirred tank, we often need only a very qualitative and primitive understanding of the reaction mechanism to make a sensible choice. For homogeneous reactions the problem has been discussed, and we have enough understanding to solve this simple problem. For heterogeneous reactions even this simple problem is not that simple, and we need either a better understanding or more data. It is, however, accessible by experimentation.

Let us however discuss some of the problems related to emulsion and precipitation polymerization. In both cases the nucleation is strongly affected by the configuration. In emulsion polymerization correct prediction of the nucleation in a continuous reactor from batch experiments has been achieved in some cases, though more experiments are needed to be conclusive. Regretfully in both cases the experiments were at low conversions where one would expect the theory to hold best since conditions are not too different from those encountered in a batch experiment. It is at very high conversions and with the type of nucleation that in batch reactors leads to uniform particle size that one would expect difficulties and where one could test the theory. However, qualitatively we know that at high conversion, nucleation rate is less than at low conversion, and therefore particle size in a continuous reactor should be larger at high conversion because of suppressed nucleation.

In precipitation polymerization our knowledge is much more qualitative. We know experimentally that in a continuous stirred tank, nucleation rates are lower and particle sizes are much larger since nucleation is suppressed by the capture of small nuclei (*26, 27*).

Particle size in itself has in both cases an effect on the uniformity of the molecular weight. Larger particle sizes might cause a high molecular weight tail because of trapping of free radicals which might offset any gains arising from more uniform polymerization conditions during the life of a particle. This can be overcome by introducing a separate nucleation stage. All these problems need further investigation before we can make even qualitative *a priori* prediction of the effect of reactor configuration on the MWD.

Other types of polymerization processes in which prediction of mixing effects on polymer properties is difficult are copolymerizations and graft polymerizations, which often depend on proper distribution of the monomers. This is a subject which has received so little theoretical consideration in the literature that one can hardly review it, but it is a subject in which fruitful work could be done. However, if one deals solely with scale-up, the considerations for safe scale-up discussed before will apply here too and will allow one at least to recognize those cases that need extra attention.

The search for best reactor configurations leads to a more refined class of optimization questions than we have encountered yet in this review. These are taken up on the next section, and the associated control questions are considered in the final section.

Optimization

Once the over-all configuration of an ordinary chemical reactor has been set, a variety of optimization problems present themselves. At the design stage, one has for example the question of minimum volume (cost) with suitable constraints on the reactor performance. However, since reactor cost is not often a really significant matter, one is more likely to encounter such questions at the operating stage, where the plant already exists, and one looks to maximize the yield.

In either case, these optima are commonly sought through variation of flow and temperature—*i.e.*, by distributing suitably the feeds and the energy inputs (or takeoffs). The mathematical methods appropriate to the problem depend on the manner of this distribution. If the distribution is discrete, as when staging the feed in a cascade of reactors, methods ranging from school calculus to modern techniques in mathematical programming can be brought to bear. If the distribution is continuous, as when studying temperature profiles in a pipeline reactor, methods of the calculus of variations are called for. For chemical reactor problems, where the situation is described by a small number of ordinary differential equations, one has resources ranging from the modern dynamic programming techniques of Bellman to the more classically grounded results of Pontrjagin. In addition there are always methods from linear algebra or from functional analysis for problems in isothermal first-order kinetics (linear problems).

One should bear in mind that at the operating stage, optimum problems are more likely to be solved experimentally than mathematically if only because of lack of knowledge of reactor kinetics. The methods here are carefully designed perturbation experiments, backed up by careful statistical analysis of the results, so as to be confident (at an assigned probability level) of detecting a real effect despite random fluctuations in the process and errors in measurement.

The foregoing discussion is for chemical reactors in general. How does it apply to polymerization reactors? At the operating stage, to maximize the production rate makes economic sense, just as it does for ordinary chemical reactors. Mathematically, it is likely to be much harder to do for polymerization reactors since even a simple homogeneous polymerization system is described by a large number of ordinary differential equations, one for each radical R_i and polymer P_i of i monomer units, or by a pair of partial differential equations if the degree of polymerization is treated as a continuous variable. Only if a small number of summary ordinary differential equations can be developed in the leading moments of the MWD—this will usually be possible when the rate constants are independent of the degree of polymerization—will the methods for conventional chemical reactors be applicable (*38*).

While to maximize the production rate always makes economic sense, to improve the quality of the product is likely to have even greater payoff in polymerization reactors. This must be done experimentally since one does not know quantitatively how the MWD defines the quality parameters of the polymer. Again, it will be especially difficult and expensive to do just because these very parameters cannot be distinguished by chemical analysis.

At the design stage the important optimum question for polymerization reactors would seem to be to come as close as possible to a desired MWD. Ideally, one would only know what MWD to ask for if one knew already how to predict product quality in terms of the MWD; in the face of competing quality demands, to characterize the desired MWD would itself be an optimum problem. In practice, if the desired MWD is broader than what can be obtained in a single reactor, one would be more likely to tailor it by blending than by implementing this or that algorithm in the calculus of variations. Still, some more immediate questions present themselves that can perhaps be answered in terms of distributed feeds and temperatures. One might, for example, aim at uniform molecular weight, or at the same MWD as an ideally mixed reactor but under conditions realizable at high viscosity. Also there is the whole question of specifying different polymer structure, such as in graft polymers or copolymers, a subject which has received little attention from the reactor engineer. One situation especially where proper staging might lead to a narrower MWD is that in which the termination rate depends strongly on the degree of polymerization. In homogeneous polymerization, this is simply the gel effect, and here temperature staging might well reduce the high molecular weight tail. In heterogeneous polymerization, this occurs when intraparticle diffusion becomes controlling (trapping), and here a separate nucleation stage operating at different conditions might have the same effect.

Even in the simplest case of a well known chemical reaction seldom are such optimization procedures used in the actual design of reactors. What this type of work leads to is an understanding of what would be gained by better design and in what direction one should go. Even if the reaction is only imperfectly understood, theoretical considerations might lead to suitable experiments. It is in this area that a better understanding of these problems would lead to real gains.

Control

In complex reactions, when product quality is important, easy controllability of the reactor should be one of the major considerations in reactor design. We often design a control system around a finished reactor instead of incorporating the control into the initial design. However these concepts are only slowly entering reactor design in general, and in that aspect polymerization is not different from other reactors, only that cases involving control difficulties are more frequent. Thus close temperature control might be more important in some polymerization reactors than in almost any other type.

Another specific aspect of continuous control in polymerization is the fact noted earlier that the desired properties of the product are not always clearly related to parameters easily measured in a continuous reactor. We therefore must try to keep as many parameters as possible constant and rely upon changes in easily measurable parameters as alarm signals. From a theoretical point of view we are dealing with a control system in which only some of the state variables are measured and used for control, and what we need is a better understanding of the region of the total state space open to us under such partial control.

Consider, for example, homogeneous polymerization of a monomer, such as in a prepolymerization reactor. Control of temperature as well as viscosity will bound the molecular weight, and from the limits of the two measured variables one can quite easily deduce bounds on the MWD of the mixed total product. If the polymerization were in solution, solvent concentration in the product stream would also have to be measured, otherwise the bounds become too large. However measurement of the molecular weight by intrinsic viscosity would suffice as a single control parameter in a stirred tank reactor.

On the other hand in a continuous emulsion or precipitation polymerization where the uniformity of the molecular weight might depend on intraparticle diffusion, measuring the intrinsic viscosity is not enough to put a bound on the MWD unless we know from experience that for this specific polymer it is a sufficient control criterion. We can easily imagine that by changing particle size and conversion the same intrinsic viscosity might relate two polymers with different MWD. For copolymers this becomes even more complex. Some theoretical work on the possible bounds of imperfectly controlled systems might be very valuable for a better understanding of such systems.

This question of partial measurement of the state vector must also be considered in any considerations of optimal control of batch systems, intentional periodic operation, or other optimal trajectory problems.

Another feature of some continuous polymerization systems is their tendency to low amplitude low frequency limit cycles (*39*) and possibly to multiple steady states (*20*). Although multiple steady states commonly arise in situations that are heat transfer controlled, they may arise as well in heterogeneous systems involving a nucleation step because of the dependence of the particle nucleation rate (and perhaps the growth rate also) on the existing particle surface. Such systems may also have a single unstable steady state with a stable limit cycle "surrounding" it, and so show sustained fluctuations very much like those shown by crystallizers (*40*). Indeed, stability analyses may be borrowed from crystallization studies and applied to these polymerization systems. However, instability in polymerization systems may arise in ways having no special analogy to crystallizer behavior, for example, because of a viscosity-dependent termination rate; this particular effect can readily be demonstrated mathematically by estimating viscosity effects in terms of the total monomer conversion.

It may also be noted that polymerization reactors are especially likely to have natural disturbances amplified by the control system. This is partly because of time delays in the measurement, which is normally an off-line laboratory procedure, and partly because of imperfect knowledge of (and fluctuations in) the reactor gains. The whole situation is made worse by the fact that the natural time scale of many disturbances in the reactor, such as catalyst batch changes, is of much the same order of magnitude as just these measurement delays, thus permitting the control system to act as a strong amplifier of these disturbances. While one has a theoretical understanding of such problems, they are in fact largely ignored in the operation of polymerization reactors, both manual and computer controlled. In this respect, the work of Box and Jenkins (*41*) can be of interest.

A deeper understanding of all such behavior is very important when sophisticated control such as direct digital control is considered. Again, the answer might often be not in a complex control but in designing the reactor system so that cyclic behavior or amplification of disturbances is avoided.

Literature Cited

(1) Chappelear, D. C., Simon, R. H., ADVAN. CHEM. SER. (1969) **91,** 1.
(2) Platzer, N., *Ind. Eng. Chem.* (Jan. 1970) **62,** 6.
(3) Lenz, R., *Ind. Eng. Chem.* (Feb. 1970) **62,** 59.
(4) Luss, D., Amundson, N. R., *J. Macromol. Sci. Rev. Macromol. Chem.* (1968) **2** (1), 145.
(5) Goppel, J. M., Van Der Vegt, A. K., *Proc. Discuss. Intern. Plast. Congr. Amsterdam Chem., 3rd, 1966,* 177.
(6) Graessley, W. W., Segal, *A.I.Ch.E. J.* (1970) **16,** 261.
(7) Ferry, J. D., "Viscoelastic Properties of Polymers," Wiley, New York, 1961 and 1970.
(8) Spencer, R. S., Dillon, R. E., *J. Colloid Sci.* (1949) **4,** 241.
(9) Zeman, R. J., *Can. Chem. Eng. Conf. 17th, Niagara Falls, N. Y., 1967.*
(10) Zeman, R. J., Amundson, N. R., *Chem. Eng. Sci.* (1965) **20,** 637.
(11) Saidel, G. M., Katz, S., *J. Polymer Sci. Pt. C* (1966) **27,** 149.
(12) Hamielec, A. E., Hodgins, J. W., Tebbens, K., *A.I.Ch.E. J.* (1967) **13,** 1087.
(13) Hui, A. W. T., Hamielec, A. E., *Proc. Natl. Meetg. A.I.Ch.E., 63rd, St. Louis, Mo., 1968.*
(14) Lee, S.-I. *et al.,* ADVAN. CHEM. SER. (1972) **109,** 78.
(15) Secor, R. M., *A.I.Ch.E. J.* (1969) **15,** 861.
(16) Nagasubramanlan, K., Graessley, W. W., ADVAN. CHEM. SER. (1972) **109,** 81.
(17) Katz, S., Shinnar, R., *Ind. Eng. Chem.* (1969) **61** (4), 60.
(18) Thies, J., Schoenemann, K., ADVAN. CHEM. SER. (1972) **109,** 87.

(19) Tukey, J. W., Blackman, R. B., *The Bell System Tech. J.* (1958) **37,** 185.
(20) Goldsmith, R. P., Amundson, N. R., *Chem. Eng. Sci.* (1965) 195, 449, 477, 501.
(21) Smith, W. V., Ewart, R. H., *J. Chem. Phys.* (1948) **16,** 592.
(22) Gardon, J. L., *J. Polymer Sci., Pt. A-1* **6,** 623, 643, 665, 687, 2853, 2859.
(23) Katz, S., Saidel, G. M., Shinnar, R., Advan. Chem. Ser. (1969) **91,** 91.
(24) Grancio, M. R., Williams, D. J., *J. Polymer Sci. Pt. C* (1969) **27,** 139; *Pt. A-1* (1970) **8,** 2617.
(25) Poehlein, G. W., DeGraff, A. W., Advan. Chem. Ser. (1972) **109,** 75.
(26) Klepfer, J., Sonshine, R., Shinnar, R., *Proc. A.I.Ch.E. Meetg., Cleveland, May 1969.*
(27) Fitch, R. M., Prenosil, M. B., Spich, K. J., *J. Polymer Sci., Pt. C* (1969) **27,** 95.
(28) Trommsdorf, E., Kohle, H., Lugally, P., *Macromol. Chem.* (1947) **1,** 169.
(29) Gehrens, H., Kocholein, E., *Z. Electrochem.* (1968) **64,** 1199.
(30) Peebles, L. H., "Copolymerization," Chap. V., Interscience, New York, 1964.
(31) Lewis, O. E., King, R. M., Advan. Chem. Ser. (1969) **91,** 25.
(32) Dunn, R. W., Hsu, C. C., Advan. Chem. Ser. (1972) **109,** 85.
(33) Church, J. M., Shinnar, R., *Ind. Eng. Chem.* (1947) **53,** 643.
(34) Biesenberger, J. A., Tadmor, Z., *Polymer Eng. Sci.* (1966) **6,** 299.
(35) Denbigh, K. G., *Trans. Faraday Soc.* (1947) **43,** 643.
(36) Biesenberger, J., *A.I.Ch.E. J.* (1965) **11,** 369.
(37) Tadmor, Z., Biesenberger, J., *Ind. Eng. Chem., Fundamentals* (1966) **5,** 336.
(38) Hicks, J., Mohan, A., Ray, W. H., *Can. J. Chem. Eng.* (1970) **47,** 590.
(39) Thomas, W. M., Mallison, W. C., *Petrol. Refiner* (1961) **5,** 211.
(40) Sherwin, M. B., Shinnar, R., Katz, S., *A.I.Ch.E. J.* (1967) **13,** 114.
(41) Box, G. E. P., Jenkins, G. M., "Time Series Analysis," Holden Day, New York, 1970.

Received November 16, 1970. Supported in part under NSF Grant GK-4131.

Contributed Papers

Emulsion Polymerization of Styrene in One Continuous Stirred-Tank Reactor

GARY W. POEHLEIN and ANDREW W. DeGRAFF, Department of Chemical Engineering, Lehigh University, Bethlehem, Pa. 18015

Emulsion polymerization studies in a single continuous stirred-tank reactor (CSTR) can be useful for two reasons. First, the experimental results can be utilized to design commercial continuous systems which are destined to become more important in the polymer industry (*1*). Secondly, the data might be more discriminatory than batch data for the elucidation of process mechanisms. Such could certainly be the case for emulsion polymerization reactions for which the transition from batch data to continuous reactor predictions is no simple matter. The three-interval reaction model proposed (*2, 3, 4*) for the batch reaction cannot have any significant meaning in a steady-state CSTR. New particles must be formed in the presence of many existing particles in the CSTR.

The variables subject to the control of the reactor designer and operator are: (1) reaction conditions, (2) emulsifier concentration, (3) initiation rate, and (4) mean residence time. The influence of these variables on the important dependent variables: (1) particle size distribution, (2) polymerization rate, (3) particle number, and (4) polymer molecular weight is of significant interest to the reactor design engineer. This paper examines the relationships between these variables for styrene polymerization.

Theory

Any satisfactory theory must account for (1) particle formation, (2) particle growth, and (3) distribution of residence times. The area of least knowledge, and thus the most questionable part of our theory, is particle formation equation. We have assumed that the particle concentration (N) in the CSTR can be expressed by Equation 1:

$$N = R_i\Theta\left(\frac{A_f}{A_f + A_a}\right) \qquad (1)$$

where R_i = rate of initiation, θ = mean residence time, A_a = adsorbed emulsifier, and A_f = unadsorbed emulsifier.

Particle growth, at least for styrene particles, can be expressed with much more confidence (*5, 6*) as follows.

$$\frac{dv}{dt} = 4\pi r^2 \frac{dr}{dt} = K_1[\mathrm{M}]n \qquad (2)$$

where v = particle volume, r = particle radius, $[M]$ = monomer concentration at the reaction site, n = average number of free radicals in the particle, K_1 = constant which depends on monomer and reaction conditions.

For very small particles $n = 1/2$, and dv/dt is a constant, but for large particles n is directly proportional to v. Stockmayer (*7*) has provided a relationship between n, particle size, and reaction conditions.

The distribution of residence times, which is also the particle age distribution in the effluent of an ideal CSTR, is given by Equation 3.

$$f(t) = \frac{1}{\Theta} e^{-t/\Theta} \qquad (3)$$

If the particles are small and $n = 1/2$, Equations 1, 2, and 3 can be solved in a straightforward manner (*8*) for the particle size distribution, polymerization rate, number of particles, and molecular weight characteristics. If the particle size distribution is broad enough so that $n > 1/2$ for the larger particles a numerical solution procedure (*8*) is required to calculate the parameters listed above.

Experimental Program

The reactor system was a simple, stirred, 1-liter vessel with two inlet flow streams: the water mixture and the monomer. The materials, concentrations, and ranges of experimental variables are listed below.

Initiator: ammonium persulfate—0.3 to 4.0 grams/100 grams H_2O.
Emulsifier: sodium lauryl sulfate—0.25 to 3.0 grams/100 grams H_2O.
Styrene: inhibitor removed by washing with NaOH solutions.
Reaction temperatures: 50° and 70°C.
Reactor mean residence time: 0.12 to 1.0 hours.

Experimental data were obtained only after eight to 10 mean residence times to assure steady operation. No cyclic behavior was observed after two to three mean residence times. Monomer conversion was always less than 25% so that a free monomer phase was present in the reactor effluent.

Results

Particle size distributions for moderate (30 minutes) to low values of mean residence times could be predicted within the accuracy of the electron microscope measurements by the ideal Smith-Ewart growth model—*i.e.*, $n = 0.5$. The Stockmayer modification was required for accurate predictions of particle size distributions for the runs at 1 hour mean residence time. Typical results for the number-average diameter (D_n) are shown below ($\theta = 1$ hour).

	Theoretical Predictions		*Experimental Value*
	Ideal Model	*Stockmayer Modification*	
D_n	858 A	1018 A	1100 A

The experimental values are apt to be slightly high because the probability of missing small particles in the electron micrographs is greater than average.

The number of particles predicted by the ideal theory is given by:

$$\frac{R_i\Theta}{N} = 1 + \frac{K_1 R_i \Theta}{(A_f + A_a)} \left\{ \frac{k_p[\mathrm{M}]\Theta}{1 - K_2[\mathrm{M}]} \right\}^{2/3} \tag{4}$$

where: K_1 and K_2 are constants which can be determined independently, $[M]$ is the monomer concentration in the particles, and k_p is the propagation rate constant.

The Stockmayer modified theory predicts the formation of fewer particles because those which grow larger than the ideal theory suggests will absorb more soap, leaving less to stabilize new particles. The experimentally measured particle concentrations were always less than predicted by the ideal theory (average deviation $\approx -35\%$). The observed differences could be caused by deviations from ideal theory, experimental counting errors, and the use of an inadequate particle formation equation.

The rate of polymerization can be related to the number of particles as follows:

where: $<n>$ is n averaged over all particles.

$$R_p = k_p[\mathrm{M}]N<n> \tag{5}$$

When deviations from the ideal model occur (large θ), N decreases, but $<n>$ increases. These effects nearly balance, and the predictions of the two models are nearly the same. R_p experimental data are more accurate than N data, and thus a much better agreement is achieved between theory and experiment. The experimental data points are scattered rather evenly above and below the theoretical curve.

Molecular weight averages were measured with a GPC, and M_w values were checked with light scattering. M_w/M_n ratios were around 2.5–3.0 for most runs. Theoretical predictions for systems of monodisperse particles suggest that M_w/M_n should be 2.0 (9). The larger values observed in our work could be caused by particle polydispersity and/or transfer reactions to monomer and polymer which were not considered in the theory (9).

(1) Platzer, N., *Ind. Eng. Chem.* (1970) **62**, 6.
(2) Harkins, W. D., *J. Am. Chem. Soc.* (1947) **69**, 1428.
(3) Smith, W. V., Ewart, R. H., *J. Chem. Phys.* (1948) **16**, 592.
(4) Gardon, J. L., *J. Polymer Sci.* (1968) **6**, 623.
(5) Roe, C. P., Brass, P. D., *J. Polymer Sci.* (1957) **24**, 401.
(6) Vanderhoff, J. W., Vitkuske, J. F., Bradford, E. B., Alfrey, Jr., T., *J. Polymer Sci.* (1956) **20**, 225.
(7) Stockmayer, W. H., *J. Polymer Sci.* (1957) **24**, 314.
(8) Poehlein, G. W., DeGraff, A. W., unpublished work.
(9) Katz, S., Shinnar, R., Saidel, G. M., ADVAN. CHEM. SER. (1969) **91**, 145.

A Generalized Dimensionless Model of Polymerization Processes

S.-I. LEE,[1] T. ISHIGE, A. E. HAMIELEC, and T. IMOTO,[2] McMaster University, Hamilton, Ontario, Canada

A generalized dimensionless model which includes gel effect and very high conversion is proposed. This is an extention of the previous ν model developed for the polymerization processes with conventional free-radical kinetics (*1*, *2*). Assuming that the radical lifetime is short compared with the reaction time and that the active polymer with chain length j, Pj, can grow with propagation probability β from P_{j-1}, one can obtain the following equations:

[1] Stevens Institute of Technology, Hoboken, N. J. 07030.
[2] Osaka City University, Osaka, Japan.

$$F(j,\nu)_{l,\ act} \equiv j^l P_j/\hat{P}_{l-1} \cong j^{l-1} \exp(-j/\nu)/(l-1)!\,\nu^l \quad (1)$$

$$\bar{P}_{l,\ act}(\nu) = \hat{P}_o/\hat{P}_{l-1} = l \cdot \nu \quad (2)$$

where $\hat{P}_l \equiv \sum_{j=1}^{\infty} j^l P_j$ and $\nu \equiv \beta/(1-\beta) \gg 1$.

Suppose initiation: K (catalyst) $+ A$ (monomer) $\xrightarrow{k_i = f \cdot kd} P_1$, and termination: $P_i + P_j \xrightarrow{k_t} M_i$ (dead polymer) $+ M_j + M_{i+j}$ and chain transfer and its reinitiation: $P_i + S$ (transfer agent) $\xrightarrow{kf} M_i + S^*$ (radical),$S^* + A \rightarrow S + P_1$. Then, the MWDs and averages of dead polymer produced at any instant can be given using Equations 1 and 2. For example, for $l = 2$,

$$f(j,\nu)_{2,\ inst} = (aF_{3,\ act} + bF_{2,\ act})/(1+\gamma) \quad (3)$$

$$\bar{p}_2(\nu) = \{(2 + 2\gamma + a)/(1+\gamma)\} \cdot \nu \quad (4)$$

where $\gamma \equiv k_f S/k_t \hat{P}_o$ and $\nu \equiv (\nu_{\gamma=o})/(1+\gamma)$.

If the reaction takes place in a HCSTR, the above MWDs are the ones of its output product at steady state. However, if it occurs in a BSTR, they should be integrated over a reaction time. Therefore, generally the ν changes with monomer conversion c (*2*). For the kinetics above, the ν–c relation can be characterized by only one dimensionless parameter α $(\equiv k_d k_t/8k_p^2 fK)^{1/2}$ if there is no gel effect. Using this together with Equations 1–4, one can eliminate many computational problems for MWDs and their averages. This ν model method can be applied to the linear dynamics of MWDs in continuous flow reactors (*2*). For example, in a HCSTR, the deviation of MWDs can be given in general as

$$\Delta F(j,\nu)_{l,\ act}/F(j,\ \nu_s)_{l,\ act} = (j - \bar{p}_{l,\ act}(\nu))\Delta\nu/\nu_s^2 \quad (5)$$

If $\Delta\nu = 0$, the reactor produces the polymer with the narrowest MWDs given in Equation 3. This implies that the same product could be obtained by keeping ν constant whether the reactor is batch or flow.

Let us propose an extended ν model (below). We assume that all the rate constants $\mathbf{k} \equiv (k_i, k_p, k_t, \ldots)$ are changeable and that these changes can be expressed by the product of the following two factors, subjective and objective to active species: chain length effect $\psi \equiv (\psi_i, \psi_p, \psi_t, \ldots)$ and reaction environmental effect $\epsilon \equiv (\epsilon_i, \epsilon_p, \epsilon_t, \ldots)$. For example, $k_t = \psi_t \cdot \epsilon_t \cdot k_{to}$. Again assuming that any active polymer is a sphere swollen with monomer, one considers the probability P that the radical end exists on the surface layer of the sphere (2). Then, $P \propto \nu^{-1/3}$. If the radical ends can react with the swollen monomers with the same reactivity outside ($\delta = 1$), $\psi_p = 1$. If their reactivity is zero inside ($\delta = 0$), $\psi_p \propto P$. As for terminations, we consider only the case $\psi_t \propto P^2 \propto \nu^{-2/3}$. The ϵ represents the reductive change of relative mobility between two reacting species. This could be related strongly to the viscosity of the over-all reaction mixture. We assume ϵ as a function of temperature T, c, $\bar{P}_l$, etc.,

$$\varepsilon = \varepsilon\,(T,\, c,\, \bar{P}_l,\, F_l(j),\, \ldots)$$

Assuming an isothermal case, we have

$$d\varepsilon/dt = \partial\varepsilon/\partial c \cdot dc/dt + \partial\varepsilon/\partial\bar{P}_l \cdot d\bar{P}_l/dt + --- \tag{6}$$

The first approximation assumes that the second and the further terms on the right would be insensitive compared with the first one. Then,

$$d\varepsilon/dt = d\varepsilon/dc \cdot dc/dt \tag{7}$$

where ϵ is regarded as a function of c only. For a numerical study it was assumed that $d\epsilon/dc = -h\epsilon^n$ for each component of ϵ with corresponding constants h and n. Denoting initial values by "o", ϵ_i/ϵ_{io}

$= k_i/k_{io} = f \cdot k_d/f_o k_{do}$, since $\psi_i = 1$. Also, $\varepsilon_p/\varepsilon_{po} = (k_p/k_{po})(\nu/\nu_o)^{1/3}$ for $\delta = 1$ and (k_p/k_{po}) for $\delta = 1$, and $\varepsilon_t/\varepsilon_{to} = (k_t/k_{to})(\nu/\nu_o)^{2/3}$.

In this way, once the apparent rate constants k_i, k_p, and k_t are evaluated, ϵ_i, ϵ_p, ϵ_t can be calculated by knowing ν which is proportional to radical lifetime.

Acknowledgment

The authors thank Imperial Oil Co., Ltd., Toronto, Canada, and Nalco Chemical Co., Chicago, Ill. for financial support of this investigation.

(1) Iwasa, Y., Lee, S.-I., Imoto, T., *Kagaku-Kogaku (Japan)* (1967) **32,** 373.
(2) Lee, S.-I., Imoto, T., "Polymerization Engineering," Nikkan Kogyo, Japan, 1970.

Continuous Reactors in Free Radical Polymerization with Branching

K. NAGASUBRAMANIAN and W. W. GRAESSLEY, Northwestern University, Evanston, Ill. 60201

The distribution of molecular weights obtained in polymerization depends on the nature of the elementary reactions that build the molecules and the reactor system. Denbigh (*1*) pointed out in 1947 that continuous stirred tank reactors (CSTR) can yield either broader or narrower distributions than batch reactors, depending on whether the molecules continue to grow during their time in the reactor or are formed in times which are short compared with the residence time. Free radical polymerization usually conforms to the latter case since radical lifetimes are typically of the order of 1 sec or less. However, if branching occurs by reactions between growing radicals and previously formed molecules, such as in the case of ethylene or vinyl acetate polymerization, the chains are reactivated periodically, and continued growth is possible. Recent calculations (*2*) have shown that substantial broadening is expected in CSTR. The purpose of the present study is to test these conclusions experimentally in the vinyl acetate system.

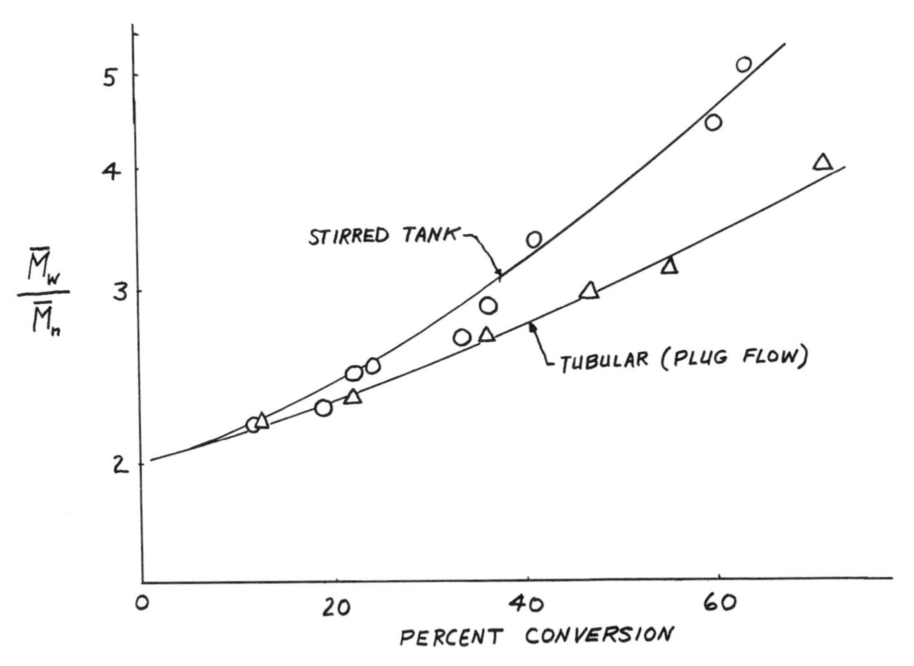

Figure 1. Solution polymerization of vinyl acetate

Solution polymerizations were conducted under argon at 60°C in batch and in a 1-liter continuous stirred reactor. The concentration of initiator (azobisisobutyronitrile) was 1.6 × 10^{-3} mole/liter, low enough to make the distribution of molecular weights independent of termination rate (controlled by propagation and chain transfer alone). The solvent was *tert*-butyl alcohol with an initial solvent–monomer ratio of 2/1. Conversions ranged from 12 to 71% in the batch system and 13 to 63% in the CSTR at steady state. Preliminary studies had shown an exponential distribution of residence times in the stirred system. Number-average molecular weights M_n were measured by osmometry; weight-average molecular weights M_w were obtained by light scattering.

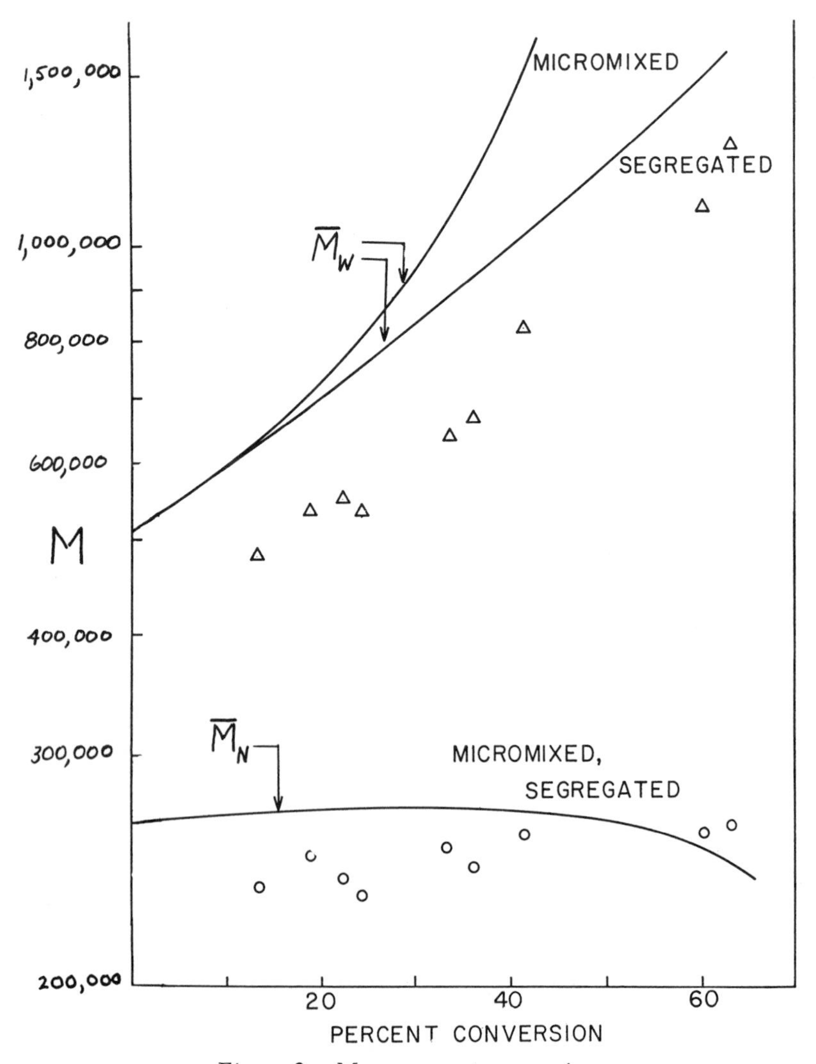

Figure 2. M vs. *percent conversion*

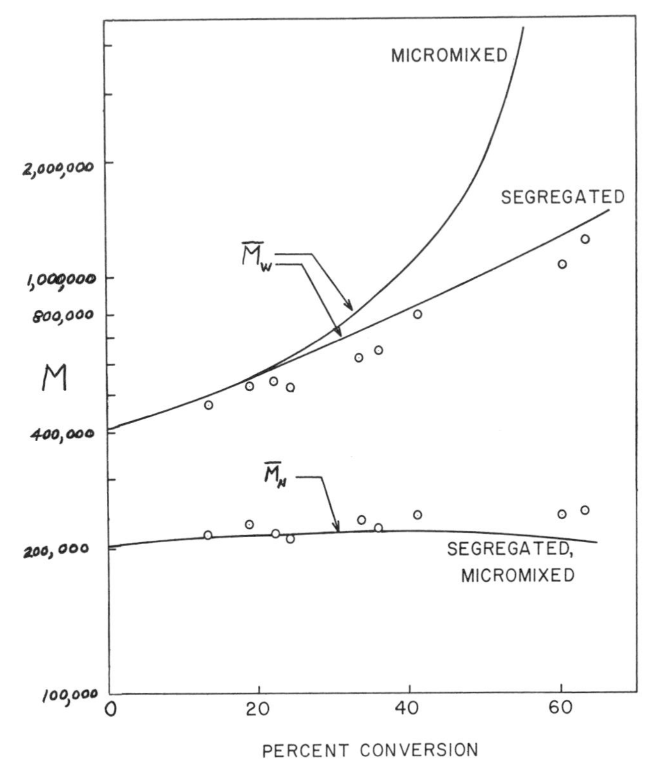

Figure 3. M vs. *percent conversion after adjusting* C_m *value*

Figure 1 shows M_w/M_n *vs.* conversion in the two reactors. As predicted, the ratio is nearly 2.0 at low conversions for both systems, but as conversion increases the CSTR products become more broadly distributed than the batch products. Branching density is also higher in the CSTR products, and this appears to be the principal broadening influence.

Branching density and molecular weight distribution are controlled by the relative rates of propagation, monomer transfer, solvent transfer, polymer transfer, and terminal bond polymerization. The kinetic constants for all reactions except solvent transfer are known from earlier batch studies on bulk polymerization of vinyl acetate at 60°C (*3*):

$$C_m = 2.43 \times 10^{-4},\ C_p = 2.36 \times 10^{-4}, \text{ and } K = 0.66$$

Differential equations were written for the population of polymer molecules and radicals and solved to obtain the first three moments of the distribution. Theoretical expressions were thereby obtained for M_n and M_w as functions of conversion x, with parameters C_m, C_p, K, and C_s. The expressions were quite different for batch and CSTR systems. Furthermore, the expressions were sensitive to the degree of segregation in the CSTR. Two cases, perfectly micromixed and perfectly segregated, were analyzed.

In the batch system excellent agreement was obtained for both M_n and M_w *vs.* x when C_m, C_p, and K from bulk polymerization were used with $C_s = 0.48 \times 10^{-4}$. However, when the same constants were used to compute M_n and M_w *vs.* x for the CSTR, the agreement was poor since experimental molecular weights were somewhat below both the micromixed and segregated curves (Figure 2). When C_m was adjusted to 3.2×10^{-4}, the values at low conversions were fitted, and the experimental data followed the segregated curve rather well (Figure 3).

The discrepancy is believed to arise from trace impurities (perhaps atmospheric oxygen) present in both reactors. An inhibition period was observed in batch polymerization, and we assume that during this time the impurity is consumed by initiator radicals. Normal polymerization subsequently takes place. In the CSTR system however, the impurity enters continuously with the feed, acting thereby to lower the molecular weight and retard the polymerization rate. Lower polymerization rates were indeed observed in the CSTR.

Thus, continuous stirred polymerization yields broader distributions than batch systems in branching polymerizations. However, although calculations for the segregated CSTR fitted the data best, one must regard the case as unproved until the inhibition anomaly is cleared up.

(1) Denbigh, K. G., *Trans. Faraday Soc.* (1947) **43**, 648; *J. Appl. Chem.* (1951) **1**, 227.
(2) Graessley, W. W., *AIChE—I. Chem. E. London Symp.* (1965) **3**, 16.
(3) Graessley, W. W., Hartung, R. D., Uy, W. C., *J. Polymer Sci. Pt. 2A* (1969) **7**, 1919.

The Use of Free Radical Homogeneous Addition Polymerization in Estimating Mixing Effects in an Imperfectly Mixed CSTR

R. W. DUNN[1] and C. C. HSU, Department of Chemical Engineering, Queen's University, Kingston, Ontario, Canada

A generalized mixing history model of an imperfectly mixed CSTR is developed where the reactor is considered to have two regions—a completely segregated entering environment and a micromixed leaving environment. It is assumed that the transfer from the entering environment to the leaving environment is instantaneous and that the distribution of times of transfer can be modelled by a mixing time distribution (MTD) which in properties is analogous to the residence time distribution (RTD).

The model requires the age distribution functions to calculate the output concentrations of the reactant from each environment. In the case of entering environment, the age distribution functions as defined by Danckwerts (*1*) can be used. These functions cannot be applied to the leaving environment because of the nature of the input. However, it can be shown that

$$E_l(t) = E_o(t) - E_e(t)$$

where E is the age frequency distribution and the subscripts o, e, and l refer to the over-all reactor system, entering environment, and leaving environment.

The average reactant concentration transferring between the environments C_m can be obtained by integrating batch reactor data with respect to the age α^* (*2*) which is defined as the maximum age of a molecule from the time it enters the system to the time it transfers to the leaving environment with a maximum life expectancy λ^*.

$$C_m = \int_o^\infty E(\alpha^*)C(\alpha^*)d\alpha^*$$

The concentration of a leaving environment plug flow reactor C_l is obtained by the following equation

$$\frac{dC_l}{d\Theta} = r(C_l) + (C_l - C_m)\frac{E_o(\Theta)}{F_e(\Theta) - F_o(\Theta)}$$

with the boundary condition

$$\frac{dC_l}{d\Theta} = 0 \text{ at } \Theta = \infty$$

[1] Present address: Rio Algom Mines, Ltd., P.O. Box 1500, Elliot Lake, Ontario, Canada.

In this equation $r(C_l)$ is the rate of reaction, θ the dimensionless time, and F the residence time distribution function. When the volume of entering environment is zero—*i.e.*, $F_e(\theta) = 1$ and $\lambda = \theta$—the above equation reduces to that derived by Zweitering (*3*).

The simplified kinetic model of a free radical homogeneous addition polymerization proposed by Tadmor and Biesenberger (*4*) was used to test the model. Experimental runs were simulated with a selected MTD and a perfectly mixed RTD by calculating the outlet conditions—*i.e.*, monomer concentration and the moments of the molecular weight distribution of polymer products. It was shown that the parameters of the MTD could be obtained from the outlet conditions of a known reactor system using the Rosenbrock pattern search method as described by Wilde (*5*) to minimize the deviation between the calculated sum of first three moments of the molecular weight distribution and that of the experimental values.

The results from this theoretical study are promising, and the method proposed here might be useful to study the imperfect mixing in a polymerization process in viscous region and to diagnose the mixing difficulties in existing industrial polymerization reactor system.

(1) Danckwerts, P. V., *Appl. Sci. Res., Sect. A* (1953) **3**, 279.
(2) Danckwerts, P. V., *Chem. Eng. Sci.* (1958) **8**, 93.
(3) Zwietering, T. N., *Chem. Eng. Sci.* (1959) **11**, 1.
(4) Tadmor, Z., Biesenberger, J. A., *Ind. Eng. Chem., Fundamentals* (1966) **5**, 336.
(5) Wilde, D. J., "Optimum Seeking Methods," Prentice-Hall, Englewood Cliffs, N. J., 1964.

Calculation of High Pressure Polyethylene Reactors by a Model Comprehending Molecular Weight Distribution and Branching of the Polymer

J. THIES and K. SCHOENEMANN, Technical University of Darmstadt, 61 Darmstadt, West Germany

This investigation of the structural kinetics of high pressure polyethylene is the continuation of the development of the industrial process. In discontinuous batch experiments the conversion of ethylene was determined from the pressure decrease (Figure 1) and described by the simplest free radical chain mechanism which sufficed for the design of a plant of 24,000 tons/year capacity (*1, 2, 3, 4*).

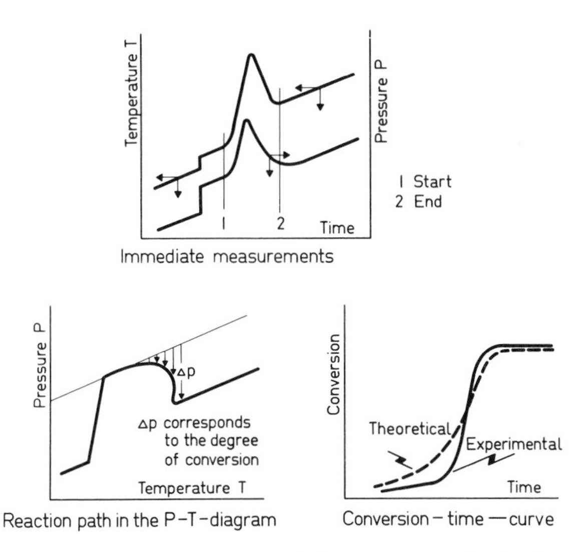

Figure 1. Determination of the reaction velocity by batch polymerization

This result was encouraging enough to proceed towards the final objective of determining by a kinetic model the molecular weight distribution and the branching of the polymer. This objective includes the fundamental problem of whether such a complicated project is feasible experimentally as well as mathematically.

To obtain the rate constants of the mass balances in the form of simple algebraic equations the laboratory experiments were carried out in a continuously operated stirred vessel microreactor (*5*). The molecular weight distribution was determined by gel permeation chromatography (*6*). The molecular weight was calculated after an indirect method based on the Flory relation of the hydrodynamic volume (*7, 8*) and on the still unknown long chain branching.

To describe the polymerization a model was postulated which included nine types of reactions: initiation, propagation, radical transfer to the monomer and to the polymers, intramolecular radical transfer, depolymerization of radicals, an undefined formation of radicals for the sake of empirical corrections, termination by disproportionation and by combination.

The mass balance of a reactant i involved in n kinds of reactions carried out in a stirred vessel in a steady state flow with the admission concentration c_{i_o}, the exit concentration c_i and the mean residence time $\tau = V/q$ is expressed by:

$$\frac{1}{\tau} \cdot (C_{i_o} - C_i) + \sum_{m=1}^{n} r_{i_m} = 0$$

As an example the mass balance for a polymer P of chain length ν can be formulated as follows:

$$\frac{1}{\tau} \cdot c_{P_\nu} = k_{UM} \cdot c_M \cdot c_{R_\nu} + k_{UP} \left(c_{R_\nu} \sum_{i=1}^{\infty} c_{P_i} - c_{P_\nu} \sum_{i=1}^{\infty} c_{R_i}\right) +$$

$$2k_D \cdot c_{R_\nu} \cdot \sum_{i=1}^{\infty} c_{R_i} + k_K \sum_{i=1}^{\nu-1} c_{R_i} \cdot c_{R(\nu-1)}$$

The quantity of P_ν is formed according to four rate terms; the first represents the formation by transfer of radicals R_ν to monomers M, the second the transfer to polymers P, the third the formation by disproportionation, and the fourth that by combination.

By means of such mass balances, which then were established for all molecules of $\nu = 1 \ldots \infty$, both formed and consumed, the nine types of reactions of the entire system are interconnected, and, therefore, the formation of the structure of the polymerizate is described.

The probability w of the appearance of a polymer P with chain length ν as a function of ν is expressed by the ratio of the respective individual concentration $c_{P\nu}$ to the total concentration of the sum of all polymers

$$w_P(\nu) = \frac{c_{P_\nu}}{\sum_{i=1}^{\infty} c_{P_i}}$$

This ratio enters the statistical zero moments μ_k of the molecular weight distribution (9)

$$\mu_{Pk} = \sum_{v=1}^{\infty} \nu^k \cdot w_P(\nu)$$

in which k represents the number of the successive statistical zero moments; thus μ is experimentally determinable.

To establish the balance equations of the statistical zero moments (which therefore represent the interdependence of the postulated nine kinds of reactions) each of the single balances for a polymer (or for a radical) of the chain length ν was multiplied by the individual factor

$e^{j\nu x}$ which is taken from the complex Fourier series from which the zero moments are derived.

$$\frac{1}{\tau} \cdot w_P(\nu) \cdot e^{j\nu x} = \left\{ \frac{k_{UM} \cdot c_M \sum_{i=1}^{\infty} c_{R_i}}{\sum_{i=1}^{\infty} c_{P_i}} \cdot w_R(\nu) + k_{UP} \sum_{i=1}^{\infty} c_{Ri}[w_R(\nu) - w_P(\nu)] + \right.$$

$$\left. \frac{2k_D \left(\sum_{i=1}^{\infty} c_{R_i}\right)^2}{\sum_{i=1}^{\infty} c_{P_i}} w_R(\nu) + \frac{k_K \left(\sum_{i=1}^{\infty} c_{R_i}\right)^2}{\sum_{i=1}^{\infty} c_{P_i}} \cdot \sum_{i=1}^{v-i} w_R(i) \cdot w_R(\nu - i) \right\} e^{j\nu x}$$

The infinite number of single balances modified in this way is always added to form a summarized balance in which the entire information of the many single balances and, therefore, of the molecular weight distribution is contained.

In these summarized balances complex Fourier series of the form

$$F_P = \sum_{v=1}^{\infty} e^{j\nu x} w_P(\nu)$$

appear which express the molecular weight distribution. Continuing the above example, the summary balance of the formation of polymers ($\nu = 1 \ldots \infty$) reads

$$\frac{1}{k_K} \left(\frac{k_{UM} \cdot c_M}{\sum c_{R_i}} + \frac{k_{UP} \sum c_{P_i}}{\sum c_{R_i}} + 2k_D \right) \cdot (F_R - F_P) + F_R^2 - F_P = 0$$

The branching of the molecules which represents the second main characteristic of structure, prevails in high pressure polyethylene in contrast to low pressure polyethylene. This is because the velocities of the various kinds of reaction depend strongly on reaction conditions. The main influence on branching is exerted by the transfer of radicals. Directed to polymers, it determines the long chain branching and thus viscosity, melt index, etc.; intramolecular radical transfer, however, causes short chain branching which influences density, crystallinity, etc. To determine this branching the material balances for the formation of the structural groups were established in a manner similar to the one as described above.

For each of the polymerizates produced at different reaction conditions, the above system of mass balances of the individual components

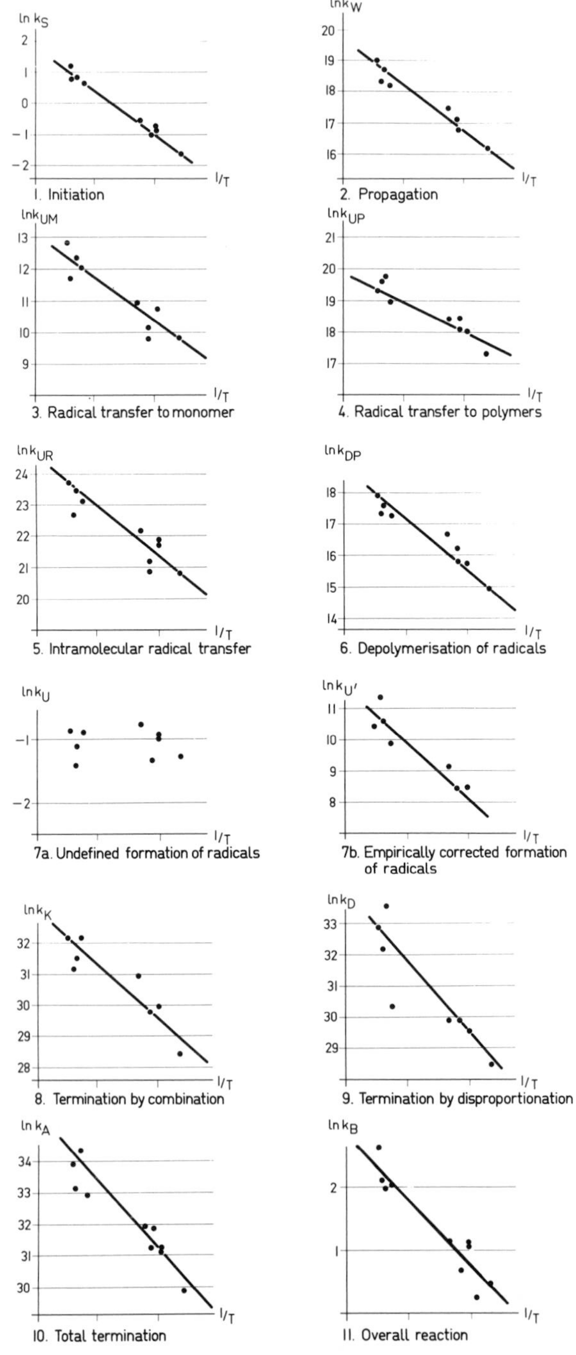

Figure 2. Arrhenius diagrams of the nine types of reactions of the structure-kinetic model

involved was solved for the unknown velocity constants by substituting 15 balance equations into one another.

The reaction velocity constants determined in the various experiments for each of the nine types of reaction were plotted in Arrhenius diagrams log k *vs.* $1/T$ (Figure 2). Since satisfactory regression lines could be drawn in all cases, the usefulness of the postulated model is proved.

The optimization by means of extensive programs (necessary for the nonlinear regression) and by the many iterations was made possible only by the good approximation values read from the Arrhenius diagrams (*10*). The internal consistency of the model was checked by comparing the conversion degrees and structure characteristics calculated with the experimental values (*11*). The mean deviation from the experimental values is not yet quite satisfactory, but it arises mainly from experimental errors in the microreactor rather than from imperfections in the model.

It is now possible to describe the structure of a polymer by a model without exceeding reasonable bounds of mathematical effort.

Acknowledgment

The authors express thanks for valuable support to W. Ring of Chemische Werke Hüls who analyzed the polymerizates and to B. Dietzer who helped in the computation.

Literature Cited

(1) Schoenemann, K., *Achema-Jahrbuch* (1962–64) **1,** 222.
(2) Schoenemann, K., *Chem. Eng. Sci.* (1963) **18,** 565.
(3) Gilles, E. D., *Chem. Ing. Tech.* (1968) **40,** 469.
(4) Schoenemann, K., Steiner, R., Luft, G., *Chem. Eng. Sci.*, in press.
(5) Hess, W., Ph.D. thesis, Darmstadt (1968).
(6) Ring, W., Chemische Werke Hüls, Marl, personal communication.
(7) Zimm, B. H., Stockmayer, W. H., *J. Chem. Phys.* (1949) **17,** 1301.
(8) Drott, E., *Intern. Seminar Gel Permeation Chromatog., 4th, Miami Beach, Fla., May 1967.*
(9) Gilles, E. D., Knöpp, U., *Regelungstechnik* (1967) **15,** 199, 261.
(10) Marquardt, W., *J. Soc. Ind. Appl. Math.* (1963) **11,** 431.
(11) Sperati, C. A., Franta, W. A., Starkweather Jr., H. W., *J. Am. Chem. Soc.* (1953) **75,** 6127.

The Effect of Imperfect Mixing on the Initiator Productivity in the High Pressure Radical Polymerization of Ethylene

TH. J. VAN DER MOLEN and C. VAN HEERDEN, DSM, Central Laboratory, Geleen, The Netherlands

The initiator consumption in the radical polymerization of ethylene was investigated in a 1-liter stirred autoclave ($h/d \sim 2$). Polymerizations were performed continuously at a constant pressure of 1600 atm. Five different types of initiators (peroxides) were needed to bridge the temperature range of 130°–270°C.

Formulation

The general formulation for the over-all polymerization rate r in moles/liter/sec is:

$$r = k_{ov}[M][I]^{1/2} = \frac{k_p}{(2k_t)^{1/2}} (n\varepsilon)^{1/2} k_i^{1/2} [M][I]^{1/2} \tag{1}$$

For use in continuous polymerization experiments this equation can be modified by a mass balance for monomer, initiator, and radicals. By means of the initiator mass balance the concentration in the reactor [I] can be converted to the concentration in the feed $[I_o]$—*viz.*, $[I] (k_i\tau + 1) = [I_o]$. For the continuous process the initiator decomposition rate is so high that always $k_i\tau >> 1$, with the consequence that k_i can be eliminated by substituting $[I] = [I_o]/k_i\tau$ in Equation 1. One then obtains:

$$r = \frac{k_p}{(2k_t)^{1/2}} (n\varepsilon)^{1/2} [M] \frac{[I_o]^{1/2}}{\tau^{1/2}} = \frac{k_p}{(2k_t)^{1/2}} (n\varepsilon)^{1/2} [M] \frac{Q_i^{1/2}}{V^{1/2}} \tag{2}$$

In developing the equations, in agreement with Symcox and Ehrlich (*1*), we used molar concentrations instead of fugacities.

The initiator consumption per unit time is not an unambiguous function of the polymerization temperature. On the one hand, theoretically the initiator consumption should decrease with increasing temperature since the propagation rate is favored more strongly than the termination rate; on the other hand, in the continuous polymerization of ethylene under high pressure a higher temperature means a higher degree of ethylene conversion and hence an increase of initiator consumption.

Thus, we use a more convenient parameter, the so-called initiator productivity: η in terms of (moles converted ethylene)/(mole initiator), both per unit time. Starting from Equation 2 the following relation for η can now be obtained simply:

$$\eta = \frac{\text{moles converted ethylene}}{\text{mole initiator}} = \frac{2k_p^2}{k_t} n\varepsilon[M] \frac{1-\xi}{\xi} \tau \tag{3}$$

In developing Equation 3 we assumed that (in theory) each molecule of initiator yields n radicals and that the efficiency amounts to ϵ. At constant pressure (1600 atm) the relationship between η and the temperature can be determined only if η is related to the same values of [M], τ, and ξ. That is why for each series of experiments calculated by Equation 3 the value of η is always corrected to standard conditions [M] = 14.0 moles/liter, τ = 120 sec, and ξ = 0.10.

If $n\epsilon$ is assumed to be constant and $k_{ov} \approx k_p/k_t^{1/2}$ can be described by an Arrhenius relation, a rectilinear relation can be expected to exist between log η and $1/T$ as long as the reactor can be considered to be perfectly mixed.

η and Polymerization Temperature

In Figure 1 the corrected values for η are plotted against $1/T$(°K). For each initiator ($n = 2$) a similar curve is found, showing the following three regions:

(1) The Lower Temperature Region. For every initiator the log η curve terminates at the low temperature side in a point where the continuous polymerization process terminates because of a relatively too small over-all rate of polymerization. For temperatures higher than this minimum, η increases owing to the increase of k_p^2/k_t. The slope of this part of the curve is determined by 2 $E_p - E_t$, which is independent of the type of initiator and amounts to 14 kcal/mole, in good agreement with the results of other investigators (2).

(2) The Transition Region. By raising the temperature further a transition region is found for each type of initiator, in which a maximum in η is reached at a critical temperature T_k. The values of η_k and T_k are different for each initiator investigated.

(3) The Upper Temperature Region. For temperatures above T_k η appears to decrease rapidly in all cases. This presumably is caused by a premature initiator decomposition before the contents of the reactor have been mixed sufficiently—in other words, radicals are wasted.

To account for these phenomena, the following characteristic times are introduced: (a) τ = reciprocal space velocity, (b) $\tau_i = 1/k_i$ = characteristic time constant for initiator decomposition, (c) τ_m = characteristic time constant for mixing in the reactor, which is of the order of the reciprocal stirrer speed. In the lower temperature region: $\tau_m < \tau_i < \tau$. Here perfect mixing in the reactor can be assumed. In the transition region at $T = T_k : \tau_m \approx \tau_i < \tau$. The half-lifetimes of the initiators, used in this investigation, are known only at low temperatures. Extrapolation of the half-life curves, as a function of the reciprocal of absolute temperature, up to the real polymerization temperatures, and taking into account the effect of high pressure on the initiator decomposition rate (a decrease of about 30% at 1600 atm relative to atmospheric pressure), results in the following values for τ_i at $T = T_k$—*viz.*, 0.07 sec (I_2), 0.04 sec (I_3) and 0.18 sec (I_4). This means that if $\tau_m \approx \tau_i$ at $T = T_k$, mixing in the reactor is accomplished after one to four rotations of the stirrer (speed = 1500 rpm), which is a reasonable result.

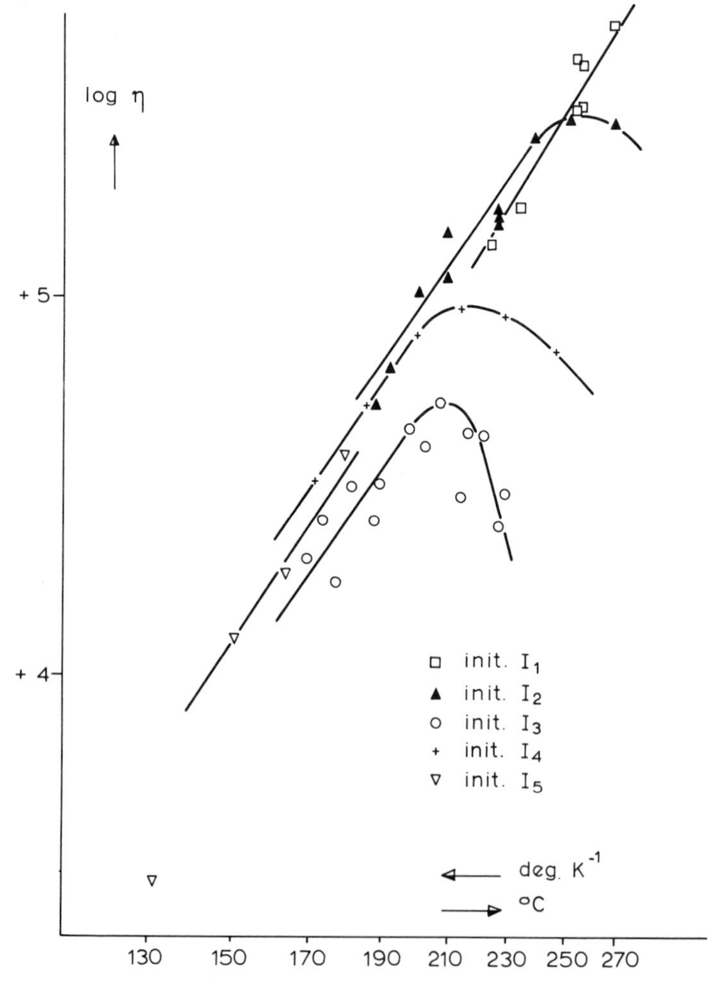

Figure 1. Polymerization temperatures and initiator productivity (corrected to standard conditions) at 1600 atm

In the upper temperature region: $\tau_i < \tau_m < \tau$. With regard to the initiator concentration the reactor can no longer be considered perfectly mixed. One may attempt to describe the rapid decrease of the log η curves at the higher temperatures proceeding from a recirculation model with τ_m as recirculation time. From this model a modified formulation for η is deduced, which also comprises the upper temperature region $T > T_k$—*viz.*:

$$\eta = \frac{k_p^2}{2k_t} n\varepsilon[M] \frac{1-\xi}{\xi} \frac{\tau}{\left[1+\left(\frac{\tau_m}{4\tau_i}\right)^2\right]^{1/2}} \quad (4)$$

The appearance of a maximum in the log η curve now can be explained simply, with the help of Equation 4. As long as $T < T_k$, $\tau_m/\tau_i < 1$ and Equation 4 becomes equivalent to Equation 3. In this lower temperature region η increases with temperature as k_p^2/k_t (apparent activation energy 14 kcal/mole) and does not depend on the choice of the initiator. However, as soon as $T > T_k$ so that $\tau_m/\tau_i > 1$ the temperature dependency of τ_m/τ_i (activation energy ~ 35 kcal/mole) will be much stronger than that of k_p^2/k_t, and the value of η will decrease more and more, going finally to a slope with an apparent "negative" activation energy of about 20 kcal/mole.

Nomenclature

E_{ov}	Activation energy (over-all rate), kcal/mole
E_p	Activation energy (propagation rate), kcal/mole
E_t	Activation energy (termination rate), kcal/mole
k_{ov}	Over-all rate constant, liter$^{1/2}$ mole$^{-1/2}$/sec
k_i	Initiator decomposition rate constant, sec^{-1}
k_p	Propagation rate constant, liter/mole/sec
k_t	Termination rate constant, liter/mole/sec
r	Ethylene conversion rate, moles/liter/sec
ξ	Degree of monomer conversion to polymer
[M]	Ethylene concentration in the reactor, moles/liter
[I]	Initiator concentration in the reactor, moles/liter
$[I_o]$	Initiator concentration in the feed, moles/liter
Q_i	Initiator feed rate, moles/sec
η	Initiator productivity $\dfrac{\text{moles converted ethylene}}{\text{mole initiator}}$
η_k	Initiator productivity at $T = T_k$
$\tau_i = 1/k_i$	Characteristic time constant for initiator decomposition, sec
τ_m	Characteristic time constant for mixing in the reactor, sec
τ	Reciprocal space velocity, sec

t	Time, sec
T	Absolute temperature, °K
T_k	Critical temperature at which η is a maximum, given the type of initiator, τ, and τ_m, °K
n	Theoretical number of radicals per molecule of initiator
n_ϵ	Real number of radicals per molecule of initiator
h	Reactor height, meters
d	Reactor diameter, meters
V	Reactor volume, liters

(1) Symcox, R. O., Ehrlich, P., *J. Am. Chem. Soc.* (1962) **84**, 531.
(2) Ehrlich, P., Mortimer, G. A., *Advan. Polymer Sci.* (1970) **7**, 386.

The Degradation of Linear and Branched Polymers: A Quantitative Study Using Simulation Techniques

P. J. MEDDINGS[1] and O. E. POTTER, Department of Chemical Engineering, Monash University, Clayton, Victoria, Australia

A simulation procedure has been developed for studying the degradation of linear and branched polymers. The simulation procedure is stochastic, requiring the repetitive scissions of the bonds to obtain the appropriate statistical features of the process. The application of the simulation procedure to linear polymers for the degradation of which analytical solutions (*1, 2*) are available has been used to test the simulation, with satisfactory results.

For linear chains the units may be defined simply by numbering bonds, but for branched structures the problem is more complex. A typical branched molecule is amylopectin, a component of starch, in which the monomeric unit is glucose and each glucose unit is potentially capable of being linked with other glucose units by an α-1,4 bond and an α-1,6 bond. Three types of chains are considered to occur in the amylopectin structure. The average structure of amylopectin is known, but it is highly improbable that each external chain would have the same length or that the number of units between branch points would always be identical. Molecules were simulated in the computer, variations of chain length and type being included by the use of probability functions.

[1] General Manager, Mauri Bros. and Thompson Ltd., Australia.

By adjusting the means and variances for the various operations performed, many different types of structure can be obtained. Figure 1 shows a typical result for the amylopectin molecule built with a mean distance between branch points of eight units with a standard deviation of 2, and the length of external chains having a mean of 12 units with a standard deviation of 2.

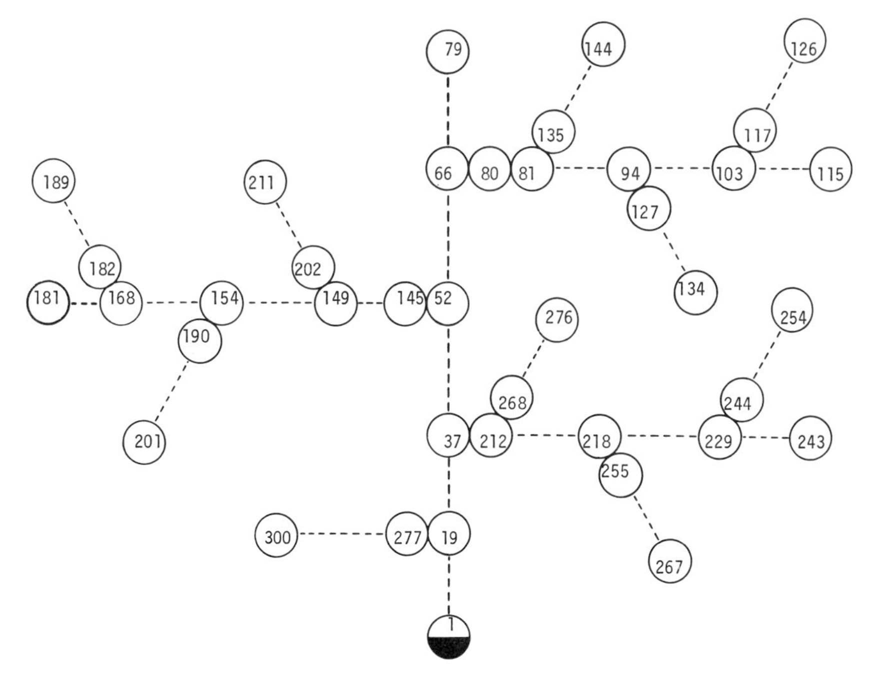

Figure 1. A simulated amylopectin molecule

The simulation procedure can be used for reactions other than first order—*e.g.*, for enzymic reactions. For such reactions involving polymers the substrate molecules available for complexing include degradation products of previous reactions. The procedure involves:

(a) The definition of a number of molecules of the substrate.

(b) Choice of one molecule of substrate which is equivalent to complexing with a single enzyme molecule.

(c) Choice of a bond in the molecule.

(d) Calculation of the interval of time involved in complexing the molecule and in the subsequent breakdown.

(e) Acceptance or rejection of the bond for scission, depending on steric qualifications etc.

(f) Redefinition of the units in the fragments of the broken molecule, including codes on the first unit and all end units to indicate their position and also the type of molecule in which they are incorporated.

To test the program for the enzymic hydrolysis of branched and linear substrates, numerical solutions of the set of nonlinear differential equations can be obtained conveniently if the molecular size of the initial substrate is small, say 12 units long. The numerical solution and the stochastic simulation procedure were compared and gave similar results —*e.g.*, at an arbitrary time the numerical solution in a particular case indicated 19.9% of bonds broken as against 20.3% for the simulation procedure and a concentration of the six-unit molecule of 6.8% as against 6.7% for the simulation procedure.

The simulation procedure provides the facility to study large and complex structures in reasonable computer time, within acceptable and predictable limits of accuracy. It is possible to allow for steric effects, a mixture of enzymes, competitive and non-competitive inhibition. If the components of a heteropolymer (*e.g.*, protein) are known, synthesis and degradation can be simulated by the method.

(1) Simha, R., *J. Applied Phys.* (1941) **12**, 569–578.
(2) Liu, S-L., Amundson, N. R., *Rubber Chem. Technol.* (1961) **34**, 995–1133.

Syntheses of Minimum-TimeTemperature Paths for Free Radical Polymerization

YOUNG D. KWON, LAWRENCE B. EVANS, and JAMES J. NOBLE, Department of Chemical Engineering, Massachusetts Institute of Technology, Cambridge, Mass. 02139

Although many papers on computing molecular weight distributions from polymerization kinetics have been published (*1*), few studies have been reported (*2, 3*) on the optimization of polymerization processes. Apparently, this is because of the complexity of the kinetic model which involves thousands of reacting species. This paper introduces a "continuous blending model" as a manageable and useful method of computing the molecular weight distribution in a free radical polymerization system and demonstrates its utility in ascertaining the system controllability and in synthesizing minimum-time–temperature paths for the thermally initiated bulk polymerization of styrene in a homeneous batch reactor.

The continuous blending model accounts for the various blending effects arising from reaction, convection, diffusion, and final mixing in polymerization systems. This model identifies the incremental portion of polymer which is formed successively at each instant of time during the batch and assigns a molecular weight distribution to it with the parameters x_n^o, x_w^o and w_x^o. Then the parameters of the final molecular weight distribution, x_{n_f}, x_{w_f} and w_{x_f}, can be computed simply by blending these incremental polymer portions. The blending processes can be represented by differential equations with x_n, x_w, and w_x as the state variables.

The continuous blending model is equivalent to use (*4*) of the moments of the molecular weight distribution if the definitions $x_n^o = dQ_1/dQ_0$, $x_w^o = dQ_2/dQ_1$, and $w_x^o = d(x[M_x]/dQ_1)$ are admitted. These definitions call for short average life of active radicals. In the styrene polymerization system, as in many other polymerization systems, x_n^o, x_w^o, and w_x^o can be represented by relatively simple empirical expressions rather than by the complex kinetic expressions. This is where the major simplification is effected with the continuous blending model.

The model is tested by applying it to the system studied by Hamielec *et al.* (*5*), who polymerized styrene in benzene with an initiator and compared the molecular weight distribution determined by gel permeation chromatography with that predicted by integrating differential equations describing the rate of change of x-mer concentration for 5000 species. Values of x_n and x_w calculated by the two blending equations and values of w_x calculated by one blending equation at intervals of $\Delta x = 100$ gave a result which is practically identical to that of Hamielec *et al.*

The continuous blending model is used to synthesize the temperature paths which polymerize a monomer to a desired conversion with desired values of x_{n_f} and x_{w_f} in the minimum possible time. Bulk polymerization of styrene by thermal initiation is taken as the example system. Because the minimum-time problem presupposes controllability, it is desirable to answer the question: is there at least one temperature path which can steer the system from the given initial state to the desired fixed end state subject to the system dynamics and the control constraint? For a nonlinear system, this is a difficult question. The continuous blending model, however, lets us ascertain the controllability of the styrene polymerization system in an explicit manner. The range of allowable temperatures required for the controllability is a function of the end value of polydispersity.

Synthesis of minimum-time–temperature paths with the continuous blending model are amenable to the application of the Lagrange multiplier method (*6*). For the case of 80% conversion the resulting computational problem involves simultaneous solution of 82 nonlinear equations. An algorithm has been developed for solving this nonlinear programming problem; its main features are a method of determining the multipliers and a method of satisfying the two-point boundary conditions during the iterations.

The resulting minimum-time–temperature paths show a gradual rise from a relatively low initial temperature to a higher terminal temperature. This pattern of the temperature path reflects the compromise between the decrease of reaction rate with the progress of conversion and the decrease in molecular weight averages for the incremental polymer portions formed at the increased system temperature.

The work reported here is a part of a comprehensive investigation (7) of the minimum-time trajectories for a distributed parameter system with the reaction and diffusion effects of blending and the regulation of these trajectories against disturbances.

Nomenclature

Q_n: nth moment of molecular weight distribution
x_n, x_w, w_x: number average, weight average chain lengths, x-mer weight fraction
$[M_x]$: Molar concentration of x-mer polymer chains
Superscript ^{o} refers to the incremental polymer, and the subscript f refers to the final values.

(1) Amundson, N. R., Luss, D., "Polymer Molecular Weight Distribution," *J. Macromol. Sci., Rev., Macromol. Chem.* (1968) **2** (1), 145.
(2) Ray, W. H., "Modeling Polymerization Reactors with Applications to Optimal Design," *Can. J. Chem. Eng.* (1967) **45**, 356.
(3) Hicks, J., Mohau, A., Ray, W. H., "The Optimal Control of Polymerization Reactors," *AIChE Natl. Meetg., 65th, Cleveland, 1969, Symp. Paper,* No. 8b.
(4) Bamford, C. H., Tompa, H., "The Calculation of Molecular Weight Distribution from Kinetic Scheme," *Trans. Faraday Soc.* (1954) **50**, 1097.
(5) Hamielec, A. E., Hodgins, J. W., Tebbens, K., "Polymer Reactors and Molecular Weight Distribution: Part II. Free Radical Polymerization in a Batch Reactor," *AIChE J.* (1967) **13** (6), 1087.
(6) Athans, M., Falb, P. L., "Optimal Control—An Introduction to the Theory and Its Applications," McGraw-Hill, New York, 1966.
(7) Kwon, Y. D., "Optimal Design and Control of Bulk Polymerization Processes," Sc.D. thesis, Department of Chemical Engineering, Massachusetts Institute of Technology (August 1970).

Some Design Models for Viscous Polymerization Reaction Systems

W. C. BRASIE, Process Systems Engineering, The Dow Chemical Co., Midland, Mich. 48640

One of the more significant challenges in polymer reaction engineering is in the design of systems to produce polymers in mass or in solution with solvents plus monomers (*1*). It is useful to provide a qualitative model of such processes whether the systems are batch or continuous and particularly for those cases where the viscosity of the system plays an important role. Models useful for these systems are described briefly.

There are several general characteristics of mass or solution polymerizations which the model must take into consideration (*1*). These include:

(a) A single-phase essentially homogeneous solution throughout the conversion range, whose viscosity may easily vary over a range of six to eight decades.

(b) The rate of polymerization involves classical mechanisms of initiation, propagation, and termination, is not necessarily single-ordered throughout the conversion range, and has typical Arrhenius temperature dependence. Exothermic heats of polymerization are in the range of 10 to 26 kcal/mole.

(c) Molecular weights produced vary inversely with temperature.

The basic data required to construct the model include the rate of conversion *vs.* temperature, monomer and solvent concentration, and concentrations of catalyst or other materials. Physical, thermal, and transport properties as a function of temperature and component concentration are needed. These include density, specific heat, thermal conductivity, diffusion coefficients, fluid viscosity, including non-Newtonian effects for continuous flow, and enthalpy and heat of reaction values.

The basic equations required to construct the model are the classical equations of continuity, momentum, energy, and reactant balances. These must be solved *via* numerical techniques on a digital computer using various finite difference methods, solution of simultaneous equations and iteration schemes which are well known but generally complex. As in most typical numerical methods, a grid of points is superimposed on the reactor control volume of interest. For flow systems, a radial grid and iteration for each related flow axis distances comprise the grid. For batch geometries, a two-dimensional system with iteration with respect to time steps and both dimensions is necessary. The use of subroutines to calculate various physical, thermal, and rate parameters allows basic building units to be used in models describing either flow or batch reaction systems. A single model can describe several different geometries by making

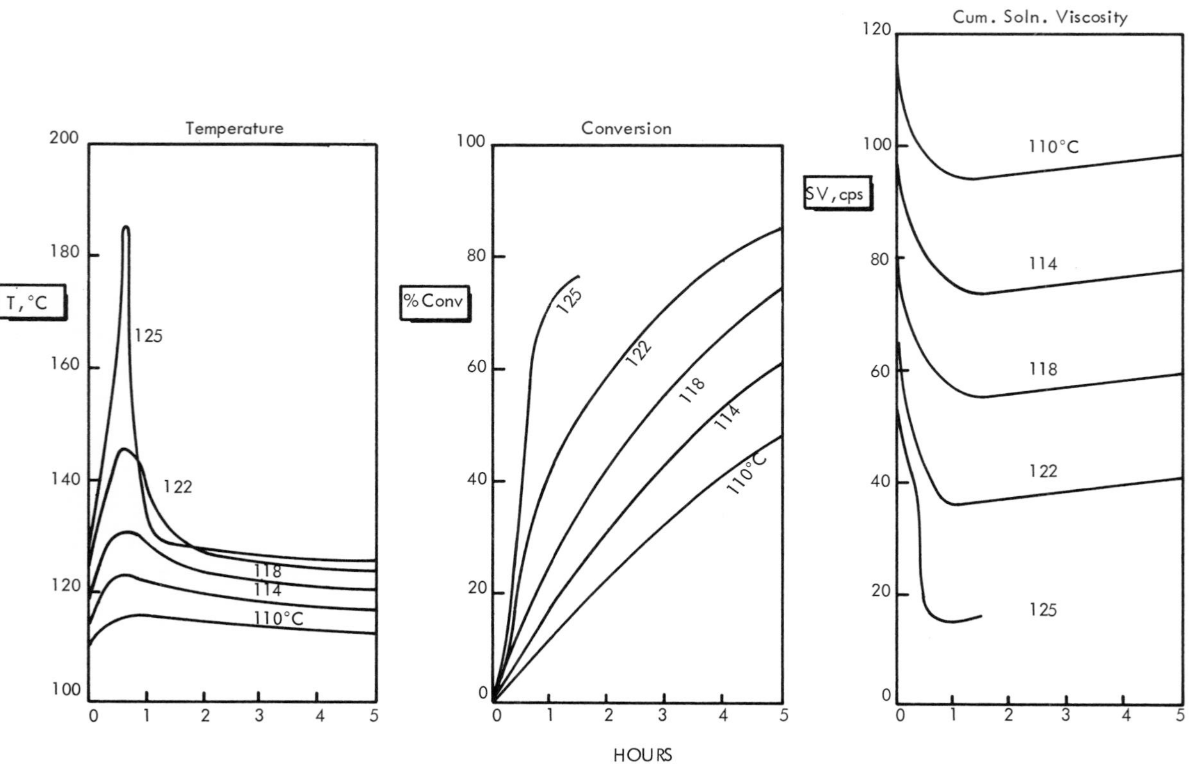

Figure 1. Time vs. *temperature, conversion and cumulative solution viscosity for a batch reactor for constant wall temperatures of 110°, 114°, 118°, 122°, and 125°C.*

small changes in boundary conditions—*e.g.*, flow inside round tubes or parallel plates or outside (parallel to) tubes or in annuli.

Accurate models must include all first-order terms and all terms of significance which affect local concentration, temperature, viscosity, or energy release including diffusion of monomers and viscous dissipation for flow systems of very high viscosity. Heat transfer is assumed to be by conduction. Natural convection is ignored, and body forces are neglected. To avoid iteration and convergence difficulties, grid spacing and time steps should be such to limit the local fractional conversion at any grid point per time step to less than 0.01.

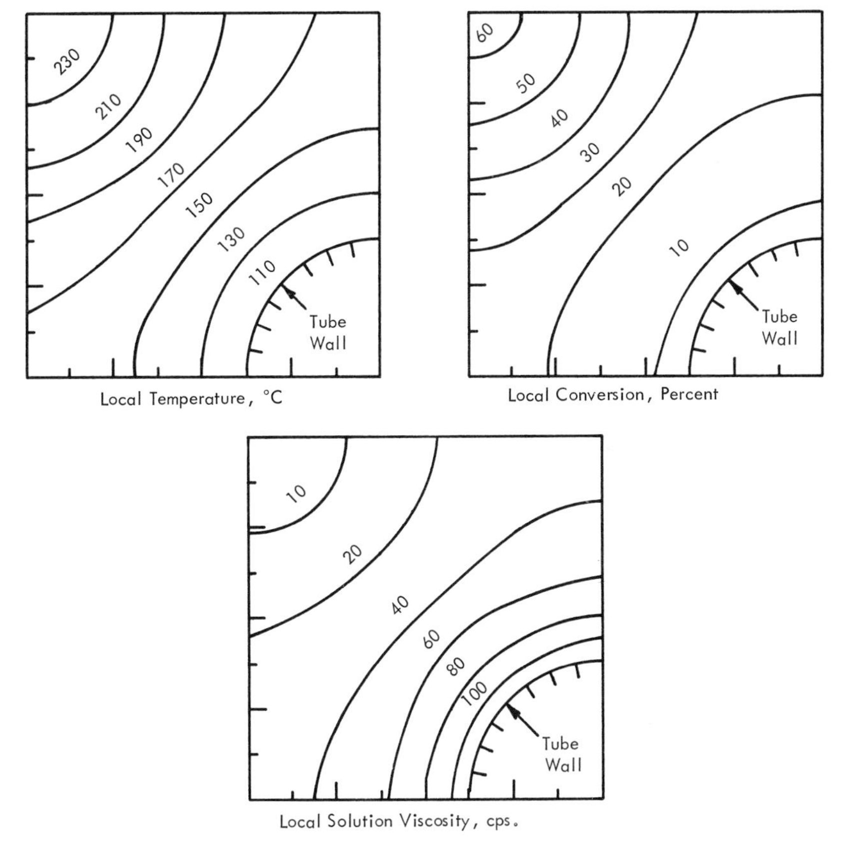

Figure 2. Local values of temperature, conversion, and solution viscosity in a 1 × 1-inch control volume of a batch reactor at 1.6 hours (T_w = *110°C)*

The exact details of the models are too voluminous to present here. Instead, it is useful to show some example results of using the models. Figure 1 shows calculations of the variation in mass average temperature, conversion, and cumulative solution viscosity (a direct measure of molecular weight) *vs.* time for a batch reactor consisting of 0.84-inch od heat transfer tubes in a 1.5-inch square pitch array. The variation of tube wall causes local runaway or hot-spot temperature conditions of increasing severity at higher wall temperature values. Correspondingly, the cumulative solution viscosity decreases, and a significant variation in terminal solution viscosity is possible by small temperature variations. Local instantaneous temperature, conversion, and solution viscosity profiles are shown in Figure 2 for a control segment of a 2-inch square pitch array of 0.75-inch od tubes at 1.6 hours after startup at 110°C and with constant wall temperature. The results are based on published data (*2*) for styrene but are illustrative of the type of results the models yield—*i.e.*, complex variations of temperature, conversion, and molecular weight.

Results for flow systems are more complex and depend strongly on individual monomer and geometric variables. The general models will predict velocity, temperature, and concentration profiles, "hot-spots," stagnant polymer layers, and similar results. Complex reactors such as Figure 3 (*3, 4*) may be modeled in principle by dividing the reactor into pieces, each of which can be described by an appropriate model with

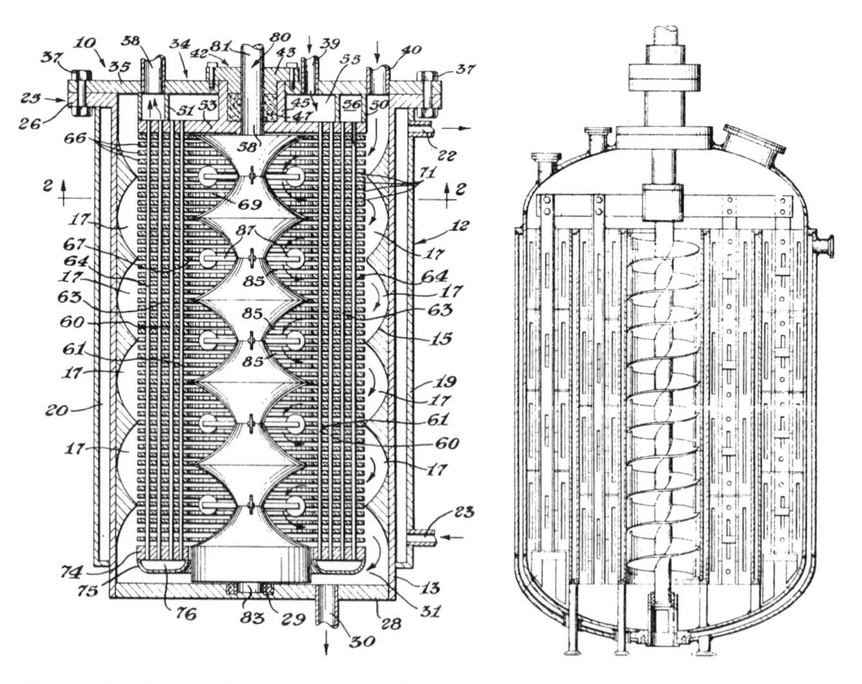

Figure 3. Patented reactors for polymeriaztion of exothermic, viscous monomer–polymer mixtures

Left: U.S. Patent 3,280,899 (3)
Right: U.S. Patent 3,206,287 (4)

qualifying assumptions. Thus, a reactor may be a series of well-mixed pots, or partially mixed pots joined by unmixed channels of finite geometry, for example. All that is required are reasonably good assumptions on the flow patterns, and good data, particularly concerning rate data and viscosity parameters. Within these limits there are few limitations to modeling viscous polymerization reaction systems, and results providing engineering accuracies are easily achievable.

(1) Brasie, W. C., "Elements of Polymer Engineering," AIChE Professors Workshop on Polymer and Monomer Engineering, Midland, Mich. (1968).
(2) Boundy, R. H., Boyer, R. F., "Styrene—Its Polymers, Copolymers, and Derivatives," Reinhold, New York, 1952.
(3) Brasie, W. C., U.S. Patent **3,280,899** (Oct. 25, 1966).
(4) Crawford, J. R., U.S. Patent **3,206,287** (May 1965).

3

Fluidized Bed Reactors: A Review

D. L. PYLE

Department of Chemical Engineering and Chemical Technology,
Imperial College, London, S.W. 7, England

Recent work relevant to the development of models for analyzing and designing fluidized bed reactors is reviewed. Special attention is placed on attempts to develop mechanistic models, and the various features of the existing models are compared and contrasted and related to other experimental and theoretical studies. In general it appears that most effort has been placed on the study of single bubbles and on attempts to correlate reactor performance in terms of these studies. In practice, as experimental studies show, many of the important characteristics of fluidized reactors can be explained only in terms of more complex models accounting for bubble interactions, coalescence, and growth. Although it is now possible to explain qualitatively, at least, many of the features of such reactors, much more theoretical and critical experimental work is needed before a priori *design methods can be used with confidence.*

The analysis of fluidized bed reactors provides one of the clearest examples we have of the complex interaction between different physical and chemical processes that typify chemical engineering. This review attempts to describe some of the characteristic features of gas fluidized beds, particularly those which seem to have the most important effect on their operation as chemical reactors. It is intended to provide a reasonable background to the field although it is not a comprehensive literature survey.

Even a superficial study of a fluidized bed in operation reveals its essential complexity. Bubbles form, at an apparently unpredictable rate, and subsequently grow—perhaps to the same dimensions as the bed itself—coalesce, and split. There is clearly a flow of gas through the bubbles and a consequent interchange with the bed. The particles move up and down and around as the bubbles pass; a fraction of the particles may be steadily elutriated from the bed. What is more, the behavior of a bed will depend strongly on many variables: particles, gas flow, bed diameter, distributor design. The questions that must ultimately be answered are (1) whether all the relevant features of bubbling fluidized beds are understood and can be built into reactor models, and (2) which features are important or critical for satisfactory reactor design and operation. This review was prepared with these questions in mind although the choice of topics for discussion was quite selective.

A number of different types of reaction are commonly carried out in fluidized reactors: gas-phase reactions, solid-catalyzed reactions, and reactions involving reaction within and between the gas and solid phases. Catalytic reactions are probably the most common in chemical engineering, but in all these cases, and particularly the last two, what is important is the contact time distribution between the two phases. For heterogeneous reactions the contact time distribution plays the same role as the residence time distribution for homogeneous reactions. The problems discussed below are those which have a particular influence on the contact time distribution.

When attempts failed to describe fluidized bed reactors in the conventional terms of a reactor with dispersion, a series of papers in the 1950's developed a model for a fluidized bed which recognized that many of the problems and contradictions in the operation, and particularly scale-up, of fluidized reactors stemmed from their two-phase character (*1–5*). It is the presence of bubbles, and their effect on gas/solid contacting and mixing, that lies at the root of the bed's behavior and gives rise to the great difference between the R.T.D. (which is easily measured) and the C.T.D. (which is not). For example, May's model (*6*), which today with a much more detailed knowledge of the fluid mechanics involved, is known to be essentially correct, described the bubbling bed as a two-phase system, characterized by an interchange between the bubbles and the emulsion or dense phase, and by mixing within the emulsion phase (Figure 1). The bubble phase, which is free of particles, is in essentially plug flow, and the gas mixing in the dense phase is characterized by a dispersion coefficient. With these assumptions, material balances on the two phases lead to:

$$\frac{\partial C_b}{\partial y} = \frac{W_s}{W_b}(C_p - C_b) \tag{1}$$

and

$$\frac{DV_p}{H^2}\frac{\partial^2 C_p}{\partial y^2} - W_e\frac{\partial C_p}{\partial y} + W_s(C_b - C_p) - KV_pC_p = 0 \tag{2}$$

Although we shall examine the mechanics of the two-phase system in more detail later, it is worth noting that May's model gives a reasonable account of the effects of gas interchange on reactor performance. What is probably the least satisfactory aspect of the theory and those preceding it is its reliance on experimental values for the various parameters in the model. The only exception is what is now known as the two-phase theory, which proposes that all the gas in excess of that required just to fluidize the bed flows through the bed as bubbles. The most significant improvement of the various bubble-type models developed since then is the ability to begin at least to forecast some of the parameters and particularly the rate of interchange between the phases. May's model was developed by van Deemter (*7*), and more recently by Mireur and Bischoff (*8*) to enable the parameters of the two-phase model to be assessed from tracer and reactor experiments; van Deemter (*9*) and Bailie (*10*) have also developed this type of model to include the presence and interchange of particles within both phases. Although these methods have probably not yet been sufficiently exploited I want to concentrate more on the developments in predicting *a priori* the performance of a particular reactor.

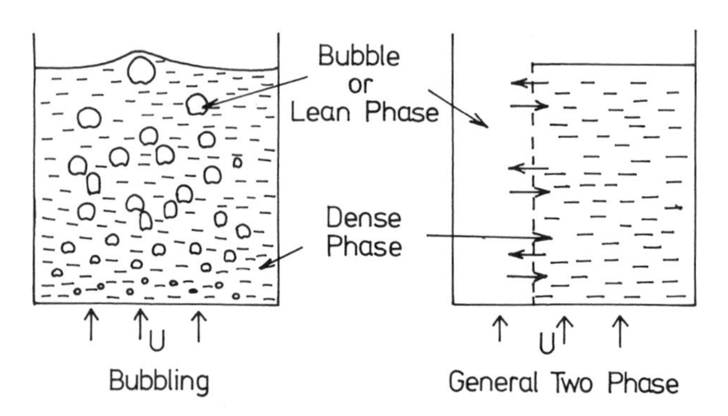

Figure 1. Two-phase models

Lanneau (*11*) attempted to measure bubble velocities and to incorporate these into a model of a chemical reactor. His model allowed for either upwards flow of gas in the dense phase or for the possibility of downwards flow and backmixing of gas (following simply from a continuity argument). The gas crossflow or interchange between the two phases was held to be largely caused by the consequence of solids interchange and was predicted to be, per unit volume of reactor,

$$\varepsilon = \frac{V_B \rho_b U_b}{V \bar{\rho} H} \, s^{-1}$$

Gas Interchange from Rising Bubbles

The work by Orcutt, Davidson, and Pigford (*12*), subsequently developed by Davidson and Harrison (*13*) was really the first to include results which had become available on the relation between bubble rising velocity and the bubble size and gas velocity. In this theory the bubble rising velocity is considered proportional to the square root of the bubble radius, which is consistent with the assumption that the particles move around the essentially spherical bubbles like an inviscid liquid. Davidson (*14*) developed a simple theory for the gas and particle motion around a bubble using this assumption and the assumption of constant voidage fraction, and this theory was used to predict the rate of gas throughflow through a bubble of diameter d_b. A diffusional flow was added to the convective term, and the basis for this theory was the assumption that resistance to diffusion resided in a gas film inside the bubble, as developed by Baird and Davidson (*15*). We return later to a comparative assessment of the various theories for interchange between the phases. For the moment we note that Davidson's theory at that time took no account of resistance to diffusion within the particulate phase nor of the interaction between the convective flow or the rate of any reaction proceeding in the dense phase and this diffusional term. The other point which should be noted is that the significance of this interchange depends on $\alpha = U_b/u_o$, which is the ratio between the bubble velocity and the interstitial gas velocity at incipient fluidization. As Davidson (*14*) showed, and subsequent experiments have confirmed, the gas flow relative to a rising bubble depends strongly on α (Figure 2). For air fluidization, systems with throughflow—*i.e.*, $\alpha < 1$—are usually confined to large particles (> 400–500 microns), or "teeter beds" to use Squires' (*16*) terminology, except possibly for conditions very close to the distributor in beds of fine par-

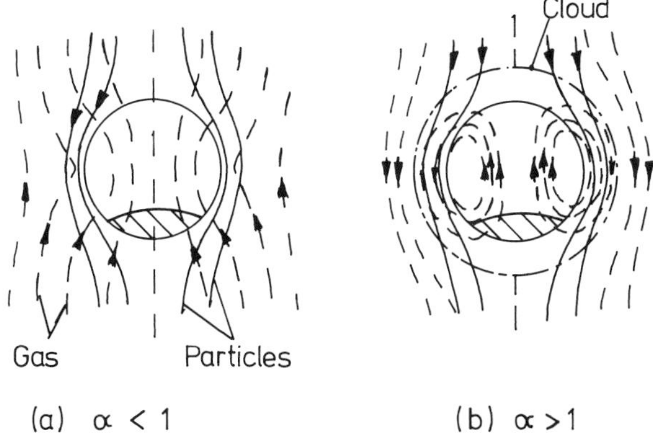

Figure 2. Flow around bubbles in fluidized beds

ticles. Many situations of industrial importance, particularly catalytic cracking units, operate with large values of α—*i.e.*, with gas clouds surrounding the bubble. In this latter case, the throughflow predicted by Davidson, and used by Davidson and Harrison, represents the rate at which gas circulates between the bubble void and the surrounding gas cloud. It is not, strictly speaking, the correct interchange between the bubble phase and the surrounding dense phase. It, or a more rigorous version of it, would appear to be the correct term to use in the analysis of the teeter bed reactor.

Since most of the subsequent work on reactor modeling has been based on this model, with the incorporation of later experimental results on bubble and wake shape or on a more refined model of the gas and particle flow (*17*) it is suitable to consider in more detail the mechanism of gas interchange between bubbles and the surrounding gas phase. For the moment let us confine ourselves to bubbles rising in an essentially infinite medium; the important and frequently encountered case of slugging is dealt with briefly later.

Considering first the case $\alpha < 1$ (Figure 2), a modification of Davidson's theory should be appropriate. It is found experimentally (*18*) that bubbles carry up with them a wake which occupies around $\frac{1}{3}$–$\frac{1}{4}$ of the bubble volume. Clearly this will contribute to any catalytic or gas–solid reaction. The effect of the wake on the gas flow in and around the bubble is as yet unknown. Assuming that Davidson's or Murray's analysis is an adequate representation, the convective contribution to the gas exchange rate is

$$q = \frac{3}{4} \pi U_o D_e^2 \tag{3a}$$

or

$$q = \frac{\pi}{4} U_o D_e^2 \tag{3b}$$

Assuming that the effects of dispersion and convection may be superposed, the total volumetric throughflow rate is (*13*):

$$Q = q + 0.975 \left(\frac{D_g^{1/2} g^{1/4}}{D_e^{1/4}} \right) \pi D_e^2 \tag{4}$$

If the effects of the diffusional resistance in the particles are included, Equation 4 becomes

$$Q = q + \frac{0.975\ \varepsilon_o}{1 + \varepsilon_o} \left(\frac{D_g^{1/2} g^{1/4}}{D_e^{1/4}} \right) \pi D_e^2 \tag{5}$$

The assessment of the diffusional contribution to this flow which allows for the interaction of the convective flow, and the effects of reaction and adsorption, as necessary, may then be found in the same way as indicated for throughflow through a slug by Hovmand and Davidson (*19*). Hovmand (*20*) has allowed for the interaction between convection and diffusion and has given an exact solution to replace Equation 4. The main conclusion is that, as in the case of slug flow, the percentage of the total throughflow caused by diffusion is less than predicted by Equation 4. For a 5-cm diameter bubble and assuming a molecular diffusion coefficient of 1 cm^2/sec, the diffusional contribution to the transfer decreases to zero from the prediction of Equation 4, over a range of particle sizes of 50–200 microns. This takes no account of the effects of a fast reaction in the dense phase which would modify the analysis further. For the particular case in question—*i.e.*, $\alpha < 1$— we can conclude that the diffusional contribution to the throughflow would in most cases be negligible.

As noted earlier, most cases of interest have $\alpha > 1$ (neglecting for the moment the situation in the immediate vicinity of the gas distributor). Referring to Figure 2b we note the following effects to be accounted for: (a) the flow pattern and mixing within the bubble, (b) the presence of the wake and its influence on contacting and gas flow, (c) the convection and diffusion between the bubble and the surrounding cloud phase (as discussed above), (d) diffusion between the cloud and the surrounding particulate phase, (e) convective interchange caused by shedding of "fragments" of the gas cloud and bubble wake. Although theory (which neglects the effects of turbulence and diffusion) has been developed for the flow within the bubble (*21–23*) and predicts a substantial proportion of the gas within the bubble to be circulating inside the void and never contacting the cloud at all, experimental tests are extremely difficult, and it seems that perfect mixing should be a fair assumption for the situation inside the bubble. Since little is known of the effect of the wake, those analyses which allow for its presence on the progress of a catalytic reaction have effectively assumed that it is contained within the bubble phase gas.

We have already considered the question of flow between the bubble and the cloud. On the basis of Davidson's theory of gas and particle motion, Hovmand's analysis should be the most appropriate. In Kunii and Levenspiel's bubbling bed model, the transfer rate given earlier by Davidson—*i.e.* Equation 4—is used. The limitations of this analysis have already been discussed.

Rowe and Partridge's (*25*) model of a fluidized reactor, of which an application is given in this volume (*27*), assumes that the bubble plus cloud ensemble (= cloud in their notation) is perfectly mixed and that a fraction of the gas within this ensemble equivalent to the volume of the

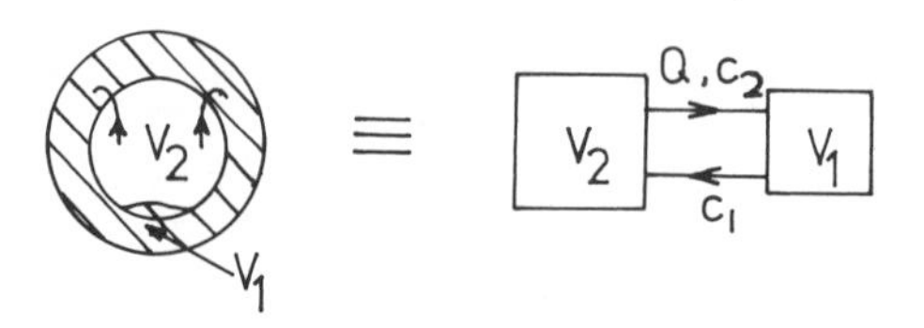

Figure 3. Reaction in bubble cloud

surrounding cloud and wake is in contact with the particles at any instant. The significance of this assumption can be grasped by the analogy shown in Figure 3. The equations which describe this situation are:

$$c_1 - c_2 = \frac{V_2}{Q}\frac{dc_2}{dt} \tag{6}$$

and

$$c_2 - c_1 = \frac{V_1}{Q}\frac{dc_1}{dt} + \frac{kV_1c_1}{Q} \tag{7}$$

so that

$$\frac{d}{dt}(V_1c_1 + V_2c_2) = -kV_1c_1 \tag{8}$$

The assumption implicit in Rowe and Partridge's model is that the time constant ($= V_2/Q$) of the empty bubble is much smaller than that of the cloud and wake where the reaction is proceeding. Since in general $V_1 < V_2$, this could only be exactly true if interchange were infinitely fast although in practice with slow reactions the assumption will be more nearly correct.

The problem of the diffusional interchange between the cloud and emulsion phase is considered in the elegant paper by Chiba and Kobayashi (*28*). Their analysis assumes that the main resistance to transfer between the bubble and the emulsion is in the transfer step between the cloud and the emulsion phase. The transfer coefficient K_B, which is defined in terms of unit volume of the bubble void, is calculated from Murray's solution for flow around the bubble and is

$$K_B = \frac{6.78}{1 - f_w}\frac{\alpha}{\alpha - 1}\left(\frac{D_G\varepsilon_oU_b}{d_b^3}\right)^{1/2} \tag{9}$$

Kunii and Levenspiel (*29*) had also previously estimated the mass transfer coefficient for transfer between cloud and emulsion (K_{ce}) by Higbie's penetration theory, obtaining

$$K_{ce} \simeq 6.78\left(\frac{\varepsilon_oD_GU_b}{d_b^3}\right)^{1/2} \tag{10}$$

The over-all transfer coefficient K_B may be calculated from

$$\frac{1}{K_B} = \frac{1}{K_{ce}} + \frac{1}{K_{bc}} \tag{11}$$

where K_{bc} is the transfer coefficient for transfer between bubble and cloud, which is, from Equation 4 above:

$$K_{bc} = 4.5 \left(\frac{U_o}{d_b}\right) + 5.85 \left(\frac{D_g^{1/2} g^{1/4}}{d_b^{5/4}}\right) \tag{12}$$

In general, as Chiba and Kobayashi point out,

$$K_B \simeq K_{ce} \tag{13}$$

These two theories are quite similar, except at low values of α (*i.e.*, with large clouds). A significant test of the theories would need to be carried out with α close to unity, which is of course difficult to achieve experimentally.

Toei and Matsuno (*30*) have attempted to account for the contribution made by shedding of fragments of the gas cloud. Their analysis needs stronger support than it has at present; they conclude that the contribution to the over-all transfer is relatively small.

Partridge and Rowe (*25*) have conjectured that the mass transfer coefficient between the cloud and the emulsion should be the same as that for transfer between a sphere of gas of the same diameter (d_c) as the cloud, moving at the same velocity as the relative velocity ($U_R = U_b - u_o$) between the cloud and the emulsion gas. The mass transfer coefficient is thus given by

$$\frac{h_m d_c}{D_G} = 2 + 0.69\ Sc^{1/3} Re^{1/2} \tag{14}$$

where $Sc = \mu/\rho D_G;\ Re = \rho U_R d_c/\mu$

Other approximate relationships have also been proposed by Kato and Wen (*31*) (an expression which is used by Yoshida and Wen (*32*) in this volume):

$$K_B = \frac{11}{d_b} \tag{15}$$

and by Toei *et al.* (*30*),

$$K_B = \frac{3}{d_b} \sim \frac{6}{d_b} \tag{16}$$

The experimental results reported by Chiba and Kobayashi (*28*) provide a useful means of testing these proposed relationships. One point concerned with Chiba and Kobayashi's results must be noted, however. It is well known that the effect of the container walls on bubbles in both liquids and fluidized beds is very significant, and in the limit the bubbles rise as slugs. Theory for the motion of slugs in fluidized beds has been given by Stewart and Davidson (*33*), and a sketch of typical gas and particle streamlines is given in Figure 4. It appears that single injected bubbles will begin to behave like slugs when the equivalent diameter of the bubble is greater than about one-third of the bed diameter [Pyle and Stewart (*34*)]. Kobayashi's ozone experiments were conducted in a 10-cm diameter bed; hence, all the results he quotes (bubble diameters greater than 5 cm) are probably taken from a slugging system or at least from systems exhibiting transitional behavior. The same is true of the helium experiments, as of a proportion of the results given by Stephens *et al.* (*35*). Exact theory exists for transfer from a slug to the cloud [Hovmand and Davidson (*19*)]. Since in this case the cloud is close to the wall, gas transferred into the cloud is presumably swept down into the slug wake and into the particulate phase. The analysis should thus give the required over-all transfer coefficient.

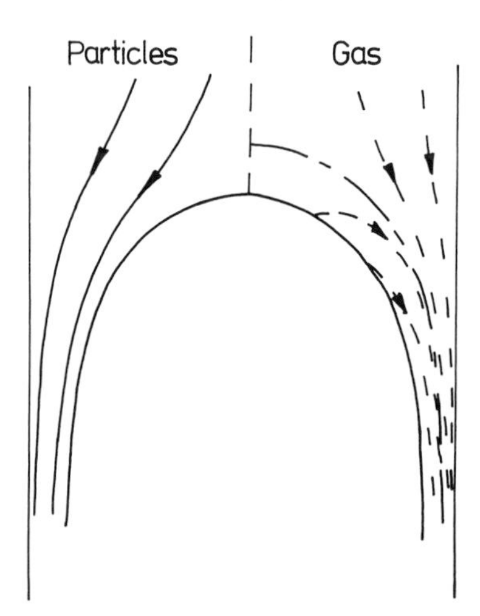

Figure 4. Flow around a slug in a fluidized bed

Experimental results (*35, 60*) for the over-all transfer coefficient defined above are compared with the available theories in Figures 5 and 6. Hovmand and Davidson's theory agrees reasonably well with the results at the slugging end of the range. Wen's approximate and convenient relationship appears to be no worse than any other theory. It cannot be

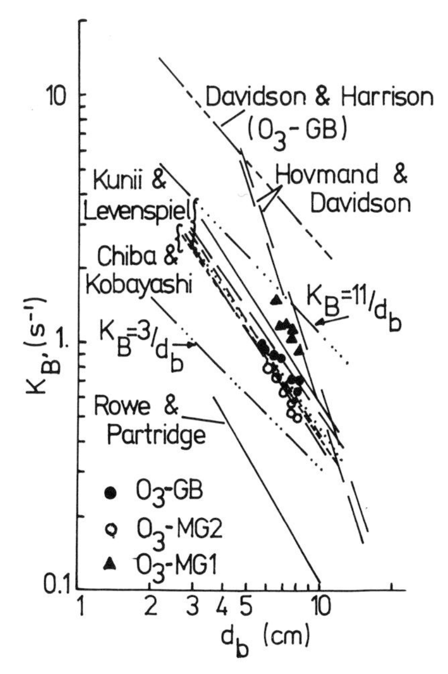

Figure 5. Gas exchange from single bubbles. Data from Chiba and Kobayashi (28).

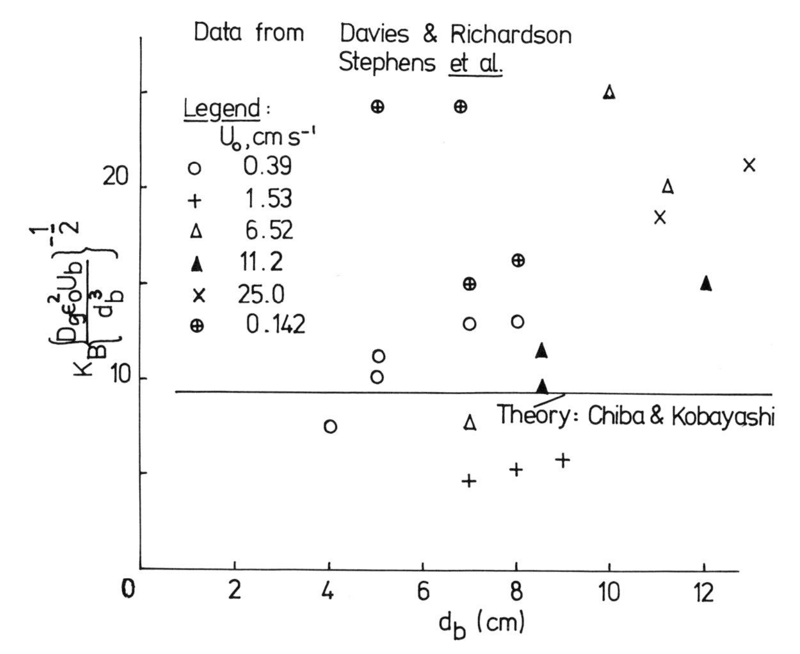

Figure 6. Gas exchange from rising bubbles

said that as yet any theory can be accepted as overwhelmingly superior to any other, and there is clearly a need for further careful experimental work in this field. Rowe and Partridge's (*25*) theory appears to underestimate the exchange rate seriously.

These expressions are valid only for bubbles with clouds; for the case where $\alpha < 1$, which may occur near the distributor, Equations 4 or 5 must be used.

Bubble Coalescence

Kobayashi's earlier results using helium (*36*) and the results quoted by Stephens *et al.* were taken in a situation where bubbles were being produced continuously and probably coalescing. Toei *et al.* (*37*) have produced a theory for transfer from the cloud based on the shedding which should result when two bubbles coalesce (Figure 7). However, on coalescence, a further effect must be included. There is reasonable evidence for the two-phase theory of fluidization—*i.e.*, that the bubble flow is the total gas flowrate minus the flow needed to fluidize the bed. When two bubbles with clouds coalesce, therefore, the resulting bubble should be such that the upwards flux of gas in the bubble phase remains constant. Thus if two bubbles of volumes V_1, V_2 and rising velocities U_{b1}, U_{b2} coalesce, the resulting bubble will have a volume V_3 and velocity U_{b3} such that

$$V_1 U_{b1} + V_2 U_{b2} = V_3 U_{b3} \tag{17}$$

Since rising velocity and volume are uniquely related, this expression is sufficient to define V_3. Now, since the volume of the resulting bubble will be different from the volumes of the two coalescing bubbles, the values of "α"—and thus the cloud dimension—will be different before and after coalescence.

Consider the coalescence of two identical circular bubbles each of radius a in a two-dimensional bed. We will assume, as usual, that the rising velocity, and thus α, is proportional to $a^{1/2}$. The resulting bubble of radius a_c will thus be defined by

$$a_c = 2^{2/5}\, a \tag{18a}$$

and

$$\alpha_c = 2^{1/5}\, \alpha \tag{18b}$$

Using Davidson's theory, the radius of the cloud A is related to the bubble radius by

$$\frac{A^2}{a^2} = \frac{\alpha + 1}{\alpha - 1} \tag{19}$$

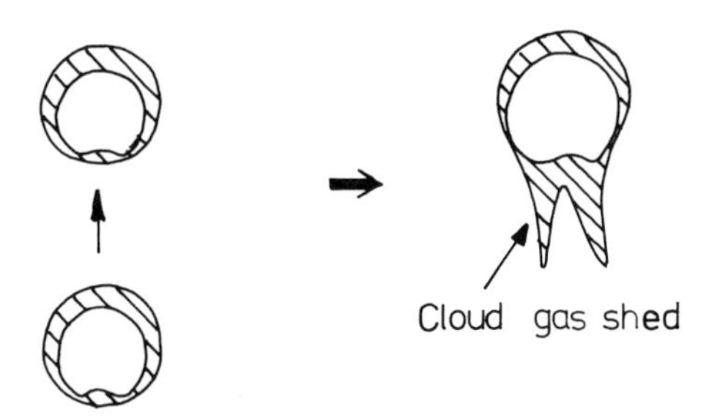

Figure 7. Coalescence, after Toei et al. (37)

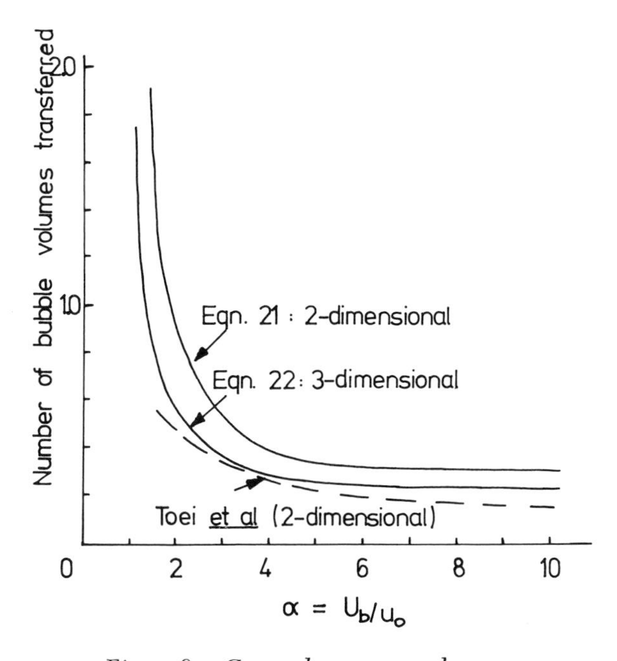

Figure 8. Gas exchange on coalescence

hence, the volume of the bubble plus cloud, per unit depth of bed, will be

$$\varepsilon(A^2 - a^2) + a^2 = \left(\frac{\alpha}{\alpha - 1}\right) a^2 \tag{20}$$

taking $\epsilon = 0.5$.

The difference between the sum of the total (ensemble) volumes of the two bubbles before coalescence and the coalesced bubble will be the volume of gas transferred from the bubble plus cloud to the dense phase. The transfer ratio K (defined as the volume transferred divided by the visible volume of a single bubble of radius a) is thus

$$K = 2\alpha\left(\frac{1}{\alpha - 1} - \frac{1}{2^{1/5}\alpha - 1}\right) \tag{21}$$

The corresponding result in three dimensions is

$$K = 2\left(\frac{\alpha - 0.5}{\alpha - 1}\right) - 2^{6/7}\left(\frac{2^{1/7}\alpha - 0.5}{2^{1/7}\alpha - 1}\right) \tag{22}$$

These transfer ratios are plotted in Figure 8, and Toei's (two-dimensional) theoretical value, based on the shed volume of cloud, but not allowing for the above affect, is also included. The predicted effect of coalescence is a substantial transfer of gas to the dense phase. If the above theory is correct, this might well account for the difficulty in interpreting the experiments referred to above and suggest a further mechanism to be included in fluid bed reactor models, especially in the region near the distributor where bubbles are smallest and coalescence most intense.

Mixing of Gas and Solids

Before discussing the effects of the gas interchange on reactor performance it is appropriate to consider the processes of mixing of gas and solids. It was soon found that a reactor-with-dispersion model gave an inadequate and misleading description of a fluidized bed. May (*6*) attempted to remedy the situation by including a dispersion term in the conservation equation for the gas phase. He suggested that since the mixing processes were intimately linked, the dispersion coefficients for gas and solids in the dense phase should be the same. We are now in a better position to assess this judgment. Let us recall briefly some of the basic physics involved. As far as the solids are concerned, mixing in bubbling beds is overwhelmingly caused by the following effects: (1) the transport of solids up the bed in the wake of the bubbles, and transfer from the wake to the bulk of the bed; (2) the movement and disturbance arising from the passage of bubbles, which is the phenomenon of "drift," first discussed by James Clerk Maxwell (*38*).

Qualitatively, the effects of both contributions is to create mixing and circulation within the bed, and a very thorough study of the basic mechanisms will be found in Rowe *et al.* (*39*). Woollard and Potter's (*40*) study broadly confirmed these mechanisms, and they found that the contribution arising from drift is the displacement of around 30–40% of the bubble volume from its original (horizontal) position. The wake occupies about one-third of the bubble volume. Rowe *et al.* (*39*) also refer to the phenomenon of wake shedding, which looks qualitatively, at any rate, to be similar to the process involved in wake shedding behind a cylinder or sphere in a liquid. No quantitative results are available, however, and it does not seem possible to assess this particular contribution to the solids interphase transfer, which is a term essential to the models proposed by Bailie (*10*) and van Deemter (*9*). The explanation proposed by Yoshida, Kunii, and Levenspiel (*41*) which envisages solids being drawn through the gas (!) cloud into the wake and then out again is as implausible as it is elegant. One only need consider a bubble with a large cloud (*i.e.* $\alpha \sim 1$) to realize the dubious character of this model. Apart from the mechanism observed by Rowe *et al.*, transfer from the wake region to the rest of the bed will occur at the top of the bed and presumably as bubbles coalesce. Indeed this last mechanism, which has not been studied to my knowledge, may well be the dominant one within the body of the bed. Latham, Hamilton, and Potter (*26*) provide a simple model similar to van Deemter's which is based on the solids movement set up by the upwards passage of solids in the bubble wakes. The comparison of Woollard and Potter's calculations of this mechanism with experimental results on the solids circulation rate is not very good.

The effect of drift can be estimated by calculating the mean distance of particle displacement. One such analysis, based on the (dubious) assumption of random bubbling (*23*) shows that the axial dispersion coefficient should be of the form:

$$\frac{D}{(U - U_o)d_b} \simeq \text{constant} \tag{23}$$

with the usual nomenclature.

Much more careful work is needed, however. It is not at all clear that a dispersion-type model is likely to be appropriate since the behavior of a bed is characterized by gross circulation patterns, differences in bubble dimensions and frequencies up the bed, and the formation of local bubble concentrations and tracks (*16, 42*). It is hardly surprising, therefore, that variations in mixing, and particularly in its scale, are found throughout the bed (*e.g.*, Refs. *43* and *44*).

It is clear that the gas movement depends heavily on the motion of the solids, and some effects of this interaction have recently been clarified. Let us first consider the bubble phase. We have already considered briefly the situation within a single bubble, which may (possibly) be taken as perfectly mixed. From an over-all point of view, however, the situation is different since it is difficult to see how backmixing could occur, on a large scale at any rate, between bubbles. Hence, it is probably in order to take the bubble phase over-all as being in plug flow.

However, Krambeck *et al.* (*45*) consider that some over-all models are not very sensitive to this assumption. For a model for control, however, it would seem important to retain this feature.

Consider next the effects of gas interchange between the bubbles and the dense phase. Latham, Hamilton, and Potter (*26*) have shown that even if the gas within the particles is assumed to be in plug flow, the effect of gas interchange is to produce a diffusion-type equation for the dense-phase gas concentration, which was also true of Davidson and Harrison's analysis of a reactor yielding as it does a second-order differential equation for the dense-phase gas. A basically similar scheme was developed further by Hiraki, Kunii, and Levenspiel (*46*). However, this method pays no attention to the other effects which operate, just as they do in the case of solids mixing. In particular there are two effects which must be considered: the gas backmixing which is caused by the gross circulation in the solids phase (discussed above) and the distribution in gas residence times caused by the presence of bubbles [*e.g.*, Rowe (*47*)].

Lanneau appears to have been one of the first to recognize the reasons for gas backmixing which is observed in practice. Local downward gas movement near a single bubble may arise simply as a consequence of the fluid mechanics; this mechanism is considered below. The approximate conditions for gross downward movement of gas in the dense phase as a consequence of the particle circulation have been devised by Kunii and Levenspiel (*29*) and by Latham, Hamilton, and Potter (*26*). Based on an argument from continuity, Latham *et al.* show that the critical gas velocity for backmixing should be given by

$$\frac{U_{\text{crit}}}{U_o} = 1 + \frac{1}{\varepsilon_o R}$$

where U_o is the incipient fluidization velocity, ϵ_o is the dense phase voidage (~ 0.5), and R is the number of bubble volumes of the dense phase which are moved, which again will be of order 0.5. It may be concluded that with fine particles especially "total" backmixing will frequently occur. Quantitative agreement is not found between theory and experiment, however, but as Kato and Wen (*31*) point out, it may often be appropriate to assume that the net upward gas velocity in the emulsion phase is zero. The implications of this and the effect of large downwards particle velocities on the bubble behavior and the distribution of gas between bubbles and dense phases seems not to have been explored sufficiently.

One effect which does not seem, as yet, to have been sufficiently recognized is the further possibility of drift in the gas phase, precisely equivalent to that in the solids phase. For a bubble with a surrounding cloud, Davidson's (*14*) theory predicts that the gas should move relative to the cloud in the same way as an inviscid liquid moves past a sphere (or cylinder, in two dimensions) of the same dimensions. The implication of this theory is that the passage of a bubble plus cloud moving at U_b through a gas interface (moving upwards at u_o), will produce a permanent distribution, or drift, of the interface. That such an effect does indeed occur is shown in Figure 9, showing the passage of a bubble through a nitrogen dioxide/air interface in a two-dimensional bed (23). Detailed comparison with the theory and its implications will be published later, but this effect will have a particularly distorting effect on the distribution of gas contact times when the clouds are large.

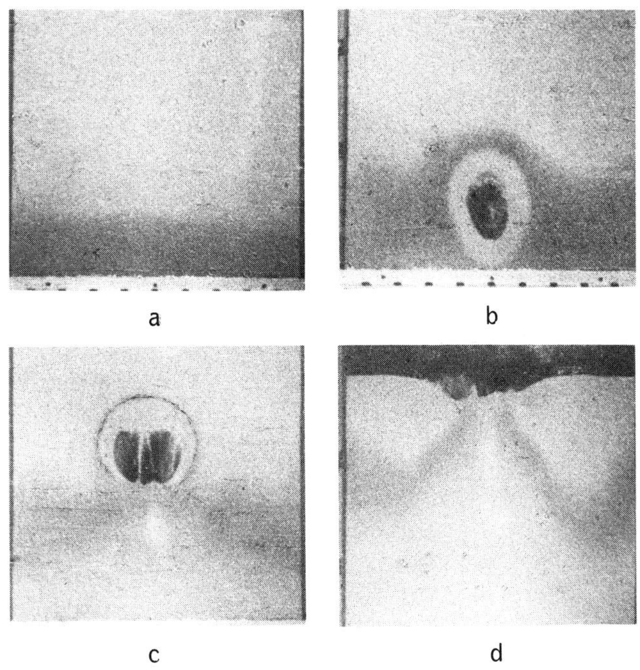

Figure 9. Gas drift around a bubble [*Anwer* (23)]

There is room for a great deal of work into mixing processes in fluidized beds, and it appears that substantial progress will only be made when we have a much more substantial understanding of the processes of bubble formation, coalescence, and distribution within the bed.

Models of Bubbling Fluidized Reactors

As indicated earlier a number of attempts have been made to characterize fluidized beds in terms of a rather general two-phase model involving mixing and transfer of both gas and solids within and between the phases [*see e.g.* van Deemter (9)]. The structure of these models is inevitably similar to the more explicitly bubble-based models, apart from the difficulty noted above in specifying *a priori* the terms within the model. Now we consider how different authors have attempted to include some of the terms discussed above into their models of bubbling fluidized beds.

The discussions above of the interphase gas exchange and mixing processes give a good indication of the different methods used by various workers. Orcutt, Davidson, and Pigford's (*12*) analysis assumed the bubbles to be characterized by a mean effective diameter D_e, and the transfer rate between bubbles and dense phase to be given by Equation 4. Assuming either perfect mixing or plug flow of gas in the dense phase gives a simple pair of equations to be solved for the performance of a catalytic reactor which are similar, in form at any rate, to Equations 1 and 2 above. Davidson and Harrison compared this model, with fair success, with the results of many studies of first-order catalytic reactions. The method of comparison used was the calculation of the bubble diameter D_e. Many of the results predicted values of D_e that were of the same order or larger than the bed, and the subsequent development of a simple theory for the onset of slugging (*33*) shows that, indeed, many of the experimental results reported in the literature are taken from slugging systems for which (noted above) the bubble model is not appropriate. The model has subsequently been developed, as indicated above, to deal with the slugging situation (*19*). In this analysis, the reasonable assumption is made that the gas transferred from the bubble goes into the main body of the dense phase.

It was indicated above that coalescence is an important factor as far as gas interchange is concerned; Hovmand and Davidson were able to measure the slug lengths by a capacitance technique, and they divided the bed into a small number of regions of (assumed) constant slug length. Each of the sections was then treated as an individual reactor and the individual reactor yields multiplied together so that the effect of the coalescence was to mix (perfectly) the two phases. It would be interesting to see if this were true; it is certainly difficult to see how it might be of a bubbling bed. One reason for investigating this—and methods of promoting coalescence—is that mixing the two phases some distance above the distributor should greatly reduce the effect of gas by-passing, which is the most serious limitation of fluidized reactors. If, for the moment, we accept Davidson and Harrison's model, and for simplicity assume perfect mixing in the dense phase, the fraction of inlet gas unconverted in a first-order catalytic reaction is given by their equation 6.9:

$$c' = \beta e^{-x} + \frac{1 - \beta e^{-x}}{1 + k' - \beta e^{-x}} \tag{24}$$

where the transfer factor $x = QH/U_AV$. Now if we consider the same bed subdivided, for simplicity, into "n" equal sections and we assume that all other conditions remain the same, then for the nth section $x_n = x/n$ and $k'_n = k'/n$. The ratio of the fraction unconverted leaving the nth section, to that given by Equation 21 is thus

$$\Phi = \frac{\beta^{n-1}\left(1 + \dfrac{1 - \beta e^{-x/n}}{\beta e^{-x/n}\,(1 + k'/n - \beta e^{-x/n})}\right)^n}{\left(1 + \dfrac{1 - \beta e^{-x}}{\beta e^{-x}(1 + k' - \beta e^{-x})}\right)} \tag{25}$$

The effect is most easily seen by considering a fast reaction where bypassing controls the yield ($c' \rightarrow \beta e^{-x}$). In this case $\Phi \neq \beta^{n-1}$, and since $\beta = (1 - U_0/U)$ is always less than 1, as expected an improvement in reactor performance should result. The same is true if plug flow is assumed in the particulate phase.

Toor and Calderbank (*48, 49*) used the same model as Orcutt *et al.* but included the effect of bubble coalescence by including a term to account for the variation in bubble frequency with height. They conclude that coalescence is unimportant, but it is doubtful if their results allow one to write off coalescence, and their conclusion is not borne out by other workers (nor does it reflect the arguments proposed earlier). This study, like most others, shows the extreme difficulty in setting up experiments which will clearly test and discriminate between models. Perhaps the largest contributory factor to this is the virtual impossibility of making simultaneous measurements of reactor performance and bubble size and spatial distribution. This is one very strong argument in favor of experimental work on slugging reactors, where at least the bubble rising velocity is determined.

In a very useful extension of this work to an 18-inch diameter bed, Toor and Calderbank (*50*) spotlighted some of the problems associated with larger scale operation and scale-up in general. There is evidence that bubble tracks and the resulting relatively stagnant regions of bed give changes in performance which are not predictable as yet. The change from a porous plate distributor to a sieve tray also caused a marked reduction in performance. It appears impossible, as we have already seen, to characterize the performance of a reactor in terms of a single effective bubble diameter because the changes in bubble diameter, and thus in exchange rate, are so marked. This result is also seen in later comparison of some theories (Figure 10). Davidson (*51*) in a review of Toor and Calderbank's work shows, however, that a simple bubble model assuming plug flow near the base of the bed followed by a bubbling (or slugging) region gives good agreement with the experiments. Clearly more work is needed if we are ever to be able to predict *a priori* the performance of industrial scale reactors. It is noteworthy that Toor and Calderbank report that a 50% change in reactor performance should theoretically result from a change in bubble size of 1.5 times.

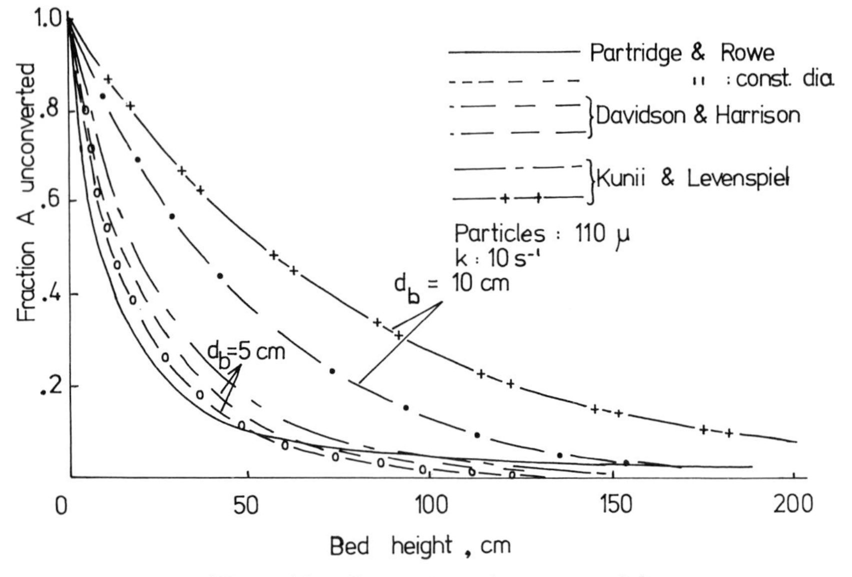

Figure 10. Comparison of reactor models

It is difficult to overemphasize the problems still remaining before adequate methods of prediction are developed. All the studies referred to reinforce the complexity of the problem; it is clear that in many situations the overwhelming influences originate in bubble interactions. The attempt to base methods of analysis and design on simple, single parameter models (*e.g.*, a mean bubble diameter) is certainly naïve and may well be seriously misleading if applied uncritically. What is clear is that far more study is needed of the effects of distributor design. Zenz (*52*) and Kelsey and Geldart (*53*) have recently shown the importance of distributor design. Kelsey and Geldart's study shows, for example, the poor performance that may result from the use of a porous plate distributor: a result that appears, incidentally, not altogether in agreement with Toor and Calderbank's finding which was noted above. Similarly the recent work of Clift and Grace (*54*) on bubble coalescence shows that substantial progress is possible along a path which looks at first sight so daunting.

One of the most striking features of a fluidized bed is the continuous disturbing effect of bubbles. The model discussed above, and those discussed later, all treat the continuously bubbling bed as an essentially steady-state system. Krambeck *et al.* (*45*) have made a significant move in treating the bed as a stochastic system since, as they point out, the time and length scale of the disturbances introduced by the bubbles is often of the same order as the operating scale of the bed itself.

Partridge and Rowe's model (*25, 27, 55*) assumes, as noted before, perfect mixing within the bubble and cloud. Allowance is made for the effect of the wake, and the mass transfer coefficient for gas interchange between the bubble and dense phases is given by Equation 11. The other feature of the model is that it incorporates bubble growth through the bed—either as an empirical relation or as an analytic function. In their model, this is certainly a critical factor, as Figure 10 shows where computed results by Partridge and Rowe are compared with predictions using their model for exactly the same operating conditions but assuming a mean bubble diameter of 5 cm. It is quite clear from this that the effect of many small bubbles near the distributor is to improve the yield over the average performance while their "exact" calculations show the serious effects of bypassing caused by large bubbles near the surface. What is needed, of course, is a sensitive experimental study of this. The work by Ellis *et al.* (*56*) does not give very good agreement between this theory and experiment (the oxydehydrogenation of butenes). Good agreement is apparently achieved between the theory and experiments on the isomerization of *n*-butenes reported in this volume (*27*) by assuming (on quite strong grounds) a linear variation in bubble diameter with bed height and calculating the coefficients in the expression for bubble diameter to give a good fit with theory.

This same study does illustrate one interesting feature of fluidized reactors with complex reaction schemes for the effect of the gas bypassing and interchange can give rise to performance very different from what might be expected and occasionally superior to a fixed bed reactor. Yates *et al.* point to the improvement in selectivity (*trans*- to *cis*-2-butene ratio) resulting from the effect of the gas bubbles as a type of distributed feed. A similar effect was previously noted by May.

Using Harrison and Davidson's model it is clear (*57*) that the assumption of plug flow in the particulate phase shows that the selectivity in the simple reaction scheme $A \begin{smallmatrix} \nearrow B \\ \searrow C \end{smallmatrix}$ depends on the interchange factor, x. The selectivity is apparently independent of x if perfectly mixed gas in the dense phase is assumed. Another effect of the "distributed feed" nature of the bubbling has recently been reported by Jones and Pyle (*58*), and the effects of bubbling on the optimum operating conditions and yield of a complex reaction scheme has been studied.

The final model to be discussed in detail is that proposed by Kunii and Levenspiel (*24, 59*). Its features will probably be fairly clear from the earlier discussion on mass transfer from bubbles. Reaction is accounted for in both wake and cloud and could be allowed for within the bubble itself, and the theory allows for transfer between both bubble and cloud and the cloud and the surrounding dense phase. The theory has been developed for the case where gas backmixing in the dense phase is assumed—*i.e.* $U/U_o > 6$, say, and there is assumed to be no flow through the dense phase. Since all the gas added to the bed at a velocity U goes through the bed as bubbles, the "continuity" addition to the bubble velocity should thus, apparently, be U rather than $U - U_o$. It is difficult to see what effect the (large) downflow of solids will have, but presumably both the bubble rising velocity and the cloud dimensions will be affected. A slow reaction system would be a good test of this theory since the possible contribution of the emulsion phase to the reaction would certainly be significant.

No adequate test of the theory has been reported, but the correlating bubble diameters reported by Kunii and Levenspiel (*59*) seem rather small. The theory would thus appear to be good at correlating existing results, but perhaps one would not be so confident in using it for design purposes from first principles (which is not to recommend overconfidence in any other model).

An alternative approach has also been taken by Latham *et al.* (*26*), but in the absence of experimental results I defer consideration of this model.

What are now needed are detailed tests of the various models against each other and experiment. In the absence of the opportunity to do the latter (and more important) job, consider Figure 10. In this graph the predictions of the models discussed are compared for one case only. The differences between the theories are particularly noticeable near the floor of the bed where the effect of the small bubbles used by Partridge and Rowe (*25*) is in evidence. Over much of the range covered, moreover, the predictions of the theories differ by up to 100% in either concentration or bed height, so perhaps experimental tests should not be too difficult. What will be recognized is the great importance of the correct prediction of the gas interchange rate under these conditions. The sensitivity of the results to this factor may be seen from the computations at a bubble diameter of 10 cm. However, reactor performance is by no means always so sensitive to bubble properties. Davidson's model shows what is generally accepted—that under conditions of fast interchange and slow reaction, the performance approaches that of a CSTR. It would be useful, however, to have a clearer definition of these conditions.

Kato and Wen (*31*) have recently reported an extension of Kunii and Levenspiel's model, in which bubble growth is accounted for by dividing the bed into a series of vertical compartments of the same height as the bubble diameter at that point. Good agreement is found between the theory and the results of a number of different workers (even though —again!—some of these results appear to be taken from slugging beds (*19*)). This same method applied to a gas–solid reaction but incorporating a model for solids interchange and assuming perfect solids mixing in each compartment is presented in this volume (*32*) and is an interesting example of the sort of situation where both gas and solids mixing is relevant.

There can be little doubt that the last decade has seen remarkable strides in the understanding and modeling of fluidized bed reactors, yet many problems still remain. There is little strong evidence or information on bubble inception at the distributor, nor are there any rational accounts of the effects of distributor design; clearly the phenomena associated with coalescence require elucidation. Similarly information and adequate theory on bubble distribution in large scale beds is vital. As I have indicated, mixing has received a good deal of attention, but no satisfactory models yet exist to deal with realistic situations, nor have we much information on the likely effects of mixing [but Toor and Calderbank (*48*) find the model rather sensitive to mixing]. There is clearly much work to be done in a difficult and challenging area.

Nomenclature

a = bubble radius
a_c = bubble radius after coalescence
A = radius of cloud
c_b = concentration in bubble phase
c_p = concentration in particulate phase
c' = ratio of concentrations at outlet and inlet
d_b = bubble diameter
d_c = cloud diameter
D = dispersion coefficient
D_e = diameter of sphere of equivalent volume
D_g = molecular diffusion coefficient
g = acceleration of gravity
h = height
h_m = mass transfer coefficient (Equation 11)
H = over-all bed height
H_o = bed height at incipient fluidization
k = reaction rate constant
k' = kH_o/U, dimensionless rate constant
K = ratio of volume transferred to bubble volume, on coalescence
K_B = over-all transfer coefficient, defined in terms of unit bubble volume
K_{bc} = transfer coefficient, from bubble to cloud
K_{ce} = transfer coefficient from cloud to emulsion phase
n = number of reactor stages or coalescences

q = convective transfer rate from bubble
Q = total transfer date from bubble
R = ratio of volume of solids moved to bubble volume
u_o = interstitial velocity at incipient fluidization
U = gas (superficial) velocity
U_A = absolute rising velocity of bubble
U_b = rising velocity of single bubble
U_{crit} = critical gas velocity for gas backmixing
U_o = superficial velocity at incipient fluidization
U_R = relative velocity between bubble and emulsion gas
V = bubble volume
V_B = bubble phase volume
V_P = volume of particulate phase
V_1, V_2 = volumes of bubble cloud and void (in Figure 3)
V_1, V_2, V_3 = volume of bubbles (Equation 14)
W_b = flow rate in bubble phase
W_e = flow rate in emulsion phase
W_s = total interchange between phases
x = QH/U_AV = transfer factor
y = dimensionless distance
α = ratio of bubble velocity to interstitial velocity = U_b/u_o
α_c = value of α for bubble after coalescence
β = $1 - U_o/U$
ϵ = voidage fraction
ϵ_o = voidage fraction at incipient fluidization
ρ = gas density
ρ_b = density of bed
Φ = ratio of concentrations (Equation 25)
μ = gas viscosity

Literature Cited

(1) Toomey, R. I., Johnstone, H. F., *Chem. Eng. Progr.* (1952) **48,** 220.
(2) Mathis, J. F., Watson, C. C., *AIChE J.* (1956) **2,** 518.
(3) Shen, C. Y., Johnstone, H. F., *AIChE J.* (1955) **1,** 349.
(4) Lewis, W. K., Gilliland, E. R., Glass, W., *AIChE J.* (1959) **5,** 419.
(5) Massimilla, L., Johnstone, H. F., *Chem. Eng. Sci.* (1961) **16,** 105.
(6) May, S. W. G., *Chem. Eng. Progr.* (1959) **55,** 49.
(7) van Deemter, J. J., *Chem. Eng. Sci.* (1961) **13,** 143.
(8) Mireur, J. P., Bischoff, K. B., *AIChE J.* (1967) **13,** 839.
(9) van Deemter, J. J., *Proc. Symp. Fluidization* (1967) 334 (Netherlands University Press, Amsterdam).
(10) Bailie, R. C., *Ibid.*, p. 322.
(11) Lanneau, K. P., *Trans. Inst. Chem. Engrs.* (1960) **38,** 125.
(12) Orcutt, J. C., Davidson, J. F., Pigford, R. L., *Chem. Eng. Progr. Symp. Ser.* (1962) **58** (38), 1.
(13) Davidson, J. F., Harrison, D., "Fluidised Particles," University Press, Cambridge, 1963.
(14) Davidson, J. F., *Trans. Inst. Chem. Engrs.* (1961) **39,** 230.
(15) Baird, M. H. I., Davidson, J. F., *Chem. Eng. Sci.* (1962) **17,** 87.
(16) Squires, A. M., *Chem. Eng. Progr. Symp. Ser.* (1962) **58** (38), 57.
(17) Murray, J. D., *J. Fluid Mech.* (1965) **22,** 57.
(18) Rowe, P. N., Partridge, B. A., *Trans. Inst. Chem. Engrs.* (1965) **43,** T157.

(19) Hovmand, S., Davidson, J. F., *Trans. Inst. Chem. Engrs.* (1968) **46,** T190.
(20) Hovmand, S., Ph.D. Dissertation, University of Cambridge (1968).
(21) Pyle, D. L., Rose, P. L., *Chem. Eng. Sci.* (1965) **20,** 25.
(22) Judd, M. R., de Kock, J. W., *Trans. Inst. Chem. Engrs.* (1966) **43,** T78.
(23) Anwer, J., Ph.D. Thesis, London University (1970).
(24) Kunii, D., Levenspiel, O., *Ind. Eng. Chem., Process Design Develop.* (1968) **7,** 481.
(25) Partridge, B. A., Rowe, P. N., *Trans. Inst. Chem. Engrs.* (1966) **44,** T335.
(26) Latham, R., Hamilton, C., Potter, O. E., *Brit. Chem. Eng.* (1968) **13** (5), 333.
(27) Yates, J. G., Rowe, P. N., Whang, S. T., *Chem. Eng. Sci.* (1970) **25,** 1387.
(28) Chiba, T., Kobayashi, H., *Chem. Eng. Sci.* (1970) **25,** 1375.
(29) Kunii, D., Levenspiel, O., *Ind. Eng. Chem. Fundamentals* (1968) **7,** 446.
(30) Toei, R., Matsuno, R., Miyakawa, H., Nishiya, K., Komagawa, Y., *Chem. Eng. (Tokyo)* (1968) **32,** 565.
(31) Kato, K., Wen, C. Y., *Chem. Eng. Sci.* (1969) **24,** 1351.
(32) Yoshida, K., Wen, C. Y., *Chem. Eng. Sci.* (1970) **25,** 1395.
(33) Stewart, P. S. B., Davidson, J. F., *Powder Technol.* (1967) **1,** 61.
(34) Pyle, D. L., Stewart, P. S. B., *Chem. Eng. Sci.* (1964) **19,** 842.
(35) Stephens, G. K., Sinclair, R. J., Potter, O. E., *Powder Technol.* (1967) **1,** 157.
(36) Kobayashi, H., Arai, F., Sunagawa, T., *Kagaku Kogaku (Chem. Eng. Japan)* (1967) **31,** 239.
(37) Toei, R., Matsuno, R., Nishitani, K., unpublished data.
(38) Maxwell, J. C., *Collected Papers* (1890) **2,** 208.
(39) Rowe, P. N., Partridge, B. A., Cheyney, A. G., Henwood, G. A., Lyall, E., *Trans. Inst. Chem. Engrs.* (1965) **43,** T271.
(40) Woollard, I. N. M., Potter, O. E., *AIChE J.* (1968) **14,** 388.
(41) Yoshida, K., Kunii, D., Levenspiel, O., *Ind. Eng. Chem. Fundamentals* (1969) **8,** 402.
(42) Grace, J. R., Harrison, D., *Proc. Tripartite Chem. Eng. Conf. (Montreal)* (1968) (Institution of Chemical Engineers, London).
(43) Morris, D. R., Gubbins, K. R., Watkins, S. B., *Trans. Inst. Chem. Engrs.* (1964) **42,** T323.
(44) Schügerl, K., *Powder Technol.* (1970) **3,** 267.
(45) Krambeck, F. J., Katz, S., Shinnar, R., *Chem. Eng. Sci.* (1969) **24,** 1497.
(46) Hiraki, I., Kunii, D., Levenspiel, O., *Powder Technol.* (1968/69) **2,** 247.
(47) Rowe, P. N., *Chem. Eng. Progr. Symp. Ser.* (1962) **58** (38), 42.
(48) Toor, F. D., Calderbank, P. H., *Proc. Symp. Fluidization* (1967) 373 (Netherlands University Press, Amsterdam).
(49) Calderbank, P. H., Toor, F. D., Lancaster, F. H., *Proc. Symp. Fluidization* (1967) 652 (Netherlands University Press, Amsterdam).
(50) Toor, F. D., Calderbank, P. H., *Proc. Tripartite Chem. Eng. Conf. (Montreal)* (1968) (Institution of Chemical Engineers, London).
(51) Davidson, J. F., *Proc. Tripartite Chem. Eng. Conf. (Montreal)* (1968) (Institution of Chemical Engineers, London).
(52) Zenz, G. A., *Proc. Tripartite Chem. Eng. Conf. (Montreal)* (1968) (Institution of Chemical Engineers, London).
(53) Kelsey, J. R., Geldart, D., *Proc. Tripartite Chem. Eng. Conf. (Montreal)* (1968) (Institution of Chemical Engineers, London).
(54) Clift, R., Grace, J. R., *Chem. Eng. Progr. Symp. Ser.* (1970) **66,** 105, 14.
(55) Rowe, P. N., Partridge, B. A., Yates, J. G., *Proc. Symp. Fluidization* (1967) 711 (Netherlands University, Press, Amsterdam).
(56) Ellis, J. E., Partridge, B. A., Lloyd, D. I., *Proc. Tripartite Chem. Eng. Conf. (Montreal)* (1968) (Institute of Chemical Engineers, London).

(57) Jones, B. R. E., Ph.D. Thesis, London University (1970).
(58) Jones, B. R. E., Pyle, D. L., *Chem. Eng. Sci.* (1970) **25**, 859.
(59) Kunii, D., Levenspiel, O., "Fluidisation Engineering," Wiley, New York, 1969.
(60) Davies, L., Richardson, J. F., *Trans. Inst. Chem. Engrs.* (1966) **44**, T293.

RECEIVED June 17, 1970.

Contributed Papers

A Model of the Mechanism of Transport in Packed, Distended, and Fluidized Beds

T. R. GALLOWAY and B. H. SAGE, California Institute of Technology, Pasadena, Calif. 91109

A boundary layer and vortex flow mechanism has been proposed for single-phase flow heat and mass transfer in packed beds made up with ordered arrays of particles, distended-ordered arrays, randomly arranged particles, and homogeneously fluidized beds. The flow is characterized by a local interstitial Reynolds number based on the actual velocity in the smallest flow passage and by a free-stream turbulence level characterizing the turbulence in the region of the particle. Both the interstitial Reynolds number and interstitial turbulence level can be derived from packing geometry and the given superficial flow velocity.

The transport is calculated from a theory based on stochastic arguments describing heat and mass transfer from single cylinders and spheres immersed in turbulently flowing liquid or gas streams by replacing the free-stream Reynolds number and turbulence level by the new interstitial Reynolds number and interstitial turbulence level appropriate for the packed bed. The applicability of this bluff body theory was verified experimentally by studying the local flow and heat transfer in a rhombohedral number six, blocked passage array of 1.5-inch spheres. The interstitial turbulence was established from previous "hot-wire" measurements, and the local interstitial velocity was measured with a pitot tube. The local heat transfer was measured with a segmented instrumented copper sphere, 1.5 inches in diameter, provided with a small calorimeter. The segment as well as the sphere could be rotated, thus positioning the calorimeter at almost any point on the surface of the sphere. The superficial Reynolds number was varied from 875 to 3618.

An analytical expression obtained from the theory has represented available transport measurements from single spheres within 8.8% involving 950 data points and packed arrays of cylindical, spherical, and commercial packing together with distended and fluidized beds of spheres within 9.8% involving 762 data points. In commercial packed columns being irrigated with liquid the height of a gas-phase transfer unit was predicted within 12% using 12 cases involving both absorption and vaporization.

Gas Exchange between the Bubble and Emulsion Phases in Gas-Solid Fluidized Beds

T. CHIBA and H. KOBAYASHI, Department of Chemical Process Engineering, Hokkaido University, Sapporo, Japan

In gas-solid fluidized beds, numerous bubbles are invariably generated near the bottom of the bed, and the gas inside the bubbles is exchanged continuously with the gas in the emulsion phase as they rise through the bed. To predict the extent of chemical reaction in a catalytic fluidized bed, it is essential to know the rate of gas exchange between the bubble and emulsion phases. Many theoretical and experimental attempts have been made to obtain reliable data on the rate of the gas exchange.

In the present study, an attempt was made to determine the gas exchange coefficient in a fluidized bed, 10 cm in diameter containing glass beads or milled glass particles. Gas bubbles containing ozone as a tracer were injected successively into the incipiently air fluidized bed. Measurements were made at four different levels in the bed on the changes of ozone concentration immediately inside the bubbles by means of an ultraviolet spectrophotometer with simultaneous measurements of the size, frequency, and rising velocity of the bubbles. Throughout this experiment the bed height was maintained within a range of 63.5 to 74.4 cm, and the heights and frequencies of bubbles measured were within a range of 4.8 to 8.1 cm and 0.37 to 0.47 sec^{-1}, respectively.

It was shown from the measurements that the concentrations of the tracer gas in the bubble decrease as the bubbles rise through the bed. From the axial variation of the concentration of the tracer gas in the bubble, gas exchange coefficients between bubble and emulsion phases were calculated with the following assumptions: (a) the concentration of ozone in a bubble is uniform, (b) bubble characteristics such as the volume, the rising velocity, and the frequency are kept unchanged throughout the bed, and (c) the shape of bubble voids and that of gas clouds are a spherical cap and a sphere, respectively. The value of gas exchange coefficient based on the unit volume of bubble void decreased with the increasing value of the height of bubble and for the same size of bubble it is larger for the larger size of particles.

A model for predicting gas exchange coefficients was also derived, based on the diffusional flow from the gas cloud surface to the emulsion phase. Assuming that the composition in the gas cloud is uniform and the mechanics of gas and particle flow around the bubble follow the analysis by Murray, the gas exchange coefficient based on the unit volume of bubble void is given as:

$$K_B = \frac{6.78}{1 - f_w} \cdot \frac{\alpha}{\alpha - 1} \cdot \left(\frac{D_G \varepsilon_{mf}^2 u_B}{D'_B{}^3} \right)^{1/2} \quad (1)$$

f_w: volume fraction of bubble wake

α : $= u_B \varepsilon_{mf}/u_{mf}$

f_w: volume fraction of bubble wake

α: $= u_B \varepsilon_{mf}/u_{mf}$

u_B, D'_B: rising velocity and diameter of bubble, respectively

ε_{mf}, u_{mf}: voidage and superficial velocity of gas at incipient fluidization respectively

D_G: diffusion coefficient

Thus, the experimental values of K_B were compared with those of Equation 1 and were in fairly good agreement. Hence, we concluded that the rate of gas exchange between the bubble and emulsion phases was largely governed by the rate of diffusion from the surface of the gas cloud.

The effect of gas adsorption on the gas exchange coefficient was also estimated, and the results of calculations were presented to demonstrate that the effect would become appreciable when the adsorption equilibrium constant becomes large.

The Isomerization of *n*-Butenes over a Fluidized Silica-Aluminia Catalyst

J. G. YATES, P. N. ROWE, and S. T. WHANG, Department of Chemical Engineering, University College, London W.C.1, England

The two-phase model of a freely bubbling gas-fluidized bed in which a chemical reaction is taking place has been described in detail by Partridge and Rowe (*1*), and its application to real systems has been investigated (*2, 3*). We investigated the model further and compared its theoretical predictions with experimental results obtained from the study of a solid catalyzed gas reaction. The reaction chosen is the isomerization of 1-butene to a mixture of *cis*- and *trans*-2-butene catalyzed by silica-alumina at 104°C. Kinetic data were obtained by a small fixed bed reactor. The reaction takes the following course:

$$\begin{array}{ccc} & A & \\ \swarrow\!\!\nearrow & & \searrow\!\!\nwarrow \\ B & \rightleftharpoons & C \end{array}$$

where A represents 1-butene, B is *trans*-2-butene, and C is *cis*-2-butene. Although the kinetics seem complex, the system may be considered to be one of first-order parallel reactions if the reactions of each individual component are examined at low conversion. All six rate constants may be derived in this manner, and the results describe accurately the interconversion up to the equilibrium condition. Gas chromatography was used for product analysis.

The fluidized bed experiments which were carried out in a 15.3 cm diameter bed were investigations of the effect on 1-butene conversion and on reaction selectivity (taken as the ratio B:C) of varying gas flow rate and bed depth. The results were expressed conveniently in terms of contact time. Comparison was made between these experimental results and two-phase model predictions; the latter were obtained on the basis of assumptions as to the bubble size distribution through the bed; further, the interphase gas exchange was assumed to be diffusional. The experimental results correlate well with the model predictions if a linear variation of bubble size with bed height is assumed. The functional dependence is given by:

$$d_B = ZH + 0.30$$

where d_B is the average bubble diameter at a distance H cm above the distributor, and Z is a constant for a fixed superficial gas velocity, u_{sup}. Good correlation is given by the following values:

$$Z = 0.066 \text{ when } u_{sup} = 1.67 \text{ cm sec}^{-1}$$

$$Z = 0.048 \text{ when } u_{sup} = 0.71 \text{ cm sec}^{-1}$$

This assumed distribution agrees substantially with that found by other investigators working with similar catalyst and with the same type of distributor (*4*) and shows a similar dependence on superficial gas velocity. Moreover recent work in this department using the x-ray technique has produced strong evidence of a linear variation of bubble diameter with bed height. Many of the salient features of a freely bubbling bed may be expressed in terms of the parameter α, the ratio of bubble to remote interstitial gas velocity. Calculation of α values from the bubble sizes given by the above equation show them to be large ($20 < \alpha < 90$) at most bed levels. This is not unexpected in view of the relatively low value of the minimum fluidization velocity (0.21 cm sec^{-1} at 104°C) of the catalyst particles forming the bed; it leads to the conclusion, however, that the cloud volume associated with the rising bubbles is small, and hence that cloud-phase reaction can be expected to contribute little to the over-all conversion. Furthermore, calculated conversions based on the incorporation of the bubble wake fraction (assumed to be 20% of the complete bubble sphere) in the cloud phase reaction zone demonstrate that this modification is insignificant as far as the predicted conversion in this case is concerned.

Calculation of the important two-phase model parameter Q_E, which gives a measure of the interphase gas exchange rate, shows clearly that although in the lower regions of the bed considerable diffusional exchange takes place, its rate decreases markedly with bed height. This is a further consequence of the rapidly increasing values of α towards the top of the bed since on the two-phase model Q_E is inversely proportional to α.

The over-all effect of fluidized bed operation on the selectivity of 1-butene isomerization was studied. For the same molar conversion of 1-butene in the fixed and fluidized beds, the *cis*- to *trans*-2-butene ratio is different. This may be accounted for on the basis of the limited solids contact of cloud phase gas coupled with the efficient contacting achieved in the emulsion phase of the fluidized bed. This is demonstrated clearly using a triangular diagram of the reacting system from which it is also apparent how the gas exchange between the two phases affects the selectivity.

Although the reaction chosen for study in this work is of little industrial importance, it provides evidence of the power of the two-phase model to predict fluidized bed performance. It should now be possible to apply the model to reactions and situations of real industrial significance.

(1) Partridge, B. A., Rowe, P. N., *Trans. Inst. Chem. Engrs.* (1966) **44,** T. 335.
(2) Rowe, P. N., Partridge, B. A., Yates, J. G., *Proc. Intern. Symp. Fluidization,* Netherlands Univ. Press, Amsterdam (1967) 711.
(3) Partridge, B. A., Ellis, J. E., Lloyd, D. I., *Tripartite Chem. Eng. Conf., Montreal, 1968.*
(4) Kunii, D., Levenspiel, O., "Fluidization Engineering," p. 237, Wiley, New York, 1969.

Three-Phase Fluidized Bed Reactors: An Application to the Production of Calcium Bisulfite Acid Solutions

GENNARO VOLPICELLI and LEOPOLDO MASSIMILLA, Istituto di Chimica Industriale e di Impianti Chimici dell'Universita, Napoli, Italy

Acid bisulfite calcium liquors are used as a leaching agent for pulp production in the paper industry. A typical liquor contains about 5% total SO_2 and 1.7% combined SO_2. Generally, the production of bisulfite acid solutions is carried out by absorbing SO_2 either in water over a limestone packing (Jenssen towers) or in milk of lime (Barker columns). The needs for increasing specific throughput capacity of the absorber, improving control of the liquor composition, preventing plugging, and adapting the process to the use of low grade limestones have suggested the application of fluidization in the production of bisulfite liquors.

Following a previous investigation on the production of bisulfite acid solutions with ordinary three-phase liquid fluidized beds (*1*), limestone from 0.6 to 10 mm in size was fluidized in a 143-mm id and 4-meter high column with water and sulfurous gas under conditions for three-phase bubble fluidization (*2*). Under these conditions limestone particles were held in suspension by transfer of momentum from the gas to the solid phase *via* the liquid medium. The supporting effect of the liquid, prevailing in ordinary three-phase fluidizations, became negligible. For a given particle size, reducing liquid and increasing gas flow rates switched the operation from bubble to plain liquid fluidization.

The three-phase bubble fluidized bed is the necessary development for a useful application of fluidization to the production of bisulfite liquor. To compete with Jenssen towers, the fluidized bed absorber should have a throughput of about 15 cu meters/hour sq meter of liquor. Considering that SO_2 content in the sulfurous gas is at most 15% (in volume), superficial gas velocities of about 50 cm/sec are required, which are well in the range for bubble fluidization, whether the liquid flows concurrently or countercurrently with respect to the gas. By comparison with the ordinary three-phase fluidization, flow conditions prevailing in a three-phase bubble fluidized bed are much less clearly defined. Slugging may develop at relatively low height-to-bed diameter ratios, but process stoichiometry requirements are met at much higher liquor specific outputs than those which could be obtained with ordinary fluidized beds.

A drawback in applying three-phase bubble fluidization might have been a solid concentration in the bed which was too low. Therefore, a parallel investigation on the behavior of three-phase bubble fluidized beds was carried out using the nonreacting system: limestone–water–air, without liquid flowing through the column. The influence of particle size, gas velocity, and distributor design on the maximum mass fraction of the solid suspended in the bed (critical solid hold up) was tested. Practically any size limestone can be handled, provided the gas is injected at sufficiently high velocity. The only limitation from a fluid dynamic standpoint is that the higher the particle size, the larger the energy loss at the distributor. For 1–2 mm limestone, critical solid holdup of about 0.45 could be obtained at a superficial gas velocity of 140 cm/sec with less than 100 mm Hg pressure drop at the distributor. For 6–10 mm limestone, holdup was 0.2 with a pressure drop of 800 mm Hg at a superficial gas velocity of 200 cm/sec.

Increase in energy lost at the distributor might reflect a change in the particle suspension mechanism. With smaller particles, the liquid motion induced by bubbles swirling in the bed is effective in suspending the solid, whereas suspension of coarser material ought to be related to the liquid motion induced by gas jetting at the bottom of the bed. Both these mechanisms occurred in three-phase fluidization with reacting system. Owing to limestone reaction a substantial amount of fines were always present in the bed even when close range coarse material was charged into the column. Stratifications occurred. Fines were intensively bubble fluidized in the upper part, whereas fresh limestone was jiggling about at the bottom.

The production capacity per unit volume of the fluid bed absorber is high. Rates of limestone dissolution of about 2 gram moles/liter hour and SO_2 absorption of 3 gram moles/liter hour were obtained at the experimental conditions investigated. However, the liquid phase in the fluidized bed absorber is mixed perfectly, and thus the two requirements for an industrial application—*i.e.*, the strength of leaching liquor and the degree of SO_2 recovery—cannot be fulfilled simultaneously with a single unit. Fractions of SO_2 taken up in the experimental absorber were about 90% in comparison with recovery of 99.9% to be reached in industrial applications. The process layout should then be based on a sequence of absorbers, with gas and liquid flowing countercurrently, in each of which limestone holdup is regulated in accordance with fluid dynamic behavior of the bed and chemical aspects of the process. Up to a certain level, limestone surface available in the bed controls the rates of SO_2 absorbed. At higher surfaces both the formation of H_2CO_3 and desorption of CO_2 may become slow steps, their relative roles on the kinetics of the over-all process depending on bed agitation and reaction conditions. On the other hand, an exceedingly high limestone surface might raise concentration of SO_3^{2-} in the liquid to induce precipitation of unsoluble calcium sulfite and plugging problems. Therefore, there is an appropriate limestone surface at which each unit of the multistage absorber should be operated. Related to the SO_2 concentration in the sulfurous gas, this surface should be large enough to allow a high fraction of SO_2 to be absorbed, but not so high as to increase the concentration of combined SO_2 in the liquor above the limits for sulfite precipitation.

From the chemical standpoint, the requirements for a given limestone surface in the bed can be fulfilled with particles of any size. However, besides considerations of energy loss at the distributor, finer particles (< 1–2 mm) are preferred to control slugging. The smaller the particle size the larger is the specific surface and the higher the solid holdup at which the bed can be operated. Correspondingly, the ratio between the height and the diameter of the bed is lower.

(1) Volpicelli, G., Massimilla, L., *Paper Trade J.* (1965) **66,** T-512.
(2) Roy, N. K., Guha, D. K., Rao, M. N., *Chem. Eng. Sci.* (1964) **19,** 215.

Noncatalytic Solid-Gas Reactions in a Fluidized Bed Reactor

KUNIO YOSHIDA and C. Y. WEN, West Virginia University, Morgantown, W. Va.

Although the performance of fluidized bed reactors has been studied by many investigators, only a few dealt with noncatalytic solid–gas reaction systems. This paper presents a mathematic model incorporating the bubble behavior and the solids movement to describe the performance of a continuously operated fluidized bed reactor in which a noncatalytic solid–gas reaction is taking place.

When a continuous flow of solids is fed into a fluidized reactor, the outgoing stream of solids is composed of particles having different ages and degrees of conversion. The average conversion of this stream depends on two factors; the rate of reaction of individual particles in the reactor environment and the flow characteristics of solids in the reactor.

Based on their "bubbling bed model" Kunii and Levenspiel showed in the book, "Fluidization Engineering," a procedure for predicting the conversion of both gas and solids leaving the bed. Their model assumes complete solids mixing, which may not be realistic under most of the fluidized bed operating conditions. Besides, it is difficult to predict the so-called "effective bubble diameter" used in their model since bubbles vary in size during the rise through the bed. Kato and Wen proposed the "bubble assemblage model" for catalytic reactions and showed the possibilities of removing this difficulty by taking into account the bubble growth and coalescence in the model.

An attempt is made to extend this model to noncatalytic solid–gas reaction systems and to demonstrate that the performance of a fluidized bed reactor can be simulated by this model. As the noncatalytic reaction we consider the case in which solid particles react with the fluidizing gas while maintaining its original size because the inert solid layer is formed around the unreacted core. The following stoichiometric equation can be used to represent this type of reaction.

$$aA \text{ (gas)} + S \text{ (solid)} \rightarrow \text{gaseous or/and solid products}$$

The proposed calculation method assumes that the reaction of solid particles can be described by the shrinking core model and that the rate of solids conversion is controlled by chemical reaction step.

The fluidized bed is assumed to be represented approximately by "N" numbers of compartments in series where the height of each compartment is equal to the size of each bubble at the corresponding bed height. Each compartment is considered to consist of the bubble phase and the emulsion phase. The change of the bubble diameter along the bed height is given by an empirical equation. The voidage distribution in the fluidized bed is assumed as follows: up to the bed height corresponding to the incipient bed height, the voidage can be considered uniform while above the incipient bed height, it increases linearly along the bed height. From these two relations, the volume of bubbles, clouds, and emulsion phase in each compartment can be computed. Also, it is considered that a part of the solids is distributed in the cloud and the wake region of the bubbles while the remainders are distributed in the emulsion phase. Since the solids are carried upward as a part of the wake of the rising bubbles, this sets up a circulation in the bed with downward movement of solids in the emulsion phase. Two methods of solids feed are considered: one in which the solids are fed to the bottom of the bed and withdrawn from the top of the bed at a constant volumetric flow rate and the other in which the flow direction is reversed. Thus, the net flow rate of solids must be obtained by either adding or subtracting the solid circulation rate to or from the solid feed rate depending on the flow direction.

From these considerations, the material balances of both gas and solid in each compartment can be established. The computational procedure is then developed and applied by using the roasting of zinc sulfide in a fluidized bed as an example.

The adequacy of the proposed procedure is demonstrated by comparing the calculated and the experimental conversions. The computational procedures and the block diagram for computer simulation of noncatalytic solid–gas reaction in fluidized beds are illustrated, and the calculations are tabulated.

The application of this model, however, should be limited to operating conditions in which the bubble velocities are much greater than u_{mf}/ϵ_{mf}. This implies that the use of the model is limited to fluidized beds of small particles, and when large particles are used, this model must be modified since clouds around the bubbles are no longer distinguishable from the emulsion phase.

4

Optimization Theory and Reactor Performance

F. HORN[1] and J. KLEIN[2]

University of Waterloo, Waterloo, Ontario, Canada

Studies of reactor optimization can be classified according to emphasis (e.g., *method of optimization, general reactor theory), according to the nature of the system being optimized* (e.g., *an invented model, a plant-data model), or according to the number of control variables* (e.g., *inlet temperature, amount of catalyst). Optimization methods can be constructive, indirect, or search routines, the last of which are used to solve most practical optimization problems. Generally, the problems in reactor optimization revolve around reactor type and control of process conditions. Continued investigation of theoretical reactor optimization problems should be encouraged since reactions or catalysts which exist could never be discovered by purely practical laboratory investigations.*

This presentation discusses only that work in which either the reactor model or the optimization method used is mathematically sophisticated. However, many important contributions to chemical reaction engineering have been made and will be made which do not meet this criterion (there are also many unimportant contributions which do meet it). A classification of relevant contributions is followed by a short review of optimization methods and by a section containing general results in reactor optimization. This is not a complete review of recent literature (for this *see* Refs. *1–3*) but an overview of the general field using typical examples.

Classification

Work done in reactor optimization can be classified according to whether the main emphasis is on (a) method of optimization, (b) gen-

[1] Present address: Department of Chemical Engineering, University of Rochester, Rochester, N. Y. 14627.

[2] Present address: Farbwerke Hoechst, Frankfurt (M), DBR.

eral reactor theory, or, (c) optimization or improvement of a special system of practical interest. In case (a) the intention is to develop a new or improved mathematical optimization method. The reactor or the reactor model is used only to demonstrate or test this method.

In case (b) the objective is to derive results of general applicability in reactor theory such as how deterioration of catalyst affects the optimum temperature policy in general. Often the usefulness of such results lies in their qualitative nature rather than in quantitative aspects. They may serve as guidance for experimental trial and error procedures or for work in category (c).

In case (c) the main objective is to optimize or to improve a particular system, either existing or planned to exist in physical reality, regardless of the elegance of the mathematics or the generality of the model used. The frequency of publications seems to decrease as one proceeds from group (a) to (c).

Another possibility of classification is based on the nature of the system subjected to optimization. This may be (1) an invented model, (2) a model derived from laboratory or plant data which incorporates general physical laws, or (3) a "black box" represented by an existing pilot plant or plants. Certain pairings between groups (a) to (c) and (1) to (3), such as (a1), (b2), (c2), and (c3) will occur more frequently than others. However, in principle, all combinations are possible. For instance, at a certain stage in optimizing a real plant, results obtained from an invented model may be helpful in suggesting a certain direction of experimental research (c1). It is also possible to try out a new search strategy with a pilot plant or laboratory reactor (a3) (*see* Ref. *4*).

The number of control variables subjected to optimization provides another distinction. The variables which in physical reality can be controlled are always finite in number though they may be numerous. Optimization methods dealing with this situation belong to the field of mathematical programming (*5–8*). An example is the optimal choice of inlet temperatures and amounts of catalyst for the stages of a multibed catalytic reactor.

In problems of a more general nature the optimum choice of functions of time or position is often considered. Then, the number of control variables is infinite and the optimization methods used belong to the field of variational calculus (*9, 10*). Examples are the optimum variation of heat input or cooling as function of time in a batch reactor or the optimum temperature profile along a tubular reactor operated in the steady state.

Optimization Methods

We classify optimization methods as follows: (A) constructive methods, (B) indirect methods, and (C) search routines. Constructive methods guarantee construction of the optimum in a finite number of steps or, at least, an arbitrarily close approach to the optimum in a sufficiently large number of steps. Such methods are, for instance, linear programming (*11–15*), convex programming (*15, 16*), dynamic programming (*17–21*), and geometric programming (*22*).

Indirect methods use necessary conditions which must be satisfied by the optimum case. Examples are the well known stationarity condition for certain programming problems, the Euler equations, and the Maximum principle of the variational calculus (*9, 10*). If the optimum is known to exist and if an algorithm is known furnishing all cases (finite in number) satisfying the necessary condition, then the optimum can be found by simple selection, and the method becomes constructive and belongs to class A.

Violation of a necessary optimum condition is sufficient for the possibility of improvement. Thus, to each necessary condition there belongs a method by which the system can be improved in any case where this condition is not satisfied. Often improvement is carried out repeatedly until either the cost (for computer time or experimentation with pilot plant or plants) becomes prohibitive or the necessary condition on which the improvements are based becomes satisfied (or very nearly so). Such algorithms are called search routines. Examples are the many variations of the method of steepest ascent (*23, 24*).

Special situations exist where it is known that a particular search routine will provide the optimum in a finite number of iterations or will approximate it closely in a sufficiently large number of iterations. Then, the method is constructive and belongs to class A.

Most practical optimization problems are solved by search routines. These methods are much more powerful than unsophisticated trial and error procedures though in general no mathematical proof is available whether or not the result finally obtained is nearly optimal. A search routine based on a strong necessary condition will yield improvements when weaker methods fail to indicate that improvements are still possible. Necessary optimum conditions are especially useful in investigating whether a system which is optimal with respect to a set of "conventional" limitations can be improved by unconventional design or operation—*i.e.*, by relaxing one or more of the "conventional" conditions. For instance, a continuous process may be optimal under the condition of stationarity, but the relaxing of this condition may lead to necessary optimal conditions which are not satisfied by the optimal steady state process. If this happens, the process can be improved by dynamic operations (*25*).

Problems with a finite number of control and process variables in most cases, can be formulated as follows:

A vector $\mathbf{y}$,

$$\mathbf{y} = (y_1, y_2, \ldots y_n) \tag{1}$$

is to be chosen such that the given real valued function f of $\mathbf{y}$ assumes a minimum (or maximum)

$$f(\mathbf{y}) \rightarrow \min \tag{2}$$

subject to the conditions

$$g_j(\mathbf{y}) \leqslant 0 \quad j = 1, 2, \ldots, m \tag{3}$$

where the g_j's denote given real valued functions. (Equality constraints are contained in this formulation since each such contraint can be represented by two inequality constraints.)

The set of vectors $\mathbf{y}$ satisfying Restrictions 3 is often called the feasible set. The function f maps each point of the feasible set into a point of the real line. The set of all such points, that is the image of the feasible set under f, is called the attainable set. The minimum (maximum) of the attainable set corresponds, if it exists, to the optimum choice (or the optimum choices) of the vector $\mathbf{y}$.

According to a well known mathematical theorem the minimum (maximum) of the attainable set exists if (a) f and the functions g_j are continuous and (b) the feasible set is not empty and bounded. In almost all practical cases the continuity requirement is satisfied. (Occasionally, regulations prescribing parameters, such as wall thickness, as function of other parameters, such as pressure, provide different formulas for different regions, which do not join continuously. Care must be taken in such cases since a computer optimization routine may be foiled by this, though, to the human mind, the matter is rather trivial.) Boundedness of the feasible set is easily achieved by additional restrictions, if necessary, preventing the coordinates of $\mathbf{y}$ to assume values *a priori* known to be unreasonably large or small.

The practically only non-trivial case of a non-existent optimum in programming problems is an empty feasible set. In a problem with many nonlinear restriction it may take great effort to find feasible vectors $\mathbf{y}$ or to establish that no such vectors exist for the restrictions chosen. The situation is different and more difficult for variational problems.

Linear programming (*11–15*) applies to the case where the functions f and g_j are linear functions of $\mathbf{y}$. This is a special case of the situation where f and the g_j's are convex functions of the vector **y**. Methods applying to this latter situation are called convex programming methods (*15, 16*). Special methods called geometric programming have been developed for the case where f and the g_j's are linear combinations with positive coefficients of power products of the coordinates of **y** (*22*).

A large class of optimization methods utilizes certain structural properties of the objective function f and the restrictions in order to decompose the problem in a set of optimization problems which can be solved sequentially and with a total effort which is smaller than the effort required without decomposition.

Figure 1 shows one of the simplest situations leading to a decomposition. A reactor is followed by a separation unit. The reactor control variables are combined to a vector **y**. Suppose that the output of the reactor can be characterized by two numbers which, of course, depend on **y**. Let $f_1(\mathbf{y})$ and $f_2(\mathbf{y})$ denote these numbers and consider a two-dimensional vector space in which points are represented by $(f_1(\mathbf{y}), f_2(\mathbf{y}))$. One can then map the feasible set into this space to obtain an attainable set representing all possible reactor outputs under the restrictions placed on **y**. Once this set has been determined, the separation unit can be optimized in a second step and without further reference to the reactor model. The attainable set together with the feasible set of the separation controls (*i.e.*, the product of these sets) form the new feasible set for the optimization of the separation unit.

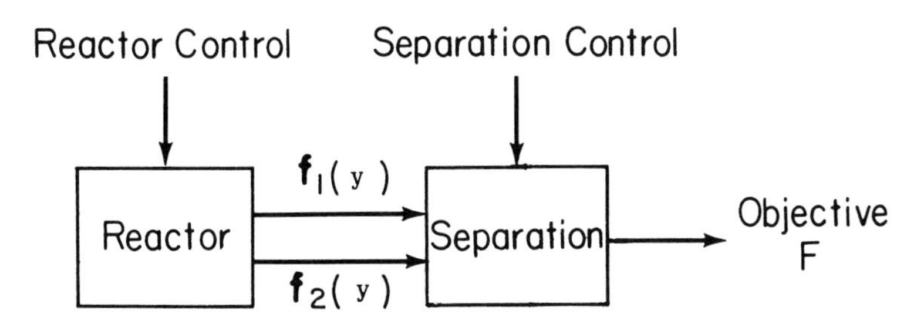

Figure 1. Simple plant decomposition

Depending on the number of control variables involved it can be shown that the effort required for the two-step optimization may be smaller than that for the one-step procedure not utilizing the attainable set (possible outputs of the reactor) and based on the feasible set equaling the product of the feasible sets for reactor and separation controls, and the relation between elements of this large feasible set and the output of the separation. Moreover, the attainable set depends only on the reactor model and, therefore, can be used again if the separation or the model for it are changed.

A further simplification arises if it can be demonstrated that the objective function (defined on the output of the separation) cannot have a minimum (maximum) at a point corresponding to an interior point of the attainable set but only at its boundary. This, for instance, is the case, if the objective function is linear with respect to the input into the separation, and in many other situations. In fact, in the example given here the case where the maximum occurs only at the boundary of the attainable set is the rule. Then, only boundary points of the attainable set need be considered, which further reduces the required effort for the optimization.

The best known example for a decomposition method is dynamic programming (*17–21*). This method is applicable to stagewise structured systems like the one in the foregoing example. However, the optimization procedure starts at the last stage not the first and involves transformations of objective functions rather than attainable sets.

A well known example of an indirect method is utilization of the condition of stationarity,

$$\partial f/\partial y_j = 0 \qquad j = 1,2, \ldots n \tag{4}$$

for a differentiable objective function at an optimum in the interior of the feasible set.

An example for a search routine associated with Equation 4 is the gradient method in which the improvement step is given by

$$y_j^* = y_j + \epsilon(\partial f/\partial y_j) \tag{5}$$

with ϵ chosen as positive (if an increase of f is desired) and sufficiently small. The improved control variables are denoted y_j^*. Many much more sophisticated search routines have been developed. These routines take restrictions into account, and some of them do not require computation of derivatives (*23, 26*).

Variational problems in reactor theory usually can be formulated as follows. Let $\mathbf{x}(t)$ be a vector the n coordinates of which characterize the state of the system at time t (or at position t). The system dynamics is represented by

$$\frac{d\mathbf{x}(t)}{dt} = \mathbf{f}(\mathbf{x}(t),\mathbf{u}(t)) \tag{6}$$

where $\mathbf{f}$ is a vector valued function of the state vector $\mathbf{x}(t)$ and the control vector $\mathbf{u}(t)$ (m coordinates). The objective is to maximize F,

$$F = \int_{t=0}^{T} f_o(\mathbf{x}(t),\mathbf{u}(t))\,dt \tag{7}$$

for either free or specified τ by choice of the control $\mathbf{u}$ where

$$\mathbf{u}(t) \epsilon U \tag{8}$$

i.e., the control, at any time, has to be taken out of a specified subset U of m space; and by choice of $\mathbf{x}(0)$ subject to

$$\mathbf{x}(0) \epsilon I$$

i.e., the initial state vector must lie in a specified initial set in n space. Furthermore, in most cases it is required that

$$\mathbf{x}(\tau) \epsilon T$$

i.e., the end state must lie within a specified target set in n space.

The methods most often used to solve this problem are Pontryagin's maximum principle (*10*), an indirect method, and search routines equivalent to the gradient method of programming, Equation 5 (*24, 27, 28*).

In variational problems, even though the problem under investigation seems well posed, the attainable set (*i.e.*, the set of realizable values of F) can be non-empty, bounded from above and yet not contain its maximum. That is, the supremum of the set is not attained though it can be approximated arbitrarily closely by realizable systems. In such cases indirect methods will fail since either no solution of the necessary condition (*i.e.*, the maximum principle) can be found or even if solutions exist and even if all such solutions are known, the corresponding F values may be far below those obtained for other realizable systems. Search routines, on the other hand, in such cases, though also failing to produce a nonexistent optimum will often clearly indicate which physical or economical aspect was left out when the problem was formulated. For instance, the control function may show more and more extreme oscillatory behavior as the number of search steps increases. This then indicates that the cost of rapid change of control must be considered to make the problem realistic. The difficulties arising sometimes with indirect methods in variational problems are demonstrated by means of a reaction example in Ref. 29.

Qualitative Results in Reactor Theory

Early problems in reactor optimization concerned the optimum mixing patterns in isothermal reactors. In addition to the limiting cases of a plug flow reactor without axial mixing and the continuous flow stirred tank reactor, various intermediates such as cascade or tube with axial dispersion have been considered.

Two optimization problems are particularly simple. In one, the problem is to maximize a given function of the output composition for a given total residence time of the system (type I problem). In the other, the total residence time becomes a control variable itself which is to be chosen together with the other control variables such that the value of the product (a given function of the output composition) is maximal (type II problem). A modification of the type I problem occurs if a cost, associated with the total residence time, is explicitly considered.

It can be shown with theorems of convex set geometry that for both types of problems a tubular reactor is superior to all other reactor types if kinetics and objective function are linear. The proof is given in Ref. *30* where isothermicity is assumed. Very likely the result is also true in the nonisothermal case (*31*). Irrespective of whether the objective function is linear or not it can be shown (*30*) that for linear kinetics a system consisting of m parallel tubes is at least as good as any other reactor type. Here m equals the number of independent reactions plus 1 in a type I problem and the number of independent reactions in a type II problem.

Another problem belonging to this general area is the optimum choice of the distribution of residence times in different tanks of a cascade. In the particular examples with linear kinetics and objective function considered early in the literature, the optimal cascade has tanks of equal size. However, it has been shown (*30*) that a corresponding general statement is false for type I problems (it is true if there is only one stoichiometrically independent reaction). As for type II problems the matter has been clarified only recently (*32*). The result is that for more than three linearly independent reactions local maxima may occur in an unequal size tank cascade. (This contradicts previous results of others (*33*) but that, save certain trivial cases, the global maximum can be achieved only if tank sizes are equal.) The possibility of improving reactor performance by delayed feed addition, recycling, or local mixing has been studied in Refs. *34–36*.

Another type of intermediate between tube and tank is the piston flow tubular reactor with axial diffusion. The system of consecutive reactions

$$\mathrm{A} \rightarrow \mathrm{B} \rightarrow \mathrm{C} \tag{11}$$

with B as desired product has been investigated for such a reactor (*37*). Obviously, for any given diffusivity D there will be an optimum residence time, for which the amount of B formed is maximal. It is well known that the optimum residence time in a tank ($D = \infty$) is larger than that in an ideal tube ($D = 0$). Interestingly, if D is varied from 0 to ∞, the optimum residence time passes through a maximum which is larger than either of the values at the boundaries ($D = 0$, $D = \infty$).

A large class of optimization problems investigated concerns the optimum control of a variable such as temperature, feed addition, or catalyst composition along a tubular reactor or as function of time in a batch reactor.

The first problems investigated in this class concern heterogeneously catalyzed exothermic reversible reactions, such as sulfur dioxide oxidation, ammonia synthesis, and the water–gas shift reaction. In a practical problem of this nature, the cost of heat exchange is important. There are, however, two idealized problems of significance which can be solved with knowledge of chemical kinetics alone. These problems arise if it is assumed that either indirect heat exchange is infinitely cheap or infinitely expensive. The truth obviously lies somewhere in between.

In the one case (low cost heat exchange) the appropriate problem is to ask for the temperature profile, which maximizes the conversion as given amount of catalyst. This "perfect indirect control" problem was solved long ago by several authors (*see* references in Ref. *38*).

In the second case (high cost heat exchange) one may ask for the distribution of cold feed gas along the tubular reactor, which maximizes the conversion η for a given amount of catalyst. This problem has been referred to as the "perfect direct control" problem (*38*).

In both cases computations can be carried out for a variety of total amounts of catalyst m. The optimum reactors can be represented by two curves in a (η, m) diagram. It is convenient to use $\log (\eta/(1 - \eta))$ and $\log m$ as coordinates since it can be shown that for ordinary kinetics the graph approaches straight lines for low and high conversions, η (*38*).

Apart from perfect control many cases of imperfect temperature control have been considered. Examples are reactors consisting of adiabatic stages with indirect and/or direct intercooling (*39*) and problems where radial heat transfer resistance within the catalyst bed is taken into account (*40*).

For a variety of complex reaction systems the problem of perfect indirect heat control and optimum temperature profiles have been determined by the maximum principle or similar methods (*29, 41–48*). Some systems with desired species underlined are listed below with the appropriate references.

$$A \rightarrow \underline{B} \rightarrow C \ (29, 41)$$

$$A \rightleftarrows \underline{B} \rightarrow C \ (41, 46)$$

$$2A \rightleftarrows \underline{B} + C, A + B \rightleftarrows D + E \ (48)$$

$$A + B \rightarrow \underline{C} \begin{array}{l} \rightarrow D \\ \rightarrow E \end{array} \qquad (50)$$

$$\begin{array}{ccccccc} A & + & B \rightarrow & C & \rightarrow & \underline{D} & (41, 49) \\ \downarrow & & & \downarrow & & \downarrow & \\ E & & & F & & G & \end{array}$$

$$A + B \begin{array}{l} \rightarrow C \\ \rightarrow \overline{D} \\ \rightarrow E \end{array} \quad (50)$$

Feed addition problems are formally similar to temperature profile problems. An interesting case regarding a parallel reaction, where both feed addition and temperature are controlled has been studied recently (*51*).

The optimum variation of catalyst compositions along the tubular reactor has been studied (*28, 52*). A typical example is given by the reaction system

$$A \rightleftarrows B \rightarrow C \tag{12}$$

where C is the desired product (*28*). It is assumed that two catalysts are available, catalyst 1 affecting the reaction $A \rightleftarrows B$ and catalyst 2 affecting the irreversible reaction $B \rightarrow C$. In such a case one should consider the possibility of using a mixture of the two catalysts, and one may ask for the variation of the fraction of catalyst 1 which maximizes for given total amount of catalyst the conversion to *C*. For the reaction given above, where the kinetics are first order, the optimum profile is piecewise constant. If the residence time is sufficiently large there are three intervals corresponding to pure catalyst 1, a particular mixture of catalysts 1 and 2, and pure catalyst 2. For sufficiently small residence times, on the other hand, there are only two sections in the optimum reactor each containing pure catalyst.

More recently problems regarding optimum control in the presence of decaying catalyst have been studied (*28, 53–61*). The objective is to maximize the integral yield of a product over a given time. In the most general model it must be assumed that the rate of decay of catalyst activity depends on activity, composition of reacting mixture, and temperature, and it may even be necessary to introduce more than one variable describing the activity of the catalyst. In the problems treated in literature special cases only have been considered.

The simplest non-trivial problems in this class arise in connection with the optimum temperature control in a well mixed batch reactor containing a catalyst (*55*) and the spatially (not temporally) isothermal fixed bed tubular reactor (*59, 60*). These two problems are not formally identical. Both have been solved under the assumptions that the activity decay is of first order in the activity, depends on temperature according to an Arrhenius law, but does not depend on the composition of the reaction of mixture.

The first of the above mentioned problems (well mixed batch reactor) is equivalent to an optimum temperature profile problem (distance replaces time) with the formal system

$$\begin{gathered} A + C \rightleftarrows B + C \\ C \rightarrow W \end{gathered} \tag{13}$$

where A is reactant, B is product, C is active catalyst, and where W represents the "inactive part" of the catalyst.

For the second problem the interesting result has been obtáined (*59, 60*) that under certain conditions, the optimum temperature control corresponds to a policy where in time intervals at the beginning and end of the operation the temperature is held at a lower and upper bound, respectively, while in the intermediate time interval it varies such that the end conversion remains time invariant.

The numerically most difficult optimization problems arise in connection with the optimum choice of temperature as a function of time and position in a fixed bed catalytic reactor with decaying catalyst. The function to be optimized depends here on two rather than one variable. This problem has been attacked by a search routine corresponding to the gradient method (*28*). Because of the great numerical effort involved only few iteration steps could be carried out, but the result nevertheless indicates qualitative features which an optimum policy should exhibit.

A somewhat less complicated problem in the optimization of reactors with decaying catalyst arises if several functions, each of one variable only, are to be chosen optimally. For instance, one may vary inlet composition and pressure as a function of time, and catalyst composition (dilution) as function of length. Such a problem has been considered in optimizing a vinyl chloride reactor with a search routine as optimization method (*61*).

Problems arising from the dynamic behavior of normally stationary processes can be described as follows. First, there are problems connected with the optimum startup, shutdown, and transition between steady states (*62, 65*). Secondly, it can be investigated whether periodic (*25, 66–72*) or stochastic (*73*) operation can improve performance over the best steady-state performance.

As an example consider a system of heterogeneously catalyzed parallel reactions

$$A \rightleftarrows P \quad A \rightleftarrows Q \tag{14}$$

Assume that the concentration of reactant A can be chosen between two limits. Then under certain conditions, involving the mechanism of P and Q formation, one can, by cycling A concentration, obtain average *P* and *Q* formation rates which are different from all realizable steady state formation rates or their mixtures (the latter are obtainable also as average rates by very slow cycling). Moreover, the average formation rates obtained by sufficiently fast cycling may correspond to a larger product value (which is reasonably defined) than the corresponding value obtainable with any steady state operation (including mixed ones) (*70*). In other words, it is possible to increase the selectivity of heterogeneous catalyst for a certain class of mechanisms by cyclic operation.

Conclusions

Most of the published work is oriented towards either optimization theory or reactor theory (cases (a) and (b) discussed earlier). Only a few publications meeting the criterion established earlier deal with the optimization of existing or planned reactors (*74–82*). In general the more practical problems of reactor optimization are solved by finite dimensional search routines. The main problem does not seem to be the selection of the optimization method but construction of a reliable model. Thus, one may ask whether continued investigation of theoretical reactor optimization problems, particularly method-oriented ones, will pay off in chemical reaction engineering.

Undoubtedly such problems have educational value since their solution requires knowledge of many areas of applied mathematics (including numerical methods and computers), physical chemistry, and engineering. Maximizing an objective function is a problem formulation which appeals to the mind of an engineer. Furthermore, qualitative results of a greater nature may suggest unconventional experiments and can provide the engineer with a deeper insight of the influence of certain variables on reactor performance. On the other hand, many engineers who have successfully designed and improved chemical reactors feel that reactor optimization theory, in particular its more sophisticated aspects, has thus far contributed little to their profession.

In the development of a chemical process, chemistry, on the one hand, and control and optimization, on the other, are separated almost completely. With very few exceptions (*e.g.*, in pyrolysis) chemical reactions are sought in the laboratory under conditions not involving a sophisticated control of physical parameters. After the main steps of the process are established, optimization and control are added to the large scale development. It is quite possible that reactions or catalysts exist which never can be discovered in this way since yield or selectivity is negligible for the controls usually applied in the laboratory. It is easy to invent plausible chemical mechanisms for which the yield for an optimum conventional control is, say only 1% of that for an optimum unconventional but in principle realizable control. Therefore, it is not unlikely that such cases really exist, and it seems worthwhile to look for them.

Literature Cited

(1) Lapidus, L., *Ind. Eng. Chem.* (1968) **60,** 43.
(2) Lapidus, L., *Ind. Eng. Chem.* (1969) **61,** 43.
(3) Denn, M. M., *Ind. Eng. Chem.* (1969) **61,** 46.
(4) Price, R. J., Rippin, D. W. T., *Chem. Eng. Sci.* (1968) **23,** 593.
(5) Wilde, D. J., Beightler, C. S., "Foundations of Optimization," Prentice-Hall, Englewood Cliffs, N. J., 1967.
(6) Beale, E. M. L., "Mathematical Programming in Practice," Wiley, New York, 1968.
(7) Cooper, L. N., Steinberg, D., "Introduction to Methods of Optimization," Saunders, Philadelphia, 1970.
(8) Williams, N., "Linear and Nonlinear Programming in Industry," Pitman, New York, 1967.
(9) Hestenes, M. R., "Calculus of Variations and Optimal Control Theory," Wiley, New York, 1966.
(10) Denn, M. M., "Optimization by Variational Methods," McGraw-Hill, New York, 1969.
(11) Dantzig, G. B., "Linear Programming and Extensions," Princeton University Press, 1963.
(12) Orchard-Hays, W., "Advanced Linear Programming Computing Techniques," McGraw-Hill, New York, 1968.
(13) Gass , S. I., "Linear Programming Methods and Applications," McGraw-Hill, New York, 1969.
(14) Sprivey, W. A., Thrall, R. M., "Linear Optimization," Holt, Rinehart, and Wilson, New York, 1970.
(15) Zukhovitsky and Avdeyeva, "Linear and Convex Programming," B. R. Gelbaum, Ed., Saunders, Philadelphia, 1966.
(16) Stoer, J., Witzgall, C., "Convexity and Optimization in Finite Dimensions," Springer, New York and Berlin, 1970.
(17) Aris, R., "The Optimal Design of Chemical Reactors, A Study in Dynamic Programming," Academic, New York, 1961.
(18) Bellman, R. E., Dreyfus, S. E., "Applied Dynamic Programming," Princeton University Press, 1962.
(19) Aris, R., "Discrete Dynamic Programming," Blaisdell, 1964.
(20) Roberts, S. M., "Dynamic Programming in Chemical Engineering and Process Control," Academic, New York, 1964.
(21) Hadley, G., "Nonlinear and Dynamic Programming," Addison-Wesley, 1964.
(22) Duffin, R. J., Peterson, E. L., Zener, C. M., "Geometric Programming," Wiley, New York, 1967.
(23) Wilde, D. J., "Optimum Seeking Methods," Prentice-Hall, Englewood Cliffs, N. J., 1964.
(24) Kelley, H. J., "Methods of Gradients" in "Optimization Techniques," G. Leitman, Ed., Academic, New York, 1962.
(25) Horn, F., Lin, R. C., *Ind. Eng. Chem. Process Design Develop.* (1967) **6,** 30.
(26) Fletcher, R., Powell, M. J. D., *Comp. J.* (1963) **6,** 163.

(27) Horn, F., Troltenier, U., *Chem. Ing. Tech.* (1960) **32**, 382.
(28) Jackson, R., *Ber. Bunsenges.* (1970) **74**, 98.
(29) Coward, I., Jackson, R., *Chem. Eng. Sci.* (1965) **20**, 911.
(30) Horn, F., *Chem. Ing. Tech.* (1970) **42**, 1185.
(31) Katz, S., Shinnar, R., private communication.
(32) Glasser, D., Horn, F., in preparation.
(33) Grutter, W. F., Messikommer, B. H., *Helv. Chim. Acta.* (1961) **54**, 285.
(34) Horn, F., Tsai, M. J., *J. Opt. Theory Appl.* (1967) **1**, 131.
(35) Jackson, R., *J. Opt. Theory Appl.* (1968) **2**, 240.
(36) Jackson, R., Senior, M. G., *Chem. Eng. Sci.* (1968) **23**, 971.
(37) Kipp, K. L., Davis, Jr., S. H., *Chem. Eng. Sci.* (1968) **23**, 833.
(38) Dyson, D. C., Horn, F., Jackson, R., Schlesinger, C. B., *Can. J. Chem. Eng.* (1967) **45**, 310.
(39) Dyson, D. C., Horn, F., *Ind. Eng. Chem., Fundamentals* (1969) **8**, 49.
(40) Paynter, J. D., Dranoff, J. S., Bankoff, S. G., ADVAN. CHEM. SER. (1972) **109**, 168.
(41) Horn, F., Troltenier, U., *Chem. Ing. Tech.* (1961) **33**, 413.
(42) Denn, M. M., Aris, R., *Ind. Eng. Chem., Fundamentals* (1965) **4**, 7.
(43) Alm, Y. K., Fan, J. T., Erickson, I. E., *A.I.Ch.E. J.* (1966) **12**, 534.
(44) Lee, E. S., *A.I.Ch.E. J.* (1967) **13**, 1043.
(45) Lee, E. S., *A.I.Ch.E. J.* (1968) **14**, 977.
(46) Zahradnik, R. L., Parkin, E. S., in "Computing Methods in Optimization Problems," L. A. Zadeh, L. W. Nenstadt, A. V. Balakrishnan, Eds., Academic, New York, 1969.
(47) Binns, D. T., Kantyka, T. A., Welland, R. C., *Trans. Inst. Chem. Eng.* (1969) **47**, T53.
(48) Jaspan, R. K., Coull, J., Anderson, T. E., ADVAN. CHEM. SER. (1972) **109**, 160.
(49) Denbigh, K. G., *Chem. Eng. Sci.* (1958) **8**, 125.
(50) Horn, F., *Ber. Bunsenges.* (1961) **65**, 209.
(51) Jackson, R., Senior, M. G., Obando, R., ADVAN. CHEM. SER. (1972) **109**, 156.
(52) Gunn, D. J., *Chem. Eng. Sci.* (1967) **22**, 963.
(53) Chou, A., Ray, W. H., Aris, R., *Trans. Inst. Chem. Eng.* (1967) **45**, T153.
(54) Seinfeld, J. H., Lapidus, L., *Ind. Eng. Chem., Process Design Develop.* (1968) **7**, 475.
(55) Szepe, S., Levenspiel, O., *Chem. Eng. Sci.* (1968) **23**, 881.
(56) Wojciechowski, B. W., *Can. J. Chem. Eng.* (1968) **46**, 48.
(57) Campbell, D. R., Wojciechowski, B. W., *Can. J. Chem. Eng.* (1969) **74**, 413.
(58) Paynter, J. D., *Chem. Eng. Sci.* (1969) **24**, 1277.
(59) Crowe, C. M., *Can. J. Chem. Eng.* (1970) **48**, 576.
(60) Lee, S. I., Crowe, C. M., *Can. J. Chem. Eng.* (1970) **48**, 192.
(61) Ogunye, A. F., Ray, W. H., ADVAN. CHEM. SER. (1972) **109**, 500.
(62) Munick, H., *A.I.Ch.E. J.* (1965) **11**, 754.
(63) Jackson, R., *Chem. Eng. Sci.* (1966) **21**, 241.
(64) Katz, S., Millmann, M. C., *Ind. Eng. Chem., Process Design Develop.* (1967) **6**, 447.
(65) Chang, K. S., Bankoff, S. G., *A.I.Ch.E. J.* (1969) **15**, 414.
(66) Douglas, J. M., Rippin, D. W. T., *Chem. Eng. Sci.* (1966) **21**, 305.
(67) Bailey, J. E., Horn, F., *Ber. Bunsenges.* (1970) **74**, 611.
(68) Douglas, J. M., Gaiton, N. Y., *Ind. Eng. Chem., Fundamentals* (1967) **6**, 265.
(69) Chang, K. S., Bankoff, S. G., *Ind. Eng. Chem., Fundamentals* (1968) **7**, 633.
(70) Horn, F., Bailey, J. E., *J. Opt. Theory Appl.* (1968) **2**, 441.
(71) Bailey, J. E., Horn, F., *Ber. Bunsenges.* (1969) **73**, 274.

(72) Pareja, G., Reilly, M. J., ADVAN. CHEM. SER. (1972) **109,** 172.
(73) Pell, Jr., J. M., Aris, R., *Ind. Eng. Chem., Fundamentals* (1969) **8,** 339.
(74) Fein, M. M., Paustian, J. E., Sarakwash, M., *Ind. Eng. Chem., Process Design Develop.* (1966) **5,** 380.
(75) Pacey, W. C., Rustin, A., *Can. J. Chem. Eng.* (1967) **45,** 305.
(76) Pan, B. Y. K., Roth, G., *Ind. Eng. Chem., Process Design Develop.* (1968) **7,** 53.
(77) Shipman, L. M., Hickman, J. B., *Chem. Eng. Progr.* (1968) **64,** 59.
(78) Brosilow, C., Nunex, E., *Can. J. Chem. Eng.* (1968) **46,** 205.
(79) Chartrand, G., Crowe, C. M., *Can. J. Chem. Eng.* (1969) **47,** 296.
(80) Sheel, J. G. P., Crowe, C. M., *Can. J. Chem. Eng.* (1969) **47,** 183.
(81) Bakemeier, H., Laurer, P., Schroder, W., *Ber. Bunsenges.* (1970) **74,** 150.
(82) Jarvan, J. E., *Ber. Bunsenges* (1970) **74,** 142.

RECEIVED May 24, 1971.

Contributed Papers

The Control of Competing Chemical Reactions

R. JACKSON and R. OBANDO, William Marsh Rice University, Houston, Tex. 77001

M. G. SENIOR,[1] University of Edinburgh, Edinburgh, Scotland

We are concerned with controlling a chemical reaction to maximize the yield of a given product, despite competing parallel reactions generating useless substances. To be specific, we consider the pair of reactions

$$A + B \xrightarrow{k_1} C$$

$$A + nB \xrightarrow{k_2} D$$

where C is the desired product and D is waste. The kinetics are assumed to be of the mass action form and to correspond to the stoichiometry of the reactions as written, so that the rate of the first reaction is proportional to the concentration of B and that of the second reaction to the nth power of this quantity. The reactions are carried out in a well-stirred, thermostatically controlled vessel into which a_0 moles of A are introduced initially, and the course of the reactions is subsequently controlled by regulating the temperature T and the rate r at which B is added to the mixture. Both variables are physically constrained to lie within specified bounds,

$$T^{\min} \leq T \leq T^{\max}; \quad 0 \leq r \leq r_m \tag{1}$$

and the total amount of B to be added during the reaction time τ is given and is denoted by Q.

The problem is to find the piecewise continuous functions of time, $T(t)$ and $r(t)$, satisfying the constraints in (1) above and maximizing

[1] Present address: Imperial Chemical Industries, Ltd., Agricultural Division, Teesside, England.

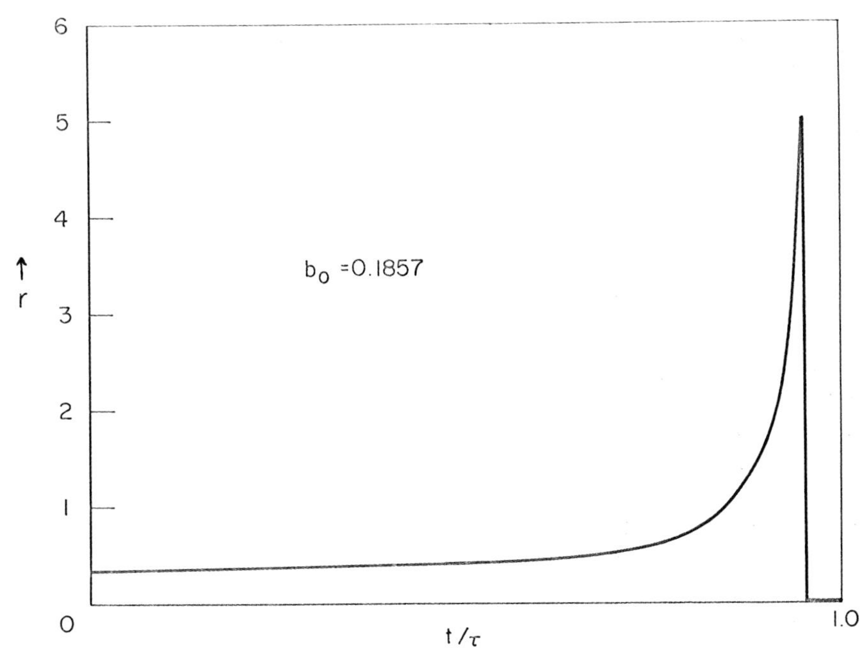

Figure 1. Optimum rate of addition of B for isothermal operation at $T = T^{max}$; $n = 2$, $a(0) = a_0 = 1.0$, $b(0) = 0.1857$, $k_1 = k_1^{max} = 1.0$, $k_2 = k_2^{max} = 2.0$, $Q = 2.5$, $\tau = 4.0$, $r_m = \infty$. *The form of* r(t) *is independent of* p.

the final concentration $c(\tau)$ of product C in the mixture. The parameter $r(t)$ must also satisfy the integral constraint

$$\int_0^\tau r\, dt = Q$$

In the interest of algebraic simplicity, it is assumed that the concentration of each substance in the mixture is proportional to the total amount of that substance present and is independent of the quantities of the remaining components in the mixture. This is true, for example, if the reactants form an ideal gas mixture and the volume is held constant, and it is approximately true at high dilution in an inert liquid solvent. The corresponding problem for a continuous tubular reactor with distributed addition of B and a controllable temperature profile is also closely related to the present one.

The problem is of such a form that Pontryagin's Maximum Principle provides necessary conditions for optimality, and it is remarkable, in that the complete form of the optimum control policy can be deduced, without detailed calculation, by reasoning based on these conditions (assuming, of course, the existence of an optimal policy).

The principal, and rather striking result is that simultaneous use of temperature and distributed addition of B for control is never optimal (with a single exception noted below) and that the choice of temperature control or control by addition of B is determined entirely by the relative values of n, the order of the side reaction with respect to B, and p, the ratio of the activation energies of the two reactions ($p = E_2/E_1$). The nature of the conclusion may be stated particularly simply in the limit as $r_m \to \infty$. When $p > n$, all available B is added at the beginning of the batch, and control is exercised entirely by temperature variation. When $p < n$, the system is operated isothermally at T^{max}, and control is exercised by regulating the rate of addition of B. The condition where $p = n$ is a singular situation where there appears to be an infinite number of optimal, and equally effective controls, some of which vary the temperature and r simultaneously.

Figure 1 shows an example of optimal control by distributed addition of B, for the values of the problem parameters indicated in the caption. This control function is optimal for $n = 2$ and for all values of p less than n. Correspondingly, the optimum value of the objective function $c(\tau)$ is independent of p, as indicated by Curve a in Figure 2.

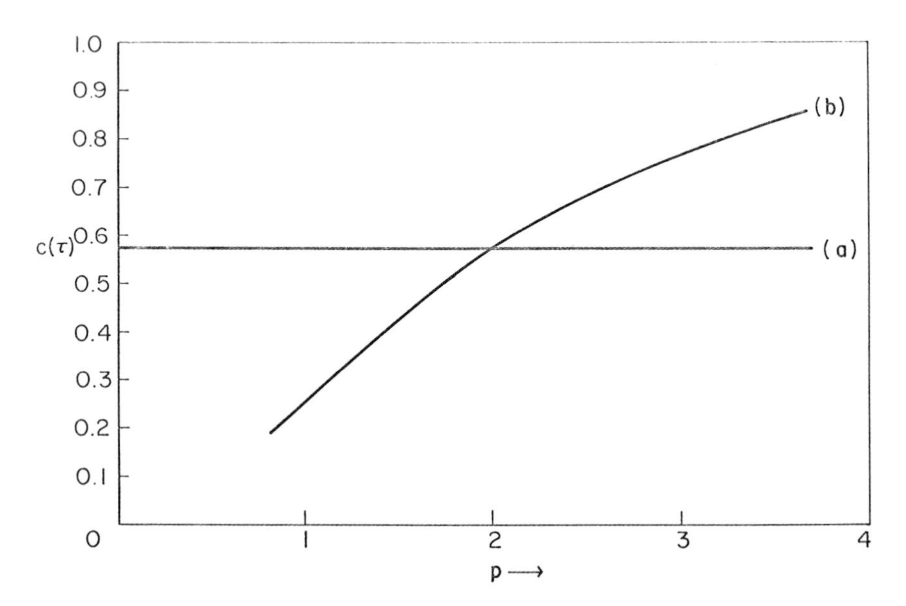

Figure 2. Optimum $c(\tau)$; $n = 2$, $a_0 = 1.0$, $Q = 2.5$, $k_1^{max} = 1.0$, $k_2^{max} = 2.0$, $k_1^{min} = k_2^{min} = 0$, $\tau = 4.0$, $r_m = \infty$

Curve a: Isothermal operation at T^{max}, *optimized with respect to distributed addition of B*

Curve b: Pure batch operation with all B added at $t = 0$, *optimized with respect to temperature variation*

When $p > n$, all B is added at $t = 0$, and Figure 3 shows the optimal temperature control policy for $p = 3.0$. The optimum values of $c(\tau)$ for temperature control, with all B added at $t = 0$, are indicated by Curve b in Figure 2. Curves a and b in this diagram cross at $p = n = 2$, as expected from the above conclusions as to the nature of the optimum control policy.

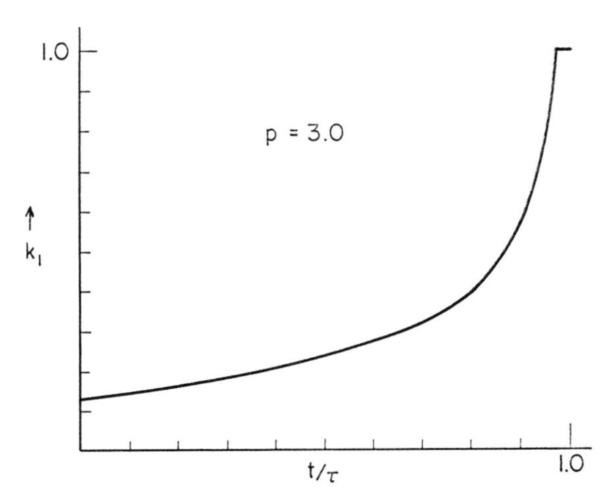

Figure 3. Optimum temperature policy with all B added at $t = 0$; $n = 2$, $a(0) = a_0 = 1.0$, $b(0) = Q = 2.5$, $k_1^{max} = 1.0$, $k_2^{max} = 2.0$, $k_1^{min} = k_2^{min} = 0$, $\tau = 4.0$, $p = 3.0$

The general reasoning leading directly to the optimal control policy is valid only for sufficiently large values of r_m and sufficiently small values of T^{min}. This is why the results presented in Figures 1 to 3 correspond to the limits $r_m \to \infty$ and $T^{min} \to 0$. For smaller values of r_m and larger values of T^{min} the optimal policies are less simple, though it is still true to say that temperature control and control by addition of B should not be used simultaneously. A number of optimal policies in this parametric region have been determined computationally.

The Optimal Thermal Design of a Reactor for Systems of Chemical Reactions

R. K. JASPAN, J. COULL and T. S. ANDERSEN, Chemical and Petroleum Engineering Department, University of Pittsburgh, Pittsburgh, Pa. 15213

The optimal thermal design of chemical reactors has been the subject of much work in recent years. The novel feature of this research is the study of a system of two bimolecular reversible reactions.

$$2B \underset{k_2}{\overset{k_1}{\rightleftarrows}} D + H$$

$$B + D \underset{k_4}{\overset{k_3}{\rightleftarrows}} T + H$$

$$f_1 = dX_1/d\tau = k_1 C_B{}^2 - k_2 C_D C_H \tag{1}$$

$$f_2 = dX_2/d\tau = k_3 C_B C_D - k_4 C_T C_H \tag{2}$$

The temperature profile which optimizes the production of D can exhibit an unusual shape, depending on inlet composition. For high purity B inlet, the upper profile in Figure 1 first falls to a local minimum, then rises to a local maximum, and finally resumes its fall to the end of the reactor. For low purity B inlet, there is no falling portion near the entrance; the profile simply rises through a maximum and then falls toward the end of the reactor.

Two Simplified Reaction Models

The first model is simply two irreversible reactions in series. This model can be expected to be valid only at very low space times and when the feed is high purity B, with little D. This simple reaction mechanism closely resembles the classic A-B-C system.

$$2B \rightarrow D + H$$

$$D \rightarrow T + H$$

$$f_1 = dX_1/d\tau = k_1 C_B{}^2 \tag{3}$$

$$f_2 = dX_2/d\tau = k_3 C_{BO} C_D \tag{4}$$

The second model also consists of two irreversible reactions but with the series-parallel feature of the original system. This model can be expected to be valid over somewhat longer space time (0.3 sec).

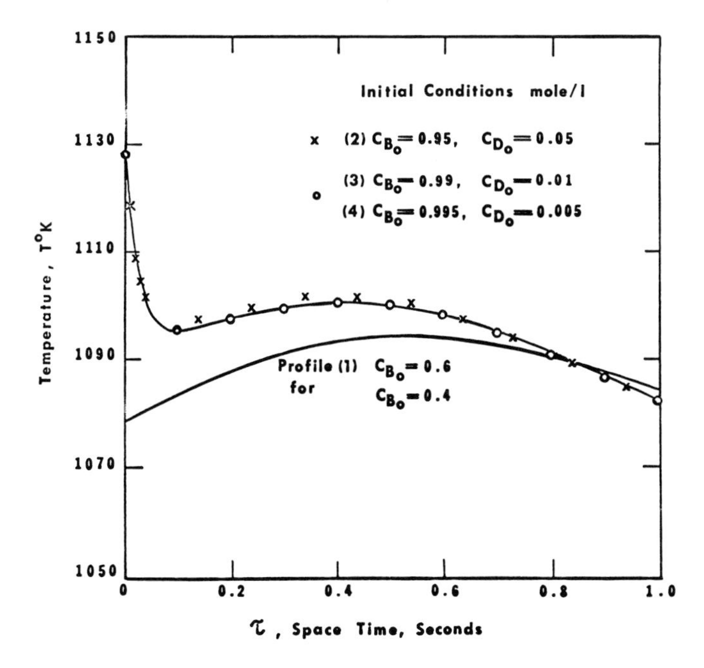

Figure 1. Optimal temperature profiles for diphenyl-type system

$$2B \rightarrow D + H$$

$$B + D \rightarrow T + H$$

$$f_1 = dX_1/d\tau = k_1 C_B{}^2 \tag{5}$$

$$f_2 = dX_2/d\tau = k_3 C_B C_D \tag{6}$$

Benzene–Diphenyl–Triphenyl

The specific reaction system chosen is typified by the benzene–diphenyl–triphenyl polymerization where B is the raw material, benzene, D is the desired product, diphenyl, T is the waste product, triphenyl, and H is the side product, hydrogen. These reactant and product values will be used to compute profit functions in Table II. Benzene costs 25¢/gal, diphenyl is worth 17¢/lb, and hydrogen is worth 20¢/MCF. Triphenyl is worthless. When the values are expressed on a molar basis, the desired products are worth more than eight times the cost of the reactant. Unfortunately the real benzene–diphenyl system results in the uninteresting trivial optimal temperature profile.

Hypothetical Kinetic Parameters

Hypothetical reaction kinetic parameters have been selected to result in a non-trivial optimum: $k_1 = 3.98 \times 10^7 \exp(-35000/RT)$, $k_2 = 1.6 \times 10^{13} \exp(-70000/RT)$, $k_3 = 1.0 \times 10^{15} \exp(-80000/RT)$, $k_4 = 2.0 \times 10^{17} \exp(-95000/RT)$ (notice that $E_1 < E_2 < E_3$).

Qualitative Optimum Temperature Profiles for Simple Reactions

Characteristics of each of three well-known optimal temperature profile systems can become dominant in the benzene–diphenyl system at different extents of reaction (compare with Figure 1).

(1) The $A \rightarrow B \rightarrow C$ system closely resembles our simple model, and the characteristics of this falling temperature profile can be expected to appear at very low space times when little D is present.

(2) The two parallel reaction system resembles our intermediate model, not for low space times, but only when a significant amount of diphenyl is present. Under these conditions the second reaction begins to compete successfully with the first for reactant B. The rising temperature profile of this simple system explains qualitatively the rising portion of the rigorous profile.

(3) Finally, the single reversible reaction can be compared with our full model after significant conversion has occurred when it is necessary to suppress or reverse the recombination of D with H to reform B. The falling temperature profile for this simple system explains the final falling portion of the profile of our rigorous model.

Optimization Procedure

The optimization procedure consists of five basic steps:

(1) Derive the differential equations for the extents of reaction from the reaction model (Equations 1–6).

(2) Apply Horn's method (*1, 2*) to find the differential equation for the optimal temperature profile in terms of the extents of reaction (state variables) and temperature (control variable). (Objective functions are discussed below.)

$$\frac{dT}{d\tau} = \left[\frac{\partial f_1}{\partial x_2} - \frac{\partial f_2}{\partial x_1} Z^2 + \left(\frac{\partial f_1}{\partial x_1} - \frac{\partial f_2}{\partial x_2} \right) Z - f_1 \frac{\partial Z}{1 \partial x_1} - f_2 \frac{\partial Z}{2 \partial x_2} \right] \left(\frac{\partial Z}{\partial T} \right)^{-1} \tag{7}$$

$$Z = (\partial f_1/\partial T)/(\partial f_2/\partial T) \tag{8}$$

(3) Guesstimate the feed temperature to enable the solution of the differential equations. First approximations for the feed temperature for the rigorous model were obtained by using the simpler models for short space times.

(4) Integrate the differential equations for extents of reaction and temperature using the Crane-Klopfenstein Predictor-Corrector method with a Runge-Rutta starting routine.

(5) Correct the feed temperature to find its global optimum value.

Objective Functions

Three objective functions for profit optimization considered in this study are shown in Table I.

Table I. Objective Functions for Profit Optimization

Function	λ_{1F}	λ_{2F}
$S_1 = D_F = X_{1F} - X_{2F}$	1.0000	−1.0000
$S_2 = .00586(B_F - B_O) + .05767(D_F - D_O)$	.04595	−.06353
$S_3 = S_2 + .0359(H_F - H_O)$	.08185	−.02763

More realistic profit functions, which were not studied would include the length of the reactor (or final space time) and a penalty cost for the separation of the worthless triphenyl from the product.

Feed Temperature Correction

The standard criterion of maximizing the final value of the Hamiltonian is applied.

$$H_f = \lambda_{1f} f_{1f} + \lambda_{2f} f_{2f}$$

$$\lambda_{if} = \partial S / \partial x_i \quad (11)$$

The final values of the adjoint variables are given in Table II. The unconstrained maximum of H_f is found when

$$\partial H_f / \partial T_0 = 0 \quad (12)$$

$$\lambda_{1f} \partial f_{1f} / \partial T_0 + \lambda_{2f} \partial f_{2f} / \partial T_0 = 0$$

A function, F, is defined from Equation 12 to be equal to zero at the optimum.

$$F = 0.1 + (\lambda_{1f} \partial f_{1f} / \partial T)/(\lambda_{2f} \partial f_{2f} / \partial T) \quad (13)$$

An adaptation of the classical root-finding technique of Newton is used to find the value of T_0 which makes $F = 0$.

$$T_0^{(j+1)} = T_0^{(j)} - \varepsilon[F/(\partial F/\partial T)]^{(j)} \tag{14}$$

$$\partial F/\partial T = (\lambda_{1f}/\lambda_{2f})[f_2' f_1'' - f_1' f_2'']/(f_2')^2 \tag{15}$$

$$f_i' = \partial f_{if}/\partial T, \quad f_i'' = \partial^2 f_{if}/\partial T^2 \tag{16}$$

Note that the derivatives are taken with respect to T instead of T_0. The damping factor ϵ was used to reduce the size of the exact correction whenever the new T_0 seemed to degrade the reactor's performance (resulting in a lower conversion to D).

Optimum Thermal Design

The results of the search for optimal initial temperatures for four inlet compositions and three objective functions are shown in Table II. Temperatures are in °K and are highest when S_3 was used as a profit function. The temperatures increase with feed purity and in fact become infinite when the feed is absolutely pure.

Table II. Optimal Initial Temperatures

Initial Composition		*Objective Function*		
B	*D*	S_1	S_2	S_3
0.60	0.40	1078.8	1073.5	1093.0
0.95	0.05	1128.4	1122.7	1140.1
0.99	0.01	1179.9	1173.4	1192.7
.995	.005	1209.5		1226.3

For feed reactant purities of 95% the profiles in Figure 1 are graphically indistinguishable. Even for an inlet purity of 60% the optimal profile resembles that for higher purity beyond the entrance region. This leads to the hypothesis that near optimal control may be retained over a band of inlet purities by simple adjustment of feed temperature.

An optimal heat load distribution was calculated from the temperature profile and appears quite simple to implement.

Summary

In this paper we have: (1) studied a system of two bimolecular reversible reactions; (2) derived and explained the optimum temperature profile which exhibits both local minimum and maximum; (3) hypothesized that near-optimal control may be retained over small perturbations in inlet composition by compensating changes in inlet temperature. Further details of this study are given by Jaspan (*3*).

(1) Horn, F., "Optimale Temperatur- und Konzentrations-verlaufe," *Chem. Eng. Sci.* (1961) **14,** 77.
(2) Horn, F., "Über die optimale Temperatur fuhring bei Kontinuerlichen chemischen Prozessen," *Z. Elektroch.* (1961) **65** (3), 209.
(3) Jaspan, R. K., "The Optimal Design of Pyrolytic Reactors," Ph.D. dissertation, School of Engineering, University of Pittsburgh (1969).

Catalytic Reforming

M. BUNDGAARD-NIELSEN and J. HENNINGSEN, Danmarks tekniske Højskole, 2800 Lyngby, Denmark

Catalytic reforming of petroleum naphthas to produce high quality gasolines is an outstanding application of heterogeneous catalysis in the petroleum industry. This paper presents a new general reaction model of the reforming process and discusses how one may optimize the performance of a catalytic reforming process unit utilizing some of the newer literature on the deactivation of dual function catalysts.

In reforming petroleum naphthas the major types of reactions are:

(a) Aromatization of C_6-ring naphthenes
(b) Isomerization of paraffins and naphthenes
(c) Cyclization (and ring opening)
(d) Hydrocracking

Reaction models of the reforming process have been suggested by Smith (*1*), Krane *et al.* (*2*), and Burnett *et al.* (*3*). Neither Smith nor Krane *et al.* take into consideration that C_5- and C_6-ring naphthenes react according to quite different patterns. The dominant reaction of the C_6-ring naphthenes is by far the direct aromatization. The C_5-ring naphthenes must undergo isomerization, which is a slow reaction and comparable with the undesired ring opening, before the much faster aromatization can take place. Burnett *et al.* present a model of the C_7 system alone.

We suggest a general model of the reforming process as outlined in Figure 1. Utilizing the proposed reaction model we can form a set of

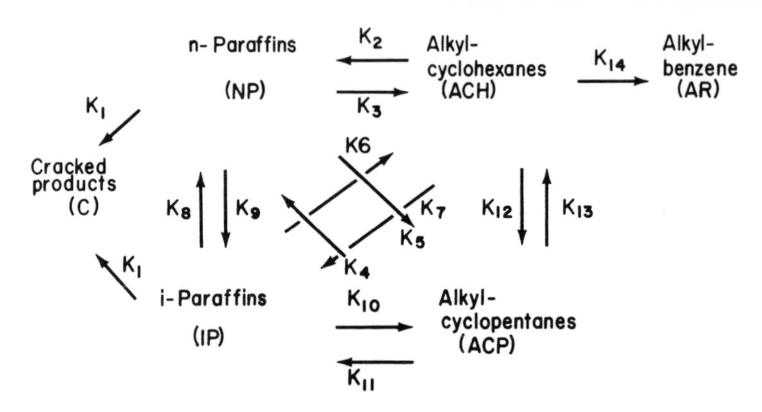

Figure 1. General model for the reforming process

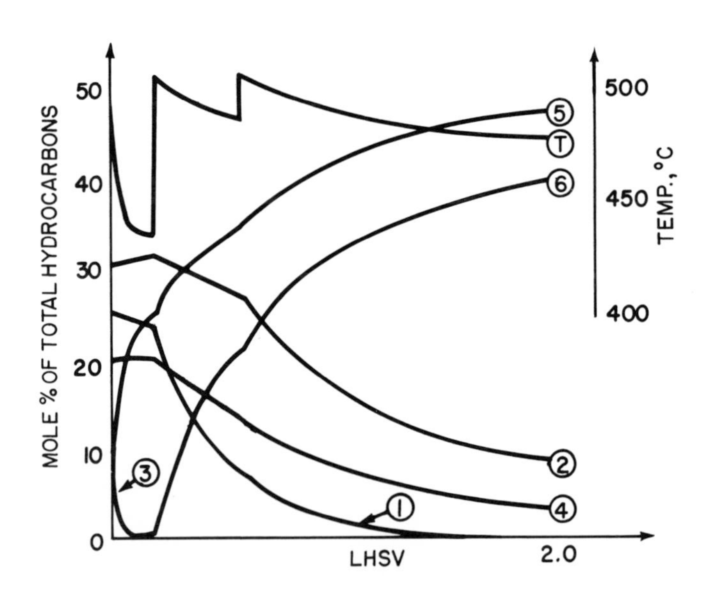

Figure 2. Catalyst distribution in the reactors: 1:2:7. Hydrogen-hydrocarbon ratio: 7. Hydrogen pressure, 30 atm.

Symbols	*Initial Value*
T: *temperature*	*500°C*
1: n-*octanes*	*25 mole %*
2: isooctanes	*30 mole %*
3: alkylcyclohexanes	*20 mole %*
4: alkylcyclopentanes	*20 mole %*
5: xylenes	*5 mole %*
6: cracked products	*0 mole %*

differential first-order equations that describe the concentration and temperature profiles in a reforming reactor system.

Based upon data available in the literature the solution of the differential equations was carried out on an IBM 7094 computer for a C_8 naphtha assuming plug flow. The result is shown in Figure 2.

The dual function catalysts used in catalytic reforming consist of a dehydrogenation–hydrogenation component, such as platinum, on an acidic component, such as halogenated alumina or silica alumina. Sinfelt (*4*), in a review on dual function catalysts, has discussed the rate-controlling steps involved in the four major types of reactions in the reforming of naphthas as outlined above.

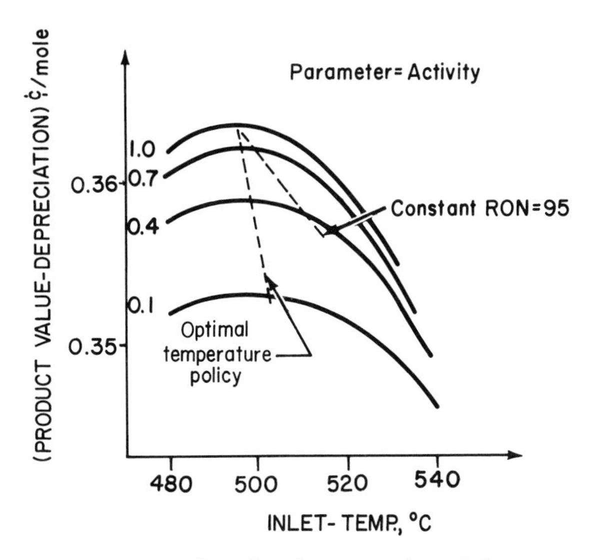

Figure 3. Catalyst distribution and naphtha composition as indicated in Figure 2. Depreciation of the catalyst at 480°C: 7 × 10⁻⁴ cent/mole processed. Inlet temperature to first and second reactor: 500°C.

The aromatization reaction is mainly transport controlled whereas the acidic function of the catalyst is rate controlling as far as the isomerization, cyclization, and cracking reactions are concerned. Thus, a reaction model where the rate constants from the isomerization, cyclization, and cracking depend upon the activity of the catalyst is a realistic one. According to Szépe and Levenspiel (*5*) we may describe the decay in activity by the equation:

$$-\frac{dA}{dt} = k \cdot \exp\left(-\varepsilon/RT\right) \cdot A^{m} \qquad (1)$$

where A = activity
t = time
ε = activation energy of catalyst decay
T = temperature (°K)
m = order of deactivation
k = a constant

We suppose that a zero- or first-order deactivation is in agreement with practical experience. Assuming a zero-order deactivation we have used the reaction model to calculate (product value) − (depreciation of the catalyst) for various inlet temperatures of the last—and controlling—reactor in a three-reactor system with the activity of the catalyst as parameter. The result is shown in Figure 3. This figure indicates that the optimal temperature policy to be followed is one which leads to a decrease in octane number of the reformate as the catalyst deactivates. This may be contrary to the way many reforming processes are operated, and we do not claim that these processes are not run profitably. It is, however, our feeling that an approach to the optimization of the reforming process as outlined here using realistic models of reaction and deactivation in connection with a simple optimization procedure may prove valuable in many cases. [This summary is based on a complete paper which has been published elsewhere (*6*).]

(1) Smith, R. B., *Chem. Eng. Progr.* (1959) **55** (6) 76.
(2) Krane, H. G., Groh, A. B., Schulman, B. L., Sinfelt, J. H., *Proc. World Petrol. Congr., 5th* (1959) Sect. III, Paper 4, p. 39.
(3) Burnett, R. L., Steinmetz, H. L., Blue, E. M., Noble, J. J., *Div. Petrol. Chem., Am. Chem. Soc.*, Preprints (April 1965) 17.
(4) Sinfelt, J. H., *Advan. Chem. Eng.* (1964) **5**, 37.
(5) Szépe, S., Levenspiel, O., *Proc. European Symp. Chem. Reaction Eng., 5th* (1968) Sect. 4.
(6) Bundgaard-Nielsen, M., Henningsen, J., *Brit. Chem. Eng.* (1970) **15** (11), 1433.

Application of a Suboptimal Design Method to a Distributed-Parameter Reactor Problem

J. D. PAYNTER,[1] J. S. DRANOFF, and S. G. BANKOFF, Department of Chemical Engineering, Northwestern University, Evanston, Ill. 60201

To fill the need for an intermediate computational method which would yield better estimates of the optimal control than the one-dimensional procedure, while requiring significantly less computer time than

[1] Present address: Esso Research Laboratories, Baton Rouge, La.

the two-dimensional approach, a partial averaging procedure has been proposed (*1*) for certain classes of nonlinear distributed-parameter, optimal control problems. In the present work, this method is applied to a realistic problem, consisting of the choice of the average temperature profile in the tubular reactor containing V_2O_5 catalyst to maximize the outlet conversion in the air oxidation of SO_2. The results demonstrate the feasibility of the suboptimal hybrid method, which uses integro-differential equations rather than ordinary differential equations to retain more information concerning the radial variations in concentration and temperature than the lumped-parameter approach.

The problem discussed is the determination of the optimal wall heat flux distribution for a packed tubular reactor of fixed length in which radial temperature and concentration gradients are significant. The objective function to be maximized is the outlet conversion of SO_2. For this exothermic reaction it is assumed that heat may be removed from the reactor but not added—*i.e.*, wall heat fluxes are restricted to non-negative values. The rate expression used here is that of Calderbank (*2, 3*) which is based on data obtained with commercial catalyst.

The two-dimensional model for the packed reactor is described by the following dimensionless material and thermal energy balances:

$$\frac{\partial \alpha}{\partial \zeta} = \beta_Y \frac{1}{x} \frac{\partial}{\partial x}\left(x \frac{\partial \alpha}{\partial x} \right) + \widetilde{R} \tag{1}$$

$$\frac{\partial \tau}{\partial \zeta} = \beta_T \frac{1}{x} \frac{\partial}{\partial x}\left(x \frac{\partial \tau}{\partial x} \right) + \widetilde{R}_T \tag{2}$$

These equations assume that the mass velocity of the gas remains uniform at any cross section of the reactor, even though temperature and composition vary with radial position. It is further assumed that the Reynolds number will be sufficiently high that the radial Peclet numbers for heat and mass transport, Pe_T and Pe, respectively, may be considered constants. Finally, these equations also assume that axial diffusion of heat and mass are negligible, that gas and solid temperatures are identical at any point, and that the thermal properties of the gas remain constant.

The boundary and initial conditions which complete the reactor model are:

$$\alpha(x, 0) = 0, \qquad \tau(x, 0) = 1.0$$

$$\frac{\partial \alpha}{\partial x}(0, \zeta) = 0, \qquad \frac{\partial \tau}{\partial x}(0, \zeta) = 0 \tag{3}$$

$$\frac{\partial \alpha}{\partial x}(1, \zeta) = 0, \qquad \frac{\partial \tau}{\partial x}(1, \zeta) = \widetilde{Q}(\zeta)$$

with

$$\widetilde{Q}_* \leq \widetilde{Q}(\zeta) \leq \widetilde{Q}^*$$

where $Q(\zeta)$ is the dimensionless heat flux.

The objective function, P, to be maximized in this problem is the average extent of reaction in the product leaving the reactor. This is defined as the usual cup-mixing value:

$$P \overset{\Delta}{=} \bar{\alpha}(1) = 2\int_0^1 \alpha(x, 1)\, x dx \tag{4}$$

Direct application of variational methods to this problem leads to the optimal policy:

$$\widetilde{Q}(\zeta) = \begin{cases} \widetilde{Q}^* \text{ when } \lambda_\tau(\zeta, 1) < 0 \\ \widetilde{Q}_* \text{ when } \lambda_\tau(\zeta, 1) > 0 \end{cases} \tag{5}$$

where the adjoint variable λ_τ is defined (along with the additional adjoint variable λ_α) by the following partial differential equations and boundary conditions:

$$\frac{\partial \lambda_\alpha}{\partial \zeta} = -\lambda_\alpha \frac{\partial \widetilde{R}}{\partial \alpha} - \lambda_\tau \frac{\partial \widetilde{R}_T}{\partial \alpha} + \beta_Y \left[\frac{\partial}{\partial x}\left(\frac{\lambda_\alpha}{x}\right) - \frac{\partial^2 \lambda_\alpha}{\partial x^2} \right] \tag{6}$$

$$\frac{\partial \lambda_\tau}{\partial \zeta} = -\lambda_\alpha \frac{\partial \widetilde{R}}{\partial \tau} - \lambda_\tau \frac{\partial \widetilde{R}_T}{\partial \tau} + \beta_T \left[\frac{\partial}{\partial x}\left(\frac{\lambda_\tau}{x}\right) - \frac{\partial^2 \lambda_\tau}{\partial x^2} \right] \tag{7}$$

$$\lambda_\alpha(x, 1) = 2x, \qquad \lambda_\tau(x, 1) = 0 \tag{8}$$

$$\lambda_\alpha(0, \zeta) = 0, \qquad \lambda_\tau(0, \zeta) = 0$$

$$\lambda_\alpha(1, \zeta) - \frac{\partial \lambda_\alpha}{\partial x}(1, \zeta) = 0, \; \lambda_\tau(1, \zeta) - \frac{\partial \lambda_\tau}{\partial x}(1, \zeta) = 0$$

Solution of the set of nonlinear partial differential Equations 1, 2, 6, and 7, together with the split boundary and initial conditions 3 and 8 will lead (hopefully) to the optimal solution. However, this is indeed a formidable computational task.

To ease the computational requirements, a hybrid method, which may be termed the partial averaging method, has been developed (*4*). Application of this approach to the problem is illustrated below briefly.

Defining the average extent of reaction and dimensionless temperature as above:

$$\bar{\alpha}(\zeta) = 2\int_0^1 \alpha(x, \zeta)\, x dx \tag{9}$$

$$\bar{\tau}(\zeta) = 2\int_0^1 \tau(x, \zeta)\ x dx \tag{10}$$

Equations 1 and 2 may be averaged similarly, yielding, upon using Equation 12:

$$\frac{d\bar{\alpha}}{d\zeta} = 2\int_0^1 \widetilde{R} x dx; \qquad \bar{\alpha}(0) = 0 \tag{11}$$

$$\frac{d\bar{\tau}}{d\zeta} = -2\beta_T \widetilde{Q}(\zeta) + 2\int_0^1 \widetilde{R}_T x dx; \qquad \bar{\tau}(0) = 1 \tag{12}$$

Since the objective function, P, depends only on $\bar{\alpha}(1)$, an effective suboptimal approach, which reduces the computational labor by a factor of almost 2, employs averaged equations in the backwards integration but the full state equations in the forward pass. The gradient in control space of a lumped-parameter Hamiltonian function, obtained by averaging the state variables obtained in the forward pass, is used to determine an improved estimate of the control function.

The result of this treatment is the suboptimal control policy:

$$\widetilde{Q}(\zeta) = \begin{cases} \widetilde{Q}^* \text{ when } \lambda_2(\zeta) < 0 \\ \widetilde{Q}_* \text{ when } \lambda_2(\zeta) > 0 \end{cases} \tag{13}$$

where the adjoint variables satisfy the following ordinary differential equations and boundary conditions:

$$\frac{d\lambda_1}{d\zeta} = -\lambda_1\left[2\int_0^1 \frac{\partial \widetilde{R}}{\partial \alpha}\ x dx\right] - \lambda_2\left[2\int_0^1 \frac{\partial \widetilde{R}_T}{\partial \alpha}\ x dx\right] \tag{14}$$

$$\frac{\partial \lambda_2}{\partial \zeta} = -\lambda_1\left[2\int_0^1 \frac{\partial \widetilde{R}}{\partial \tau}\ x dx\right] - \lambda_2\left[2\int_0^1 \frac{\partial \widetilde{R}_T}{\partial \tau}\ x dx\right] \tag{15}$$

$$\lambda_2(1) = 1 \qquad \lambda_2(1) = 0 \tag{16}$$

Comparison of the above equations with those of the direct approach, Equations 6, 7, and 8, shows the great simplification which results from this suboptimal method. This method takes advantage of the unique features of the present problem in which the objective function involves only the average extent of reaction and the control variable $Q(\zeta)$, is a function only of axial position. The essence of this method is that the reactor is still simulated by the two-dimensional model while the optimization calculations utilize the simplified average variable equations. Ap-

Table I. Variation of Optimal Exit Extent of Reaction with Wall Heat Flux Profile Using Two-Dimensional Reactor State Equations

Heat Flux Profile	$\bar{\alpha}(1)$
Suboptimal profile 1 (*4*)	0.9492
Suboptimal profile 2 (*4*)	0.9438
Optimal profile for plug-flow (one-dimensional) model	0.9073
(Optimal plug-flow profile in plug-flow model)	(0.9657)

plication of this technique to problems where the objective function, or control variable, is a function of both space variables would require additional approximations.

Despite the apparently flat optimum, there is a significant difference between the final profiles determined from the suboptimal procedure for the two-dimensional reactor and that corresponding to the optimal one-dimensional reactor. The results are summarized in Table I. Ignoring the radial gradients gives an estimate of the optimal control which apparently results in $\bar{\alpha}(1) = 0.966$, whereas actually it would be 0.907 because of the radial effects. The estimates obtained with the partial averaging technique are significantly higher (0.949 and 0.944), and do not appear to depend upon the choice of initial profile. The importance of the more accurate calculation is therefore evident.

(1) Paynter, J. D., Dranoff, J. S., Bankoff, S. G., *Ind. Eng. Chem., Process Design Develop.* (1970) **9,** 303; (1971) **10,** 244.
(2) Calderbank, P. H., *Chem. Eng. Progr.* (1953) **49,** 585.
(3) Calderbank, P. H., *J. Appl. Chem. (London)* (1952) **2,** 482.
(4) Paynter, J. D., Ph.D. Thesis, Northwestern University, Evanston, Ill. (1968).

Periodic Operation of Reactor-Recycle Systems

GUILLERMO PAREJA[1] and MATTHEW J. REILLY, Department of Chemical Engineering, Carnegie-Mellon University, Pittsburgh, Pa. 15213

It has recently been established, both theoretically and experimentally, that the periodic operation of chemical reactors can be superior to the conventional steady-state design (*1–9*). Douglas and co-workers (*1–7*) studied extensively the periodic behavior of continuous stirred tank reactors and demonstrated that a positive feedback control system can be used to improve reactor performance. The inherent feedback present in the tubular reactor–recycle system suggests that it might be improved by periodic operation. Such an improvement is taken to mean that the

[1] Present address: Monsanto Co., Sauget, Ill. 62201.

yield of a desired product has been increased or, for multiple reactions, that the ratio of the product yield to the yield of a waste product has been increased. The yield and selectivity under periodic operation are compared with yield and selectivity under steady operation, both with and without recycle. The basis for comparison is that all three types of operation produce the same average quantity of feed in the same sized reactor.

Periodic Recycle with One Feed Stream

The system contains a mixing tank with volume V_t followed by a plug-flow tubular reactor of volume V (Figure 1). The reactor's effluent stream is split into recycle and output streams. The recycle stream flows through

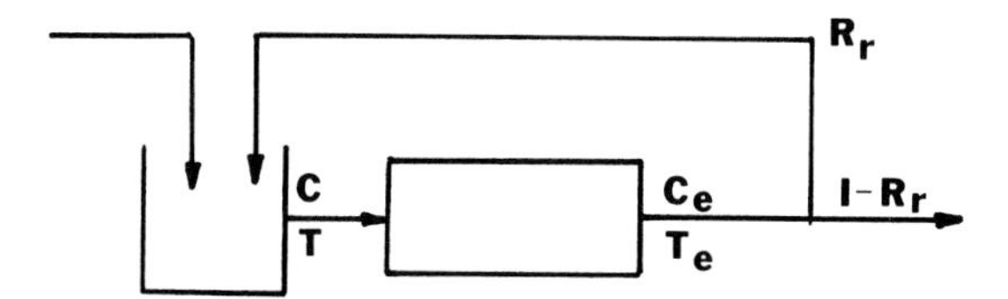

Figure 1. Reactor-recycle system

the recycle line of negligible volume back to the mixing tank where it is mixed with the feed stream. The volumetric flow rate of the recycle stream is taken to be:

$$R(t) = \overline{R}(1 + a_r \sin \omega t) \tag{1}$$

The feed flow rate is chosen to maintain a constant velocity within the tubular reactor. Thus

$$F(t) = \overline{F} - a_r \overline{R} \sin \omega t \tag{2}$$

Consider a single, irreversible, first-order reaction

$$\mathrm{A} \longrightarrow \mathrm{P}$$

which occurs isothermally only within the reactor. For a plug flow reactor, the effluent reactant concentration is given by

$$C_{A,e}(t) = e^{-k\tau} C_A(t) \tag{3}$$

where $C_A(t)$ is reactant concentration in the mixing tank and hence in the reactor input stream. A material balance for the reactant around the mixing tank yields

$$V_t \frac{dC_A}{dt} = R(t)\, e^{-k\tau}\, C_A(t - \tau) + F(t)\, C_{A,f}(t) - (\bar{F} + \bar{R})\, C_A(t) \qquad (4)$$

We wish to consider two modes of periodic operation:

(1) The complementary mode in which the reactant feed concentration is constant. Since the feed flow rate $F(t)$ is complementary to the recycle flow rate, the molar reactant feed rate $F(t)C_{A,f}$ varies with time in the complementary mode.

(2) The balanced mode in which any oscillations in the feed flow rate are balanced by oscillations in the reactant feed concentration. Thus the balanced mode provides for a constant molar reactant feed rate.

Some latter examples involving multiple feed streams, some of which are constant, may be interpreted as combining these two modes of operation.

Perturbation Analysis for the Complementary Mode

The reactant material balance, Equation 4 may be rewritten as:

$$\tau_t \frac{dC_A}{dt} = R_r(1 + a_r \sin \omega t)\, e^{-k\tau}\, C_A(t - \tau) \qquad (5)$$

$$+ [1 - R_r(1 + a_r \sin \omega t)]\, C_{A,f} - C_A(t);$$

$$R_r = \bar{R}/(\bar{F} + \bar{R})$$

Equation 5 is a linear differential-difference equation with periodic coefficients. A perturbation technique may be used to solve this equation, and we may write

$$C_A(t) = C_A^{(0)}(t) + a_r C_A^{(1)}(t) + a_r^2 C_A^{(2)}(t) + \ldots \qquad (6)$$

Substitution of this expression into Equation 5 and subsequent collection of the coefficients of a_r yield the set of differential-difference equations

$$\tau_t \frac{dC_A^{(0)}}{dt} = R_r\, e^{-k\tau} C_A^{(0)}(t - \tau) - C_A^{(0)}(t) + (1 - R_r)\, C_{A,f} \qquad (7)$$

$$\tau_t \frac{dC_A^{(1)}}{dt} = R_r\, e^{-k\tau} C_A^{(1)}(t - \tau) - C_A^{(1)}(t)$$

$$+ R_r(e^{-k\tau}\, C_A^{(0)}(t - \tau) - C_{A,f}) \sin \omega t \qquad (8)$$

$$\tau_t \frac{dC_A^{(2)}}{dt} = R_r\, e^{-k\tau} C_A^{(2)}(t - \tau) - C_A^{(2)}(t)$$

$$+ R_r\, e^{-k\tau} C_A^{(1)}(t - \tau) \sin \omega t \qquad (9)$$

Our primary interest is in the asymptotic periodic solution of these equations. It is verified readily that these solutions are given by

$$C_A^{(0)} = C_{A,f}(1 - R_r)/(1 - R_r\, e^{-k\tau}) \tag{10}$$

$$C_A^{(1)} = E_c(A \sin \omega t + B \cos \omega t) \tag{11}$$

and

$$C_A^{(2)} = \left(\frac{R_r\, e^{-k\tau}}{1 - R_r\, e^{-k\tau}}\right)\frac{E_c A_e}{2} + \psi_1(t) \tag{12}$$

$\psi_1(t)$ is a linear combination of sin $(2\,\omega t)$ and cos $(2\,\omega t)$ where

$$E_c = R_r(e^{-k\tau}\, C_A^{(0)} - C_{A,f}) \tag{13}$$

$$A = \frac{1 - \omega_c\, R_r\, e^{-k\tau}}{(1 - \omega_c\, R_r\, e^{-k\tau})^2 + (\omega\tau_t + \omega_s\, R_r\, e^{-k\tau})^2} \tag{14}$$

$$B = \frac{\omega\tau t + \omega_s\, R_r\, e^{-k\tau}}{(1 - \omega_c\, R_r\, e^{-k\tau})^2 + (\omega\tau_t + \omega_s\, R_r\, e^{-k\tau})^2} \tag{15}$$

with $\omega_c = \cos \omega\tau$, $\omega_s = \sin \omega\tau$, and $A_e = A\omega_c + B\omega_s$. $C_A^{(0)}$ is the steady-state solution for the case of constant recycle flow.

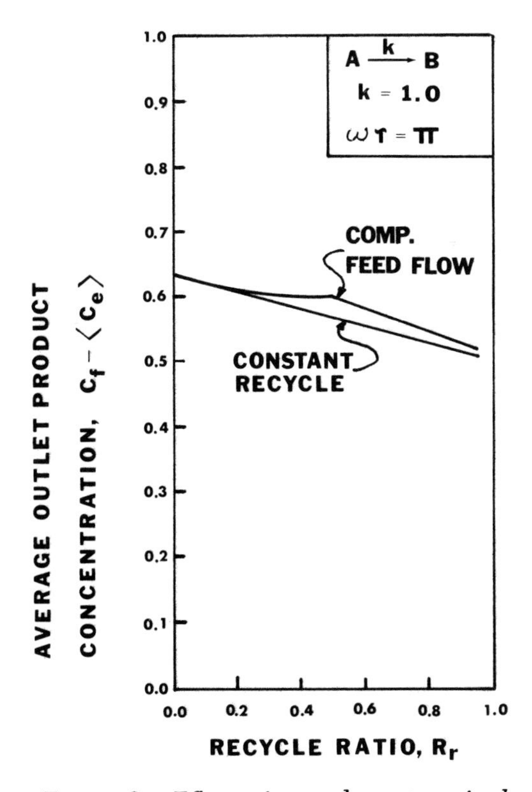

Figure 2. Effect of complementary feed flow

The flow-averaged effluent concentration is defined as

$$<C_{A,e}> = \frac{\omega/2\pi}{F} \int_0^{2\pi/\omega} C_{A,e}(t)\, F(t)\, dt \tag{16}$$

Substitution of $C_{A,e}(t)$ from Equation 3 and of $F(t)$ from Equation 2 yields

$$<C_{A,e}> = e^{-k\tau}\left[C_A^{(0)} - \frac{a_r^2}{2}\left(\frac{R_r}{1 - R_r}\right)\left(\frac{1 - e^{-k\tau}}{1 - R_r e^{-k\tau}}\right) E_c A_e\right] \tag{17}$$

The flow-averaged effluent concentration is the sum of the steady state effluent concentration plus a correction term, which is proportional to the square of the amplitude of the recycle oscillations. The correction term is also proportional to A_e which can be positive or negative, depending upon the frequency ω. With $\omega_c = -1$, $\tau_t = 0$, $R_r = 0.75$ and $e^{-k\tau} = 1/9$, Equation 17 becomes

$$<C_{A,e}> = e^{-k\tau}\, C_A^{(0)}(1 - 3.4a_r^2) \tag{18}$$

Thus, if $a_r^2 = 0.1$, the effluent reactant concentration is 34% lower for periodic operation than for the nonoscillatory case. It is not necessary to have a nonlinear reaction to produce a shift in the average conversion. The predictions of the perturbation analysis, Equation 17, coincide within 0.1% with the numerical results. Figure 2 presents the effect of complementary feed flow. The amplitude of the oscillations is the maximum allowable—*i.e.*, $a_r = 1.0$ for $R_r \leqslant 0.5$, $a_r = \dfrac{1 - R_r}{R_r}$ for $R_r \geqslant 0.5$.

Perturbation Analysis for the Balanced Mode

For this mode the material balance, Equation 4, becomes

$$\tau_t \frac{dC_A}{dt} = R_r(1 + a_r \sin \omega t)\, e^{-k\tau}\, C_A(t - \tau) + (1 - R_r)<C_{A,F}> - C_A(t) \tag{19}$$

A perturbation analysis similar to that for the complementary mode yields equations identical to Equations 10, 11, 12, and 17 except that E_c is replaced by E_B with

$$E_B = R_r e^{-k\tau}\, C_A^{(0)} \tag{20}$$

While E_c is negative, E_B is positive. Hence the two modes produce opposite shifts in the average conversion.

Periodic Recycle with Two Feed Streams

The opposing effects of the two modes of operation can be used, in a multiple reaction system, to increase the yield of the desired product while decreasing the yield of the waste product.

Consider the following reactions taking place isothermally in the tubular reactor

$$2A \xrightarrow{k_1} U \qquad (21)$$

$$A + 2B \xrightarrow{k_2} P$$

with flow rates given by

$$\begin{aligned} F_A(t) &= \overline{F}_A \\ F_B(t) &= \overline{F}_B - \overline{R}\, a_r \sin \omega t \\ R(t) &= \overline{R}(1 + a_r \sin \omega t) \end{aligned} \qquad (22)$$

Reactant A has a constant molar feed rate and is under conditions similar to the balanced feed flow mode. On the other hand, reactant B is under the same conditions as those in the complementary feed flow mode. Hence we can expect the yield of P to increase and the yield of U to decrease. These effects are shown by the following examples:

For $R_r = 0.5$, $k_1 = 2.0$, $k_2 = 1.0$

	No Recycle	*Constant Recycle*	*Sel. Per Feed*
S	0.979	0.961	1.049
Y	0.328	0.325	0.241

$$Y = \frac{\langle C_{p,e} \rangle}{\left(\dfrac{\overline{F}_A}{\overline{F}_A + \overline{F}_B}\right) C_{A,F} - \langle C_{A,e} \rangle}; \qquad S = \frac{\langle C_P, e \rangle}{\langle C_U, e \rangle}$$

For $k_1 = 4.0$, $k_2 = 1.0$

	No Recycle	*Constant Recycle*	*Sel. Per Feed*
S	0.688	0.697	0.764
Y	0.257	0.258	0.273

In the first example no recycle is superior to constant recycle, and in the second example constant recycle is superior to no recycle. However, in both cases the periodically operated system gives the best performance of the three.

(1) Douglas, J. M., Rippin, D. W. T., *Chem. Eng. Sci.* (1966) **21**, 305.
(2) Douglas, J. M., *Ind. Eng. Chem., Process Design Develop.* (1967) **6**, 34.
(3) Douglas, J. M., Gaitonde, N. Y., *Ind. Eng. Chem., Fundamentals* (1967) **6**, 265.
(4) Gaitonde, N. Y., Douglas, J. M., *AIChE J.* (1969) **15** (6), 902.
(5) Baccaro, G. P., Gaitonde, N. Y., Douglas, J. M., *AIChE J.* (1970) **16** (2), 249.
(6) Dorawala, T. G., Douglas, J. M., unpublished data.
(7) Ritter, A. B., Douglas, J. M., unpublished data.
(8) Horn, F. J. M., Lin, R. C., *Ind. Eng. Chem., Process Design Develop.* (1967) **6**, 21.
(9) Chang, K. S., Bankoff, S. G., *Ind. Eng. Chem., Fundamentals* (1968) **7**, 633.

Cyclic Operation of Reaction Systems: The Influence of Diffusion on Catalyst Selectivity

J. E. BAILEY[1] and F. J. M. HORN,[2] Department of Chemical Engineering, William Marsh Rice University, Houston, Tex. 77001

In several earlier papers (*1, 2, 3*) it has been shown that cycling can increase catalytic selectivity relative to steady-state operation when mass transfer resistance is neglected. Therefore, it is natural to inquire whether the beneficial effects of periodic operation persist in the presence of diffusional resistance to mass transport.

Although a similar query has been answered in the affirmative in another paper (*4*), the model there was relatively simple. All mass transfer resistance was lumped into a stagnant fluid boundary layer around the catalyst pellet, and intraparticle concentration gradients were neglected. This study reveals that such a model is inadequate when reaction is sufficiently fast compared with diffusive transport and/or when changes in conditions at the external catalyst surface occur too rapidly. The need for a distributed-parameter description of adsorbent particles in Rolke and Wilhelm's recuperative parametric pumping model is dictated by similar considerations (*5*). Although the distributed-parameter model used here is still idealized, it is a significant improvement over the lumped model.

[1] Present address: Department of Chemical Engineering, University of Houston, Houston, Tex. 77004.
[2] Present address: Department of Chemical Engineering, University of Rochester, Rochester, N. Y. 14627.

In the detailed discussion of results, frequent use is made of the recently proposed set-theoretic definition of selectivity (*1*, *3*). Briefly, selectivities corresponding to various classes of operations (*e.g.*, steady-state operations) are attainable sets in a space spanned by the average rates of product formation. This definition is especially well suited for discussing the relative merits of periodic and steady-state operation.

Mathematical Model of Reaction and Diffusion within a Catalyst Slab

It is assumed that reactions take place in a slab of catalyst of thickness $2L$ and that the local mass transport of a component of the fluid phase, say reactant R, within this slab can be described by

$$N_R = -D_R \frac{\partial c_R}{\partial z} \quad (1)$$

Thus, any bulk flow contributions to the molar flux of R in the positive z direction, N_R, are assumed negligible, and both the effective diffusivity of R and total concentration are considered constant. This simplified picture of the complex multicomponent diffusion process is commonly employed in analyses primarily concerned with reaction–diffusion interaction. Also, any mass transfer resistance between the bulk fluid phase and the exterior catalyst surface is neglected, and the catalyst is assumed isothermal. Justification of the latter assumptions for gas–porous solid catalyst systems has been discussed recently (*6*). Finally, the control variable which is manipulated to alter selectivity is the reactant concentration at the exterior catalyst surface. It is assumed that there is a specified upper bound $c_{R\max}$ on boundary reactant concentration.

The reaction mechanism which is analyzed is the following:

$$R \underset{2}{\overset{1}{\rightleftarrows}} A \quad (2a)$$

$$R + A \xrightarrow{3} P_1 \quad (2b)$$

$$R \xrightarrow{4} P_2 \quad (2c)$$

A denotes a chemisorbed species, and the over-all reaction (2c) is assumed second order in the concentration of reactant R. This mechanism has been examined within several different contexts in earlier papers. First, it was shown that in the absence of mass transfer effects, cycling R concentration gave much larger selectivity than steady-state operations (*1*, *3*). Favorable effects of cycling remained when a lumped mass transfer resistance element was added to the model, at least if the relative rate of mass transfer to chemical reaction was sufficiently large (*4*). Therefore, Reaction 2 is of interest for two reasons. First, cyclic operation may be favorable even with mass diffusion. Also, the results allow a comparison among three different models of mass transport.

It is said that cycling improves selectivity if any of the time-average rates for periodic operation are not contained in the convex hull of the steady-state selectivity, co S_{ss} (*see* Ref. *3*). In this situation there exist average rates of production which can be obtained by cycling but which are not realizable by steady-state operation or by mixing the outputs of several steady-state reactors operated in parallel. Hence, improved catalyst performance might be available under dynamic operation.

Cycling with Large Diffusional Resistance

Upon first consideration the prospects for improving selectivity when mass transfer resistance is large may seem slight. This skepticism is based on a previous analysis which indicated that rapid switching of reactant concentration at all points of the catalyst surface led to the desired improvement (*3*). In that investigation switching was assumed so rapid that chemisorbed intermediates could not follow and approached a time-invariant condition called a relaxed steady state. Clearly, if diffusional resistance is large, only very slow changes in the reactant concentration near the center of the catalyst slab can be effected by switching at the exterior catalyst surface. Hence there is no possibility of approximating anything resembling a relaxed steady state for chemisorbed species far away from the catalyst exterior.

However, when there is a large resistance to mass transport, reactant concentrations—and hence reaction rates—are relatively large near the exterior surface of the catalyst and decrease rapidly to near zero just inside the slab. Thus, the major contribution to the average rates of production comes from the neighborhood of the exterior slab surface. Since this is also the region where switching of the control has the largest and most immediate effect, it may be possible to obtain average rates by cycling which are not in co S_{ss} even in the "diffusion limited" regime.

Qualitative physical criteria for improvement by cycling may be formulated as follows. Owing to the damping effect of diffusion on switches in the fluid phase composition, for a given period τ there will be a characteristic penetration depth L_{τ} for the effects of switching. There is also a characteristic penetration depth L_R for reaction which, like L_{τ}, decreases as diffusional resistance increases. Finally, there is a characteristic time t_{ss} for the chemisorbed species A to reach steady state; t_{ss} depends only on the reaction kinetics.

For cycling to have a significant effect on selectivity, L_{τ} must be on the same order as L_R or larger, and τ must be on the same order as t_{ss} or smaller:

$$L_\tau \gtrsim L_R \tag{3}$$

$$\tau \lesssim t_{ss} \tag{4}$$

Condition 3 means that switching affects the reaction zone, and Condition 4 indicates that switching is sufficiently rapid so that chemisorbed intermediates are not always at steady state. If cycling is so slow that the steady-state approximation is valid for chemisorbed species, the average rates obtained will be in co S_{ss} and therefore of little interest (*2*).

The results of several calculations for cases of large mass transfer resistance are consistent with the simple arguments given above. There is a wide range of kinetic parameters for which the criteria above are satisfied for some cycling frequency. In some cases the ratio of time-average P_2 production to time-average formation of P_1 could be made nearly twice as large by cycling as by any steady-state operation.

Asymptotic Dependence of Average Rates on a Thiele Modulus

By an argument similar to that given by Petersen (*7*), it is shown that time-average production rates for a given boundary concentration control are inversely proportional to a Thiele modulus η for sufficiently large values of the modulus. η is defined here by

$$\eta = L\sqrt{\frac{k_3 c_{R\max}}{D_R}}. \tag{5}$$

This is a useful result since it allows extrapolation of results calculated for one case of large diffusional resistance to all other asymptotically large values of the Thiele modulus. Also, this conclusion means that comparisons of ratios of average rates for one large η apply to all other η corresponding to the diffusion limited regime.

Discussion

The results above are particularly interesting when compared with the analysis of a lumped-parameter mass transfer resistance model (*4*). For the simpler lumped-parameter model, no improvement by periodic operation is possible for reaction system 2 when mass transport is very slow relative to reaction velocity. This is not the case for the distributed-parameter model in the same situation—*i.e.*, very large η. However, as is obvious by comparing the steep internal concentration profiles in the diffusion-limited case with the spatially uniform concentration distributions assumed for the lumped-parameter model, the lumped-parameter model is a poor representation of even steady-state catalyst behavior when mass transfer resistance is large. This inadequacy is amplified when çycling with a period smaller than the characteristic response time of the system is considered.

Thus, while the simplest model is acceptable for a rough preliminary investigation, a more accurate representation of mass transfer resistance should be analyzed before any final conclusions regarding the possibility or impossibility of improving catalyst selectivity by periodic operation are made. There may be cases where the ability to increase the performance of a catalyst by dynamic operation depends on the presence of large intraparticle composition and/or temperature gradients.

Acknowledgments

This work was supported by the National Science Foundation Grant No. GU-1153. One of the authors (J.E.B.) was also supported by a NASA Traineeship.

(1) Bailey, J. E., Horn, F., *Ber. Bunsenges. Phys. Chem.* (1969) **73,** 274.
(2) Bailey, J. E., Horn, F., *Ber. Bunsenges. Phys. Chem.* (1970) **74,** 611.
(3) Horn, F. J. M., Bailey, J. E., *J. Optimization Theory Applications* (1968) **2,** 441.
(4) Bailey, J. E., Horn, F. J. M., Lin, R. C., *Proc. Natl. AIChE Meetg., 63rd, St. Louis, 1968.*
(5) Rolke, R. W., Wilhelm, R. H., *Ind. Eng. Chem., Fundamentals* (1969) **8,** 235.
(6) Carberry, J. J., White, D., *Ind. Eng. Chem.* (1969) **61,** 27.
(7) Petersen, E. E., "Chemical Reaction Analysis," Prentice-Hall, Englewood Cliffs, N. J., 1965.

5

Physical Phenomena in Catalysis and in Gas–Solid Surface Reactions

E. WICKE

Institut für Physikalische Chemie, Westfälische Wilhelms-Universität, 4400 Münster, Germany

Physical phenomena are understood as intraparticle and interparticle transport processes of mass and of heat in fixed bed reactors. Starting with measurements of temperature and of concentration profiles along short combustion zones in packed beds, our present knowledge about these processes and their influence on chemical reactions at solid surfaces is summarized. The main topics are the isothermal in-pore diffusion—the interior of porous catalyst pellets usually can be taken as isothermal—the gas–solid heat and mass transfer, and the intensity of the dispersion effects in packed beds. Particular emphasis is given to the discussion of experimental methods. Special attention is called to the differences between heat and mass interparticle transport processes, brought about in the range of medium and low Reynolds numbers by solid heat conduction and by radiation.

Physical phenomena and catalysis is a broad field which has developed rapidly in recent years. Its increasing importance can be judged by the new extended and revised editions of the excellent book by Frank-Kamenetzky (*1*) and of the useful monograph by Satterfield (*2*). This review is restricted to prominent directions in this field. Within this framework it should be possible not only to review the knowledge accumulated, but to point to some difficulties and problems which have remained unsolved. In line with personal preferences, special emphasis is given to methods and to results of experimental investigations.

Some interesting temperature measurements and concentration profiles in catalyst beds have been done recently by Fieguth (*3*). The experimental device used was a tubular reactor suitable for work under adiabatic conditions (Figure 1) [described in detail elsewhere (*4*)], with a packed bed of catalyst particles. These were cylindrical pellets from alumina, about 3 mm in diameter and height, containing 0.3 wt % platinum as active component. The oxidation of carbon monoxide in air was chosen as the test reaction. One of the catalyst pellets was equipped with two thermocouples for temperature measurements in the center and at the surface (Figure 1, right). This pellet was imbedded in the catalyst layer at the same level as the open end of a suction tube. A thermocouple in the orifice of this tube measured the gas temperature, and the

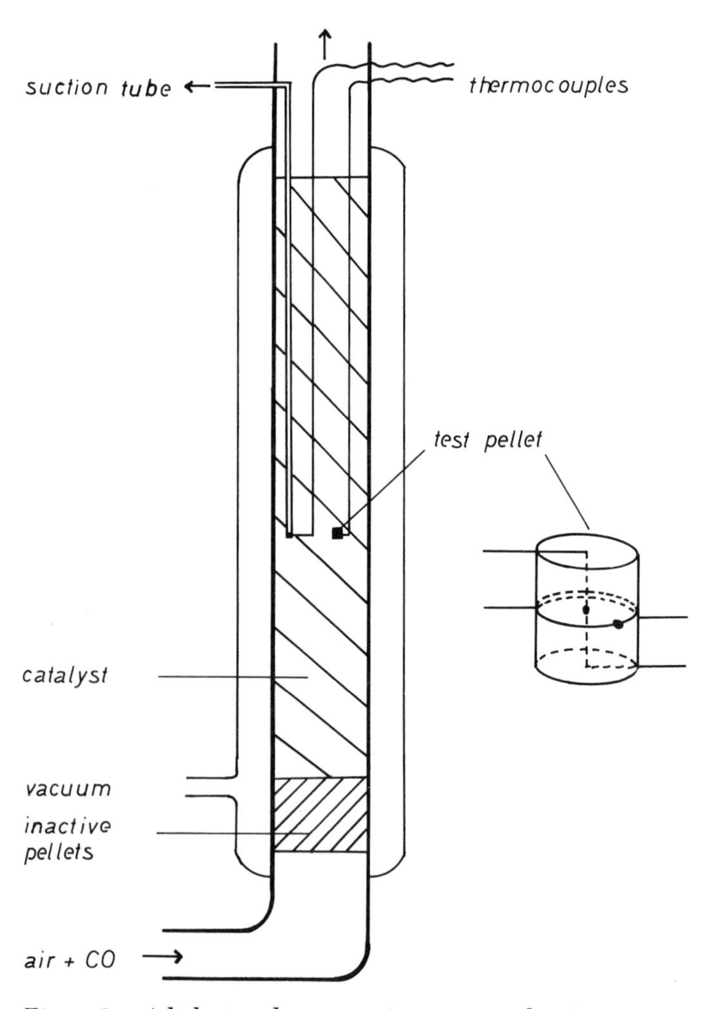

Figure 1. Adiabatic tube reactor from quartz glass for measuring concentration and temperature profiles. Right side: 3 × 3 mm catalyst pellet with central and equatorial thermocouple.

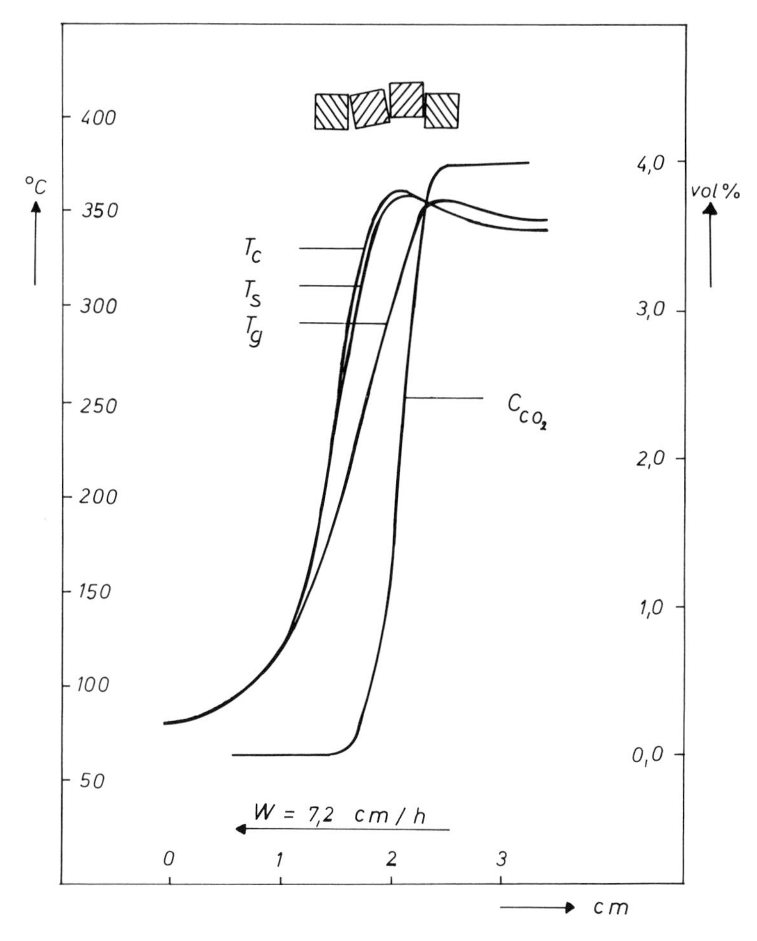

Figure 2. Profiles of CO_2 concentration, gas temperature, T_g, and pellet temperature (surface T_s, center T_c) along moving combustion zone (3). Top: catalyst pellets, schematic. 4% CO in air, linear flow velocity $u_o = 3.7$ cm/sec.

gas flow withdrawn through the tube was analyzed continuously for CO_2. The inlet air flow contained a few percent CO; if its temperature was raised to about 120°C, a steep temperature increase built up in the entrance of the catalyst bed, and the CO was oxidized completely along a distance of a few particle diameters. By lowering the inlet gas temperature, the combustion zone could be pushed back from the entrance and could then be shifted back and forth through the catalyst bed by changing the gas flow rate (4). A short reaction zone of this type was moved through the section of the bed where the measuring devices had been fixed; thus, the concentration and temperature profiles along the zone were obtained. Figure 2 gives an example with an inlet gas composed of 4% CO and air preheated to 70°C. The reaction zone moves at 7.2 cm/hr opposite the direction of the gas flow. The concentration profile needs about 8 minutes to pass by the orifice of the suction tube; it extends therefore along about 1 cm or three particle diameters. The gas temperature profile is markedly broader, extending along five to six particle diameters.

Regarding these profiles the following statements can be made:

(1) The pellet interior is nearly isothermal; the temperature in the center, T_c, exceeds the surface temperature T_s by only a few degrees.

(2) The pellet as a whole is superheated appreciably above the gas flow; the excess temperature $T_s - T_g$ attains values up to 100°C.

(3) The steepness of the concentration profile indicates that the axial dispersion of mass is very small under these conditions (modified Reynolds number $Re_p \approx 10$); in particular there is no back mixing of mass.

(4) The axial dispersion of heat is appreciably larger, as shown by the gas temperature profile, T_g. There is also back mixing of heat against the gas flow; the temperature profile extends upstream to particles which have no chemical heat production as yet.

(5) The first increase of CO_2 concentration, indicating the onset of reaction and chemical heat production, should coincide with the start of the pellets' superheating over the gas flow. Contrary to this, the CO_2 concentration in Figure 2 starts to increase about one particle diameter later. Such shifts in the profiles are characteristic for measurements of this type because the distance of one particle diameter represents the general uncertainty in fixing the position of a reaction zone in a packed bed.

(6) In view of the heterogeneous nature of the packed bed the question arises as to how far the smooth profiles are meaningful, or if stepwise functions would not be better approximations. Fundamentally, would it not be more advantageous to work with difference equations and with cell methods instead of differential equations? Along with this is the problem of the behavior of catalyst particles in steep gradients of concentration and temperature which has been studied recently by several groups of authors (5–9).

The background of these temperature and concentration profiles in theory and experience contains the whole field of physical phenomena in catalysis and in gas–solid reactions; a compilation of these phenomena is shown in Table I.

Intraparticle Problems—Isothermal

The predominant internal problem of porous solids in catalysis and in surface reaction with gases is in-pore diffusion. Measurements as well as theoretical considerations are made usually with countercurrent diffusion through porous plates. The general transport equation for stationary multicomponent diffusion at constant temperature and pressure is:

$$-c \text{ grad } x_i = \sum_k \frac{J_i x_k - J_k x_i}{\psi D_{ik}} + \frac{J_i}{D_i{}^{K}{}_{eff}} \quad (1)$$

This represents the balance between the driving force for the molecular species i and the resistances to transport, as they are induced by intermolecular collisions (normal gas diffusion according to Stefan-Maxwell) and by wall collisions (Knudsen flow). Here x_i and x_k are the mole fractions, J_i and J_k are the molar fluxes of the species i and k; c is the total molar concentration in the gas phase. The D_{ik} represent the binary bulk diffusion coefficients, the factor ψ accounts for the restrictions of bulk diffusion by the porous network (structural factor), and $D_i{}^{K}{}_{eff}$ is the effective Knudsen diffusion coefficient of species i. For counter diffusion

Table I. Physical Phenomena

Intraparticle (internal problems)

Isothermal:	*Non-isothermal:*
in–pore diffusion, normal gas diffusion, Knudsen diffusion, stoichiometry	internal superheating and stability

Interparticle (external problems)

interphase transport

dispersion effects

of mass:	of heat:
diffusion, convection (gas phase only)	conducton, convection, irradiation (gas and solid phase)

of two components, with equal pressures on both sides of the porous plate (open system), the molar fluxes are inversely proportional to the roots of the molecular masses:

$$J_2/J_1 = -\sqrt{M_1/M_2} \equiv \alpha - 1 \quad (2)$$

This holds for both Knudsen flow and in the pressure range where normal gas diffusion occurs in the porous medium, as was shown first by Hoogschagen in 1953 (*10, 11*). For binary counter diffusion (Components 1,2):

$$J_1 = -D_{1\,eff}\, c \text{ grad } x_1 \quad (3)$$

with
$$\frac{1}{D_{1\,\mathrm{eff}}} = \frac{1 - \alpha x_1}{\psi D_{12}} + \frac{1}{D_1{}^{\mathrm{K}}{}_{\mathrm{eff}}} \tag{3a}$$

This, and its integrated form:

$$J_1 = \psi D_{12} \frac{c}{\alpha L} \cdot \ln \left(\frac{1 - \alpha x_{1L} + \psi D_{12}/D_1{}^{\mathrm{K}}{}_{\mathrm{eff}}}{1 - \alpha x_{1o} + \psi D_{12}/D_1{}^{\mathrm{K}}{}_{\mathrm{eff}}} \right) \tag{3b}$$

are the expressions used most often in diffusion studies (*12, 13, 14, 15, 16, 17, 18*). Here x_{1o}, x_{1L} are the mole fractions of Component 1 at the two faces of the porous plate from thickness L.

For self-diffusion Equation 3a reduces to:

$$\frac{1}{D_{\mathrm{eff}}} = \frac{1}{\psi D^G} + \frac{1}{D^{\mathrm{K}}{}_{\mathrm{eff}}} = \frac{1}{\psi D_o{}^G} \frac{P}{P_o} + \frac{1}{D^{\mathrm{K}}{}_{\mathrm{eff}}} \tag{4}$$

where $D_o{}^G$ is the bulk diffusion coefficient at gas pressure P_o. Equation 4 represents a generalization of a formula proposed in 1944 by Bosanquet (*19*) for self-diffusion in a cylindrical capillary to account for the transition range between bulk diffusion and Knudsen flow. Calculations by kinetic gas theory agree closely with the results of Bosanquet's additive resistance formula, as Pollard and Present (*20*) later showed. Experimental proof was attempted by Mingle *et al.* (*21*) by measuring self-diffusion of CH_4 and CO_2, tagged by carbon-14, through 1 mm glass capillary at pressures around 0.1 torr. The results scattered appreciably and cannot be considered conclusive. As a matter of fact, a convincing verification of the Bosanquet formula at low pressures, running from the transition range to pure Knudsen flow, has not yet been performed.

If the system of transport pores under consideration has a fairly well defined mean radius r, the effective Knudsen diffusion coefficient also can be taken as the product of a structural factor and a transport quantity—*i.e.*, a single capillary coefficient: $D^{\mathrm{K}}{}_{\mathrm{eff}} = \psi^{\mathrm{K}} D^{\mathrm{K}}(r)$ with $D^{\mathrm{K}}(r) = \frac{2}{3} w r$, where w is the mean molecular speed. The structural factor ψ^{K} usually has a value similar to ψ in the normal gas diffusion term of Equation 3a or 4; $\psi \approx \psi^{\mathrm{K}}$ is known as the "permeability" of the pore system in question.

To describe binary countercurrent diffusion through porous media with arbitrary pore size distribution, two different methods have been developed: (1) the method of random pore distribution, and (2) the method of structural factors or of permeabilities. The first method was developed by Wakao and Smith (*16*) for bidispersed solids. These are made by compressing porous powder particles and thereby contain micro- and macropores. For this method the void fractions ϵ_n and the mean pore radii r_n of the different groups of pores must be known from pore size distribution data. The probability that the macro- and/or

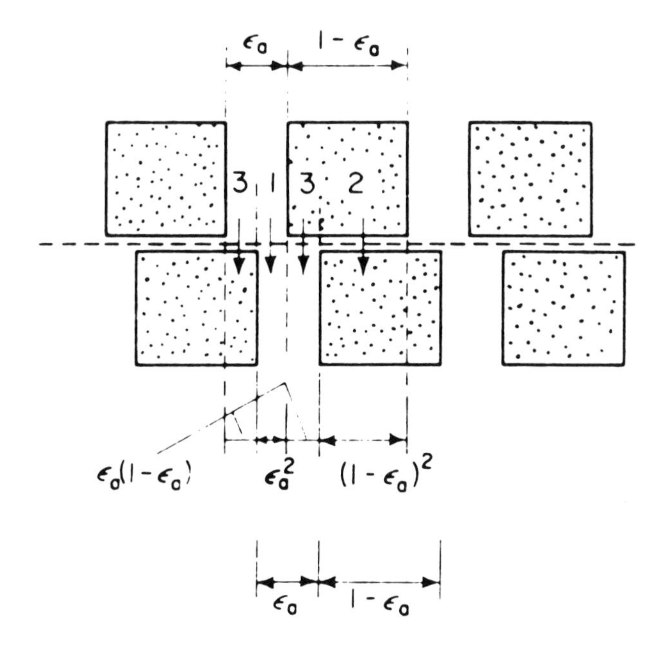

Figure 3. Combinations of pore alignments in bidispersed porous materials (16)

micropores line up along the diffusion path is assumed proportional to the product of the volume fractions in question. Figure 3 demonstrates the three combinations of pore alignment in this random pore model; ϵ_a is the void fraction of the macropores between the particles, $1 - \epsilon_a$ the volume fraction of the particles containing the micropores. For these three types of alignment the diffusivities are set up according to the model of straight cylindrical pores and are then combined in parallel flow to the transport equation. With a number of such terms, however, the transport equation becomes cumbersome, even more so if the method is extended to three different groups of pores (*18*). The method of structural factors, on the other hand, is based on diffusion measurements only and on the use of Equation 3 and 3a or, if self-diffusion is concerned, of Equation 4. Its application to a material, wherein a uniform group of pores dominates in diffusional transport, needs two diffusion measurements, one at high gas pressures to determine the structural factor ψ, the other at low pressures to determine the effective Knudsen coefficient $D_1{}^{K}{}_{eff}$. With these two empirical terms Equation 3a can then be applied as an approximation through the whole pressure range. If two groups of pores, I and II, are engaged equally in the diffusion process, the transport equation must be composed of two terms, for instance in case of self-diffusion:

$$D_{\text{eff}} = \frac{\psi_I \, D^G}{1 + D^G/D^K\,(r_I)} + \frac{\psi_{II} \, D^G}{1 + D^G/D^K\,(r_{II})} \tag{5}$$

A two-term-expression like this is sufficient to describe even complicated diffusional behavior (*22, 23*).

Figure 4 represents data of countercurrent diffusion of N_2 against He measured by Cunningham and Geankoplis (*18*) with three porous tablets. The tablets were prepared by compressing a porous powder of alumina to samples of different density, and contained a macropore and a micropore system. In the low density sample (1) the macropores with $r \approx 10^{-4}$ cm predominate in contributing to diffusion flux; the same is true for the medium sample A, whereas in the high density sample (2) the micropores with $r \approx 6 \cdot 10^{-7}$ cm predominate. The authors set up the transport equation as:

$$J_1 = -\frac{D_e}{1 - \alpha x_1}\, c \text{ grad } x_1 \tag{6}$$

instead of Equation 3; thus a diffusion coefficient D_e was defined according to:

$$\frac{1}{D_e} = \frac{1}{\psi D_{12}} + \frac{1}{(1 - \alpha x_1)\, D_1{}^K{}_{\text{eff}}} \tag{7}$$

The mean values of the mole fractions along the diffusion path were about the same in all experimental runs; for nitrogen: $\bar{x}_1 \approx 0.4$ to 0.5. The influence of composition could therefore be neglected in first approximation compared with the large influence of total gas pressure which was changed in the range 1:1000. To evaluate the measurements Equation 6 was integrated along the diffusion path at constant gas pressure P and temperature T to:

$$J_1 = \frac{D_e}{\alpha L}\,\frac{P}{RT}\,\ln\left(\frac{1 - \alpha x_{1L}}{1 - \alpha x_{1o}}\right) \tag{6a}$$

whereby D_e was taken as a constant mean value. With J_1, x_{10} and x_{1L} measured for N_2 this mean value was calculated from Equation 6a, and the product D_eP was plotted *vs.* P (Figure 4). The straight lines represent Knudsen flow, the bent parts of the curves are the transition range to normal gas diffusion. The dashed lines give the results of random pore calculations, based on the model of Figure 3, with the micro- and the macropore system listed in the caption to Figure 4. The dotted lines were obtained by the method of structural factors assuming one uniform group of pores only in each case. Although the samples had a bidispersed porous structure, this simple description agrees remarkably well with the measurements, thereby confirming that in each run one group of pores predominated in diffusional transport.

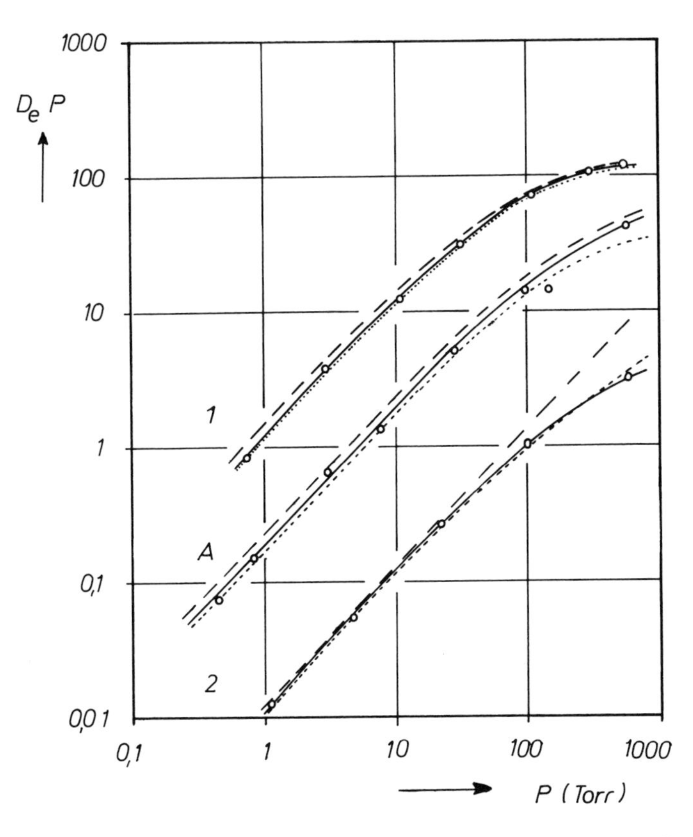

Figure 4. Experimental and predicted diffusivities for several bidispersed materials (18)

– – – – – *Method of random pores*
• • • • • *Method of structural factors*

Porosities (ϵ) *and mean pore radii* (r, *A units*):

N°	micro		macro	
1	ϵ_i = *0.326*,	r = *76*	ϵ_a = *0.521*,	r = *9800*
A	*0.468*	*85*	*0.303*	*4300*
2	*0.537*	*58*	*0.052*	*1500*

If in the porous solid a chemical reaction proceeds: $\nu_1 A_1 + \nu_2 A_2 \rightarrow \nu_3 A_3$, the molar fluxes of the gas components under steady-state conditions must be in the ratios of the stoichiometric numbers ν_i (< 0 for reactants, > 0 for products):

$$\frac{J_1}{\nu_1} = \frac{J}{\nu_2} = \frac{J_3}{\nu_3} \tag{8}$$

The reaction rate per unit volume, dependent on temperature and concentrations, is then:

$$\text{div } J_i = \nu_i \cdot r_v(T, c_1, c_2 \ldots) \tag{9}$$

The stoichiometric ratios (Equation 8) generally do not agree with the inverse ratios of the roots of the molecular masses, Equation 2; this means that: $\nu_i/\nu_k \neq \sqrt{M_k/M_i}$ is generally true. In the case of Knudsen flow within the porous pellet this contradiction is solved by pressure gradients, which build up along the diffusion paths to such values, that the fluxes of the components are in stoichiometric ratio. If there is normal gas diffusion, however, the total pressure remains constant throughout the pellet; the stoichiometric fluxes are then adjusted by proper gradients of the mole fractions which are established in the stationary run of reaction. These problems have been treated in detail recently by Hugo (*24, 25, 26*), who developed stoichiometric flux diagrams of the reaction components for the three transport processes: bulk aerodynamic flow, Knudsen flow, and normal gas diffusion. In the latter case the flux of a component *i* during a steady-state reaction can be described by a generalized diffusion coefficient:

$$D_i{}^G{}_{\text{eff}} = \nu_i \frac{\psi D_{12}}{\sum_k \frac{D_{12}}{D_{ik}} \cdot (\nu_i x_k - \nu_k x_i)} \tag{10}$$

in the transport equation: $J_i = D_i{}^G{}_{\text{eff}} c \text{ grad } x_i$, as shown earlier (*27*). These generalized coefficients contain the influence of the different values of the binary coefficients D_{ik} of all pair combinations i,k and the influence of the volume change by reaction. If all binary coefficients have the same value (D_{12}), the expression reduces to:

$$D_i{}^G{}_{\text{eff}} = \frac{\psi D_{12}}{1 - (x_i/\nu_i) \cdot \sum_k \nu_k} \tag{11}$$

and contains nothing more than the influence of the increase in the mole number $\sum_k \nu_k$. If there are only two components, it follows from Equations 2 and 8: $\sum_k \nu_k/\nu_i = \alpha$, and Equation 11 simplifies to the countercurrent expression, Equation 3a.

In the transition range between Knudsen flow and normal gas diffusion the Bosanquet formula (Equation 4) can be used to obtain the approximate relation:

$$\frac{1}{D_{i\,\mathrm{eff}}} \approx \frac{1}{D_i{}^G{}_{\mathrm{eff}}} + \frac{1}{D_i{}^K{}_{\mathrm{eff}}} \tag{12}$$

This approximation is restricted principally to cases where the total gas pressure can still be taken as constant along the diffusion path—*i.e.*, where the influence of Knudsen diffusion is small. The more general problem—that pressure gradients are built up to satisfy the stoichiometric flow conditions in the transition range—has not been solved until now.

To evaluate the influence of in-pore diffusion on reaction rate, Equation 9 must be integrated. This requires, in addition to knowledge of the transport equation, knowledge of the reaction rate law. With an empirical n^{th} order law the integration yields the following expression for the Thiele modulus:

$$\varphi = \frac{V_p}{S_p}\left[\frac{n+1}{2} \cdot \frac{1}{\overline{D}_{i\,\mathrm{eff}}} \cdot \frac{(-\nu_i) r_{vs}}{c_{is} - c_{i\mathrm{eq}}}\right]^{1/2} \tag{13}$$

V_p and S_p are the volume and the external surface area of the porous pellet, r_{vs} is the reaction rate, and $c_{is} - c_{i\,\mathrm{eq}}$ is the distance of concentration of component i from equilibrium at the external pellet surface. In the Knudsen range $D_{i\,\mathrm{eff}}$ is to be substituted by the effective Knudsen coefficient of component i, in the range of normal gas diffusion $D_{i\,\mathrm{eff}}$ represents a mean value of the generalized diffusion coefficient, Equation 10, taken over the concentration field in the pellets interior (for details *see* Hugo (*26*)). In the scope of this approximation the values of the diffusion coefficients in the transition range can be estimated by the modified Bosanquet formula (Equation 12), although in principle it is insufficient for stoichiometric flow.

To use this formula the structural factor ψ must be known. Its value is usually determined by measuring steady-state countercurrent diffusion by two gas streams, which bypass the two faces of a cylindrical or plate-shaped probe of the porous material. The method was developed by Kallenbach (*29*) in 1941 and has been applied frequently since then (*14, 15, 16, 17, 18, 30, 31*) at normal and reduced pressures. It has also been used with remarkable success at high pressures by Paratella (*32*) to measure the binary diffusion coefficient in CO/N_2 mixtures up to 150 atm by using a parallel pore plate with known permeability. As a matter of fact this method was being used in the early 1900's in the U. S. Department of Agriculture by Buckingham (*33*), when he measured the diffusivity of gases in soils.

Recently a new steady-state method has been worked out by Hugo *et al.* (*34, 35, 36*) in connection with permeability measurements on nuclear graphites. It uses the *p,o*-hydrogen conversion and deals with the simple case of self-diffusion. The principle of the method is shown in

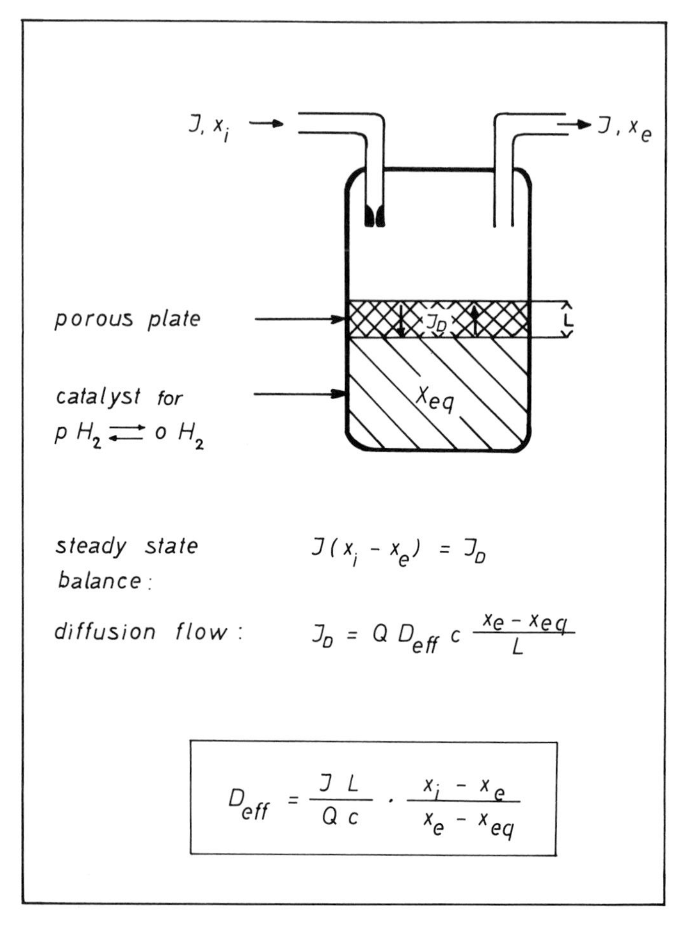

Figure 5. Reaction diffusion cell for diffusivity measurements in porous materials by the p-*hydrogen method after Hugo* et al. (34, 35, 36, 37)

Figure 5. A stream of hydrogen, enriched with the para species of mole fraction x_i, enters a mixing chamber which is separated by the porous sample from a platinum catalyst. The para molecules diffuse to the catalyst—diffusion flow J_D—where the equilibrium mole fraction x_{eq} of p-hydrogen is established, and the ortho molecules diffuse back through the porous plate. The stationary flow balance of the mixing chamber and the diffusion equation give a simple relation for the effective self-diffusion coefficient of hydrogen in the probe (*see* Figure 5). To evaluate this relation, the mole fractions of p-hydrogen in the inlet and outlet stream, x_i and x_e, are measured continuously by thermal conductivity. The method can be used only for porous materials which do not catalyze the conversion. A catalytic activity for o,p-H_2 conversion must be poisoned beforehand, or the measurements must be made with inactive catalyst support material.

It is practical to run measurements at different hydrogen pressures, and to plot $1/D_{eff}$ *vs.* pressure. In this way Figure 6 demonstrates the values that have been measured by Hugo and Beyer (*36, 37*) with a nuclear graphite in its original condition and after increasing amounts of

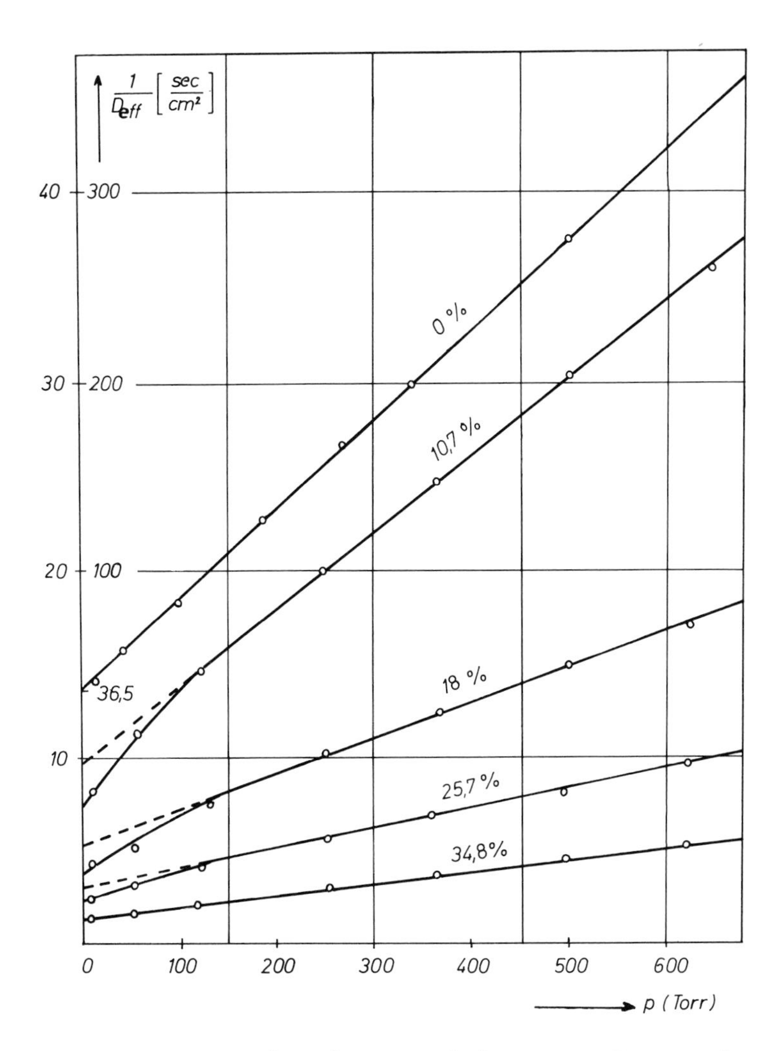

Figure 6. Pressure dependence of self-diffusion coefficient of hydrogen in a nuclear graphite at room temperature (Bosanquet diagram) after different burnoffs. The line for 0% burnoff has been shifted upward and drawn with a 10-fold reduced scale (36, 37).

burnoff (with CO_2 at 1000°C). Figure 7 represents pore size distributions of some of the probes obtained by Hg porosimetry. The measurements on the original material give a straight line in the $1/D_{eff}$ *vs.* P plot as expected for a uniform group of transport pores. From the slope of the line, according to Equation 4, the structural factor (permeability) can be calculated: $\psi = 2 \times 10^{-3}$, from the intercept on the ordinate the effective Knudsen diffusion coefficient, and from this the mean radius of the transport pores: $r = 1, 2 \times 10^{-4}$ cm. [The self-diffusion coefficient of hydrogen at 20°C was taken as $D_{H_2} = 1.43$ cm²/sec.] This value agrees well with the macropore peak in the pore size distribution (Figure 7) for 0% burnoff. With increasing burnoff the micropore peak shifts from very small radii to larger values, obviously because narrow connections (micropores) between macropore voids burn out to larger diameters and finally merge in the group of macropores. The simultaneous influence of two different pore groups on the diffusional transport gives rise to deviations from straight lines in the Bosanquet diagram (Figure 6). The deviations at low pressures are induced by the influence of the macropores; to account for this a two term relation like Equation 5 has to be used (*22, 23*). It is interesting that the measurements of Cunningham and Geankoplis (*18*) on Sample 1 in Figure 4 also give a straight

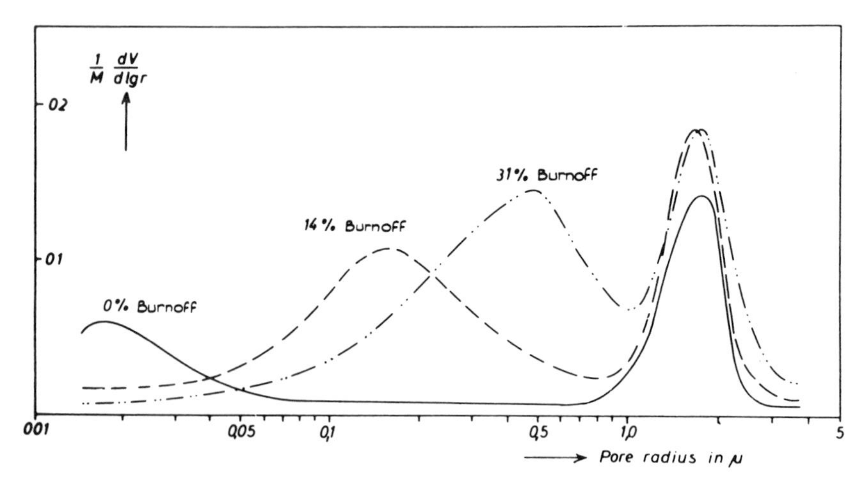

Figure 7. Pore size distributions of the nuclear graphite after different burnoffs (36, 37)

line when plotted in a $1/D_e$ *vs.* P diagram as demonstrated in Figure 8. The values of the structural factor and the mean pore radius, derived from the slope and intercept of this line, agree well with the values given by the authors, thereby confirming that in this bidispersed system the macropores dominate in diffusional transport. [The evaluation was based on

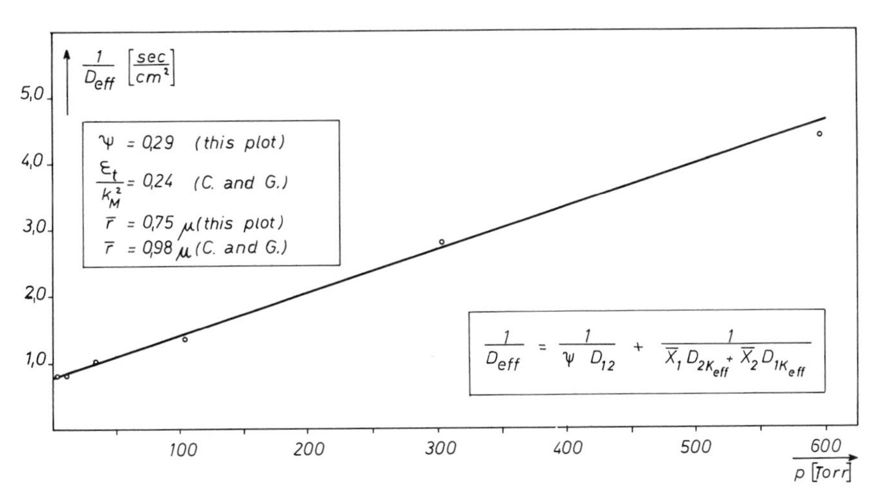

Figure 8. $1/D_e$ vs. P plot of the measurements of Cunningham and Geankoplis (18) on Sample 1 (see Figure 4)

Equation 7; by substituting α by $1 - \sqrt{M_1/M_2}$ according to Equation 2, and x_1, $1 - x_1$ by $\bar{x}_1$, $\bar{x}_2$ respectively, the equation given in the figure is obtained.]

The steady-state diffusion methods, useful and reliable as they are in basic research, have some disadvantages for direct application to catalyst pellets in industrial practice. Single samples only can be used after special preparation, and erroneous results must be expected if the material contains blind pores or anisotropic pore distributions. Unsteady state methods are often more generally applicable; among these the pulse technique, developed first by Deisler and Wilhelm (*38*) in 1953, seems to be of particular importance. Supported by experiences in gas chromatography this method is now in progress at several places (*39*, *40*, *41*). From the pulse technique readily applicable test methods are to be expected, suitable for supervising catalysts in practice, because their results will be statistical mean values from a number of catalyst pellets.

Intraparticle Problems (Non-isothermal)

The problem of internal superheating of catalyst particles by chemical heat production has received much interest in recent years. The coupling of reaction rate as a strongly nonlinear function of temperature with the essentially linear transport processes of heat and mass may give rise to multiple stationary solutions of the balance equations—*i.e.*, to multiple steady states of the reaction system. Much theoretical work has been done to investigate the conditions under which multiple steady-state profiles of concentration and temperature occur in the interior of a catalyst pellet and to determine the criteria of stability of the steady states. A survey of these problems and of the extended literature has been given by Aris (*42*) (*see also* V. Hlaváček (*8*)). The boundary conditions at the external surface of the pellet are usually choosen in such a way (*43*) that the surface temperature T_s and the surface concentration c_s remain fixed. In addition to the Thiele modulus φ, the characteristic parameters are the maximum relative excess temperature in the pellet center, β, and the dimensionless activation energy, γ:

$$\beta = \frac{(-\Delta H)\, D_{eff}\, c_s}{k_e\, T_s} = \frac{\Delta T_{max}}{T_s}; \quad \gamma = \frac{E}{RT_s}$$

(ΔH = reaction enthalpy; k_e = heat conductivity of the pellet material). For multiple steady states to occur in the interior of the pellet—with fixed surface temperature and with a first-order reaction—the product $\gamma\beta$ must exceed a value of about 5. Recently Hlaváček *et al.* (*44*) have accumulated data of γ, β, and φ from industrial catalytic processes and from experimental studies (*see* Table II). In no case is the limit $\gamma\beta \geqslant 5$ attained; in most industrial processes the values are much smaller. As a

Table II. Parameters of Some Exothermic Catalytic Reactions (*44*)

Reaction	γ	$\gamma\beta$	φ
Synthesis of NH_3 (*45*)	29.4	0.0018	1.2
Synthesis of higher alcohols from CO and H_2 (*45*)	28.4	0.024	—
Oxidation of CH_3OH to CH_2O (*45*)	16.0	0.175	1.1
Synthesis of vinyl chloride from C_2H_2 and HCl (*45*)	6.5	1.65	0.27
Hydrogenation of C_2H_4 (*46*)	23–27	1.0–2.7	0.2–2.8
Oxidation of H_2 (*47*)	6.75–7.52	0.21–2.3	0.8–2.0
Oxidation of C_2H_4 to ethylene oxide (*45*)	13.4	1.76	0.08
Decomposition of N_2O (*49*)	22.0	1.0–2.0	1.0–5.0
Hydrogenation of C_6H_6 (*50*)	14–16	1.7–2.0	0.05–1.9
Oxidation of SO_2 (*45*)	14.8	0.175	0.9

matter of fact, multiple steady states in the interior of catalyst pellets have not been observed experimentally, even in cases where high internal superheating was obtained, as in studies of the hydrogenation of ethylene [(*46*), $\Delta T = T_c - T_s$ up to 37°C], the oxidation of hydrogen [(*47, 48*), ΔT more than 100°C], and the decomposition of N_2O [(*49*), ΔT up to 36°C]. The reason for the failure to observe multiple steady states is that with increasing temperature in the pellets' interior the concentration of the reactants, within a small temperature interval, recedes towards the external surface. Only a thin reaction zone remains, with steep concentration gradients. Hence, induced by the small diffusivities in the porous structure, the interior of the catalyst particles usually can be taken as isothermal.

Interphase Transport

When overheating by an exothermic reaction occurs, the catalyst particles usually overheat as the whole, as shown in Figure 2. The temperature gradients appear predominantly in front of the external pellet surface, where the resistance of heat transfer to the bypassing gas flow must be overcome. It is the limitation in the intensity of this external heat transfer compared with the chemical heat production within the particle which really gives rise to multiple steady states of the system. In this *external problem* three steady states are possible, and the medium one is unstable (*51, 52*). The particles in these states have temperatures more or less different from the gas flow. Their interiors, however, are essentially isothermal, for the reasons stated above [(*see also* Hugo and Wicke (*22*)]. In this case also it seems unlikely that the internal problem should have multiple solutions. The phenomenon of more than three steady states, as calculated by combining the external and internal problem (*53*), should scarcely be observed experimentally. An instructive discussion of the possible situations has been given by Cresswell (*54*) (Figure 9). Here the Thiele modulus is defined by the reaction rate under the conditions of the bypassing gas flow as a reference value; an increasing Thiele modulus therefore means increasing temperatures of gas flow.

Cresswell distinguishes four regions with different behavior of the system. In the kinetic region 1 the reaction parameters are the same throughout the porous pellet. In Region 2 the temperature in the pellet increases, but the interior remains essentially isothermal; the reaction is confined more and more to a thin shell at the pellet surface. This is why the effectiveness factor normally does not increase in this range but decreases. In Region 3 the limitation of external heat transfer gives rise to appreciable superheating of the pellet; this causes a dramatic increase of the effectiveness factor, connected with three steady states. In Region 4 the pellet is in the upper stable state where the external mass transfer controls the rate of conversion.

Mass and heat transfer between particles and gas flow in packed beds has been studied increasingly in the last years. This is especially true for the range of low Reynolds numbers, below about 100, which is most important for gas–solid reactions and catalysis. The traditional method

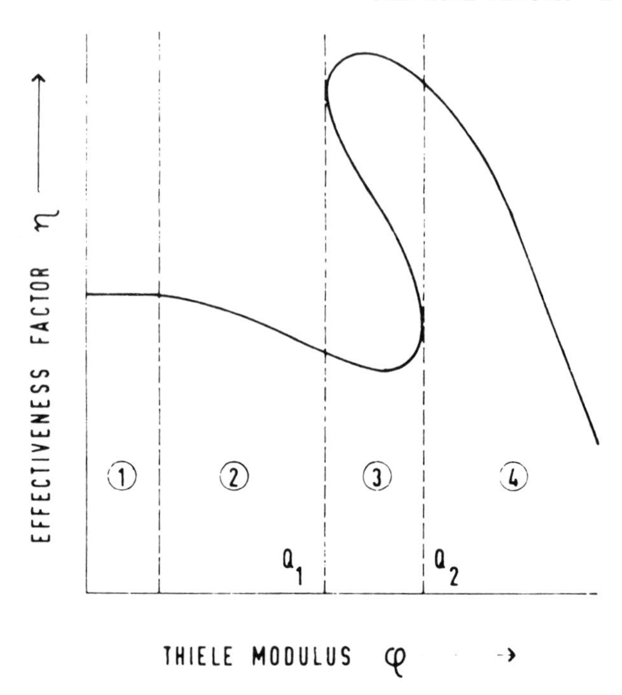

Figure 9. Main rate-controlling regions of an exothermic reaction at a catalyst pellet. Region 1: kinetic; Region 2: in-pore diffusion, non-isothermal; Region 3: multiple steady states; Region 4: interphase mass transfer, after Cresswell (54).

of measurement (the adiabatic evaporation of liquids from wetted porous particles) is difficult to perform in this range because the gas stream at the outlet of the test section will be nearly saturated with the vapor. The test section, therefore, must be short, two or three layers of particles only, suitably sandwiched between beds of solid dry pellets of the same size. Such a short arrangement, of course, has strong entrance and exit effects, whose influence can be observed in the fact that the wetted particles approach the wet-bulb temperature of the inlet air only at high air velocities. The same is true when the wetted particles are interspersed randomly in a bed of solid dry pellets ("expanded" or "dispersed" bed of wetted particles) (*55*). The temperatures of the wet surfaces, therefore, must be measured carefully if reliable results of mass transfer are to be obtained. Petrovic and Thodos (*56*) recently performed such measurements with uniformly packed and with dispersed beds at low Reynolds numbers: $3 < Re < 230$. They correlated the values found for the mass transfer factor j_d by:

$$\varepsilon j_d = 0.357\, Re^{-0.359} \tag{14}$$

where ϵ is the fraction of voids in the fixed bed. The factor ϵ in Equation 14 originates from the work of McConnachie (*57*), who measured mass and heat transfer into air from spheres held fixed in space by wires in a body-centered cubic arrangement, a "distended" bed. His results, obtained in the range $0.416 < \epsilon < 0.778$; $100 < Re < 2000$, could be approximately related to: $j_d \sim 1/\epsilon$ at Re = constant, as Sen Gupta and Thodos (*58*) have shown. Petrovic and Thodos, in deriving Equation 14 from their measurements, tried to account for the influence of axial mixing by using a mixing cell model instead of the usual method of logarithmic mean driving force. With the same model they recalculated results of earlier investigations (*55, 59, 60*) on packed and dispersed beds and thereby extended Equation 14 to the range $3 < Re < 1000$.

On the other hand, however, important parameters are missing in this correlation. For uniformly packed beds, for example, the dimensionless length of the test section—*i.e.*, the number of wetted particle layers used—is expected as parameter to account for end effects. For dispersed bed arrangements the mean value of shortest distance (or the number density) of the wetted particles should be implicit in Equation 14. Neglect of these parameters may be the reason for the broad scattering of experimental data with respect to Equation 14; hence, the correlation will have to be refined in future, especially in the low Reynolds number range.

Interphase heat transfer in this range is characterized by the fact that influences of heat conduction and of radiation between the particles in question and their surroundings can no longer be neglected, as is normally done at high Reynolds numbers. These effects favor the rate of interphase heat transport, but there are no comparable effects in mass transfer. The ratio of heat to mass transfer rates, therefore, should increase with decreasing Re number. De Acetis and Thodos in their investigations (*55*) of heat and mass transfer with wetted particles in short test sections and in dispersed beds, obtained values of $j_h/j_d \approx 1.5$ for the ratio of heat to mass transfer factor (especially at low Re numbers) compared with only 1.07 in the classical work of Hougen *et al.* (*59, 60*) at higher Reynolds numbers, $Re > 350$ [(*see also* Sen Gupta and Thodos (*58*)]. Satterfield and Cortez (*61*) have reported on differencies between mass and heat transfer data obtained with woven-wire screens in gas flow, (they call it "discrepancies") which can be attributed to longitudinal heat conduction.

In 1950 Wilhelm (*62*) compiled the effective mechanisms for heat transport in packed beds, and together with Singer he worked out the scheme shown in Figure 10. It considers the heat conduction from particle to particle through the contact areas and through the fluid fillets around them, the interphase heat transfer, and the molecular and eddy conductivity in the fluid phase. The diagram can be taken as a program for research in this field; until today this program has not been completely worked out. Remarkable progress has been achieved in recent years by Schlünder (63) in investigations of combined heat and mass transfer between gas flow and single bodies enclosed in packed beds. From his numerous important results the one shown in Figure 11 seems

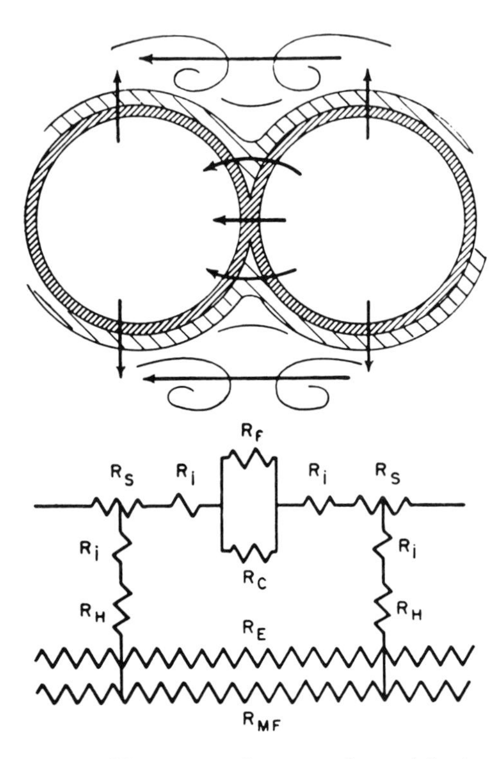

R_s = Thermal resistance of particle itself.
R_f = Thermal resistance of fluid fillet between particles.
R_c = Thermal resistance of contact area.
R_i = Thermal resistance of impurities: oxides, grease, etc.
R_H = Thermal resistance for heat flow from particle to main body of fluid.
R_{MF} = Thermal resistance of fluid at rest (molecular conductive effect).
R_E = Thermal resistance of turbulent flow.

Figure 10. Scheme of mechanisms for heat flow in packed beds, after Singer and Wilhelm (62)

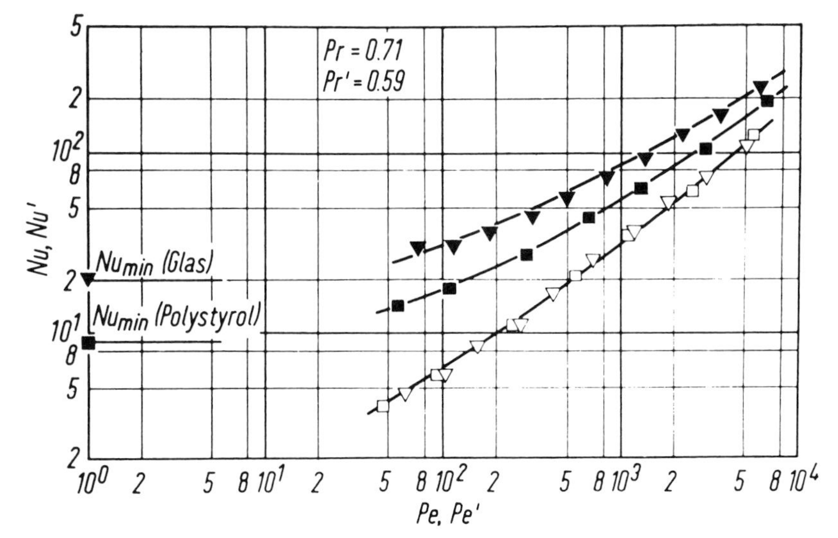

Figure 11. Heat and mass transfer (Nusselt numbers Nu and Nu') between a wetted porous cylinder, 30 × 30 mm, and air flow in packed beds of 2-mm pellets from glass (▼, ▽) and from polystyrol (■, □) after Schlünder (63)

especially interesting. A cylindrical cell from porous alumina, 30 × 30 mm, was enclosed in a packed bed of spheres, 2 mm in diameter, from glass and from polystyrol, and was wetted continuously with water by thin glass tubes. From the steady-state water influx to the evaporation cell and from the temperatures measured at the surface of this cell, and in the air flow through the packed bed, the mass and heat transfer data were calculated. They are shown in Figure 11 as Nu *vs.* Pe for heat transfer and Nu′ *vs.* Pe′ for mass transfer respectively. In the definition of Nusselt and Peclet numbers half the circumference of the evaporation cell, $d_p \times \pi/2$, has been taken as the characteristic length. The range of Pe numbers, $50 < \mathrm{Pe} < 7500$ in Figure 11, corresponds to the range $20 < \mathrm{Re} < 3000$ of modified Reynolds numbers. As Figure 11 demonstrates, the rate of mass transfer from the evaporation cell does not depend on the material of the spheres in the packed bed surrounding the cell, but the rate of mass transfer from the evaporation cell does not depend on the material of the spheres in the packed bed surrounding the cell, but the rate of heat transfer is appreciably higher in the bed of glass spheres than in the bed of polystyrol spheres. The reason for this is the higher heat conductivity of the glass spheres' packing at zero gas flow rate, as is obvious from the values of $\mathrm{Nu_{min}}$ on the ordinate (Figure 11). With increasing Pe number (or Re number) the ratio of heat transfer to mass transfer rate decreases continually. This behavior can be explained physically by a boundary layer model, wherein the packing is treated like a continuous phase, as Schlünder (63) has shown. From the results in Figure

11 it can be expected that for exothermic reactions in catalyst beds the tendency of single catalyst pellets to overheat and to produce hot spots in the packed bed increases with increasing gas flow rate.

Dispersion Effects

A discussion of dispersion effects in packed beds requires an appreciation of the basic research in this field by Wilhelm and co-workers. In 1962 he presented an excellent review (*64*) on transport phenomena and chemical reactor design. From this paper the famous survey of the characteristic regions for *dispersion of mass* in packed beds has been reproduced in Figure 12. Most important for catalysis and gas–solid reactions is the fact that the dispersion behaves non-isotropically. At Re $\geqslant$ 10 the measurements for axial dispersion accumulate in the range of Pe numbers between 1 and 2. The theory, regarding the voids between adjacent

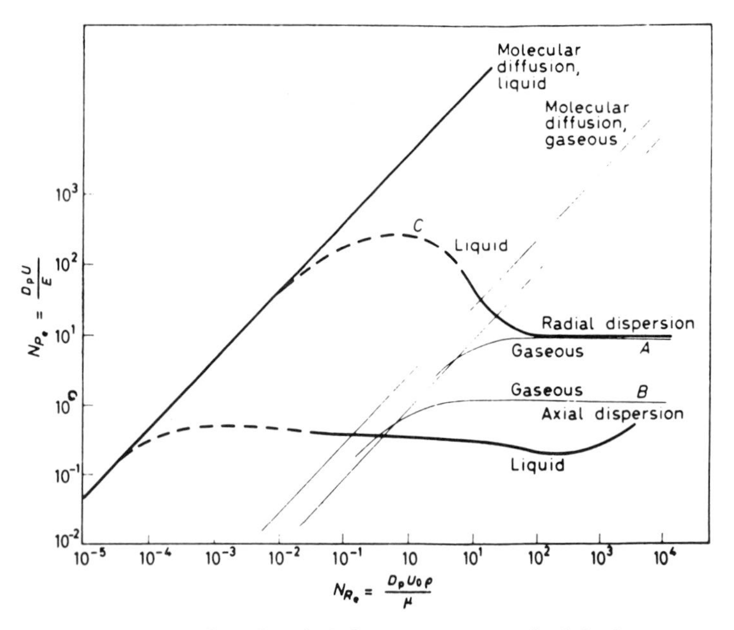

Figure 12. Axial and radial dispersion in packed beds; interlocking hydrodynamic regions after Wilhelm (64)

particles as mixing stages, yields $Pe_{ax} = 2$ in the limit of infinitely deep beds. For radial dispersion the experimental values of Pe accumulate between 8 and 12; the theory, based upon a lateral random walk model, yields: $Pe_r = 8$. In measurements of the local radial dispersion of CO_2 in air at different distances from the tube wall, Fahien and Smith (*65*) found minimum values of $Pe_r = 8$ near the axis of the packed bed; approaching the tube wall, values of Pe_r increased—*i.e.*, the dispersivity decreased. This behavior can be attributed to the increasing fraction of voids between the particles when approaching the wall; the effect diminishes with increasing ratio of tube to particle diameter.

The *dispersivity of heat* in packed beds is larger (as expected) than the dispersivity of mass because the dispersion of heat is supported by the conduction and by the radiation from particle to particle, whereas no analogous processes are effective in mass transport. Wilhelm was one of the first to draw attention to these differences; in a comparison of radial mass dispersion data with results of heat dispersion measurements (*66*) he demonstrated in 1950 the excess of heat dispersivity over mass dispersivity at low Reynolds numbers.

About 10 years later Yagi *et al.* showed in detailed investigations (*67, 68*) that the effective thermal conductivity in packed beds with gas flow can be described by a two-term relationship of the following type:

$$k_{eff} = k_o + \frac{u\rho c_p d_p}{Pe^*} \qquad (15)$$

The first term, k_o, is the heat conductivity of the packed bed without convection, the second term represents the contribution of the gas flow to the effective thermal conductivity (u = mean flow velocity between the particles, ρ, c_p = density and heat capacity of the gas). The dimensionless quantity Pe* was found to be about 10 for radial effective conductivity and about 3 for conduction parallel to the axis. Hence, the second term in Equation (15) is non-isotropic; it corresponds obviously to the dispersivity of mass in the range Re $\geqslant$ 10 in Figure 12.

Equation 15 can be rearranged to give:

$$\frac{u\rho c_p d_p}{k_{eff}} \equiv \mathrm{Pe} = \frac{\mathrm{Pe}^*}{1 + \mathrm{Pe}^* \dfrac{k_o}{u\rho c_p d_p}} \qquad (15a)$$

for the Peclet number of heat dispersion. This expression depends on gas flow rate (or on Re number) in a similar way as Figure 12 shows for the Peclet number of mass dispersion. The transition ranges, however, are shifted to larger values of Re (10- to 100-fold) according to the ratio of the thermal diffusivity, $k_o/\rho c_p$, in the packed bed without convection to the mass diffusivity D_o. For this reason the axial heat dispersion is, at low Re numbers, appreciably larger than the axial mass dispersion, as demonstrated by the different slopes of the gas temperature and the concentration profiles in Figure 2.

This difference in quantity between axial dispersion of heat and of mass is connected with a more fundamental difference. The axial dispersion of mass is predominantly a forward dispersion; only little back mixing occurs against the direction of gas flow because the pellets are practically impenetrable for the gas molecules. The dispersivity of mass, therefore, must be considered principally by mixing cell models. The dispersivity of heat, on the other hand, up to Reynolds numbers of several hundred is composed of one part which represents a real two-sided diffusivity (k_o in Equation 15), and a second part which represents a forward dispersion. The effective conductivity, therefore, must be represented principally by a model, which combines a continuous with a mixing cell character. Recently Amundson (*69, 70*) conceived such a model (Figure 13). It consists of mixing cells, 1 and 2, which are separated by particles; these particles represent with their heat conduction the continuous part of the model. At higher temperatures radiational heat transport also must be considered; if there are steep gradients, as in Figure 2, this requires calculations with stepwise temperature profiles, similar to mixing cell methods (*71*).

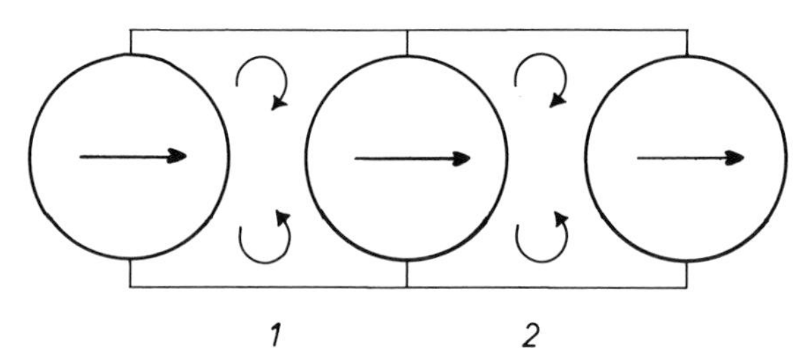

Figure 13. Combined model for heat dispersion in packed beds after Amundson (69, 70). *Mixing cells (1 and 2) connected by heat conduction through the adjacent particles.*

In view of the complicated nature of the effective axial heat conductivity, analysis of the adiabatic packed bed reactor with different values of dispersivities of mass and of heat has not yet been worked out in theory; this remains a problem for the future. Meanwhile investigations of transport processes concentrate more and more on the range of conditions which are most important for catalysis and for gas–solid reactions, first of all the range of low Reynolds numbers. We hope to learn more in near future about the manifold and interesting interactions of physical phenomena with chemical reaction rates in this range.

Literature Cited

(1) Frank-Kamentzky, David A., "Diffusion and Heat Transfer in Chemical Kinetics," 2nd ed., Plenum Press, New York, London, 1969.
(2) Satterfield, Charles N., "Mass Transfer in Heterogeneous Catalysis," M.I.T. Press, Cambridge, Mass., 1970.
(3) Fieguth, P., Ph.D. Thesis, Münster, 1971.

(4) Padberg, G., Wicke, E., *Chem. Eng. Sci.* (1967) **22,** 1035.
(5) Petersen, E. E. *et al., Chem. Eng. Sci.* (1964) **19,** 683, 783.
(6) Pismen, L. M., *Dokl. Akad. Nauk. SSSR* (1966) **167,** 151.
(7) Bischoff, K. B., *Chem. Eng. Sci.* (1968) **23,** 451.
(8) Hlaváceck, V., "Diffusion und Reaktion in porösen Kontakten," *Fortschr. Chem. Forschg.* (1970) **13,** 455.
(9) Copelowitz, I., Aris, R., *Chem. Eng. Sci.* (1970) **25,** 885.
(10) Hoogschagen, J., *J. Chem. Phys.* (1953) **21,** 2096.
(11) Hoogschagen, J., *Ind. Eng. Chem.* (1955) **47,** 906.
(12) Evans, R. B., Watson, G. M., Mason, E. A., *J. Chem. Phys.* (1961) **35,** 2076.
(13) Evans, R. B., Watson, G. M., Mason, E. A., *J. Chem. Phys.* (1962) **36,** 1894.
(14) Scott, D. S., Dullien, F. A. L., *A.I.Ch.E. J.* (1962) **8,** 113.
(15) Scott, D. S., Dullien, F. A. L., *Chem. Eng. Sci.* (1962) **17,** 771.
(16) Wakao, N., Smith, J. M., *Chem. Eng. Sci.* (1962) **17,** 825.
(17) Rothfeld, L. B., *A.I.Ch.E. J.* (1963) **9,** 19.
(18) Cunningham, R. S., Geankoplis, C. J., *Ind. Eng. Chem., Fundamentals* (1968) **7,** 535.
(19) Bosanquet, C. H., *British TA Rept.* **BR-507** (1944).
(20) Pollard, W. G., Present, R. D., *Phys. Rev.* (1948) **73,** 762.
(21) Mistler, T. E., Correl, G. R., Mingle, J. O., *A.I.Ch.E. J.* (1970) **16,** 32.
(22) Hugo, P., Wicke, E., *Chem.-Ing.-Techn.* (1968) **40,** 1133.
(23) Hugo, P., "The Influence of Pore Structure on the Pressure and Temperature Dependence of the Effective Diffusion Coefficient," *Symp. Fundamentals Transport Phenomena Porous Media,* Haifa, Israel, Feb. 1969.
(24) Hugo, P., *Chem. Eng. Sci.* (1965) **20,** 187.
(25) *Ibid.,* p. 385.
(26) *Ibid.,* p. 975.
(27) Wicke, E., "Alta Tecnologia Chimica," p. 225, Academia Nazionale Dei Lincei (Fond. Donegani), Rome, 1961.
(28) Wicke, E., Kallenbach, R., *Kolloid-Z.* (1941) **97,** 135.
(29) Kallenbach, R., Ph.D. Thesis, Göttingen, 1941.
(30) Weisz, P. B., *Z. physik. Chem.* (1957) **11,** 1.
(31) Weisz, P. B., Schwarz, A. B., *J. Catalysis* (1962) **1,** 399.
(32) Paratella, A., Bortolozzo, G., "Consiglio Nazionale Delle Ricerche," Vol. 5, p. 293, Rome, 1967.
(33) Buckingham, E., *U.S. Dept. Agr., Bur. Soils Bull.* **25** (1904).
(34) Hugo, P., Clemons, D. B., unpublished data, 1964.
(35) Hugo, P., Paratella, A., "Consiglio Nazionale Delle Ricerche," Vol. 5, p. 283, Rome, 1967.
(36) Beyer, H. D., Hugo, P., *DRAGON Project Rept.* **437** (1966).
(37) Beyer, H.-D., Diploma Thesis, Münster, 1966.
(38) Deisler, P. F., Wilhelm, R. H., *Ind. Eng. Chem.* (1953) **45,** 1219.
(39) Loffler, A. J., *J. Catalysis* (1966) **5,** 22.
(40) Davis, B. R., Scott, D. S., "Measurements of the Effective Diffusivity of Porous Pellets," *Novosibirsk Symp. Congr. Catalysis,* July 1968.
(41) Schneider, P., Smith, J. M., "Studies on Intraparticle Transport Processes by Chromatography," *Novosibirsk Symp. Congr. Catalysis,* July 1968.
(42) Aris, R., *Chem. Eng. Sci.* (1969) **24,** 149.
(43) Weisz, P. B., Hicks, J. S., *Chem. Eng. Sci.* (1962) **17,** 265.
(44) Hlaváček, V., Kubičeck, M., Marek, M., *J. Catalysis* (1969) **15,** 17, 31.
(45) Slinko, M. G., Malinovskaja, O. A., Beskov, V. S., *Chim. Prom.* (1967) **42,** 641.
(46) Cunningham, R. A., Carberry, J. J., Smith, J. M., *A.I.Ch.E. J.* (1965) **11,** 636.
(47) Maymo, J. A., Smith, J. M., *A.I.Ch.E. J.* (1966) **12,** 845.

(48) Wurzbacher, G., *J. Catalysis* (1966) **5,** 476.
(49) Hugo, P., Müller, R., *Chem. Eng. Sci.* (1967) **22,** 901.
(50) Butt, J. B., Irving, J. P., *Chem. Eng. Sci.* (1967) **22,** 1859.
(51) Frank-Kamenetzky, D. A., Z. *Tech. Fiz.* (1939) **9,** 1457.
(52) Wagner, C., *Chem. Tech. (Berlin)* (1945) **18,** 1, 28.
(53) Hatfield, B., Aris, R., *Chem. Eng. Sci.* (1969) **24,** 1213.
(54) Cresswell, D. L., *Chem. Eng. Sci.* (1970) **25,** 267.
(55) De Acetis, J., Thodos, G., *Ind. Eng. Chem.* (1960) **52,** 1003.
(56) Petrovic, L. J., Thodos, G., *Ind. Eng. Chem., Fundamentals* (1968) **7,** 274.
(57) McConnachie, J. T. L., M.S. Thesis, Northwestern University, Evanston, Ill. (1961).
(58) Sen Gupta, A., Thodos, G., *A.I.Ch.E. J.* (1962) **8,** 608.
(59) Gamson, B. W., Thodos, G., Hougen, O. A., *Trans. Am. Inst. Chem. Eng.* (1943) **39,** 1.
(60) Hougen, O. A., Watson, K. W., "Chemical Process Principles," Part III, p. 985, Wiley, New York, 1947.
(61) Satterfield, C. N., Cortez, D. H., ADVAN. CHEM. SER. (1972) **109,** 209.
(62) Singer, E., Wilhelm, R. H., *Chem. Eng. Progr.* (1950) **46,** 343.
(63) Schlünder, E. U., *Chem.-Ing.-Tech.* (1966) **38,** 967.
(64) Wilhelm, R. H., *Pure Appl. Chem.* (1962) **5,** 403.
(65) Fahien, R. W., Smith, J. M., *A.I.Ch.E. J.* (1955) **1,** 28.
(66) Bernard, R. A., Wilhelm, R. H., *Chem. Eng. Progr., Symp. Ser.* (1950) **46,** 233.
(67) Yagi, S., Wakao, N., *A.I.Ch.E. J.* (1959) **5,** 79.
(68) Yagi, S., Kunii, D., Wakao, N., *A.I.Ch.E. J.* (1960) **6,** 543.
(69) Vanderveen, J. W., Luss, D., Amundson, N. R., *A.I.Ch.E. J.* (1968) **14,** 636.
(70) Amundson, N. R., *Ber. Bunsenges. Physik. Chem.* (1970) **74,** 90.
(71) Vortmeyer, D., ADVAN. CHEM. SER. (1972) **109,** 43.

RECEIVED September 14, 1970.

Contributed Papers

Mass Transfer Characteristics of Woven-Wire Screen Catalysts

CHARLES N. SATTERFIELD and DOUGLAS H. CORTEZ,[1] Department of Chemical Engineering, Massachusetts Institute of Technology, Cambridge, Mass. 02139

The objectives of this work were to determine the mass transfer characteristics of woven-wire screen catalysts and to investigate the possibility that the catalytic oxidation of hydrocarbons in air might under some circumstances proceed by initiation of reaction on a catalyst surface followed by propagation of the reaction into the surrounding gas, so-called hetero-homogeneous catalysis (*1*).

The catalytic oxidation of either 1-hexene or toluene in excess air on one to three platinum gauzes in series was carried out in one of two tubular flow reactors. Hydrocarbon concentrations were varied from 600 to 4000 ppm. To measure whether homogeneous reaction downstream from the catalyst was significant, a movable water-cooled probe was mounted vertically below the catalyst. Gas samples were analyzed by gas chromatography. The screens consisted of 52 or 45 mesh platinum gauzes or 40 or 80 mesh platinum–10% Rh gauzes.

The rate of reaction increased rapidly with increased temperature, but above about 350°C it became almost temperature independent, which is strongly suggestive of a mass transfer-controlled process. Assuming that the reaction rates at high temperatures were indeed controlled by mass transfer, mass transfer coefficients were calculated, converted to j_D factors, and correlated with the Reynolds number by the relationship

$$j_D = CN^{-m}{}_{\mathrm{Re},d}$$

in which the Reynolds number is based on wire diameter and interstitial velocity. A least-squares analysis gave $C = .865$ and $m = .648$ with a mean deviation of $\pm 12.5\%$. There was no significant difference between mass transfer coefficients for two or three screens in series or for degree of separation of two screens, nor was there any effect of probe position on the results. It was concluded that there was no positive evidence that

[1] Present address: TRW Systems Group, One Space Park, Redondo Beach, Calif.

any of the hydrocarbon oxidation occurred as a homogeneous process.

The results agree fairly closely with those of B. Gay and R. Maughan (2) obtained by measuring the mass transfer rates of mercury from single screens into flowing nitrogen, but they disagree substantially with values of j_H as determined from heat transfer rates to a gas from stacks of woven screens as measured by J. E. Coppage and A. L. London (3). We concluded that this was caused by the neglect of axial heat transfer in the work of Coppage and London. Their data after adjustment for this effect, that of Gay and Maughan, and the data of P. H. Vogtlander and A. P. Bakker (4) for liquid-phase mass transfer to screens, plus our data were recorrelated in terms of flow past a single cylinder, and the transport characteristics of screen catalysts were shown to be very similar to that of infinite cylinders. Comparison of all these data with the ammonia oxidation studies of Dixon and Longfield (5) showed that their rates of reaction never exceeded that which would be predicted for a mass transfer-controlled process. There is thus no evidence that the ammonia oxidation reaction in their case occurred as a hetero-homogeneous process.

(1) Satterfield, C. N., Cortez, D. H., "Mass Transfer Characteristics of Woven-Wire Screen Catalysts," *Ind. Eng. Chem., Fundamentals* (1970) **9**, 613–620.
(2) Gay, B., Maughan, R., *Intern. J. Heat Mass Transfer* (1963) **6**, 277–287.
(3) Coppage, J. E., London, A. L., *Chem. Eng. Progr.* (1956) **52** (2), 57F–63F.
(4) Vogtlander, P. H., Bakker, A. P., *Chem. Eng. Sci.* (1963) **18**, 583–589.
(5) Dixon, J. K., Longfield, J. E., "Catalysis," P. H. Emmett, Ed., Vol. 7, pp. 281–304, Reinhold, New York, 1960.

Experimental Study of Laboratory Flow Reactors

JOHN C. ZAHNER, Mobil Research and Development Corp., Central Research Division, P.O. Box 1025, Princeton, N. J. 08540

In studying heterogeneously catalyzed reactions of gases, the laboratory integral conversion flow reactor has a number of desirable features. One of the more desirable aspects is that simply by changing the flow rate, one can obtain rather quickly a reaction path over a wide range of conversions. Of course, for the data to be interpreted unambiguously, it should be free of significant heat and mass transfer effects such that the plug flow description of an isothermal reactor is valid.

Axial diffusion can be a significant factor in affecting adversely the results of flow reactors, especially for high conversions at low flow rates in shallow beds. If the residence time becomes too long, molecular diffusion becomes important since the residence time can become comparable with the diffusion time, $(\text{length})^2$/effective diffusion coefficient; the reactor then tends to behave more like an ideal stirred tank reactor where the concentrations become uniform throughout the length of the bed. In the intermediate regions, interpretation of experimental data for kinetic purposes becomes difficult.

For gas flow rates of 8–110 scc/min and conversions up to 90%, negligible deviations from the plug flow description were obtained in the Pt-catalyzed exchange of deuterium for hydrogen in neopentane as long as the axial diffusion modulus was appropriately low. Further, all deviations from the plug flow description were in agreement quantitatively with the theory of axial diffusion. In some experiments the catalyst was diluted about 10:1 with inert chips; no new mass transfer effects resulted.

One can arrive at a conservative design for a laboratory flow reactor operating around atmospheric presusre, 230°C, 10–200 scc of gas/min, and catalyst activities as measured by a first-order rate constant of 0.1 to 2 (cc of gas)/(sec, cc of catalyst).

Considering heat effects first, two major temperature differences are important in removing the heat of reaction: (1) ΔT between the catalyst particle and bulk fluid temperature, and (2) ΔT between the center and reactor wall.

The second difference is minimized by keeping the heat generated (catalyst) per unit of length small. Based on the following considerations:

(a) A limiting Nusselt number of 4
(b) Effective bed conductivity of 1.4×10^{-3} cal/(sec, cm, °C)
(c) Only enough catalyst to obtain 90% conversion at 10 scc/min
(d) Remaining space will be filled with inerts
(e) Approximate T, p, and k as above

one then arrives at:

$$\Delta T(°\text{C}) = \frac{x\Delta H(\text{kcal/gm-mole})}{L(\text{reactor length, cm})}$$

Thus, for a reactor 20 cm long, a reaction having an effective heat of reaction (mole fraction · ΔH) of 20 kcal/gm-mole, the maximum estimate for the ΔT is 1°C.

A catalyst particle of 80 mesh and activity of 2 cc of gas/(sec, cc of catalyst) will generate a temperature difference between the particle and an infinite gas of 0.5°C for an effective $x\Delta H$ of 20 kcal/gm-mole. Thus, use of 80 mesh particles or smaller will keep this ΔT acceptable.

Furthermore, 80 mesh will give satisfactory intraparticle diffusion even for effective diffusion coefficients of 10^{-3} cm^2/sec. For a reactor 20 cm long, a diameter of 1.1 cm or less gives an excellent axial diffusion modulus of 0.05 or less and an "effectiveness factor for axial diffusion" of 0.95 or better over the entire conversion range.

The experiments that made us more aware of axial diffusion are outlined below. These experiments were not performed to test the particular reactor scheme above but to resolve some irregularities from previous experiments.

Table I. Intrinsic Kinetics

Run No.	*Time, min.*	*Deuteride Composition,[a] Mole %*							
		d_0	d_1	d_2	d_3	d_4	d_5	d_6	d_7
514	89.9	89.21	9.75	.87	.18	—	—	—	—
	180.	79.79	17.87	2.09	.27	—	—	—	—
	300.	67.12	26.58	5.24	.86	.21	—	—	—
	574.	42.90	36.48	15.29	4.09	1.06	.18	—	—
	746.	31.78	37.55	21.20	7.14	1.90	.45	—	—
	1306.	9.72	24.81	29.63	21.56	10.05	3.15	.92	.18
515	8.3	93.24	6.14	.57	.05	—	—	—	—
	24.2	81.76	15.87	2.07	.32	—	—	—	—
	45.9	68.46	25.50	4.97	.90	.18	—	—	—
	68.6	56.20	32.38	9.18	1.89	.38	—	—	—
	89.1	46.31	36.27	13.24	3.35	.66	.18	—	—
	113.8	36.89	37.26	18.31	5.85	1.42	.28	—	—
524	18.5	96.19	3.55	.24	.02	—	—	—	—
	45.2	90.62	8.56	.71	.11	—	—	—	—
	100.	80.33	17.03	2.19	.41	.04	—	—	—
	310.	40.47	37.11	16.35	4.67	1.12	.23	.06	—
	366.	32.30	37.68	20.56	7.18	1.90	.34	.03	.02

[a] No values obtained for d_8 and d_9.

Table II. Flow

Run No.	*Q, scc/min*	*ADM*	d_0	d_1
520[a]	8.25	1.5	56.6	26.4
	Plug Flow	(1.7)	(56.6)	(32.2)
	13.4	0.9	70.0	22.1
	Plug Flow	(1.1)	(70.0)	(24.7)
533[b]	8.94	0.5	24.6	26.6
	Plug Flow	(0.5)	(24.6)	(35.9)
	14.7	0.3	34.7	32.1
	Plug Flow	(0.3)	(34.7)	(37.7)

[a] Catalyst = 350 mg, 100–200 mesh, 11 mm id, L = 0.5 cm.
[b] Catalyst = 90 mg, 100–200 mesh, 11 mm id, L = 1.4 cm, dilution = 10.1.

The intrinsic kinetics were established in a glass batch reactor of about 330 cc volume. It was constructed from 55 mm od borosilicate tube and was 15 cm long with a flattened base which was 2.5 cm wide for dispersing the catalyst. The reactor had a glass stirrer along the long axis with two fans attached and a glass-encased piece of iron; the stirrer was operated magnetically.

The catalyst used was Baker-Sinclair 0.6% Pt-alumina prepared by Englehard Industries. It was pretreated in flowing hydrogen for about 1 hour at 300°C. The reaction mixture was prepared by passing the deuterium at 5 psig over neopentane at solid CO_2 temperatures, charged to the reactor, and isolated. The mole ratio was about 120:1 at those conditions.

A previous report concerning the deuterium exchange reaction with neopentane (*1*) showed that the whole exchange process could be discussed and modeled as an hypothetical exchange with the *tert*-butyl ions which are the predominant peaks from the various deuterated neopentanes. This procedure is followed here; the various amounts of d_i (i = 0, 1, 2, . . . , 9), refer to the fraction of *tert*-butyl ions in the mass spectrometer containing i atoms of deuterium.

Three experiments performed in the batch reactor at 160°C, which establish the intrinsic kinetics are presented in Table I. The data deviate from the theoretical binomial curves to the extent that the maximum amount of d_1 is 37.5% *vs.* 38.8% for the binomial calculation. Data from all three runs fall on the same curves, indicating that the same relative intrinsic kinetics are common to all three experiments. When these experiments were performed, activity control was not established, and the specific activity varied from 0.7 to 2.0 (cc of gas)/gm of catalyst-sec) and varied appreciably during two of the runs. However, in contradistinction to previous work (*1*) with supported palladium, the selectivity

Experiments

Deuteride Composition, Mole %							
d_2	d_3	d_4	d_5	d_6	d_7	d_8	d_9
10.4	4.5	1.5	.4	.1	—	—	—
(9.0)	(1.8)	(0.4)	—	—	—	—	—
5.7	0.7	.5	—	—	—	—	—
(4.4)	(0.7)	—	—	—	—	—	—
20.5	13.3	7.8	4.2	1.9	.8	.2	.1
(25.0)	(10.3)	(3.3)	(0.8)	—	—	—	—
18.8	8.8	3.7	1.3	0.5	0.1	—	—
(19.4)	(6.2)	(2.7)	(0.3)	—	—	—	—

did not change with activity in these experiments. In terms of the theory for the effects concerning surface catalytic rate to desorption rate competition (*1*), a value of 0.05 for β, the ratio of surface to desorption rate, would fit the data quite well. We conclude that for the catalyst pretreated in hydrogen at 300°C for times on the order of an hour, the intrinsic kinetics at 160°C are almost binomial.

The same materials were used in the flow experiments as in the batch experiments for the intrinsic kinetics. Neopentane was introduced in the same manner. Two different reactors were used. One was a standard sealing tube containing a 1-cm course frit, id = 1.1 cm. Catalyst was placed directly on top of the frit. Another reactor was a section of 1/4-inch glass tubing, id = 0.40 cm. Catalyst was placed on top of about 1/2 inch of glass wool; the glass wool was compacted by tapping with two glass rods, one from each end. The reactors were mounted vertically, and the reactant gases flowed from top down. The reaction temperature was *ca.* 160°C.

The results of the flow experiments are summarized in the introductory remarks. All deviations from the plug flow description (same as intrinsic kinetics) are predictable from axial diffusion considerations. In Table II some observed compositions where axial diffusion was a factor are compared with what the composition would be for plug flow at the same amount of unreacted d_0. Axial diffusion gives a less selective resultless main product, d_1 and more secondary products, d_2, d_3, . . . , etc. In fact, one observes compounds that are not indicated by the intrinsic kinetics. Table II also gives the estimated values for ADM, the axial diffusion modulus; the ADM's in parenthesis are the values that must be assumed to calculate the observed product distribution. The agreement is excellent.

The theory of axial diffusion is included in most texts with some differences in nomenclature. To relate our work with these texts we use the following mass balance:

$$-\ (\mathrm{ADM})\frac{d^2x}{dy^2} + \frac{dx}{dy} + \frac{mk}{Q}\,x = 0$$

where x = mole fraction, and y = fractional reactor length.

One can make a conservative close approximation to the ADM by ignoring the Reynolds number:

$$\mathrm{ADM} \cong \frac{1}{L}\left(\frac{\varepsilon 2D}{(3-\varepsilon)}\frac{A}{Q} + \frac{d_p}{2}\right)$$

where

Q/A is superficial velocity

ε is void fraction

d_p is the particle diameter

D is the ordinary molecular diffusion coefficient

(1) Dwyer, F. G., Eagleton, L. C., Wei, J., Zahner, J. C., *Proc. Roy. Soc. A* (1968) **302**, 253–270.

Gas Pressure Buildup Within a Porous Catalyst Particle Which is Wet by a Liquid Reactant

J. J. SANGIOVANNI and A. S. KESTEN, United Aircraft Research Laboratories, East Hartford, Conn. 06108

Catalytic reactors which use porous catalyst particles to promote the exothermic decomposition of liquid reactants to form gaseous products can exhibit unusual startup characteristics. Liquid entering the reactor can surround and wet particles near the inlet of the bed and wick into the pores by capillary flow. Catalytic decomposition of this liquid, or of the vapor in equilibrium with the liquid, within the porous structure of the particles results in gaseous reaction products which are temporarily trapped within the particles by the liquid blocking the pores. Liquid penetration into the pores, which is accompanied by gas pressure buildup arising from reaction, continues until the gas pressure exceeds the capillary pressure. For particles with pores of radii of the order of a few hundred Angstroms, this capillary pressure can be as high as 1000 psia for typical liquids. Pressure buildup continues as the depth of liquid penetration decreases until the liquid finally is expelled from the catalyst particle pores. Liquid penetration and subsequent expulsion would be evidenced in the interstitial (bulk fluid) phase of the reactor by a long induction period, during which decomposition products would not appear, followed by a sudden pressure excursion.

An analysis of the gas pressure buildup within a porous catalyst particle which is wet by liquid reactant is presented in detail in Ref. *1;* the analysis considers the simultaneous processes of mass transfer, heat transfer, and chemical reaction within the particle. Integral equations are used to relate the temperature and species concentration profiles within the porous structure, and a computational scheme is presented for solving these integral equations. This analysis has been used to compute concentration and temperature profiles, pressure buildup, and liquid penetration depth as a function of time within a porous catalyst particle used to promote the decomposition of hydrazine by the exothermic reaction

$$2\ N_2H_4 \rightarrow 2\ NH_3 + N_2 + H_2$$

The calculations pertain to a Shell 405 catalyst particle for which estimates have been made of the kinetics of the catalytic decomposition of hydrazine. The rate of heterogeneous chemical reaction can be approximated by

$$r_A = 10^{10}\rho_A e^{-2500/T}$$

where ρ_A is the hydrazine concentration and T is the temperature. An illustrative example is considered which is typical of conditions existing for the startup of a hydrazine catalytic reactor. The gas pressure buildup for this case is shown in Figure 1 for three radial positions within the catalyst particle. The corresponding liquid hydrazine penetration depth is presented in Figure 2 for a range of pore radii since there is considerable uncertainty regarding the pore radius which is most typical for this analysis; whereas the average pore radius for the Shell 405 catalyst is less than 100 A, the pores near the particle surface where capillary flow of liquid reactant takes place would be expected to be considerably larger. In

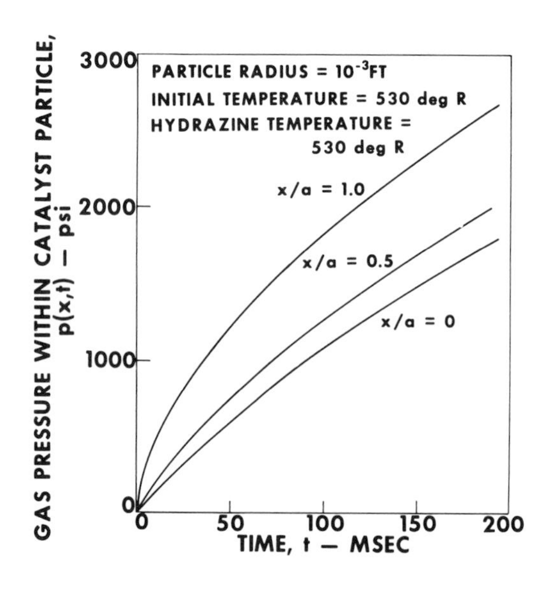

Figure 1. Gas pressure buildup within a porous catalyst particle

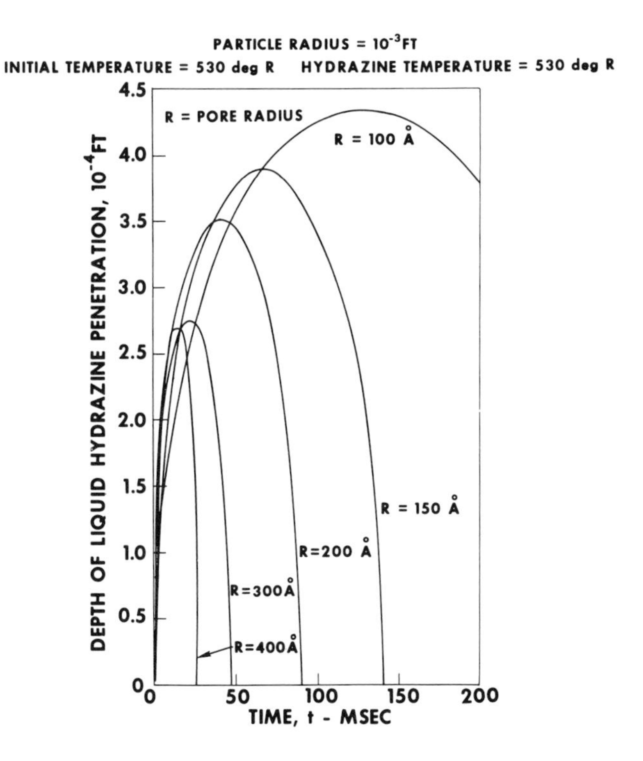

Figure 2. Liquid hydrazine penetration depth into catalyst pores

addition, most of the gas products will probably escape from the largest accessible pores. The liquid hydrazine residence time is obtained from Figure 2 as the period of time during which the liquid occupies catalyst pores.

Similar calculations were performed for liquid hydrazine temperatures and initial catalyst particle temperatures between 495° and 580°R; all other parameters were fixed at the values used in computing the results shown in Figures 1 and 2. The liquid hydrazine residence time is plotted in Figure 3 as a function of temperature for a number of pore radii. A marked change in liquid hydrazine residence time is shown for only a modest change in the hydrazine temperature. This effect results primarily from the large variation in hydrazine vapor pressure over the temperature range 495°–580°R, which affects the rate of reaction directly by virtue of the hydrazine concentration.

Computed liquid pore residence times can be compared with observed induction periods between liquid injection and gas pressure excursions in hydrazine catalytic reactors. In a series of startup tests conducted

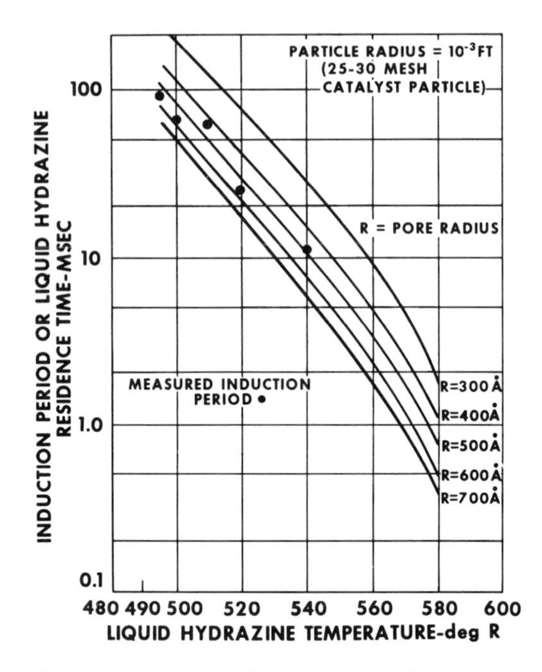

Figure 3. Liquid hydrazine residence time and comparison with measured induction period

with a catalytic reactor which uses Shell 405 catalyst to promote the decomposition of liquid hydrazine to form gaseous products for spacecraft thrusters, the temperature of liquid hydrazine surrounding the catalyst particles had a predominant effect on the initial startup transient; the other parameters exhibited a secondary influence by what appeared to be slight, temporary changes in the catalyst activity. In general, decreasing the liquid hydrazine temperature resulted in a pronounced increase in the time elapsed between the arrival of liquid hydrazine in the reactor and the first indication of an increase in reactor gas pressure. This induction period exhibited the same trends as the computed liquid hydrazine residence times, as shown in Figure 3. The nearly identical dependence of the induction period and the liquid residence time on liquid hydrazine temperature lends credence to the liquid penetration model as an explanation for the unusually long ignition delays observed during startup of certain liquid-feed catalytic reactors.

(1) Sangiovanni, J. J., Kesten, A. S., "Analysis of Gas Pressure Buildup within a Porous Catalyst Particle Which is wet by a Liquid Reactant," *Chem. Eng. Sci.*, in press.

Carbon Deposition on Catalysts during Conversion of Oxygenated Organic Compounds

IB DYBKJAER, Fa. Haldor Topsoe, 2950 Vedbaek, Denmark

ANDERS BJÖRKMAN, Instituttet for Kemiindustri, Technical University of Denmark, 2800 Lyngby, Denmark

This study is aimed at a better understanding of the mechanism of carbon formation in a specific case by studying the rate of carbon formation together with careful identification of simultaneously occurring intermediates and by-products. The reactants were ethanol and acetaldehyde, alone or mixed with steam. The catalysts were oxides with special emphasis on mixtures which were able to produce acetone (commercially) by the following over-all reaction:

$$2\ C_2H_5OH + H_2O \rightarrow (CH_3)_2\ CO + CO_2 + 4\ H_2$$

Acetaldehyde and acetic acid appear as intermediates, and a variety of by-products can be found in minor quantities.

An initial series of long range experiments (*1*) indicated a possible connection between carbon formation and the presence of certain by-products. The best catalyst for the reaction with respect to initial activity was Fecronz, a mixture of Fe_2O_3, Cr_2O_3 and ZnO (1:1:3 by weight), but this catalyst was deactivated rapidly by carbon formation. The catalyst Cafecronz (Fecronz + 10% CaO) had a lower initial activity, but after partial deactivation, its activity was fairly constant for a considerable time (Figure 1). The rate of carbon formation was consistently high on Fecronz; on Cafecronz it was lower, and after the initial phase of deacti-

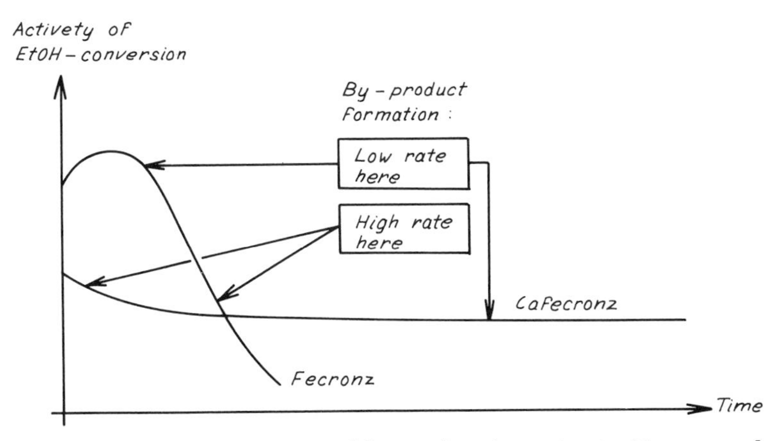

Figure 1. Activity vs. *time and by-product formation for Fecronz and Cafecronz*

vation, it stopped almost completely. Concerning the formation of by-products, remarkable differences were observed in the behavior of the two catalysts. With Fecronz, the amount of higher boiling by-products was low during the initial phase, but as the activity was reduced, increasing amounts of mainly methyl *n*-propyl ketone and mesityl oxide were formed. With Cafecronz the formation of by-products was most pronounced in the initial phase, and together with the above mentioned aliphatic compounds significant amounts of isophorone and aromatic by-products were formed, mainly *m*-cresol and 3,5-dimethylphenol. During the constant-activity period, by-product formation was low. These observations are indicated qualitatively in Figure 1.

These observations gave rise to the assumption that the course of carbon formation may have some connection with the by-product species, seemingly so that the formation of aromatic by-products inhibits the carbon formation and subsequent catalyst deactivation.

In a later study (2) this possibility was studied in more detail. This study was to be a general approach, but acetone formation was the most rewarding aspect. The reaction was studied in a recycle reactor with a stainless steel loop and a glass reactor. The catalysts used were Fecronz, Cafecronz, and Fecronz-K (Fecronz + 1% K_2O). Only the initial phases of deactivation were studied (duration of experiments 12 hours or less). The amount of carbon on the catalyst was measured as a function of time, and the reaction products and by-products were determined carefully. Only the most important conclusions are given here.

The kinetics of the carbon formation can be described by the well-known equation:

$$W_c = k\ \theta^n$$

where W_c = amount of carbon and θ = time. The values of n were higher than those normally found (*ca.* 0.5) as shown in Table I.
The high value of n indicates that the widely accepted mechanism of carbon formation—diffusion through a layer of previously formed carbon—may not be valid in this case. This is also indicated by the high value of the apparent activation energy on Fecronz and Fecronz-K. It was

Table I. Values of k and n

°C	Const.	Fecronz	Fecronz-K	Cafecronz
450	k	0.20	0.042	0.10
	n	0.75	1.7	1.0
480	k	0.35	0.072	0.10
	n	0.75	1.7	1.0
E_a, kcal/mole		20	19	0

shown that the rate of carbon formation is directly related to the acetaldehyde concentration. The apparent temperature independence of carbon formation on Cafecronz can be explained from the significantly reduced acetaldehyde concentration at the higher temperature.

It was not possible to derive a conclusive quantitative relationship between the amount of high boiling by-products and various parameters. There are indications that the formation of these products is mainly a function of residence time when both acetaldehyde and acetone are present.

Based on the amount of acetone formed, the amounts of both high boiling by-products and carbon are highest on the base-containing catalyst during the initial period (up to 12 hrs). Further, the rate of carbon formation was highest from acetaldehyde alone or from acetone alone and lowest when both compounds were present. On the other hand, aromatic by-products were formed only when both acetaldehyde and acetone were present in the reaction mixture.

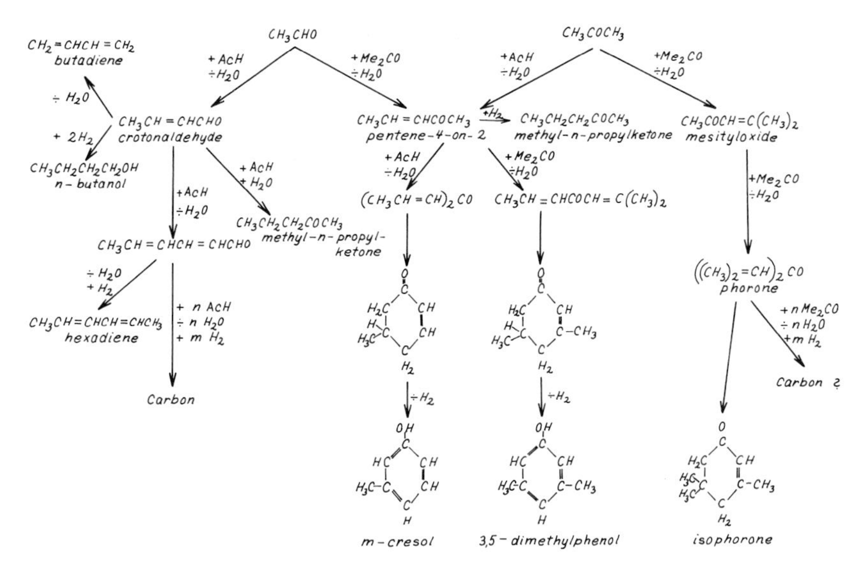

Figure 2. By-product formation from acetaldehyde and acetone

These observations might be explained by the simplified reaction paths shown in Figure 2. At the left, the mechanism of carbon formation by condensation of acetaldehyde is shown. At the right, a similar series of reactions shows the formation of carbon from acetone. In this reaction path mesityl oxide and isophorone are formed as intermediates.

The mechanism of the formation of *m*-cresol and 3,5-dimethylphenol is shown in the middle of the figure. These compounds are derived from mixed condensations between acetaldehyde and acetone with subsequent ring formation and aromatization.

The experimental results indicate that the behavior of the three catalysts with respect to formation of carbon and by-products may be explained by assuming that they all catalyze condensation reactions but that the addition of basic compounds to Fecronz promotes condensation reactions and especially the formation of isophorone and aromatic compounds. In the initial phase all reactions take place, but after some time, the direct condensation to carbon on the base-containing catalysts is inhibited, and the activity remains constant with formation of predominantly aromatic by-products.

Formation of acetonyl acetone and phenol was observed. Acetonyl acetone is probably the source of phenol—*e.g.*, *via* a five-carbon ring as an intermediate. Other identified compounds were ethyl acetate, acetaldehyde diethylacetal, and ethyl vinyl ether; they should be of little or no interest with regard to carbon formation.

The results of this investigation may be rather uncertain. Some aspects of the discussion and conclusions are only tentative. Further work will be needed to relate the different results, but it seems likely that one might arrive at a fairly thorough understanding of catalyst deactivation.

(1) Björkman, A., unpublished work.
(2) Dybkjaer, I., Licentiat thesis, The Technical University of Denmark, 1969.

6

Two-Phase and Slurry Reactors

GIANNI ASTARITA

Istituto di Principi di Ingegneria Chimica, Università di Napoli, Naples, Italy

The rational analysis of two-phase and slurry reactors is based on an understanding of both the interaction of mass transfer and chemical kinetics on a microscopic scale, and the influence of two-phase fluid mechanics on the over-all performance. Knowledge of both aspects has been increasing rapidly in recent years, and even difficult problems such as the prediction of selectivity for complex reaction schemes have been approached successfully. Superposition of research work in apparently unrelated fields is decreasing as the general theory is understood better. The most interesting and promising field of application of the body of research results available in the field is the area of biological processes, where two-phase reactions occur often.

The analysis of the performance, and ultimately the rational design, of two-phase and slurry reactors is a challenging task for the chemical engineer. Fluid mechanics, transport phenomena, and chemical kinetics interact on both a microscopic and a macroscopic scale in such reactors, making a rigorous analysis of even the simplest two-phase reactor an almost hopeless task. Fortunately, simplifications are often possible, and modeling of multiphase reactors suggests satisfactory procedures for their rational design.

The analysis of chemical reactors—homogeneous or multiphase—requires an understanding of the phenomena involved in both the local (or differential) and the over-all (or integral) situation. When dealing with integral reactor problems, some simple model for the local conditions is assumed, and attention is focused on the integration of the pertinent rate equations over the reactor volume. Procedures for modeling the local situation have been discussed in a recent review (*1*); procedures for attacking the integral reactor design problem, based on such modeling of the local conditions, have been discussed in a recent series of papers by Russell and co-workers (*2, 3, 4, 5, 6, 7*). Russell gives (*8*) an analysis of the product distribution obtainable in integral two-phase reactors.

The major difficulty encountered in analyzing integral two-phase reactors is connected with the unsatisfactory state of the art as far as fluid mechanics of two-phase systems is concerned. In the work quoted above (*2, 3, 4, 5, 6, 7*), Russell reviews the abundant literature on gas–liquid flow and analyzes the applicability of published results to the design of two-phase reactors. Knowledge on liquid–liquid systems is much more limited; some indication on the fluid mechanics involved, which is relevant to the design of liquid–liquid reactors, is given by Wijffels and Rietema (*9*). The work of Corrigan and Miller (*10*), and the paper by Nelson (*11*) discuss the design of staged two-phase reactors, while general procedures for cocurrent and countercurrent packed column continuous reactors are discussed by Astarita (*12*).

This review is concerned mainly with the other aspects of the problem—*i.e.*, understanding the local conditions and the microscopic interaction of transport phenomena and chemical reaction. Since available reviews (*1, 12*) are about four years old, this paper covers the literature from 1966 on. Although an effort has been made to cover all the pertinent chemical engineering literature, the survey does not claim to be complete.

Local Rate Equations

In multiphase reactors, the chemical reactions involved take place mainly in only one of the phases, though this is not generally true (*13*). The rate equations for the reactive phase are formulations of the energy balance and of the component mass balances which take into account the generation arising from the chemical reaction. The differential rate equations, in a general form, are:

Energy balance:

$$(\alpha\nabla - \mathbf{v})\,\nabla T = \frac{\partial T}{\partial t} + \frac{\Delta H\ r(\mathbf{c})e^{-\Delta E/RT}}{\rho c_p} \tag{1}$$

Component "*i*" mass balance:

$$(D_i\nabla - \mathbf{v})\nabla c_i = \frac{\partial c_i}{\partial t} + \nu_i r(\mathbf{c})e^{-\Delta E/RT} \tag{2}$$

The symbols are defined in the Nomenclature section.

Of course, Equations 1 and 2 also govern the local behavior of homogeneous reactors. The additional complication which arises for multiphase reactors lies in the fact that through boundary conditions stating the assumption that physical equilibrium prevails at the interface between any two phases, Equations 1 and 2 are coupled with the rate equations for the nonreactive phases.

Integration of Equations 1 and 2 requires, in principle, a detailed knowledge of the fluid mechanics involved because of the appearance of the convective terms $\mathbf{v}\nabla T$ and $\mathbf{v}\nabla c_i$, where $\mathbf{v}$ is the instantaneous local velocity vector. Such a detailed knowledge of the fluid mechanics of two-phase systems is almost never available, and some modeling of the fluid mechanics involved is thus required to handle the problem. Fluid mechanics modeling is discussed next.

In two-phase and slurry reactors the reactive phase is often subjected to vigorous mixing. Consequently, concentration and temperature tend to be uniform, except in a comparatively thin region near the interface with the nonreactive phase. When the chemical reactions involved are sufficiently slow, they occur mainly in the bulk of the reactive phase and therefore do not interfere microscopically with the transport phenomena: the reaction term can then be dropped altogether from Equations 1 and 2, though of course one needs to take into account the reactions when writing the boundary conditions and the integral balance equations. Under these conditions, the over-all process can be analyzed in terms of classical results from transport phenomena theory and chemical kinetics. This is the situation encountered in the so-called "slow reaction regime" (*1, 12*); one may encounter systems where the reaction rate is controlling (the kinetic sub-regime) and systems where the rate of heat or mass transfer is controlling (the diffusional sub-regime).

A more complicated situation arises when the reaction rate is sufficiently large so that the reaction takes place appreciably even in the interface region, altering the concentration and/or temperature profiles which govern the rate of transfer of heat and/or mass. Under these conditions, the interaction occurs on a microscopic scale, and the two individual phenomena—*viz.*, reaction and transport—must be analyzed simultaneously on the basis of Equations 1 and 2. The typical approach is to solve these equations for some appropriate model of the process considered and to compare the solution with that of either the classical transport equations or the classical homogeneous reactor equations written for the same model. The differences observed are a measure of the theoretically predicted influence of reaction on transport rate or of transport on reaction rate and can be compared with experimental results to assess the validity of the model used.

Simultaneous solution of the coupled heat and mass transfer equations is seldom possible. Fortunately, the isothermal approximation is often justified; this corresponds to assuming that the right side of Equation 1 is small enough to yield a small value of the temperature gradient in the interface region. Under these conditions, the temperature does not appear as a variable in Equation 2, which can be solved independently. Most of the research work discussed below is concerned with the isothermal case; corrections for heat effects, which were first considered by Danckwerts (*14*), have been discussed recently by Carberry (*15*), Clegg and Mann (*16*), and Danckwerts (*17*).

Modeling of the Fluid Mechanics

When dealing with the term $\mathbf{v}\nabla c_i$ in Equation 2, some simple model for the fluid mechanics in the region of interference of reaction and transport is required. Fortunately, the concentration gradient ∇c_i is approximately orthogonal to the interface, while the velocity vector is approximately parallel to the interface; therefore, the scalar product of the two vectors is small, and under some conditions the term $\mathbf{v}\nabla c_i$ can be dropped from the rate equations, thus resulting in a major simplification.

In the well-known film theory and penetration theory models, as well as in numerous derived models, it is assumed that the velocity is parallel to the interface and also that it is constant throughout the thin layer near the interface over which the concentration is not uniform—*i.e.*, where ∇c_i is different from zero. This is a justified assumption whenever the reactive phase is much more viscous than the adjoining phase, so that the velocity gradient near the interface can be neglected. By a proper choice of the coordinate system, the constant velocity $\mathbf{v}$ can be reduced to zero, and the term $\mathbf{v}\nabla c_i$ drops out of the rate equations.

A more complicated situation arises when the viscosity of the reactive phase is of the same order of magnitude or smaller than the viscosity of the adjoining phase, as well as in any other situation where the velocity gradient is not negligible near the interface. The theory of simultaneous mass transfer and chemical reaction in a boundary layer (*12, 18*) is developed for such problems: the concentration gradient has a non-zero component parallel to the interface, and therefore even if the velocity is parallel to the interface, the term $\mathbf{v}\nabla c_i$ cannot be dropped. The article by Sedriks and Kenney (*13*), which analyzes a trickle bed reactor where the packing is the catalyst, deals with a problem of this type. Much research work has recently been done (*19–27*) on the problem of simultaneous mass transfer and chemical reaction in the boundary layer surrounding a solid or a fluid sphere.

The case where the velocity vector has a non-zero component normal to the interface has been analyzed recently by Szekely (*28*). This problem arises when the mass transfer rates are so large that the bulk flow normal to the interface cannot be neglected.

A major problem arises when turbulence is significant in the region where the rate equations need to be integrated. In fact, in the term $\mathbf{v}\nabla c_i$ the instantaneous values of both $\mathbf{v}$ and ∇c_i are to be understood; therefore, even if the scalar product of the average values is zero, the average value of the scalar product may not be zero. Simple models of turbulence have been used to analyze problems of this type (*29–33*).

Apart from the special problems discussed above, most research on simultaneous mass transfer and chemical reaction is based on the film and penetration theory models. Although for these models the term $\mathbf{v}\nabla c_i$ does not appear explicitly in the rate equations, modeling of the fluid mechanics is still required. For the film theory approach, the thickness of the film appears as an independent parameter; for the penetration theory, both the average life of surface elements and the distribution of life spans appear as parameters. Of course, different models yield different results; how important these differences are as far as the predictive ability of the models is concerned is still a debated point (*34, 35*).

The simplest case, as far as modeling of the fluid mechanics is involved, is the one where the reactive phase is a solid; in that case $\mathbf{v} = 0$, and there are no adjustable parameters to be considered. Several papers concerning the analysis of gas–solid noncatalytic reacting systems have been published (*31–42*).

The remainder of this review discusses three subjects within the general field of simultaneous mass transfer and chemical reaction which seem to be the most interesting and promising. These are: the analysis of specific reacting systems of direct pragmatic interest; the measurement of interface areas in two-phase fluid–fluid systems by the so-called chemical method; and, possibly most interesting of all, the analysis of selectivity in two-phase and slurry reactors. Most of the recently published literature in the field is concerned with one or more of these subjects, though some theoretical research on more traditional problems in the field has also been carried out (*43–54*).

Recently Studied Reacting Systems

The rate equation for mass transfer with chemical reaction, Equation 2, is really a set of equations, one for each of the components considered; furthermore, the term $r(\mathbf{c})$ is essentially nonlinear and in general couples together all the equations of the set. Consequently, although the general philosophy of approach to problems of simultaneous mass transfer and chemical reaction is well understood, the theory needs to be reformulated in a different way for every special reacting system one wishes to consider, each one of which gives rise to a different form for the $r(\mathbf{c})$ function (*i.e.*, is characterized by its own kinetic mechanism).

Indeed, it is a general feature of chemical kinetics that every reacting system must be studied separately, although general rules of qualitative character can be formulated.

Table I gives a summary of some of the reacting systems which have been studied recently in some detail and with some effort at a scientific approach. Carbon dioxide absorption in monoethanolamine and diethanolamine solutions has been the subject of numerous investigations (*55–62*), and owing to the large amount of available information, some subtle problems have been analyzed. In particular, interfacial turbulence, possibly initiated by the mass transfer phenomenon itself, may be responsible for some discrepancies among experimental results and theoretical predictions (*56*), although a recent series of papers by Thomas (*60, 61, 62*) disputes this conclusion.

Table I. Two-Phase Reacting Systems Investigated in Recent Years

System	*Reference*
CO_2 absorption in ethanolamines	*55–63*
CS_2 absorption in monoethanolamine	*64*
Cl_2 absorption in water	*26, 65*
Sodium dithionite oxidation	*66*
H_2S absorption in alkaline solutions	*67*
Cuprous chloride oxidation	*49, 68*
Isobutylene hydration	*69, 79*
Hydrolysis of esters	*71*
Polycondensation	*72*
CO_2 absorption in Na_2SO_4	*73*
Oxidation of cumene	*74*
Hydrogenation of crotonaldehyde	*13*
Homogeneously catalyzed gas–liquid reactions	*75, 76*
Gas–solid reactions	*37–42*
Liquid–liquid systems	*77*

Some reacting systems present an interesting challenge to the designer; in fact, the requirements of a large reacting capacity and of a fast reaction rate often conflict. The typical example is the absorption of carbon dioxide in hydroxide and in ethanolamine solutions. As long as the carbonation ratio—*i.e.*, moles of CO_2 absorbed per mole of liquid-phase reactant—is kept low, the reaction rate is very fast, the reaction product being the carbonate for hydroxide solutions and the carbamate for amine solutions:

$$CO_2 + 2MOH \rightarrow M_2CO_3 + H_2O$$

$$CO_2 + 2RNH \rightarrow RN\ COOH_2NR$$

Unfortunately, the stoichiometry of the reactions involved does not allow a carbonation ratio higher than 0.5. When this limit is reached, reversal to the bicarbonate form sets in (*12*):

$$CO_2 + M_2CO_3 + H_2O \rightarrow 2MHCO_3$$

$$CO_2 + RN\ COO\ H_2\ NR + 2H_2O \rightarrow 2\ RNH_2\ HCO_3$$

Unfortunately, the rate of bicarbonate formation is generally very slow. It can be enhanced by catalyzing homogeneously the direct attack of carbon dioxide on water:

$$CO_2 + H_2O \rightarrow HCO_3^- + H^+$$

which proceeds parallel to the attack of carbon dioxide on hydroxyl ions. Studies of systems of this type have been published by Danckwerts and McNeil (*75*) and by Pohorecki (*76*).

Absorption of hydrogen sulfide in several aqueous alkaline solutions has been studied systematically (*61*), thus laying the groundwork for work on selectivity during simultaneous absorption of two gases to be discussed later. Hydrogen sulfide undergoes extremely fast proton transfer reactions in alkaline solutions. These reactions are often reversible, and this creates an interesting problem of mass transfer accompanied by reversible infinitely fast reaction; the rate of reaction r is no longer an explicit function of the concentration vector $\mathbf{c}$ but depends on the rate by which diffusion brings the reactants together. The theory for this case, which was discussed for the hydrogen sulfide system by Gioia and Astarita (*67*), was formulated later in more general terms by Danckwerts (*78*) and by Ulanowicz and Frazier (*54*).

A very interesting system has been considered by Hoftyzer and Kreveler (*72*)—*viz.*, polycondensation reactions. In the late stages of the reaction the rate is controlled by the desorption of the volatile product, and thus a theory for desorption accompanied by chemical reaction is required. Such a theory is, of course, also useful in analyzing the regeneration stage of a chemical absorption unit.

The work of Beek, Marrucci, and Davis on the sulfidation of alkali cellulose (*37*) and the work of Sedriks and Kenney on hydrogenation of crotonaldehyde (*13*) are also particularly interesting because special effects are important; heat transfer in the sulfidation of alkali cellulose and the parallel gas-phase reaction on non-wetted parts of the catalyst in the hydrogenation of crotonaldehyde.

Chemical Method for Determining Interface Area

The measurement of interfacial area in fluid–fluid systems presents great difficulties, and all available methods are applicable only in a limited range of conditions, often giving doubtful results. The chemical method, which is based on measurements of reaction rates, has been used widely in recent research on fluid–fluid systems.

The principle on which the method is based can easily be illustrated on the basis of the penetration theory or the film theory model. It can be shown (*1, 12*) that, whenever some characteristic time scale of flow is much larger than the time scale of the chemical reaction, Equation 2 reduces to:

$$D_i \frac{d^2 c_i}{dx^2} = r(\mathbf{c}) \tag{3}$$

with boundary conditions of the following type: at interface,

$$c_i = c_i^o \text{ for volatile components} \tag{4}$$

$$\frac{dc_i}{dx} = 0 \text{ for nonvolatile components} \tag{5}$$

where $c_i = C_{eq}$,

$$\frac{dc_i}{dx} = 0 \text{ for all components} \tag{6}$$

The solution of Equations 3–6 is clearly independent of the fluid mechanics involved, which influence neither the differential equation nor the boundary conditions. Apart from the actual analytical solution, it may be stated directly that the total transfer rate, which at steady state equals the total reaction rate, will be given by an equation of the following form:

$$\phi = - DA \frac{dc_i}{dx}/\text{interface} = AF \tag{7}$$

where A is the total interface area, and F is a parameter which does not depend on the fluid mechanics involved. Hence, it is possible to determine A from the transfer rate ϕ_i. F can be obtained by calibration experiments with the same system under conditions where A is known, such as in a laminar jet or a wetted wall column.

This general idea has been used in various situations, and interesting results have been obtained. Table II gives a summary of fluid–fluid systems which have been investigated recently by this technique.

For gas–liquid systems, the reacting system which has been used more frequently is the oxidation of sodium sulfite, catalyzed by cobalt or manganese ions present in the liquid phase. This system presents a

Table II. Systems For Which the Interface Area Has Been Determined by the Chemical Method

System	*Reference*
Plate columns	*81, 82*
Packed columns	*83, 84*
Sieve trays	*85, 86*
Sparged stirred tanks	*87, 88*
Cocurrent two-phase flow	*89, 90, 91*
Liquid–liquid systems	*92, 93*

complex kinetic mechanism (*79*), with the order of the reaction with respect to oxygen ranging from zero to 2.

The paper by Van den Berg (*80*) is an interesting example of application of the method to a complex system—*i.e.*, a three-phase stirred tank reactor. Once the interface area has been determined, mass transfer coefficients can be calculated from measured over-all transfer rates. In this way Van den Berg obtains an interesting correlation of the transfer coefficients based on Kolmogoroff's theory of isotropic turbulence.

The chemical method for measuring interface areas meets with some major difficulties. The first is that the effective interfacial area for mass transfer may not coincide with the geometrical one, and the chemical method may actually be measuring neither (*57, 79*). In particular, the chemical method tends to give equal weight to quasi-stagnant regions and to continuously renewed ones while the former are much less effective than the latter in most processes.

The second difficulty is more important. The ease by which the coalescence phenomena occur is strongly influenced by the presence of even minor amounts of solutes (*94, 95*), which are always present when the chemical method is used (the reactants themselves). Therefore, the interface area in any given system may be quite different when the chemical method is used from what it is under other conditions. In particular, in the case of sparged reactors the average bubble size when reactants are added to the liquid phase may drop by an order of magnitude; hence, the chemical method can grossly overestimate the interface area which would be obtained in the absence of the reaction. Despite this, the chemical method has interesting possibilities, and in some cases it is the only available method for determining interface areas.

Selectivity

The analysis of complex reaction mechanisms for which problems of selectivity need to be considered has been extended only recently to include multiphase systems. This is an important and diversified field where both theoretical and experimental work is badly needed and where the groundwork is just being laid. Research in this field can be of major pragmatic importance by offering a new tool to control selectivity in industrial operations, beyond the traditional ones of temperature and mixing conditions.

Problems involving selectivity are in a sense unique among engineering problems because an incomplete understanding cannot be obviated by increasing the cost of operation but must be paid for in terms of a technically poor performance. If the conversion obtainable from a given reactor cannot be predicted accurately, a somewhat oversized reactor will solve the problem, but if the residence time for maximum selectivity is not predicted accurately, both a low and a high prediction will result in an actually lower selectivity and thus in a technically poor result.

Selectivity is influenced strongly by mass transfer when one or more of the reactants and/or products can be transferred to or from a nonreactive phase. As a limit, were it possible to withdraw the desired product from the reactive phase as soon as it is formed, the selectivity for a consecutive reaction scheme would reach 100%. The rate of transfer itself is influenced by the chemical reactions, so that very complex situations may be encountered.

The pioneer work in this area has been done by Van de Vusse (*96, 97, 98*), who has given a thorough analysis of the consecutive reaction scheme:

$$A + B \rightarrow C$$

$$A + C \rightarrow D$$

where C is the desired product. In his first paper (*96*), Van de Vusse analyzes the case where A is transferred from a nonreactive phase, and both reactions are first order; the film theory model is used, and experimental results on chlorination of *p*-cresol are given. The predicted and the observed selectivity are lower than would be obtained in the homogeneous case—*i.e.*, diffusion has a negative effect. In contrast with the homogeneous case, when there is diffusion control, the selectivity may not approach unity when the conversion approaches zero.

In his second paper (*97*) Van de Vusse extends the analysis to the case where the reactions are not both first order, with similar results. His third paper (*98*) attacks the problem where the desired product can be transferred from the reactive phase, and for this problem the mass transfer causes an increase of selectivity over the homogeneous case, and 100% selectivity can be obtained if the mass transfer is infinitely fast. Different modes of operation—*i.e.*, cocurrent and countercurrent plug flow of both phases as well as possible combinations of plug flow and perfectly mixed flow—are analyzed in detail. The same problem has been analyzed also by Bridgwater (*99*).

Szekely and Bridgwater (*100*) analyzed the influence on predicted selectivity of different models of the fluid mechanics involved, such as the film theory and the penetrating theory. In contrast with the situation encountered in predicting transfer rates, where the model assumed is largely immaterial, predicted selectivities may change by as much as 20% from one model to another. Physically, this implies that details of the fluid mechanics involved may have a major influence on the selectivity obtainable in two-phase reactors.

Bridgwater and Carberry (*101*) have discussed the selectivity for a parallel reaction mechanism in two-phase reactors. If the two reactions are of different order, mass transfer may alter the selectivity with respect to that possible in a homogeneous reactor.

Russell discussed product distribution obtainable in gas–liquid integral reactors (*8*); no great differences are reported for the homogeneous reactor case. Corrigan and Miller (*10*) have analyzed the performance of a staged reactor with reflux for a consecutive reaction scheme. If the volatilities of all components are equal, the staged reactor behaves as a plug-flow reactor. If the volatilities are quite different, all the reactants are present simultaneously only on the feed stage, and the staged reactor performs as a stirred-tank reactor. It is interesting that for intermediate volatilities the staged reactor may, in terms of selectivity, outperform both the plug-flow and the stirred-tank reactor.

A somewhat different problem of selectivity has been analyzed by Gioia (*102–105*)—the simultaneous absorption of hydrogen sulfide and carbon dioxide. Selectivity for hydrogen sulfide is often a desirable feature; here selectivity is defined as:

$$S = \frac{\phi_{H_2S}/\phi_{CO_2}}{\phi^{\circ}_{H_2S}/\phi^{\circ}_{CO_2}}$$

where the ϕ°'s are absorption rates possible in the absence of chemical reaction. The chemical reaction always causes S values to be larger than 1, and for some systems extremely large values are obtained; in fact, under some conditions practically no carbon dioxide is absorbed. Experimental results are in good agreement with theoretical predictions.

Conclusions

Research on heterogeneous reacting systems has reached a stage of development where essentially any pragmatically interesting problem of coupling between transport phenomena and chemical reaction can be attacked with reasonable confidence in obtaining a significant solution. As in most problems in chemical engineering, the major difficulty is encountered in handling the fluid mechanics involved, and future major breakthroughs will probably be related to an improved understanding of the fluid dynamics of heterogeneous systems.

Coupling of mass transfer and chemical reactions is a feature of many biological processes taking place in living systems. It seems therefore that the scientific background accumulated by chemical engineers in this field could be applied usefully to a rational analysis of some interesting biological problems.

Such interdisciplinary research work is often hindered by a lack of communication among scientists working in seemingly unrelated fields, a problem whose solution is unfortunately not simply a technical one. Scientists should make a consistent effort to find an efficient way to eliminate this communication gap.

Nomenclature

- A total interface area, cm^2
- c_i concentration of component i, gram mole/cm^3
- c_i^o value of c_i at interface, gram mole/cm^3
- c_{eq} equilibrium concentration, gram mole/cm^3
- c concentration vector, $(c_1, \ldots c_n)$, gram mole/cm^3
- c_p specific heat, cal/gram, °K
- D_i diffusivity of component i, cm^2/sec
- ΔE activation energy, cal/gram mole
- ΔH heat of reaction, cal/gram mole
- r rate of reaction, gram mole/cm^3, sec
- R gas constant, cal/gram mole, °K
- t time, sec
- T temperature, °K
- $\mathbf{v}$ velocity vector, cm/sec
- α heat diffusivity, cm^2/sec
- ν_i stoichiometric coefficient, dimensionless
- ρ density, grams/cm^3
- ϕ mass flux, gram mole/sec
- ∇ gradient operator, cm^{-1}

Literature Cited

(1) Astarita, G., *Ind. Eng. Chem.* (1966) **58** (8), 18.
(2) Anderson, R. J., Russell, T. W. F., *Chem. Eng.* (1965) **72** (25), 134.
(3) *Ibid.*, (1965) **72** (26), 99.
(4) *Ibid.*, (1966) **73** (1), 87.
(5) Cichy, P. T., Russell, T. W. F., *Ind. Eng. Chem.* (1969) **61** (8), 15.
(6) Cichy, P. T., Ultman, T. S., Russell, T. W. F., *Ind. Eng. Chem.* (1969) **61** (8), 6.
(7) Schaftlein, R. W., Russell, T. W. F., *Ind. Eng. Chem.* (1968) **60** (5), 12.
(8) Russell, T. W. F., Rothenberger, R., ADVAN. CHEM. SER. (1972) **109,** 255.
(9) Wijffels, J. B., Rietema, K., ADVAN. CHEM. SER. (1972) **109,** 245.
(10) Corrigan, T. E., Miller, J. H., *A.I.Ch.E. J.* (1967) **13,** 809.
(11) Nelson, P. A., ADVAN. CHEM. SER. (1972) **109,** 237.
(12) Astarita, G., "Mass Transfer with Chemical Reaction," Elsevier, Amsterdam, 1967.

(13) Sedriks, W., Kenney, C. N., ADVAN. CHEM. SER. (1972) **109,** 251.
(14) Danckwerts, P. V., *Appl. Sci. Res.* (1952) **A3,** 385.
(15) Carberry, J. J., *Chem. Eng. Sci.* (1966) **21,** 951.
(16) Clegg, G. T., Mann, R., *Chem. Eng. Sci.* (1969) **24,** 321.
(17) Danckwerts, P. V., *Chem. Eng. Sci.* (1967) **22,** 472.
(18) Astarita, G., *Chim. Ind. (Milan)* (1963) **45,** 441.
(19) Estrin, J., Schmidt, H. E., *A.I.Ch.E. J.* (1968) **14,** 678.
(20) Goddard, J. D., Acrivos, A., *Quart. J. Mech. Appl. Math.* (1967) **20,** 471.
(21) Ishi, T., Johnson, A. I., *Can. J. Chem. Eng.* (1969) **47,** 272.
(22) Johnson, A. I., Akehata, T., *Can. J. Chem. Eng.* (1969) **47,** 88.
(23) Johnson, A. I., Hamielec, A. E., Houghton, W. T., *A.I.Ch.E. J.* (1967) **13,** 379.
(24) Johnson, A. I., Hamielec, A. E., Houghton, W. T., *Can. J. Chem. Eng.* (1967) **45,** 140.
(25) Rutland, L., Pfeffer, R., *A.I.Ch.E. J.* (1967) **13,** 182.
(26) Takahashi, T., Hatanaka, M., Konaka, R., *Can. J. Chem. Eng.* (1967) **45,** 145.
(27) Wu, C. C., Pfeffer, R., *Ind. Eng. Chem., Fundamentals* (1970) **9,** 101.
(28) Szekely, J., *Chem. Eng. Sci.* (1967) **22,** 777.
(29) Lee, J., *Phys. Fluids* (1966) **9,** 1753.
(30) Mao, K. W., Toor, H. L., *A.I.Ch.E. J.* (1970) **16,** 49.
(31) O'Brien, E. E., *Phys. Fluids* (1966) **9,** 215.
(32) *Ibid.*, p. 1561.
(33) Toor, H. L., *A.I.Ch.E. J.* (1962) **8,** 70.
(34) Huang, C. J., Kuo, C. H., *A.I.Ch.E. J.* (1965) **11,** 901.
(35) Porter, K. E., Roberts, D., *Chem. Eng. Sci.* (1969) **24,** 695.
(36) Smith, J. L., Winnick, J., *A.I.Ch.E. J.* (1967) **13,** 1207.
(37) Beek, W. J., Marrucci, G., Davis, S. H., *Chem. Eng. Sci.* (1968) **23,** 1347.
(38) Beveridge, G. S. G., Goldie, P. J., *Chem. Eng. Sci.* (1968) **23,** 913.
(39) Ishida, M., Wen, C. V., *A.I.Ch.E. J.* (1968) **14,** 311.
(40) Novosad, Z., Rehakova', M., *European Symp. Chem. Reaction Eng., 4th, Bruxelles, 1968.*
(41) Ross, L. W., *A.I.Ch.E. J.* (1969) **15,** 136.
(42) Wang, S. C., *A.I.Ch.E. J.* (1969) **15,** 624.
(43) Boerma, H., Lankester, J. H., *Chem. Eng. Sci.* (1968) **23,** 799.
(44) Coeuret, F., Jamet, B., Ronco, J. J., *Chem. Eng. Sci.* (1970) **25,** 17.
(45) Gunn, D. J., *Chem. Eng. Sci.* (1967) **22,** 1439.
(46) Jhaveri, A. S., *Chem. Eng. Sci.* (1969) **24,** 1738.
(47) Jhaveri, A. S., Sharma, M. M., *Chem. Eng. Sci.* (1969) **24,** 189.
(48) Karanth, P. K., Mohan Rao, N. A., *Trans. Faraday Soc.* (1967) **63,** 355.
(49) Ramachandran, P. A., Sharma, M. M., *Chem. Eng. Sci.* (1966) **24,** 1681.
(50) Ronco, J. J., Coeuret, F., *Chem. Eng. Sci.* (1969) **24,** 423.
(51) Standardt, G., *Chem. Eng. Sci.* (1969) **22,** 1655.
(52) Stewart, W. E., *Chem. Eng. Sci.* (1968) **23,** 483.
(53) Tavlarides, L. L., Gal-or, B., *Chem. Eng. Sci.* (1969) **24,** 553.
(54) Ulanowicz, R. E., Frazier, G. C., *Chem. Eng. Sci.* (1968) **23,** 1335.
(55) Barrere, C. A., Deans, M. A., *A.I.Ch.E. J.* (1968) **14,** 280.
(56) Brian, P. L. T., Vivian, J. E., Matiatos, D. C., *A.I.Ch.E. J.* (1967) **13,** 28.
(57) Danckwerts, P. V., McNeil, K. M., *Chem. Eng. Sci.* (1967) **22,** 925.
(58) Danckwerts, P. V., Tavares da Silva, A., *Chem. Eng. Sci.* (1967) **22,** 1513.
(59) Gupta, V. P., Douglas, W. J. M., *A.I.Ch.E. J.* (1967) **13,** 883.
(60) Thomas, W. J., *A.I.Ch.E. J.* (1966) **12,** 1051.
(61) Thomas, W. J., Mc. Nicholl, E., *Chem. Eng. Sci.* (1967) **22,** 1877.
(62) Thomas, W. J., Mc. Nicholl, E., *Trans. Inst. Chem. Engrs. (London)* (1969) **47,** 325.

(63) Danckwerts, P. V., Sharma, M. M., *Trans. Inst. Chem. Engrs. (London)* (1966) **44**, CE244.
(64) Kothari, P. J., Sharma, M. M., *Chem. Eng. Sci.* (1966) **21**, 391.
(65) Brian, P. L. T., Vivian, J. E., Piazza, C., *Chem. Eng. Sci.* (1966) **21**, 551.
(66) Jhaveri, A. S., Sharma, M. M., *Chem. Eng. Sci.* (1968) **23**, 1.
(67) Gioia, F., Astarita, G., *Ind. Eng. Chem., Fundamentals* (1967) **6**, 370.
(68) Jhaveri, A. S., Sharma, M. M., *Chem. Eng. Sci.* (1967) **22**, 1.
(69) Gehlawat, J. K., Sharma, M. M., *Chem. Eng. Sci.* (1968) **23**, 1173.
(70) Solbrig, C. W., Gidaspow, D., *Can. J. Chem. Eng.* (1967) **45**, 35.
(71) Nanda, A. K., Sharma, M. M., *Chem. Eng. Sci.* (1967) **22**, 769.
(72) Hoftyzer, P. J., van Krevelen, D. W., *European Symp. Chem. Reaction Eng., 4th, Bruxelles, 1968.*
(73) Navratil, J., Nylvt, J., *Chem. Prum.* (1967) **17**, 235.
(74) Low, D. I. R., *Can. J. Chem. Eng.* (1967) **45**, 166.
(75) Danckwerts, P. V., McNeil, K. M., *Trans. Inst. Chem. Engrs. (London)* (1967) **45**, 32.
(76) Pohorecki, R., *Chem. Eng. Sci.* (1968) **23**, 1447.
(77) Sharma, M. M., Kanda, A. K., *Trans. Inst. Chem. Engrs. (London)* (1968) **46**, 44.
(78) Danckwerts, P. V., *Chem. Eng. Sci.* (1968) **23**, 1045.
(79) Astarita, G., Marrucci, G., Coleti, L., *Chim. Ind. (Milan)* (1964) **46**, 1021.
(80) Van den Berg, H. J., Advan. Chem. Ser. (1972) **109**, 240.
(81) Porter, K. E., King, M. B., Varshey, K. C., *Trans. Inst. Chem. Engrs. (London)* (1966) **44**, 274.
(82) Sharma, M. M., Gupta, R. K., *Trans. Inst. Chem. Engrs. (London)* (1967) **45**, 169.
(83) De Waal, K. J. A., Beek, W. J., *Chem. Eng. Sci.* (1967) **22**, 585.
(84) Jhaveri, A. S., Sharma, M. M., *Chem. Eng. Sci.* (1968) **23**, 669.
(85) Eben, C. D., Pigford, R. L., *Chem. Eng. Sci.* (1965) **20**, 803.
(86) Pasiuk-Bromikowska, W., *Chem. Eng. Sci.* (1969) **24**, 1139.
(87) Linek, V., Mayrhoferova, J., *Chem. Eng. Sci.* (1969) **24**, 481.
(88) Reith, T., Beek, W. J., *European Symp. Chem. Reaction Eng., 4th, Bruxelles, 1968.*
(89) Gregory, G. A., Scott, D. S., *Intern. Symp. Cocurrent Gas-Liquid Flow, Waterloo, Ont., 1968.*
(90) Scott, D. S., Haydruk, W., *Can. J. Chem. Eng.* (1966) **44**, 130.
(91) Wales, C. E., *A.I.Ch.E. J.* (1966) **12**, 1166.
(92) Fernandes, J. B., Sharma, M. M., *Chem. Eng. Sci.* (1967) **22**, 1267.
(93) Nanda, A. K., Sharma, M. M., *Chem. Eng. Sci.* (1966) **21**, 707.
(94) Marrucci, G., Nicodemo, L., *Chem. Eng. Sci.* (1967) **22**, 1257.
(95) Zieminski, S., Caron, M. M., Blackmore, R. B., *Ind. Eng. Chem., Fundamentals* (1967) **6**, 233.
(96) Van de Vusse, J. G., *Chem. Eng. Sci.* (1966) **21**, 631.
(97) *Ibid.*, p. 645.
(98) *Ibid.*, p. 1239.
(99) Bridgwater, J., *Chem. Eng. Sci.* (1967) **22**, 185.
(100) Szekely, J., Bridgwater, J., *Chem. Eng. Sci.* (1967) **22**, 711.
(101) Bridgwater, J., Carberry, J. J., *Brit. Chem. Eng.* (1967) **12**, 217.
(102) Gioia, F., *Chim. Ind. (Milan)* (1967) **49**, 921.
(103) *Ibid.*, p. 1287.
(104) Gioia, F., Astarita, G., Marrucci, G., *European Symp. Chem. Reaction Eng., 4th, Bruxelles, 1968.*
(105) Gioia, F., Marrucci, G., *A.I.Ch.E. Natl. Mtg., New Orleans, 1969.*

Received September 8, 1970.

Contributed Papers

Countercurrent Equilibrium Stage Separation with Reaction

PAUL A. NELSON, Shell Development Co., Emeryville, Calif.

It has been shown (*1*) that the classical methods of solving equilibrium stage problems—*viz.*, Thiele-Geddes, Lewis-Matheson, etc.—are not sufficiently powerful to solve all problems of this type. The relationships which describe such a problem may be viewed as a high dimensional set of nonlinear algebraic equations. Tierney and his collaborators (*2*, *3*) have applied a standard mathematical tool, the Newton-Raphson method, to formulate a powerful algorithm for solving a wider class of problems than had been possible with any single technique. In a study of the use of equilibrium stage devices as reactors, the author found that this problem had not been treated in general. Extensions of the classical algorithms to include a nonlinear reaction rate expression were found not to converge. The simplest reliable technique has proved to be a non-trivial extension of the work of Tierney *et al.*

In formulating the problem we restrict our consideration to countercurrent flow between equilibrium stages and use the terminology of distillation, remembering that the extension to more general staged systems is readily made. It is assumed that each stage is a perfectly mixed reactor. Also the vapor leaving each stage is in physical equilibrium with the liquid leaving that stage. It can be shown (*2*) that the material balance equations for component j can be written

$$\mathbf{s}^{(j)} \equiv Z^{(j)}\mathbf{x}^{(j)} + \mathbf{f}^{(j)} + \mathbf{r}^{(j)} = 0 \tag{1}$$

[As a notation convention, italic capital letters symbolize matrices, and bold lower case letters indicate column vectors.] Equation 1 differs from Equation 6 of Ref. 2 only in the inclusion of the column vector $\mathbf{r}^{(j)}$ representing the rate of creation of component j by reaction, which in general has nonlinear dependence on the composition vector $\mathbf{x}^{(j)}$. When the system is thermodynamically nonideal, the flow matrix $Z^{(j)} \equiv L + VK^{(j)}$ also depends on $\mathbf{x}^{(j)}$ since $K^{(j)}$, the diagonal matrix of vapor–liquid equilibrium coefficients, is composition dependent. L and V are flow connection matrices defined in Ref. 2.

The vectors $\mathbf{t}$ and $\mathbf{v}$, describing the stage temperatures and vapor flow rates, must be chosen such that when Equation 1 is solved for each component j, the sum of mole fractions in each phase on each tray is unity, and the energy flow is in balance. These two criteria are formulated respectively as:

$$\mathbf{d}_m \equiv \sum_{j=1}^{M} (I - K^{(j)})\mathbf{x}^{(j)} = 0 \tag{2}$$

$$\mathbf{d}_e \equiv L\mathbf{h} + V\mathbf{g} + \mathbf{q} = 0 \tag{3}$$

where I represents an identity matrix of appropriate dimension, $\mathbf{h}$ and $\mathbf{g}$ are vectors representing the enthalpy of the liquid and vapor streams leaving each stage, and $\mathbf{q}$ represents the feed stream enthalpies.

We begin by generating an iterative solution to Equation 1. We assume that at any stage of the calculation we have estimated values for all of the composition vectors. For some $\mathbf{t}$ and $\mathbf{v}$ we wish to calculate improved values, $\mathbf{x}'^{(j)}$. A suitable method is to use a Newton-Raphson algorithm based on Equation 1:

$$\mathbf{x}'^{(j)} = \mathbf{x}^{(j)} - \left[\frac{\partial \mathbf{s}^{(j)}}{\partial \mathbf{x}^{(j)}}\right]^{-1} \mathbf{s}^{(j)};\ j = 1,2, \ldots, M \tag{4}$$

This equation is only approximate in the sense that the various components j have been decoupled by neglecting all terms of the form $\frac{\partial \mathbf{x}^{(k)}}{\partial \mathbf{r}^{(j)}}$ where $k \neq j$. Equation 4 can be written

$$\mathbf{x}'^{(j)} = \left[Z^{(j)} + V\frac{\partial K^{(j)}}{\partial \mathbf{x}^{(j)}}\mathbf{x}^{(j)} + \frac{\partial \mathbf{r}^{(j)}}{\partial \mathbf{x}^{(j)}}\right]^{-1} \tag{5}$$

$$\left[V\frac{\partial K^{(j)}}{\partial \mathbf{x}^{(j)}}\mathbf{x}^{(j)}\mathbf{x}^{(j)} + \frac{\partial \mathbf{r}^{(j)}}{\partial \mathbf{x}^{(j)}}\mathbf{x}^{(j)} - \mathbf{f}^{(j)} - \mathbf{r}^{(j)}\right]$$

The corresponding values of $\mathbf{d}'_m$ and $\mathbf{d}'_e$ are readily calculated from the definitions in Equations 2 and 3. Now we go back and ask how we should have chosen $\mathbf{t}$ and $\mathbf{v}$ in such a way that the compositions $\mathbf{x}'^{(j)}$ would cause Equations 2 and 3 to be satisfied. We can construct a Newton-Raphson method based on these latter equations. In partitioned matrix form, the method gives:

$$\begin{pmatrix} \mathbf{v} \\ - \\ \mathbf{t} \end{pmatrix}_{new} = \begin{pmatrix} \mathbf{v} \\ - \\ \mathbf{t} \end{pmatrix}_{old} - \left(\begin{array}{c|c} \dfrac{\partial \mathbf{d}'_m}{\partial \mathbf{v}} & \dfrac{\partial \mathbf{d}'_m}{\partial \mathbf{t}} \\ \hline \dfrac{\partial \mathbf{d}'_e}{\partial \mathbf{v}} & \dfrac{\partial \mathbf{d}'_e}{\partial \mathbf{t}} \end{array} \right)^{-1} \begin{pmatrix} \mathbf{d}'_m \\ --- \\ \mathbf{d}'_e \end{pmatrix} \tag{6}$$

To use Equation 6 we must develop expressions for the elements of the matrix of partial derivatives. We begin by differentiating Equation 5 with respect to $\mathbf{t}$ and $\mathbf{v}$. This yields

$$\frac{\partial \mathbf{x}'^{(j)}}{\partial \mathbf{t}} = -\left[Z^{(j)} + V \frac{\partial K^{(j)}}{\partial \mathbf{x}^{(j)}} \mathbf{x}^{(j)} + \frac{\partial \mathbf{r}^{(j)}}{\partial \mathbf{x}^{(j)}}\right]^{-1} \left[V \frac{\partial K^{(j)}}{\partial \mathbf{t}} \mathbf{x}'^{(j)} + \frac{\partial \mathbf{r}^{(j)}}{\partial \mathbf{t}} + \sum_{k=1}^{j-1} \frac{\partial \mathbf{r}^{(j)}}{\partial \mathbf{x}'^{(k)}} \frac{\partial \mathbf{x}'^{(k)}}{\partial \mathbf{t}}\right] \tag{7}$$

$$\frac{\partial \mathbf{x}'^{(j)}}{\partial \mathbf{v}} = -\left[Z^{(j)} + V \frac{\partial K^{(j)}}{\partial \mathbf{x}^{(j)}} \mathbf{x}^{(j)} + \frac{\partial \mathbf{r}^{(j)}}{\partial \mathbf{x}^{(j)}}\right]^{-1} \left[\frac{\partial Z^{(j)}}{\partial \mathbf{v}} \mathbf{x}'^{(j)} + \sum_{k=1}^{j-1} \frac{\partial \mathbf{r}^{(j)}}{\partial \mathbf{x}'^{(k)}} \frac{\partial \mathbf{x}'^{(k)}}{\partial \mathbf{v}}\right] \tag{8}$$

The submatrices of the Jacobian matrix in Equation 6 are then given by

$$\frac{\partial \mathbf{d}'_m}{\partial \mathbf{v}} = \sum_{j=1}^{M} (I - K^{(j)}) \frac{\partial \mathbf{x}'^{(j)}}{\partial \mathbf{v}} \tag{9}$$

$$\frac{\partial \mathbf{d}'_m}{\partial \mathbf{t}} = \sum_{j=1}^{M} \left[(I - K^{(j)}) \frac{\partial \mathbf{x}'^{(j)}}{\partial \mathbf{t}} - \frac{\partial K^{(j)}}{\partial \mathbf{t}} \mathbf{x}'^{(j)}\right] \tag{10}$$

$$\frac{\partial \mathbf{d}'_e}{\partial \mathbf{v}} = \sum_{j=1}^{M} \left[L\hat{H}^{(j)} + VK^{(j)}\hat{G}^{(j)}\right] \frac{\partial \mathbf{x}'^{(j)}}{\partial \mathbf{v}} + \Xi \tag{11}$$

$$\frac{\partial \mathbf{d}'_e}{\partial \mathbf{t}} = \sum_{j=1}^{M} \left[(L\hat{H}^{(j)} + VK^{(j)}\hat{G}^{(j)}) \frac{\partial \mathbf{x}'^{(j)}}{\partial \mathbf{t}} + V\hat{G}^{(j)} \frac{\partial K^{(j)}}{\partial \mathbf{t}} \mathbf{x}'^{(j)}\right] + L\Gamma_{(liq)} + V\Gamma_{(vap)} \tag{12}$$

In these equations $H^{(j)}$ and $G^{(j)}$ are diagonal matrices of partial molar enthalpy, $\Gamma_{(liq)}$ and $\Gamma_{(vap)}$ are diagonal matrices representing total stream heat capacities, and the matrix Ξ is defined as:

$$\Xi = \begin{pmatrix} h_1 - g_1 & g_2 - h_1 & & & & \\ h_2 - g_1 & h_1 - g_2 & g_3 - h_2 & & & \\ h_3 - h_2 & & h_2 - g_3 & \cdot & & \\ \cdot & & \cdot & \cdot & \cdot & \\ \cdot & & & \cdot & \cdot & \\ \cdot & & & \cdot & \cdot & \\ \cdot & & & & \cdot & g_n - h_{n-1} \\ h_n - h_{n-1} & & & & \cdot & h_{n-1} - g_n \end{pmatrix} \quad (13)$$

In an obvious simplification of notation, Equation 6 may be rewritten

$$\Delta \mathbf{w} = - J^{-1}\mathbf{d} \quad (14)$$

Experience has shown that when Equation 14 is used, computational instabilities may arise which make it impossible to converge to a solution. A modification of the algorithm embodied in Equation 14 known as the method of "damped least squares" has proven to be quite stable and reliable. This is achieved by replacing Equation 14 by

$$\Delta \mathbf{w} = - (J^T J + \alpha I)^{-1} J^T \mathbf{d} \quad (15)$$

where α is called the "damping factor" (*4*). When $\alpha = 0$, Equation 15 reduces to Equation 14. As α is increased, the step size $\Delta \mathbf{w}$ decreases, and for suitably large α convergence can be achieved.

(1) Friday, J. R., Smith, B. D., *A.I.Ch.E. J.* (1964) **10**, 698.
(2) Tierney, J. W., Bruno, J. A., *A.I.Ch.E. J.* (1967) **13**, 556.
(3) Tierney, J. W., Yanosik, J. L., *A.I.Ch.E. J.* (1969) **15**, 897.
(4) Feder, D. P., *Appl. Optics* (1963) **2**, 1209.

Physical Aspects of a Three-Phase System in a Stirred Tank Reactor

H. J. VAN DEN BERG, Unilever Research, Vlaardingen, The Netherlands

The principal physical aspects involved in a stirred tank reactor in which particles are suspended and gas is blown through are:

(a) Mass transfer across the gas–liquid interface, determined by the gas–liquid interfacial area and the mass transfer coefficient.

(b) Mass transfer between liquid and suspended particles.

(c) Power consumption and gas holdup.

The results of interfacial area measurements and those of mass transfer experiments with suspended particles are reported.

Methods

The gas–liquid interfacial area was measured using the oxidation of sodium sulfite solutions. The theoretical background of this method and the study of the mechanism and kinetics of the reaction have been the subject of several publications (*1, 2, 3*). We used the recent results of Reith, who employed $CoSO_4$ as a catalyst, and showed that the reaction is second order in oxygen and zero order in sulfite for sulfite concentrations higher than 0.4 kmole/m^3. Mass transfer between liquid and suspended particles was determined by the ion-exchange bead method used by Calderbank and Jones (*4*) and Harriott (*5*).

Experimental

The experiments were done in a Perspex vessel (diameter, 28.8 cm) provided with four vertical baffles (width, 2.88 cm) and a six-blade turbine stirrer. In all experiments the liquid filling height was equal to the vessel diameter, and the stirrer was placed at half the liquid filling height. The temperature in the vessel was controlled *via* an external cooling circuit, a conductivity cell being taken up in this circuit just below the vessel for mass transfer experiments. The vessel was further provided with pH electrodes, thermometers, and sampling valves. Speeds from zero to 15 rps could be reached with a hydraulic variator stirrer. Between the variator and the stirrer, a torque meter (Dr. Staiger, Mohilo and Co.) was taken up in the stirrer axis. The gas supply was metered with rotameters, the gas being saturated with water before being passed through the vessel. For interfacial area measurements the whole vessel was kept under slight pressure to force a continuous sample flow of outlet gas through a drier and an oxygen analyzer (Servomex, type OA 137).

Results

Interfacial Area. Because of the poor reproducibility of the measurements (the spread was about 15%), we performed several experiments in which we added particles to the sulfite–air system that had attained the stationary state. In these experiments no differences in outlet oxygen concentration were observed. The results are shown in Figure 1. Because of the large number of experimental points, only the average value and the spread are given. At low stirrer speeds, the interfacial area depends

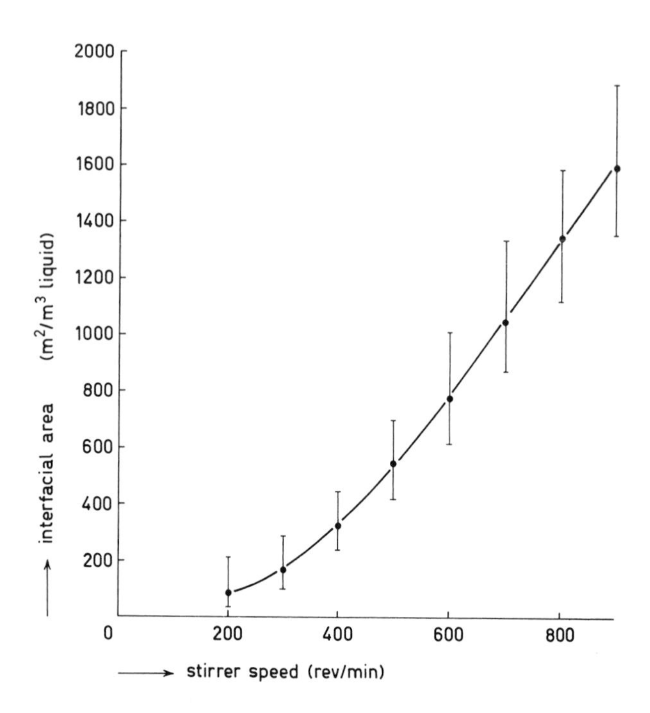

Figure 1. Interfacial area as a function of stirrer speed

on the gas flow rate. For stirrer speeds $\geqslant$ 500 rpm, the interfacial area depends only on the stirrer speed. The results compare fairly well with those of Reith (*1*).

Mass-Transfer Coefficients. Several authors have used Kolmogoroff's theory of isotropic turbulence to describe their results on mass transfer from suspended particles. They calculate a slip velocity for a particle in suspension and use this velocity in an equation for forced convection mass transfer from a stationary particle.

The expression for the slip velocity involves the power dissipation per unit mass of liquid. Since the power dissipation is influenced strongly by the presence of gas bubbles in the vessel, we thought that a correlation based on power dissipation could be attractive for a three-phase system.

For particles larger than the scale of the energy-dissipating eddies, the expression for the slip velocity reads (*6*):

$$v_s = C_1 \varepsilon^{1/3} d_p^{1/3}$$

If we use this expression in the Frössling equation, we get:

$$Sh = 2 + C_2 Re_s^{1/2} Sc^{1/3}$$

where

$$Re_s = \frac{\varepsilon^{1/3} d_p^{4/3}}{\nu}$$

The scale of the dissipating eddies η is given (7) by

$$\eta = 0.5\, \nu^{3/4} \varepsilon^{-1/4}$$

For water at room temperature and a power dissipation of 1 watt/kg we find for η about 15μm. This is well below the size of the smallest particles in our experiments.

The experimental results for one particle size are given in Figure 2. Since the influence of the gas velocity seems to be expressed well in terms of power input, we have given the mass transfer results for three particle sizes in Figure 3. All these results can be correlated according to:

$$Sh = 2 + 0.75\, Re_s^{1/2} Sc^{1/3}$$

with a spread of about 15%.

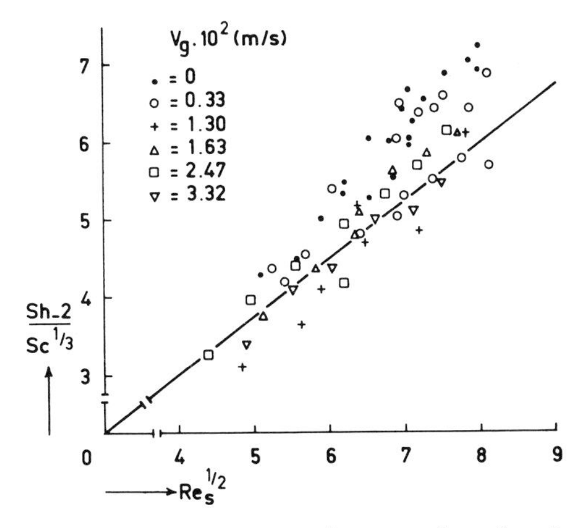

Figure 2. Mass transfer between liquid and particles of a diameter of 384 μm at various gas velocities

To calculate the *Sh* and *Sc* number, we needed a value for the diffusion coefficient. The so-called Nerst coefficient as discussed by Helfferich (8) was chosen.

Conclusions

The interfacial area between gas and liquid in a stirred vessel is not influenced by the presence of particles of diameters from 75 to 600 μm and in concentrations up to 4 wt %. The mass transfer between liquid

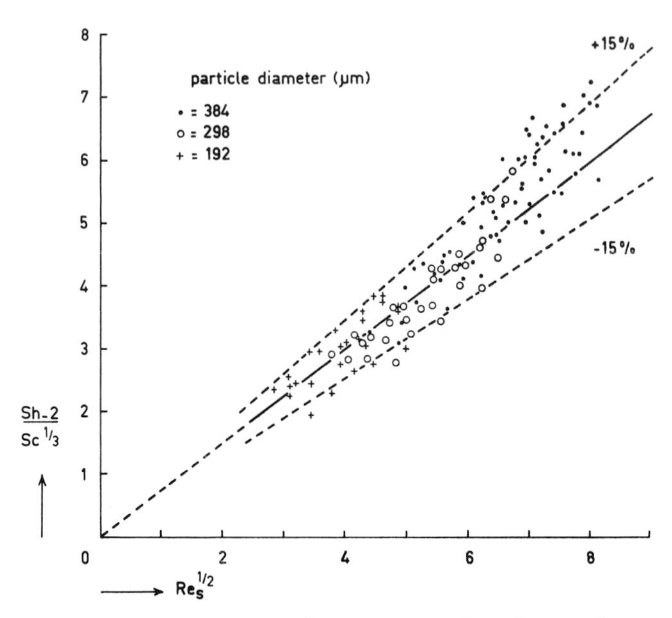

Figure 3. Mass transfer between liquid and particles of three different sizes

and suspended particles is influenced by the presence of gas. A good description of the results can be given on the basis of Kolmogoroff's theory of isotropic turbulence.

Nomenclature

d_s	diameter of stirrer, meters
d_p	diameter of particle, meters
C_1 C_2	empirical constants
k_1	mass transfer coefficients in the liquid phase, meters/sec
v_s	slip velocity of a particle, meters/sec
v_g	superficial gas velocity, meters/sec
n	stirrer speed, sec^{-1}
D	diffusion coefficient, $meters^2/sec$
ϵ	power dissipation per unit mass, watts/kg
η	microscale of turbulence
ν	Kinematic viscosity, $meters^2/sec$
$Sh = k_1 d_p / D$	Sherwood number
$Re_s = v_s d_p / \nu$	Reynolds number
$Sc = \nu / D$	Schmidt number

(1) Reith, T., Ph.D. Thesis, Delft, 1968.
(2) De Waal, K. J. A., Okeson, J. C., *Chem. Eng. Sci.* (1966) **21**, 559.

(3) Westerterp, K. A., Van Dierendonck, L. L., De Kraa, J. A., *Chem. Eng. Sci.* (1963) **18,** 157.
(4) Calderbank, P. H., Jones, S. J. R., *Trans. Inst. Chem. Eng.* (1961) **39,** 363.
(5) Harriott, P., *A.I.Ch.E. J.* (1962) **8,** 93.
(6) Shinnar, R., Church, J. M., *Ind. Eng. Chem.* (1960) **52,** 253.
(7) Towsend, A. A., *Proc. Roy. Soc. (London)* (1951) **A208,** 534.
(8) Hellerich, F. J., *Phys. Chem.* (1965) **69,** 1178.

Axial Mixing in Liquid-Liquid Spray Columns

J.-B. WIJFFELS[1] and K. RIETEMA, Eindhoven University of Technology, The Netherlands

Three 20-meter long operational spray columns of 70, 85, and 100 cm diameter and two 6-meter long laboratory columns of 15 and 45 cm diameter have been investigated. All measurements were carried out for density differences between continuous and dispersed phases smaller than 200 kg/m^3, dispersed phase holdup less than 8% and droplet Reynolds numbers greater than 100. The effective axial mixing coefficient of the continuous phase has been calculated from residence time distribution measurements. The circulatory behavior of spray columns has been studied in the laboratory equipment.

For all cases that have been subject of this research the motion of droplets through the continuous phase of the spray column is irregular. The spray column is treated as a turbulent system. The dispersion is considered to be a single fluid. Quantities that appear in the momentum balance such as pressure p and shear stress τ are considered to be averages over a volume element large compared with the dimension of the droplets. The relevant velocity that appears in the balance equation is the mass average velocity w. Accordingly, the shear stress is assumed to be proportional to the negative gradient of the mass average velocity. The dispersed phase is introduced at one end of the column and travels countercurrently to the continuous phase through the central region. For the sake of the model the column is roughly divided into a droplet free layer near the wall and a core of turbulent dispersion of constant dispersed phase holdup $(1 - \epsilon_k)$. A solution is sought for the steady-state operation of slender columns. No end effects are considered. It is assumed that the

[1] Present address: Koninklijke/Shell-Laboratory, Amsterdam, The Netherlands.

flow pattern is axisymmetrical, while the velocity is aligned with the axial direction, and that the pressure is constant over the cross section of the column. The model contains the core radius b as an undetermined constant which can be found by applying the condition that the pressure gradient is stationary. The special case of small density differences between the phases and low values of dispersed phase holdup is considered in detail. The simplified model for the case when the thickness of the wall layer is small compared with the radius of the column is given here; a more general treatment has been given in the Ph.D. thesis of J.-B. Wijffels.

The momentum balance for the axial direction reads:

$$\frac{1}{r}\frac{d}{dr}(r\tau) = -\frac{dp}{dz} - \rho g \tag{1}$$

where r is the distance from the axis of the column, z the distance in axial direction, ρ the density, and g the acceleration of gravity. Since the density of the core $\rho_k = \epsilon_k\rho_c + (1 - \epsilon_k)\rho_d$ is assumed constant, integration over the core yields:

$$2\tau_k/b = -\frac{dp}{dz} - \rho_k g \tag{2}$$

where τ_k is the shear stress at the core boundary. For the shear stress in the core,

$$\tau = \frac{r}{b}\tau_k \tag{3}$$

With the condition that the shear stress is continuous at the core boundary, integration over the column yields:

$$2\tau_w/R = -\frac{dp}{dz} - \bar{\rho}g \tag{4}$$

where τ_w denotes the shear stress at the wall. The average density in the column $\bar{\rho} = \epsilon\rho_c + (1 - \epsilon)\rho_d$, and the average dispersed phase holdup $(1 - \epsilon) = b^2(1 - \epsilon_k)/R^2$. To find a solution for the flow pattern it is assumed that:

$$\tau = -\rho\nu_t\frac{dw}{dr} \tag{5}$$

This may be regarded as a definition of the turbulent kinematic viscosity ν_t. The mass average velocity $w = x_c v + x_d v_d$ in which x_c and x_d are the mass fractions, and v and v_d are the linear velocities of the continuous and dispersed phase respectively. The major part of the wall layer is assumed to be in turbulent motion also, and since the wall layer is assumed to be small, its velocity is considered constant. The flow pattern is then given by:

$$w = 2(v_w - w_0)\left(\frac{r^2}{b^2} - \frac{1}{2}\right) + w_0 \text{ for } 0 < r < b \tag{6}$$

$$w = v = v_w \qquad \text{for } b < r < R$$

in which w_0 is the mass velocity averaged over the core. Directly adjacent to the wall, however, a turbulence-free layer exists with molecular viscosity, in which the velocity decreases rapidly to zero. This laminar sublayer is assumed to be very thin so that its contribution to the flow rate of the continuous phase may be neglected. The shear stress at the wall τ_w is assumed to be related to the velocity outside the laminar sublayer v_w according to

$$\tau_w = f\rho_c v_s v_w \tag{7}$$

in which f is the friction factor and the slip velocity $v_s = v - v_d$. By subtracting Equation 2 from 4 the momentum balance for the wall layer may now be written as

$$2f\rho_c v_s v_w \frac{b^2}{R} + 8\rho_k \nu_t (v_w - w_0) = (R^2 - b^2)(1 - \varepsilon)(\rho_d - \rho_c)g \tag{8}$$

Since $v_d = w - x_c v_s$, the flow rate of the dispersed phase, Q_d, equals:

$$Q_d = \pi b^2(1 - \varepsilon_k)(w_0 - x_c v_s) \tag{9}$$

The total mass flow rate is equated to zero or:

$$\pi b^2 \rho_k w_0 + \pi(R^2 - b^2)\rho_c v_w = 0 \tag{10}$$

To enable us to write the equations in a dimensionless form, a reciprocal Reynolds number, n, and a gravitational acceleration number, a, are introduced as

$$n = 4\nu_t/v_s R \tag{11}$$

$$a = (\rho_d - \rho_c)gR/2\rho_c v_s^2 = \frac{3}{8}\frac{R}{d}C_D \tag{12}$$

in which C_D is the so defined droplet drag coefficient, which is approximately equal to 1, and d is the droplet diameter. With the notation $u = w/v_s$ and $y = b/R$, Equations 10, 8, 9, and 4 are simplified, neglecting density differences and second-order terms $(y \sim 1)$ whenever they are immaterial to the discussion, to give respectively:

$$u_0 = -u_w(1 - y^2) \tag{13}$$

$$(n + f)u_w = a(1 - \varepsilon)(1 - y^2) \tag{14}$$

$$F = Q_d/\pi R^2 v_s = -(1 - \varepsilon)(1 + f_c) \tag{15}$$

and

$$G = R(-\frac{dp}{dz} - \rho_c g)/2\rho_c v_s^2 = a(1 - \varepsilon)(1 + g_c) \tag{16}$$

in which

$$f_c = a(1 - \varepsilon)(1 - y^2)^2/(n + f) - (1 - \varepsilon) \tag{17}$$

and

$$g_c = f(1 - y^2)/(n + f) \tag{18}$$

The pressure gradient is assumed to be stationary:

$$\delta G = 0 \tag{19}$$

or

$$\delta(1 - \varepsilon) = -(1 - \varepsilon)\frac{f}{n + f}\delta(1 - y^2) \tag{20}$$

The dispersed flow rate F is constant. Therefore,

$$\delta(1 - \varepsilon) = -2a(1 - \varepsilon)^2\frac{1 - y^2}{n + f}\delta(1 - y^2) \tag{21}$$

Because f_c and g_c are small compared with 1, their contribution to the left side of Equations 20 and 21 has been neglected. Furthermore the variations of v_s, f, n, and a have been assumed negligibly small. The solution is now easily found to be

$$1 - y^2 = f/2a(1 - \varepsilon) \tag{22}$$

and by means of Equation 14 it follows that

$$u_w = f/2(n + f) \tag{23}$$

For spray columns, to which the foregoing treatment of the circulatory behavior is applicable, an expression for the effective axial mixing coefficient can be derived. The fraction of the continuous phase β that is free to move independently of the droplets is distinguished from the fraction $(1 - \beta)$ that is kept in the wake of the droplets. If the length of the column L is very large compared with its radius and the time necessary for convection to make an appreciable change in concentration $L/2v_w^*$ is very large compared with $R^2/16\beta E_1$, the effective axial mixing coefficient is:

$$E_{ax,eff} = \frac{v_w^{*2}R^2}{48\beta E_1} + \frac{(1-\beta)}{U}\left(v_s^2 + \frac{1}{3}v_w^{*2}\right) + E_1 \tag{24}$$

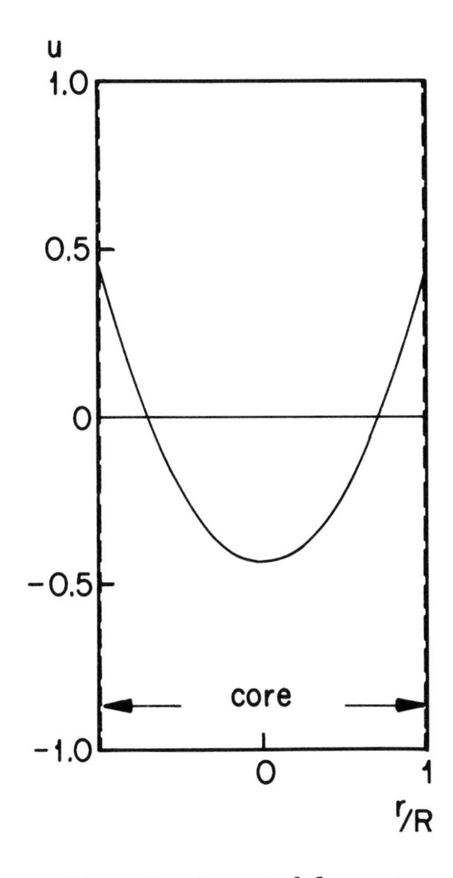

Figure 1. Computed flow pattern in heavy dispersed phase, R/d = *200,* $(1 - \varepsilon) = 0.04$. *Conditions are similar to those in the largest of the spray towers investigated.*

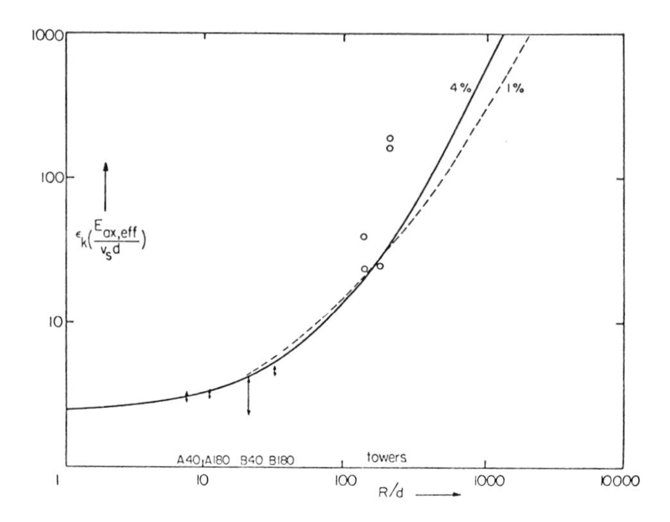

Figure 2. A general representation of the results for 4% dispersed phase holdup. Experimental results for the spray tower fat splitters are indicated by circles. The range of experimental values found for the laboratory columns A and B is indicated by arrows for density differences of 40 and 180 kg/m³. The continuous line is computed from Equation 24. For comparison the computed values for 1% dispersed phase holdup are represented by the dotted lines.

in which $v_w^* = v_w - w_0$, E_1 is the intrinsic diffusion coefficient of the free continuous phase, and U is the exchange coefficient that represents the fraction wake volume exchanged per unit time. The terms on the right side of Equation 24 describe the effects of the non-uniform flow pattern in the continuous phase, the translation of continuous phase carried in the wake of the droplets, and the intrinsic diffusion respectively.

Discussion

The flow patterns have been made visible by means of ink injections in two 6-meter long laboratory columns: A with a diameter of 15 cm and B with a diameter of 45 cm. The dispersed phase was a kerosene–trichloroethylene mixture; the continuous phase was fresh tap water. The flow patterns were measured by a movie camera for dispersed phase densities of 820 and 960 kg/m³. By comparison with the experimental results f was found to be 0.06. The thickness of the wall layer d_w may be calculated from Equation 22 to give $d_w/d = 1/25(1 - \epsilon)$. For small columns for which the hydraulic diameter of the dispersion $d_h = 2\epsilon d/$

$3(1 - \epsilon)$ is much smaller than the column radius, the kinematic viscosity could be estimated: $\nu_t = 0.0165\ v_w R$. For large columns it is expected that $\nu_t = v_s d$. In the general model a combined expression has been used. The circulation velocity was one-fourth of the slip velocity in the laboratory columns and should reach its maximum value of half the slip velocity in large columns. The computed flow pattern is shown in Figure 1.

The residence time distribution in the continuous phase was determined in the laboratory columns, using a saturated NH_4Cl solution as tracer material, as well as in three operational fat splitters by effecting a complete change in fat feed. The model is compared with the result in Figure 2. The exchange coefficient was estimated to be $Ud_h/v_s = 0.27$, and the intrinsic diffusion coefficient to be $E_1/v_s R = n$.

The predominant cause for axial mixing in small columns appears to be the translation by wake transport and in large columns the translation by circulatory flow. This treatment is valid only if the thickness of the wall layer is small compared with the radius of the column.

Acknowledgment

The permission granted by Unilever-Emery, Gouda N.V. to obtain the experimental data reported from fat splitters of this company is acknowledged.

Partial Wetting in Trickle Bed Reactors

W. SEDRIKS[1] and C. N. KENNEY, Department of Chemical Engineering, Cambridge University, Cambridge, England

Three-phase systems, in which reaction occurs on a solid catalyst in the presence of both gas- and liquid-phase reactants, have been used to carry out hydrogenation and oxidation reactions, either in a trickle bed, where liquid trickles over a fixed bed of catalyst in the presence of a gas, or in a slurry reactor, where the catalyst is suspended in liquid. The use of three-phase and in particular trickle bed reactors has been partly influenced by the limited understanding of not only the complex hydro-

[1] Present address: Shell Development Co., Emeryville, Calif.

dynamics, but also the mass transfer and diffusional effects which arise in such systems. To obtain further information on the influence of liquid-phase pore diffusion and of the importance of partial catalyst wetting in trickle beds, we recently studied the hydrogenation of crotonaldehyde to *n*-butyraldehyde at near ambient conditions over a catalyst consisting of palladium deposited on a porous pelleted alumina. The investigations consisted of three separate studies.

The apparent intrinsic kinetics of the reaction were first determined. The catalyst was ground to a fine powder. Reaction rates were measured for a semibatch stirred-cell reactor in which the catalyst was suspended in the liquid reactant. The absence of mass transfer limitation was verified.

The apparent intrinsic kinetics were found to be first order with respect to the concentration of hydrogen in the liquid and zero order with respect to the crotonaldehyde concentration. The apparent activation energy was approximately 11 kcal/gram mole at atmospheric pressure and at temperatures in the range 30°–70°C.

Liquid-phase pore diffusion in the pelleted catalyst under reaction conditions was investigated next. The stirred cell was again used, whole catalyst pellets being placed in wire mesh baskets attached to the impeller and swept through the liquid. With this arrangement resistance to mass transfer to the outside of the pellets could be essentially eliminated. Rates of reaction were measured and compared with the intrinsic rates.

Even though the palladium was deposited only within a relatively thin shell on the outside of the pellets, diffusion of hydrogen through the liquid-filled pores limited the rate of reaction markedly. Effectiveness factors of approximately 0.1 were measured. Consistent with these, the apparent activation energy was 7 kcal/gram mole, and the reaction remained first order with respect to hydrogen concentration in the liquid. Assuming that transfer in the pores took place by bulk diffusion, a tortuosity factor of 1.6 was calculated for the catalyst.

Lastly, the effects of partial wetting of the catalyst were examined in a trickle bed reactor (4.3 cm, id) in which liquid flowed over a shallow bed of catalyst with co-current flow of gas. Partial wetting was not difficult to achieve. In fact, with 3/16-inch pellets in the column (P.T.I. runs) wetting appeared far from complete even at the highest value of liquid flow rate. Under these conditions approximately 50–60% of the pellets were dry, and the degree of wetting increased only slightly with flow rate. Wetting was much more extensive when the catalyst pellets were interspersed with inert 8–16 mesh α-alumina granules, being effectively complete with preflooding and *ca.* 90% without preflooding (P.T.2 runs). Measured rates are shown in Figure 1, where, for comparison, the results obtained with the pellets and powder are also given.

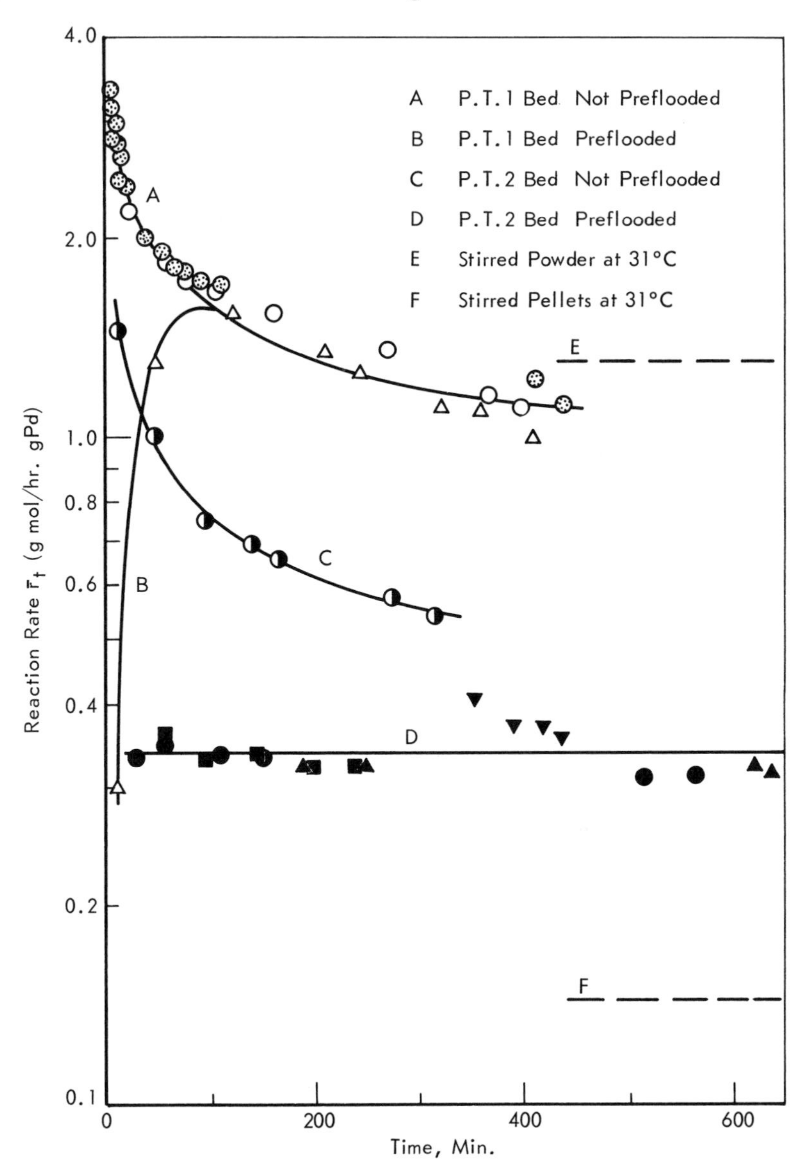

Figure 1. Reaction rates in pellet trickle beds

The most striking observation is that a higher reaction rate (gram moles/hr(grams Pd)) is obtained when the catalyst pellets are not completely covered with liquid; hence, the existence of a gas-phase reaction on the solid catalyst occurring simultaneously with the liquid-phase reaction seemed probable. This was substantiated by measuring the rate of hydrogenation in a dry bed with a single active pellet in which the feed consisted of hydrogen saturated with crotonaldehyde vapor.

Provided the species required for the reaction on the solid catalyst are the same in both liquid and gas phases, and the phases are in equilibrium, then since the chemical potentials must be the same in both phases, the reaction rates on wet and dry catalyst must be identical. Any

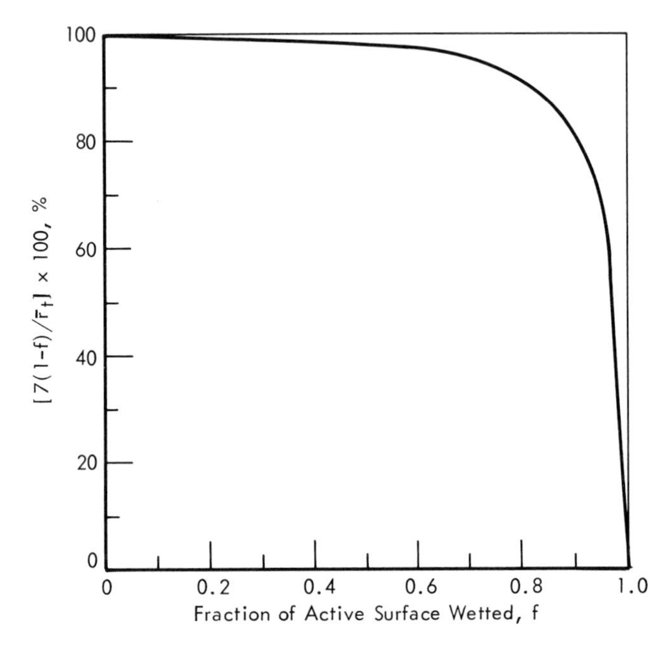

Figure 2. Gas-phase contribution (%) to over-all reaction rate in partially wetted trickle bed

differences must be caused by lack of equilibrium—*i.e.*, by differences in mass transport rates or temperature. Since mass transport in the gas phase will usually be much faster, in general the reaction rate in a trickle bed must therefore be expressed in some form that acknowledges the separate contributions of the wet and dry catalyst—*e.g.*, a simplified expression that could serve as a first approximation would be:

$$r_t = fr_L + (1 - f)r_G$$

where f is the total fraction of active catalyst surface covered by liquid, and r_L and r_G are the rates on wetted and dry catalyst respectively. Only rarely can the r_G contribution be neglected.

To illustrate this from the present work, the rate measured with the single dry pellet was *ca.* 7 gram moles/hr(grams Pd) for a bulk bed temperature of 31°C. Calculations showed that no pore diffusion limitation was present but that the interior of the pellet may well have been at a somewhat higher temperature. The comparable rate for the submerged pellets where pore diffusion was shown to be the limiting step was 0.15 gram mole/hr(gram Pd). As a rough approximation, the above values may be taken as the values for r_G and r_L in the trickle bed, and the fraction of the gas-phase contribution to the total rate plotted *vs.* the wetted fraction f from the above equation. This plot is shown in Figure 2 and illustrates the conclusion that except at very high values of f, the gas-phase contribution completely dominates the total rate in the trickle beds. In the light of this, the slow but marked decline in reaction rate with time observed in the trickle bed experiments is explained by gradual seepage of liquid into the pores of initially unwetted pellets. The fact that the gas-phase contribution has almost always been discounted in trickle bed studies can explain some of the discrepancies observed.

In theory, because evaporation is a relatively fast process, one should be able to design partially wetted beds to operate closer to intrinsic rates without losing the advantages that operation at "liquid phase" temperatures sometimes offers—*i.e.*, such a reactor would combine the functions of a gas–solid reactor and a gas–liquid contactor, the liquid acting mainly as the source of reactants and sink for the products. The problems associated with the requirement of good liquid distribution and control of the reactor may, however, normally require operation at close to total wetting in any practical reactor design.

Product Distribution Problems in Gas-Liquid Reactors

T. W. F. RUSSELL and R. ROTHENBERGER, University of Delaware, Newark, Del. 19711

The design and analysis of gas–liquid reactors is, at best, a complex problem of considerable difficulty. In the most general case, one must deal with simultaneous energy, mass, and momentum transport. Rigorous design calculations require consideration of these phenomena on both the microscopic and the macroscopic levels. Solution of the resulting coupled equations is invariably an awesome task.

In practice, one introduces a number of assumptions to make the design more tractable. The concept of the so-called "ideal" reactor is one such simplification whose utility is well established. Analysis of such reactors allows one to establish limits on the performance of an actual process. In addition, conclusions drawn from such analyses are also applicable to "non-ideal" reactors. These points are discussed in considerable detail by Russell and co-workers (*1, 2, 3*).

Even these simplified models contain numerous unknown parameters. A meaningful design, therefore, begins with an experimental program to obtain kinetic and mass transfer data and information concerning the basic fluid mechanics. Knowledge of the gross fluid mechanics in the particular piece of equipment is essential for rational design. Unfortunately, it is this area which is probably the least understood in gas–liquid reactors.

It is possible, however, to distinguish various configurations which may exist in ideal reactors. Two-phase reactors, then, can be adequately designed by analyzing reactors displaying different behavior in terms of the gross fluid mechanics of each phase. The application of this approach to tank-type and tubular reactor design has been discussed by Schaftlein and Russell (*3*) and Cichy *et al.* (*1*). To consider the question of product distribution in gas–liquid reactors, we restrict our attention to the (a) plug-flow gas–well-mixed liquid and the (b) well-mixed gas–well-mixed liquid continuous flow tank type (CFTR) configurations. In the case of tubular reactors we consider (a) continuous fluid phases with a well-defined interface and (b) discrete gas-phase units in a continuous liquid phase. The reactions are of the following types which are representative of a large class of industrial reactions of pragmatic interest: Competitive-consecutive:

$$A + B \rightarrow R$$

$$R + B \rightarrow S$$

$$S + B \rightarrow T$$

Parallel:

$$A + B \rightarrow R$$

$$A + B \rightarrow S$$

B is the volatile reactant and A, R, S, and T exist in the liquid phase only.

For each reactor configuration described above the appropriate form of the reaction rate expressions appearing in the liquid-phase component mass balances depends upon whether the reaction occurs in the bulk liquid or in a narrow region adjacent to the gas–liquid interface. To determine where the reaction occurs, it is necessary to refer to Astarita's (*4*) classification of reaction regimes.

Astarita (*4*) defines two characteristic times for the reacting systems —*i.e.*, the time for reaction to proceed to an appreciable extent and the time a liquid volume element is exposed to the interface. Values of these two parameters and the concentrations of reactants in the liquid phase allow determination of the chemical absorption regime. There are three major regimes: slow, fast, and instantaneous. Several characteristics of each regime are relevant to the product distribution problem. In the slow regime reaction occurs in the bulk liquid, and in the fast regime reaction occurs in a thin film adjacent to the interface. The important point is that in both cases there is no concentration gradient of the liquid-phase reactant through the film. In the instantaneous regime, on the other hand, the liquid-phase reactant and all products of the reaction are no longer considered to have uniform concentration profiles near the interface.

In the slow reaction regime the reaction rate expression is written in terms of bulk concentrations. In the fast regime, however, it is convenient to define the rate of reaction on the basis of interfacial area. In so doing, we use a surface concentration for the volatile reactant and are essentially modeling the reaction as if it occurred at the interface. This type of rate expression is useful for experimental studies in the fast reaction regime.

If we consider all the reactions discussed previously as being elementary, the product distribution expressions are independent of the volatile reactant concentration. This means that the product distributions for a given reacting system and reactor configuration will be identical for both the slow and fast reaction regimes.

Solution of the component mass balance relations for both the competitive-consecutive and parallel reactions yields expressions for the product concentrations which are identical to those obtained for homogeneous reactions (*2*). This is a consequence of the condition that for the slow and fast reaction regimes the composition of the liquid phase is uniform except for the volatile reactant. Therefore, for reactions of the type considered here occurring in the slow and fast reaction regimes, the homogeneous and heterogeneous product distributions are identical for the same conversion of liquid phase reactant.

For reactions occurring in the instantaneous regime or in the region between the fast and instantaneous regimes, the single-phase and two-phase selectivities will differ; however, an examination of industrial reactions whose primary purpose is the production of a saleable product indicates that practically all these reactions occur in the slow or fast reaction regimes. Instantaneous reactions are frequently acid–base type and are encountered most often in recovery operations rather than product manufacturing. Computational studies of some reactions of pragmatic interest indicate that for a large class of reactions equilibrium between the gas and liquid is readily achieved.

This simplifies the process analysis considerably and allows comparisons between tubular and tank systems on the basis of reactor size and the effect of reactor type on the attendant process equipment. Although most gas–liquid reactions are carried out industrially in tank-type devices, analysis of the total process synthesis indicates some distinct advantages of tubular gas–liquid systems.

(1) Cichy, P. T., Ultman, J. J., Russell, T. W. F., *Ind. Eng. Chem.* (Aug. 1969) **61,** 7.
(2) Russell, T. W. F., Buzzelli, D. T., *Ind. Eng. Chem., Process Design Develop.* (Jan. 1969) **8,** 2.
(3) Schaftlein, R. W., Russell, T. W. F., *Ind. Eng. Chem.* (May 1968) **60,** 12.
(4) Astarita, G., "Mass Transfer with Chemical Reaction," Elsevier, Amsterdam, 1967.

7

Catalyst Deactivation

JOHN B. BUTT

Department of Chemical Engineering, Northwestern University,
Evanston, Ill. 60201

Research on catalyst deactivation has been concerned with three generally distinguishable areas: (1) investigation of the fundamental mechanisms and the kinetics of deactivation processes, (2) determination of deactivation rates in particulate catalysts, and (3) study of deactivation effects on the operation of real reactor systems. These areas range from the microscopic to the macroscopic and involve a large range of interests and techniques. Catalyst deactivation can be described kinetically by the same forms as conventional kinetics so that chemical reaction engineering problems associated with deactivation can normally be thought of as involving time variant systems in which the interactions of rates of reaction, rates of transport (diffusive and convective), and rates of deactivation determine system behavior. Obviously compromise will be an intimate part of catalytic processes faced with time-dependent activity; hence, a number of significant optimization problems arise in analyzing large scale process systems.

While the phenomenon of catalyst deactivation has been recognized for most of the history of catalysis, our concern with it in terms of systematic observation and interpretation extends back only about 30 years. In 1951 Maxted (*1*) presented a landmark review of work on the poisoning of metallic catalysts to that time, and it was only a few years earlier that Voorhies (2) had reported his now-famous correlation of coke deposition rates on natural and synthetic cracking catalysts. Since then there have been many studies of various deactivation problems, but one cannot escape the general feeling that the bulk of such problems are still imperfectly understood or not yet explored. In such a sense, then, catalyst deactivation appears to be one of the darker areas of an art (catalysis) which, more often than not, assumes various magical (and always mysterious) characteristics.

The following review presents a representative selection (no attempt has been made to be comprehensive) of work on various aspects of deactivation. Four general sections are discussed which correspond generally to the natural divisions of the subject matter. First considered are some of the basic chemical and/or physical mechanisms of deactivation together with their probable intrusions into the natural activity or selectivity of the catalyst. Several specific chemical systems are used to illustrate these mechanisms, and the observed results are formulated in terms of generalized reaction schemes which allow exposition of the formal kinetics of heterogeneous reactions subject to catalyst deactivation. Secondly, observed kinetics and theoretical studies dealing with the deactivation of single particles are considered, using the reaction models derived in the first section. Such analysis corresponds to the determination of point rate and selectivities, which would then be used in reactor design in a manner similar to analogous studies of mass and heat transfer rate limitations. In the third section the results of intrinsic and individual particle deactivation behavior are incorporated into macroscopic models for chemical reactor design and analysis, and the composite behavior of process systems is examined. Parametric effects on activity and selectivity are of particular interest, and the implementation of various operational possibilities in this regard are discussed. This leads naturally to the final section, which considers a number of optimization problems concerning chemical reactor design or operation in the face of catalyst decay. This area is particularly active in chemical reaction engineering at present.

Microscopic Events and their Models: Kinetics of Reactions with Catalyst Deactivation

While it may not seem much of a problem at first glance, one of the more difficult things in catalysis is to define a useful and representative measure of activity. In general this requires some comparison among various states of the same catalyst if deactivation problems are to be considered as well. Some measures of activity are:

(a) Temperature required for a given conversion
(b) Temperature for a given product quality
(c) Conversion achieved
(d) Space velocity required for a given conversion at set temperature
(e) Reaction rate
(f) Rate constants or parameters extracted from kinetics studies.

Most of these measures are related to data obtainable from laboratory reactor investigation, and completely different conclusions regarding activity can be reached from several different measures. The second measure, for example, really may depend more on the selectivity than the activity of a catalyst while for the last two careful and tedious experimentation is required, and uncertainties of interpretation are involved in reaction kinetics studies. Indeed, when catalyst deactivation is occurring, the task of the kineticist is even more formidable since under these conditions the activity (and other properties) of the catalyst are functions of the entire history of the catalyst, including preparation, handling, storing, pretreatment, regeneration, and specific poisoning processes. When the reaction rate functional can be expressed as the product of two terms, kinetic dependencies which are time independent and activity dependencies which are not, the rate equation is termed separable (*3*), and one actually requires two equations to describe the over-all kinetics: the instantaneous rate equation and some expression relating the time dependency of activity. Fortunately, many real catalytic deactivation mechanisms, at least those which may be described chemically, do lead to separable rate equations, for there would be little one could do in analysis if this were not so. The separable form, thus, allows the expression of an instantaneous rate of reaction as the product of individual factors which are independent of one another. For example, one might write the product of three terms, one containing a concentration dependence, one a temperature dependence, and one the activity (previous history) dependence:

$$r_T = r_1([C])r_2(T)r_3(s) \tag{1}$$

in which C represent concentration, T temperature, and s the activity. Equation 1 must be supplemented with some description of the time variation of s:

$$r_s = r_4([C]r_5(T)r_6(s) \tag{2}$$

where the individual rate factors are again taken to be separable. The forms which the individual rate factors in Equations 1 and 2 assume are well known; these generally would involve some power-law dependence on concentration (r_1 and r_4), an Arrhenius temperature dependence (r_2 and r_5), and perhaps a power law dependence on activity (r_3 and r_6). If more complicated kinetics are involved, such as Langmuir-Hinshelwood forms, the separable form of equation can still be maintained as long as adsorption coefficients are not affected by the deactivation.

At this point it might be well to look further into some detailed mechanisms of typical deactivation processes, with particular attention to determination of whether these separable rate/deactivation equations are reasonable forms in view of the microscopic chemical or physical events occurring. We will in general adhere to a broad division of deactivation processes into three general classes. These three are:

(a) Poisoning: loss of activity caused by strong chemisorption of some impurity, normally contained in the reacting mixture.

(b) Fouling: loss of activity caused by reactant or product degradation on the catalyst surface; coke formation is the most important example.

(c) Aging: loss of activity caused by sintering; decrease of active surface.

The first two are chemical in nature, and particularly in the first case detailed and elegant studies of specific systems have been carried out. Aging is usually thought of as a physical process, though our approach to the analysis of this problem will be analogous to the chemical mechanisms, and we will see that sintering may not always be independent of the chemical events occurring.

Catalyst Poisoning. Perhaps the most elaborate and comprehensive work in the general field of catalyst deactivation has been concerned with studies of individual systems, catalyst–reaction–poison, which can be well characterized chemically. Many of these investigations have not only dealt with specific problems associated with poisoning but have revealed much information concerning the catalytic chemistry of the reaction/catalyst system involved—*i.e.*, what specific chemical properties of a given substance make it a good catalyst for a certain type of reaction.

Four such studies are reviewed here, chosen either because of the diversity of chemistry involved, the importance of the reaction involved, the depth of the information provided concerning the catalytic properties of the material involved, or all three. The first is the work of Maxted (*1*) concerning the poisoning of metallic catalysts by typical metallic and nonmetallic substances, and the importance of *d* electrons in the formation of these strongly chemisorbed poisons. Second is the work of Mills, Boedecker, and Oblad (*4*) on determination of the acid nature of silica–alumina cracking catalysts and its relationship to their activity *via* poisoning experiments with alkali metals and basic nitrogen compounds (with note of the many subsequent investigations of similar catalysts by these techniques). Third are the results of Pines and Haag (*5*) demonstrating the variations possible in acid site strength on alumina and the dependence of the catalytic properties of a given alumina on this site acidity distribution by measurement of alkali metal, ammonia, water, and trimethylamine poisoning effects on several test reactions. Finally we consider briefly the elucidation of the role of surface oxide on the activity of nickel oxide in oxidation reactions by Parravano (*6*) through inhibition of the activity of fresh oxide surface with CO chemisorption.

Metallic Catalysts. Most of the basic information compiled by Maxted is an identification of metals susceptible to poisoning and the primary materials responsible for such poisoning. The bulk of the background here is experience with metals used as hydrogenation catalysts, though with some added information from Fischer-Tropsch synthesis catalysts (cobalt) and ammonia synthesis (iron). Accordingly, the range of metallic catalysts considered is somewhat limited, but the results are impressive nonetheless. Figure 1 lists those metals judged most susceptible to poisoning, arranged as they appear in the periodic series. These consist predominantly of group VIII metals with their group Ib relatives. Silver and gold are included provisionally since they have very small activity for ordinary hydrogenation.

The common poisons for these metals can be classified into three groups:

(a) Molecules containing elements of groups Vb and VIb, as listed in Figure 1, or the free elements (Not N_2)

(b) Compounds of catalytically toxic metals and metallic ions.

(c) Molecules containing multiple bonds such as CO or strong sorbates such as benzene.

In many cases the effects listed under group (c) above can be accounted for by proper adsorption inhibition terms in the rate equation, and thus do not count as true poisoning effects.

For molecules containing group Vb or VIb elements, the degree of toxicity is related directly to the state of the toxic element in the molecule. If the potential poison has its normal valency orbitals saturated by stable bonding to other elements in the molecule, there is no toxic activity. If there are, however, unshared electron pairs or empty valency orbitals, then chemisorptive bonding to the metal is possible and the material is a catalyst poison. These contrasting states are illustrated in Figure 1 for some typical examples showing the electronic configuration of the toxic element. Most of these results pertain to observations made in hydrogenations with Pd, Pt, or Ni; under severe reducing conditions some of the nontoxic structures can be converted to toxic forms; particularly susceptible in this regard are arsenic and antimony compounds which are generally easily transformed into toxic arsine and stibine under hydrogenating conditions. Under other conditions such transformations may not occur, so it is quite possible for a substance to poison a catalyst for one reaction but not another. Maxted cites the example of arsenic compounds, which are general poisons for platinum in hydrogenation reactions, *via* arsine

$$\left[\begin{array}{c} \mathrm{H:\overset{\cdot\cdot}{\underset{\cdot\cdot}{As}}:H} \\ \mathrm{H} \end{array}\right]$$

formation, yet have no effect on catalytic activity in the decomposition of hydrogen peroxide, presumably by remaining in a "shielded" or saturated form such as the arsenate under strong oxidizing conditions

$$\left[\begin{array}{c} \mathrm{O} \\ \mathrm{O:\overset{\cdot\cdot}{\underset{\cdot\cdot}{As}}:O} \\ \mathrm{O} \end{array}\right]^{3-}$$

Susceptible Metals – Hydrogenation

Catalysts Arrayed in Periodic Series

Fe (26)	Co (27)	Ni (28)	Cu (29)
Ru (44)	Rh (45)	Pd (46)	[Ag] (47)
Os (76)	Ir (77)	Pt (78)	[Au] (79)

Nonmetallic Poisons

Group Vb	Group VIb
N	O
P	S
As	Se
Sb	Te

Summary of Toxicity of Metallic Ions

Metal ions tested					Electronic occupation of external orbitals	Toxicity towards platinum
Li^+	Be^{2+}				No *d* shell.	Nontoxic
Na^+	Mg^{2+}	Al^{3+}			No internal *d* shell.	Nontoxic
K^+	Ca^{2+}				3*d* ○ ○ ○ ○ ○ | 4*s* ○	Nontoxic
Rb^+	Sr^{2+}		Zr^{4+}		4*d* ○ ○ ○ ○ ○ | 5*s* ○	Nontoxic
Cs^+	Ba^{2+}	La^{3+}			5*d* ○ ○ ○ ○ ○ | 6*s* ○	Nontoxic
		Ce^{3+}				
			Th^{4+}		6*d* ○ ○ ○ ○ ○ | 7*s* ○	Nontoxic
Cu^+	Zn^{2+}				3*d* (:) (:) (:) (:) (:) | 4*s* ○	Toxic
Cu^{2+}					3*d* (:) (:) (:) (:) ⊙ | 4*s* ○	Toxic
Ag^+	Cd^{2+}	In^{3+}			4*d* (:) (:) (:) (:) (:) | 5*s* ○	Toxic
			Sn^{2+}		4*d* (:) (:) (:) (:) (:) | 5*s* (:)	Toxic
Au^+	Hg^{2+}				5*d* (:) (:) (:) (:) (:) | 6*s* ○	Toxic
	Hg^+				5*d* (:) (:) (:) (:) (:) | 6*s* ⊙	Toxic
		Tl^+	Pb^{2+}	Bi^{3+}	5*d* (:) (:) (:) (:) (:) | 6*s* (:)	Toxic
Cr^{2+}					3*d* ⊙ ⊙ ⊙ ⊙ ○ | 4*s* ○	Nontoxic
Cr^{3+}					3*d* ⊙ ⊙ ⊙ ○ ○ | 4*s* ○	Nontoxic
Mn^{2+}					3*d* ⊙ ⊙ ⊙ ⊙ ⊙ | 4*s* ○	Toxic
Fe^{2+}					3*d* (:) ⊙ ⊙ ⊙ ⊙ | 4*s* ○	Toxic
Co^{2+}					3*d* (:) (:) ⊙ ⊙ ⊙ | 4*s* ○	Toxic
Ni^{2+}					3*d* (:) (:) (:) ⊙ ⊙ | 4*s* ○	Toxic

Figure 1. Poisons for

Influence of Electronic Configuration on Toxicity

Toxic types	Nontoxic type (shielded structure)
H : S : H — Hydrogen sulfide; H : P : H (with H) — Phosphine	$[O : P : O (O, O)]^{3-}$ — Phosphate ion
$[O : S : O (O)]^{2-}$ — Sulfite ion (also selenite and tellurite)	$[O : S : O (O, O)]^{2-}$ — Sulfate ion (also selenate and tellurate)
(R)C : S : H — Organic thiol	(R)C : S : OH (O, O) — Sulfonic acid
(R)C : S : C(R′) — Organic sulfide	(R)C : S : C(R′) (O, O) — Sulfone
Pyridine (toxic)	Pyridinium ion (nontoxic, shielded structure)
Piperidine (toxic)	Piperidinium ion (nontoxic, shielded structure)

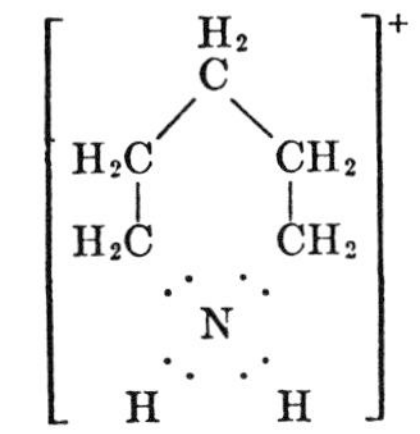

Piperidinium ion
(nontoxic, shielded structure)

Advances in Catalysis

metallic catalysts (1)

The toxicity of nitrogen compounds for hydrogenations has also been interpreted in the same manner, although these (as the example pyridine in Figure 1) are not as pronounced in their effect as materials such as arsine or thiophene. The shielding theory is well supported also by results with dry ammonia

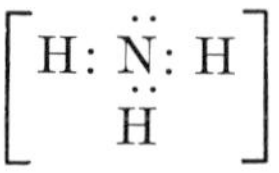

$$\left[\begin{array}{c} \text{H} : \ddot{\underset{\cdot\cdot}{\text{N}}} : \text{H} \\ \text{H} \end{array}\right]$$

which was found to poison the hydrogenation of cyclohexene in a non-aqueous environment where the nontoxic ammonium ion could not be formed.

The toxicity of metals or metallic ions towards the catalytic activity of metals has been thoroughly surveyed by Maxted and Marsden (7) with respect to the hydrogenation activity of platinum as a test reaction. The interpretation which they suggest, also summarized in Figure 1, emphasizes the existence of some connection between the toxicity of a metallic ion and the structure of its *d* band. Theories invoking the nature of the *d*-electrons in explanation of various facets of chemisorption on metals are nothing new, and it is probably correct to view them as providing one with valuable qualitative insight into the nature of the bond without necessarily being quantitatively correct. From the results of Maxted and Marsden one concludes that those materials which are toxic all have occupied *d* orbitals; when this does not occur, there is no poisoning effect. These observations also hold for other derivatives of the metals. Although in the latter case it cannot be ruled out that occupied *s* or *p* levels might take part in chemisorptive bonding (these levels being vacant in the metallic ion), metals are found in compounds with such higher levels occupied in stable bond formation, yet the compounds are quite toxic. A good example is that cited of tetramethyllead, where the Pb configuration corresponding to Figure 1 is:

Pb | ⦂ ⦂ ⦂ ⦂ ⦂ || ⊙ ⊙ ⊙ ⊙

$5d$ Hybridized $6sp^3$

and the hybridized $6sp^3$ levels are bonded to carbon in the molecule. This is a strong poison for hydrogenation with platinum, yet there is no opportunity for the *s* and *p* orbitals to take part in chemisorptive bonding.

The conclusion is that such metallic compounds or the ions themselves are involved *via* the *d*-shell in the formation of chemisorbed complexes resembling intermetallic species and that the nature of this bonding (and consequently poisoning behavior) is thus fundamentally different

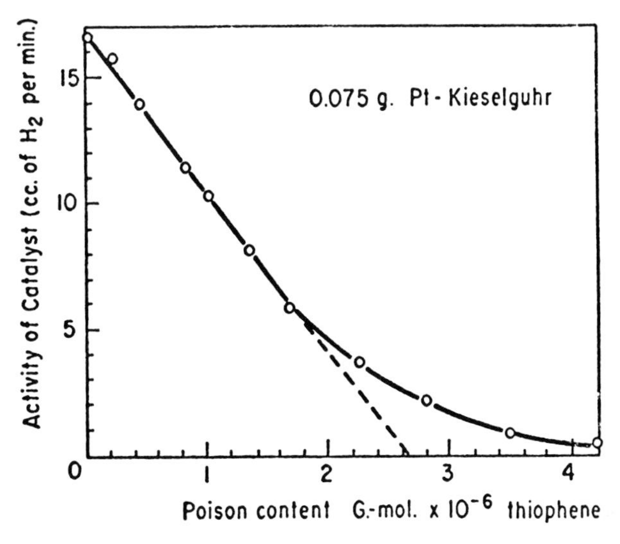

Figure 2a. Form of a typical poisoning curve. The figure shows the depression of the activity of a supported platinum catalyst, by increasing amounts of thiophene, in the liquid-phase hydrogenation of crotonic acid (1).

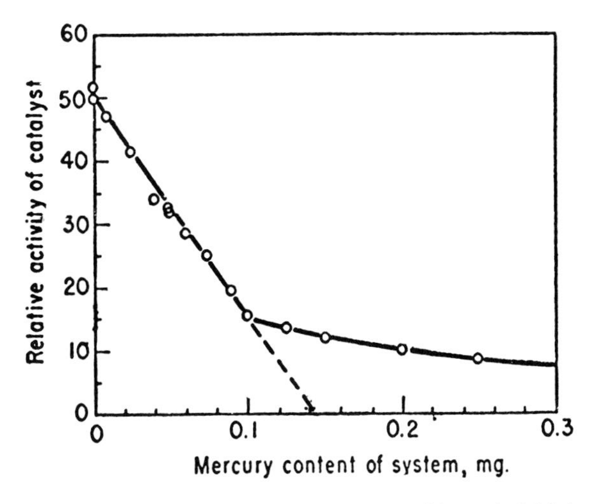

Figure 2b. Poisoning curve for a platinum catalyst, poisoned by mercury ions, in the decomposition of hydrogen peroxide (1)

from the group Vb–VIb poisons which can reasonably be viewed as bonding *via* coordination with *s* or *p* valence electrons.

While further discussion of the details of surface bonds involved in chemisorption would be desirable at this point, it is a luxury which space will not permit. An introduction to some of the theory and much of the available data is given by Hayward and Trapnell (*8*) regarding chemisorption on metals. More recently, also, molecular orbital descriptions have been used to analyze bonding at metallic surfaces; an excellent example, directed toward ethylene and CO bonding on Ni (related to category *c* of the poisons listed by Maxted), is the study of Bond (*9*).

Quantitative relationships between the amount of deactivation and the catalytic activity are relatively easily obtained when poisoning (as contrasted with coking or sintering) is the mechanism of deactivation. Maxted presented a number of these correlations, relating activity to

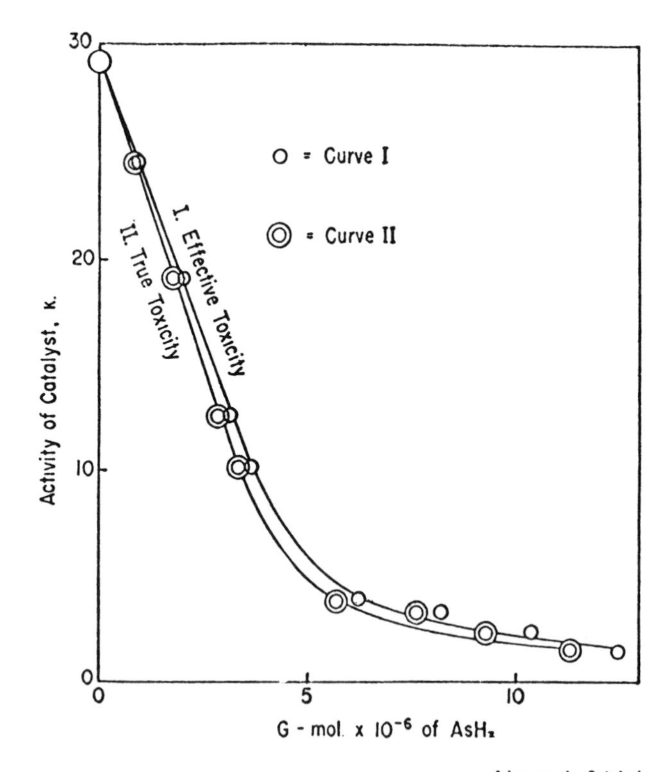

Figure 2c. Effective and true poisoning graphs for a platinum catalyst (0.05 g) poisoned with AsH_3. Curve I (effective toxicity) is based on the total poison present in the system. Curve II (true toxicity) is based on the amount of poison actually adsorbed on the catalyst (1).

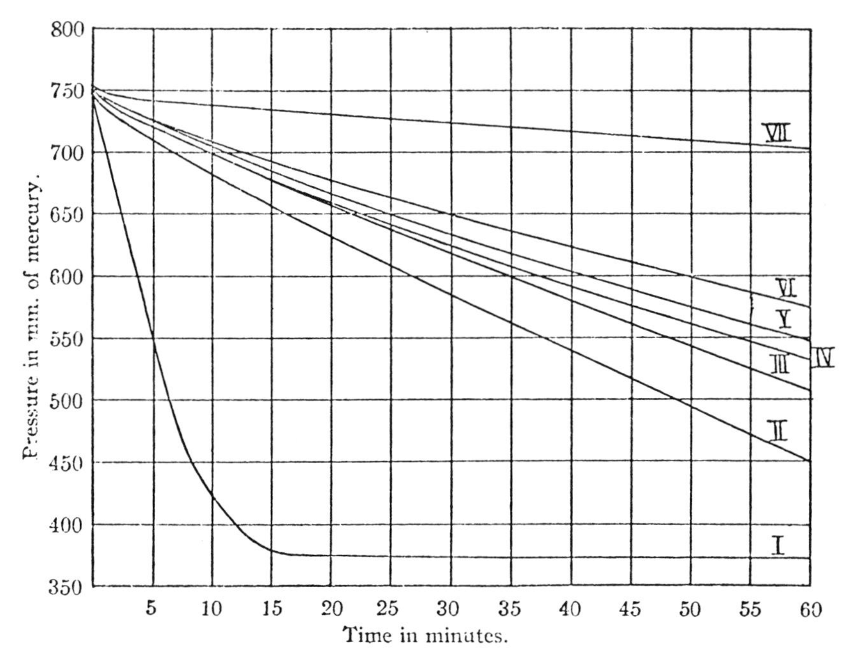

Journal of the American Chemical Society

Figure 3a. Velocity measurement on ethylene-hydrogen mixtures at 0° in the presence of carbon monoxide. Curves I—no CO; II—0.05 cc CO; III—0.08 cc CO; IV—0.33 cc CO; V—0.69 cc CO; VI—1.97 cc CO; VII—9.14 cc CO (10).

amount of poison present, as illustrated on Figure 2. This form of correlation, however, had been used for quite a while before the work of Maxted; Figure 3 gives a much earlier correlation of this type, describing the inhibition of ethylene hydrogenation on Cu by CO chemisorption (*10*). Obviously in some cases the activity is affected strongly by only small amounts of poison—*e.g.*, AsH_3–Pt (Figure 2) or CO–Cu (Figure 3) —with the effect tapering off at higher poison concentrations. In several of these cases the effect is linear over a substantial portion of the range so that one might represent activity as:

$$s = s_o - \alpha_1 C_p \tag{3}$$

in which s_o is the original activity, α_1 a poisoning coefficient representative of the effect on activity per unit of poison, and C_p the concentration of poison. Such a linear relationship implies a homogeneous or nonspecific removal of activity from the system, and such poisoning has been occasionally referred to as "nonselective." A fine example of a completely

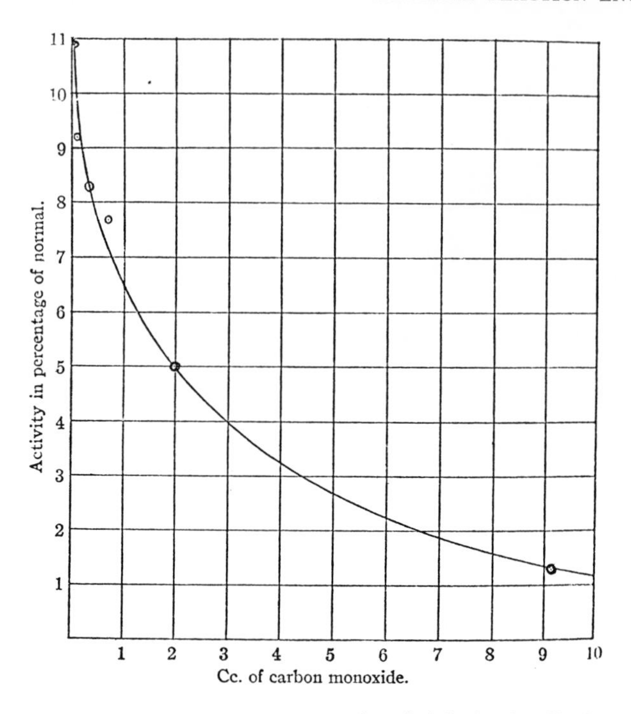

Journal of the American Chemical Society

Figure 3b. Activity of the catalyst as a function of the volume of carbon monoxide introduced (10)

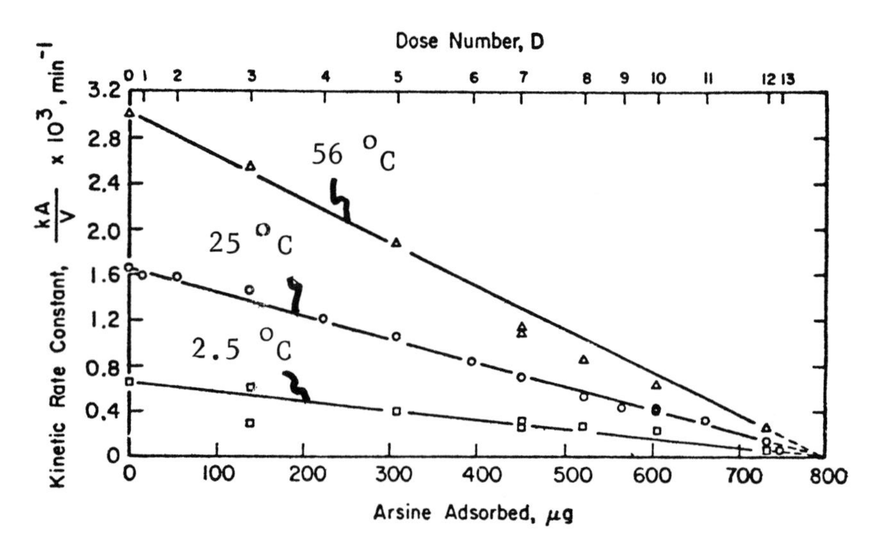

Journal of Catalysis

Figure 4a. Kinetic rate constants as a function of the quantity of arsine adsorbed on film B (11)

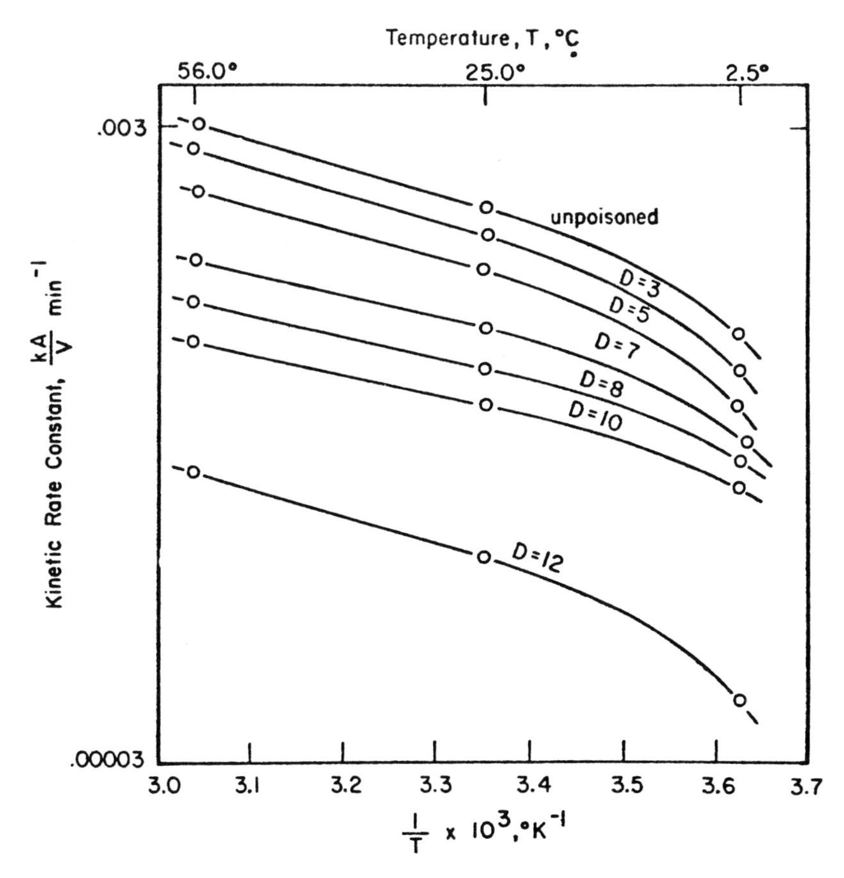

Figure 4b. Kinetic rate constants for a progressively poisoned film (11)

nonselective poisoning system is shown in Figure 4 from the results of Clay and Petersen (*11*) on arsine poisoning of Pt-catalyzed cyclopropane hydrogenolysis. The nonlinear, or selective, deactivation behavior has been represented empirically by a variety of expressions such as:

$$
\begin{aligned}
s &= s_o \exp(-\alpha_2 C_p) \\
1/s &= 1/s_o + \alpha_3 C_p \qquad (4) \\
s &= \alpha_4 C_p{}^{\Omega}
\end{aligned}
$$

There is no fundamental significance associated with any of these forms, *per se,* although we shall see later that since they express the integral of the activity decay, such forms may be used to infer rate equations for deactivation by poisoning and (as the reaction engineer is occasionally

Table I. Catalyst

	Description	Physical Properties: Area sq m/g	Physical Properties: Bulk dens., g/cc	CAT-A: Gaso. vol %	CAT-A: Coke, wt %
I A	Houdry type S, SiO_2–12.5% Al_2O_3	273	0.61	45.1	3.2
II A	Same	196	.60	35.6	1.7
III A	Same	—	.62	32.1	1.6
IV A	Same	111	.67	28.2	1.0
V A	Same	44	.96	16.2	0.5
VI A	SiO_2–1% Al_2O_3	390	.73	27.2	1.3
VII A	SiO_2	330	.74	7.0	0.3
VIII Z	SiO_2–9.5% ZrO_2	251	.63	35.7	2.0
IX M	SiO_2–32% MgO	426	.65	47.0	3.7
X C	Filtrol clay	41	.91	25.9	1.4
—	Empty reactor	—	—	5.0	0.1

[a] Data of Mills *et al.* (*4*).

tempted to do against his better judgment) perhaps indicate something concerning the mechanism of the poisoning. Points of initial deviation from linearity, as in the three examples of Figure 2, are often taken to indicate some change in the nature of the deactivation or of the surface being deactivated. Given the precision of most kinetic data, this would appear to be a rather risky procedure without supplemental chemical information supporting the interpretation.

Cracking Catalysts and Nitrogen Compounds. It is now well known that basic organic compounds are effective poisons for acid-catalyzed reactions such as isomerization, cracking, double bond migration, etc. Effective catalysts for such reactions are acidic oxides, notably silica-alumina, and one of the earlier studies of such poisoning (at a time when it was not at all clear that these catalysts owed their activity to their surface acid function) is that of Mills, Boedeker, and Oblad (*4*) on the deactivation of cracking catalysts by nitrogen-containing organic compounds. Some of the specific catalysts which they investigated, with typical activity and selectivity patterns for the nondeactivated material, are given in Table I. The catalytic properties of these materials for cumene cracking were studied in the presence of a number of basic organic nitrogen compounds such as quinoline, pyridine, piperidine, and aniline. One interesting result is illustrated on Figure 5a for quinoline

Properties [a]

Catalytic Properties						
CAT-A		*Cumene Cracking*				*Ability to Chemisorb*
Gas, wt %	*Gas grav. (air = 1.0)*	*Benzene wt %*	*Coke, wt %*	*Gas, wt %*	*Gas grav. (air = 1.0)*	*Quinoline at 315° m/g*
10.1	1.58	39.7	0.62	17.2	1.41	0.06
6.1	1.55	40.7	0.39	15.9	1.40	.044
6.2	1.57	—	—	—	—	.027
3.1	1.50	27.3	0.1	10.1	1.40	.021
1.2	1.23	19.5	.05	6.9	1.37	.009
4.3	1.49	—	—	—	—	.020
—	—	1.7	—	—	—	.001
4.2	1.41	26.3	.16	8.9	1.33	.033
5.5	1.49	—	—	—	—	.09
2.7	1.13	—	—	—	—	.018
0.6	1.1	—	0.1	0.7	1.0	—

adsorption on SiO_2 *vs.* SiO_2–Al_2O_3; this material is completely reversibly adsorbed on the silica catalyst, but on adding alumina, two types of adsorption may be identified: a reversible mode analogous to that on pure SiO_2 and an irreversible (not desorbed at 315°C) mode presumably associated with the introduction of Al_2O_3 into the catalyst. Obviously the chemistry of the silica–alumina with respect to the nitrogen compound is different from that of the pure silica.

Partial poisoning studies, in effect giving poisoning correlations such as those of Figure 2, were carried out with catalyst IIIA (SiO_2–Al_2O_3) of Table I using CAT-A (425°C, 1.5 LHSV, 10 min with cumene feed). The results are shown in Figure 5b for various nitrogen compounds of interest; these all appear to be highly selective poisons with activity depending almost exponentially on poison concentration. The levels of surface coverage are not shown explicitly in the figure, but as an example for catalyst IA, at 4% surface coverage (assuming 36 A^2 as the area of the flat lying quinoline molecule) its activity for cumene cracking at 425°C is decreased by a factor of about seven. Thus, as Mills *et al.* concluded: ". . . by far the major part of the surface does not contribute to the cracking activity of the catalyst. . . ."

The basic nitrogen compounds effective in poisoning would be ranked according to their effectiveness as quinaldine >quinoline >pyrrole > piperidine >decylamine >aniline, which is not in the same ratio as their basicity, piperidine being the strongest base. This discrepancy, however, is resolved by looking at the susceptibility of the nitrogen compounds themselves to cracking. These results, shown in Table II, indicate that the most basic, piperidine, is about 54% cracked itself in the poisoning tests and presumably is proportionately decreased in its poisoning effectiveness. None of the other materials, except for decylamine (which rates low on the basicity scale anyway) are cracked to a significant extent.

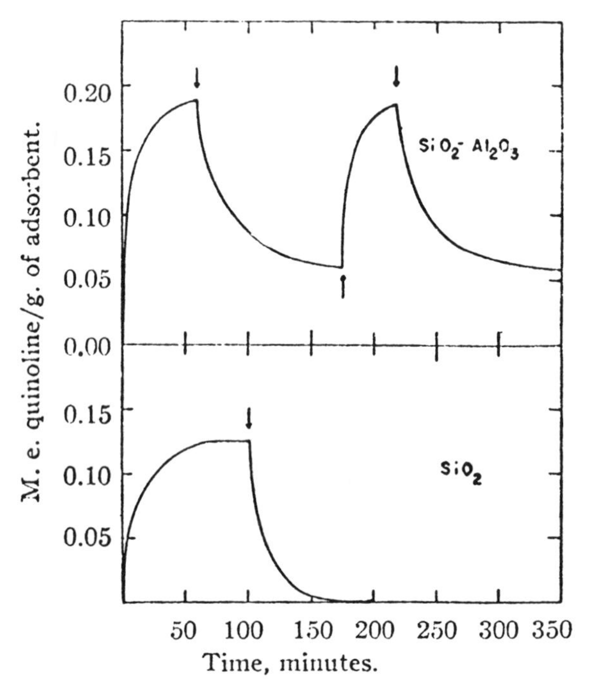

Journal of the American Chemical Society

Figure 5a. Sorption and desorption of quinoline at 315° on calcined silica gel VIIA or on catalyst IA (SiO_2–12.5% Al_2O_3). Quinoline partial pressure in flowing nitrogen stream was 73 mm. Arrows indicate start of sorption or desorption (4).

In spite of the apparent differences in quinoline chemisorption between alumina and silica–alumina shown in Figure 5a, various types of catalysts appear to be affected similarly in their cracking activity by basic nitrogen poisoning, as shown in Figure 5c. This indicates that the amount of quinoline chemisorption ". . . thus measures a fundamental property of the catalyst which is related to its ability to act as a catalyst." The property, of course, is the acidity of the catalyst, and the mechanism of poisoning is postulated to occur by chemisorption of the poison on incompletely coordinated aluminum ions or silicon ions which on the surface form Lewis acids:

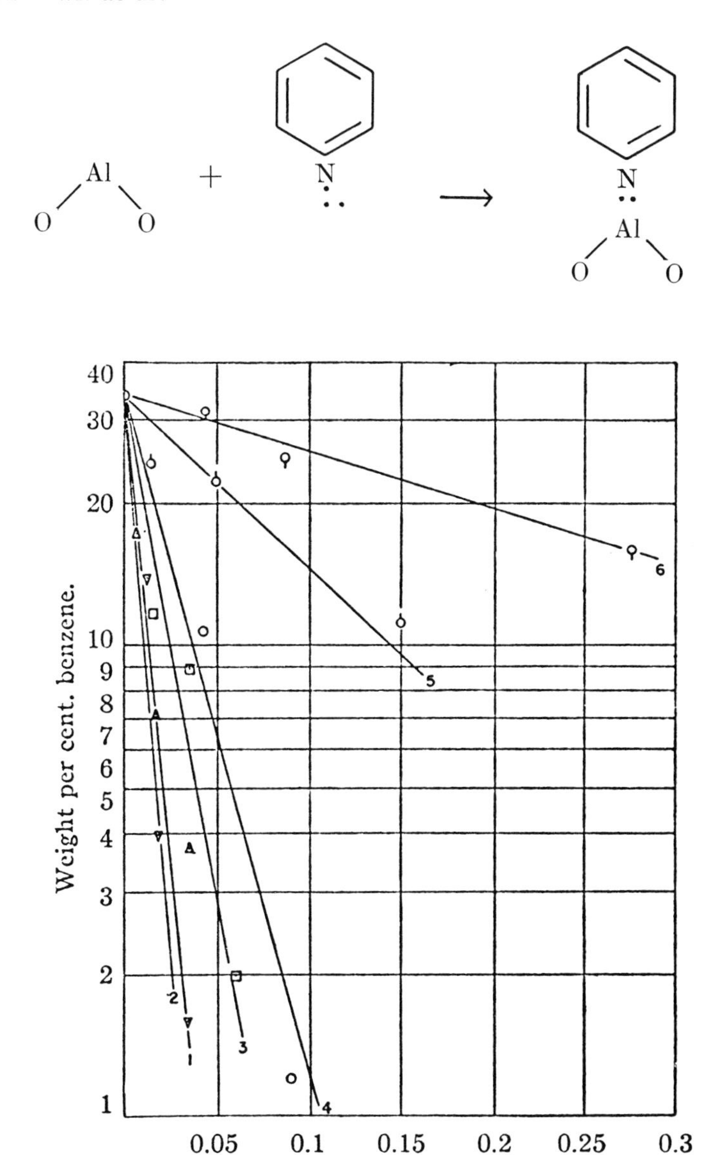

Journal of the American Chemical Society

Figure 5b. Poisoning effect of organic nitrogen compounds on the catalytic dealkylation of cumene catalyst IIIA: 1, quinoline; 2, quinaldine; 3, pyrrole; 4, piperidine; 5, decylamine; 6, aniline (4)

Several interesting points related to the behavior of such oxides as catalysts are left unanswered by these experiments. In particular, we refer to questions concerning the nature of the acidity of these surfaces, which has direct bearing on their chemisorptive and poisoning behavior. Experiments shown in Figure 5a indicated that there might be two types of adsorption sites on a silica–alumina, one related to the SiO_2 and the second (stronger) related to the Al_2O_3. There is not space here to reiterate the mountains of published research concerning the nature of the acidity (Lewis or Brønsted) or the strength of acid-acting sites on Al_2O_3 or SiO_2–Al_2O_3; however, to continue our examples of specific poisoning studies we look at one careful study directed toward the elucidation of the strength and properties of the acid sites on alumina.

Table II. Cracking of Nitrogen Compounds[d]

CAT-A Conditions: 425°C, 1.5 LHSV, 10 min on stream, 50 ml liquid charged; catalyst IIIA

	Vol, ml of liquid recovered					
Compound cracked	*BP below charge*	*BP of charge*	*BP above charge*	*Coke, wt %*	*Gas, wt %*	*Gas gravity*[b]
Piperidine	none	23.2	18.1	2.7	3.1	0.5
Pyridine	none	48.4[a]	none	1.2	0.3	1.6
Quinaldine	none	47.8[a]	none	1.3	0.1	0.6
Quinoline	none	48.0[a]	none	1.4	0.3	0.8
Aniline	none	48.4[a]	none	1.5	0.2	1.1
Decylamine[c]	23.7	19.3[a]	none	4.5	4.9	0.8

[a] Calculated.
[b] Caused by small amount of gas formed; gas gravity values are not accurate.
[c] Based on the ammonia absorbed by standard HCl, 57.4 wt % of the decylamine cracked.
[d] Data of Mills *et al.* (*4*).

Catalytic Activity and Intrinsic Acidity of Alumina. The approach which was taken in this fine study by Pines and Haag (*5*) is that of using model reactions for the evaluation of relative activities of a series of aluminas. The aluminas studied were prepared by varying techniques designed to produce differing surface acidities. The major types were as follows (*see* Table III):

(a) Pure—prepared from aluminum isopropoxide or from aluminum hydroxide and aluminum nitrate (Method A or B)

(b) Impregnated—pure catalyst impregnated with NaCl or NaOH

(c) Aluminate–alkali containing alumina precipited from KOH solution (Method C)

The incorporation of alkali in the impregnated or aluminate catalysts

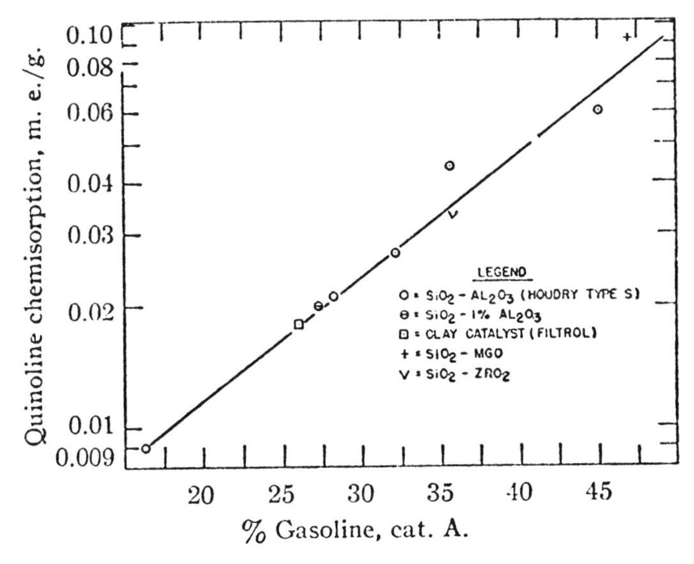

Journal of the American Chemical Society

Figure 5c. Quinoline chemisorption at 315°C as a function of activity for cracking light East Texas gas oil (4)

Table III. Methods of Preparing and Designating Alumina Catalysts[b]

Catalyst	*Method of prepn.*	*Temp., °C*	*Conditions, hr atm*	*Wt % Na or (K)*	*Area, m^2/g*
1	A (pure)	400	4 (10 mm)	—	246
2	A	600	4 (10 mm)	—	147
3	A	700	4 (10 mm)	—	152
9-1	A impreg. with NaOH	600	4 N_2	0.11	280
9-2	A impreg. with NaOH	600	4 N_2	0.2	—
9-3	A impreg. with NaOH	600	4 N_2	0.4	—
10-1	A impreg. with NaCl	600	4 N_2	0.2	—
10-2	A impreg. with NaCl	600	4 N_2	0.6	215
10-3	A impreg. with NaCl	600	4 N_2	1.5	—
11(2)[a]	C (in NaOH)	600	4 N_2	0.65	371
11(4)[a]	C (in NaOH)	600	4 N_2	—	—
11(6)[a]	C (in NaOH)	600	4 N_2	—	—
18	C (in KOH)	360	16 N_2	—	—
19	C(7)[a] (in KOH)	360	16 N_2	(0.09)	384
19[a]	C(7)[a] (in KOH)	700	4 N_2	—	298

[a] Number of washings after precipitation.
[b] Data of Pines and Haag (*5*).

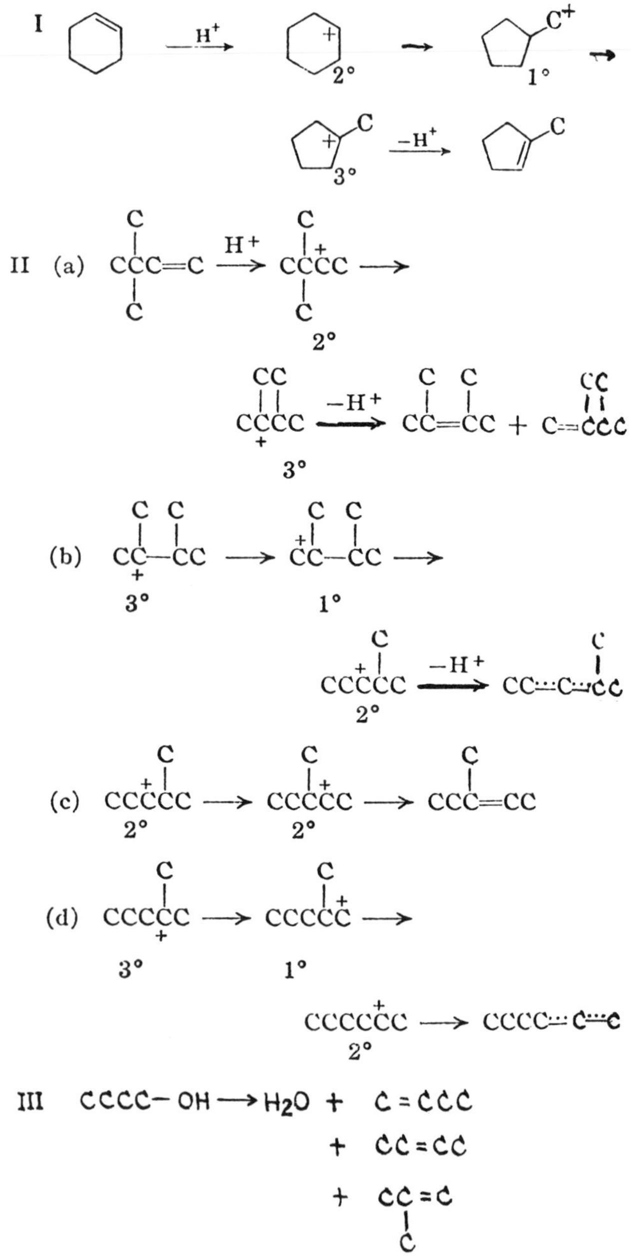

I cyclohexene $\xrightarrow{H^+}$ cyclohexyl cation (2°) → cyclopentylmethyl cation C^+ (1°) →

methylcyclopentyl cation (3°) $\xrightarrow{-H^+}$ methylcyclopentene

II (a) $CC(C)(C)C{=}C \xrightarrow{H^+} CC(C)(C)\overset{+}{C}C$ → (2°)

$CC(C)\overset{+}{C}(C)C$ (3°) $\xrightarrow{-H^+}$ $CC(C){=}C(C)C$ + $C{=}C(C)C(C)C$

(b) $C\overset{+}{C}(C)C(C)C$ (3°) → $\overset{+}{C}C(C)C(C)C$ (1°) →

$CC\overset{+}{C}C(C)C$ (2°) $\xrightarrow{-H^+}$ $CC{=}C{=}C(C)C$

(c) $CC\overset{+}{C}C(C)C$ (2°) → $CCC\overset{+}{C}(C)C$ (2°) → $CCC(C){=}CC$

(d) $CCC\overset{+}{C}(C)C$ (3°) → $CCCC(C)\overset{+}{C}$ (1°) →

$CCCC\overset{+}{C}C$ (2°) → $CCCC{=}C{=}C$

III $CCCC{-}OH \longrightarrow H_2O$ + $C{=}CCC$
+ $CC{=}CC$
+ $CC(C){=}C$

Figure 6a. Model reactions used to evaluate relative activities of aluminas

decreases the acidity of these materials relative to the isopropoxide preparations, and the model reactions used were chosen to provide some detail about these changes. These reactions, shown in Figure 6a, are (I) the isomerization of cyclohexane (CH), (II) isomerization of 3,3-dimethyl-1-butene (3,3-DMB), and (III) the dehydration of 1-butanol (BuOH). Carbonium ion mechanisms are involved in all these reactions, and the relative stability of these ions ($3° > 2° > 1°$) can be used as a measure of the acid strength required to catalyze its rearrangement. Hence the CH isomerization ($2° \rightarrow 1°$) should be more difficult to carry out on a given surface than 3,3-DMB (IIa) isomerization, where the $2° \rightarrow 3°$ rearrangement is involved. Subsequent steps in the 3,3-DMB sequence of II are also of interest. Step IIb, yielding 2-methylpentene from 2,3-DMB, is a $3° \rightarrow 1°$ transformation and would proceed more slowly than the first step. Similar deductions can be made concerning IIc and IId as shown; the point is that it is not only the total reaction of 3,3-DMB which can be used to indicate some measure of total acidity but also the selectivity between the individual steps which can be used to give some indication of the distribution of acid strength. Rearrangements *via* 1° carbonium ions would occur only on relatively strong acid sites at measurable rates, 2° and 3° on both strong and weak sites. Finally, alcohol dehydration should also occur on both strong and weak acid sites.

The results (relative catalytic activity here based on conversion under set conditions rather than directly determined rates) for CH isomerization are in good agreement with expectations based on catalyst acidity. The pure isopropoxide catalysts demonstrated high activity for the isomerization, while the alkali doped types had very little activity for CH isomerization. The 3,3-DMB isomerizations were also in line with expectation, although the results here were somewhat more complex. The over-all degree of isomerization was higher than for CH for all catalysts, indicating that some acidity was effective in this reaction which did not participate in the CH isomerization, although the product distributions did vary considerably for the differing materials. Butanol dehydration also occurs on all types at conversion levels above the CH isomerization, again with a wide variation in product distribution (between selective 1-butene or an *n*-butene mixture).

Generally, all the experimental results are in accord with the expectation that there exists a distribution of acid site strength on alumina, but this is far from all the information available to us from the study. What are the factors which affect this acid strength which so markedly affects the catalytic properties of an alumina? Surface hydration is one choice since cracking catalysts require a small concentration of water for maximum activity. The CH isomerization was conducted with a series of isopropoxide catalysts heated to differing levels before use, and the meas-

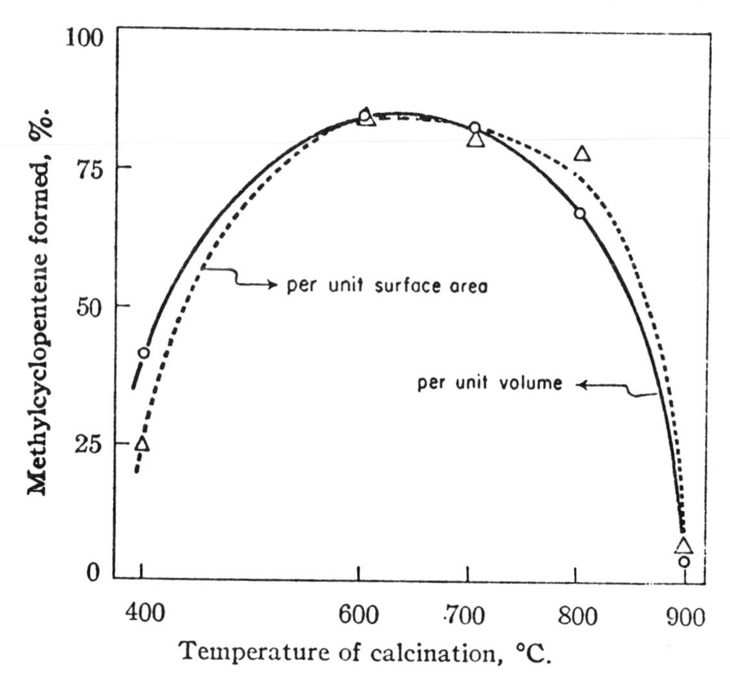

Journal of the American Chemical Society

Figure 6b. Isomerization of cyclohexene (410°, HLSV 2.0): as a function of calcination temperature of alumina. Solid line, isomerization activity per unit volume; dashed line, isomerization activity per unit surface area (5).

ured specific isomerization activity is determined as a function of calcination temperature, as shown in Figure 6b. Sintering resulting from the treatment at different temperatures is accounted for in the specific activities, so there appears to be a quite strong dependence of the catalytic properties (*i.e.*, strongly acidic properties) on the calcination temperature. Since the latter would determine equilibrium surface water content, one infers that acidity, activity, and surface hydration are all related. Subsequent experiments of this sort gave similar results for the 3,3-DMB isomerization; hence the hydration effect appears related to total acidity. It is also interesting to note that very large amounts of water on the surface (lower calcination temperatures) act as a poison for the isomerization activity while at the opposite end of the scale some water is required for catalytic activity. These notions of surface hydration are consistent with dissociative adsorption of water *via* hydroxylation of the surface aluminum and protonation of the oxide; the acidity of the surface arises from sites corresponding to partial hydration. Peri (*12*) has been able to use this as the basis for a Monte Carlo simulation of the surface hydration of alumina which results in a reasonable interpretation of observed infrared spectra in terms of particular groupings of partially hydrated sites, and the approach has recently been extended successfully to the prediction of reaction selectivity (between olefin and ether formation) for ethanol dehydration as a function of surface hydration (*13*).

A second factor affecting the acidity of the catalysts of Pines and Haag is certainly the amount of alkali incorporated into the impregnated or aluminate samples. Figure 6c gives some of the results of tests conducted on both CH and 3,3-DMB isomerizations, and the butanol dehydration. In all cases the activity is diminished by the presence of the alkali, and the similarity in behavior for the various reactions again suggests that similar types of sites are responsible for the catalytic properties of the alumina. The aluminate catalysts are poisoned by alkali contents an order of magnitude smaller than those shown for the impregnated material; presumably this is related to the distribution of poison on the internal surface of the alumina as well as the mechanism of poisoning. The question of poison distribution is one to which we shall return later in more detail with regard to simultaneous diffusion and deactivation phenomena.

In summary high activity can result either from a large number of acid sites or strong acidity of the sites. The CH isomerization is a measure

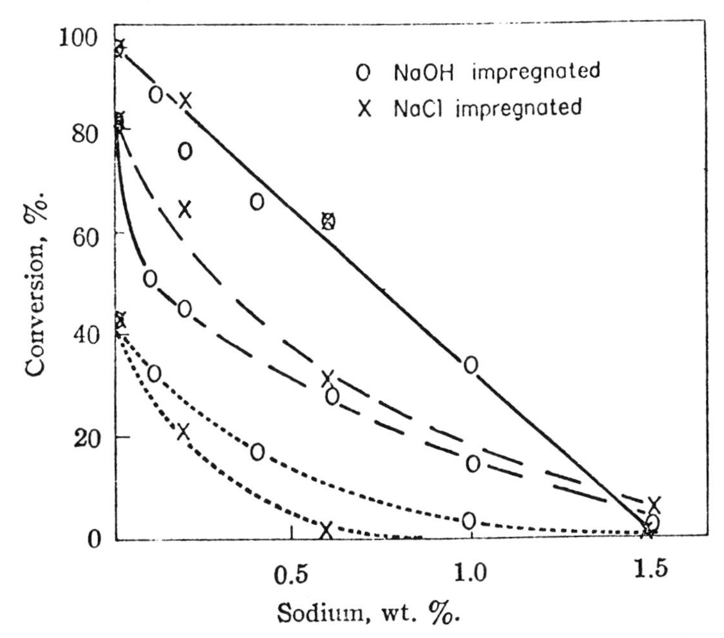

Journal of the American Chemical Society

Figure 6c. Isomerization of cyclohexene (dotted line), of 3,3-dimethly-1-butene (solid line), and dehydration of 1-butanol (dashed line) over impregnated aluminas (350°, HLSV 2.0) (5)

of strong acid sites, dehydration a measure of total acidity. Similarly the extent and selectivity of the 3,3-DMB isomerization is a measure of total acidity (over-all conversion), weak acidity (production of primary product 2,3-DMB) and strong acidity (further isomerizations to 2- and 3-MP). Aluminas without alkali poisons have a strongly acidic surface, those prepared from the aluminate have no strong acid sites, but are weakly acidic, while impregnation seems to deactivate both strong and weak sites indiscriminately. From further experiments on the adsorption of alkali, Pines and Haag obtain values of about 10^{13} strong acid sites per cm^2 (aluminate measurements), and about 7.5×10^{13} total sites involved in 1-butanol dehydration (Figure 7c). Thus pure alumina has roughly 10^{14} acidic sites per cm^2, with about 10% of this number effective in isomerization.

We recall that Mills *et al.* correlated the activity of alumina and silica–alumina with quinoline adsorption, and the amine index is a well known means for determination of the acidity of solid surfaces. An interesting further result obtained by Pines and Haag is the relationship between amine index and catalytic activity. Apparently the index measures total acidity, and hence one can determine relative activities for a

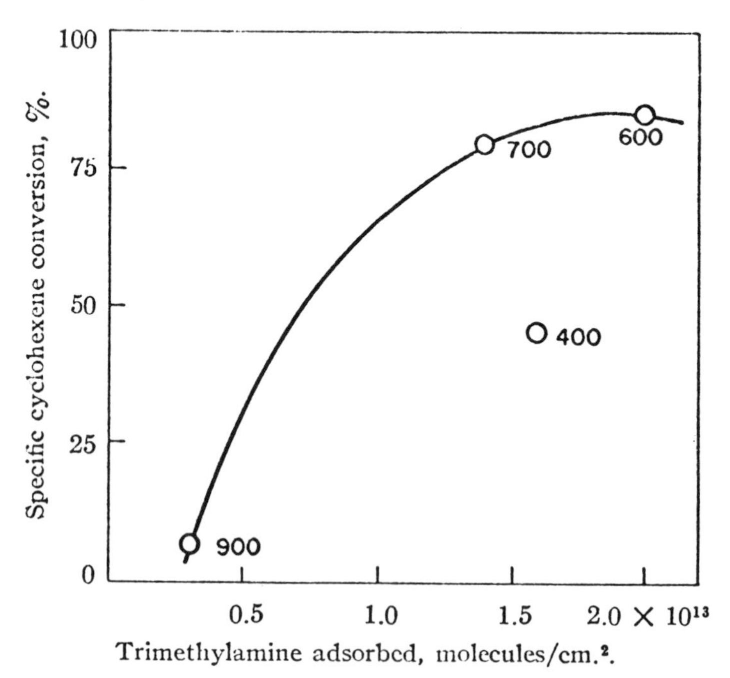

Journal of the American Chemical Society

Figure 7a. Comparison between trimethylamine adsorption and isomerization of cyclohexene (410°, HLSV 0.5) (5)

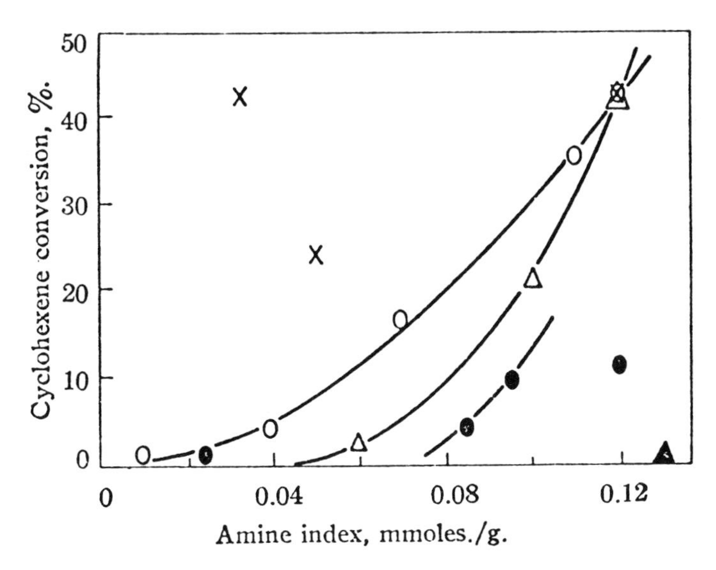

Figure 7b. Cyclohexene conversion (350°, HLSV 2.0) vs. *acidity; aluminas:* ○, *NaOH impregnated;* ●, *from potassium aluminate;* △, *NaCl impregnated;* ▲, *from sodium aluminate;* ×, *from isopropoxide* (5)

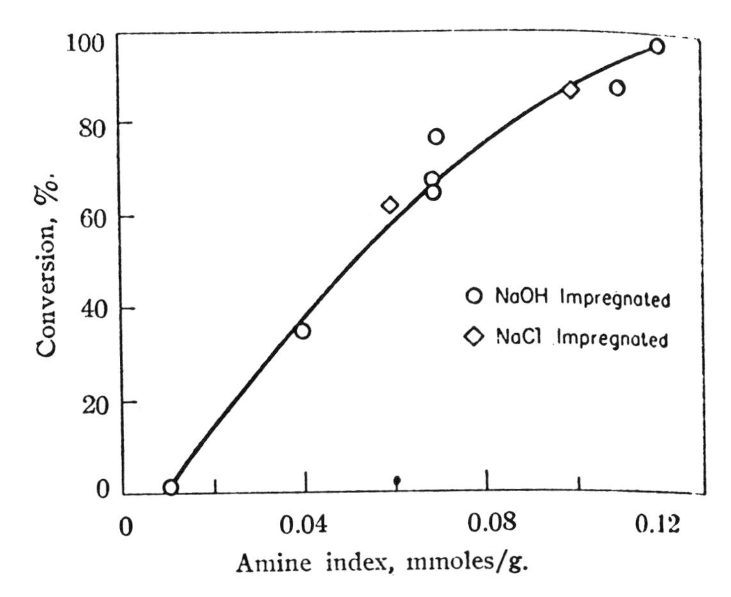

Figure 7c. Conversion of 3,3-dimethyl-1-butene (350°, HLSV 2.0) vs. *acidity; alumina from isopropoxide:* ○, *NaOH impregnated;* ◇, *NaCl impregnated* (5)

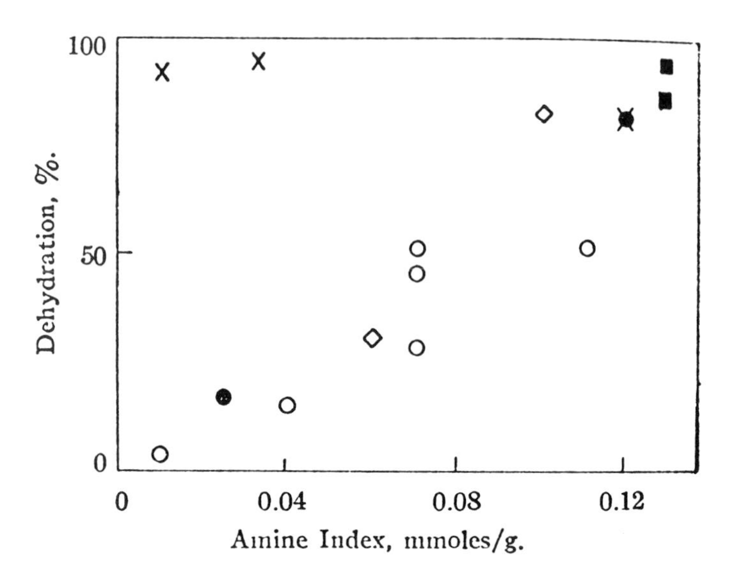

Journal of the American Chemical Society

Figure 7d. Dehydration of 1-butanol (350°, HLSV 2.0) vs. *acidity; alumina: ×, from isopropoxide; ●, from potassium aluminate; ■, from sodium aluminate; ○, NaOH impregnated; ◇, NaCl impregnated* (5)

"family" of related aluminas, as shown on Figure 7a–c. However, if one attempts to compare activity for aluminas from different sources, no correlation is obtained, as shown by the differing curves on Figure 7b for the three types of alumina in a strong acid reaction, or the shotgun pattern of Figure 7d for a total acid reaction. The fact that amine index is generally a satisfactory measure of cracking catalyst activity is at least indirect evidence of a relatively homogeneous acidity of the sites on silica–alumina. Since two types of sites are indicated in Figure 5a, it might be concluded that only one is effective in determination of the cracking activity.

The results with trimethylamine chemisorption on alumina are thus in good agreement with the reaction experiments. While elucidation of the detailed catalytic chemistry of alumina is not the object of the present discussion, it is interesting to note that much of what was learned in this study was accomplished by specific studies of the poisoning behavior of the catalyst.

SURFACE OXIDATION EFFECTS ON CATALYTIC ACTIVITY. This example study by Parravano (*6*) illustrates a common problem in many oxidation catalysts where the "active" or lattice oxygen of the solid material is involved in the reaction. Such systems appear particularly confusing on first sight since it seems impossible to reproduce activity or to obtain a consistent interpretation of the kinetics. The reduction of the surface and the accompanying change in activity may not be poisoning in the normal sense of the term, but the effects are similar. For the CO oxidation on NiO investigated here, initial rate of reaction data were not correlated by any standard kinetic expression, rather the initial rates of reaction appeared to be represented by an Elovich equation:

$$r_T = ae^{-\alpha' x} \tag{5}$$

in which r is the rate in terms of fraction converted per time, α' and a are the Elovich constants, and x is the fraction converted. Figure 8a shows the initial rate data obtained in a constant volume experiment, plotted according to the integrated form of Equation 5. At longer reaction times, large deviations from the Elovich correlation were noted, which indicated that after this initial period of rapid activity decrease, stabilization of the catalyst occurred and higher conversion data could be correlated by a conventional first-order plot at the lower temperatures or with an apparent order of 1.25 at the higher temperatures. The first-order results are shown in Figure 8b. The apparent activation energies determined were 9.7 kcal/mole for the Elovich region and 2.2 kcal/mole for the low temperature–high conversion experiments, indicating a considerable difference in the mechanism of oxidation in the two regions. Pretreatment of the catalyst with carbon monoxide (no oxygen) eliminated the Elovich

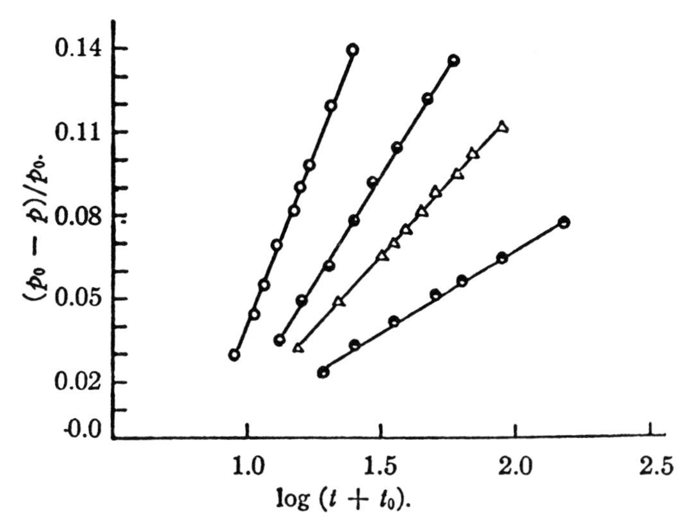

Figure 8a. Initial rates of carbon monoxide oxidation on nickel oxide: (0.483 g); p_{CO}, 186 mm.; p_{O_2}, 94 mm.; ◓, 161°; △, 181°; ◒, 200°; ○, 212° (6)

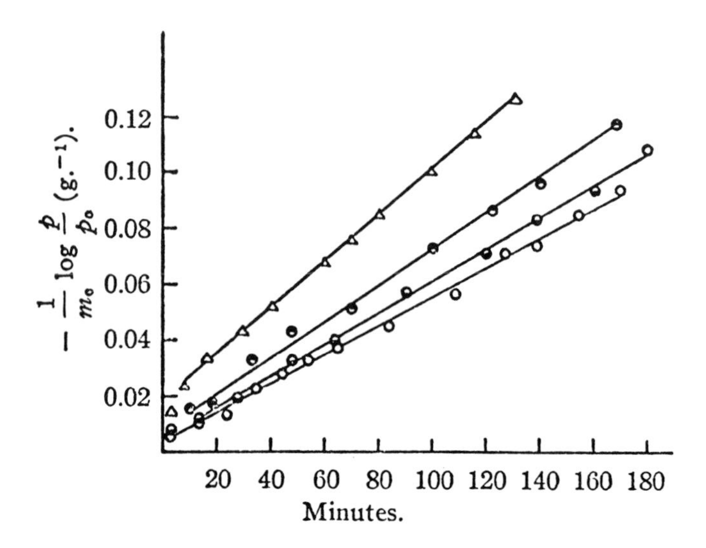

Journal of the American Chemical Society

Figure 8b. Carbon monoxide oxidation on nickel oxide: p_{CO}, *186 mm.;* p_{O_2}, *94 mm.; low temperature range:* ○, *106°;* ◒, *125°;* ◓, *147°;* △, *174°* (6)

region from observed behavior, but O_2 pretreatment did not. This gives strong evidence that the initial process on fresh catalyst consists in the removal of CO by excess O_2 on the surface. After this initial period of operation the surface establishes an equilibrium with the reaction mixture; hence the reaction sequence is explained by the following:

$$O \cdot S + CO \rightarrow S + CO_2 \qquad \text{(fast)}$$

$$S + 1/2\, O_2 \rightleftarrows O \cdot S \qquad \text{(slower)}$$

The initial stabilization occurs through the predominance of the first reaction, with the steady-state activity level attained when the [O · S] adjusts to equilibrium with the O_2 partial pressure.

Much effort has been expended in various phases of oxidation catalysis to relate activity properties to the oxidation state of the surface. In addition, one may wish to know the selectivity of the catalyst as well in hydrocarbon oxidations. The reader is referred to an interesting study by Callahan and Grasselli (*14*) for an example of one method of analysis of the selectivity problem.

Coke Formation in Hydrocarbon Reactions. Over the years it has become apparent that the chemical details of coke-formation mechanisms are not as amenable to inspection nor as obvious as the mechanisms of chemical poisoning. On the other hand, the kinetics of coke formation have been relatively thoroughly investigated (albeit indirectly in most cases). On most catalysts used in hydrocarbon reactions where coking is observed, highly unsaturated species of high molecular weight are adsorbed preferentially, and in particular, polyring aromatics have been associated with coke formation. After adsorption, these aromatics undergo further condensation and hydrogen elimination on the surface to form hydrogen-deficient coke. In this section we present a more or less chronological development of information in this area.

The starting point for developing the technology of catalyst fouling would be the early, empirical correlations of carbon formation in fixed bed cracking, such as those of Voorhies (*2*). The basic correlation is shown in Figure 9a, in which each point represents a run on a fresh or regenerated catalyst for the time indicated. The result obtained, for differing catalysts, feed materials, and feed rates at fixed temperature is a correlation of the form:

$$C_c = A\theta^n \tag{6}$$

in which C_c is the concentration of carbon on the catalyst (which may be related to the activity s as used in Equation 1), A and n are the correlation constants, and θ is the length of the process period involved.

One of the problems involved in the specific results obtained by Voorhies was the apparent independence of the coking correlation of

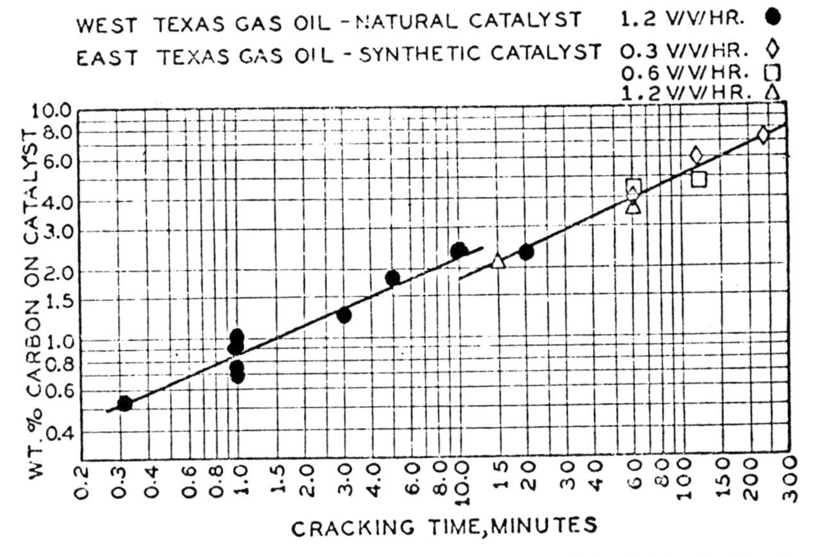

Figure 9a. Carbon formation vs. *cracking time in fixed-bed catalytic cracking* (2)

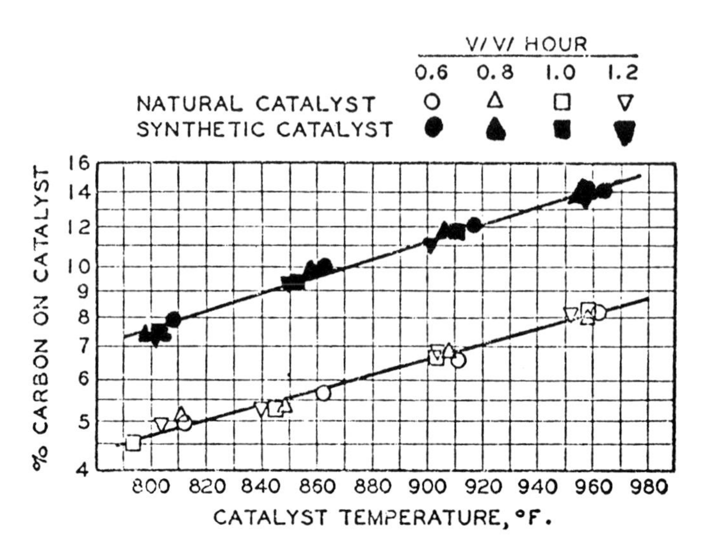

Industrial and Engineering Chemistry

Figure 9b. Carbon formation vs. *cracking temperature for 2-hour cracking periods* (2)

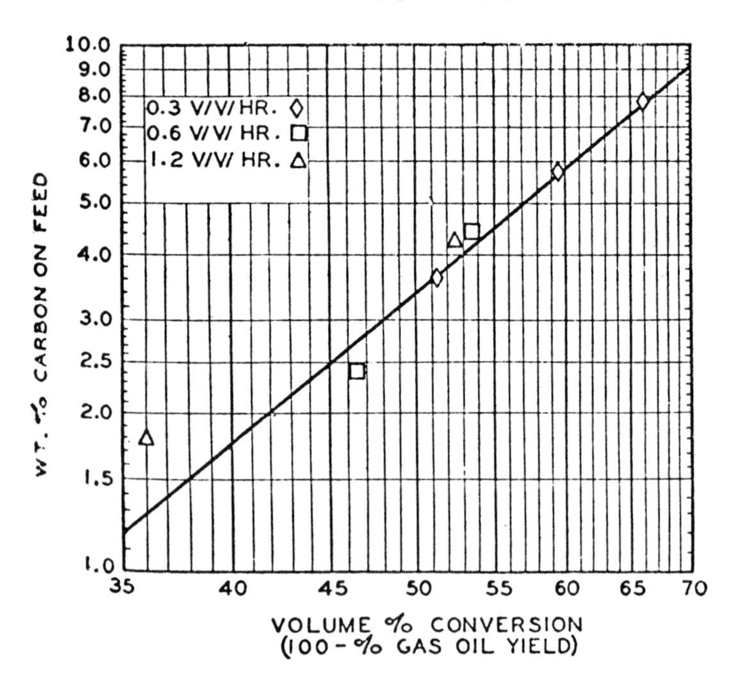

Industrial and Engineering Chemistry

Figure 9c. Carbon yield vs. *feed conversion for fixed-bed catalytic cracking of an East Texas gas oil with synthetic catalyst* (2) feed rate; this point has been restudied by several workers, and the consensus is that feed rate is important in determining the constants of Equation 6. We shall see some more detailed explanations later.

The temperature dependence of carbon deposition is not unlike those characteristic of chemical reaction (*i.e.*, log rate *vs.* $1/T$) although temperature rather than its reciprocal is plotted on Figure 9b. It would, however, probably be incorrect to pursue the analogy too far, other than to remark that an activation energy corresponding to these data is rather low compared with many chemical reactions and may indicate diffusion control.

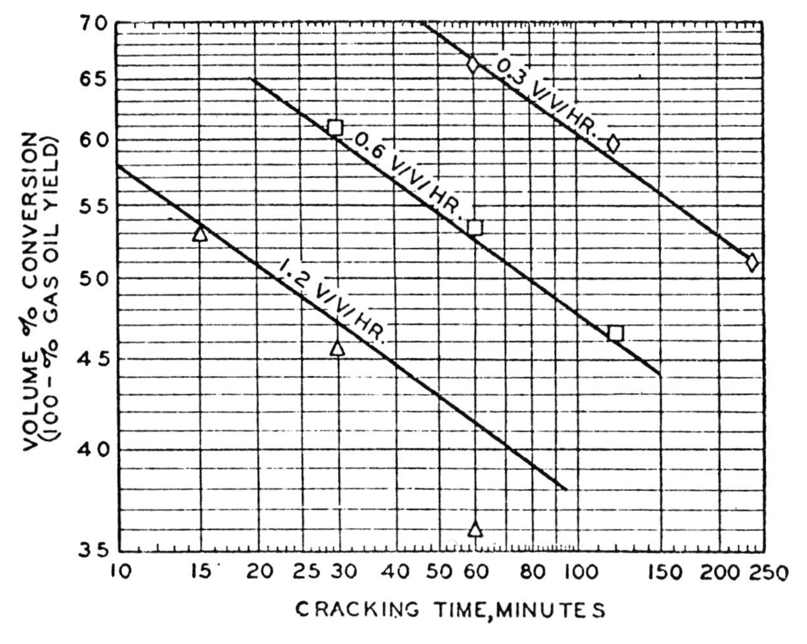

Industrial and Engineering Chemistry

Figure 9d. Feed conversion vs. *cracking time and feed rate for fixed-bed catalytic cracking of an East Texas gas oil with synthetic catalyst* (2)

A correlation was also obtained by Voorhies between carbon yield and conversion, depicted in Figure 9c, which suggests the relationship of the coking reactions to the other reactions occurring. Further, since carbon yield is a function of process time (Figure 9a), and conversion is a function of carbon yield (Figure 9c), then one should obtain some correlation between conversion and process time, as indeed is shown in Figure 9d. These results correspond to an equation of the form:

$$x_v = B\theta^{m}U^{p} \tag{7}$$

in which x_v is the volumetric conversion, U is the feed rate, and B, m, and p the correlation constants.

The general forms of the Voorhies correlations, Equations 6 and 7 (and particularly the former), fit a very large amount of catalyst deactivation data, not only coking, and are encountered throughout the literature. The only difficulty with such results is that they are, after all, only empiricisms (though very useful ones) and leave one with no clear picture of more fundamental information concerning kinetics or possible routes for the formation of coke. In addition, a correlation such as Equation 6 incorporates catalyst deactivation into the formulation of the over-all kinetics by way of Equation 1 in an *ex post facto* fashion. The separable form and the fundamental variables of concentration, temperature, and time are retained, but Equation 1 becomes:

$$r_T = r_1([C])r_2(T)s(\theta) \tag{1a}$$

Despite these shortcomings, it is quite striking that any correlation at all could be obtained from such different catalysts, operating conditions, and complex feeds. Very extensive applications of correlations of this type

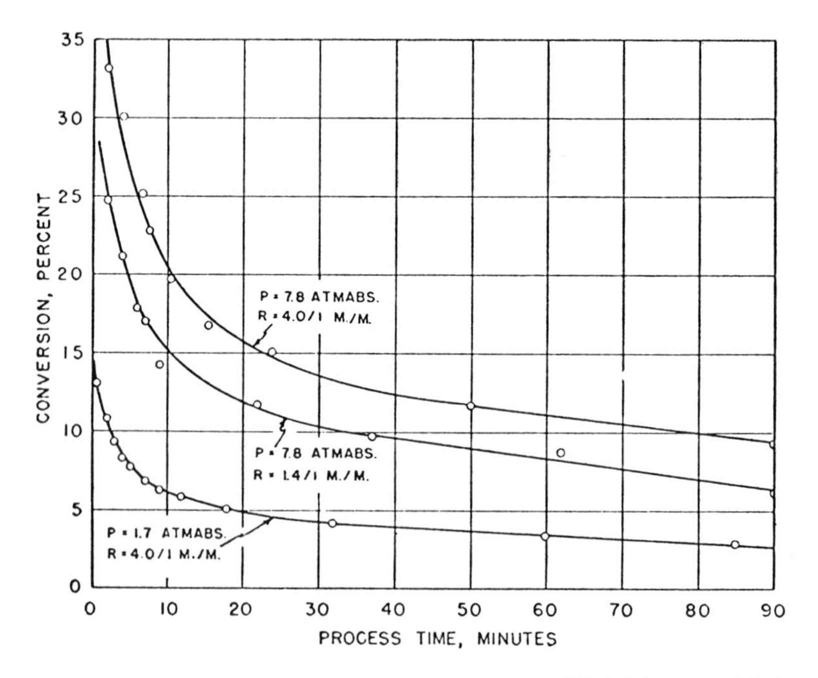

Chemical Engineering Science

Figure 10a. Conversion vs. process time at various conditions of total pressure and H_2/C_6H_{12} molar feed ratio (all other conditions "standard") (16)

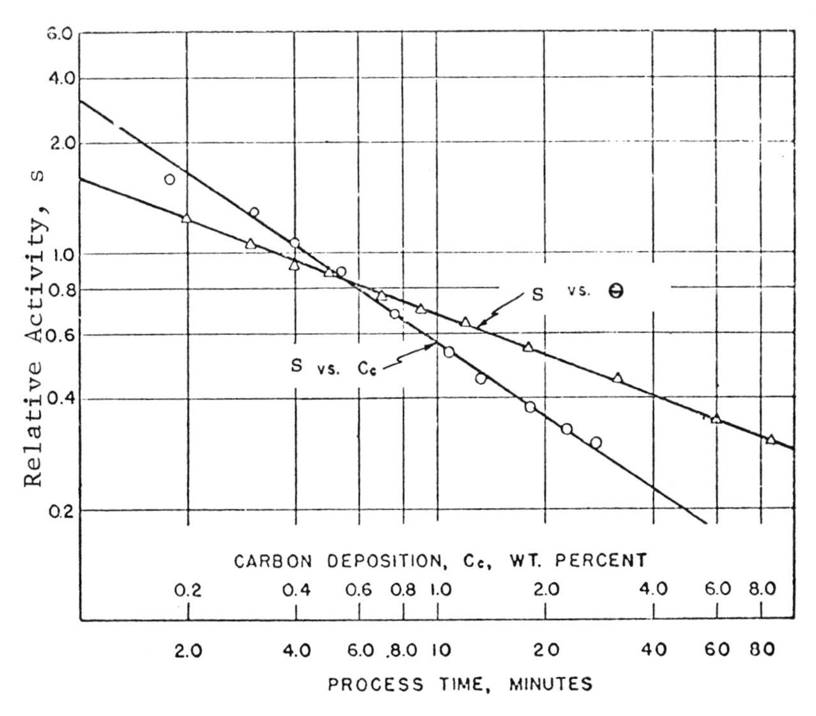

Chemical Engineering Science

Figure 10b. Decline in activity with process time and with carbon deposition ("standard" conditions, except process time) (16)

to cracking reactions are reported by Blanding (*15*). Some of the subsequent studies we use as illustrations were concerned with experiments in more well characterized systems to eliminate some of these uncertainties.

Rudershausen and Watson (*16*) studied the aromatization of cyclohexane on a commercial molybdena-alumina gel catalyst (10% molybdena as MoO_3) with surface area of 90 m^2/gram. The "standard" conditions of the studies reported in Figure 10 refer to fixed bed reactor, 31.2 grams catalyst, (W/F) of 10.4 grams catalyst/(gram mole C_6H_{12}/hour), pressure of 1.7 atm, T of 940°F and a H_2/C_6H_{12} molar feed ratio of 4:1. Behavior similar to that noted by Voorhies was observed in this study. The initial very rapid decline in catalyst activity with time of operation shown in Figure 10a can be correlated as an exponential function of either process time or carbon deposition, Figure 10b. This suggests kinetics of the form:

$$\frac{dC_c}{d\theta} = \frac{K}{C_c} \tag{8}$$

where K is some rate coefficient. In contrast to the results of Voorhies, however, Rudershausen and Watson observed a strong temperature dependence of the coking rate, indicating that their K was related to the actual coking kinetics. The form of Equation 8 can be related to more fundamental events by the following argument. Assume that the reaction involved is the addition of an adsorbed molecule of reactant to an adjacently adsorbed large coke molecule, to produce a yet larger adsorbed coke molecule. Hence:

$$A_n + A \rightarrow A_{n+1}$$

where A_n is the adsorbed coke molecule, A the addition molecule and A_{n+1} the product coke. From a conventional Langmuir-Hinshelwood

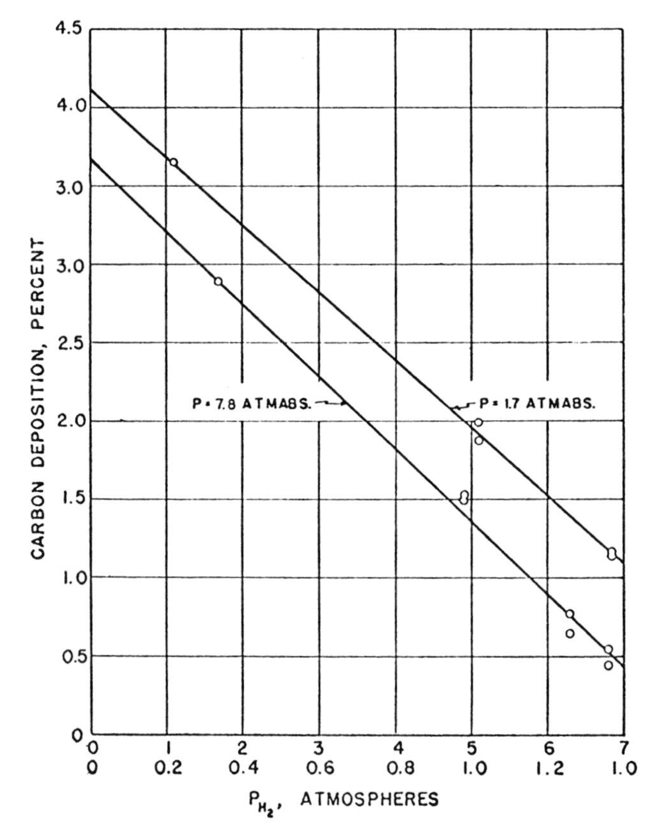

Chemical Engineering Science

Figure 10c. Effect of partial pressure of hydrogen on carbon deposition ("standard" conditions, except total pressure and H_2/C_6H_{12} molar feed ratio) (16)

analysis for surface reaction between adjacent sites rate determining, the rate of coking is:

$$r_c = \frac{K' C_{A_n} \ P_A}{(1 + K_A P_A + K_R P_R + C_{A_n})^2} \tag{9}$$

where C_{A_n} is the concentration of coke on the surface, P_A and P_R the partial pressures of coke precursor and reactants, respectively, and K_A, K_R, and K' are the adsorption and rate constants involved in the LH description. Under conditions such that the strong adsorption of coke makes the last term in the denominator of Equation 9 predominate, then:

$$r_c = \frac{K' C_{A_n} P_A}{C_{A_n}^2} = \frac{K' P_A}{C_{A_n}} \tag{10}$$

For a given run under the conditions of the experiments of Rudershausen and Watson, P_A represents some average value which appears constant; hence Equations 10 and 8 are identical, with $K'P_A = K$ and $C_{A_n} = C_c$. Another experimental fact supporting the general form of Equation 9 is the suppression of coke formation by increasing hydrogen partial pressure, as shown in Figure 10c. This would presumably be not only the result of increased adsorption competition, represented by the $K_R P_R$ term but also of some reversal or competition with the coking reaction.

The dual-site view of the coking mechanism has been supported, at least in part, by more recent evidence (*17*), which we discuss subsequently. Although the coking certainly affects aromatization activity, it is not clear that the identical catalytic functions involved in coking are also involved in the main reactions. Prater and Lago (*18*) studied this aspect of the mechanism of coking by looking at the formation of a particular type of coke product formed from cumene hydroperoxide on silica–alumina (90 wt % SiO_2, 10 wt % Al_2O_3, 350 m^2/gram) which did not lead to deactivation of the catalyst. Even though the catalyst was not deactivated by this type of coke, however, one could not say that the sites responsible for coke formation were different from those active in the main reaction since it is quite possible to visualize the kinetic scheme as:

$$R + S \rightleftarrows R \cdot S \rightarrow P + S \qquad \text{(main)}$$

$$A + S \rightleftarrows A \cdot S \rightarrow \text{coke} + S$$

where R and P are reactants and products in the main cracking reaction, and the scheme does not include coke removing active sites from the system in this particular instance. This type of scheme leads to a rate equation for coke appearance of the form:

$$r_c = \frac{K''P_A}{(1 + K_AP_A + K_RP_R)} \tag{11}$$

The experimental data on coke formation are given in Figures 11a and b in terms of coke formation as a function of hydroperoxide concentration (A in the above reaction scheme) and time of operation. The linear relationship with concentration (actually with mole fraction, x_A; Prater and Lago reported an independence of rate on total pressure level, but we

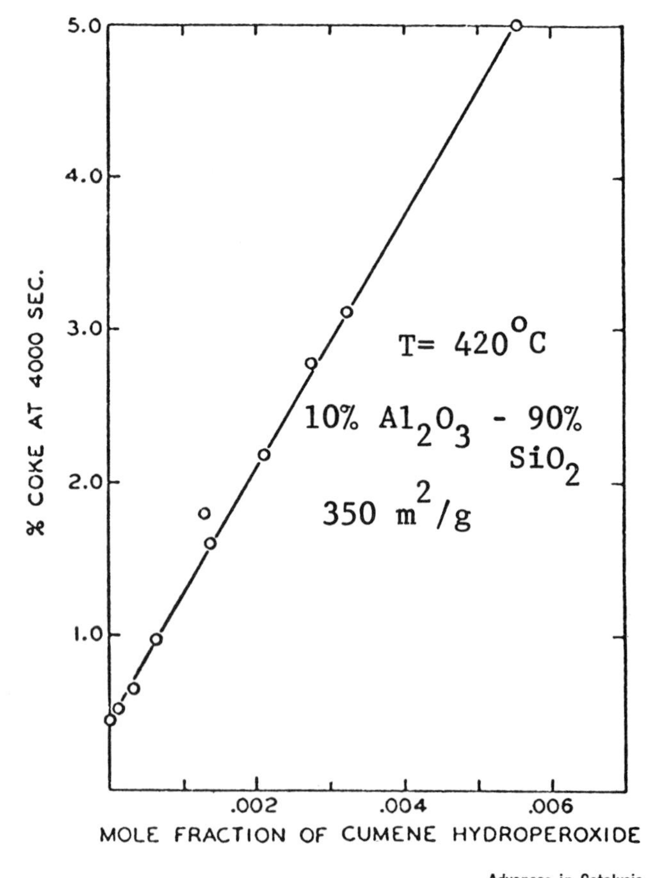

Advances in Catalysis

Figure 11a. Percentage of coke by weight on catalyst at 4000 sec as a function of mole fraction of cumene hydroperoxide in cumene (18)

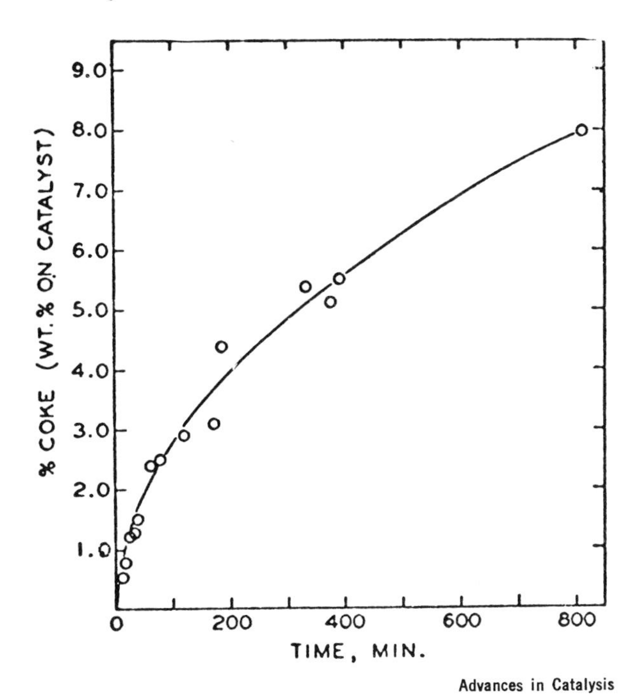

Figure 11b. Percentage of coke by weight on catalyst as a function of time for cumene containing 0.28 mole % cumene hydroperoxide (18)

will not concern ourselves with that for the present purpose), and the nearly square root dependence on time can thus be written as:

$$C_c = (2K\theta)^{1/2}P_A \tag{12}$$

which is obtained from the rate equation:

$$r_c = \frac{dC_c}{d\theta} = \frac{KP_A{}^2}{C_c} \tag{13}$$

Obviously, the forms of Equations 11 and 13 are not compatible, and again the indication is that coke is not formed by direct action of cracking sites but possibly by some other activity which, in the case of coke production which deactivates the catalyst, yields a product that subsequently is adsorbed strongly on the active cracking sites. Figures 11c and 11d indicate that this type of behavior is also observed with more complex feed materials which also contain inhibitors whose cracking reactions are similar to those of cumene hydroperoxide. Results such as those of Rudershausen and Watson and Prater and Lago, thus, establish at least a primitive case for the existence of a kind of bifunctionality in the catalysis of coke formation in cracking reactions, even on different types of catalysts. Mechanistic speculation aside, however, our real interest in these results at this point of the discussion is the demonstration of the existence of Voorhies-type relationships in differing reactions and differing catalysts by different investigators, and the fact that such results can be related to plausible kinetic schemes.

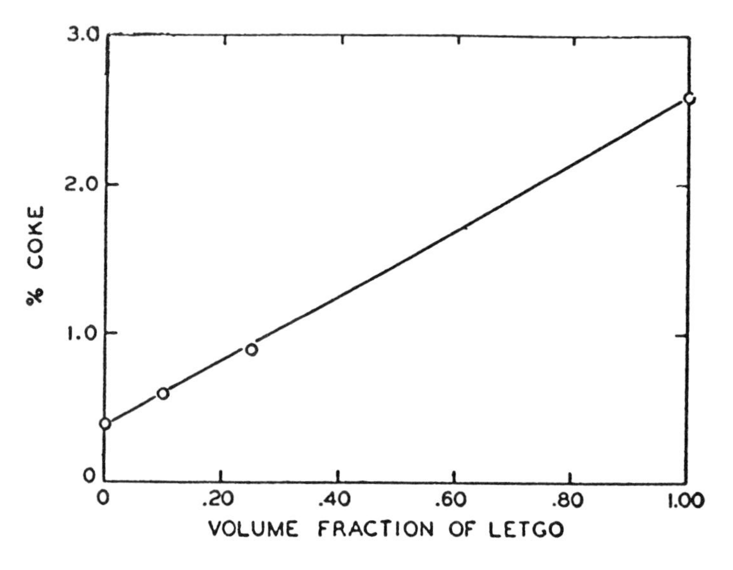

Advances in Catalysis

Figure 11c. Percentage of coke by weight as a function of volume fraction of light East Texas gas oil present in cumene (18)

More recent and detailed experimentation on the kinetics of coke deposition by Eberly *et al.* (*17*) has shown that the correlations are not quite as simple as those originally indicated by Voorhies. Their work was conducted with *n*-hexadecane and LETGO feeds, cracked at 500°C by a 13% Al_2O_3–87% SiO_2 catalyst of 382 m^2/gram surface area. One of the purposes was to test the validity of the Voorhies form over a wider range of test conditions than reported previously, and some of the primary results are given in Table IV. Excellent correlation by the form of Equation 6 is obtained in all cases, but the data in the table indicate that the values of the correlation constants are definite functions of reaction conditions and feed composition. Of particular interest are two factors, (a) variation of the correlation with feed rate, which was not found by Voorhies, and (b) the range of values of *n*.

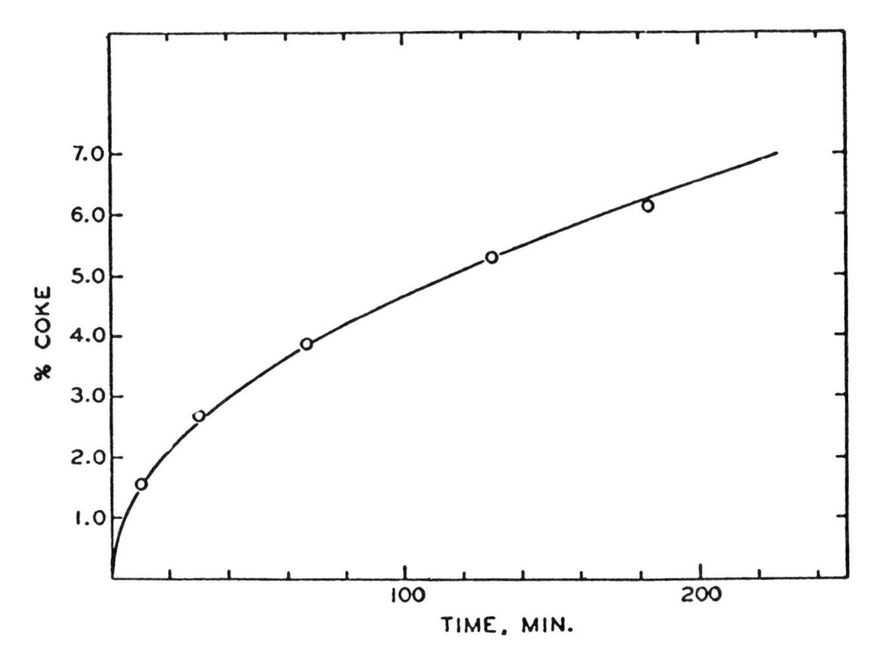

Figure 11d. Effect of amount of coke deposited on catalyst as a function of time for cracking of light East Texas gas oil. The solid line represents the relationship: % coke = $0.47\sqrt{t\,(min.)}$ (18).

Table IV. Constants in $C_c = A\theta^n$

Fixed bed of 13% Al_2O_3–87% SiO_2 at 500°C[a]

Feed	*Vol/Vol/Hour*	*A, wt %*	n
n-Hexadecane	0.2	0.049	0.97
	0.5	0.11	0.86
	1	0.16	0.78
	2	0.22	0.70
	5	0.29	0.60
	10	0.31	0.52
	20	0.29	0.44
East Texas light gas oil	0.5	0.25	0.82
	2	0.52	0.70
	5	1.05	0.42
	10	0.90	0.41

[a] Data of Eberly *et al.* (*17*).

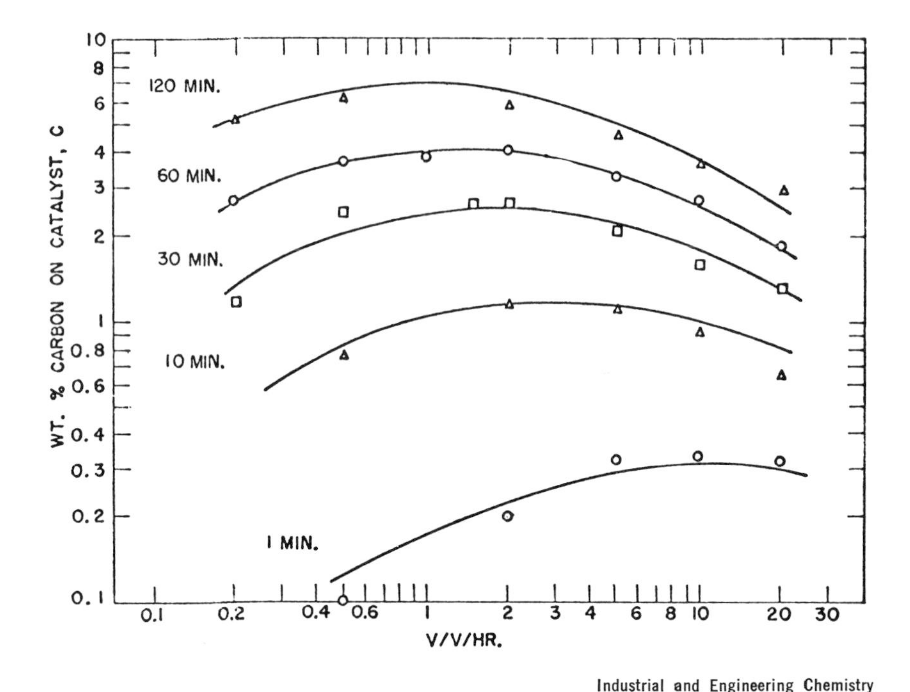

Industrial and Engineering Chemistry

Figure 12a. Carbon formation on 13% Al_2O_3*–87%* SiO_2 *from cracking of* n-*hexadecane at 500°C* (17)

Considering the latter point first, a number of workers (*18*) had obtained correlations with the square root of time, and this had been interpreted in various quarters as being indicative of a diffusion limitation in coke formation. We already know this is not necessarily so since the $n = 1/2$ could be obtained from perfectly respectable kinetics as in the case of Rudershausen and Watson; however, the range of n values in Table IV provides additional support to the hypothesis that coking is not a universally diffusion-limited process. In addition, Eberly *et al.* investigated the effect of particle size on net carbon deposition, finding no dependence on size in the range 75–2400 microns. This again is evidence for the absence of diffusional effects.

The experimental results demonstrating dependence of coking kinetics on space velocity are shown in Figure 12a, where the solid lines represent a regression fit to the experimental data. The form of dependence varies somewhat with time of operation but over limited ranges of space velocity (Voorhies observed vol/vol/hour from 0.3 to 1.2), particularly at longer reaction times, the carbon deposition is nearly independent of feed rate. A cross plot of the regression lines in Figure 12a can be used to obtain carbon deposition profiles in a fixed bed, Figure 12b; these are very strong functions of feed rate and in some ranges indicate a maximum. Maximum carbon deposition corresponds to minimum catalytic activity, and it is worth noting that even with such variable and spatially distributed bed activity very similar characteristics in terms of conversion *vs.* time for the reactor over-all are observed. Figure 12c illustrates the similarity of results obtained with *n*-hexadecane to a cracking feed more representative of reality.

Another kind of question can be asked regarding the interrelation between diffusion and coking, this being whether the formation of coke has any effect on the catalyst diffusivity. In such an event, one would visualize the coke formed as building up in layers, shutting off some pores of the catalyst structure entirely and reducing the dimensions of others. Ozawa and Bischoff (*19*) have conducted experiments with ethylene cracking at 350°–500°C and 1 atm on a commercial silica–alumina catalyst (Mobil Durabead, 90 wt % SiO_2, 9.7 wt % Al_2O_3, ~200 m^2/gram) to investigate this point. Some of their typical results are shown in Table V. The effective diffusivity data indicate that except for some decline

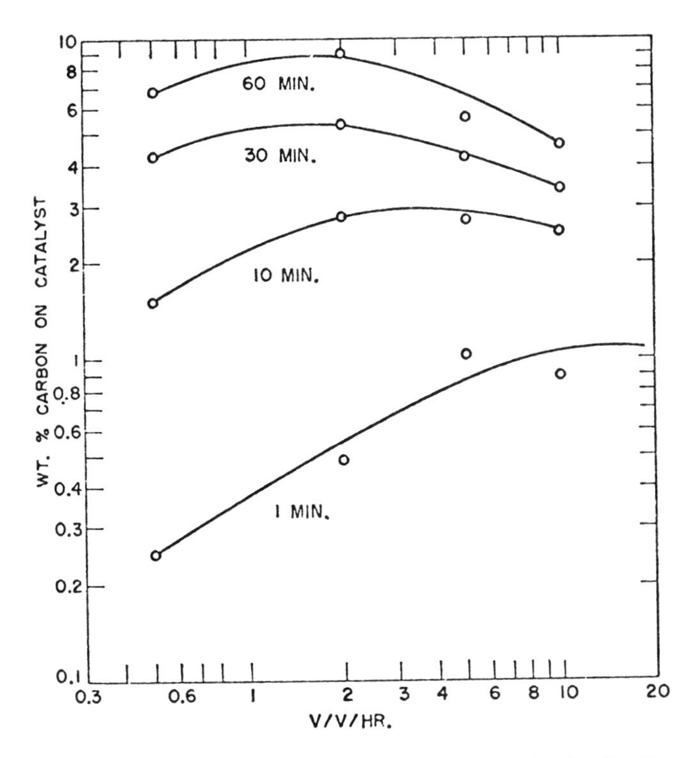

Industrial and Engineering Chemistry

Figure 12b. Carbon formation on 13% Al_2O_3–87% SiO_2 from cracking of East Texas light gas oil at 550°C (17)

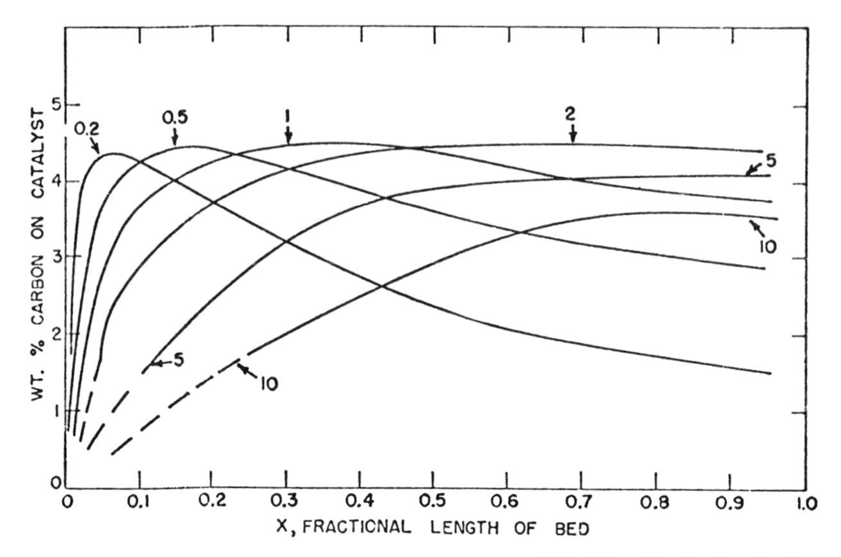

Industrial and Engineering Chemistry

Figure 12c. Weight per cent carbon on catalyst as a function of bed length. n-*Hexadecane at 500°C and 60-minute cycle time. Numbers indicate v/v/hr* (17).

Table V. Effect of Coke on Diffusivity[a, b]

Sample	*Wt % Coke on CAT*	*Before Regeneration Diffusivity sq cm/sec*	*After Regeneration at 450°C Diffusivity sq cm/sec*
CAT-C	0.395	0.0104	0.0094
CAT-C	0.521	0.0104	0.0104
CAT-C	0.954	0.0090	0.0086
CAT-B (fresh)	0.0	0.0217	—
CAT-B	0.201	0.0175	0.0192
CAT-B	0.584	0.0174	0.0196
CAT-B	0.274	0.0188	0.0188

Surface Area of Fresh and Coked Catalyst

Wt % Coke on Catalyst	*Surface Area, m^2/g*
0.0	211
0.20	208
0.39	202
0.58	200
0.95	213

[a] Hydrogen at room temperature.
[b] Data of Ozawa and Bischoff. CAT-B—Particle diameters between 4.76 and 3.36 mm; CAT-C—Particle diameters between 3.36 and 2.38 mm.

which is associated with the initial lay-down of coke on the surface, the mass transfer characteristics of the catalyst are unaffected by carbon deposition. (The self-consistency of diffusion results is such that the experimental values of diffusivity are the same before and after regeneration within experimental error.) In addition, measurements of the total surface area (BET) indicate essentially no variation with coke formation, in support of the diffusivity results.

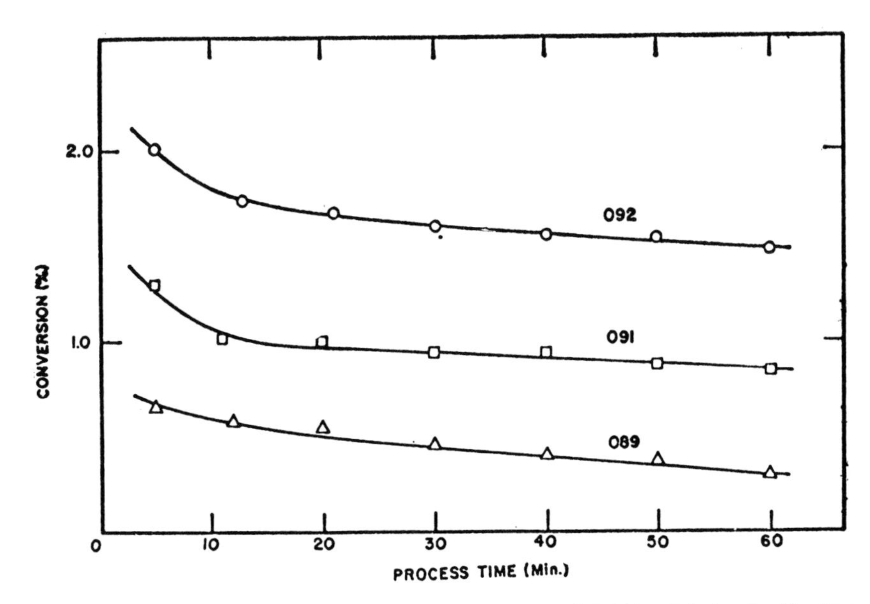

Industrial and Engineering Chemistry

Figure 13a. Relation of conversion to process time (catalyst deactivation) (19)

In their experimental measurements of conversion *vs.* process time, shown in Figure 13a, Ozawa and Bischoff noted an initial rapid decay in activity, followed by a slower, long term transient. Similar behavior had been noted by several prior workers, such as Rudershausen and Watson (Figure 10a) with cyclohexane aromatization on molybdena-alumina and Pozzi and Rase (*20*) with isobutylene hydrogenation on Ni-kieselguhr; hence, it was postulated that the coking rate could arbitrarily be associated with one of two regimes, depending on the level of coke deposition or process time. They were able to check this experimentally, using a direct and continuous gravimetric measurement of the total catalyst weight, the variation with time being attributed to coke deposition. The results for CAT-B and CAT-C are shown in Figures 13b and c; these precise measurements clearly indicate that two straight lines on the log-log correlation, intersecting at a time of about 10 minutes, are required

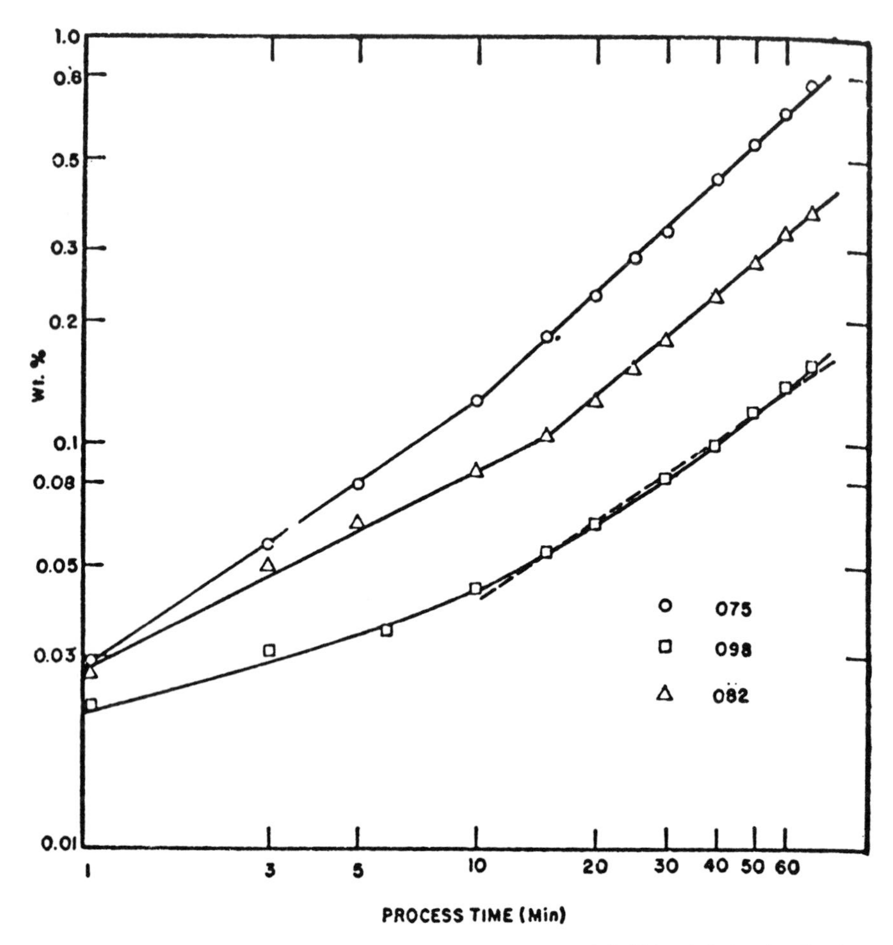

Industrial and Engineering Chemistry

Figure 13b. Relation of weight per cent coke on catalyst to process time (Catalyst B) (19)

to represent the coke concentration on catalyst *via* a Voorhies correlation. We shall see later that such behavior is not inconsistent with a type of diffusion-limited coke deposition in which a coke layer is progressively deposited within the catalyst structure. The values of the Voorhies exponent for the coking at $\theta > 10$ minutes obtained by Ozawa and Bischoff for differing flow rates and temperatures, are given in Table VI. Again, a range of values for n is obtained, generally increasing with increasing temperature which is, of course, opposite to what would be expected if coking rates were diffusion limited.

There are surprisingly few studies in detail of the possible chemical mechanisms of coke formation, but the work which has been done is quite informative in some aspects and does not conform exactly to the expectations one might have as a result of the investigations we have just discussed. One relatively recent and detailed study of the mechanism of coke formation is that of Appleby *et al.* (*21*) who were concerned with determination of the importance of aromatic content on the extent of coke formation, the mechanism of coke formation *via* aromatic reactions, and the contribution of very reactive compounds such as olefins to the over-all coke formation. Although we have seen that noncyclic materials can lead to significant coke formation, as in the studies of Rudershausen and Watson and Ozawa and Bischoff, the bulk of evidence is that the aromatic content of process streams is primarily responsible for coking. Typical

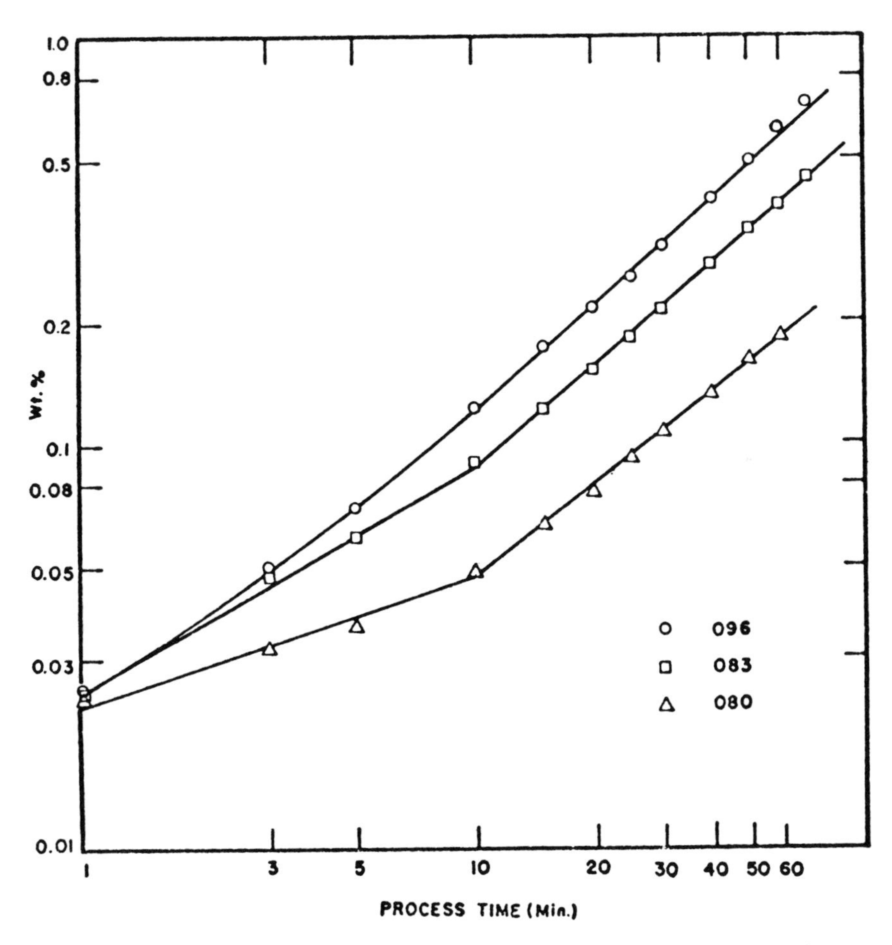

Figure 13c. Relation of weight per cent coke on catalyst to process time (Catalyst C) (19)

of such structures are benzene, naphthalene, phenanthrene, chrysene, and pyrene rings with various alkyl substituents. In cracking reactions, the side chains are effectively removed first, so the bare ring structures seem to be the important intermediates in coke formation. An indication of the importance of aromatic content is shown in Table VII, where the coke yield of a hydrogenated cycle oil is about one-third that of the nonhydrogenated material. The monoaromatics here are mostly Tetralin and indane types, diaromatics are naphthalenes, triaromatics phenanthrenes, and higher pyrene, perylene, etc.

The experiments reported by Appleby *et al.* consisted of cracking a series of aromatic, naphthenic, heterocyclic, and paraffinic hydrocarbons on either a commercial silica–zirconia–alumina catalyst (86.2/9.4/4.3 wt %; UOP Type B) or a synthetic silica–alumina (90/10 wt %; American Cyanamid Co.). The reaction conditions generally were 500°C, 1 atm reactant pressure, and flow rates on the order of 10 moles/liter/hour. One series of experiments was conducted with *n*-butene and 1,3-butadiene to investigate the coking properties of olefins, which although not of the structure of the final coke product, are nonetheless very reactive compounds which could interact with already adsorbed species and undergo condensation and hydrogen elimination. At 500°C a substantial amount of higher boiling products was produced, 22% by wt in the case of *n*-butene, of which half was aromatic. The coke formation was about 15 wt % of charge and contained 4.9 wt % H_2, compared with 14.3% H_2 in the feed. Since similar results were obtained with the butadiene, it was concluded that even with low molecular weight straight-chain

Table VI. Exponent in Voorhies Correlation[a]

Run No.	*Reaction Temp., °C*	*C_2H_4 Flow Rate, cc/min*	n, *$\theta > 10$*
		Reactor A	
077	499	100	0.902
096	499	75	0.885
075	498	50	0.913
094	500	30	0.881
097	450	75	0.742
082	449	50	0.820
083	450	30	0.841
098	400	75	0.625
080	399	50	0.736
081	399	30	0.605
099	350	50	0.554

[a] Data of Ozawa and Bischoff (*19*).

Table VII. Aromatics Composition and Coke Formation from a Recycle Oil before and after Hydrogenation[a]

	Recycle Oil	Hydrogenated
H_2 added, cu ft/bbl	0	745
Aromatics content, mmoles/100 grams		
monoaromatics	27	102
diaromatics	39	15
triaromatics	73	8
higher	23	5
Coke yield in catalytic cracking, wt % of feed, at 50% conversion	13	4

[a] Data of Appleby *et al.* (21).

hydrocarbons (excluding paraffins, which were found generally unreactive at these conditions) polymerization and cyclization reactions occur which produce significant quantities of aromatics and low hydrogen coke.

Experiments conducted with polycyclic aromatics and some heterocyclic compounds containing S, N_2, and O_2 determined the coke formation on catalyst at the conditions as shown in Figure 14a. The general characteristics of coking behavior are indicated by the sequences shown in the figure. Coke formation is more pronounced in condensed rings (naphthalene → anthracene, etc.) than in linked rings (biphenyl → terphenyl, etc.), while with the heterocyclic materials the hydrocarbon analog was always most productive of coke. The conversion of nitrogen bases such as quinoline is in general accord with the results of Mills *et al.*, discussed previously.

In the cracking of both olefins and aromatics, substantial reductions in surface area were obtained: *n*-butene and phenanthrene giving similar results of 22 and 33% loss in surface area at 2.8 and 10.4 wt % coke on catalyst, respectively. This is, of course, in considerable disagreement with the results of Ozawa and Bischoff, although the levels of coke deposition were somewhat lower in their study. This point seems to be the focus of considerable disagreement among various experimental results reported in the literature. Other conflicting studies, for example, are those of Ramser and Hill (22) who found a 27% decrease in surface on formation of 2.2 wt % carbon, and Haldeman and Botty (23) who found very little effect of coke deposit on either surface area or pore size distribution at levels on the order of several wt %. Since the catalysts used in the various studies were somewhat different, however, one must conclude that the effect of coking on the mass transfer characteristics of cracking catalysts is structure sensitive, and each case must be examined individually.

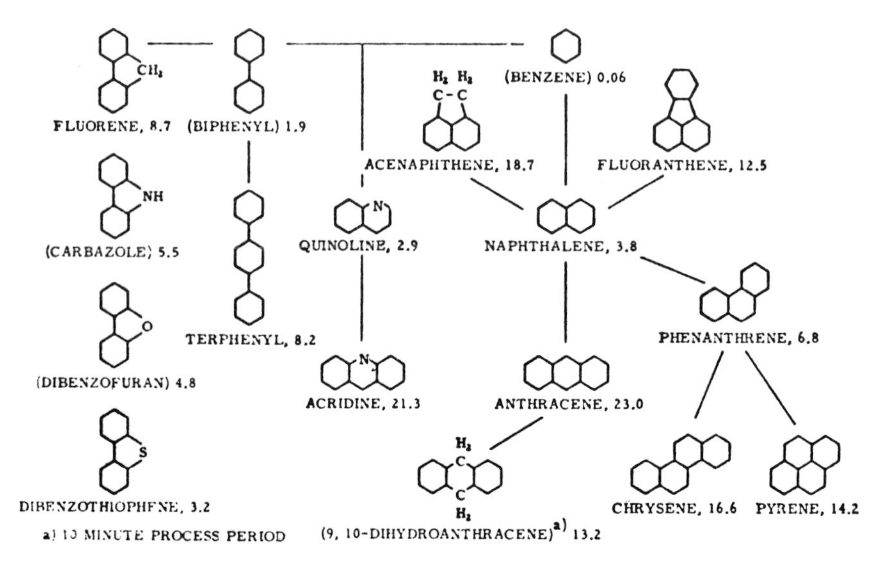

Industrial and Engineering Chemistry

Figure 14a. Coke formation from polycyclic and heterocyclic aromatics over fresh silica–alumina catalyst. Temperature = 500°C; process period = 15 minutes; pressure = atmospheric; flow rates = ca. *10 to 13 moles/liter/hour, except for compounds named in parentheses; numbers indicate coke as weight per cent of catalyst* (21).

The point has been considered in detail by Levinter *et al.* (*24*) in a series of coking experiments on silica–alumina catalysts, the most important of which involved styrene–benzene mixtures at 500°C and LHSV of 5 cc/hour-gram catalyst. In all cases a limiting degree of coking was observed, which varied with styrene concentration (benzene being essentially inert for coke formation under the conditions employed), with catalyst average pore radius and with flow rate and average catalyst grain size. This limiting degree of coke formation, however, did not in general correspond (in terms of wt % coke on catalyst) to the maximum possible amount of coke deposition computed from known catalyst porosity and coke density, about 68% for the experimental materials used. The limiting amount varied from 10 to 50 wt % on catalyst but generally was associated with a lower limiting specific area of around $2m^2$/gram (initial surface areas were not reported), which indicates strongly that the limiting value of coke formation corresponds to pore blocking and consequent elimination of accessible surface. This pore blockage may occur more or less uniformly throughout the catalyst structure, or, if temperatures are high and coking rates large, in a discontinuous reaction zone or shell (discussed later in detail).

The quantitative results of Levinter *et al.* are probably of less interest than the demonstration that pore blocking can occur on coke deposition to various extents depending strongly on catalyst properties and reaction conditions. Hence, under extreme conditions it might be possible to eliminate completely accessible surface at very low total levels of coke on catalyst while under more moderate conditions near-theoretical limiting amounts would be required before surface access was denied. The results of various workers concerning changes in diffusivity and surface area which we have discussed, then, are not necessarily in conflict; one can conclude that coke formation can, under proper conditions, substantially alter physical and transport properties of catalysts as well as their activity.

Table VIII. Reaction of a Benzene–Naphthalene Mixture[d]

Composition, 78 mole % benzene + 22 mole % naphthalene
Temperature, 500°C

Process period, 1 hr
Pressure, atmospheric
Flow rate, 6.0 moles/liter/hr
Catalyst, fresh silica–alumina

		Distillate[a]				
Fraction	*Gas*	*Below 161°C*	*161–261°C*	*261–492°C*	*Residue above 492°C*	*Coke*
Charge, wt %	0.01	69.52	23.79	0.18	2.26	2.63
Analysis						
carbon	—	—	93.7	93.4	95.2	85.2
hydrogen	—	—	6.3	6.6	4.8	2.9
ash[b]	—	—	—	—	—	11.9
Loss on heating[c]	—	—	—	—	—	1.5

Some Polycyclic Aromatics Identified in Residual Oil from Benzene–Naphthalene Mixture

Polycyclic	*Relative Molar Quantities*
Fluoranthene	6
Perylene	4
Benzofluoranthene	1
1,2-Benzanthracene	3
Chrysene	3
Pyrene	1

[a] An effort was made to separate roughly the benzene, combined naphthalene, and methylnaphthalene fractions, and higher boiling distillate. The temperatures listed are of the vapor in a Claisen flask. The quantities of material involved were too small to avoid some remixing in the take-off manifold.
[b] By combustion of coke remaining after NaOH dissolution of catalyst.
[c] In nitrogen at 200°C.
[d] Data of Appleby *et al.* (*21*).

The structure of the coke layer deposited by various hydrocarbons was examined by Appleby *et al.* *via* x-ray diffraction; it was concluded that all deposits were quite similar, involving a turbostratic layer structure suggestive of condensed aromatic nuclei of high molecular weight serving as the reaction intermediates in coke formation.

One particularly interesting result shown in Figure 14a is the very high activity of anthracene and other condensed aromatics in coke formation, which led Appleby *et al.* to postulate that interactions involving benzene and naphthalene might be taken as representative of important reaction paths in coke formation. The results of some experiments along this line are given in Table VIII. The product fraction above the boiling point of the feed mixture is of particular interest since it would contain those materials most closely related to coke formation, most particularly the high boiling residue ($>492°C$) which is formed in amounts about equal to the coke formation, assuming the figure given for the latter refers to total feed. The spectra of that portion of the residue soluble in benzene showed the compounds listed in the second part of Table VIII to be present in the indicated relative quantities, in total amounting to about 10% of the entire high boiling residue. The first three listed are direct addition products, while the others demonstrate the occurrence of possible complex dehydrogenation or polymerization steps. Nonetheless, it seems clear that an entire spectrum of condensed aromatics can be formed by reactions of this type—that is, characterized in general by the simple type of addition visualized by Rudershausen and Watson.

Further experiments were conducted to test the effect of side chains on coke formation using alkyl benzenes and naphthalenes. Coke increases with the total number of substituted alkyl carbon atoms, rather than with the number of alkyl substituents—viz.:

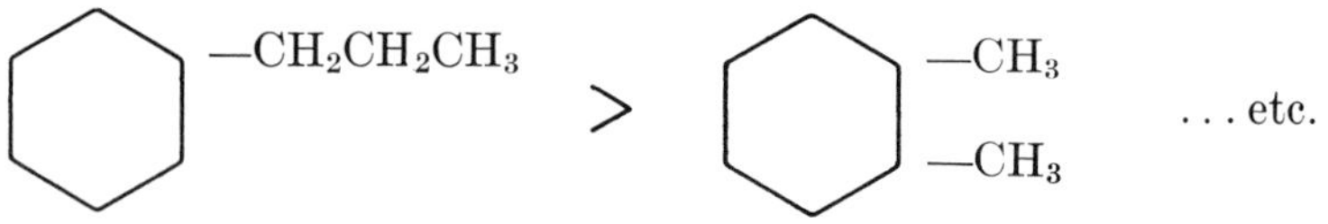

$$C_6H_5\text{—}CH_2CH_2CH_3 > C_6H_4(\text{—}CH_3)(\text{—}CH_3) \quad \ldots \text{etc.}$$

Structural factors such as this are far more important in determining coke formation tendencies than is molecular weight, and over-all reactivity would, as in the case of cracking activity, be expected to be related to the acid–base properties of the catalyst–reactant involved. A rough correlation of this sort is shown on Figure 14b, in which a property related to the basicity of the aromatic is plotted *vs.* carbon deposition for the cracking experiments of Appleby *et al.* There is certainly some evidence here that the acidic property of the catalyst is responsible for coke formation, presumably *via* the formation of carbonium ions generated from various aromatics. For the benzene–naphthalene reaction, then, one could write:

I ⟶ II

+ II ⟶ III or IV

III ⟶ V (ring closure of III)

IV ⟶ VI (proton elimination)

V ⟶ VII (proton elimination)

where the original H atoms and double bonds have been eliminated for simplicity. Dehydrogenation of VII would lead to the principal fluoranthene product listed in Table VIII. It is also worth noting that, depending on the details of the benzene interaction with the surface and the

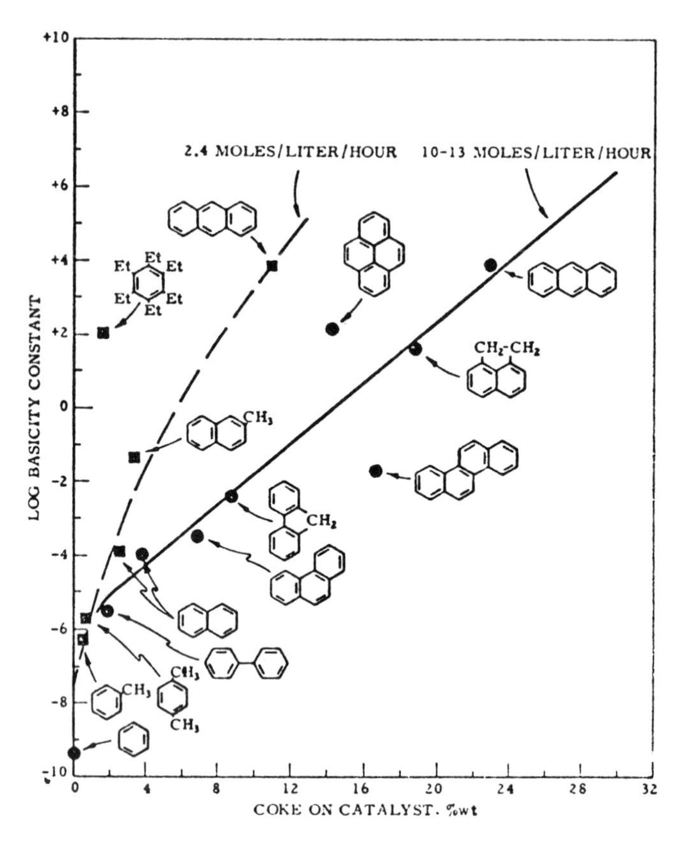

Industrial and Engineering Chemistry

Figure 14b. Coke formation in catalytic cracking and hydrocarbon basicity (21). [*Basicity as defined by E. L. Mackor, A. Hofstra, and J. H. van der Waals,* Trans. Faraday Soc. *(1958)* **54,** *66, 186*].

speed of intermediate steps, such a scheme is not inconsistent with the "dual site" analysis used by prior workers. However, the implication is that similar catalytic functions are involved in coking and cracking, which is not exactly the picture provided by some of the data we have discussed.

A final study we consider dealing with the structure and mechanism of the formation of coke is the infrared work of Eberly *et al.* (*17*). They determined the spectra of the carbon deposit resulting from the cracking of various materials on a SiO_2–Al_2O_3 catalyst, described previously. The substance of their results is given in Figures 15a and b. The particular bands of interest here are those of the C–H stretch region, 3100–2800 cm^{-1}, with aromatic, methylene, and methyl groups as shown in Figure 15a. A second feature of the spectra of interest is the region 700–1800

cm^{-1}, shown in Figure 15b for the 1-methylnaphthalene cracking. These frequencies can be assigned to aromatic skeletal vibrations, and since the intensities in this region are strong compared with the various C–H frequencies, it was concluded that the coke deposits were largely composed of highly condensed aromatic structures of low hydrogen content, observations which are quite in agreement with those of Appleby *et al.* Additional evidence for the low hydrogen content of the coke is provided by the data on olefin formation from *n*-hexadecane cracking shown in Figure 15c. These indicate that conversion to olefins decreases at high feed conversion, which Eberly *et al.* interpret as being caused by hydrogenation of the unsaturates by hydrogen transfer from the adsorbed molecules.

Most of the information available in the literature concerning the kinetics and mechanism of coke formation deals with silica–alumina catalysts, a situation which is reflected in this review. The advent of

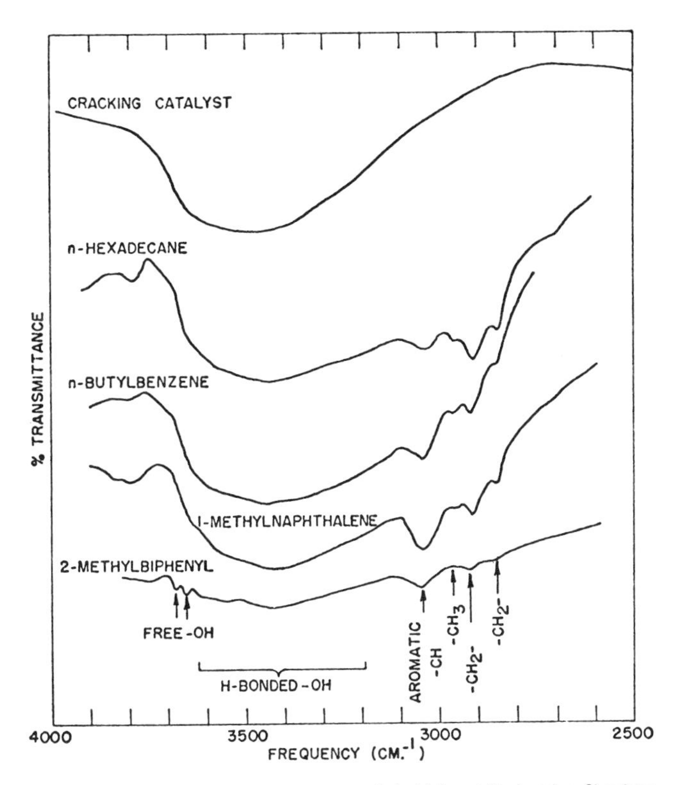

Industrial and Engineering Chemistry

Figure 15a. Infrared spectra of coke deposits from 13% Al_2O_3–87% SiO_2. Obtained by cracking various compounds at 445°C and 0.4 v/v/hr (17).

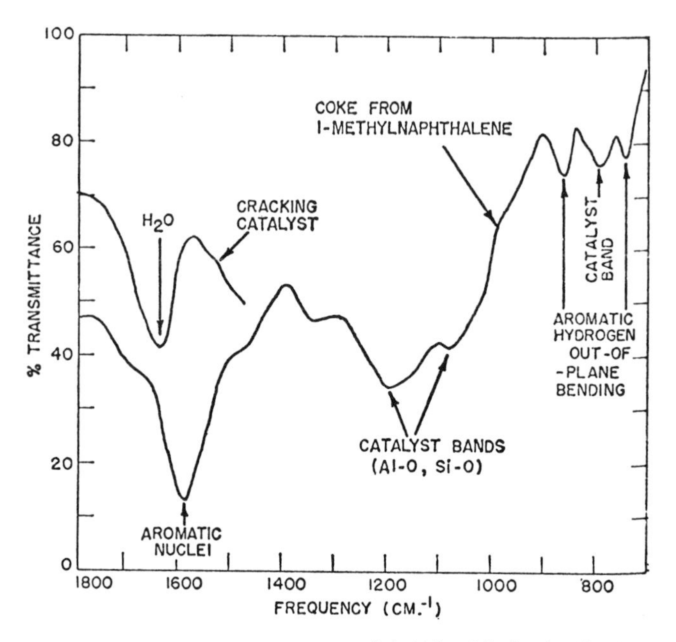

Industrial and Engineering Chemistry

Figure 15b. Infrared spectrum of 1-methylnaphthalene coke deposit (17)

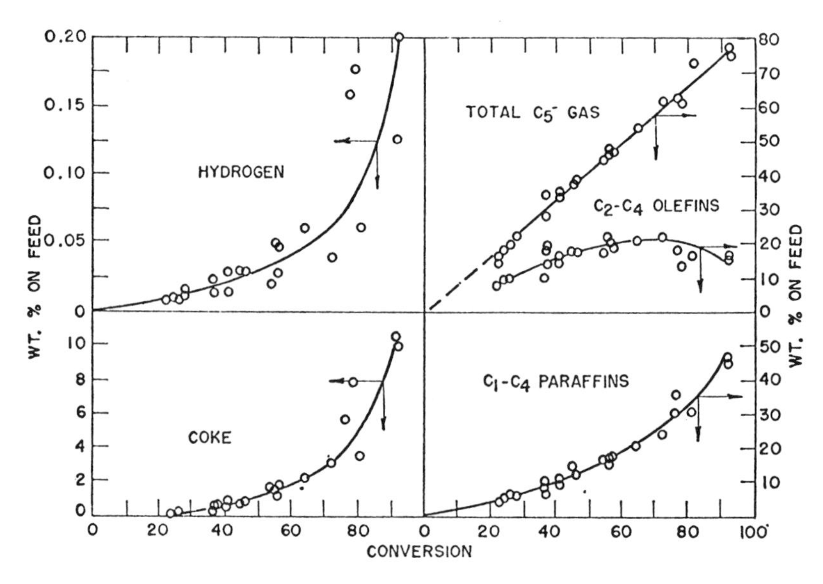

Industrial and Engineering Chemistry

Figure 15c. Product distribution from cracking n-*hexadecane on 13% Al_2O_3–87% SiO_2 at 500°C* (17)

commercial zeolitic catalysts has produced a certain amount of similar information with respect to these materials. Coke formation on zeolites is the concern of a recent experimental study by Eberly and Kimberlin (*25*), and some typical recent data on zeolitic deactivation kinetics are given by Weekman (*26, 27*) and by Weekman and Nace (*28*).

Sintering of Active Surface Area. Cohesive information or analysis of catalyst deactivation by sintering, particularly the kinetics thereof, is not widely available in the literature, and the discussion here is correspondingly limited. Indeed, we treat only two examples, the sintering of a Pt–Al_2O_3 catalyst and a "chemically-assisted" type of sintering of Pt film which has been described recently.

Sintering is normally thought of as a thermal phenomenon in which crystallite agglomeration and growth occur by several possible mechanisms (*i.e.*, surface or grain boundary self-diffusion) which can be considered activated (*29*); hence, long periods at high temperature lead to the loss of internal surface areas in porous structures typical of catalysts or supports. A typical example of this is shown in Figure 16a, giving the data of MacIver, Tobin, and Barth (*30*) for η- and γ-alumina surface area

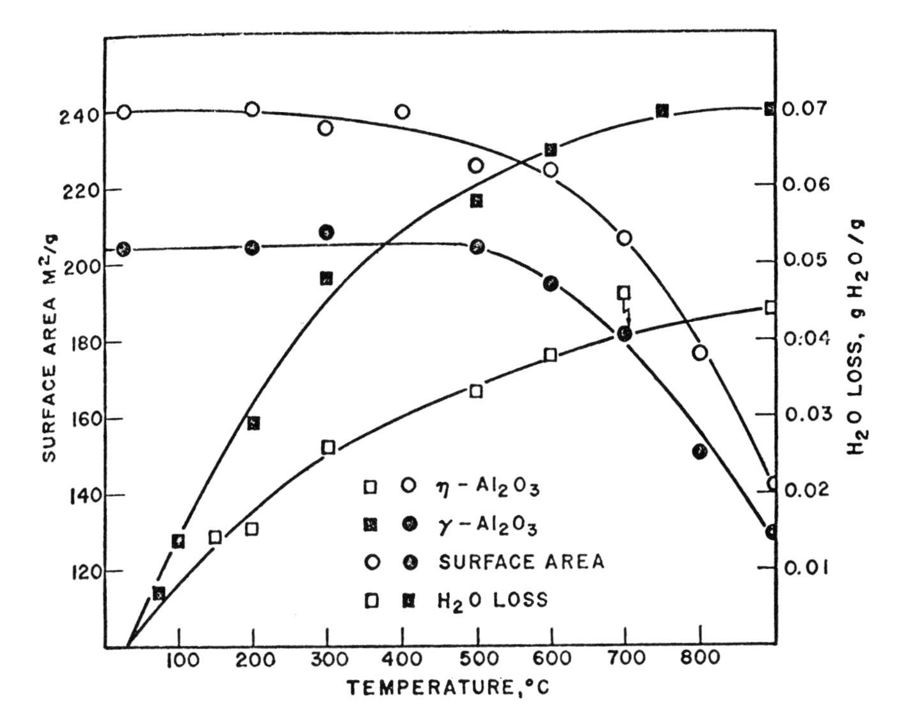

Figure 16a. Surface area and water loss of eta- and gamma-alumina (30)

Table IX. Platinum Size Measurements[a]

Sample	*hrs at 780°C*	*cc H_2 (STP)/ g cat*	*m^2/ g cat*	*Area, m^2/g Pt*	d_s *average Pt size (A)*
1	0	0.282	1.397	233	10
2	2	0.245	1.213	202	12
3	4	0.087	0.431	72	33
4	10	0.039	0.193	32	73
5	17	0.018	0.039	15	158
6	72	0.0062	0.031	5	452

n-Heptane Reforming Experiments

Sample	*1*	*2*	*3*	*4*	*5*
Area, m^2/g Pt	*233*	*202*	*72*	*32*	*15*
d_s*(A)*	*10*	*12*	*33*	*73*	*158*
MSHSV, mole/m^2 Pt hr $\times 10^{-2}$	*1.75*	*2.02*	*5.8*	*12.7*	*27.1*
mole % *n*-heptane remaining	3.0	3.5	4.6	7.0	9.8
mole % dehydrocyclization	37.4	32.8	26.8	21.6	17.7
mole % isomerization	9.0	10.6	14.2	21.7	24.3
mole % hydrocracking	50.6	53.1	54.4	49.7	48.2

[a] Data of Maat and Moscou (*33*).

and water content. These equilibrium values illustrate the magnitude of surface (and activity) loss possible; from 500° to 800°C the η-alumina loses almost 30% of its original surface. Similar magnitudes are typical of other materials also, as those who have encouuntered sintering problems can attest only too well.

The incorporation of sintering kinetics into expressions such as Equation 2 would be desirable since this type of deactivation could then be handled in analysis in a manner similar to poisoning or coking deactivations. As indicated, it is possible to represent sintering as an activated process; for example, the surface and grain boundary self-diffusion coefficients of nickel have activation energies of 14.3 and 28 kcal/mole, respectively (*31*). The system of catalytic interest for which the most information is available is supported platinum. The original work on Pt sintering kinetics was reported by Herrman *et al.* (*32*), who correlated the rate of decrease in surface area of a Pt–Al_2O_3 catalyst as second order in remaining surface area. This approach forms the basis of the study of Maat and Moscou (*33*), who were interested in sintering effects both on activity and selectivity in Pt-catalyzed reforming of naphtha. They used a commercial powerforming catalyst, 0.6% by wt Pt on Al_2O_3 containing approximately 0.5 wt % chloride. Experimental results on the kinetics of the sintering of this catalyst are given in Table IX. These experiments consisted of heating the original sample for the time indicated in an inert atmosphere, then determining surface area of the Pt by hydrogen chemisorption, B–E in Table IX (*34*) and by electron micrography, F. As indicated in column F, the crystallite size distribution is very wide, and the average sizes obtained from chemisorption are perhaps more useful in kinetic interpretation. If sintering kinetics are taken as second order in Pt area, we have:

$$\frac{1}{A_s} = \frac{1}{A_{s_o}} + k_A\theta \qquad (14)$$

where A_s and A_{s_o} are current and initial surface areas, respectively, and k_A is some sintering rate constant. A plot of the data of Maat and Moscou according to Equation 14 is shown in Figure 16b; a good correlation is obtained. Previous data of Herrmann *et al.* indicated an activation energy of about 70 kcal/mole for the sintering of Pt; this value, while high, is not inconsistent with those of volume diffusion coefficients in some fcc metals.

The effects of sintering on reactivity and activity in *n*-heptane reforming at 200 psia, 500°C, 2.44 gram/hr-gram catalyst and H_2: *n*-heptane of 5.3 mole/mole are also summarized in Table IX. Significant effects on activity are seen although not as large as might have been expected; change from 233 to 5 m^2/gram Pt, a 98% reduction in area, gives a change in conversion from 97 to 76%, only a 25% reduction. This

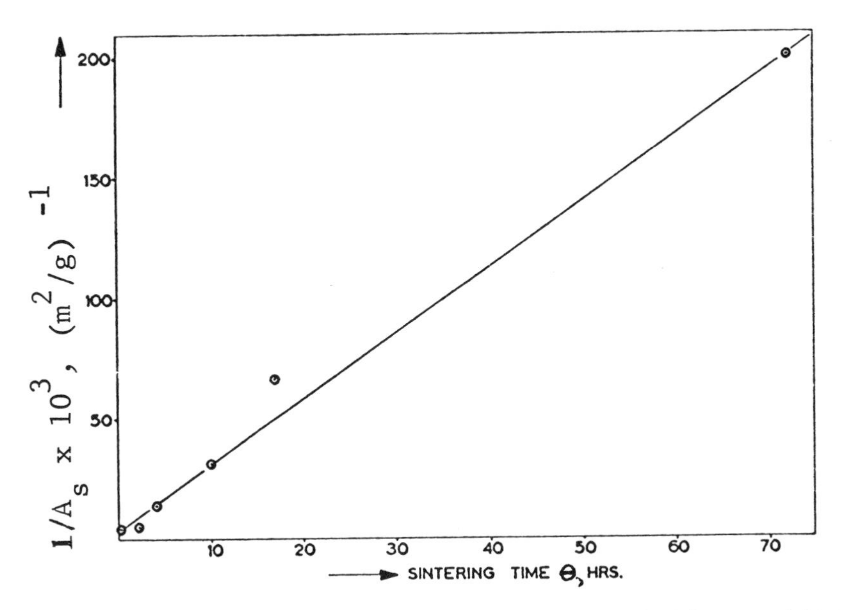

Figure 16b. Second-order sintering reaction (33)

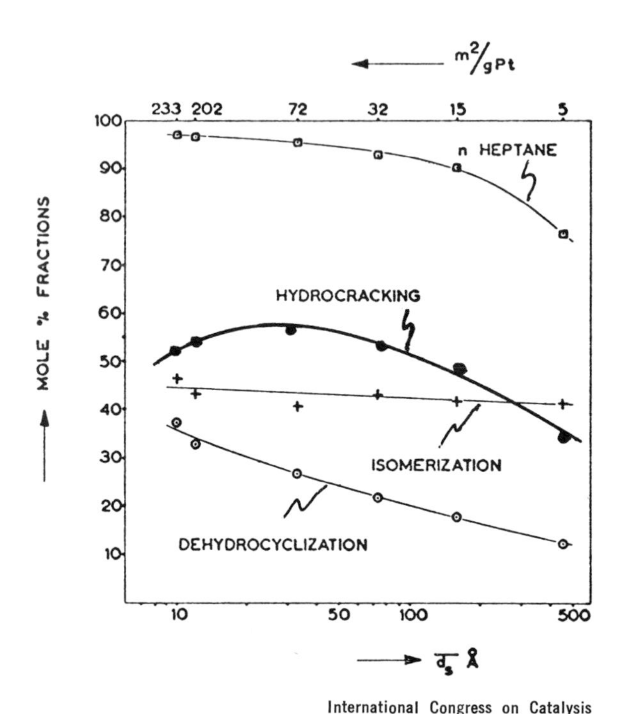

Figure 16c. Product distribution of n-*heptane reforming* (33)

is in part caused by differing intrinsic activity associated with different crystallite sizes, and is a sufficiently complex problem to be well beyond the present discussion. A further important message is conveyed by these results, however, and is illustrated in Figure 16c. The selectivity between the various *n*-heptane reforming products changes markedly with differing degrees of sintering since changes in activity seem primarily to be reflected in decreased cyclization reactions; the sum of cyclization and isomerization remains approximately constant. Since the aromatics produced in dehydrocyclization reactions would contribute more to an octane rating of the product than any of the other components, this decrease represents a real change in product quality resulting from a selectivity alteration caused by deactivation. It is important, then, to keep in mind that catalyst deactivations of all sorts may affect not only the activity, but the selectivity properties of a catalyst as well.

Another mechanism of sintering of metallic surfaces has been reported by a number of investigators, of which the study by Clay and Petersen (*11*) is a recent example. In these cases, the chemisorption of a poison on the metal surface causes crystallite growth, such that one can regard the phenomenon as a chemically assisted type of sintering. Clay and Petersen observed the sintering of Pt films on chemisorption of arsine; the resulting non-selective decrease in activity of the film for cyclopropane hydrogenolysis, which we have already shown in Figure 4, was demonstrated to be correlated directly with a decrease in the surface area of the film. The net sintering effect was probably the primary result of the formation of $PtAs_2$ on the surface on chemisorption of AsH_3 according to:

$$Pt + 2AsH_3 \rightarrow PtAs_2 + 3H_2$$

$PtAs_2$ has a much lower melting point than the metal; hence its formation would tend to decrease considerably the thermal stability of the catalyst at a given temperature level. Moreover, there is approximately a 35% increase in the lattice parameter as one goes from Pt to $PtAs_2$, and the heat of chemisorption involved in the Pt–AsH_3 interaction was estimated by Clay and Petersen to be in the range of 90–120 kcal/mole Pt. Thus, the combination of altered thermal and geometric properties of the surface plus significant heat effects on chemisorption was thought to induce the loss of surface area which was directly reflected in loss of activity.

Observations of directly induced thermal sintering are relatively widely available for cracking catalysts. Not much rate analysis has been done, with the exception of a recent engineering study by Gwyn (*35*), and since the over-all problem is closely bound up with the question of the development of thermal gradients within catalyst particles on coke burnoff (*36*), our discussion of this particular type of sintering is postponed to a later section. Further detail on the physical mechanisms of sintering and experimental methods for their investigation can be found in the paper by Kingery and Berg (*29*).

It seems clear from the foregoing examples on poisoning, fouling, and sintering that whatever the microscopic mechanism of the process, one may consider it in terms of an additional chemical process to that of the main catalysis which is occurring. This has important implications in chemical reaction engineering since it means that the separable form of rate equation, as in Equations 1 and 2, does correspond to physical (or chemical) reality, and one generally should be able to isolate and identify catalyst deactivation effects on the behavior of macroscopic systems. For reaction modeling purposes, then, it is perfectly adequate to represent general reaction schemes as indicated under "Mechanisms" in Figure 17. For example, in the type I system [nomenclature employed by analogy to the discussion of selectivity by Wheeler (*37*)] the poisoning of active catalytic sites by impurity chemisorption is represented by the step $L + S \rightarrow L \cdot S$. Type II represents a reactant or product inhibition such as $A + S \rightarrow A \cdot S$, which would be associated with coke formation. Obviously these simple models can be expanded in their form to represent various situations such as poisoning or fouling of polyfunctional catalysts, reversible chemisorptions, etc. Some work of this sort has been done in reactor modeling and is discussed later; the important point here is that deactivation can be analyzed in terms of normal reaction kinetics with associated rate constants and activation energies, even when an apparently physical process such as sintering is concerned.

Type I $A+S \rightarrow B+S$
$L+S \rightarrow L \cdot S$

Type II $A+S \rightarrow B+S$
$A+S \rightarrow A \cdot S$

$A+S \rightarrow B+S$
$B+S \rightarrow B \cdot S.$

Chemical Engineering Science

Figure 17. Type I and type II mechanisms (77)

In fact, there is a bit of divergence in the manner in which kinetics of deactivation have been handled in the literature. Szépe and Levenspiel (3) have tabulated the forms of activity correlation noted in studies such as those we have discussed, as shown in Figure 17a and b, many of which are variants of the forms of Equation 4. We have demonstrated an example of almost every type shown here, and in each case it is possible to relate the observed activity correlation to a rate equation based on activity only, as shown in Table X. Thus, while the exponential form of activity correlation would arise from the rate equation for deactivation:

$$\frac{ds}{d\theta} = -\beta_2 s \qquad (15)$$

the corresponding form according to a type I mechanism would be:

$$\frac{ds}{d\theta} = -\beta_2 s l \qquad (16)$$

where the appearance on the RHS of the concentration of poison, l, prevents the rate of deactivation from being a function only of current activity, s. The distinction between Equations 15 and 16 is important to keep in mind since the literature we discuss subsequently treats deactivation kinetics in both ways.

As shown in Table X, Szépe and Levenspiel propose a general power law form for the kinetics of catalyst deactivation, by whatever means.

Form of activity decay	Author	System	Equation
Linear	Maxted †	Liquid-phase hydrogenation of crotonic acid on platinum (poisoned by thiophene), and other hydrogenations	$s' = s'_o - \beta_1\theta$
	Eley–Rideal †	Para-hydrogen conversion on tungsten (poisoned by oxygen)	
Exponential	Pease–Stewart	Hydrogenation of ethylene on copper (poisoned by carbon monoxide)	
	Herington–Rideal	Dehydrogenation of paraffins on chromia-alumina	$s' = s'_o \exp(-\beta_2\theta)$
Hyperbolic	Germain–Maurel ‡ Maat - Moscou	Dehydrogenation of 1-1-3 cyclohexane on platinum-alumina; Pt - Al_2O_3 sintering	$1/s' = 1/s'_o + \beta_3\theta$
	Pozzi–Rase	Hydrogenation of iso-butylene on nickel	$x_o - x_\infty / x - x_\infty = 1 + \beta_4\theta$
Reciprocal power function	See b for a list of some observations of this form		$s = A\theta^{-\beta_5}$ $s' = A'\theta^{-\beta_6}$
Elovich-equation	Parravano ††	Oxidation of carbon monoxide on NiO	$dx/d\theta = B\exp(-\beta_7 x)$

† Good linear fit only in the initial range of deactivation.
‡ The initial data are not well fitted.
Data obtained in low conversion region.
†† Data obtained in batch system, at low conversion levels.

Proceedings of the 4th European Federation on Chemical Reaction Engineering

Figure 17a. Experimentally found deactivation equations (3); *s' = conversion definition; s = rate definition*

Author	System	Value of n $C_c = A\theta^n$	Value of β_5 $s = A\theta^{-\beta_5}$	Value of β_6 $s' = A\theta^{-\beta_6}$
Voorhies †	Gas oil cracking on natural and synthetic catalysts	0.38–0.53	—	0.19
Blanding	Catalytic cracking of light East Texas gas oil on natural clay	—	0.57	—
Ruderhaus Watson	Aromatization of cyclohexane on MO-Al hydroforming catalyst	0.55	0.36	—
Crawford–Cunningham	Cracking of East Texas gas oil on natural and synthetic catalysts	0.44	—	—
Prater–Lago	Cracking of cumene on silica-alumina	0.5	—	—

† Both fixed and fluidized beds. Derivation is given for n= 0.5.

Proceedings of the 4th European Federation on Chemical Reaction Engineering

Figure 17b. Coke formation and deactivation in various hydrocarbon reactions (3)

While it is felt that forms such as Equation 16 may be more fundamental than pure activity expressions, this distinction is a small one. The important point is that it is possible to obtain a description of the kinetics of deactivation which, for the purposes of the reaction analyst, may be incorporated into the basic equations of conservation for the microscopic system and not tacked on *a postiori via* the device of time-dependent correlation and time averaging.

Table X. Summary of Deactivation Equations in Terms of the Power Law Form[a]

Type of Decay	*Integrated Form*	*Differential Form*	*Exponent in* $-ds/d\theta = k's^w$
Linear	$s = s_o - \beta_1\theta$	$-ds/d\theta = \beta_1$	0
Exponential	$s = s_o \exp(-\beta_2\theta)$	$-ds/d\theta = \beta_2 s$	1
Hyperbolic	$1/s = 1/s_o + \beta_3\theta$	$-ds/d\theta = \beta_3 s^2$	2
Voorhies	$s = A\theta^{-\beta_5}$	$\begin{cases} ds/d\theta = \\ \beta_5 A^{1/\beta_5} s^{(\beta_5+1)/\beta_5} \end{cases}$	$\beta_5 + 1/\beta_5$
Elovich	$dx/d\theta = B \exp(-\beta_7 x)$		

[a] Adapted from Szépe and Levenspiel (*3*).

Deactivation Effects in Intermediate Systems: The Analysis of Individual Particles

When one attempts to incorporate the details of deactivation kinetics into macroscopic or semimacroscopic catalytic systems, a number of problems arise which are reminiscent of those associated with the analysis of transport-limited catalytic reactions. Indeed, in some cases the two problems are not entirely unrelated, and one of the first studies of deactivation effects on the behavior of individual catalyst particles was that by Wheeler (*37*), which demonstrated the apparently selective poisoning of a catalyst by homogeneous adsorption of poison under conditions of diffusion limitation.

This section discusses that work and some of the more general analyses which have followed it concerning deactivation and diffusion effects, both singly and combined, on the action of individual catalyst particles—the intermediate region between the microscopic events of deactivation which we considered in the previous section and the macroscopic world of catalytic reactor behavior which we discuss subsequently. It is assumed that the reader is familiar with the elements of transport-limited catalysis; for a review, *see* Satterfield (*38*).

Homogeneous and Pore Mouth Poisoning. In his famous study of reaction rates and selectivity in catalysis Wheeler (*37*) carried out some of the early work on the intermediate system. His concern was twofold:

(a) what is the nature of uniform catalyst poisoning in situations where the catalysis is diffusion limited, and (b) what is the effect of nonuniform poisoning (selective in the sense of Figures 2 and 3 in representation of the intrinsic poisoning) in such a situation?

Wheeler visualized the porous catalytic structure as a composite of intersecting capillaries of average radius $\bar{r}$ and showed that for a particle of arbitrary shape the Thiele modulus could be defined on the basis of a length parameter computed by:

$$W = \frac{V_p}{S_x}\sqrt{2} \tag{17}$$

where the $\sqrt{2}$ is related to the tortuosity of the porous structure. The Thiele modulus so defined, for a first-order, irreversible reaction, is:

$$h = W\sqrt{\frac{2k_a}{\bar{r}D}} \tag{18}$$

where k_a is the rate constant per unit surface area, D is the effective diffusivity of diffusing species in the porous structure, V_p is the particle volume, and S_x is its external surface area. The effectiveness factor for slab geometry, or an equivalent single pore model, is:

$$\eta = \frac{\tanh h}{h} \tag{19}$$

and the rate of reaction is proportional to $h \tanh h$. If we assume that the nonselective activity expression, Equation 3, represents the poisoning which is occurring, then the intrinsic activity of the pore (or pellet) at any time is given by $k_a s$, and the Thiele modulus is:

$$h = W\sqrt{\frac{2k_a s}{\bar{r}D}} = h_o\sqrt{s} \tag{18a}$$

where h_o is the modulus for the unpoisoned pellet. The ratio of rates in the poisoned and nonpoisoned cases is:

$$F = \frac{\text{Rate poisoned}}{\text{Rate nonpoisoned}} = \frac{h_o\sqrt{s}\tanh h_o\sqrt{s}}{h_o \tanh h_o} = \sqrt{s}\,\frac{\tanh h_o\sqrt{s}}{\tanh h_o} \tag{20}$$

For small h_o (small diffusion effect), $F \rightarrow s$, and the intrinsic nonselective nature of the poisoning is retained. However, when transport limitations are large, the tanh terms approach unity and $F \rightarrow \sqrt{s}$. Wheeler termed this "anti-selective" poisoning since it is slower than the non-selective case, as shown in Figure 18a, Curves A and B.

This sort of antiselective result is not particularly exciting, however, in that none of the experimental data discussed really demonstrate such a trend. The reason, as proposed by Wheeler, is that the physical concepts involved in postulating Equation 18a are faulty in most cases of poisoning. For effective poisons, rapid and strong chemisorption occurs, such that if a poison molecule is diffusing down the length of a catalytically active pore, not many collisions with the surface will occur before it is chemisorbed. The result after a period of operation is a very uneven distribution of activity through the pore, with a dead zone growing progressively from the pore mouth to the interior of the pore as the

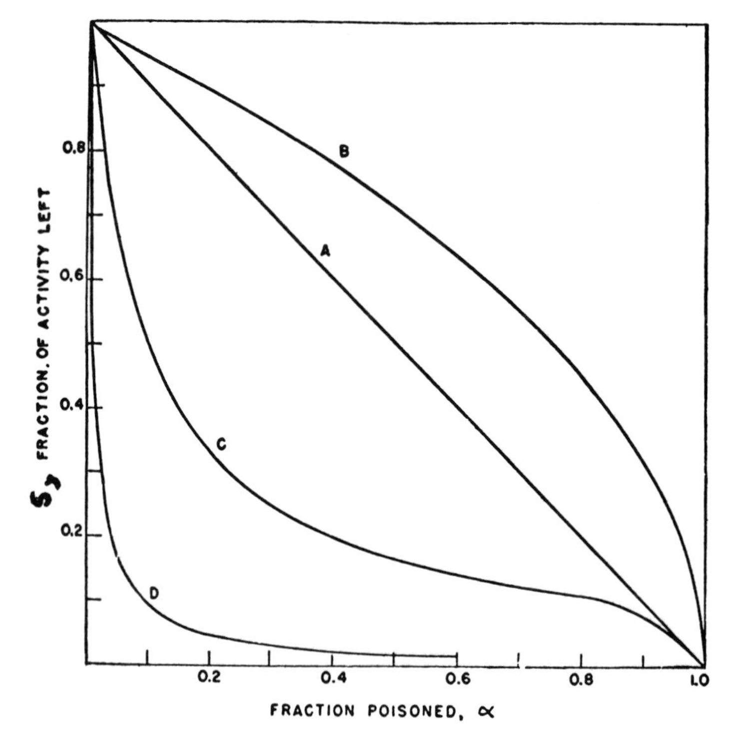

Figure 18a. Types of poisoning curves to be expected for porous catalysts. Curve A is for a nonporous catalyst or for a porous catalyst in which h_0 is very small and poison is distributed homogeneously. Curve B is for homogeneous adsorption of poison with h_0 large. Curves C and D are for preferential adsorption of poison near the pore mouth. For curve C, $h_0 = 10$ and for curve D, $h_0 = 100$ (37).

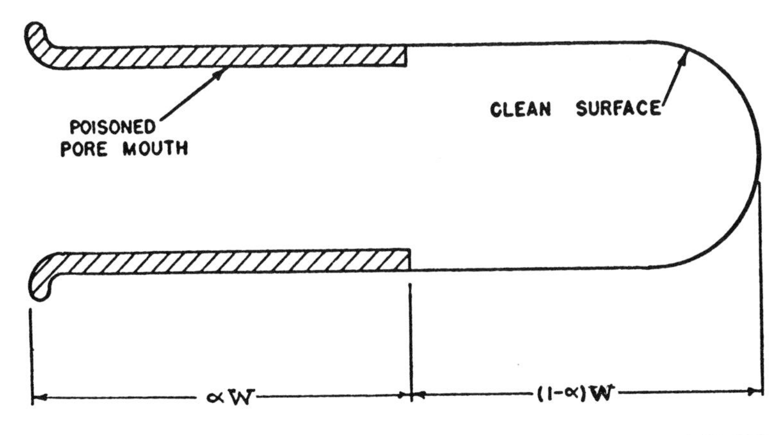

A. Wheeler, "Catalysis," Reinhold, 1955

Figure 18b. Schematic representation of the preferential adsorption of poison near the mouth of a pore. α is the fraction of the pore of length W *covered with poison* (37).

time of utilization increases. In the limit of very fast adsorption one would approach the situation of Figure 18b, which depicts a pore of length W with sufficient poison added to cover the fraction α of its length (and surface). For diffusion-limited reactions, this means that reactants must first traverse the inactive length αW before reaching catalytic surface active for the transformation. Transport in the nonactive section would occur with linear concentration gradient $(C_o - C)/\alpha W$, where C_o and C are concentrations of reactant at the pore entrance and at point αW, respectively. This transport is sufficient at the steady state to supply the quantity of material which has reacted in the active part of the pore; hence the following balance may be written:

$$\frac{\pi \bar{r}^2 D(C_o - C)}{\alpha W} = \pi \bar{r} C \sqrt{2\bar{r} k_a D} \tanh [h_o(1-\alpha)]$$

in which the diffusion rate is given on the LHS and the chemical reaction rate on the RHS, derived from $k_{\text{actual}} = k_a \eta$. One eventually obtains for the rate of reaction:

$$r_T = \frac{\pi \bar{r} \sqrt{2\bar{r} k_o D} \tanh [h_o(1-\alpha)] C_o}{1 + \alpha h_o} \tag{21}$$

whence:

$$F = \left[\frac{\tanh [h_o(1-\alpha)]}{\tanh h_o}\right] \left(\frac{1}{1+\alpha h_o}\right)$$

$$= \left[\frac{\tanh h_o \sqrt{s}}{\tanh h_o}\right] \cdot \left(\frac{1}{1 + (1-\sqrt{s}) h_o}\right) \tag{22}$$

This relationship is also shown in Figure 18a, in curves C and D for representative values of the Thiele modulus and fraction poisoned. Marked selectivity is shown here, even though that portion of the catalyst which is poisoned is poisoned uniformly (*i.e.*, completely). The reason for such extreme effects as those shown for the strongly diffusion-limited example of curve D is that the outer layer of the catalyst particle (or initial portion of the pores) which normally would constitute the major active surface available is precisely that surface which is rendered inactive by the pore mouth poisoning. Not considered in this analysis is the possibility that preferential poisoning at the pore mouth can also lead to decreasing cross section available for mass transport and eventual plugging off of the pore. Such a mechanism may be the cause of changes in surface areas (and presumably mass transfer characteristics) on coking of catalysts such as observed by Appleby *et al.* and Levinter *et al.*

The temperature effects of such poisoning are also notable. Curve D of Figure 18c shows the following progression as the temperature is increased from 200°K. At the lower range the effect of both poisoning and diffusion will not appear since the rate of reaction is sufficiently slow that all the unpoisoned surface is available for reaction and the rate of diffusion through the poisoned portion of the structure is also not limiting. (Note, however, that absolute rate measurements would still be distorted by the factor $\sqrt{s}$ from their true value on the unpoisoned catalyst). As the temperature is increased, the diffusion rate through the pore mouth will begin to become slower than the reaction on the unpoisoned pore, and the active portion of the pore will show some decrease in the availability of surface itself. Both of these effects are reflected by a decrease in apparent activation energy and would correspond to the intermediate range of diffusion-poisoning effects (*i.e.*, $h_o\sqrt{s}$ on the order of unity). At sufficiently high temperatures the process becomes completely diffusion limited, and the apparent activation energy decreases even further. (The limit of zero activation energy as shown in Figure 18 is very different from severely diffusion-limited reactions which approach an activation energy approximately half that of the true value, providing external boundary layer effects around the catalyst particle are not important in transport limitation. Such behavior is predicted by Equation 21 for large values of $h_o\alpha$.) Since both the pore mouth poisoning and the decrease in availability of active surface occur simultaneously and lead to increasing diffusion limitation as temperature is increased, the two effects are not separately identifiable in an analysis such as that of Figure 18c. One can see, however, that the effect of pore mouth poisoning is to intensify the effects of diffusion limitation and broaden the region of operating conditions in which it affects the rate and temperature dependence of reaction.

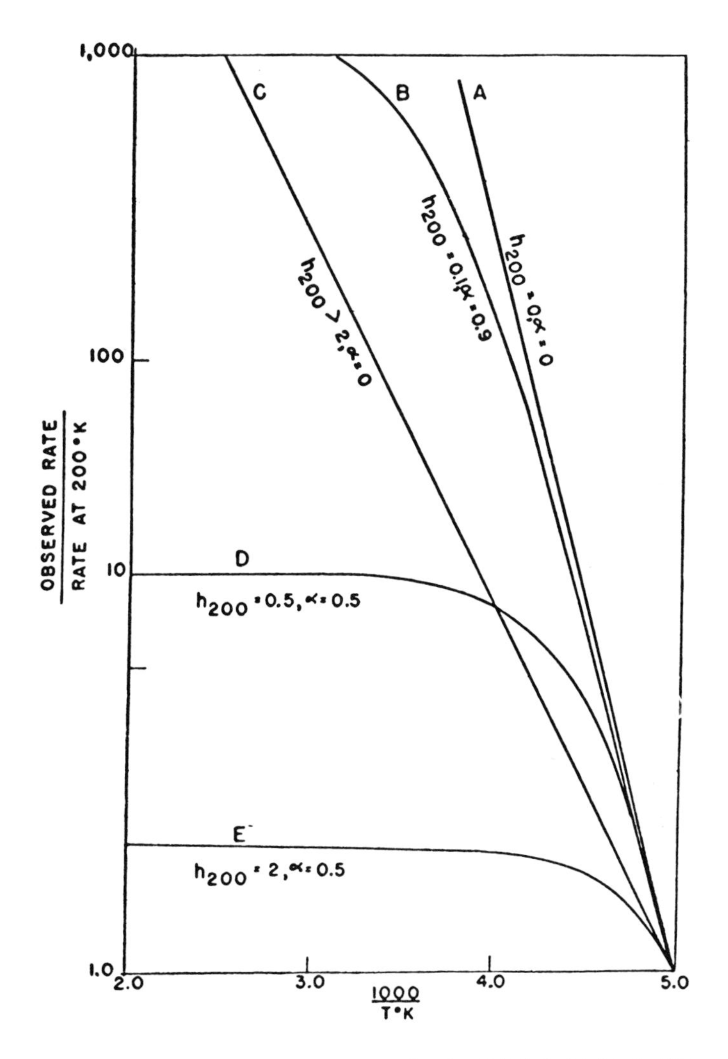

A. Wheeler, "Catalysis," Reinhold, 1955

Figure 18c. "Effect of poison and pore size on apparent activation energy. Plots of observed reaction rate vs. 1/T *for a hypothetical catalyst having 11,000 kcal intrinsic activation energy (e.g., nickel in ethylene hydrogenation) but prepared with different pore sizes and poisoned to varying extent with poison preferentially adsorbed near the pore mouth. Curve A: large pores, no poison. Curve B: fairly large pores, 90% poisoned* ($h_{200} - 0.1$, $\alpha = 0.9$). *Curve C: small pores, no poison. Curve D: moderate size pores, 50% poisoned* ($h_{200} = 2$, $\alpha = 0.5$). *The horizontal portions of D and E correspond to diffusion controlled reaction"* (37).

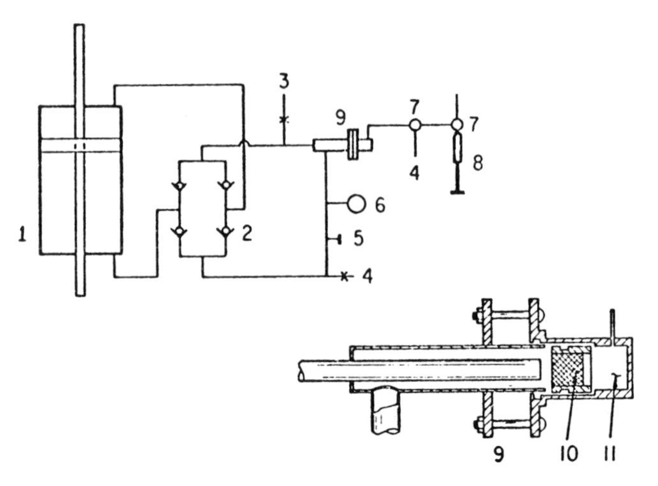

Figure 19a. Reaction apparatus: 1. Reciprocating piston pump; 2. ball check valves; 3. reactant feed inlet; 4. to vacuum pump; 5. septum for syringe sampling; 6. Bourdon-tube pressure gauge; 7. three-way valve; 8. 1 cm³ gas-tight syringe; 9. Roiter-type reactor; 10. sample catalyst pellet; 11. center-plane chamber (39)

Some work has been reported by Balder and Petersen (*39*) and by Dougharty (*40*) in which some of the concepts proposed by Wheeler were examined experimentally. The heart of these experimental studies is a single pellet reactor developed by Balder and Petersen, which permits simultaneous measurements of over-all reaction rates and centerplane concentrations of reactants which can be related to an interpretation of poisoning mechanisms. The elements of the reactor are shown in Figure 19a; the catalyst pellet is contained in the reactor in the form of a cylindrical disc, and transport through the disc occurs parallel to the axis of the cylinder. Composition sampling may be done on either side of the disc, and rates may be measured by change in reactant concentration from the inlet to outlet sides of the reactor. The following equations, which correspond to the development of Wheeler, may be written for the single pellet reactor.

For uniform poisoning:

$$r_T = \frac{\tanh h_o}{h_o} kC_o\pi\bar{r}^2W \tag{23}$$

$$\psi(1) = \frac{C_1}{C_o} = \frac{1}{\cosh h_o} \tag{24}$$

where C_1 is the center line concentration, k is a first-order rate constant (not based on unit surface area), and the Thiele modulus is defined as:

$$h_o = W\sqrt{\frac{k}{D}}$$

For pore-mouth poisoning:

$$r_T = \frac{\tanh h_o\sqrt{s}}{h_o\sqrt{s}}\, kC\overline{\pi r^2}W\sqrt{s} \tag{25}$$

$$\psi(\sqrt{s}) = \frac{1}{1 + (1-\sqrt{s})h_o \tanh(h_o\sqrt{s})}$$

or:

$$r_T = \left[\frac{\tanh(h_o\sqrt{s})}{h_o[1 + (1-\sqrt{s})h_o \tanh(h_o\sqrt{s})]}\right] kC_o\overline{\pi r^2}W \tag{26}$$

and:

$$\psi(1) = \frac{C_1}{C_o} = \frac{1}{\cosh(h_o\sqrt{s})\{1 + (1-\sqrt{s})h_o \tanh(h_o\sqrt{s})\}} \tag{27}$$

Equations 23, 24, 26, and 27 can be used to interpret poisoning data. The shapes of the curves obtained from the two sets of equations are the same (Figure 19b), but their positions for equal values of the Thiele modulus are different. Thus, if one knows or can measure the value of the Thiele modulus for the unpoisoned catalyst, h_o, an interpretation of poisoning data according to the type calculations indicated in Figure 19b can be carried out. This can be accomplished from data at initial conditions easily since a measurement of centerplane composition at that time can be used with Equation 24 to calculate h_o directly.

The experiments Balder and Petersen conducted involved the hydrogenolysis of cyclopropane at 35°C, 780 torr H_2 and 58 torr C_3 partial pressures, on a 0.75% Pt on Al_2O_3 catalyst which had been reduced in H_2 at 600°C and treated under 10^{-5} torr vacuum at 80°C for 16 hours before use. Some experimental results for rate and centerplane composition of cyclopropane (under reaction conditions the kinetics are approximately first order in cyclopropane) are shown in Figures 19c and d. From extrapolation of the centerplane composition to initial conditions, the value of h_o is determined to be 3.2. Calculated rate–concentration variations for uniform poisoning theory and pore mouth poisoning theory, using this value of h_o are shown in Figure 19e, in comparison with the experimental data. Neither theory fits the results obtained, indicating that in this system the mechanism of poisoning must result in a

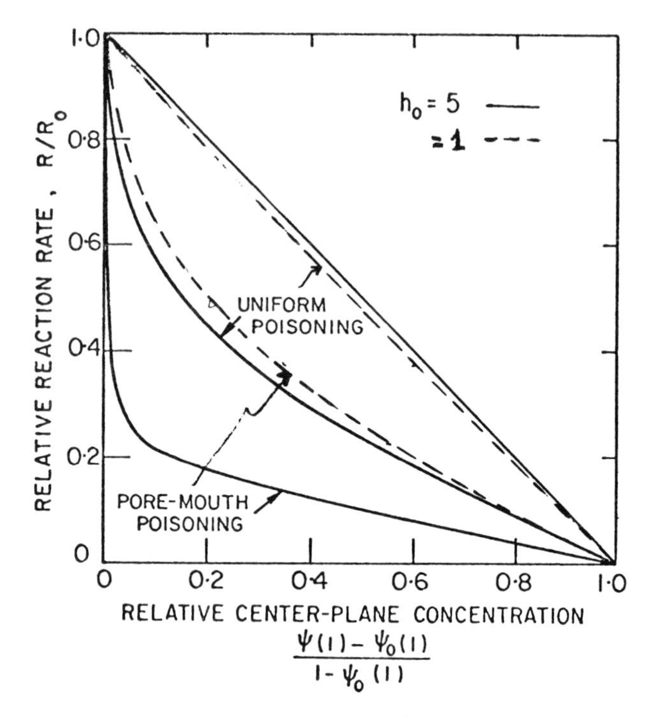

Figure 19b. Comparison of uniform and pore-mouth poisoning; $h_0 = 5,1$ (39)

situation intermediate to uniform or pore mouth results—*i.e.*, there is a gradient of poison concentration within the particle resulting in a continuously non-uniform surface.

The nature of the poison in these experiments of Balder and Petersen is not identified, but the Wheeler development implies separate component chemisorption such as the type I mechanism described earlier. Dougharty (*40*) has recently extended the single pellet reactor analysis to other cases of deactivation. The pore mouth poisoning–homogeneous poisoning comparison involved in the Wheeler-Balder and Petersen work, unfortunately, does not give a unique result. Thus Dougharty showed that one could solve the one dimensional diffusion–reaction problem, assuming various models for the deactivation kinetics and quasi-steady state, and obtain satisfactory interpretation of the experimental results. This problem is formulated as:

$$D \frac{\partial^2 C}{\partial w^2} - kc = 0 \qquad (28)$$

where:

$$C = C_o \text{ at } w = 0$$

$$\frac{\partial C}{\partial w} = 0 \text{ at } w = W$$

$$\frac{\partial k}{\partial \theta} = f(k,C)$$

It was shown that $f(k,C) = -\beta k^2 C$ (where β was arbitrary constant) fit the experimental results of Balder and Petersen. This form of deactivation rate equation indicates some type of self-poisoning mechanism, and a non-uniform distribution of activity throughout the pellet. Detailed mechanistic interpretation is risky, as usual, but the second-order dependency on activity could be taken as indicative of some type of sintering (where activity is directly proportional to surface area). Of course, non-uniqueness of experimental results in terms of mechanistic interpretation is nothing new to reaction kinetics or catalysis, so the variants seen between Wheeler, Balder and Petersen, and Dougharty can be understood as representative of a normal state of affairs.

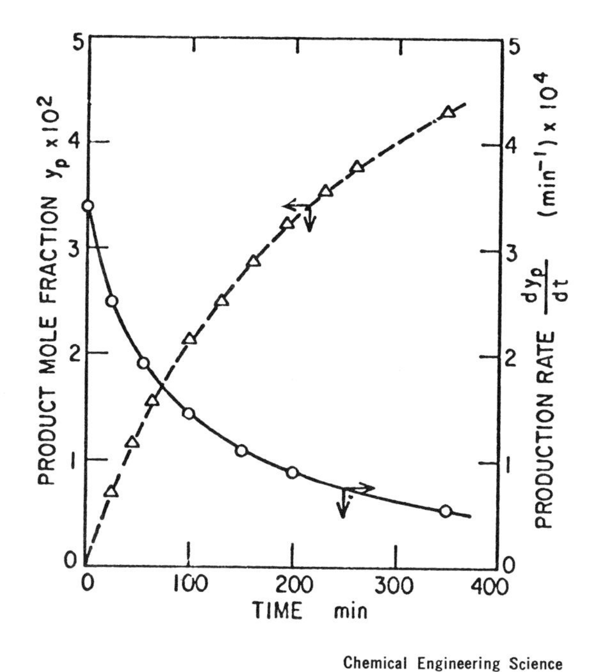

Chemical Engineering Science

Figure 19c. Bulk rate results during poisoning (39)

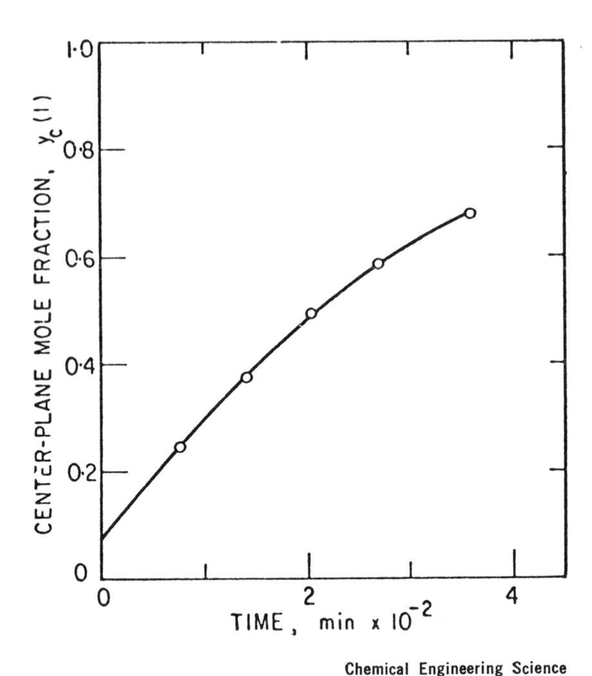

Figure 19d. Center-plane results during poisoning (39)

General Analysis of Intraparticle Deactivation Processes and Their Effects. Somewhat more general treatments of intraparticle deactivation effects have been given by Masamune and Smith (*41*), Ozawa and Bischoff (*42*), and Murakami *et al.* (*43*), and the latter two papers present experimental data for deactivation from coking rates in cracking reactions.

The analysis of Masamune and Smith considered both independent, parallel, and series deactivation mechanisms, which we have previously denoted as type I (independent) or type II (parallel or series reactant or product inhibition). Intraparticle diffusion resistances are considered in the analysis, so that the extremes of uniform poisoning or shell (pore mouth) deposition are possible, depending on relative rates of poisoning and mass transport. We thus have a complex interrelation between the nature of the deactivation—dictated by the mechanism of deactivation and the relative rates of deactivation, mass transfer, and the main reaction—and the ordinary mass transfer–chemical reaction problem. It is convenient to use the concept of particle effectiveness factor for the main reaction, which in this case will depend on time as well as the normal reaction parameters. The limiting assumptions in the Masamune and Smith treatment were isothermal conditions within the particle and the kinetics of deactivation dependent on a linear function of the concentration of poison or coke on the surface, which corresponds to the form of Equation 16. Thus we have for type II systems:

$$A + S \rightarrow B + S \qquad (k_A;\ \text{main})$$

$$A + S \rightarrow A \cdot S \qquad (k_{A,d};\ \text{fouling})$$

or,

$$B + S \rightarrow B \cdot S \qquad (k_{B,d};\ \text{fouling})$$

The effect of coke ($A \cdot S$ or $B \cdot S$) on the kinetics of deactivation are taken to be linearly related, as stated, by:

$$r_T = \rho k_A [A][s] \tag{29}$$

$$\frac{d[s]}{d\theta} = \rho k_{A,d} [A][s] \tag{30}$$

or

$$\frac{d[s]}{d\theta} = -\rho k_{B,d} [B][s]$$

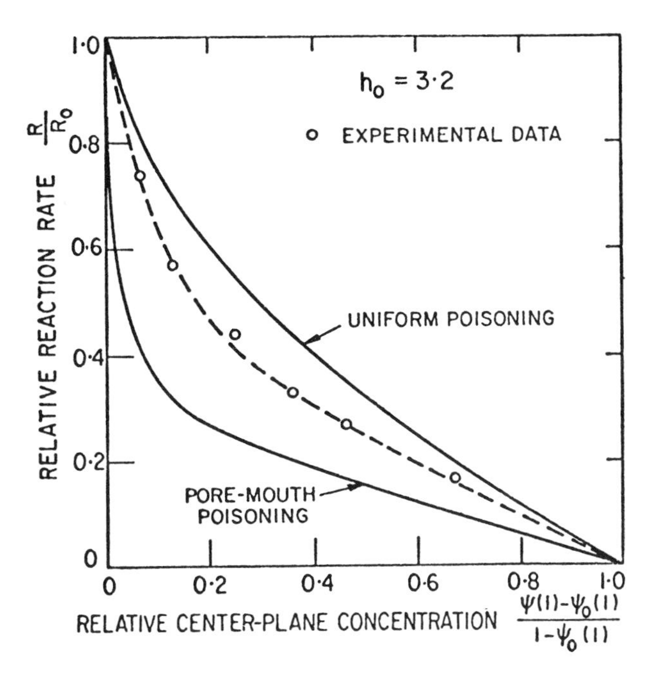

Chemical Engineering Science

Figure 19e. Comparison of experimental poisoning behavior with uniform and pore-mouth poisoning (39)

This type of formulation for deactivation kinetics has been used more recently in a considerable amount of work on reactor modeling; hence, the work of Masamune and Smith is convenient in relating intermediate events to the macroscopic scale. While Equations 29 and 30 refer on the microscopic scale to deactivations which are nonselective, this will be completely obscured by the time the final results of analysis are obtained. The mass conservation equations are:

$$D_A \nabla^2[A] - \epsilon_p \frac{\partial [A]}{\partial \theta} - \rho k_A [A][s] = 0 \tag{31}$$

$$D_B \nabla^2[B] - \epsilon_p \frac{\partial [B]}{\partial \theta} + \rho k_A [A][s] = 0 \tag{32}$$

with initial and boundary conditions:

$$[s] = 1, \theta = 0, R \geqslant r \geqslant 0 \tag{33}$$

$$[A] = [A]_o, [B] = [B]_o, \theta \geqslant 0, r = R \tag{34}$$

$$\frac{\partial [A]}{\partial r} = \frac{\partial [B]}{\partial r} = 0; \theta \geqslant 0, r = R \tag{35}$$

$$\begin{aligned} \nabla^2[A] - \frac{\rho k_A}{D_A}[A] = 0; \theta = 0, R \geqslant r \geqslant 0 \\ \nabla^2[B] + \frac{\rho k_A}{D_B}[A] = 0; \theta = 0, R \geqslant r \geqslant 0 \end{aligned} \tag{36}$$

where r is the spherical coordinate, R the particle radius, ρ the particle density, k_A the rate constant in vol/wt-time units, ϵ_p the porosity, and D_A and D_B the appropriate effective diffusivities (which are considered independent of both time and position). Equations 36 give the initial concentration distributions within the particle, assuming that relaxation times for initial mass distribution are rapid with respect to catalyst activity changes. The rate of poisoning is taken to be slow compared with the main reaction; hence the kinetics of the deactivation reaction do not appear directly in the major conservation equations (31 and 32). The problem is solved by simultaneous solution of Equations 31–36 with the boundary conditions shown. Obviously this is a complex procedure; numerical methods are required, and the details are given by the original authors.

Some primary results for the product inhibition (series fouling) case of type II are given in Figure 20a–c in terms of the dimensionless vari-

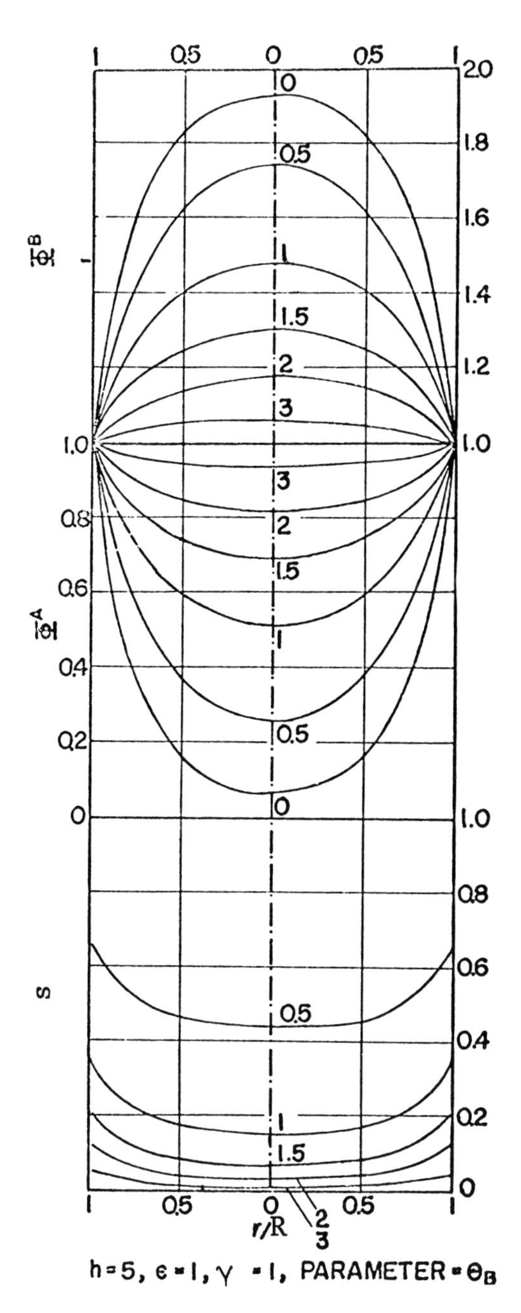

Figure 20a. Profiles for self-fouling, series mechanism, h = 5 (41)

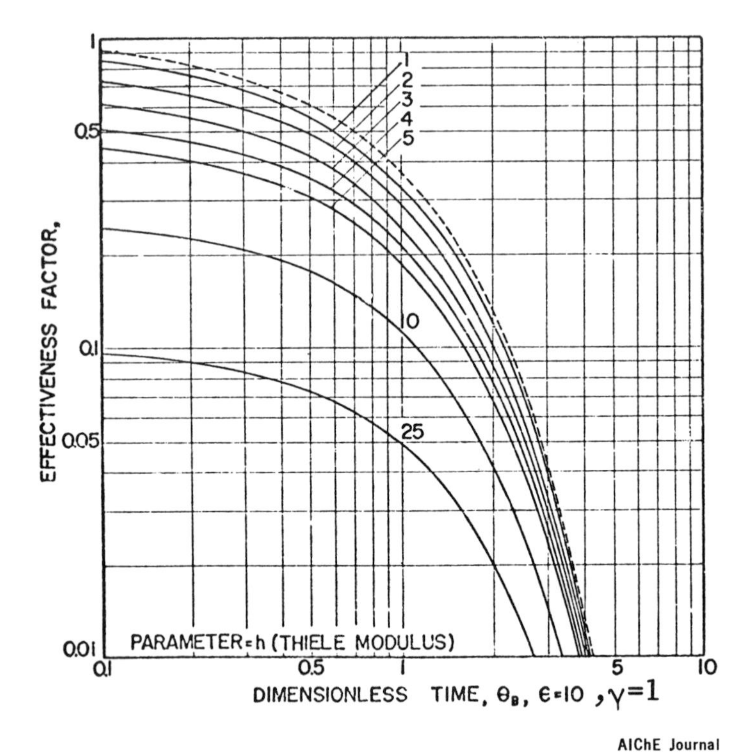

Figure 20b. Effectiveness factor for series fouling, $\varepsilon = 10$ (41)

ables arising from Equations 31–36. Figure 20a shows representations of the profiles within the particle as a function of the time of operation for a specified intermediate value of Thiele modulus (under initial conditions; $h = R(\rho k_A/D_A)^{1/2}$), diffusivity ratio ($\gamma = D_A/D_B$) and conversion ($\epsilon = [B]_o/[A]_o$). The concentration of B is at a maximum at the particle center, that of A a minimum, and these profiles are symmetric (the amount of B which forms coke is small compared with the total product). Since the coke is formed from product B, its concentration is also a maximum at the center of the particle although the profile becomes rather flat in this example as time of operation proceeds. Figure 20b shows the effectiveness factors *vs.* time of operation for differing degrees of diffusional retardation and at a given conversion specified by ϵ. In this case the greater the diffusion resistance (larger h), the larger is the extent of deactivation regardless of time of operation or extent of fouling. Figure 20c is essentially a cross-plot of 20b, in which the influence of diffusional resistance on effectiveness is shown directly for varying times of operation. The limiting curve (dotted) indicates initial conditions (no deactivation) and is actually the normal isothermal solution for effectiveness factor in spherical coordinates:

$$\eta = \frac{3}{h^2}(h \coth h - 1) \tag{37}$$

Time-average behavior in any of the cases discussed can be obtained *via* integration along the appropriate curve.

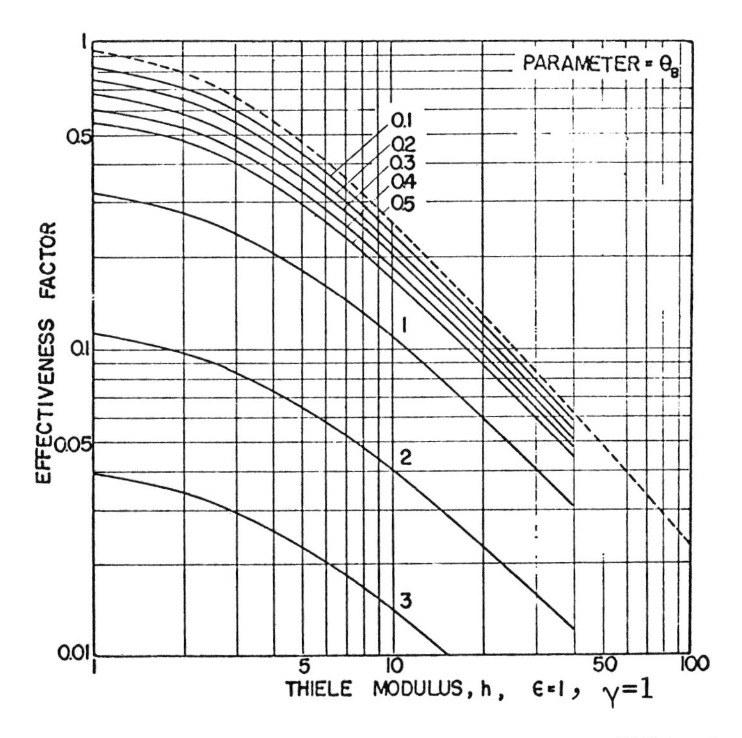

AIChE Journal

Figure 20c. Effect of diffusion resistance on series fouling, $\varepsilon = 1$ (41)

For reactant inhibition, type II, analogous results are given in Figure 21a and b. Again the profiles become fairly flat after extended periods of operation, although in contrast to product inhibition here the maximum coking effect is found near the outer surface. The effectiveness factor as a function of time of operation is shown in Figure 21b; there is a crossover demonstrated which indicates that for shorter times of operation maximum effectiveness is attained with minimum diffusion resistance while at long terms of operation the more severely diffusion-limited particle is actually the more active. The reason seems to be that the sharply decreased activity which occurs near the surface after some time of operation tends to compensate for the sharp gradients of reactant concentration which would be encountered in the absence of deactivation.

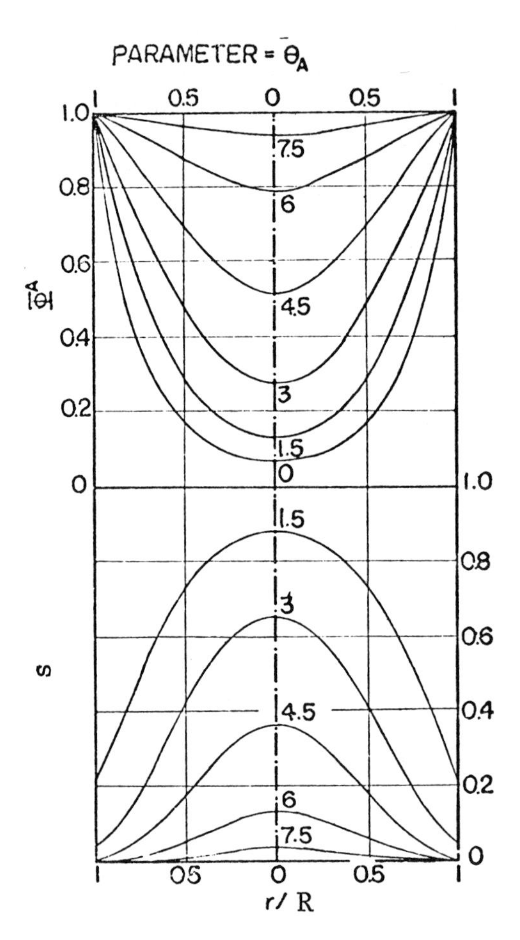

AIChE Journal

Figure 21a. Profiles for self-fouling, parallel mechanism, h = 5 (41)

Notation for Figures 21a, b, and c:

Φ^A, Φ^B = $[A]/[A]_o$, $[B]/[B]_o$
ϵ = $[B]_o/[A]_o$
γ = D_A/D_B
θ_B = $[B]_o k_{B,d}\theta/[s]_\infty$, *time variable*
h = $R(\rho k_A/D_A)^{1/2}$
$[s]_\infty$ = *total concentration of active sites*

For type I deactivation, the rate equation analog of Equations 30 is:

$$\frac{d[\mathrm{s}]}{d\theta} = -k_{\mathrm{L,d}}[\mathrm{L}]\,[\mathrm{s}] \tag{38}$$

for the deactivation step:

$$\mathrm{L} + \mathrm{S} \rightarrow \mathrm{L} \cdot \mathrm{S}$$

Mathematical details pertinent to the solution of the type I problem are similar to those for type II; the second conservation equation, Equation 32, however, would now refer to the impurity poison L. One must, then, introduce into the parameters of the solution a second Thiele modulus which is related to the diffusion limitation of the poisoning reaction. Thus:

$$h_{\mathrm{L}} = R\sqrt{\frac{k_{\mathrm{L,d}}\rho}{D_{\mathrm{L}}}}$$

Results for the example shown in Figure 21c ($h = 5$, $h_{\mathrm{L}} = 10$) are similar to type II reactant inhibition in terms of the activity profiles within the

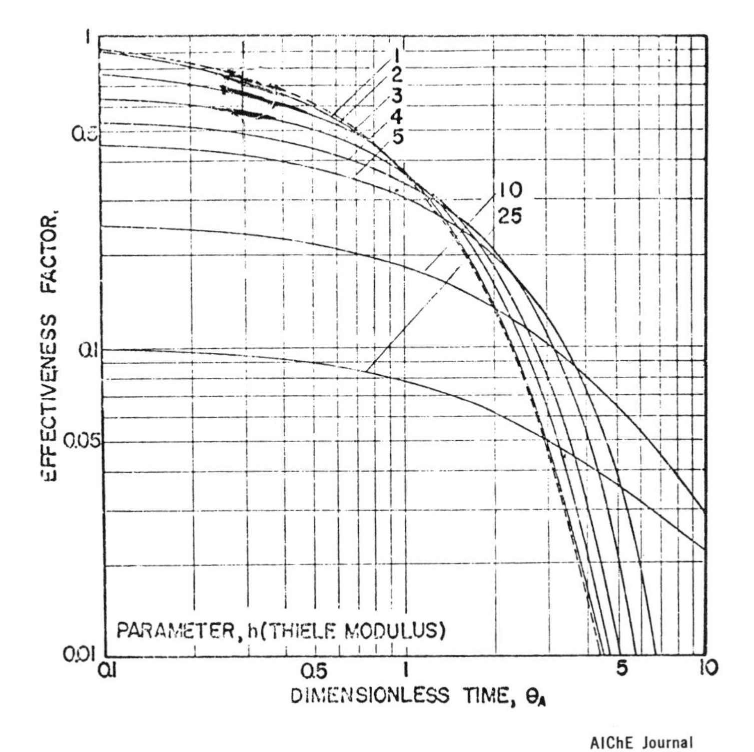

Figure 21b. Effectiveness factor for parallel fouling (41)

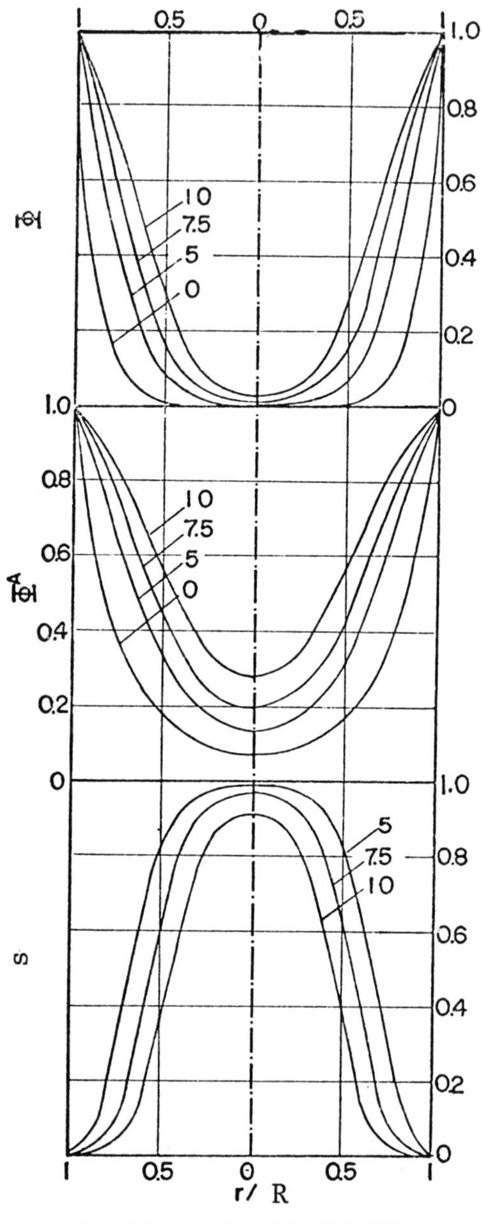

Figure 21c. Profiles for independent fouling, h = *5,* h_L = *10* (41)

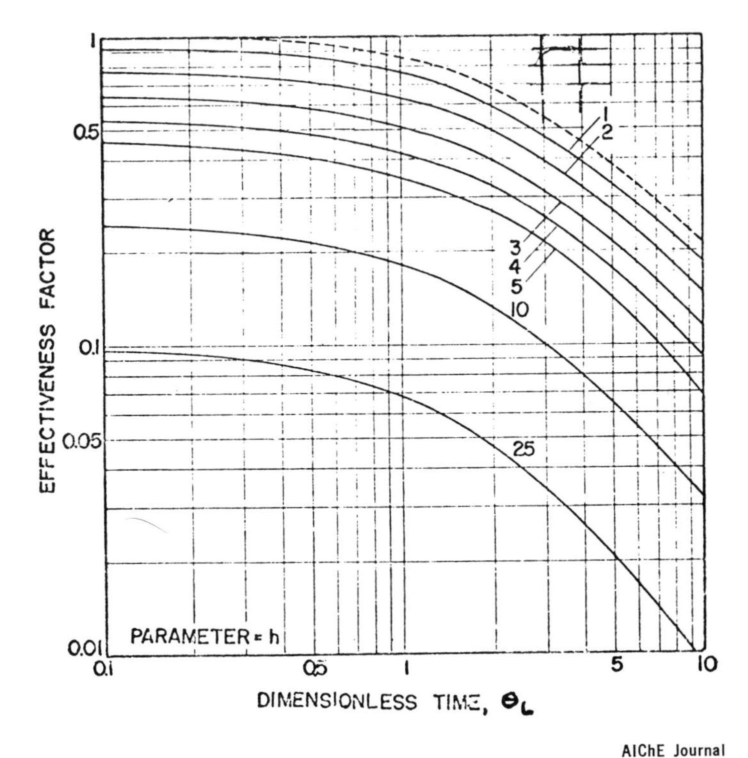

Figure 21d. Effectiveness factor for independent fouling, $h_L = 10$ (41)

particle, and intraparticle reactant and poison concentrations. Note, however, that for the value of h_L chosen, one sees even more selective poisoning of the outer layer of the particle even at short times (bottom of Figure 21c); this begins to approach the shell progressive or pore mouth result, and we discuss this situation in more detail subsequently. Figure 21d gives the effectives–time of operation result as a function of the diffusional resistance of the main reaction. The main point here is that while the intraparticle profiles look like those for type II reactant fouling, the time behavior of activity here looks like type II product fouling. Finally, the effect of the diffusion resistance of the fouling reaction is shown on Figure 21e. The upper bound here represents no deactivation (*i.e.*, it is impossible for the poison to penetrate into the pore structure) and is thus independent of time, and η is given by Equation 37. The lower limit can be obtained analytically, and is:

$$\eta = e^{-\theta_L} \frac{3\ (h \coth h - 1)}{h^2} \tag{39}$$

A diffusion-limited poisoning reaction is seen to have less deleterious effects on activity than a nonlimited one, and the effect can be substantial. When type I deactivation occurs, maximum activity is retained by the catalyst which exhibits high diffusional resistance for the poisoning reaction and low resistance for the main reaction. In practical cases, as exemplified by organic base poisoning of cracking catalysts, for example, this would represent a nearly impossible task.

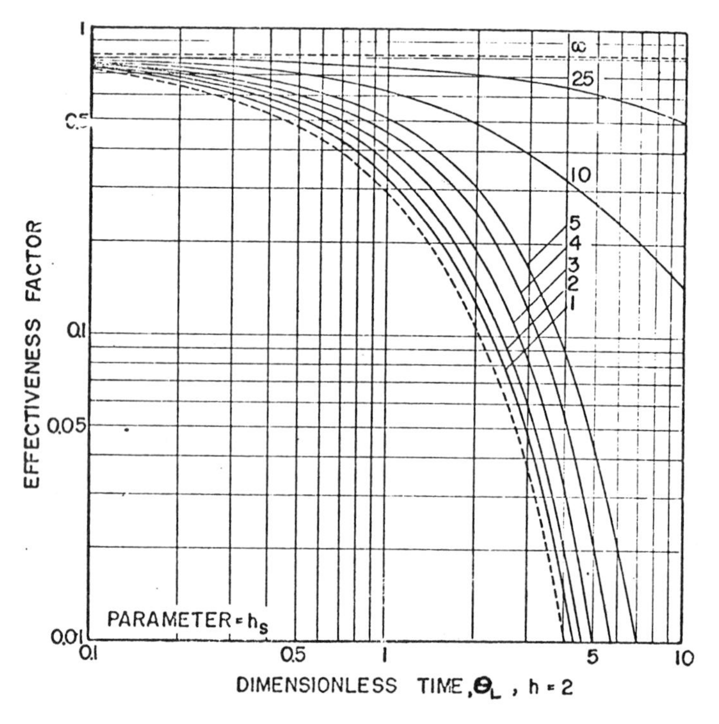

Figure 21e. Effect of diffusion resistance for fouling reaction on effectiveness factor, h = 2 (41)

Ozawa and Bischoff (*42*) have taken a very similar approach to that of Masamune and Smith in the detailed analysis of their experimental data on ethylene cracking, which we have discussed in part previously. Since they measured both kinetics and diffusivities, they were able to estimate a Thiele modulus for their catalyst and reaction conditions, coming to the conclusion on this basis that diffusional resistance was negligible. The parallel mechanism (type II, reactant) was found to be in best agreement with results for the ethylene cracking. Since the parallel mechanism is used, it is thus implied that the absence of diffusion limitation pertains to both cracking and fouling reactions. Their experiments

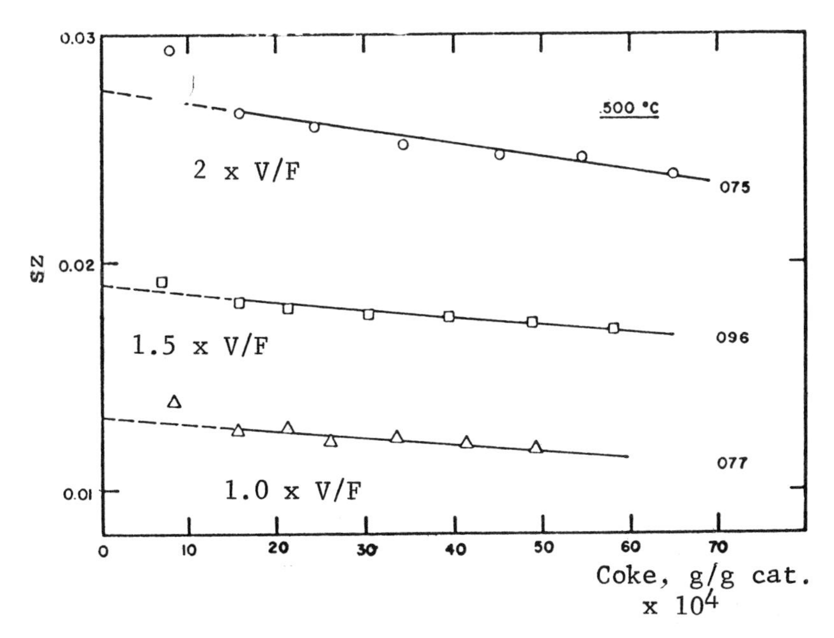

Industrial and Engineering Chemistry

Figure 22a. Determination of fouling function at 500°C (42)

Notation for Figures 22a,b,c,d,e:
h_L — $R(\rho k_{L,d}/D_L)^{1/2}$
θ_L — $k_{L,d}[L]_o\theta'/[s]_\infty$

were run in a differential reactor, so direct measurements of catalyst activity *vs.* the amount of coke on catalyst were possible, thus affording a direct test of the form of coking kinetics (*i.e.*, linear) assumed in the study of Masamune and Smith. Some of the experimental results of Ozawa and Bischoff are given in Figure 22a, where the activity function, $sz(z \equiv (k_A + k_{A,d})v/F$, and v/F is the space velocity in g. catalyst/cm³feed/min) is plotted as a function of coke on catalyst. Some deviations from linearity are noted at low coke loadings, which would correspond roughly to the initial segment of the Voorhies correlation reported in Figure 13, but in general the activity function can be correlated over the range of space velocity investigated as a linear function of coke deposition, and consequently of activity remaining (*i.e.*, [s]) as assumed in the formulation of Equations 29 and 30. The data of Ozawa and Bischoff also afford a measure of the parameters involved in the linear activity correlation of Equation 3. Rewriting that equation in terms of fractional activity:

$$\frac{s}{s_o} = 1 - \left(\frac{\alpha_1}{s_o}\right) C_p \tag{3a}$$

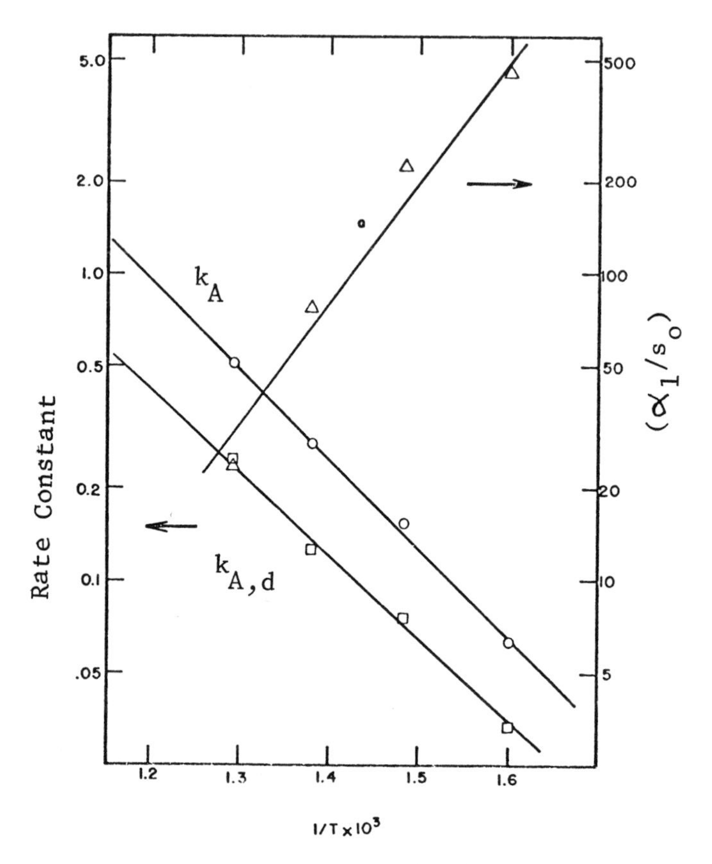

Industrial and Engineering Chemistry

Figure 22b. Arrhenius plot of rate constants and fouling parameter (42)

Figure 22b shows a plot of the values of ethane cracking and coking rate constants, as well as the activity parameter (α_1/s_o) from Equation 3a. A nicely self-consistent picture of the simultaneous kinetics of cracking and coking is obtained from this interpretation, and the values of the kinetic parameters obtained reproduce the experimentally determined coking rate of the catalyst quite well, as shown in Figure 22c. The activation energies indicated by this interpretation were 13.1 kcal/mole for the main cracking reaction and 12.5 kcal/mole for the coking reaction—approximately equal temperature sensitivity for the two processes.

A final test of the linear activity correlation was attempted by Ozawa and Bischoff, by correlating the data of Eberly *et al.* on *n*-hexadecane cracking, some of which we have discussed before. The details of the analysis are deferred to a later section in which the problem of chemical reactor analysis is discussed (the experiments of Eberly *et al.* involved finite conversions in fixed beds of catalyst); however, the average weight of coke on catalyst in the bed can be obtained, assuming the linear activity relationship, as:

$$\overline{C}_p = (\alpha_1/s_o)^{-1} + \frac{(\alpha_1/s_o)^{-1}}{\alpha} \ln \Phi_{A,\text{exit}} \tag{40}$$

in which:

$$\Phi_{A,\text{exit}} = \frac{1}{1 - e^{-\alpha\beta\delta}(1 - e^{-\alpha})} \tag{41}$$

$$\alpha = S_R W \rho_B (k_A + k_{A,d})/F$$

$$\beta = S_R W/F$$

$$\delta = (F\theta/S_R \epsilon W) - 1$$

and S_R is the bed cross section, W is bed length, ρ_B is bed density, θ is time of operation, ϵ is bed porosity, and F is flow rate. Equation 40 indicates $\overline{C}_p$ *vs.* $\ln \Phi_{A,\text{exit}}$ should be linear, and the data of Eberly *et al.* plotted

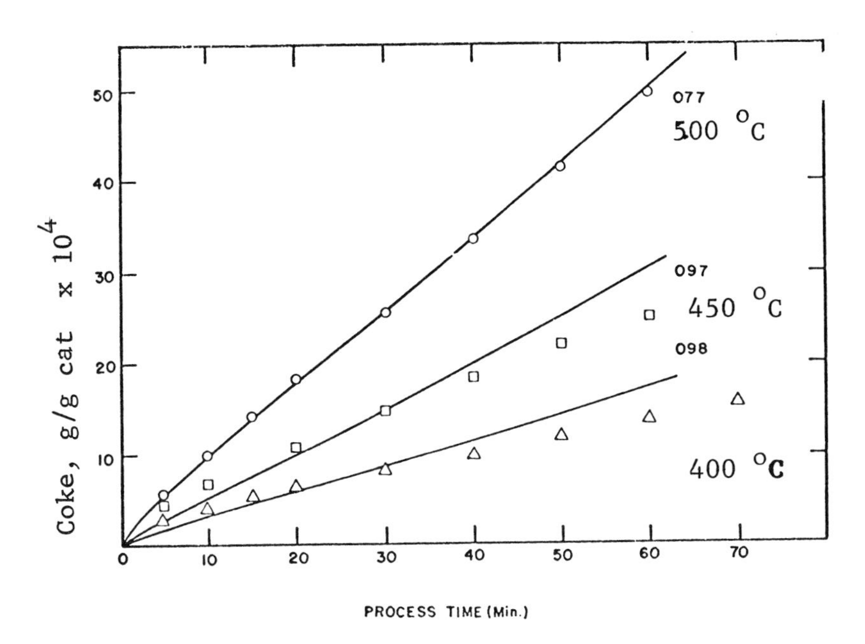

Industrial and Engineering Chemistry

Figure 22c. Comparison of calculated and experimental results (42)

according to this suggestion are shown on Figure 22d. The linear representation is subject to some deviations at higher coke contents, but below about 5% is obeyed very well.

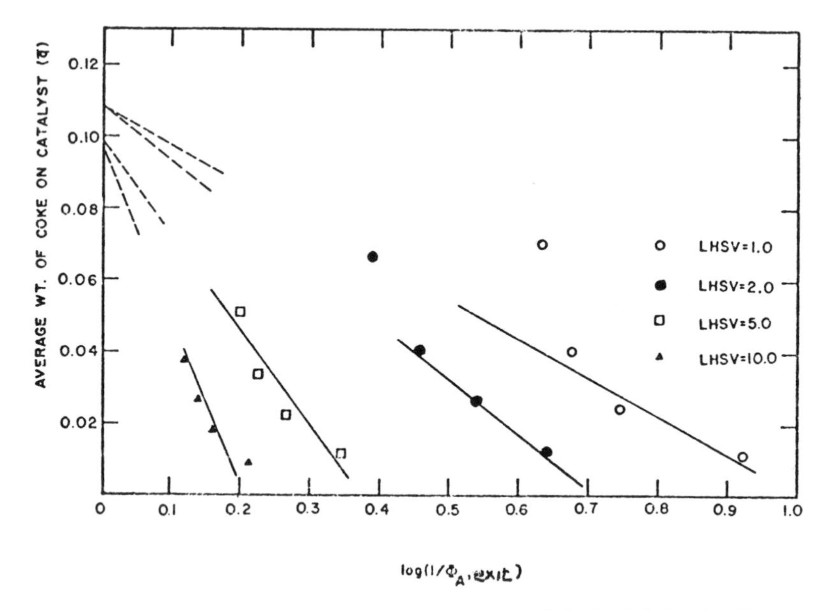

Figure 22d. Relation of (1 − conversion) and weight of coke for cracking of n-*hexadecane. Data of Eberly* et al. (42).

Hence, it seems from these data of Ozawa and Bischoff and Eberly *et al.*, that the linear relationship of activity to coke content is reasonable, and expectations based on the type II mechanism analysis of Masamune and Smith should be at least qualitatively correct insofar as the general format of the deactivation kinetics is concerned. These are, of course, net second-order reactions, so the picture is not inconsistent with the dual site mechanistic proposals which we discussed previously.

Richardson (*44*) has recently reported an experimental method for measuring carbon profiles in catalyst particles using a microcombustion technique. Coked catalyst is regenerated under diffusion-limited conditions, such that shell progressive removal of carbon is observed. The amounts of reaction products (CO_2 and H_2O) of the combustion are measured quantitatively for a series of experiments in which successively longer regeneration times are used; in each experiment the coked "core" radius is measured after combustion by visual inspection of the pellet. These data allow the determination of a carbon concentration *vs.* radius profile, which can then be used in view of the results of Masamune and

Smith to discriminate between series and parallel fouling mechanisms. Richardson has demonstrated this technique for a cobalt molybdate–alumina catalyst coked under hydrotreating conditions (700°–800°F, 1000–1500 psig) with a full boiling range coal-derived feed, and with a 600°F + petroleum stock. Carbon profiles in the former case exhibited a minimum at the pellet center, characteristic of parallel fouling, and just the opposite, characteristic of series fouling, in the latter experiment. The confidence with which one can make such an interpretation depends on the relative rates of the cracking (or hydrotreating) reactions to those reactions responsible for coke formation, however.

Murakami *et al.* (*43*) extended the analysis of Masamune and Smith to consider the situation in which poisoning or fouling (actually they consider only type II mechanisms) are not slow compared with the main reaction. This requires inclusion of the deactivation kinetics in the mass conservation equations; thus for type II reactant inhibition we have, for example:

$$D_A \nabla^2[A] - \epsilon_p \frac{\partial[A]}{\partial\theta} - \rho k_A[A][s] - \rho k_{A,d}[A][s] = 0 \qquad (31a)$$

Details of the analysis and results are generally similar to those of Masamune and Smith, with reactant and coke profiles corresponding for the most part to those shown in Figures 20 and 21. One significant difference did appear in the analysis for non-negligible deactivation rates, however, and that was associated with intraparticle coke profiles from type II product inhibition at high values of Thiele modulus. These showed a reversal at high h, changing from a maximum centerline value (Figure 20a) at low Thiele modulus to a minimum centerline value at high h. Somewhat different parameter definitions are used by Murakami *et al.* than by Masamune and Smith; thus, a direct comparison with Figure 20 is not afforded by their results; however, a qualitative idea of the reversal in profile shape which their analysis predicts is given by Figure 23a. The implication of this reversal is also that at some intermediate value of Thiele modulus the coke profile will be nearly flat and the fouling will appear to be uniform. Perhaps the most valuable part of the work of Murakami *et al.* was the experimentation on two representative type II schemes, the disproportionation of toluene (parallel reaction):

toluene (A) → xylene (B) + benzene

toluene (A) → coke (A · S)

and the dehydrogenation of *n*-butyl alcohol (series reaction):

alcohol (A) → aldehyde (B) → coke (B · S)

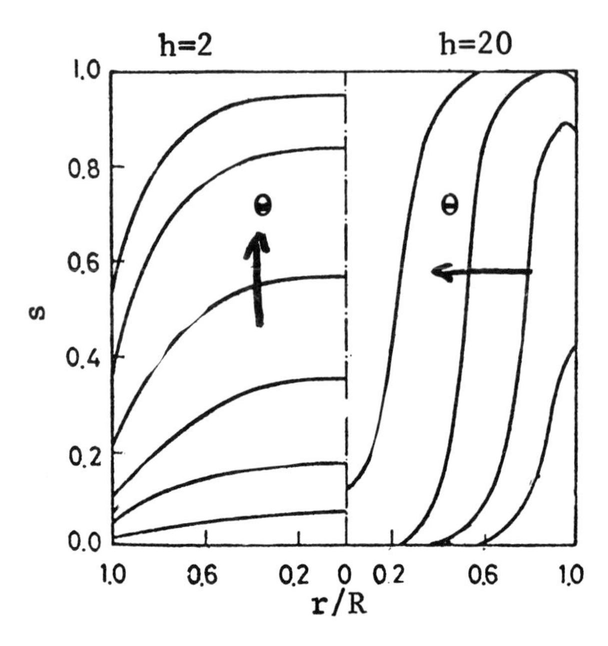

Industrial and Engineering Chemistry

Figure 23a. Example of coke profile reversal at differing h *for type II product inhibition;* $(k_A/k_{A,d}) = 1$; $\gamma = 1$ (43)

Catalysts were an alumina–boria (10% boria) for the disproportionation reaction and a commercial alumina (10% wt sodium) for the dehydration. Catalyst pretreatment consisted of nitrogen flow at reaction temperature, and both disproportionation and dehydrogenation reactions were carried out in flow reactors under essentially differential conversion conditions. The major results of interest are shown in Figure 23b and c. In b, the lower temperature pellet is relatively uniformly coked; activity at this condition is approximately 85% that of the uncoked catalyst. The higher temperature pellet verifies the effect predicted by theory as the Thiele modulus becomes larger and diffusion limitation of the coking reaction occurs (Figure 21a); the coke precursor, toluene, has a higher concentration near the particle surface and, under conditions of strong diffusion limitation, a large gradient in this region. This is seen for the 530°C particle; activity at this condition is approximately 80% that of the uncoked catalyst. In c, the prediction of coke profile reversal illustrated on Figure 23a is verified for the series alcohol dehydrogenation reaction. The aldehyde product has a maximum concentration in the center of the pellet for small h (Figure 23c, low temperature), and at the outer surface for high h since the reactant cannot penetrate far into the structure and coke is formed rapidly from the dehydrogenation prod-

Industrial and Engineering Chemistry

Figure 23b. Cross section of catalyst pellet after 10 minutes' reaction of disproportionation of toluene. Left, 440°C. Right, 530°C. Black part represents coke deposited (43)

uct. Activity decreases are 30% for the 400°C catalyst and about 50% for the 480°C catalyst. Such behavior is wholly a function of the relative magnitudes of main reaction and coking rates, or, in terms of the parameters of the problem, the selectivity factor ($k_{A,d}/k_A$), set at unity in Figure 23a.

Another recent study of intraparticle poisoning profiles with some accompanying experimental data is that of Gioia, Gibilaro, and Greco (*45*). Their analysis involves a type I reversible chemisorption specified by Langmuir isotherm; competition for adsorption sites with the reactant is not considered, so the analysis actually is for nothing more than time dependent intraparticle diffusion and sorption. Experimental data were obtained for the water vapor-poisoned hydrogenation of ethylene on a copper–magnesia catalyst; poison adsorption was said to be very rapid, and correspondence with the theoretical model checked by comparison of the total fraction of area poisoned determined experimentally (by decrease in conversion) with those determined from integration of the sorption profiles. The authors claim good agreement between theory and

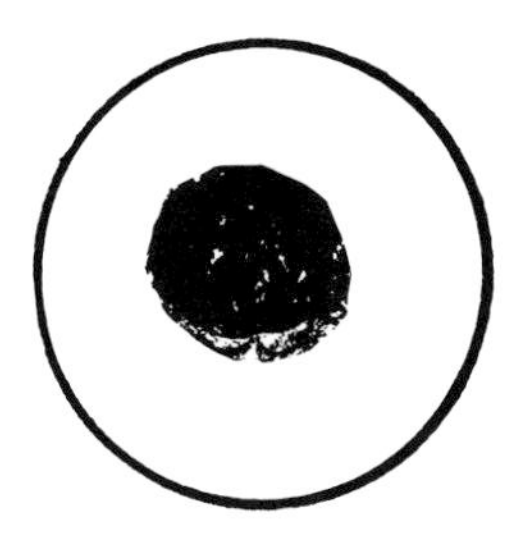

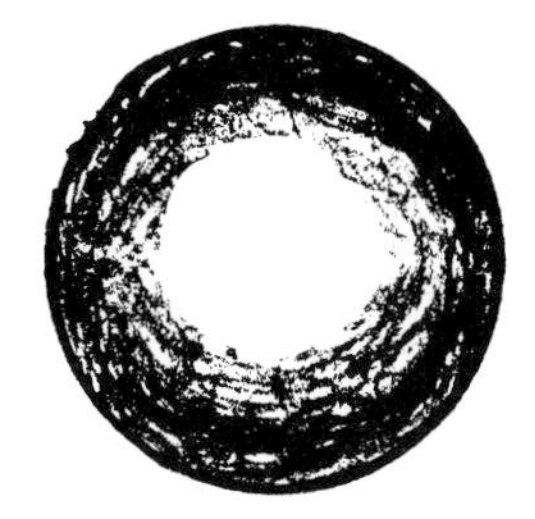

Industrial and Engineering Chemistry

Figure 23c. Cross section of catalyst pellet after dehydrogenation of n-*butyl alcohol. Left, 400°C after 50 minutes. Right, 480°C after 30 minutes. Black part represents coke deposited* (43).

experiment; however their enthusiasm is not shared by the writer; there is considerable scatter in the experimental–theoretical correlation, and it is not clear that the type of "independent" poisoning postulated is a true deactivation effect or is more closely related to sorption competition. One is left with the feeling that such a model probably oversimplifies events and a consideration of some direct interaction with reactants in the analysis would have been more satisfactory. A direct extension of the Masamune and Smith analysis, using Langmuir-Hinshelwood rate forms, has also recently been given by Chu (*46*).

Regeneration Kinetics and the Shell Progressive Mechanism. The previous discussion of deactivation effects on the intermediate scale has shown the close interconnection between the mass transport properties of a given catalyst/reaction system and its corresponding deactivation behavior. Particularly, starting with the pore mouth-poisoning model, it has been shown that very sharp concentration gradients of poison can be developed within the porous material, as is directly shown by the results of Murakami *et al.* in Figure 23. Such very sharp, almost discontinuous, gradients have come to be recognized as particularly important when coke is removed from a catalyst during regeneration. Although this review is primarily concerned with the gloomy side of catalysis, as embodied in deactivation, and a discussion of detoxification or regeneration is generally beyond the present scope, the question of coke removal is so intimately bound up with that of coke formation in the literature that an exception must be made here. At the expense of some repetition of portions of the material already discussed, we consider some problems associated with regeneration of coked catalysts.

The first problem to consider in coke removal is the kinetics of carbon burning. A good example of what is observed is given by the data of Rudershausen and Watson (*16*) on coke removal from the molybdena–alumina catalyst they used in their aromatization studies. These data are given in Figure 24a; it is clear that two different types of carbon burning are involved. The first occurs at a high rate, passes through a maximum, and then decreases rapidly until an apparent second type, occurring at a low but relatively steady state, is observed. It was suggested that the first period corresponded to removal of carbon from the external surface of the catalyst particles, and the low rate period corresponded to the combustion of carbon in the internal pore structure of the catalyst. The problem of carbon burning kinetics and the apparent duality of rate periods was subsequently treated in detail by Ausman and Watson (*47*), the essentials of which are presented here [*see also* Bondi *et al.* (*36*)].

Ausman and Watson assumed in their analysis that the external burning of carbon (termed the "constant" rate period) would continue

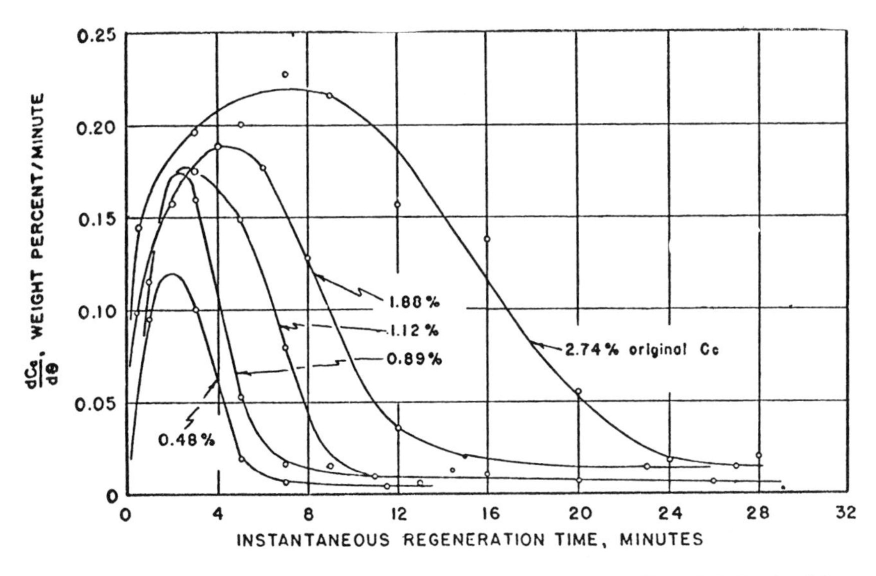

Chemical Engineering Science

Figure 24a. Carbon evolution rates at various levels of original total carbon deposition (16)

until all the coke on the external surface had been removed; then the second period (the "falling" rate period) commenced with burning of the interior coke deposit. Two separate analytical formulations of the problem are thus required, corresponding to each of the burning periods. An oxygen balance on a spherical particle for the constant rate period is, in dimensionless form:

$$\frac{1}{\xi^2}\frac{d}{d\xi}\left(\xi^2\frac{d\psi}{d\xi}\right) = N_r{}^2\psi \tag{42}$$

where

$$\xi = r/R$$

$$\psi = y_{O_2}/y_{O_2,g}$$

$$N_r{}^2 = \frac{kPR^2}{D_{O_2}CN}$$

in which y_{O_2} and $y_{O_2,g}$ are oxygen mole fractions in the pellet and the bulk gas, respectively, D_{O_2} is the oxygen intraparticle diffusivity, k is the carbon burning rate constant, C is the total (gas) molal concentration, P is the total pressure, and N is the molal ratio of carbon to oxygen in the reaction. The carbon burning rate is taken to depend only on the oxygen partial pressure in the gas; hence the kinetics of oxygen disappearance are given by:

$$r_{O_2} = -kPy_{O_2}N \text{ (moles } O_2/\text{hr-ft}^3) \quad (43)$$

This form of carbon burning kinetics is disputed by Bondi, Miller, and Schlaffer and by the later work of Weisz and Goodwin (*48*), both of whom indicate first order dependence on both oxygen and carbon:

$$r_c = -kPy_{O_2}C_c \quad (43a)$$

This would seem to be fairly well established and experimentally supported, so the kinetics of Equation 43 represent a limitation on the analysis. The kinetics of coke oxidation are discussed in more detail later.

The boundary conditions used by Ausman and Watson to solve Equation 42 were:

$$\xi = 1, \frac{d\psi}{d\xi} = \frac{1}{\text{Sh}} (1 - \psi) \quad (44)$$

$$\xi = 0, \psi \text{ finite}$$

The Sherwood number defined in Equation 44 expresses the ratio of external to internal mass transport rate coefficients:

$$\text{Sh} = D_{O_2}/R_g T_g k_g R \quad (45)$$

where R_g is the gas constant, T_g is the bulk gas temperature, and k_g is the external mass transfer coefficient. It is interesting that Ausman and Watson chose this form for the boundary conditions of the problem, since the signal importance of external boundary layer transport in various types of diffusion-limited catalytic problems has only recently been rediscovered by those active in the field. The solution to Equations 42 and 44 is:

$$\psi = \frac{1}{\beta_m \xi} \left[\frac{\sinh (N_r \xi)}{\sinh (N_r)} \right] \quad (46)$$

where

$$\beta_m = 1 + \text{Sh} \left(\frac{N_r}{\tanh N_r} - 1 \right)$$

During the constant rate period, the O_2 profile of Equation 46 is time invariant, so the carbon burning rate at any particular point will be constant (though, of course it will vary throughout the structure). After θ time of regeneration, the carbon concentration at any point is:

$$C_c = C_c^\circ - (kPy_{O_2})\left(\frac{M_c}{\rho_s}\right)\theta \tag{47}$$

in which M_c is the molecular weight of carbon, ρ_s is the catalyst particle density and C_c, C_c° are in units of lb carbon per lb catalyst. In dimensionless form, one obtains:

$$\frac{C_c}{C_c^\circ} = X = 1 - \frac{N_r^2\tau}{\beta_m\xi}\left[\frac{\sinh(N_r\xi)}{\sinh(N_r)}\right] \tag{48}$$

where

$$\tau = \frac{ND_{O_2}Cy_{O_2,g}N_c\theta}{\rho_s C_c^\circ R^2}$$

The fraction of initial carbon remaining in the pellet at τ is:

$$X_F = \frac{\int_0^1 X\xi^2 d\xi}{\int_0^1 \xi^2 d\xi} = 1 - \left(\frac{3}{\beta_m}\right)\left(\frac{N_r}{\tanh N_r} - 1\right)\tau \tag{49}$$

and the length of the constant rate period ($X = 0$ at $\xi = 1$ when the constant rate period ends) is from Equation 48:

$$\tau_{CRP} = \frac{\beta_m}{N_\tau^2} \tag{50}$$

We are now in a position to see what these results indicate as far as the over-all carbon burning rate is concerned—*i.e.*, what the general form of the kinetics of observed carbon disappearance would be during the constant rate period. Let us assume, as Ausman and Watson did, that such a rate would be given by:

$$r'_c, \text{over-all} = k'_c f(y_{O_2}) f(X_F) \text{ lb mole/hr-ft}^3 \tag{51}$$

where $f(X_F)$ indicates some functional dependence with respect to the "average" carbon concentration. The over-all rate of change of carbon content with time is obtained from Equation 49 as:

$$\frac{dX_F}{d\tau} = 1 - \left(\frac{3}{\beta_m}\right)\left(\frac{N_r}{\tanh N_r} - 1\right) \tag{52}$$

and terms of Equation 51 can be shown by comparison to be:

$$f(y_{O_2}) = y_{O_2,g}$$

$$f(X_F) = 1 \tag{53}$$

$$k'_c = \frac{3ND_{O_2}C}{R}\left\{\frac{[(N_r/\tanh N_r) - 1]}{1 + \mathrm{Sh}[(N_r/\tanh N_r) - 1]}\right\}$$

In the falling rate period, the analysis is somewhat more complex since as this period progresses, a carbon-free zone develops within the particle where the only process occurring is the diffusion of oxygen to that portion of the particle still containing coke. This is illustrated generally in Figure 24b, where for values of $\tau > \tau_{CRP}$ one sees that the carbon profiles fall to zero at radial positions within the external surface of the particle.

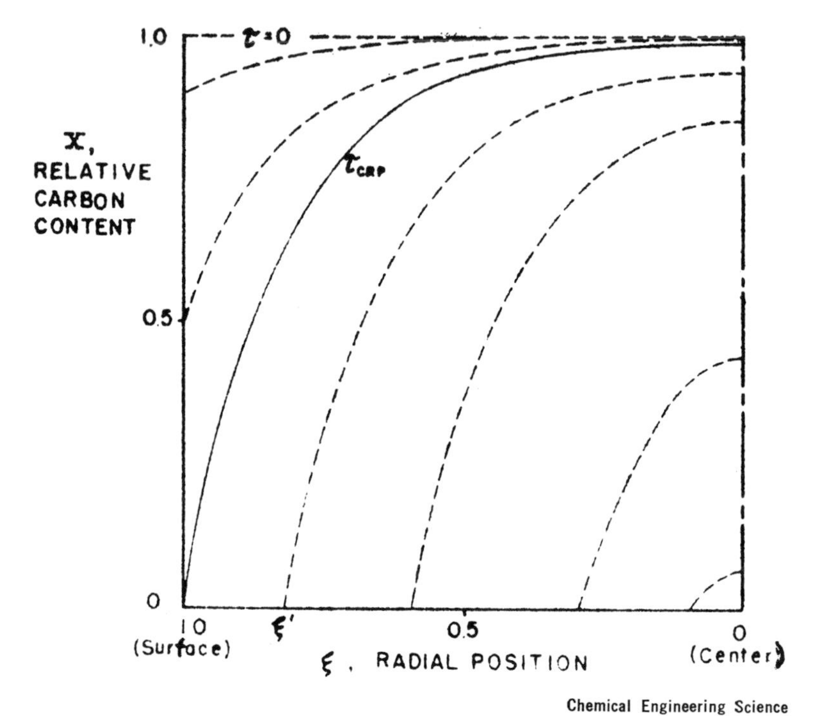

Figure 24b. Carbon profiles for $N_r = 6.80$ (47)

In the diffusion zone, $\xi > \xi'$ on the figure, we have:

$$\frac{1}{\xi^2}\frac{d}{\xi^2}\left(\xi^2\frac{d\psi}{d\xi}\right) = 0$$

$$\xi = 1 \qquad \frac{d\psi}{d\xi} = \frac{1}{\mathrm{Sh}}(1-\psi) \tag{54}$$

$$\xi = \xi' \qquad \psi = \psi'$$

whence:

$$\psi = \psi'\left\{\frac{\mathrm{Sh} + [(1/\xi) - 1]}{\mathrm{Sh} + [(1/\xi') - 1]}\right\} + \frac{(1/\xi') - (1/\xi)}{\mathrm{Sh} + [(1/\xi') - 1]} \tag{55}$$

In the diffusion–reaction zone, Equation 42 applies with the first boundary condition of Equation 44 changed to $\psi = \psi'$ at $\xi = \xi'$. This result is:

$$\psi = \psi' \left(\frac{\xi' \sinh (N_r \xi)}{\xi \sinh (N_r \xi')} \right) \tag{56}$$

The interfacial oxygen concentration ψ' can be obtained by equating the gradients of ψ and ψ' at ξ':

$$\psi' = \frac{1}{1 + [(1 - \xi') + \mathrm{Sh}\, \xi'] \{ [N_r \xi' / \tanh (N_r \xi')] - 1 \}}$$

The carbon concentration at any point in the reaction zone is:

$$X = X_{CRP} - N_r^2 \int_{\tau_{CRP}}^{\tau} \psi d\tau \tag{57}$$

with:

$$X_{CRP} = 1 - \left(\frac{1}{\xi} \right) \frac{\sinh (N_r \xi)}{\sinh (N_r)}$$

which allows one to compute profiles as shown on Figure 24b. If one attempts to cast the kinetics of the falling rate period into the form of Equation 51, using the same procedures as before, the result is a very messy expression in which $f(y_{O_2})$ is again given by $y_{O_{2},g}$ but $k'_c f(X_F)$ is a nonseparable implicit expression involving N_r, ξ', Sh, D_{O_2}, N, and C; in fact, a horrible mess. Thus, aside from the valuable analysis which Ausman and Watson provided concerning the details of regeneration mechanisms and coke profiles during regeneration, they also showed that general, over-all burning rates are very complex kinetic functions, particularly in the falling rate period, which may not be amenable to simple interpretation in terms of standard rate equation forms.

In practice, however, the kinetics and mechanism of coke burnoff may not always be quite as complex as these results would indicate. It was assumed in the development by Ausman and Watson that a combined mass transport-chemical reaction rate process determined the over-all kinetics of carbon burning. From information such as that in Figure 23, however, we know that rate-controlling regimes can change as temperature changes and, by analogy to deactivation mechanisms, one would expect a transition from uniform regeneration at lower temperatures (kinetics of carbon burning control), through a transition such as that analyzed by Ausman and Watson, to a point at which rates are completely diffusion-limited, and the burnoff of coke occurs at a sharp and moving

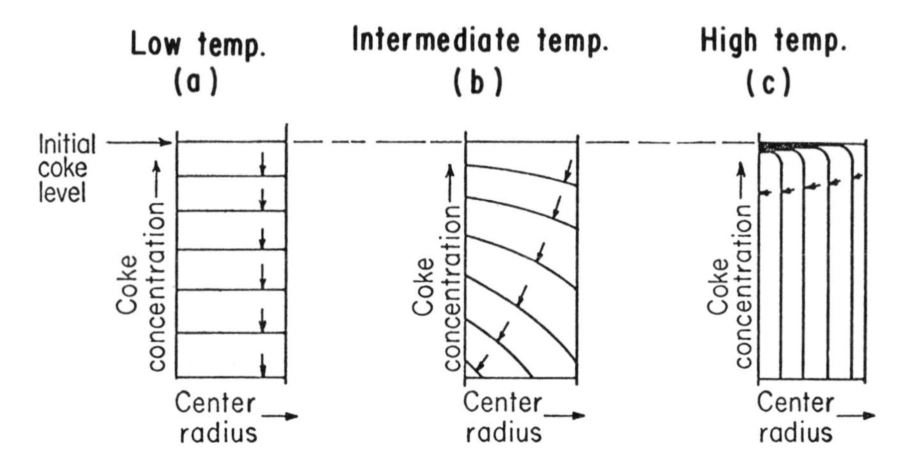

Figure 25a. Coke concentration vs. *radius in beads for successive stages of burnoff for three temperature regions* (49)

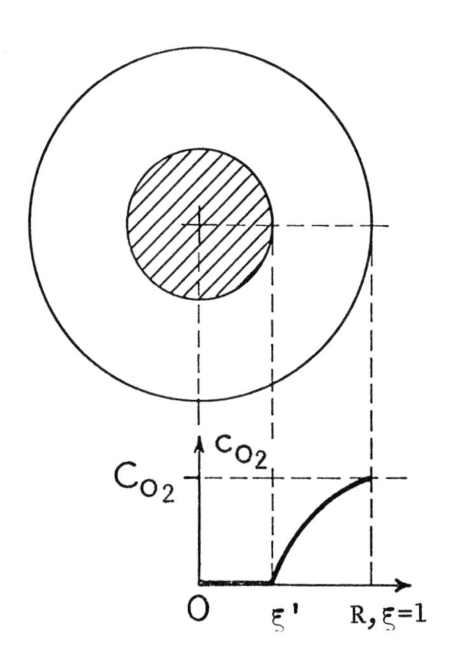

Figure 25b. Model of shell progressive combustion (49)

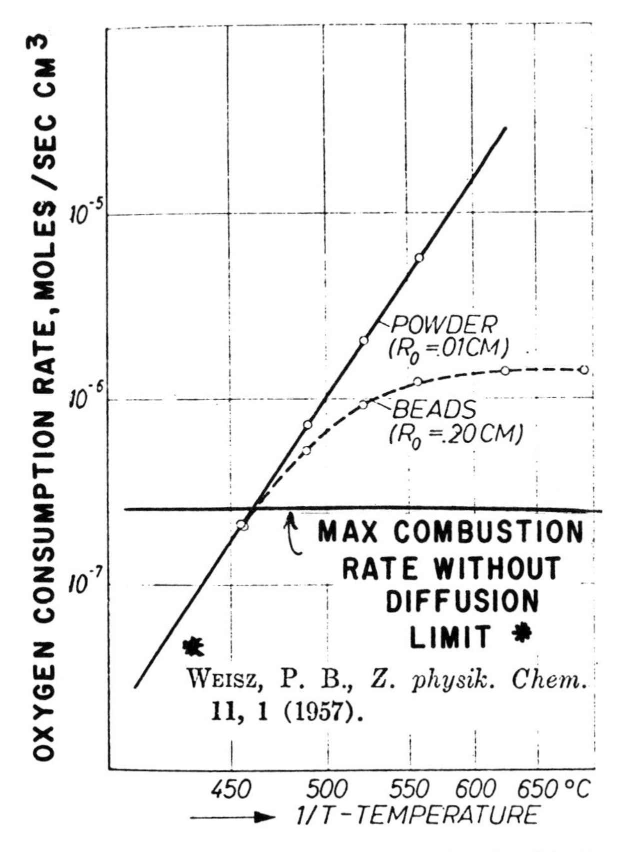

Figure 25c. Average observed burning rates of conventional silica–alumina cracking catalyst. Initial carbon contents, 3.4 wt %. Beads (dashed line) and ground up catalyst (full curve) (49).

interface in the particle. The latter event has been termed a "shell progressive" regeneration and has been studied experimentally by Weisz and Goodwin (*49*), whose illustrations of the concepts just described are shown in Figures 25a and b. The results of some experiments along these lines are shown in Figure 25c. One can estimate, using the criterion proposed by Weisz (*50*) for the absence of significant diffusional limitations on reactions within porous media, at what point such limitations (referring specifically to diffusion of oxygen) might become apparent in regenerations. For regeneration kinetics this criterion requires

$$\left(\frac{R^2}{D_{O_2}}\right)\left(\frac{r_{O_2}}{C_{O_2,g}}\right) < 1 \qquad (56)$$

for no appreciable oxygen concentration gradients, where r_{O_2} is the *observed* rate of consumption of oxygen per unit volume of catalyst, and $C_{O_2,g}$ is the oxygen concentration in the bulk gas phase. The analysis of Weisz and Goodwin is based on Equation 56, using typical values for conventional cracking catalysts. For this: $R = 0.2$ cm, $D_{O_2} \sim 5 \times 10^{-3}$ cm²/sec, $C_{O_2,g} \sim 3 \times 10^{-6}$ mole/cm³, and one has:

$$r_{O_2} < 4 \times 10^{-7} \text{ moles } O_2/\text{cm}^3 \text{ sec}$$

Above this level of oxygen consumption, appreciable gradients of O_2 concentration will occur, and the removal of coke will be nonuniform. The data in Figure 25c represent average burning rates (to 85% removal) for carbon removal from a conventional cracking catalyst containing an initial coke concentration of 3.4 wt %. The maximum combustion limit computed above is indicated on this plot, and deviations from the Arrhenius form, typical of diffusional intrusions, are observed for the beaded catalysts while the powder results indicate that nothing funny is going on with the intrinsic kinetics. Weisz and Goodwin also present visual evidence of coke profiles from beads partially regenerated at 450°, 515°, and 625°C, indicating the presence of the three regions depicted in Figure 25a.

Analysis of the shell progressive region of regeneration kinetics is relatively simple and represents a limiting case of the Ausman and Watson analysis. The rate of reaction of oxygen is equal to the diffusion rate at the particle surface in the shell progression region, so:

$$r_{O_2} = D_{O_2} \left(\frac{dC_{O_2}}{dr} \right)_R 4\pi R^2 = \frac{4\pi R D_{O_2} P}{R' y_{O_2,g}} \left(\frac{d\psi}{d\xi} \right)_{\xi=1} \tag{57}$$

and the carbon removal rate is N times this amount. The fraction of initial carbon remaining at any time is given simply by the ratio of the volumes of two spheres (assuming that geometry), one with $\xi = 1$ and one with $\xi = \xi'$:

$$X_F = (\xi')^3 \tag{58}$$

The gradient $(d\psi/d\xi)_{\xi=1}$ in Equation 57 is obtained from Equation 54 with the boundary conditions:

$$\xi = 1 \qquad \psi = 1$$

$$\xi = \xi' \qquad \psi = 0$$

The result obtained by Weisz and Goodwin for the fraction of initial carbon remaining according to the shell progressive mechanism is:

$$f(X_F) = \frac{1}{2}(1 - X_F^{2/3}) - \frac{1}{3}(1 - X_F) = \tau' \qquad (59)$$

where

$$\tau' \equiv \frac{ND_{O_2}Cy_{O_2.g}}{R^2C_c^\circ}\theta$$

Experimental burnoff results at 700°C plotted *vs.* time according to Equation 59 are shown on Figure 26a for 350 m^2/gram silica–alumina catalysts differing in initial coke content and diameter. The linear function is obeyed well over the entire range of burnoff times studied. Because of this linearity, the burnoff rate for a given catalyst can be characterized by the time required for a certain percent of total burnoff; Weisz and Goodwin employ the "85% burnoff time" as such a characteristic value which is experimentally convenient to measure. From Equation 59, with $X_F = 0.15$:

$$\theta_{85} = (0.076)\frac{R^2C_c^\circ}{ND_{O_2}Cy_{O_2.g}} \quad \text{minutes} \qquad (60)$$

The various proportionalities indicated in Equation 60 are subject to simple, direct experimental test; the results of a number of such tests are shown in Figure 26b, c, and d.

The dependence of burnoff on C_c°, initial carbon concentration, is a direct proportionality, as indicated by the agreement of experiment

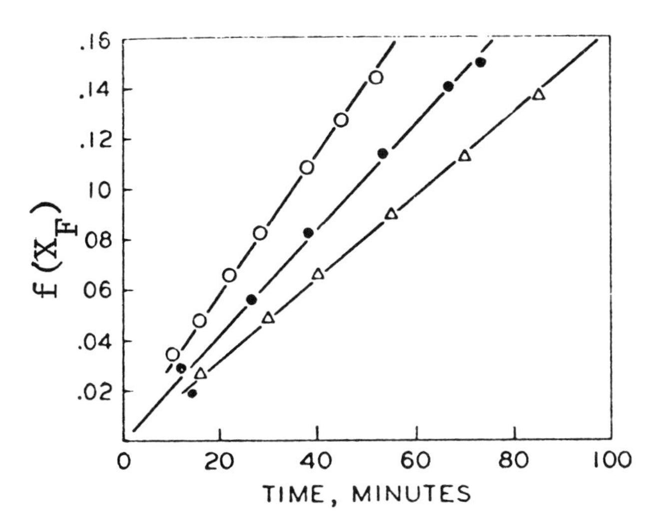

Figure 26a. Burnoff function vs. time on three different beads. X_F = *fractional amount carbon remaining* (49).

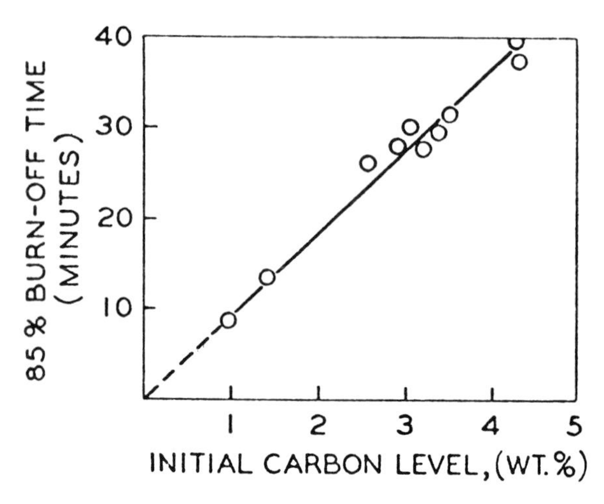

Journal of Catalysis

Figure 26b. Dependence of burnoff time on initial carbon level, for diffusion controlled combustion (silica–alumina cracking catalyst, 700°C) (49)

with Equation 60 shown in Figure 26b. The experiments represent burnoff measurements with a commercial silica–alumina catalyst (10 wt % Al_2O_3, 0.15 wt % Cr_2O_3), uniform beads of 0.4 cm diameter, coked to initial levels between one and 5 wt % carbon. The dependence on particle size was tested with beads of the same catalyst of varying diameter coked to an initial level of 3.0 ± 0.2 wt %. Again, the correlation with theoretical burnoff expectations ($\theta_{85} \alpha R^2$) is excellent, as shown in Figure 26c. The dependence on diffusivity—*i.e.*, the dependence of burnoff on the structure of the catalyst—was investigated using uniform size samples of widely different pore structure. The diffusivity was determined with hydrogen by a modified Wicke-Kallenbach (*51*) counterdiffusion measurement; since oxygen diffusivity at constant temperature should be proportional to hydrogen diffusivity, it serves to correlate θ_{85} with D_{H_2} for the various materials studied. Again, good agreement is obtained (Figure 26d). The major characteristics of the various materials used in the studies of Figure 26d are summarized in Table XI. A final test, that of the inverse proportionality of the burnoff time to oxygen partial pressure, was also conducted by Weisz and Goodwin, once again with excellent agreement with the theoretical prediction.

From these extensive experiments it seems clear that the regeneration of coked catalysts at high temperatures is governed by diffusion limits, and the shell progressive mechanism provides a satisfactory representation of the over-all process. Further, we have seen that coke deposition (as well as general poisoning by reaction impurities) can also be gov-

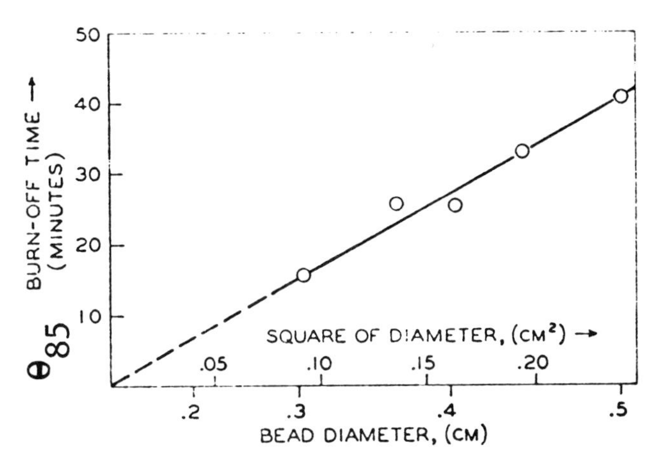

Journal of Catalysis

Figure 26c. Dependence of burnoff time on bead size for diffusion controlled combustion (49)

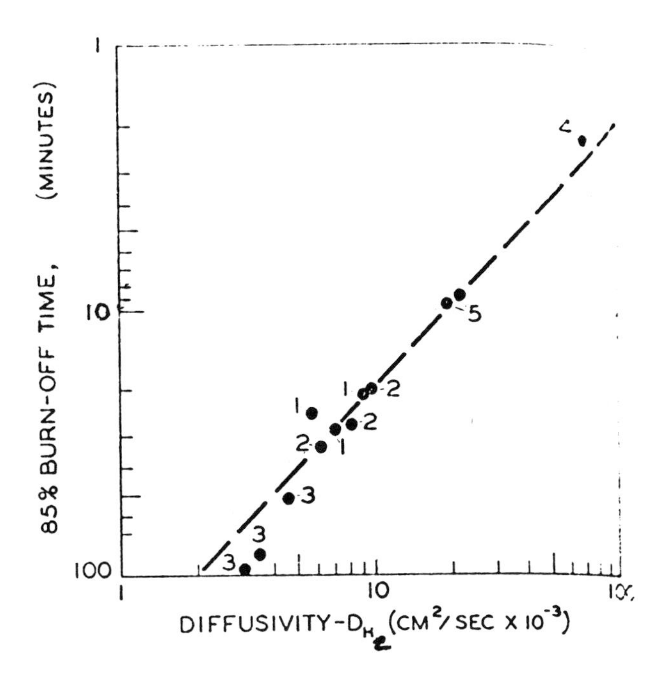

Journal of Catalysis

Figure 26d. Dependence of burnoff time on structural diffusivity, of various types of spherical particles, for diffusion controlled combustion region (49)

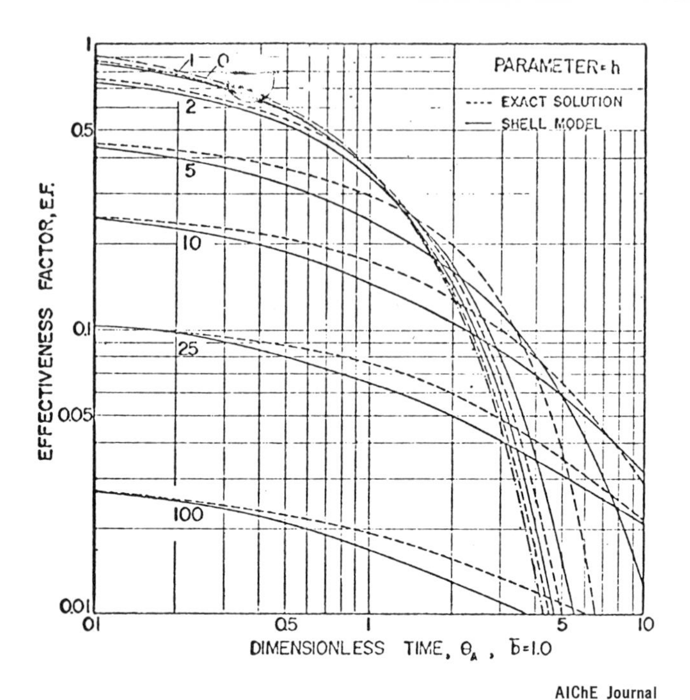

AIChE Journal

Figure 27a. Comparison of shell model and numerical solution for parallel fouling, b = *1.0* (41)

erned by diffusion limits under certain conditions (*viz.*, Murakami *et al.*), so that shell progressive processes occupy an important position in interpreting and analyzing both deactivation and regeneration. Other studies of regeneration kinetics include the earlier investigation of Haggerhammer

Table XI. Catalysts Used in Diffusivity Experiments[a]

Number (from Figure 26d)	*Description*
1	Laboratory-prepared silica–alumina; surface areas from 270 to 400 m^2/g.
2	Commercial silica–alumina with 0.15 wt % chromia.
3	Similar to sample 2, but after commercial service and exhibiting some thermal damage.
4	High diffusivity silica–alumina with a bimodal pore size distribution. Prepared according to method of Weisz and Schwartz (*52*).
5	Chromia–alumina with high coke combustion activity (*52*).

[a] Data of Weisz and Goodwin (*49*).

and Lee (*53*) on coke combustion on cracking catalysts in moving beds.

A number of further studies of shell progressive or related mechanisms have been reported, representing for the most part extension of the basic theme of Ausman and Watson and Weisz and Goodwin which we have chosen for illustration. Masamune and Smith (*41*) could, of course, easily simplify their general solutions to express the limiting case of the shell progressive analysis. In general, they found that over-all calculations according to the shell progressive simplification give results within about 15% of the detailed numerical calculations for type I and type II reactant inhibition, even in those cases where the diffusion through the deactivated (or regenerated, as the case may be) portion of the particle is not the rate-determining step nor close to it. For type II product inhibition, the shell progressive model was found to be much more restricted in application, with large discrepancies noted between the model and numerical calculations unless the shell-diffusion step was rate limiting.

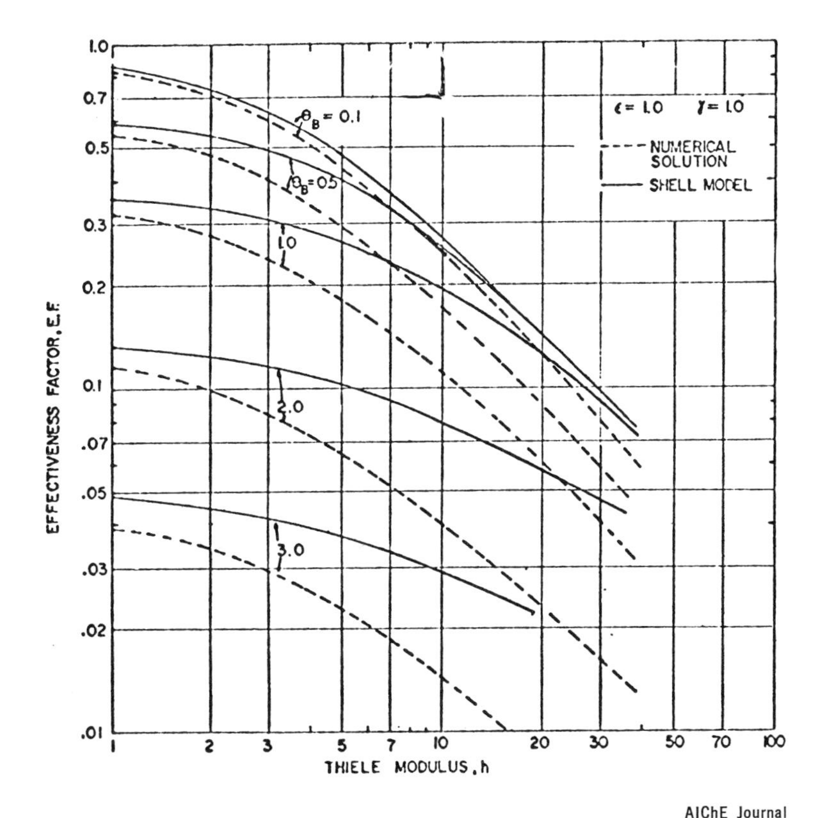

Figure 27b. Comparison of shell model and numerical solution for series fouling, $\varepsilon = 1.0$, $\gamma = 1.0$ (41)

The formal solution procedure used by Masamune and Smith is similar to that of Ausman and Watson; a general illustration of the comparison between shell progressive and detailed calculation results, in terms of the particle effectiveness factor referred to initial conditions is given in Figure 27a-c (*cf.*, Figures 20 and 21).

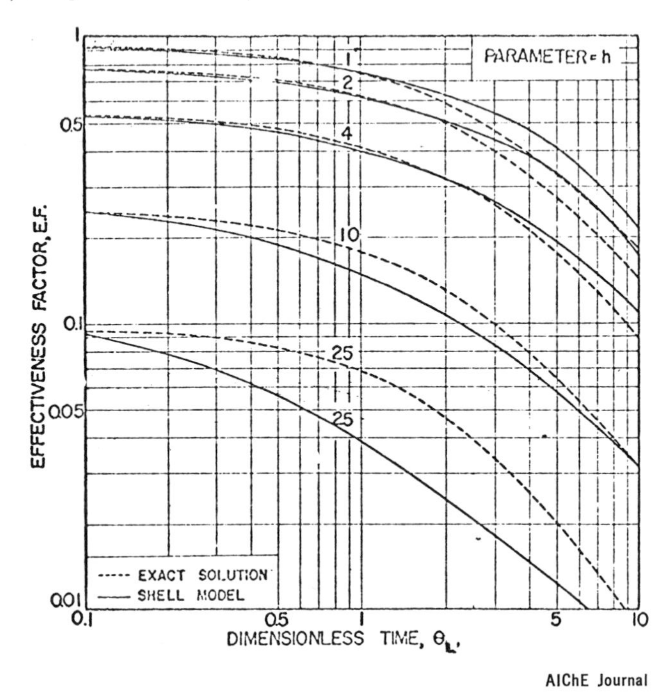

Figure 27c. Comparison of shell model and numerical solution for independent fouling, $h_L = 10$*,* $b_L = 1.0$ (41)

Further significant exploration of the shell progressive analysis has also been given by Carberry and Gorring (*54*). While their analysis presumes no single step to be rate controlling from among bulk gas phase transport, diffusion through the regenerated or poisoned region with no reaction, or reaction in the central core, some of the generality of the other studies mentioned here is missing since diffusion limitation in the reaction zone was not considered. Nonetheless, the interconnection between any of the three possible modes of rate limitation, either in poisoning or in regeneration, was clearly demonstrated in terms of three functionally variant relationships between process time and fraction deactivated (or regenerated). Pure shell progressive poisoning or regeneration is claimed to be expected only when the Thiele modulus (assuming first order type I or type II reactant inhibition) is greater than approximately 200, although we can see from the results of Masamune and Smith that this is overly restrictive, at least for type I and type II (parallel) deactivation. The discrepancy is more apparent than real; the value of $h > 200$ ensures near-discontinuous concentration profiles at the interface between active and inactive portions of the particle; the results in Figures 27a and c indicate only that similar behavior is obtained even when the profiles near this interface are not discontinuous—*i.e.*, there is an interfacial zone between active and inactive portions of the catalyst.

A primary objective of Carberry and Gorring was the relationship of over-all fouling or regeneration to time of operation in order to provide an analytical basis for the formulation of time-dependent activity expressions such as those of Voorhies which had been derived from experimental observation. The characteristic time-deactivation relations associated with external mass transfer, shell progressive diffusion, or chemical reaction controlling in a flat slab catalyst were:

$$t_m = \rho L(1 - X_F)/Nk_g C_o$$
$$t_d = \rho L^2(1 - X_F)^2/2NDC_o \qquad (61)$$
$$t_c = \rho L(1 - X_F)/NkC_o$$

in which t_m, t_d, and t_c are the characteristic times in the order listed, L is the half-thickness, ρ is the saturation concentration of poison or coke, k_g is an external mass transfer coefficient, and k is a first-order rate constant per unit area. The results are readily generalized to spherical coordinates and written in nondimensional form, where the over-all time conversion relationship is expressed as the sum of the individual terms of Equation 61:

$$\theta = t_m + t_d + t_c = \frac{\rho R^2}{NDC_o}\left[\frac{(1 - X_F)}{3}\left(\frac{1}{Nu} - 1\right) + \frac{(1 - X_F{}^{2/3})}{2} + \frac{(1 - X_F)^{1/3}}{Da}\right] \qquad (62)$$

where Nu is (k_gR/D), and $Da = (kR/D)$. The relationship of Equation 62 is shown in Figure 27d for a range of Nu and Da. The plot of activity *vs.* $\theta^{1/2}$ is linear over a considerable portion of the range included for all the parameters and not far from linear over even a larger range. Significant curvature does occur at small values of θ, which would be in general accord with observations such as those of Ozawa and Bischoff who had to use two line segments to fit activity data at short process times; however, it is apparent that shell progressive mechanisms (or variants on that theme) do give rise to integral time-activity relationships which are in accord with existing correlations. Some additional comments on the analysis are given by White and Carberry (55).

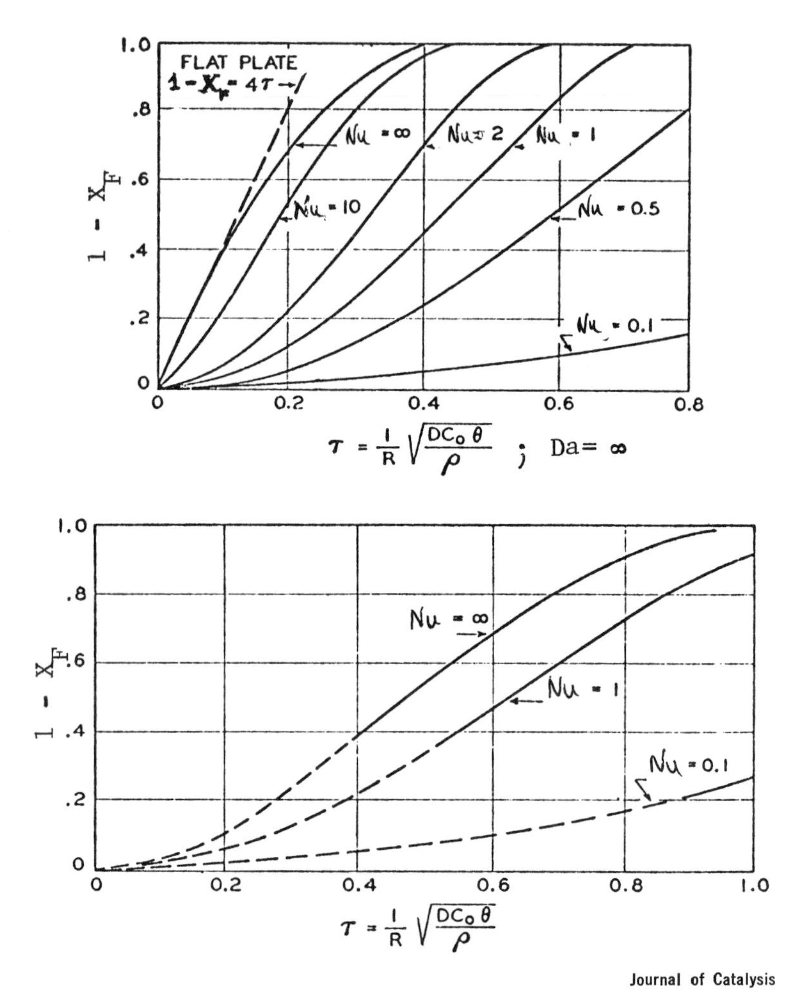

Figure 27d. Fraction of spherical catalyst poisoned or regenerated vs. *dimensionless time,* Da = *1,* ∞ (54)

A major assumption involved in all these studies of shell progressive poisoning or regeneration is that of pseudo-steady-state, in which the interface is taken to be stationary and the steady-state diffusion–reaction problem solved. The validity of this assumption has been examined in detail by Bischoff (*56*), who showed that the accuracy of the approximation depends on the ratio of reactant concentration in the bulk gas phase to the bulk density of the solid, C_o/ρ_s, being a small number. Since this is the case in most gas–solid systems, the pseudo-steady-state assumption would not appear to constitute a serious limitation to the shell progressive analysis.

For further details regarding shell progressive or "shrinking core" type problems in various applications, the reader is referred to the recent, extensive surveys by Wen (*57, 58*).

Nonisothermal Analysis. All the studies discussed to this point have been based on isothermal models of the reaction and deactivation process, although it is well known that particularly when diffusional limits are present, nonisothermal effects can become dominant in determining the over-all behavior. There is, surprisingly, relatively little literature dealing with such possibilities. The analysis of Masamune and Smith has been extended by Sagara, Masamune, and Smith (*59*) for type II deactivation only with a heat generation term for the main reaction (but not the fouling reaction) added. Such inclusion introduces three more parameters into an already parameter-laden analysis, but unfortunately this cannot be avoided. The new quantities are β, the heat of reaction parameter, and γ_A, $\gamma_{A,d}$, the Arrhenius or temperature sensitivity parameters as defined by Weisz and Hicks (*60*). In addition to the mass conservation relations of Equations 31 and 32, the energy balance must be included:

$$k_e \nabla^2 T + (-\Delta H) \rho k_A [A][s] = 0 \tag{63}$$

which is solved by numerical means simultaneously with the pseudo-steady-state mass conservation equations to obtain net catalytic behavior in terms of a time-dependent effectiveness factor as before. The boundary conditions of temperature used in this solution are analogous to those of Equations 33–36 for composition.

Some interesting example results for type II reactant inhibition are shown in Figure 28. The case shown is representative of a moderately exothermic reaction, with activation energy for the main reaction about 20 kcal/mole (based on reference temperature of 500°K) and for the deactivation reaction of 30 kcal/mole. The detailed profiles in Figure 28a are obtained for an intermediate level of diffusional resistance, $h = 5$, and the strong interaction between the internal temperature and reactant concentration leads to very complex activity profiles, at least for a limited period of operation. Ultimately, long term behavior results in nearly uniform activity profiles, however, as were found in the isothermal case (Figure 21). The over-all effectiveness as compared with isothermal operation is shown for this example in Figure 28b. For short lived processing periods the catalyst approximates an ordinary nonisothermal diffusion limit (in spite of the strange activity profiles) which for the illustration results in an increase in effectiveness relative to isothermal operation. As time progresses, though, the amount of deactivation proceeds at an accelerated rate until finally it more than offsets the advantage in rate arising from the thermal gradient, and nonisothermality eventually results

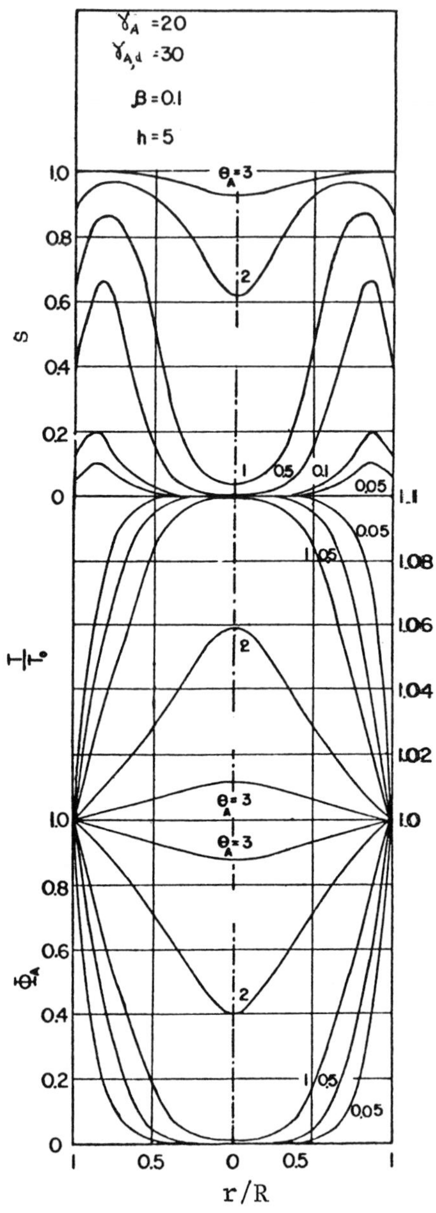

Figure 28a. Radial profiles for parallel fouling (59)

Notation on Figures 28a,b,c (c.f., *Figure 20*):

$\beta = (-\Delta H)D_A[A]_o/k_eT_o$

$\gamma_A = \Delta E_A/RT_o$

$\gamma_{A,d} = \Delta E_{A,d}/RT_o$

in poorer performance. The characteristic crossover of effectiveness *vs.* time of operation curves (h as the parameter) noted previously for type II parallel deactivation (Figure 21b) is also a characteristic of nonisothermal conditions, so that the seemingly more diffusion-limited catalyst under initial conditions of operation will actually be the best over the long run.

A few additional remarks concerning the activity profiles shown in Figure 28a are warranted. Sagara *et al.* show that the specific shape depends strongly on the magnitude of the Thiele modulus. For a value of $h = 10$, activity profiles are nearly discontinuous, and nonisothermal shell progressive behavior is obtained. A value of $h = 1$ gives almost flat activity profiles for the parameter employed. In this light, then, it appears that changes by a factor of, say, 2 in the Thiele modulus can drastically modify catalyst behavior. If the formation of coke can affect the mass transport properties of some catalysts as much as the results of some investigators we have discussed previously seem to indicate, then the catalyst itself becomes an adaptive system with time-dependent h which could exhibit quantitatively different types of behavior depending on its age. Such variability could have important implications in the determination of macroscopic reactor system stability. A second point involved in these sample results of Sagara *et al.* concerns the importance of the parameters. In the illustration, $\gamma_{A,d} > \gamma_A$, which means that as temperature is increased the deactivation is accelerated faster than the main reaction.

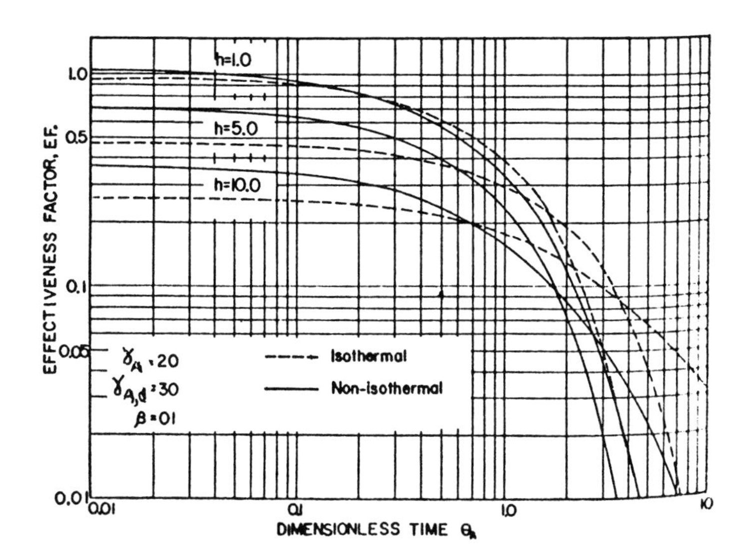

AIChE Journal

Figure 28b. Effectiveness factors for parallel fouling (59)

We see later that in deactivation systems where nonisothermalities are involved the question of whether the deactivation reaction is more or less sensitive to temperature than the main reaction is of primary importance in determining over-all behavior. The interactions are coupled and highly nonlinear, of course, so that it is almost impossible to estimate reliably what the net result of a change in temperature sensitivity will be. For present purposes, however, it is only necessary to realize that those results illustrated in Figure 28 are specific to $\gamma_{A,d} > \gamma_A$ and the analysis could differ considerably for $\gamma_A > \gamma_{A,d}$.

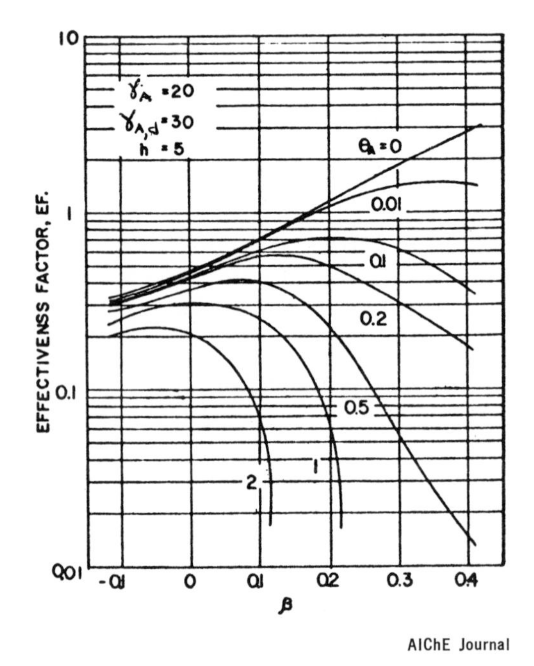

Figure 28c. Effect of heat of reaction for parallel fouling (59)

For reasonable values of activation energies, regardless of their relative magnitudes, observed behavior is also going to be quite sensitive to the heat of reaction. This is shown in Figure 28c, which shows the extreme dependence of effective operating life on the magnitude of the heat generation parameter, β. Once again, the question of time-dependent transport properties might be important since the ratio of effective diffusivity to effective thermal conductivity is involved in the definition of β, and, once again, modifications in behavior of an order of magnitude may be at stake.

Sagara *et al.* also present results for type II product inhibition under nonisothermal reaction conditions. For this type of deactivation mechanism the opposing effects of reactant concentration and temperature which resulted in the odd activity profiles of Figure 28a are absent; coke is formed from product and both product concentration and temperature levels increase with distance from the pellet surface. Qualitative behavior is again similar to that of the isothermal system (*cf.*, Figure 20b), and the comparison of effectiveness in the nonisothermal to the isothermal case follows that for reactant inhibition: at short times effectiveness is greater than the isothermal case, decreasing as time of operation proceeds more rapidly than the isothermal value and ultimately falling far short of that case.

This general analysis of Sagara *et al.* provides a sufficient consideration of nonisothermal effects, per se, for our purposes; however, there is an additional temperature-related problem of a more specific sort having to do with sintering. It was mentioned before that sintering in many cases could be regarded as a thermally activated process, and certainly in cases of deactivation or regeneration where large temperature gradients are encountered it would be possible for physical modification of the catalyst to occur as a result of thermal aging or sintering. One situation in which such a phenomenon is well known is that of addition of fresh cracking catalyst to a fluid-bed reactor-regenerator recycle. Wilson *et al.* (*61*) have showed that approximately 80% of the surface area of a fresh silica–alumina cracking catalyst is lost in the first few hours of utilization —somewhat of a puzzle because the normal temperature levels encountered in reactor–regenerator systems are not sufficiently high nor are coke levels on spent catalyst from the reactor in themselves sufficiently large to produce the temperature gradients necessary to yield the observed sintering. The problem has been fairly well resolved in the intervening years by an analysis based on temperature gradients resulting from the combustion of an oil-rich phase in the nonequilibrium catalyst leaving the reactor (*35, 36*). As a result of this work some very interesting information on nonisothermalities in coke burning has emerged. The fresh catalyst apparently contains sufficient surface area and pore volume to produce coke contents much larger than those of the equilibrium catalyst. In addition, hydrocarbon sorption on fresh catalyst is much stronger than on the aged surface, as shown by the data of Bondi *et al.* in Figure 29a for two model compounds. Thus, when the fresh catalyst is exposed to high oxygen concentrations, burning rates are observed which are many orders of magnitude greater than comparable coke burning rates, resulting from the combustion of the strongly adsorbed hydrocarbons. Assuming that the burning rates under such conditions are determined by oxygen diffusion into the catalyst particle, one derives a mass conservation equation analogous to that of Ausman and Watson (Equation 54). Bondi *et al.* used a slab geometry approximation, presuming the major portion of the

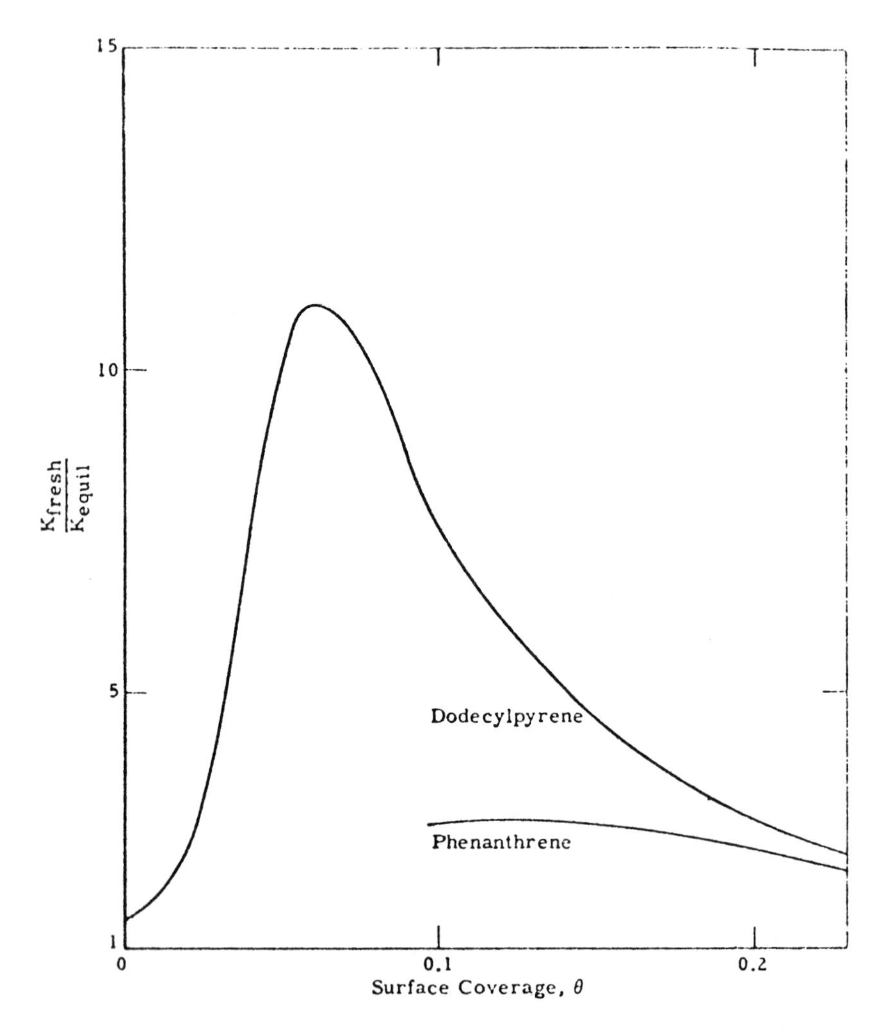

Industrial and Engineering Chemistry

Figure 29a. Relative adsorption on fresh and equilibrium catalysts of dodecylpyrene and phenanthrene (36)

Parameter values for Figures 29a,b,c,d:
Average heat transfer coeff. = 400 Btu/hr-ft²-°F
Heat capacity of catalyst = 0.27 Btu/lb-°F
Thermal conductivity of catalyst = 0.3 Btu/hr-ft-°F
Heats of combustion: carbon, −14000 Btu/lb; hydrogen, −55000 Btu/lb
Activation energy = 62500 cal/mole; $r_c = -kW_cP_{O_2}$ *where* W_c *= coke wt on catalyst*
Particle diameter = 80 microns; pore volume = 0.95 cm³/g

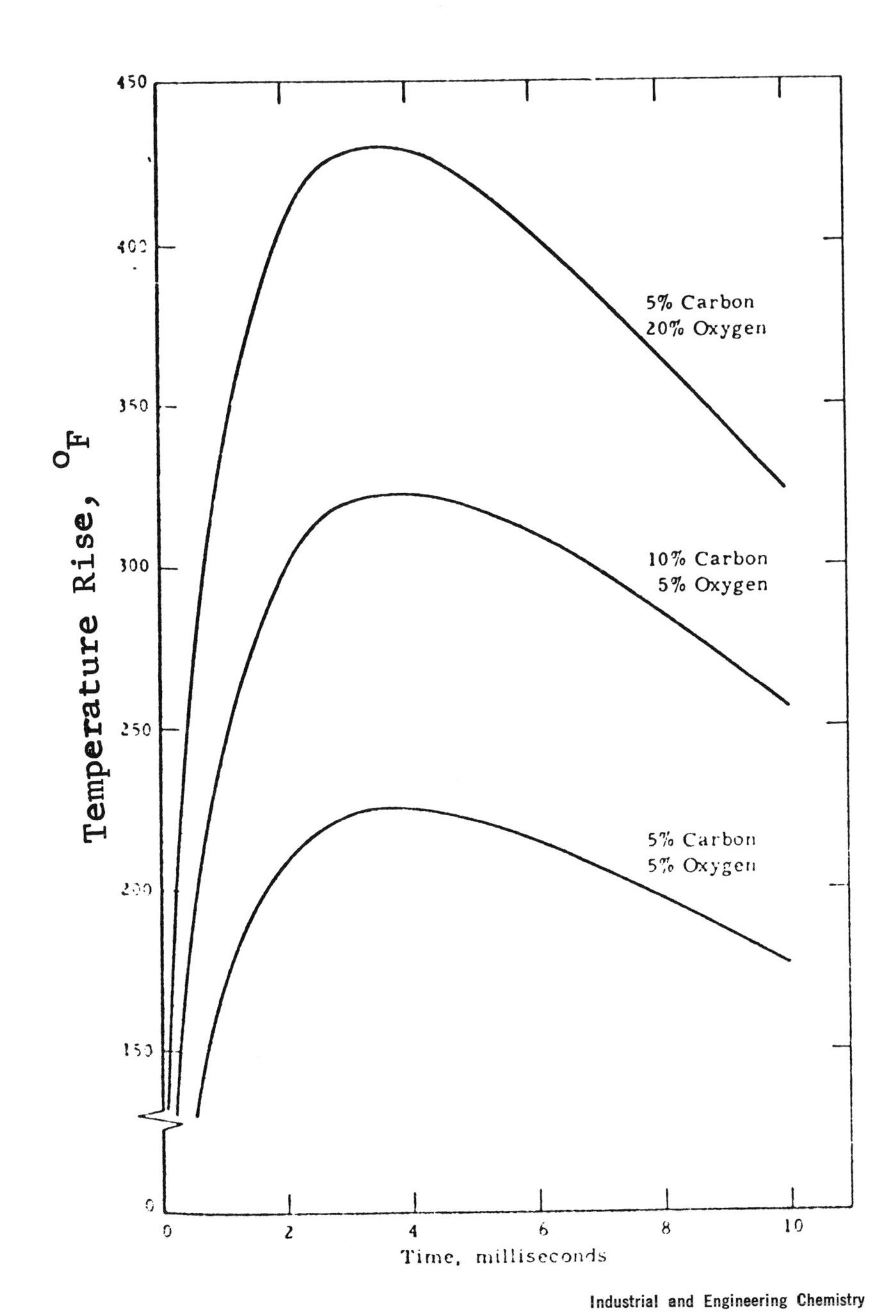

Industrial and Engineering Chemistry

Figure 29b. Diffusion controlled burning, no gradient in particle (36)

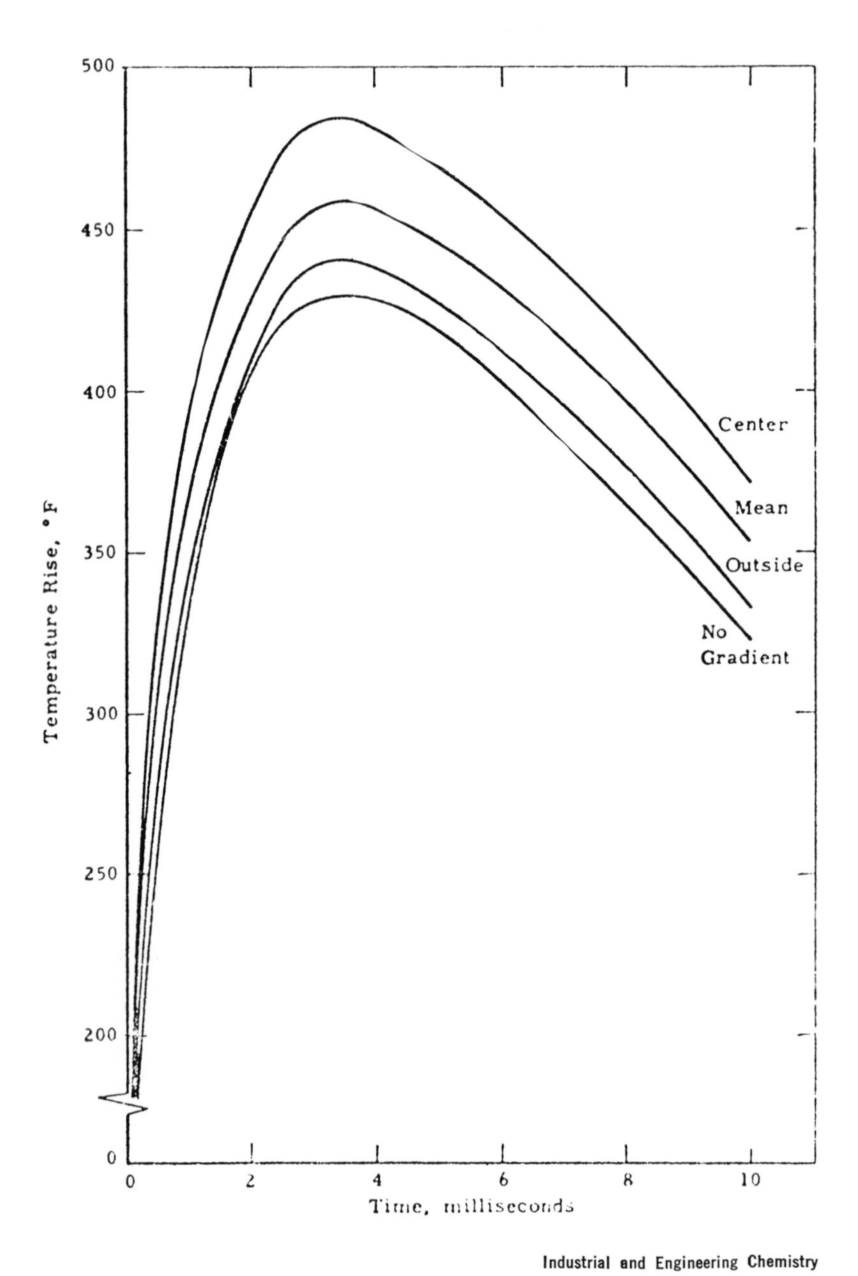

Industrial and Engineering Chemistry

Figure 29c. Diffusion controlled burning, gradient in particle. Carbon 5%. Oxygen 20% (36).

temperature rise to occur before combustion has penetrated very far into the particle. Two calculations were reported for the associated temperature rise—one assuming isothermal conditions within the particle and the second allowing for the existence of temperature gradients. Details of the mathematical procedures are not presented here, but the major results of such calculations and associated parameter values are shown in Figure 29b-d: The temperature histories shown support the general picture of a very rapid temperature rise in the particle on contact with the oxygen-rich phase, with subsequent rapid decay. The magnitudes of temperature

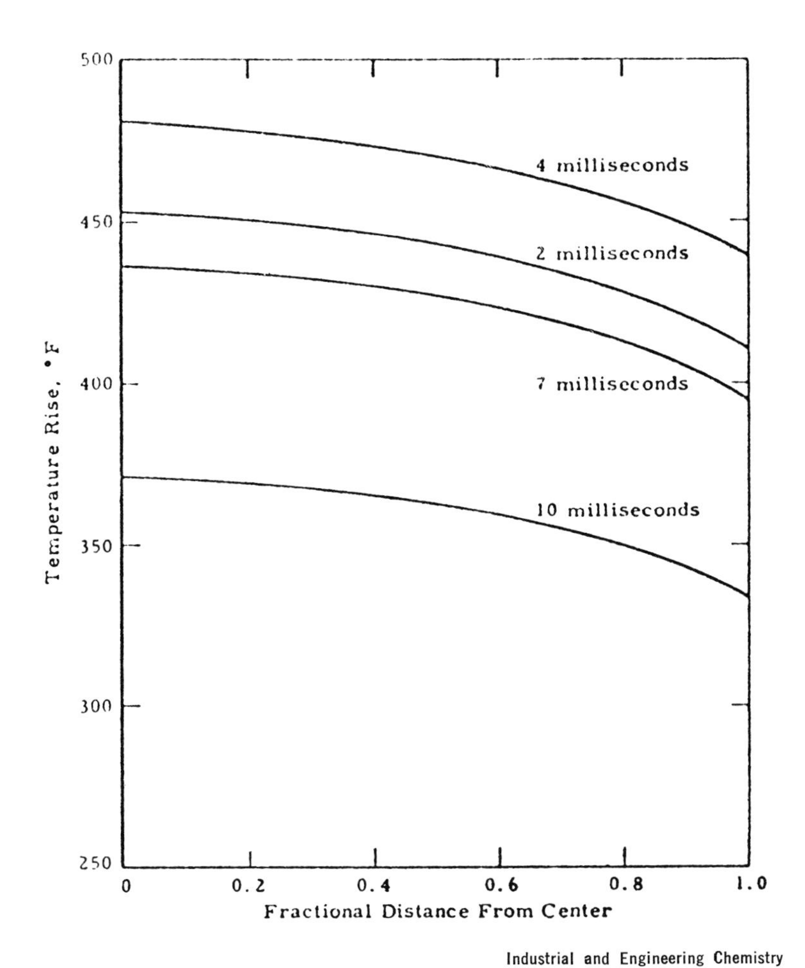

Figure 29d. Diffusion controlled burning. Carbon 5%. Oxygen 20% (36).

rise indicated for the typical parameter values given in the figure are of the order of several hundreds of degrees Farenheit, more than adequate to produce the temperature levels required for rapid thermal sintering. There is no qualitative difference in the results of the calculations assuming no gradients (Figure 28b) and those accounting for possible nonisothermality within the particle (Figure 28c). The indicated mean temperatures in the latter case run about 30°F higher than the corresponding isothermal value, and the total gradient within the particle is approximately 40°F. Uncertainties in the parameter values used to determine these results are sufficient to indicate that the isothermal calculation provides an adequate estimate. Thus the rapid loss of surface area of fresh cracking catalyst appears to be well explained in terms of thermal sintering caused by large temperature rises occasioned by combustion of relatively hydrogen-rich coke contained preferentially and in large quantity in the pore volume of the material.

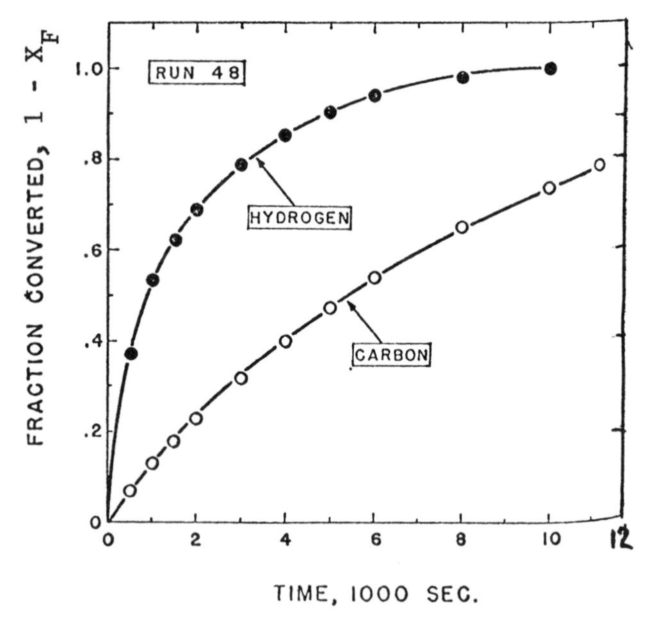

Figure 30a. Carbon and hydrogen conversion with time (62)

Conditions for Figures 30a,b:
13.3 mg. C sample (9.6% by wt C on catalyst). 400 cc. STP/ min. gas flow. 800 mm. approximate total pressure.

Run No.	Temp., °C	O_2 Press., Atm.
48	*480*	*0.062*

Some subsequent, detailed studies of the oxidation of coked silica–alumina catalysts have been reported by Massoth (*62*) and by Massoth and Menon (*63*). The former study reports experimental evidence indicating that the carbon and hydrogen components of coke are oxidized at differing rates. Under conditions corresponding to chemical control of the coke oxidation (*49, see* Figure 25), below 475°C with 2-mm diameter beads, the data shown in Figure 30a were reported by Massoth. Clearly the carbon and hydrogen are being oxidized at different rates, and one is

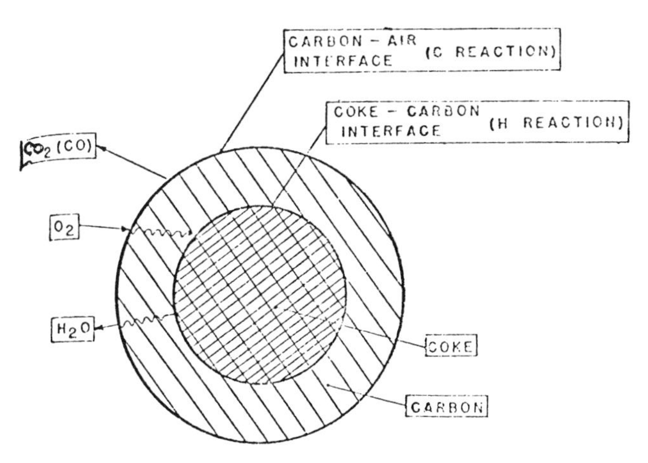

Industrial and Engineering Chemistry

Figure 30b. Model for simultaneous carbon and hydrogen reaction (62)

therefore involved with two separate, parallel gas–solid reactions which may or may not be coupled. Massoth was able to obtain a conventional chemical rate correlation for the coke oxidation data; however the hydrogen did not fit a surface reaction model. The final analysis given proposed a double core model for the over-all oxidation (Figure 30b) in which two interfaces develop as the reaction proceeds, the outer that of the carbon–oxygen and the inner that of hydrogen–oxygen. The hydrogen burning rates are taken to be controlled by oxygen diffusion through the unreacted carbon layer to the inner core, and the carbon burning rates chemically controlled. Quantitative development of this double core model resulted in a satisfactory correlation of data on both burning rates and product distribution variations with time.

Massoth and Menon (*63*) have further investigated the initial temperature transients involved in the oxidation of coked catalyst, and identify such temperature excursions with the oxidation of a strongly adsorbed hydrocarbon species containing a higher ratio of hydrogen than normal coke. This active species appears to be formed normally during the coking.

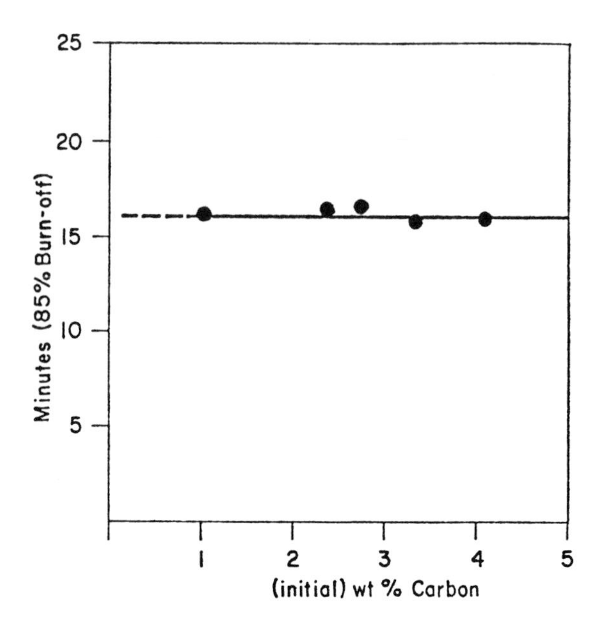

Journal of Catalysis

Figure 31a. Independence of the intrinsic burning rate from initial coke level (silica–alumina at 527°C). For 10% Al_2O_3; 100–500 m^2/g (48).

Such reactivity and associated temperature transients on initial oxidation should not be confused with the reactivity problems of fresh catalyst discussed previously; the studies of Massoth and Menon presumably involve equilibrated silica–alumina catalysts, and the temperature rises involved in this experimentation are on the order of 100°F. Nonetheless, substantial thermal gradients are involved in regenerating even equilibrium catalyst, so that long term thermal sintering is a primary event in the over-all deactivation of fluid-bed silica–alumina catalysts. Further, the disparity between carbon oxidation and hydrogen oxidation kinetics discovered by Massoth indicates the true complexity of the regeneration process and the difficulties of determination of the true kinetics of the coke oxidation reaction.

Kinetics of Coke Oxidation. While this section generally deals with deactivation effects in intermediate systems, we have seen that the kinetics and mechanisms of coke deposition and regeneration form the starting point for many of these problems. Discussion of the question of what the intrinsic kinetics of coke oxidation are at this point may seem somewhat out of place, yet the question has arisen at several places in the prevision discussion (as with Equation 43) so it is appropriate now to consider the kinetic problem in more detail.

We have seen from the early results of Weisz and Goodwin (*49*) the range of conditions over which coke oxidation is governed by oxygen diffusion, leading in the limiting case to a shell progressive burning process. Intrinsic kinetics of the carbon burning process were reported by Bondi, Miller, and Schlaffer (*36*) to be first order in carbon and first order in oxygen, as given in Equation 43a:

$$r_c = -kPy_{O_2}C_c \tag{43a}$$

No detailed data were given by Bondi *et al.* to support this expression; however the subsequent study of Weisz and Goodwin (*48*) clearly demonstrated the validity of the first-order dependence on carbon concentration and explored a number of associated factors in the oxidation reaction.

The reasonableness of first order kinetics with respect to coke concentration is suggested by the prior work of Haldeman and Botty (*23*), who measured the dispersion of coke deposits with an electron microscope. They found these deposits to be highly dispersed, which suggested that a large weight fraction of coke could exist on a catalyst in a state such that the entire amount is accessible to the oxygen reactant. For example, the carbon atom in graphite occupies an area of about 4 A^2,

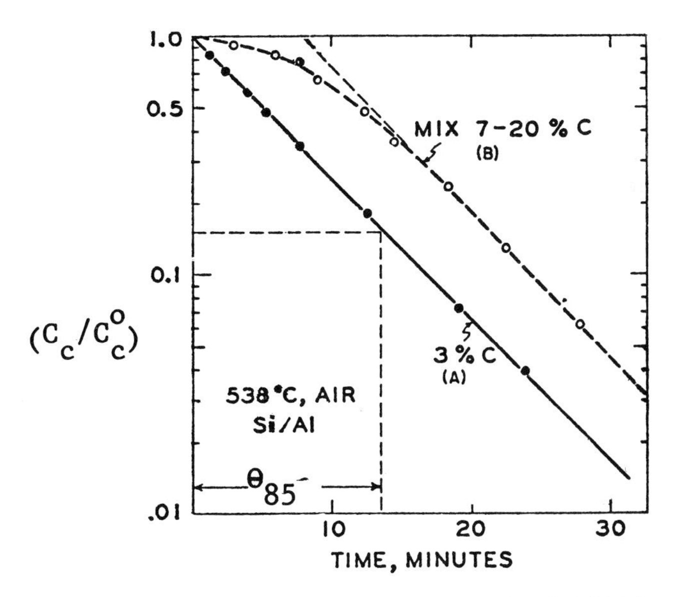

Journal of Catalysis

Figure 31b. Typical examples of rate plots of carbon remaining vs. burning time. Curve A: Normal sample. Curve B: Initial flattening due to carbon overload (partial inaccessibility) (48).

and if this is taken to be the area of a coke unit, then at one monolayer coverage on a catalyst of 250 m^2/gram area there is 12.5 wt % carbon. Experimentally, Weisz and Goodwin confirm the uniform accessibility of coke on a 250 m^2/gram catalyst up to about 6 wt % carbon; thus for a given temperature and oxygen partial pressure within this uniformly accessible region:

$$\frac{dC_c}{d\theta} = -kC_c \tag{64}$$

Verification of this, expressed in terms of the time required for 85% coke burnoff, θ_{85} ($k = 1.9/\theta_{85}$) is shown on Figure 31a for burning kinetics as a function of initial carbon content and on Figure 31b for the conversion as a function of time. The initial flattening on curve B of Figure 31b indicates the region in which total carbon surface is not uniformly available to the reactant oxygen.

Combustion behavior was found to be independent of the structural properties or history of the SiO_2–Al_2O_3 catalysts investigated, and inde-

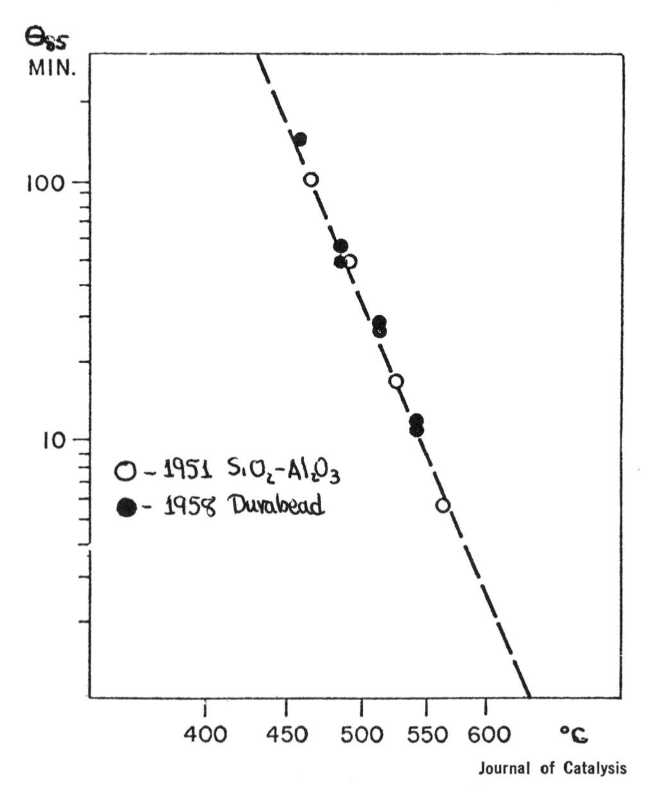

Figure 32a. Burning rate constant on different SiO_2–Al_2O_3 catalysts (48)

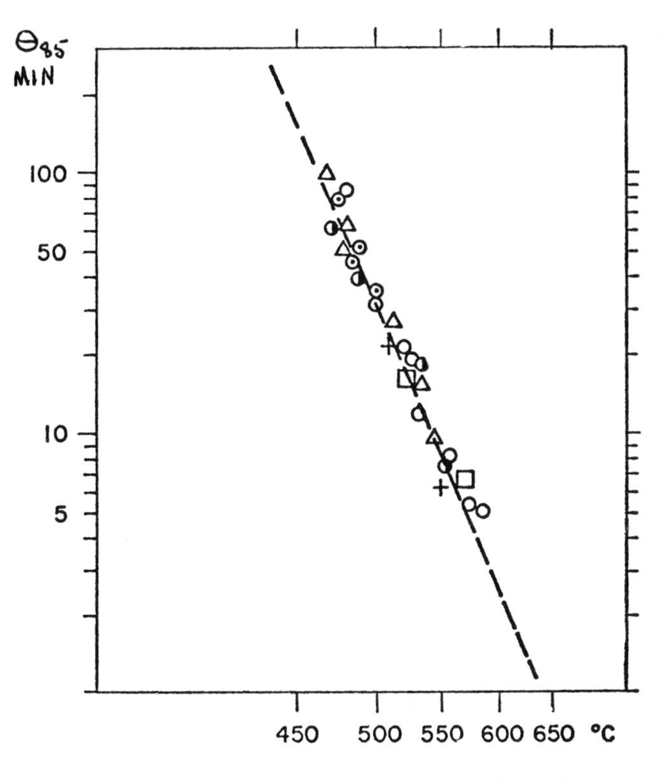

Journal of Catalysis

Figure 32b. Burning rate independence from origin of the coke (48)

Legend: (a) laboratory cracking of a (light East Texas) gas oil at 470°C (○) and (⊙); (b) commercial refinery cracking in adiabatic TCC moving bed reactor (◑); (c) cracking of cumene in a differential reactor at 420°C (□); (d) cracking of a C_6 to C_{10} naphtha mixture at 535°C (△); (e) reaction of and exposure to propylene at 420°C (+)

pendent as well of the source of the coke, including laboratory cracking of LETGO, commercial cracking in a TCC moving-bed unit, etc., as shown in Figure 32 a,b. Further, it was concluded that the combustion reaction as observed on silica–alumina was uncatalyzed, by comparison of the intrinsic rate constant on SiO_2–Al_2O_3 with those measured on a variety of other oxides, as shown in Figure 32c. The activation energy, corresponding to the lines drawn in Figure 32, is 37.6 ± 1.6 kcal/mole, and the intrinsic burning rate constant in all these cases is:

$$k = (4 \times 10^7) \exp [-(37{,}600 \pm 1{,}600)/RT] \text{ sec}^{-1} \qquad (65)$$

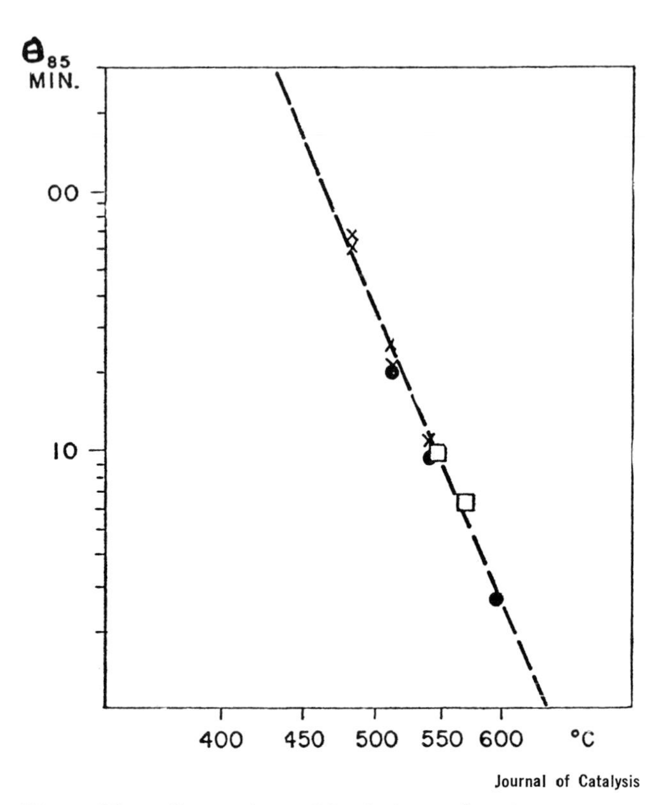

Figure 32c. Comparison of intrinsic combustion rate constant on various noncatalyzing oxide bases: ×, *Filtrol 110;* □, *silica–magnesia;* ●, *Fuller's earth. Dashed line is standard noncatalyzed kinetics* (48)

Initially enhanced burning rates were noted by Weisz and Goodwin, as we have discussed previously with respect to the results of Massoth and Menon, although in this work it was concluded that such transients were associated with the combustion of residual oil since the effect could be eliminated by prepurging test samples at high temperatures. The reported activation energy is in good agreement with the value of 34.5 kcal/mole previously reported by Haldeman and Botty (*23*). Further comparison of the value of the specific rate constant determined by Weisz and Goodwin with that for graphite oxidation determined by Gulbransen and Andrew (*64*) (determined for conditions of $Py_{O_2} = 0.21$ and 4.1×10^{-16} cm² per graphite atom) gives:

$$k \text{ (Weisz and Goodwin)} = 4 \times 10^7 \exp(-37{,}600 \pm 1600/RT)$$

$$k \text{ (Gulbransen and Andrew)} = 3 \times 10^7 \exp(-36{,}700/RT)$$

This comparison provides additional support for the contention that the coke oxidation is independent of carbon structure and is not catalyzed by the silica–alumina.

A very striking example of the catalysis of coke oxidation, however, was provided by incorporating a transition metal oxide such as chromia into the SiO_2–Al_2O_3. Similar activation energies were observed, but the absolute rate was approximately a factor of four greater under identical conditions on a catalyst containing only 0.15 wt % Cr_2O_3. When one deals with other types of transition oxide containing catalysts, such as chromia–alumina, the enhancement of burning rate is even more pronounced; differences by orders of magnitude from those on conventional SiO_2–Al_2O_3 are to be expected.

Detailed mechanistic explanation of the observations of Weisz and Goodwin is complicated, however, particularly by the extreme sensitivity of rate to very small amounts of transition metal incorporated into the SiO_2–Al_2O_3 matrix and the similarity of activation energies between the catalyzed and non-catalyzed oxidation. They concluded that the observed temperature coefficient did not reflect the activation energy of a simple kinetic step, but was the result of a mechanism proceeding via a mobile transition state possessing an unfavorable free energy of formation, or of a mechanism producing a product strongly inhibiting the oxidation by removing carbon sites from the reaction. This result again corroborates the message concerning coke oxidation and catalyst regeneration that we (by now) have seen from the results of each group of workers discussed so far: the over-all kinetics of coke combustion are comparatively simple, but the mechanism is apparently not.

The general first-order dependency of burning rate on carbon tested so thoroughly by Weisz and Goodwin appears well established, and has been investigated, in addition to the works cited, by Massoth (*62*) and Walker *et al.* (*65*). Kinetics with respect to oxygen have also been well established as first order by Walker *et al.* and corroborated in several subsequent studies, notably that of Massoth. It was indicated previously that when the combustion of hydrogen is considered separately from that of carbon, a difference in rates is observed such that the analytical treatment of both reactions would involve a model of independent and parallel gas–solid interactions. The combustion of carbon is the slow reaction, however, and for purposes of regeneration analysis or design this complication may be safely neglected. Hence, Equation 43a with rate parameters and activation energy as reported by Weisz and Goodwin provides an adequate representation of the intrinsic kinetics of coke oxidation.

Deactivation Effects in Macroscopic Systems: Analysis of Chemical Reactor Performance

The consequences of catalyst deactivation on reactor behavior and performance almost deserve the title of the forgotten effect in chemical reaction engineering. While our review to this point has been, in its content, necessarily selective it would almost be possible from here on to include all the pertinent studies of catalyst poisoning and reactor systems since there have been so few. It will be apparent early in the discussion of reactor systems subject to catalyst decay that one is never very far from discussing pure optimization problems, and, indeed, it is in this area that much of the interest and effort in the area is directed at present. Nonetheless, we make a rough division in this discussion of macroscopic effects between those studies directed predominantly toward modeling, simulation, or parametric identification on the one hand and those more purely concerned with optimization problems on the other. Needless to say, the distinction between the two is apt to become blurred at times. Further, while fluidized beds are certainly of considerable importance in catalytic process, the bulk of literature information deals with fixed beds, and hence the primary concerns of this review are in that area. We shall see, however, that some general relations can be established between certain types of deactivation effects in fixed beds and their analogs in moving or fluidized beds. For a detailed analysis of activity levels in fluidized reactor–regenerator systems, using integral time-dependent activity functions of the form of Equation 4, the works of Petersen (*66*) and of Gwyn and Colthart (*67*) are comprehensive examples.

Identification of the Behavior of Reactors Subject to Deactivation. Serious analytical studies of catalyst deactivation date from the last decade. In 1961 Anderson and Whitehouse (*68*) published one of the first such studies, dealing with the effect of specified relative activity equations and poison or activity distributions within the reactor on the over-all observed performance. This is not a reactor modeling study per se but rather one in which concentration and activity profiles are arbitrarily set; hence, the relation between microscopic and macroscopic poisoning effects is specified only indirectly. They consider various forms of the activity relationships similar to those of Equations 3 and 4:

$$\frac{s}{s_o} = 1 - \frac{\alpha_1}{s_o} C_p$$

$$\frac{s}{s_o} = (1 + s_o\alpha_3 C_p)^{-1} \tag{4a}$$

$$\frac{s}{s_o} = \exp(-\alpha_2 C_p)$$

Table XII. Activity Correlations and Poison Distribution Equations (68)

Case	*Effectiveness of Poison per Increment of Bed*	*Tendency of Poison to Accumulate near Inlet*
I	moderate to very strong	very strong
II	moderate: $[s/s_o = 1 - (\alpha_1/s_o)C_p]$	independent of distribution
III	strong: $[s/s_o = (1 + s_o\alpha_3 C_p)^{-1}]$	very strong: $[f(C_p/C_p^o) = \exp(-b'w)]$
IV	strong (III)	strong: $[f(C_p/C_p^o) = (1 + b'w)^{-1}]$
V	very strong: $[s/s_o = \exp(-\alpha_2/C_p)]$	very strong (III)
VI	very strong (V)	strong (IV)
VII	very strong (V)	very strong: $[f(C_p/C_p^o) = (1 - b'w)]$
VIII	strong (III)	very strong (VII)
IX	weak: $[s/s_o = (1 - s_o\alpha_3 C_p)^{-1/2}]$	very strong (III)

and various types of poison or activity distribution equations with reactor length, *w:*

$$f(C_p/C_p^o) = (1 - b'w) \qquad b'w \leqslant 1$$

$$f(C_p/C_p^o) = \exp(-b'w) \tag{66}$$

$$f(C_p/C_p^o) = (1 + b'w)^{-1}$$

where b' is a constant and $f(C_p/C_p^o)$ is the poison distribution function. Characteristics of the over-all reactor behavior and/or performance can be obtained from Equations 4a and 66 by suitable averaging. Thus:

$$\overline{F} = \frac{1}{W}\int_0^W (s/s_o)\,dw$$

where $\overline{F}$ is the average relative activity of the reactor with respect to initial conditions, and:

$$\overline{(C_p/C_p^o)} = \frac{1}{W}\int_0^W f(C_p/C_p^o)\,dw \tag{67}$$

where $\overline{(C_p/C_p^o)}$ is the average poison concentration in the reactor. Integration of Equation 67 for the various poison distribution functions yields expressions such as the following for the exponential function:

$$\overline{C_p} = [C_p{}^o(1 - C_p{}^W/C_p{}^o)]/\ln(C_p{}^o/C_p{}^W)$$

Anderson and Whitehouse examined nine combinations of poisoning and distribution equations, as detailed in Table XII. Some results of these calculations are shown in Figure 33 for the particular examples of Cases III and IV with their associated activity and profile expressions. The solid lines in Figure 33a represent operation for varying conditions of inlet poison concentration (parameter $(\alpha_1/s_o)C_p{}^o$) and the dashed lines are for varying exit poison concentrations (parameter $C_p{}^W/C_p{}^o$). The operation is not very sensitive, in terms of average activity *vs.* average poison concentration, to this second parameter, but it depends strongly on inlet conditions. Some further detail on poisoning effectiveness is given in Figure 33b, which shows, in crossplot of Figure 33a, how the effectiveness of a poison increases with $(\alpha_1/s_o)C_p{}^o$—essentially a measure of the impact on over-all behavior of the magnitude of the activity coefficients in the "microscopic" relationships of Equations 3 and 4—and also demonstrates that the manner of distribution of poison within the bed affects its average behavior. Perhaps the most interesting result of Anderson and Whitehouse is, unfortunately, not shown on the figure. That is, the results in terms of average values (*i.e.*, average activities, average poison concentration) or corresponding observables (conversion might be a rough measure of average activity, for instance) look alike for all the combinations studied. We note on Figure 33a that $(1 - \overline{F})$ and $(\alpha_1/S_o)\overline{C_p}$ are linear on logarithmic coordinates at lower values of activity. This behavior is characteristic of all systems, the sense of the parametric

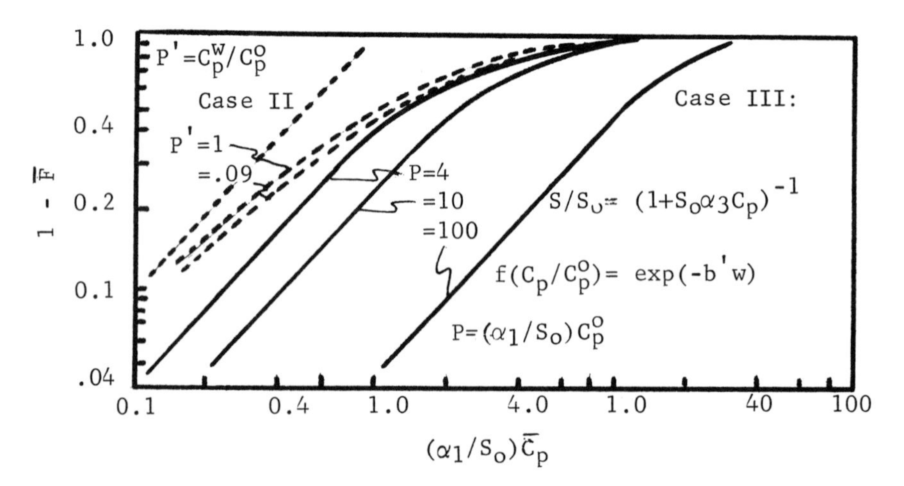

Industrial and Engineering Chemistry

Figure 33a. Loss in average relative activity as a function of $\overline{C_p}$*.* ---*, constant values of* f'*.* ———*, constant values of* aS_o (68).

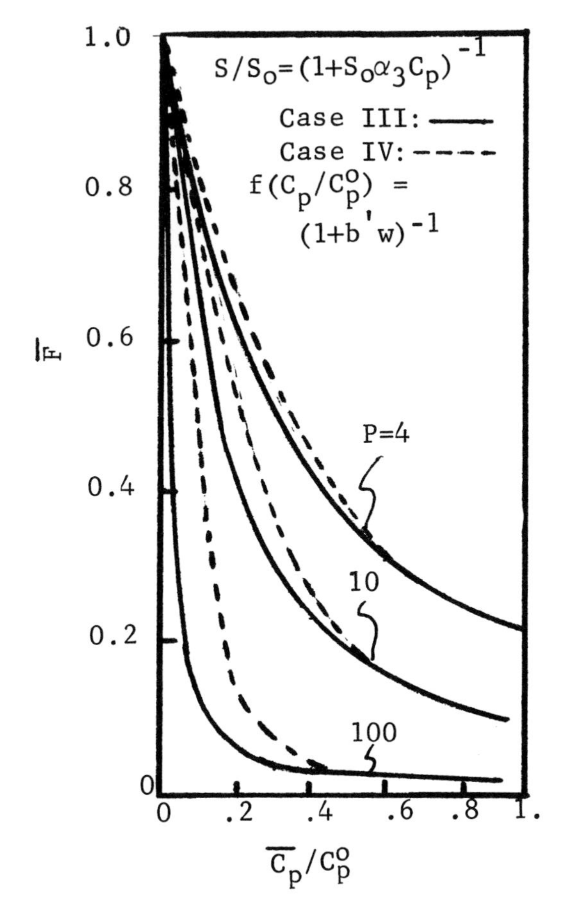

Industrial and Engineering Chemistry

Figure 33b. Relative activity is dependent upon both the poisoning equation and the distribution of poison (68)

dependence on $(\alpha_3/S_o)C_p^o$ is characteristic of all systems, and the insensitivity to (C_p^w/C_p^o) is characteristic of all systems. Hence, in application to activity problems all mechanisms of catalyst poisoning and modes of reactor performance turn out to look very much the same, and thus no wonder the seeming universal nature of the Voorhies form correlation. We see subsequently that all poisonings do not look the same when it comes to problems of selectivity variation, but it is well to keep in mind that the indistinguishability of activity behavior is really another instance of the difficulty in inferring detail concerning kinetics or mechanistics from integral reaction data.

A more detailed study of reactor behavior, using the general approach of Anderson and Whitehouse, has recently been given by Wheeler and Robell (*69*). Rather than obtaining average activities *via* Equation 67, however, they compute conversions for the type I poisoning of an isothermal, irreversible first-order reaction in a plug flow reactor where the rate constant is taken to be a function of poison concentration and the poison concentration is distributed in the bed according to fixed bed adsorption theory. Thus we have:

$$\ln\left(\frac{C_w}{C_o}\right) = \frac{1}{W}\int_0^W k\, dw$$

Where k is the first-order rate constant, C_o and C_w are the inlet and outlet concentrations of reactant, v is the superficial velocity, k_o is the initial rate constant, and W is the reactor length. The rate constant is a function of poison concentration, which in turn is a function of position and time of operation. The relationship of k to poison concentration is taken from the previous work of Wheeler (*37*) on selective poisoning. The family of curves relating activity to poison concentration shown in Figure 18a can be represented in analytical form by:

$$\frac{k}{k_o} = s = \left[\frac{1}{1 + h_o C_p / C_{p\infty}} - \frac{C_p / C_{p\infty}}{1 + h_o}\right] \qquad (68)$$

where h_o is the Thiele modulus for the unpoisoned catalyst and, in the context of Equation 68, is the parameter which determines the degree of selectivity of the poisoning, C_p is the poison concentration, and $C_{p\infty}$ the saturation poison concentration ($s = 0$ when $C_p = C_{p\infty}$).

The distribution of poison in the bed is determined from the fixed bed adsorption theory of Bohart and Adams (*70*), in which adsorption kinetics are given by:

$$\frac{\partial C_p}{\partial \theta} = r_{\text{ads}} = k_{\text{ads}} C_{p,g}(1 - C_p/C_{p}\infty)$$

and the conservation equation, presuming isothermality and plug flow with no dispersion, is:

$$\frac{\partial C_{p,g}}{\partial w} = r_{\text{ads}}$$

where k_{ads} is the adsorption rate constant, and $C_{p,g}$ is the concentration of poison in the gas phase. The adequacy of kinetics such as these in describing fixed bed adsorption has been an enduring topic of research over several decades; for present purposes we need only consider, as noted by Wheeler and Robell, that the kinetic form is of a semiempirical nature which can be used to reproduce a number of physically reasonable situations *via* variation of the parameters k_{ads} and $C_{p\infty}$. The solution to this adsorption problem for a bed initially free of contamination is:

$$\frac{C_p}{C_{p\infty}} = \frac{1 - \exp(-N_t\theta/\theta_\infty)}{1 + \exp(-N_t\theta/\theta_\infty)\,[\exp(N_t w/W) - 1]} \qquad (69)$$

where $N_t \equiv k_{\text{ads}}W/v$ is the number of adsorption transfer units in the reactor and θ_∞ is the ratio of the total capacity of the catalyst for poison adsorption to the rate at which poison is introduced to the reactor.

$$\theta_\infty = \rho_B C_{p\infty} W / M v C^o{}_{p,g}$$

Here ρ_B is the bulk density of catalyst, M is the molecular weight of poison, and $C^o{}_{p,g}$ is the inlet concentration of poison.

Equations 68 and 69 can be combined and the reactor equation integral evaluated analytically. The general solution for exit conversion as a function of time of operation is:

$$\ln\left(\frac{C_w}{C_o}\right) = -\frac{k_o/k_{\text{ads}}}{1 + h_o}\left\{\ln\left[1 + \exp\left(-N_t\frac{\theta}{\theta_\infty}\right)(\exp N_t - 1)\right]\right.$$

$$\left. + h_o \ln\left[1 + \frac{\exp\left(-N_t\frac{\theta}{\theta_\infty}\right)(\exp N_t - 1)}{1 + h_o\left[1 - \exp\left(-N_t\frac{\theta}{\theta_\infty}\right)\right]}\right]\right\} \qquad (70)$$

For completely nonselective poisoning, $h_o = 0$ and Equation 70 becomes:

$$\ln\left(\frac{C_w}{C_o}\right) = -\left(\frac{k_o}{k_{\text{ads}}}\right)\ln\left\{1 - \exp\left(-N_t\frac{\theta}{\theta_\infty}\right)\right.$$

$$\left. + \exp\left[N_t\left(1 - \frac{\theta}{\theta_\infty}\right)\right]\right\} \qquad (70a)$$

Wheeler and Robell (69) discuss various limiting cases of Equation 70; those important in parametric evaluation (of which there are four: θ_∞, N_t, (k_o/k_{ads}) and h_o) are for zero time, for conversion before breakthrough of poison from the bed, conversion at long time ($\theta/\theta_\infty >> 1$), and conversion at $\theta = \theta_\infty$.

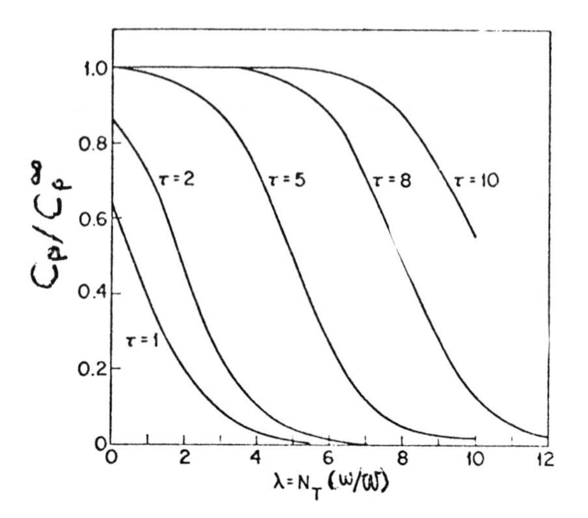

Journal of Catalysis

Figure 33c. Bohart-Adams generalized poisoning wave profiles. Plots of C_D/C_D^∞ vs. $\lambda = N_T(w/W)$ *for various reduced times,* $\tau = N_T\theta/\theta_\infty$ (69).

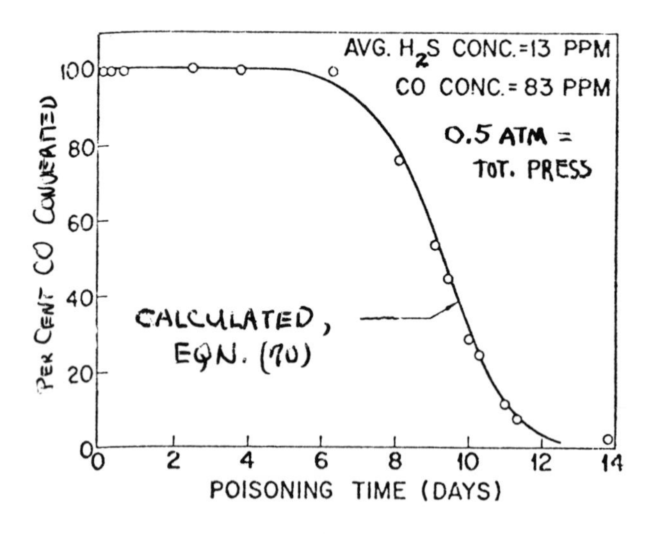

Journal of Catalysis

Figure 33d. Comparison between calculated and experimental catalytic activity history for CO oxidation at 25°C. Curve is calculated, and points are experimental. Catalyst 1% Pt–1% Pd supported on alumina (69).

Some typical poison profiles obtained from Equation 69 are shown in Figure 33c. One can see clearly one characteristic of the fixed-bed adsorption solution used; after an initial period of transient operation there is established a "wave" of poison concentration which passes through the bed at a constant velocity and with a fixed shape. This type of behavior, involving a constant zone or band of adsorption is, of course, well known in various fixed bed problems, and later it will be apparent that such fixed waves are also a prominent characteristic of the deactivation of fixed-bed reactors—even where the analysis is carried out from a somwhat more fundamental point of view than that used here.

The approach of Wheeler and Robell for correlating poisoning data and for estimating the effects of changes in design factors such as v or W (at least in the sense of qualitative trends), would seem to be most useful. Some experimental verification for the procedure is illustrated in Figure 33d for the H_2S poisoning of 1% Pt–1% Pd–Al_2O_3 catalyst in the oxidation of CO by O_2 at 0.5 atm. The parameters of the model were estimated from the experimental data after assuming that $h_o = 0$ and poisoning was nonselective (curve A of Figure 18a). The resulting computed fit from Equation 70 to experimental data is very good, particularly in matching the time of breakthrough and in the approach to long time behavior. Although four parameters may seem like many when it comes to self-congratuation on agreement between theory (model) and experiment, we shall soon learn just exactly how complicated the problem really is.

Some Analytical Reactor Studies. The first extended analysis of reactor behavior as a result of catalyst deactivation was given by Froment and Bischoff (*71, 72*). This study focused on the nature of reactor transients and activity profiles resulting from the solution of the appropriate continuity equations and kinetics of deactivation formulated according to mechanistic schemes such as types I and II, rather than in terms of time-dependent correlation. The procedures used by Bischoff and Froment are similar to those used by a number of subsequent workers, so we shall follow their development in some detail. The reactor models used here and in other studies we discuss are, for the most part, simple one-dimensional formulations in order to focus primary attention on the basic deactivation effects. Froment and Bischoff in addition assume flat velocity profiles, constant density and total molal concentration, and isothermal conditions. For the main reaction in either type I or type II mechanisms, we have for reactant A, mole fraction x:

$$\frac{\partial x}{\partial \tau} + \frac{\partial x}{\partial z} = - \frac{\Omega \rho_B d_p}{F} r_A \tag{71}$$

in which the nondimensional variables τ and z are defined as:

$$\tau = \frac{F}{\epsilon \rho_A \Omega d_p} \theta$$

$$z = w/d_p, Z = W/d_p$$

where F is feed rate in mass/time, ϵ is void fraction, d_p is the catalyst particle diameter, Ω the reactor (total) cross section, ρ_A the mass density of reactant A, ρ_B the bulk density of catalyst, w reactor length, θ time, and r_A the rate of disappearance of A. For the rate of deactivation, which will be measured by the rate of accumulation of coke or poison on the catalyst we have:

$$\frac{\partial C_c}{\partial \tau} = \frac{\epsilon \rho_A \Omega d_p}{F} r_c \tag{72}$$

where C_c is the carbon content of the catalyst in wt/total weight. By changing variables, Equations 71 and 72 can be simplified to:

$$\frac{\partial x}{\partial z} = -\frac{\Omega \rho_B d_p}{F} r_A \tag{73}$$

$$\frac{\partial C_c}{\partial \eta'} = \frac{\Omega \rho_A \epsilon d_p}{F} r_c \tag{74}$$

in which $\eta' \equiv \tau - z$, where η' is the variable along a characteristic of Equation 71.

Bischoff and Froment consider both cases of type II deactivation, so that for the reactant inhibition:

$$r_A = k'_A P x + k'_{A,d} P x \tag{75}$$

$$r_c = k'_{A,d} P x \tag{76}$$

and for product inhibition:

$$r_A = k'_A P x \tag{77}$$

$$r_c = k'_{A,d} P (1 - x) \tag{78}$$

where P is the total pressure. The form of these rate equations does not correspond exactly to that of Equations 29 and 30, written as examples for the type II system, since the proportionality of both main and fouling reactions to instantaneous catalytic activity is not given. This functionality was not included in the analysis of Froment and Bischoff as a direct proportionality; rather the catalytic activity function used was the integrated form of Table X. Two cases were considered:

$$k'_A = k'_{A_o}s \quad \text{or} \quad k'_{A,d} = k'_{A,d_o}s$$
$$s = \exp(-\alpha_2 C_c); \quad s = \frac{1}{1 + K_c C_c} \tag{79}$$

corresponding to the strong and very strong activity relationships explored by Anderson and Whitehouse. Of the four possible combinations of deactivation mechanism and activity relationship, we consider in detail first the development for type II reactant inhibition with an exponential activity function. Froment and Bischoff consider here only the activity effect on k'_A. The conservation equations become:

$$\frac{\partial x}{\partial z} = -a[\exp(-\alpha_2 C_c) + \nu]x \tag{80}$$

$$\frac{\partial C_c}{\partial \eta'} = bx \tag{81}$$

in which

$$a = \frac{\Omega \rho_B d_p P}{F} k'_{A_o}$$

$$b = \frac{\Omega \rho_A \epsilon d_p P}{F} k'_{A,d_o}$$

$$\nu = k'_{A,d_o}/k'_{A_o}$$

with the boundary conditions:

$$x(0,\eta') = 1$$
$$C_c(z,0) = 0 \tag{82}$$

expressing the uniformity of initial carbon content at zero time and the introduction of a pure feed material into the reactor. For the case of $\nu << 1$, Froment and Bischoff are able to obtain an analytical solution to Equations 75–82 as:

$$x = \{1 + \exp(-\alpha_2 b\eta')[\exp(az) - 1]\}^{-1} \tag{83}$$

$$\exp(-\alpha_2 C_c) = \{1 + \exp(-az)[\exp(\alpha_2 b\eta') - 1]\}^{-1} \tag{84}$$

The results for reactant composition and carbon profiles according to these two equations are given in Figures 34a and b as a function of the time of operation of the reactor. In general these profiles are non-uniform,

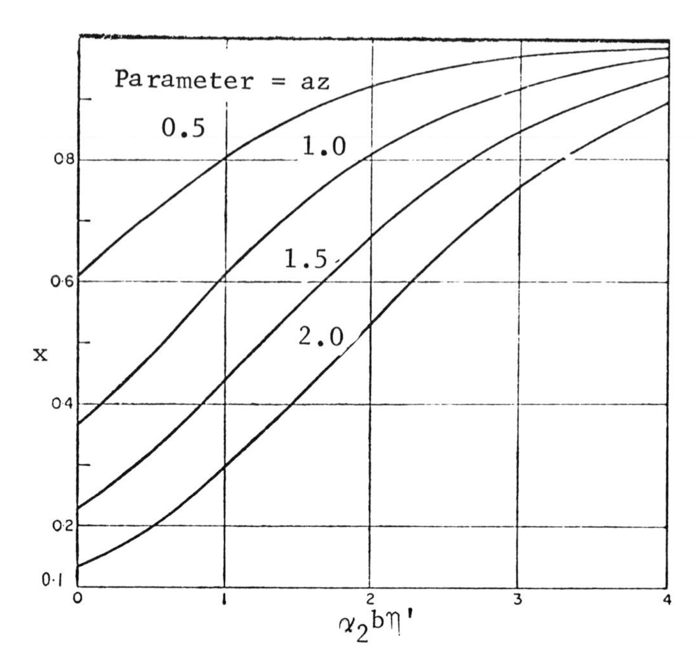

Chemical Engineering Science

Figure 34a. Reactant mole fraction vs. *time group for parallel reaction mechanism with exponential activity function* (71)

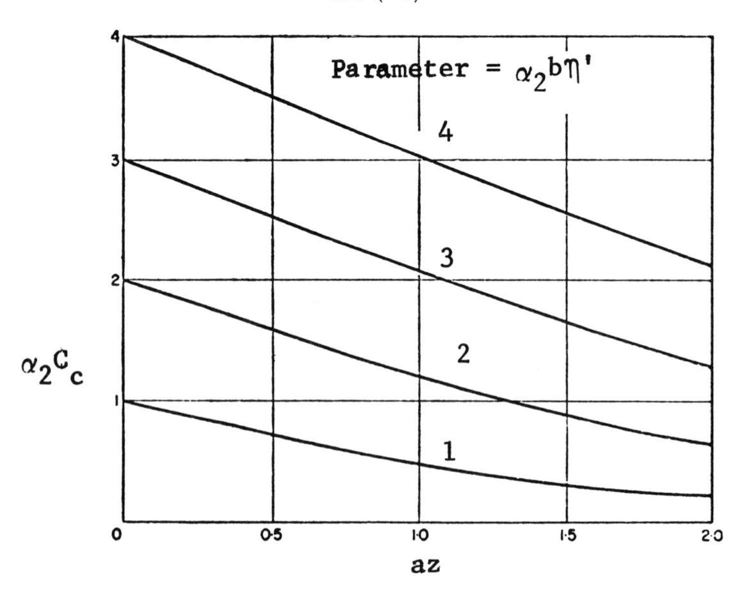

Chemical Engineering Science

Figure 34b. Carbon profiles for parallel reaction mechanism with exponential activity function (71)

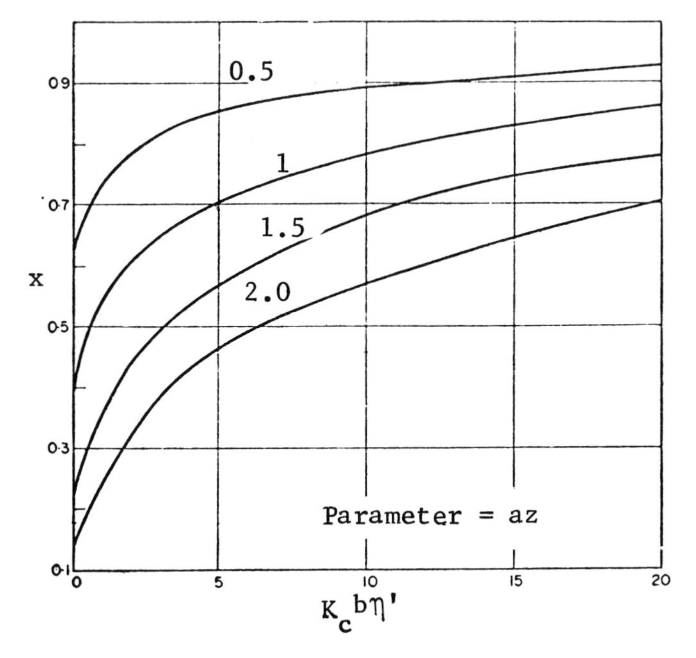

Chemical Engineering Science

Figure 34c. Reactant mole fraction vs. *time group for parallel reaction mechanism with hyperbolic activity function* (71)

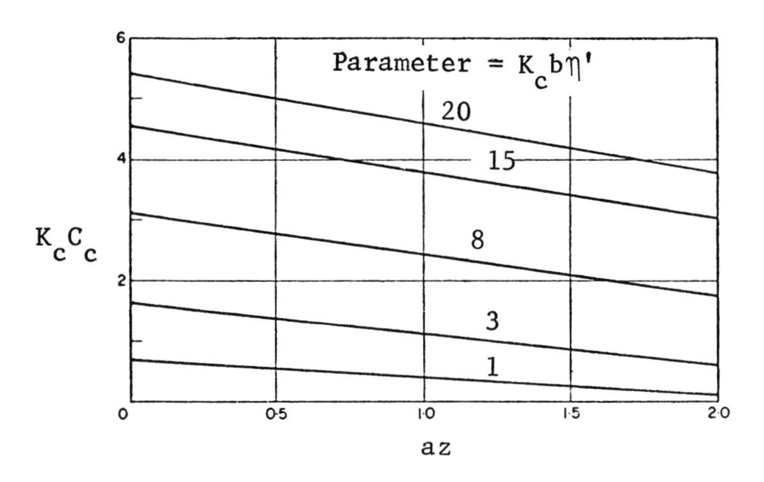

Chemical Engineering Science

Figure 34d. Carbon profiles for parallel reaction mechanism with hyperbolic activity function (71)

nonlinear, and time variant in shape as well as magnitude; simplifications based on assumed profile shapes representing average behavior may be seriously misleading.

For the Langmuir form of activity function the conservation equations are

$$\frac{\partial x}{\partial z} = -\frac{ax}{1 + K_c C_c} \tag{85}$$

$$\frac{\partial C_c}{\partial \eta'} = \frac{bx}{1 + K_c C_c} \tag{86}$$

where a and b are as defined before except k'_{A_o} is written as $(k'_{A_o} + k'_{A,d_o})$ in a. These equations may also be solved for the same boundary conditions (Equation 84) to give:

$$x = \exp[-az + (1 + 2K_c b\eta')^{1/2} - 1 - K_c C_c] \tag{87}$$

$$K_c C_c \exp[K_c C_c] = [(1 + 2K_c b\eta')^{1/2} - 1] \exp[-az + (1 + 2K_c b\eta')^{1/2} - 1] \tag{88}$$

Reactant and activity profiles computed from these expressions for the Langmuir-type activity relation, type II reactant inhibition, are shown in Figures 34c and d.

Similar solutions can be obtained for type II, product inhibition with the various activity relationships. Figures 35a and b show these results, where all parameters are as defined for Equations 80 and 81 and only activity effects on k'_A are considered. The major distinction, which is one of mechanism of deactivation rather than the activity relationship, is that coke deposition increases with reactor length in this case, while decreasing with reactor length in the event of reactant fouling. No results were reported for the Langmuir activity relation.

Bischoff and Froment were also able to cast their computed results into the Voorhies power law form and thus to compare it with macroscopic reactor behavior. Since the over-all activity or conversion, on which the observed correlation is based, is determined by some approximate function of the average bed activity (coke concentration), they obtained a mean value for C_c over the bed as a function of time of operation. Thus, from the results of Figure 34b and Figure 35b, for various values of the parameter $\alpha_2 b\eta'$ one has:

$$\overline{\alpha_2 C_c} = \frac{1}{W}\int_0^W \alpha_2 C_c dw \tag{89}$$

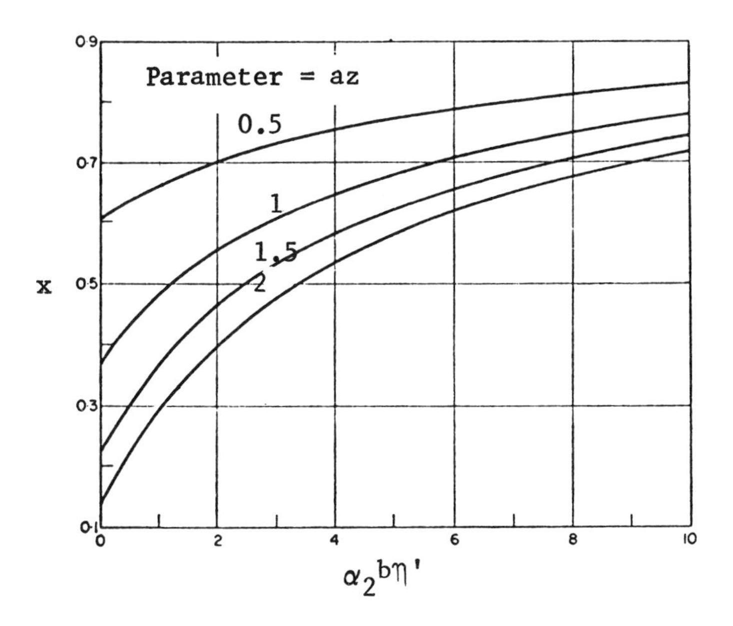

Chemical Engineering Science

Figure 35a. Reactant mole fraction vs. time group for consecutive reaction mechanism with exponential activity (71)

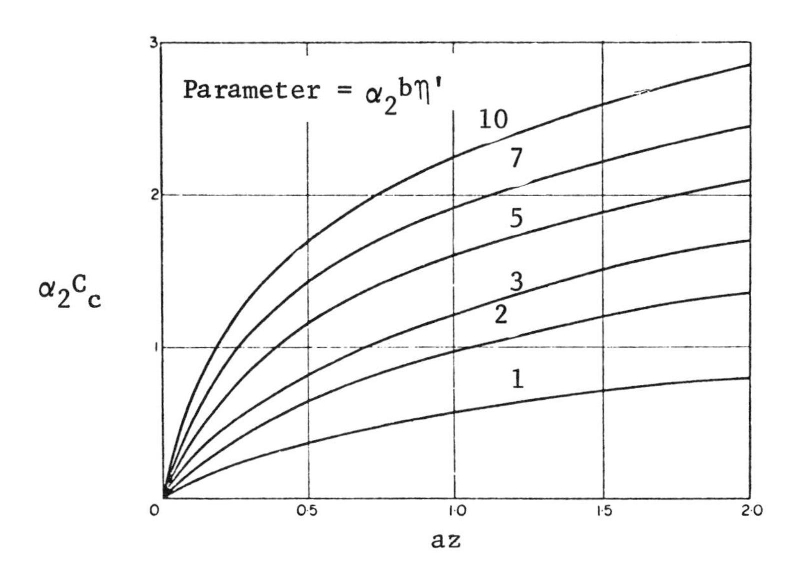

Chemical Engineering Science

Figure 35b. Carbon profiles for consecutive reaction mechanism with exponential activity (71)

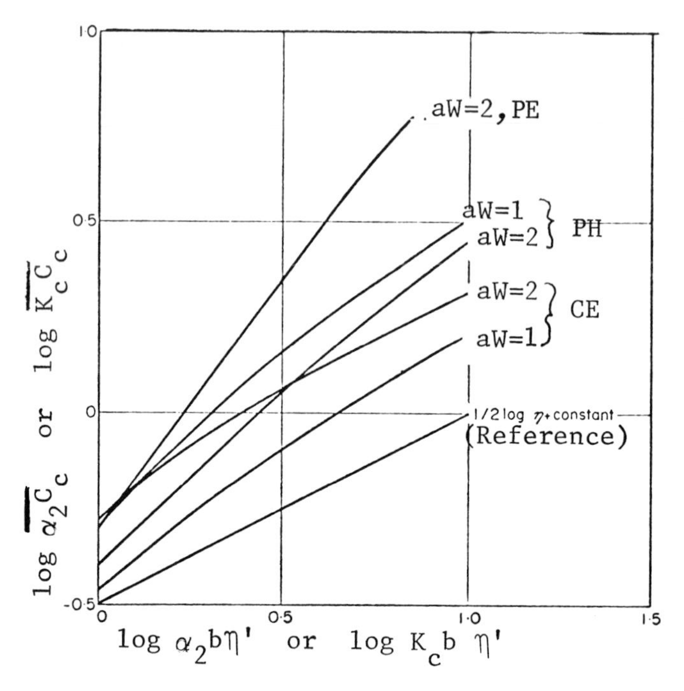

Chemical Engineering Science

Figure 35c. Distance averaged carbon content vs. *time groups. PE—parallel reaction mechanism with exponential activity function. PH—parallel reaction mechanism with hyperbolic activity function. CE—consecutive reaction mechanism with exponential activity function* (71).

These average values, plotted *vs.* time of operation according to the form:

$$\overline{\alpha_2 C_c} = A'\eta'^n \approx A\theta^n \tag{90}$$

($\eta' \sim \tau$ for extended periods of operation) are shown in Figure 35c. The slopes of these curves are of primary interest since their relative positions may be changed by parametric variation. Similar behavior is indicated for type II reactant fouling-Langmuir activity (PH) and type II product fouling-exponential activity (CE), in which cases n decreases from unity to about 0.5 as the time of utilization becomes large. The two curves plotted for these two mechanisms correspond to differing space velocities (given by the aW parameter). One observes an opposition of behavior; both point and average C_c increase with space velocity for the product fouling but decrease with space velocity for the reactant mechanism. It is possible that variations in carbon formation with space velocity such as those measured by Eberly *et al.* (*cf.*, Figure 12a), where both increases and decreases are noted in different ranges, reflect a mixed mechanism of coke formation, involving both reactant and product degradation. The type II reactant fouling with exponential deactivation (PE) seems to correlate linearly over a very wide range of utilization times with unity exponent. Hence, all the variants of mechanism and activity function examined here can be correlated satisfactorally on the Voorhies-type plot, at least over certain ranges of operating conditions or catalyst life, again giving at least partial explanation for the seeming universal applicability of the correlation.

A final, important consequence of this analysis of Froment and Bischoff is shown in Figure 35d. Under conditions of deactivation, nonuniform distributions of catalytic activity will be encountered which arise

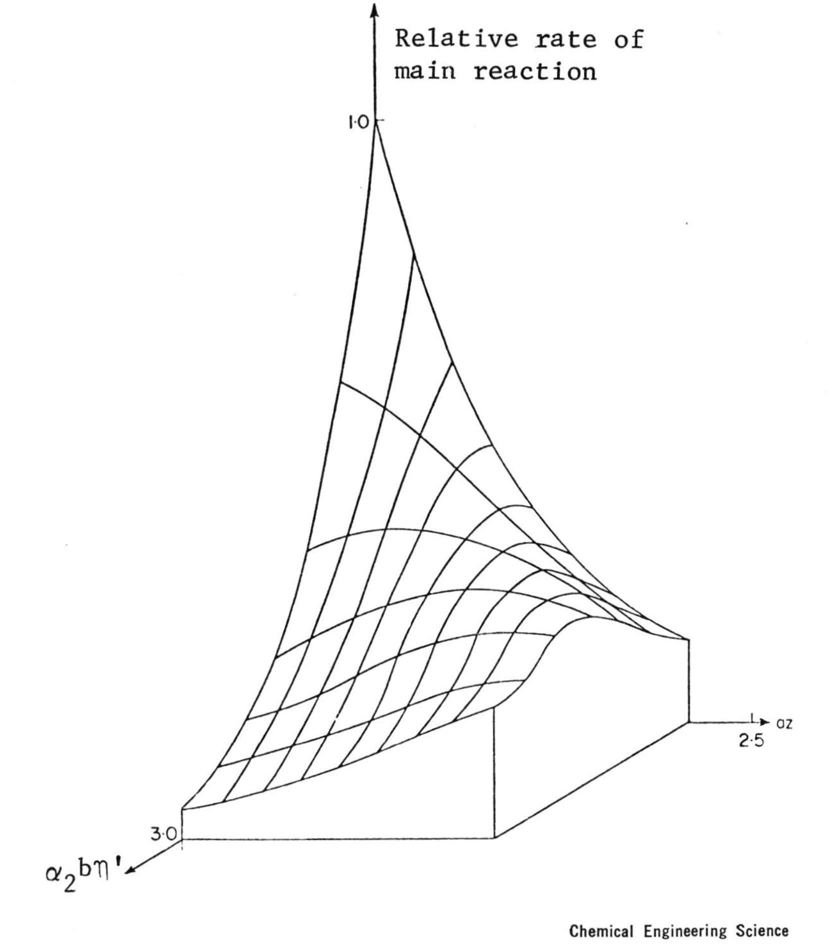

Chemical Engineering Science

Figure 35d. Rate surface for parallel reaction mechanism with exponential activity function (71)

from the nonuniform carbon deposition such as we have seen in Figures 34b and 35b. Particularly for parallel reaction, where the coke concentration decreases with reactor length, this implies that the locus of maximum reactivity may be altered since the reactor inlet—normally the most active portion of the bed because of the high concentration of reactants —has been deactivated to a greater extent than other portions of the reactor. For the type II reactant fouling with exponential activity function, whose coke profiles are given on Figure 34b, we may write for the relative rate of reaction (relative to that with no deactivation):

$$\frac{r_A}{r_A{}^o} = \frac{k'_A P_x}{k_A{}^o P} = x \exp(-\alpha_2 C_c) \tag{91}$$

where the contribution of the deactivation term in Equation 75 has been ignored. Substitution for x and $\exp(-\alpha_2 C_c)$ from Equations 83 and 84 allows direct evaluation of the relative rate in terms of distance and time parameters, and the resulting rate "surface" is shown on Figure 35d. The development of a maximum in the reaction rate is clearly demonstrated here; the maximum passes along the length of the bed and eventually out the end. One thus has an activity wave within the reactor, and if the operation is not isothermal (as has been assumed here) this activity wave will appear as a thermal one in the form of continuously varying temperature profiles within the bed. This point is discussed in full detail in the following section on reactor regeneration, but for a preview of what such temperature waves look like in practice, look at Figure 50b (p. 450).

The activity wave observed by Froment and Bischoff is a result of the opposing trends of reactant concentration profile and carbon concentration profile in the reactor on the rate of reaction. It would be observed, then, only for cases in which the carbon concentration decreases with bed length. This situation is not entirely analogous to reactor regeneration since in the present case we are dealing with profiles which are continuous throughout the bed, whereas in regeneration (at least at higher temperatures) one is concerned with a narrow zone of carbon oxidation which moves through the bed generally at a fixed rate and with fixed dimensions.

In a subsequent study Froment and Bischoff pointed out the dangers of kinetic analysis when catalyst deactivation leads to activity profiles within the reactor. In a normal integral reactor experiment, rate constants are derived from conversion data, as for the first-order example:

$$\frac{W}{F} = -\frac{1}{k'_A P} \int_{x_o}^{x} \frac{dx}{x} \tag{92}$$

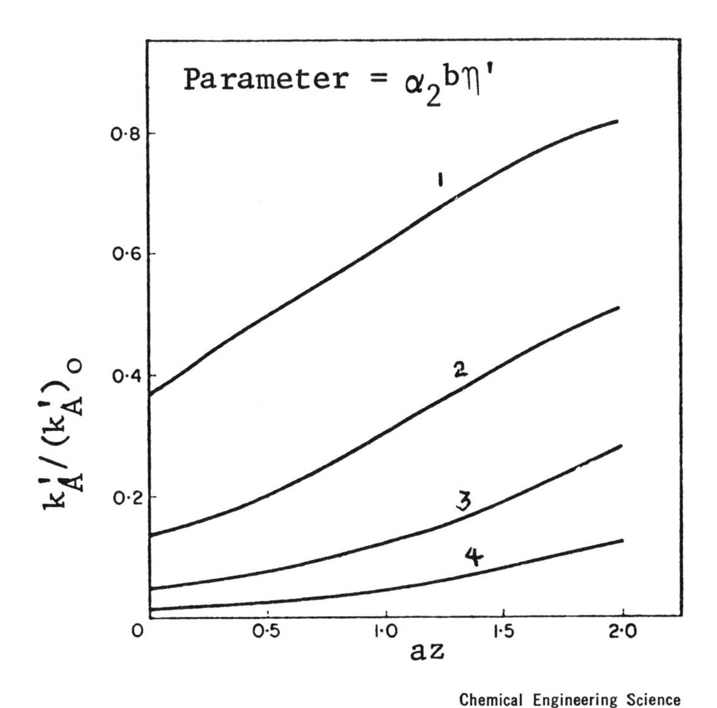

Chemical Engineering Science

Figure 36a. Rate coefficient vs. *dimensionless position in reactor* (72)

If activity changes with time, one either obtains several values of $k'_A = k'_A(\theta)$ and extrapolates back to zero time, or attempts to correct values of rate constants obtained under varying conditions back to a common carbon level. One problem, of course, is that Equation 92 presumes k'_A to be constant over the reactor which, in the face of carbon concentration profiles, it cannot be. So one actually measures an average k'_A:

$$\overline{k'_A} = \frac{1}{W}\int_0^W k'_A(w,\theta)\,dw \tag{93}$$

and relates it in terms of carbon concentration also to an average value

$$\overline{C_c} = \frac{1}{W}\int_0^W C_c(w,\theta)\,dw \tag{94}$$

However, the implied relationship, that

$$\overline{k'_A} = k'_A(\overline{C_c}) \tag{95}$$

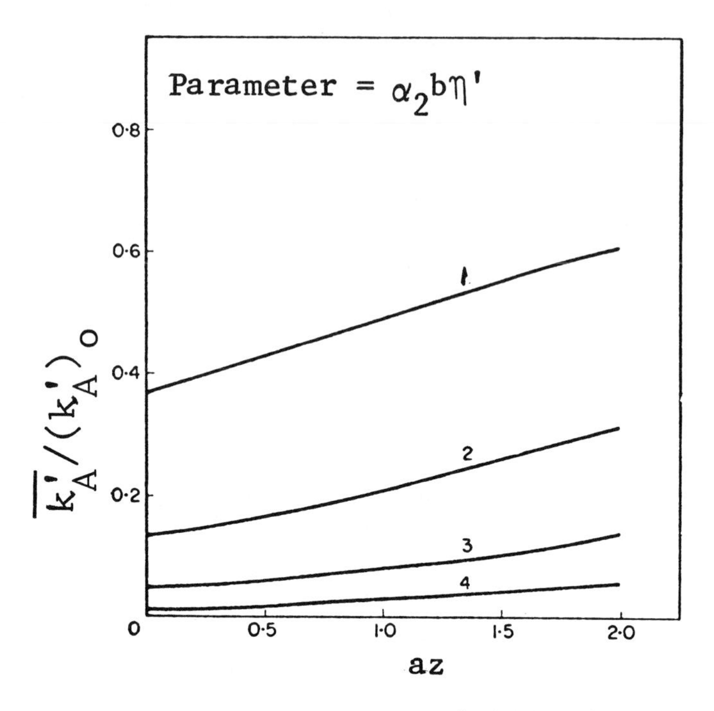

Figure 36b. Average rate coefficient vs. *reactor length group. Parallel reaction mechanism with exponential activity function* (72).

is not true when there are carbon profiles present. Consider again the example of type II reactant fouling with an exponential activity function. Carbon profiles for this case are given in Figure 34b. Now if:

$$k'_A = (k'_A)_o \exp(-\alpha_2 C_c) \tag{96}$$

where $(k'_A)_o$ is the rate constant at the bed entrance, then the relative rate constant, $k'_A/(k'_A)_o$ is given by Equation 84, as shown in Figure 36a. If we average this relative constant, according to Equation 93, the result is:

$$\frac{\overline{k'_A}}{(k'_A)_o} = \frac{1}{aZ}\left[\alpha_2 C_c{}^W - \alpha_2 C_c{}^o + \ln\frac{\exp(\alpha_2 C_c{}^o) - 1}{\exp(\alpha_2 C_c{}^W) - 1}\right] \tag{97}$$

where W refers to bed exit values. This result is shown in Figure 36b. The rate constant at the average carbon concentration $k'_A(\overline{C}_c)$ can be obtained from the average, $\overline{\alpha_2 C_c}$, of Equation 89 and the definition of Equation 96. The ratio so calculated:

$$\frac{\overline{k'_A}}{k'_A(\overline{C}_c)} = \frac{\exp\overline{(\alpha_2 C_c)}}{aZ}\left[\alpha_2 C_c{}^W - \alpha_2 C_c{}^o + \ln\frac{\exp(\alpha_2 C_c{}^o) - 1}{\exp(\alpha_2 C_c{}^W) - 1}\right] \quad (98)$$

is shown in Figure 36c. There are large deviations from unity, thus directly illustrating the falseness of Equation 95. To obtain unequivocal kinetic results from such experiments, it is necessary to know the carbon profile or to be able to extrapolate reliably to zero time. A Voorhies correlation may be useful for the latter although the dangers concomitant with the utilization of extrapolated kinetics hardly need to be cataloged here.

Some quite nice experimental data are available concerning reactor behavior and the formation of coke profiles corresponding generally to the treatment of Froment and Bischoff. These experiments by van Zoonen (73), dealt with the hydroisomerization of olefins (conversion to a high iso to normal paraffinic product) on a silica–alumina nickel sulfide catalyst in the temperature range 300°–400°C and hydrogen pressures *ca.* 40 atm. The reaction proceeds by a complex mechanism on the silica–alumina function to yield isoparaffins and diolefins; the latter are hydrogenated by the NiS function back to monoolefins. Coke is formed from the olefins *via* the diolefin intermediate; hence the coking rate is proportional to the olefin partial pressure. Both the main reaction and the coke-forming

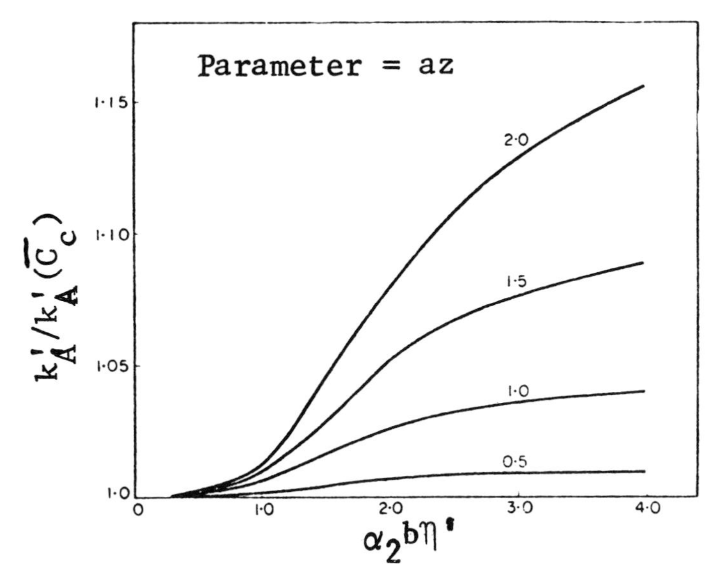

Chemical Engineering Science

Figure 36c. Comparison of average rate coefficient with rate coefficient evaluated at average carbon content. Parallel reaction mechanism with exponential activity function (72).

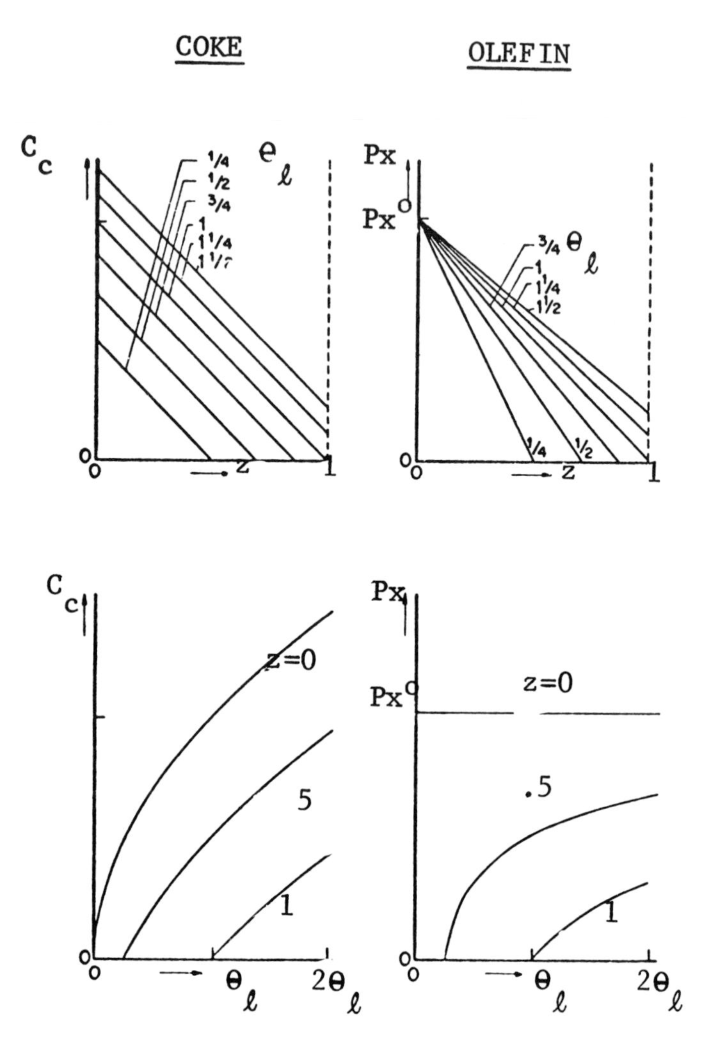

International Congress on Catalysis

Figure 37a. Fixed bed hydroisomerization. Theoretical relations between coke content of catalyst, C_c *or olefin partial pressure,* Px, *and location* z *or run time* θ_1 (73).

reaction were taken to be first order in olefin and inversely proportional to coke concentration. The result of these assumptions, together with the normal ones concerning isothermality and plug flow, is a set of equations quite similar to those of Froment and Bischoff, except for slightly different kinetics. Thus:

$$r_A = -\frac{C_c}{k'_A Px}$$

and (99)

$$r_c = \frac{k'_{A,d} Px}{C_c}$$

for van Zoonen, where Px represents the olefin partial pressure. In spite of the different kinetics, solutions for the coke content of the catalyst as a function of reactor length are similar to those of Froment and Bischoff, decreasing almost linearly with length as characteristic of the type II reactant fouling postulated. It is an interesting consequence of the kinetics that r_A, the rate of reaction of olefin, is infinite where $C_c = 0$. That is, one would not detect olefin in the gas phase where there is fresh catalyst.

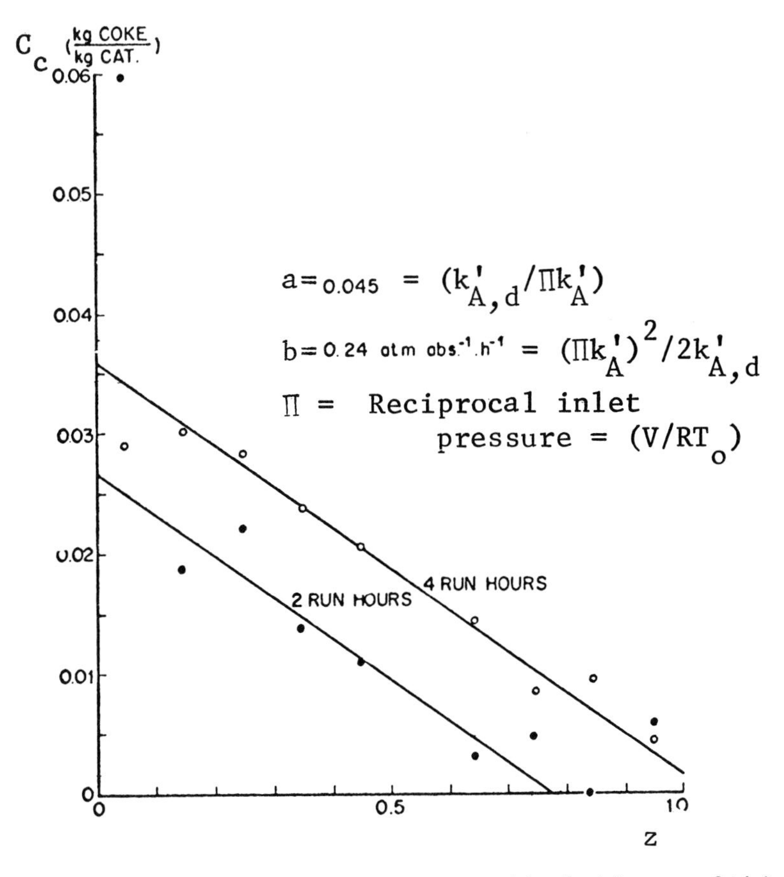

International Congress on Catalysis

Figure 37b. Relation between coke content of catalyst C_c *and location z in catalyst bed. 1-Hexene, atm pressure,* $S_v = 0.11$ *kg · kg*$^{-1}$ *· h*$^{-1}$*,* H_2*/olefin mole ratio = 4.7* (73).

While such may not be precisely true in the limit, this general behavior leads to the reactor system exhibiting "breakthrough" of the reactant (olefin) when the coke front reaches the exit of the bed. These aspects of the qualitative nature of reactor behavior are shown in Figure 37a; the coexistent coke and olefin profiles are shown at the top for various times of operation, and at the bottom is shown the olefin breakthrough at the reactor exit when the coke concentration at that point becomes finite. Experimental confirmation of some of these results is shown in Figure 37b for the hydroisomerization of 1-hexene. The lines on these plots are computed results reflecting the best fit using the rate constants of Equation 99 as adjustable parameters as shown, and reflect good agreement with theory as far as the expected coke profiles are concerned. Other aspects of the theory are correlated with experimental results on Figures 38a and b. The first of these presents data on the catalyst coke content as a function of position in the bed, corresponding to the lower left panel of Figure 37a. This plot is essentially a cross plot of Figure 34b as presented by Froment and Bischoff. Again the trend is correct for the parallel

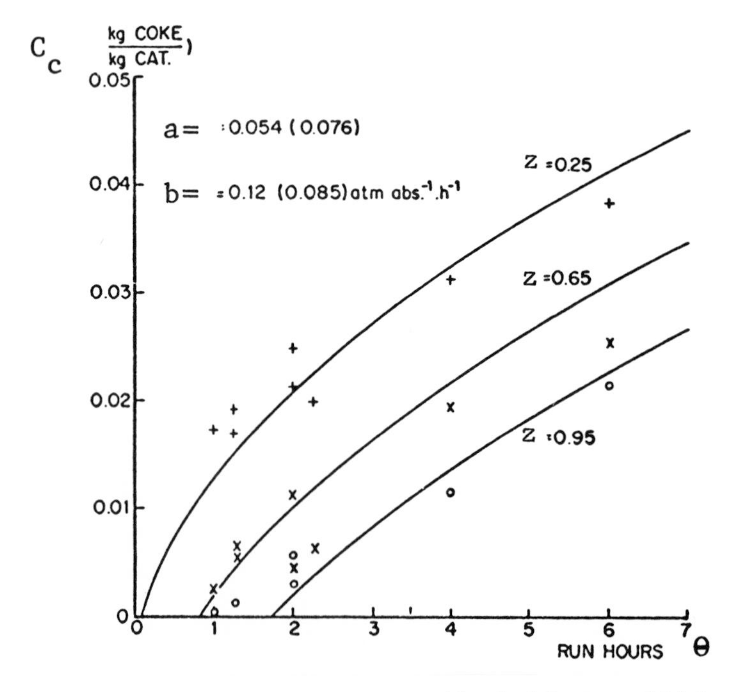

International Congress on Catalysis

Figure 38a. Relation between coke content of catalyst (at three locations) and duration of run. 1-Pentene, atm pressure, $S_v = 0.14$ *kg · kg*$^{-1}$ *·* h^{-1}*,* H_2*/olefin mole ratio = 2.4. Catalyst* 55359 (73).

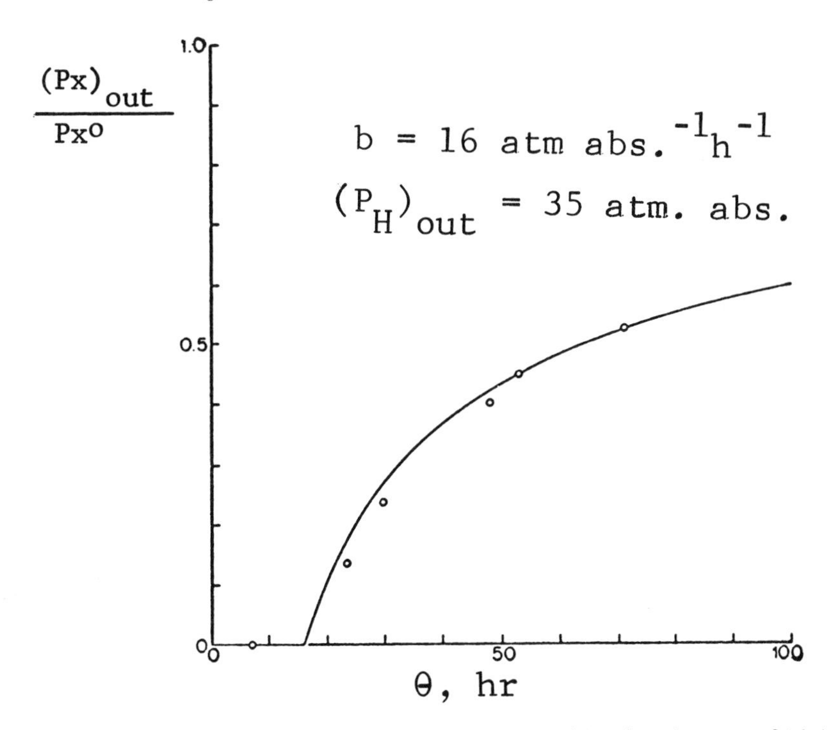

International Congress on Catalysis

Figure 38b. Partial pressure of unconverted normal hexenes at reactor outlet as a function of run duration. Pressure = 46 atm abs., H_2/olefin mole ratio = 3, S_v = 3.1 kg · kg^{-1} · h^{-1}. Catalyst 54652 (73).

reaction mechanism assumed and, except for some scatter at low processing times, the agreement with calculation using the two rate constants as adjustable parameters is satisfactory. The olefin breakthrough indicated by the postulated kinetics is shown on Figure 38b. Clearly there is a period during which exit concentration of olefin is zero, and if one wishes not to lose reactant olefin to the product stream, the point of breakthrough defines the useful catalyst life rather than some absolute level of activity reduction or some temperature limitation on reactor operation.

In all, the simple reactor model and quasi-steady state assumptions regarding the relative rates of main and deactivation reactions first detailed by Froment and Bischoff and tested to some length experimentally by van Zoonen provide a reasonable representation of reactor performance under conditions of deactivation, at least for coking mechanisms. The successful confrontation of theory with experiment here provides some *a postiori* justification for the model and assumptions and, as we have hinted at the start of this section, subsequent workers concerned with either reactor/regenerator modeling or optimization have not ventured far from this comfortable base. A somewhat similar analysis has been reported also by Weekman (*27*), but since his primary concern was comparison of the efficiencies of various reactor types we defer discussion of that work to the section on optimization.

Olson (*74*) has studied the deactivating reactor problem for the general case in which a shell progressive mechanism for the deactivation is applicable. The reactor model used is similar to those of previous studies, but both intra- and interphase mass transport resistances are included in the poisoning kinetics, and three parameters are important

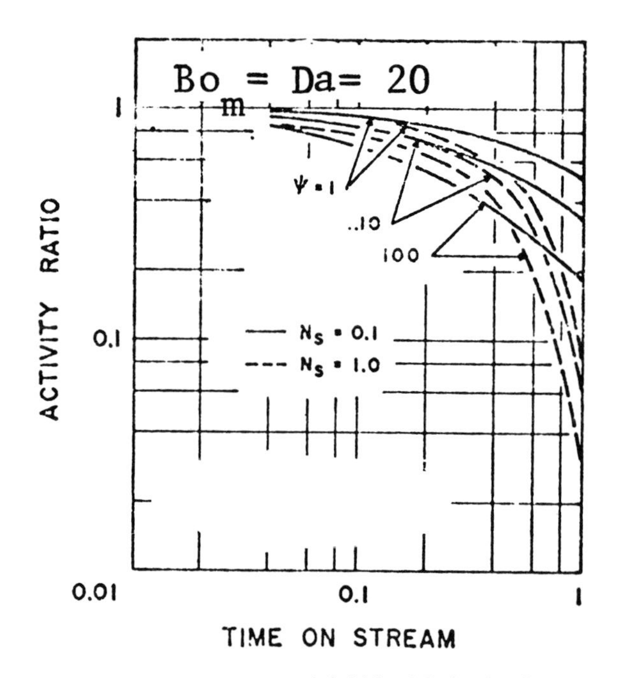

Industrial and Engineering Chemistry

Figure 39a. Activity of bed. The fraction of the initial activity of the bed is plotted as a function of the time on stream. At time equal to unity, the reactor has been supplied with enough poison to sature the bed completely. The parameter N *describes the transport rate of poison into the pellet while the Thiele parameter refers to the major chemical reaction* (74).

For Figures 39a,b,c:

$$T = \left(\frac{v'\theta - w}{W}\right)\{\epsilon a C_A{}^0/[(1 - \epsilon)q^\infty]\}$$

$C_A{}^0$ = *Inlet concentration*
a = *Stoichiometric coefficient, moles(solid)/moles(fluid)*
q^∞ = *Equilibrium concentration on solid, moles/cm³*

in determining the over-all reactor behavior. These are mass Biot number, $Bo_m = (k_g R/D_K)$, the Damköhler number for the poisoning reaction, $Da = (k'_{A,d} R/D_K)$ and a solid diffusion transfer unit parameter, $N_s = (3D_K W/R^2 v')$ where k_g is a fluid phase mass transfer coefficient, R the catalyst pellet (spherical) radius, D_K the Knudsen diffusivity of poisoning reactant within the catalyst, W the reactor length and v' the interstitial linear velocity in the bed. Olson detailed some typical estimates for these parameters in his study. For a first-order, irreversible reaction in the catalyst, the effectiveness factor corresponding to a particle in which the unpoisoned core radius is R_p is:

$$\eta = 3\psi^{-2}\left[Bo'_m + \frac{1 - R_p}{R_p} + \frac{1}{R_p(R_p\psi \coth(R_p\psi) - 1)} \right]^{-1} \tag{100}$$

where the prime on Bo'_m refers to the Biot number for the catalytic system as distinguished from the poisoning reaction, and ψ is a Thiele modulus, $(k'_A R^2/D_K)^{1/2}$ for the main reaction. The effectiveness of Equation 100 can be averaged over the bed length and expressed in terms of an activity ratio:

$$A(T) = \frac{\bar{\eta}}{\eta_o} \tag{101}$$

where η_o is the effectiveness factor of the unpoisoned bed. Some computed results are shown in Figure 39a as a function of the fractional time required to saturate the bed completely. The larger values of N_s, reflecting a larger D_K and thus more efficient penetration of poison into the catalyst, produce rapid bed deactivation at fractional times approaching unity. There is a crossover for the curves, at high ψ, between low and high N_s. Low N_s, meaning slow adsorption of poison, generally improves reactor performance, but if the Thiele modulus is large, the near uniformly poisoned reactor is less active than one in which a sharp poison front is present (owing to the nonlinear nature of the relationship between ψ and effectiveness). Thus an *a priori* estimate of relative activity for different cases can be risky if the main reaction is severely diffusion limited. Figure 39b shows the poison profiles in the reactor for various values of N_s and times of operation. The point illustrated is, again, the extreme sensitivity of the results to N_s, this time in the sense of the sharpness of the poisoning front ($R_p = 0$ means a completely deactivated catalyst); also the constant bandwidth or profile pattern assumption which is popular in problems of this sort (*see*, for detail, the following discussion of fixed bed regeneration) may or may not be applicable depending on values of these parameters. In Figure 39b, the assumption is not applicable within the time period illustrated for the two lower values of N_s, as

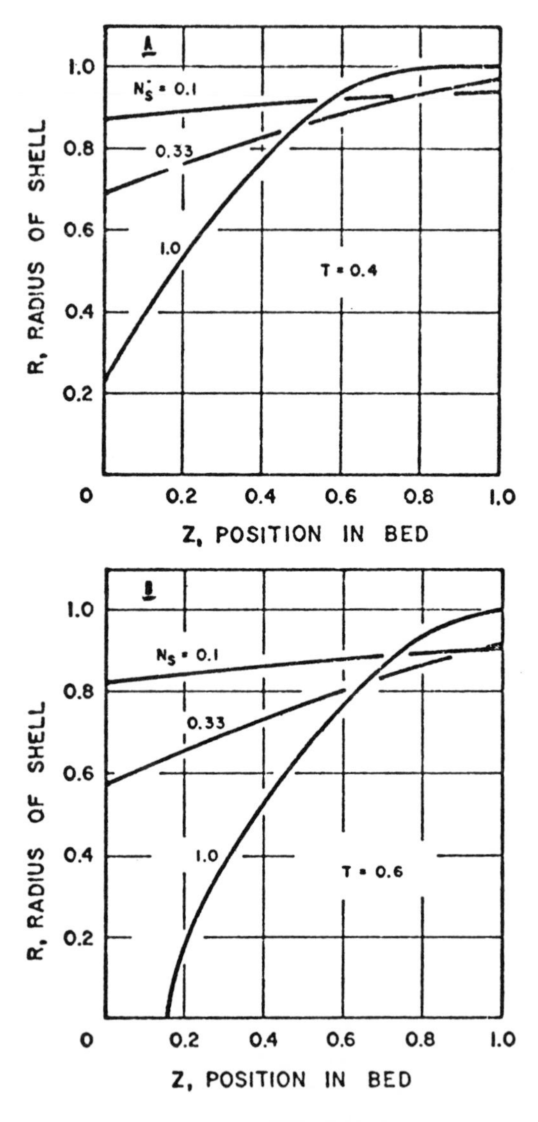

Industrial and Engineering Chemistry

Figure 39b. Interfacial position of poisoned shell. A. The radial position of the poisoned shell is shown as a function of axial position at dimensionless time, T = 0.4. The catalyst at the entrance to the bed is not completely poisoned for any value of the parameter N_s. Therefore, the "constant band width" assumption is not valid. B. Results displayed are similar to A except that the time has advanced to T = 0.6. The constant band width model is valid only for the system in which N_s = 1.0 (74).

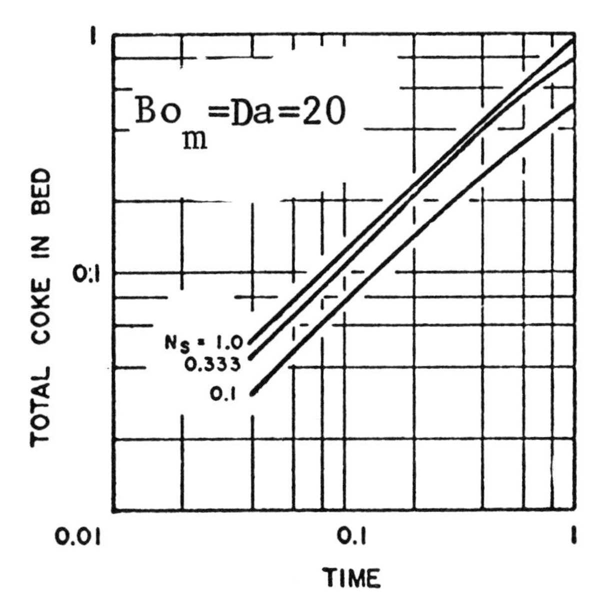

Industrial and Engineering Chemistry

Figure 39c. Coke buildup. The total amount of coke adsorbed in the bed is very nearly a linear function of time. As the parameter N_s *increases, the bed tends toward piston-flow adsorption* (74).

we see essentially a continuous activity profile through the bed. A final, interesting result of Olson is the calculation of reactor behavior for the deposition of coke by a shell progressive mechanism. The results shown in Figure 39c show the average coke concentrations, obtained by an averaging as per Equation 94 as a function of time of operation. The power law correlation is obeyed to a good approximation, the slope (exponent in correlation) in the illustration being approximately 0.85. In all these results, the sensitivity of reactor performance to N_s is demonstrated; it is thus quite obvious (and seemingly almost too simple) that by decreasing the value of D_K and hence N_s, one can effect considerable improvement in reactor performance. Stated conversely, the catalyst pore structure and mass transport characteristics are predominant in determining the reactor operation, for set conditions (which determine Bo_m and Da). Olson cites this example: changing the diffusivity from 0.333 to 0.1 increased the on-stream time by 50% for any total coke level below 30%.

Numerical methods obviously are required to solve problems with complex kinetics or mechanisms of deactivation, such as that just discussed; however, some relatively general methods have been developed for analytical solution of the equations representing fixed bed deactivation by Bischoff (*75*) and Ozawa (*76*). The conservation equations we have been involved with are generally of the form:

$$\frac{\partial u}{\partial z} = -g(v)u \tag{101}$$

$$\frac{\partial v}{\partial \tau} = g(v)u \tag{102}$$

where u and v are dimensionless reactant and poison concentration variables, z a dimensionless position variable and τ a dimensionless process time variable. The function $g(v)$ describes the effect of poison on the reaction rate constants (taken to be the same for both), and the normal boundary conditions are:

$$\begin{aligned} u(0,\tau) &= u_o(\tau) \\ v(z,0) &= v_o(z) \end{aligned} \tag{103}$$

we assume a solution of the form:

$$u(z,\tau) = \frac{v(z,\tau) - v_o(z)}{v(0,\tau) - v_o(z)} \cdot u_o(\tau) \tag{104}$$

Substituting Equation 102 in 104:

$$\frac{\partial v(z,\tau)}{\partial \tau} = g(v)\, \frac{v(z,\tau) - v_o(z)}{v(0,\tau) - v_o(z)}\, u_o(\tau)$$

For $z = 0$ we have from this:

$$\frac{\partial v(0,\tau)}{\partial \tau} = g[v(0,\tau)]u_o(\tau) \tag{105}$$

Equation 105 can be integrated with respect to τ:

$$\int_0^{\tau} u_o(\tau')d\tau' = \int_{v(0,0)}^{v(0,\tau)} \frac{dv'}{g(v')} = \int_{v_o}^{v(0,\tau)} \frac{dv'}{g(v')} \tag{106}$$

Equation 106 then gives a relationship between $v(0,\tau)$ and τ, appearing in the limits of both sides of the equation. Repeating the procedure with Equations 101 and 104, we have:

$$\frac{\partial u(z,\tau)}{\partial z} = -g[v(z,\tau)]\, \frac{v(z,\tau) - v_o(z)}{v(0,\tau) - v_o(z)}\, u_o(y) \tag{107}$$

which can be integrated, for $v_o(z) = v_o$, to:

$$-z = \int_{v(0,\tau)}^{v(z,\tau)} \frac{dv'}{(v' - v_o)g(v')} \tag{108}$$

Equation 108 defines the relationship between $v(z,\tau)$, z, and $v(0,y)$. The over-all solution to the problem is then given by Equations 104, 106, and 108. Bischoff (*75*) describes the technique in detail and gives example solutions for the studies of both Froment and Bischoff (*71*) and Olson (*74*).

The method proposed by Ozawa (*76*) uses a Legendre transformation to convert the set of partial differential equations (Equations 101–103) into a corresponding set of ordinary differential equations. We write the RHS of these two equations in general form as $= f(u,v)$. In such a case, we can define a potential function ϕ, such that:

$$u = \frac{\partial \phi}{\partial \tau} = \phi_\tau \tag{109}$$

$$v = \frac{\partial \phi}{\partial z} = \phi_z \tag{110}$$

so the original equations become:

$$\phi_{z\tau} = f(\phi_z, \phi_\tau) \tag{111}$$

Let us assume that the kinetics involved in $f(u,v)$ are separable and at least one portion of the kinetic expression is linear in the dependent variable—*i.e.*, $f(u,v) = g(u) \cdot h(v)$ where, say, $g(u)$ is a linear function of u. Introducing the transformation:

$$\Phi = u\tau + vz - \phi \tag{112}$$

so:

$$\Phi_v = z, \; \Phi_u = \tau$$

one can eventually reduce the original set of equations to the following:

$$\frac{du}{u} = \frac{dv}{(v - v_o) + [(\partial v_o/\partial z)/h(v_o)]} \tag{113}$$

$$dz = \frac{dv}{h(v)\,[(v - v_o) + (\partial v_o/\partial z)/h(v_o)]} \tag{114}$$

and

$$\tau = \Phi_u = \frac{\partial}{\partial u}\left[\int z \cdot dv + \text{const}\right] \tag{115}$$

Equations 113–115 with the boundary conditions given in Equation 103 provide an implicit solution to the problem. Ozawa also has presented detail on solutions when the linearity restriction on $f(u,v)$ is not met; the method has been used to solve a problem on fixed bed regeneration, as discussed next.

The final type of deactivating reactor analysis problem which has been considered in any detail is concerned with behavior when the conversion is maintained constant (*i.e.*, time invariant). Such operation is often dictated when large units are subject to long term deactivation effects and constant conversion is required not to upset subsequent processing units. In this event, the temperature level of the reactor is used to compensate for the catalyst deactivation, and the thermal parameters of the main reaction and the deactivation, particularly activation energy, have a great influence on the operation. The first study of this problem (*77*) dealt with nonselective poisoning according to type I and II mechanisms (Figure 17), and the analysis has subsequently been extended (*78, 79*) to problems involving polyfunctional catalysts with parallel modes of deactivation as well as reaction occurring. Since the reactor temperature is a variable in this type problem, all the equations become nonlinear *via* the Arrhenius temperature dependence of rate constants. The solution to these problems has been carried out using the mixing cells in series approximation (*81*) of the plug flow, one dimensional model of the fixed bed reactor. For the example of type II reactant fouling which we have discussed in previous examples:

$$A + S \rightarrow B + S; \qquad k_A \text{ (main reaction)}$$

$$A + S \rightarrow A \quad S; \qquad k_{A,d} \text{ (poisoning reaction)}$$

The concentration profiles through the bed are given at any time by:

$$[a]_n = \prod^{n} [1 + (k^o{}_A\Phi_{A,n} + k^o{}_{A,d}\Phi_{d,n})S_o\theta(s)_n]^{-1} \tag{116}$$

where $\Phi_{i,n} = \exp[\gamma_i(1 + 1/\phi_n)]$, n is the cell number, θ holding time per cell, the parameter γ_i is the temperature sensitivity of reaction i, defined as E_i/RT_o, and $k^\circ{}_A S_o$, and $k^\circ{}_{A,d} S_o$ refer to the specific rate constants for unpoisoned catalyst. T_o is a reference temperature which can be taken as the initial operating temperature. In each cell deactivation occurs at a rate which is small compared with the residence time, so:

$$r = \frac{d[s]_n}{d\beta} = -\tau \left(\frac{A_o}{S_o}\right) k^o{}_{A,d} S_o \Phi_{d,n} [a]_n [s]_n \qquad (117)$$

where τ is the total residence time in the reactor and β the total time of operation in terms of number of residence times. The kinetics of deactivation expressed in Equation 117 are first order, and the term "nonselective" has been used to describe this linear behavior in contrast to nonlinear kinetics such as those describing the sintering result of Maat and Moscou. This is not the same sense in which the term nonselective, meaning linear activity–poison concentration relationship, has been used in some other sources. If the temperature is constant through the reactor or if there is a constant temperature profile, then:

$$\frac{dT_{n-1}}{dt} = \frac{dT_n}{dt} = \frac{dT_{n+1}}{dt} \cdots \qquad (118)$$

Equations 116–118 can be solved by a numerical procedure described by Butt and Rohan, subject to the constant conversion constraint, $(d[a]_N/d\beta) = 0$. Some characteristics of reactor behavior are shown in Figure 40a, where reactor temperature *vs.* time of operation is illustrated for operation with a constant conversion of 60% for the type II reactant fouling case. The parameter X shown on the figure is the residence-time-concentration grouping appearing in the deactivation rate equation $X \equiv \tau \left(\frac{A_o}{S_o}\right) k^o{}_{A,d} S_o$. The exponential nature of the increase is evident, representing the "activated" reaction process, though one obtains a near-linear correlation over a certain range of process time, depending on the parameters of the reaction system—principal of which are the intrinsic selectivity, ψ, between main and deactivation reactions, given by $(k^o{}_A/k^o{}_{A,d})$, and the relative temperature sensitivity of the two reactions, given by $(\gamma_A/\gamma_{A,d})$. The initial temperature of operation indicated in Figure 40a is determined wholly by the intrinsic activity of the catalyst for the main reaction; deactivation has no influence on initial operation. In these examples the reduced temperature ϕ is defined with respect to the reference temperature T_o involved in the definition of γ; $\phi = T/T_o$. Operation at $\phi = 1$ then implies catalyst activity at a level just sufficient to give the specified conversion at an actual value of temperature, $T = T_o$.

Constant conversion operation also leads to a particular form of linear activity correlation which is roughly comparable with the Voorhies form. This is shown in Figure 40b, which is essentially a cross plot of the results on Figure 40a. The shapes of the temperature–time curves turn out to be independent of X, depending only on the value of $(\gamma_A/\gamma_{A,d})$; hence a characteristic activity plot, log β *vs.* log X at parametric values of the reactor temperature level, is linear, leading to at least a qualitatively valuable means for estimating catalyst life (time of operation to a specified maximum temperature level) as a function of the intrinsic activity of the deactivation reaction.

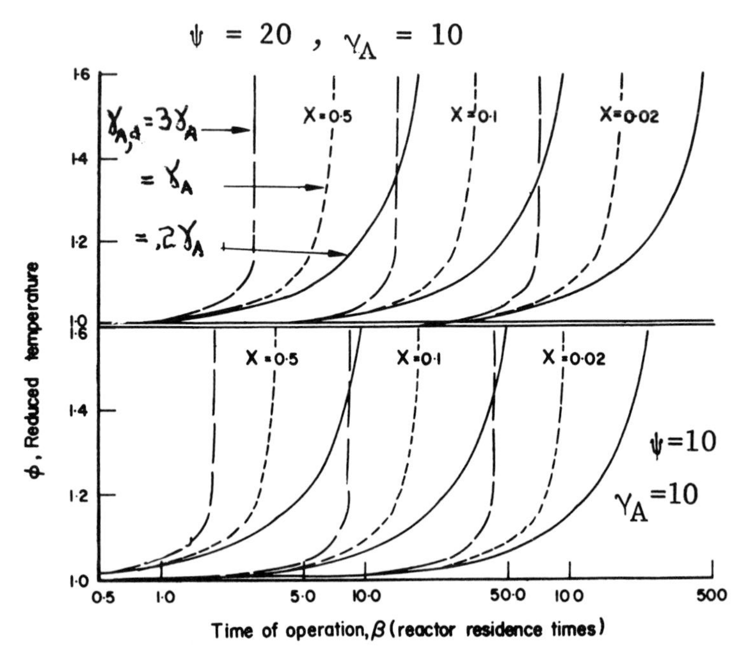

Chemical Engineering Science

Figure 40a. Effect of catalyst deactivation on reactor operation: temperature increase at constant conversion for type II poisoning (77)

A major qualitative result to remember from the examples of Figure 40 is the very strong dependence of catalyst life here on the activation energy (*i.e.*, temperature sensitivity) of the deactivation reaction. It is not surprising, at least in hindsight, to see that when $(\gamma_A/\gamma_{A,d})$ becomes on the order of or smaller than unity the deactivation rates become very large; in the limit this would really represent an unstable process since the control variable (temperature) which is being used to contain the effects of deactivation actually preferentially promotes that process, leading to a kind of kinetic positive feedback which can be modified only by allowing a certain drift to lower conversion levels with time or by restricting cycle times. The deactivation characteristics illustrated on Figure 40 for the type II reactant deactivation are also qualitatively the same for other mechanisms of deactivation including type I (feed impurity) and selective sintering. From the point of observed behavior

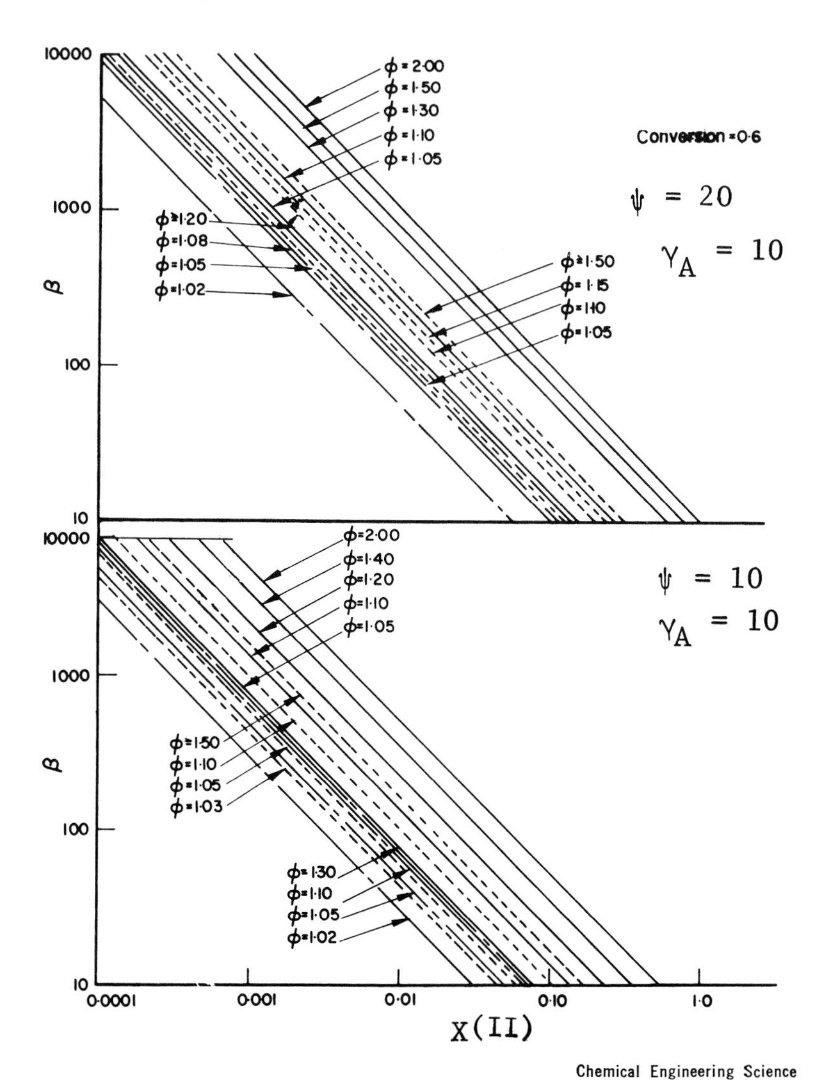

Figure 40b. Catalyst activity characteristics for type II poisoning over a range of conditions (77)

then the constant conversion operation is similar to other modes of operation; the reactor acts as an efficient information filter, and no real detail concerning mechanisms or kinetic schemes of deactivation is available from such observation.

The use of temperature as a control variable, so far illustrated by the temporal variation of conditions in an isothermal reactor, suggests that various nonisothermal modes of operation might possibly be used to reduce deactivation rates in certain situations. Again, this problem borders on that of optimal operation or control, which will be treated in more detail subsequently; however, the question was examined briefly (and somewhat arbitrarily) by Butt and Rohan, and it is convenient to summarize their main results at this point. Two alternative modes of operation were considered—one which required the rate of reaction to be maintained at its initial value (rates under initial conditions) at each point in the bed and a second which called for a uniform deactivation of

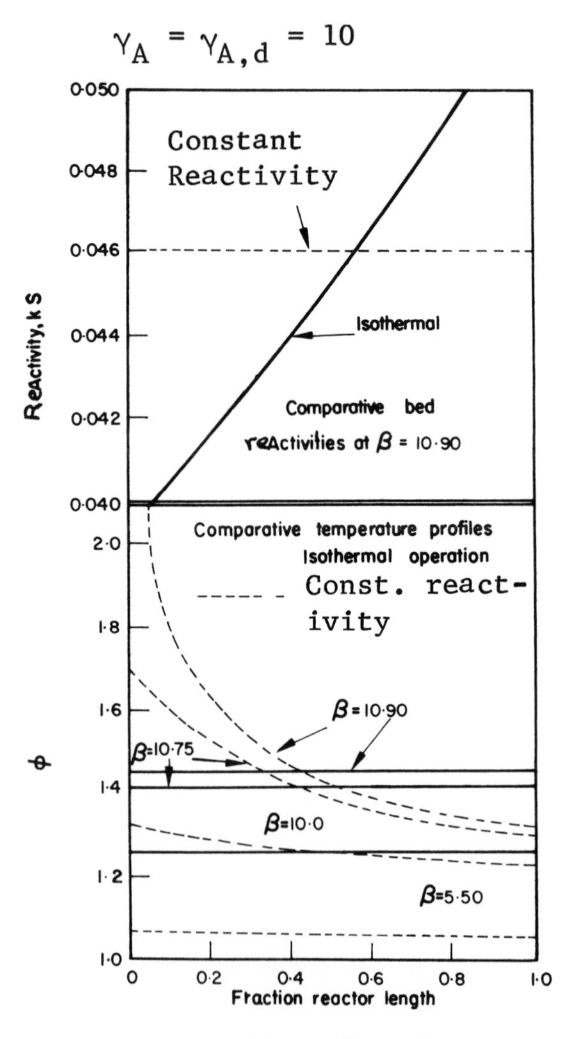

Chemical Engineering Science

Figure 41a. Example comparison of constant reactivity policy with isothermal operation for X = *0.1,* ψ = *10, conversion* = *0.6* (77)

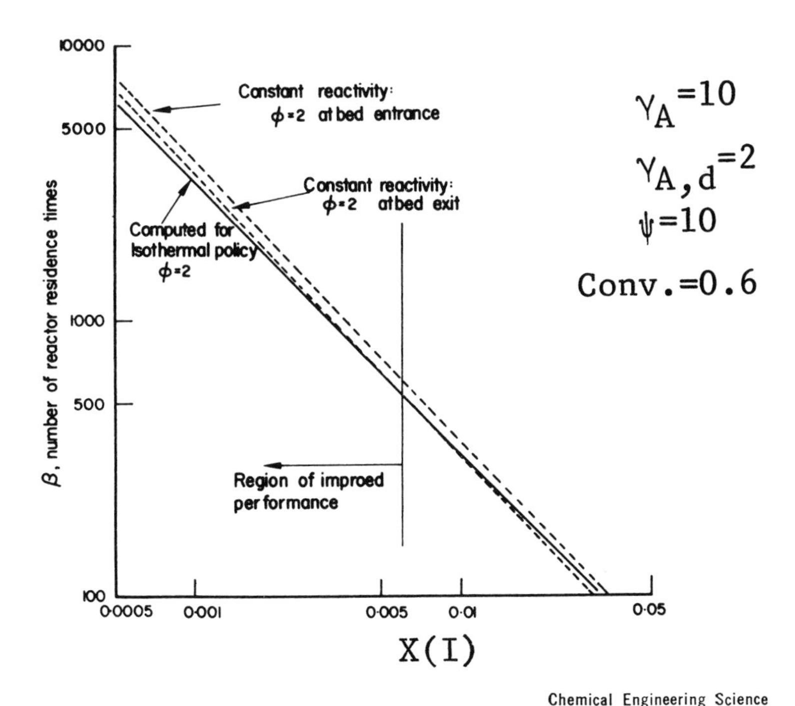

Chemical Engineering Science

Figure 41b. Example of catalyst life increase resulting from utilization of constant reactivity policy (77)

the bed such that, at any time, the catalyst activity did not vary from point to point. It is convenient to refer to these as constant reactivity and constant activity policies, respectively. (Unfortunately, this terminology was reversed in the original paper.) The former mode might be desirable in the sense that a well defined operation is maintained throughout the cycle time. Heat generation rates and conversions would be invariant with both time and position here. The latter mode would lead to a uniform utilization of the catalyst, possibly desirable in cases where the economics of processing or regeneration were such that it would be desirable to eliminate bed activity profiles. In each case one can use the specification of the mode of operation to compute the required temperature profile through the reactor as a function of the time of operation. For constant rate of reaction:

$$\frac{d}{d\beta}\{k^{o}{}_{A}S_{o}\tau\Phi_{A,n}[s]_{n}\} = 0 \tag{119}$$

where $[s]$ is obtained by simultaneous solution of Equation 117. For the uniform deactivation of the bed:

$$[s]_{n-1} = [s]_n = [s]_{n+1}$$

and

$$\frac{d[s]_{n-1}}{d\beta} = \frac{d[s]_n}{d\beta} = \frac{d[s]_{n+1}}{d\beta} \tag{120}$$

The required temperature profiles as a function of time of operation can be calculated for the two modes of operation from Equations 119 and 120. The results of these example calculations are given in Figure 41. The temperature profiles in both cases are time dependent since either activity or point rates vary with time. In general, the constant reactivity policy yields catalyst life results which differ little from isothermal operation. An example of one situation, however, in which this policy leads to substantial reduction in catalyst life is shown in Figure 41a. The parametric values here are quite normal, except that $\gamma_A \approx \gamma_{A,d}$. Application of a constant reactivity policy in the face of temperature-sensitive deactivation leads to the development of pronounced temperature profiles, *viz.*, $\beta = 10.75$ and $\beta = 10.90$, which lead to violation of temperature constraints near the entrance of the reactor after relatively short periods of operation. On the other hand, some indication was obtained that the constant activity policy could indeed lead to increased life relative to isothermal operation. Some of these results are shown in Figures 41b and c. The major improvement seems to occur in systems where the de-

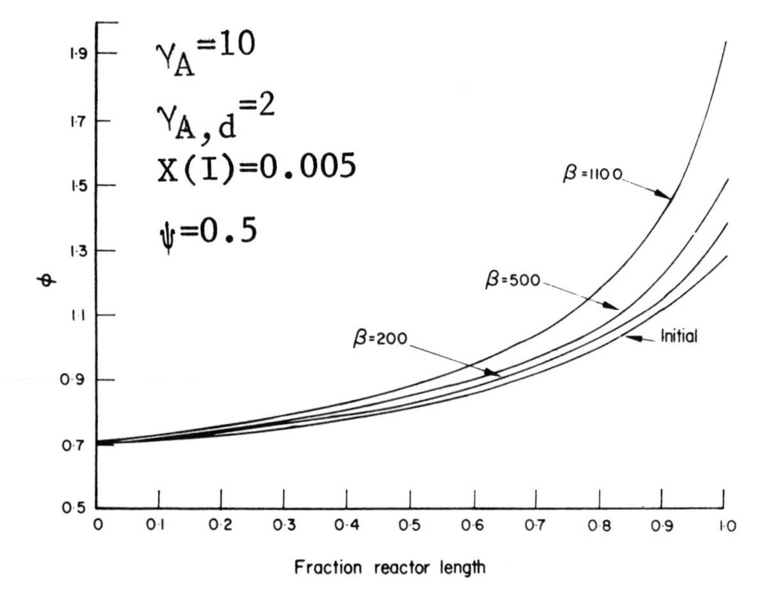

Chemical Engineering Science

Figure 41c. Example of reactor temperature profiles required for implementation of constant reactivity policy (77)

activation reaction, although less temperature sensitive than the main reaction ($\gamma_A > \gamma_{A,d}$), is subject either to only very low poison concentrations or is under conditions of initial operation preferentially catalyzed ($k^o{}_A < k^o{}_{A,d}$). The latter would be the case possibly in certain instances of initial rapid poisoning or coking of fresh catalyst such as those observed by Ozawa and Bischoff or Rudershausen and Watson. The computation example for a type I deactivation shown, giving both the temperature profile development as a function of time of operation (b), and correlation of β *vs.* X compared for limiting $\phi = 2$ (c), displays this behavior. The indications and comparisons are only qualitative. One might legitimately retain the impression that temperature profile manipulation may lead to some improvement in performance under restricted conditions when temperature increases with bed length, but the magnitude of such improvements appears to be relatively modest; decreasing profiles in certain circumstances can lead to substantial reductions in catalyst life. Subsequent, more detailed studies of the problem agree with this view.

The extension of this analysis to polyfunctional catalysts introduces, aside from the multiplication of parameters inherent with the additional complexity of the reaction deactivation system, the important new variable of catalyst composition. The problem has been studied for a series reaction scheme, A → B → C, in which the A–B reaction and the B–C reaction are catalyzed by different functions, and recently extended to a more realistic, nontrivial (*82*) case A ⇄ B → C. Since the results for the two are generally similar, we shall consider the simpler scheme in detail here. Denoting the two catalyst functions by U and V, we have for a type I example:

$$\left.\begin{array}{ll} U: & A + S_1 \rightarrow B + S_1(k_A) \\ V: & B + S_2 \rightarrow C + S_2(k_B) \end{array}\right| \quad \text{(main reaction)}$$

$$\left.\begin{array}{ll} U: & L + S_1 \rightarrow L \cdot S_1(k_{A,d}) \\ V: & M + S_2 \rightarrow M \cdot S_2(k_{B,d}) \end{array}\right| \quad \text{(deactivation reaction)}$$

Details of the calculation procedure follow those for the single reaction example just discussed. Some approximate limits on the parameters which appear in the analysis are given in Table XIIIA, together with some comment on their function and significance. Further detail on parametric values is given by Weisz and Hicks (*60*). The relative activation energies for various main and deactivation reactions are of signal importance in determining long term deactivation rates and reactor performance during

Table XIIIA. Some Approximate Limits on Parameters Associated with Nonselective Catalyst Poisoning

Parameter	*Description*	*Range*	*Comment*
ψ	Ratio of activity for main to deactivation reaction	>1 minimum. generally >10	A good catalyst would not be more active for poisoning than the main reaction step
γ	Activation energy or temperature sensitivity	$5 > \gamma > 35$	This corresponds to the range of activation energies for a large number of catalytic reactions
$\gamma_A/\gamma_{A,d}$	Temperature sensitivity ratio	>1 is norm	This range is of the most interest for desirable catalysts
(L_o/S_1^o), (M_o/S_2^o)	Relative feed impurity concentration	$<<1$	Feed impurity poisons are normally encountered in small or trace amounts
ϵ	Fraction of total of U function	$0 < \epsilon < 1$	Catalyst composition; variable at will.
$k^o{}_A/k^o{}_{B,r}$	Equilibrium constant; A $\rightleftarrows$ B	$0 < K < \infty$	$K = \infty$ corresponds to an irreversible sequence
$\gamma_A : \gamma_B :$ $\gamma_{A,d}$, etc.	Relative value of γ for the various reactions	All combinations	Some restrictions as noted above: $\gamma_{B,r} > \gamma_A$ may not be desirable in some cases.

constant conversion operation. The major results of the bifunctional analysis are obtained from calculations on a set of five different systems, differing in their relative activation energies and listed in Table XIIIB. Numerical values for the other parameters, notably the catalyst function activities, are shown as they appear on the figures. Startup conditions as a function of catalyst composition are shown in Figure 42a. There are well defined minima in the startup temperature, listed in Table XIIIB for the various systems, which depend only on the relative values of γ_A and γ_B. The location of these minima reflects the imbalance between the efficiency of the two functions in attaining the required conversion, and such results correspond generally to those of an earlier study by Gunn and Thomas (*83*) concerned with the optimal formulation of bifunctional catalysts.

For reactor operation either with or without the optimal catalyst composition required for initial operation, the temperature *vs.* time of operation or the β *vs.* X crossplots appear very much for the bifunctional case as illustrated for the single function catalyst in Figure 40. The temperature *vs.* time of operation characteristic is indicated more or less by the requirements of the less temperature-sensitive function for meeting conversion specification—a kind of rate limiting step if one wants to think of it in that way. Hence, systems such as 4 and 5 of Table XIIIB, limited by $\gamma_{A,d} = 5$ and $\gamma_{B,d} = 5$, respectively, and identical in main reaction sensitivities, exhibit the same temperature *vs.* time of operation behavior when initial optimum catalyst composition is used (similar comments apply to the pair 2-3). Product distribution (*i.e.*, catalytic selectivity), however, is another matter; indeed, the sensitivity of selectivity to deactivation is striking (though, again, perhaps not so in hindsight) and in many cases may well be more important than simple activity variation. Figure 42b shows at the top that for $\gamma_A > \gamma_B$, the production of intermediate B is favored such that after extended operation only B and C

$$k_A^o \, S_1^o \, \tau = k_B^o \, S_2^o \, \tau = \xi$$

$$k_{A,d}^o \, S_1^o \, \tau = k_{B,d}^o \, S_2^o \, \tau = 0.1$$

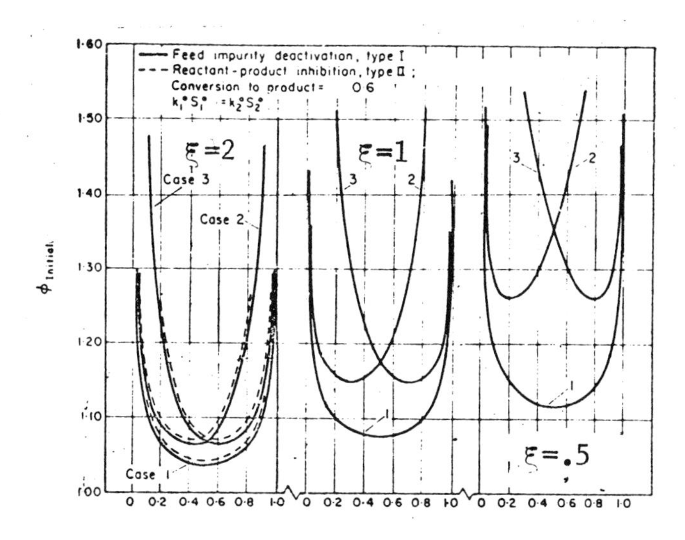

Symposium on Chemical Reaction Engineering

Figure 42a. Conditions of initial operation as a function of catalyst composition (78)

Table XIIIB. Sample Bifunctional Reaction System Temperature Sensitivities (*79*)

Case	γ_A	γ_B	$\gamma_{A,d}$	$\gamma_{B,d}$	ϵ[a]	*Selectivity*[b]
1	20	20	5	5	0.5	None
2	20	5	5	5	0.4	B & C
3	5	20	5	5	0.6	A & C
4	20	20	5	10	0.5	B & C
4a	20	20	5	10	0.3[c]	B & C
5	20	20	10	5	0.5	A & C

[a] Fraction of U function giving minimum initial temperature for set conversion.
[b] Intrinsic selectivity: products appearing in effluent as $\beta \rightarrow \infty$.
[c] Not optimum initial catalyst composition.

appear in the product; hence the designation of intrinsic selectivity for B and C in Table XIIIB for the illustrative case 2. Correspondingly, just the opposite holds for case 3 where $\gamma_A < \gamma_B$—both dealing with equivalent deactivation reactions. If $\gamma_{A,d} \neq \gamma_{B,d}$, there are two possibilities, using case 2 as an example:

(1) $\gamma_A > \gamma_B$, $\gamma_{A,d} < \gamma_{B,d}$: the V function deactivates rapidly, reinforcing the main reaction selectivity for the formation of B, so such a system would exhibit very strong intrinsic B-C selectivity;

(2) $\gamma_A > \gamma_B$, $\gamma_{A,d} > \gamma_{B,d}$: the deactivation is contrary to main reaction selectivity.

Further detail on selectivity is given in the lower portion of Figure 42b. Illustrated first is the identity of ϕ–β behavior for systems with differing rate limiting functions, as described for cases 4 and 5 and shown in the middle panel, and the exactly opposite selectivity behavior as shown in the bottom panel. Second is the effect of catalyst composition change on temperature and selectivity results. For case 4, $\gamma_{B,d} > \gamma_{A,d}$ and the V function controls the behavior of the system if one uses the initial optimum catalyst composition, $\epsilon = 0.5$. It is reasonable to expect that by increasing the amount of this function one might be able to counter the inherent selectivity of the system for intermediate product and at the same time not substantially affect the net catalyst deactivation rate as given by the ϕ–β curve. The results of changing ϵ accordingly, from 0.5 to 0.3, are also shown; the latter hope apparently can be realized, but the selectivity problem remains intractable. For a given time the reaction system selectivity is altered, but the trend of product distribution with time of operation is not changed. One can see that when deactivation occurs in these systems, the initial selectivity of the reaction cannot be maintained, nor can it be permanently altered *via* catalyst formulation but tends to a limit of reactant/product or intermediate/product composition only. The deactivation reactions, thus, completely dominate product distribution in the long run.

Catalyst life as a function of catalyst composition also demonstrates interesting behavior (Figure 42c). In case 1, where equivalent functions are identical, the deactivation of both is the same, and the optimum catalyst composition for reactor operation to any temperature level corresponds to the initial condition optimum. One also notes, on Figure 42b, that for this case since no imbalance between the two functions can develop, there is no trend in selectivity with time of operation. Cases 2 and 4, though, clearly indicate that in general catalyst formulations which

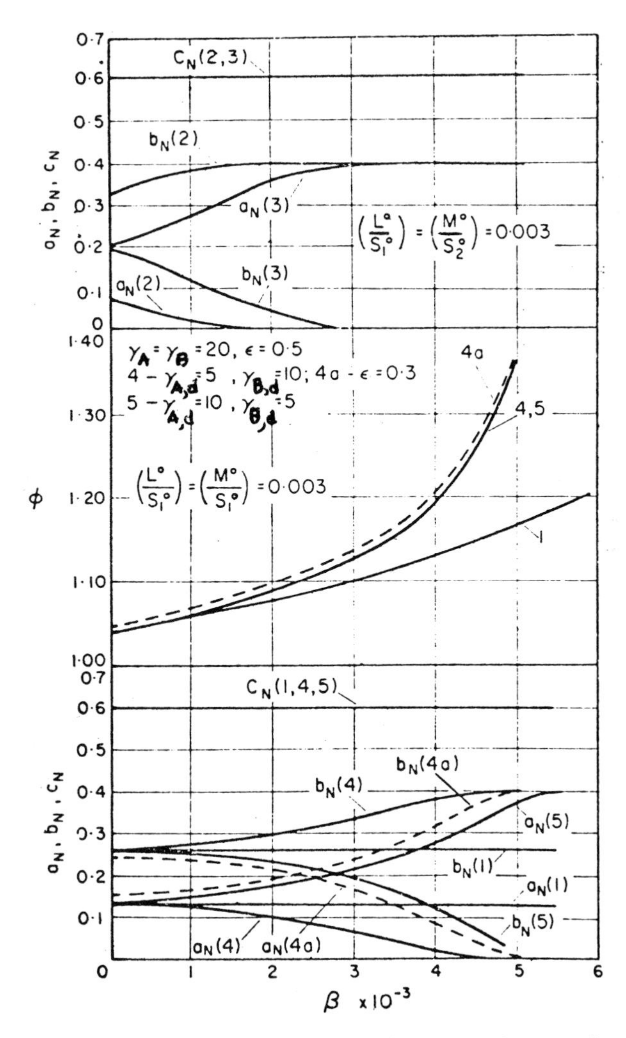

Symposium on Chemical Reaction Engineering

Figure 42b. Deactivation effects on product distribution (78)

Conversion = 0.6

$$k_A^o S_1^o \tau = k_B^o S_2^o \tau = 2.0$$

$$\left(\frac{L^o}{S_1^o}\right) = \left(\frac{M^o}{S_2^o}\right) = 0.003$$

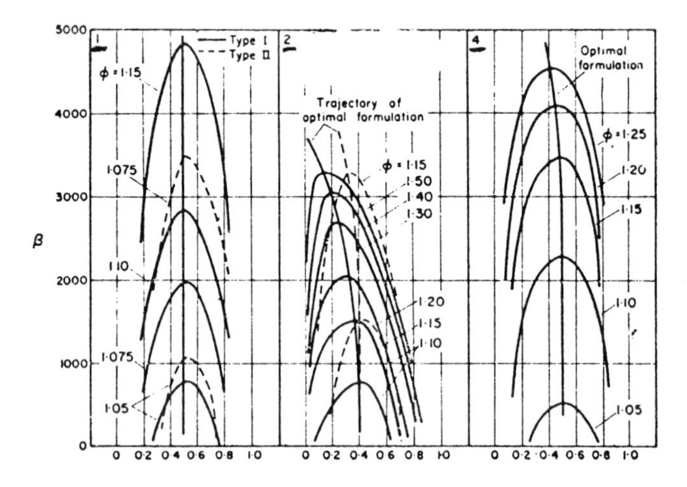

Symposium on Chemical Reaction Engineering

Figure 42c. Catalyst life as a function of composition for some example reaction systems (78)

minimize starting temperature are not optimal for extended operation if deactivation occurs. For case 2, where $\gamma_{A,d} = \gamma_{B,d}$ but $\gamma_A > \gamma_B$, the preferable formulation for extended operation includes less of the more temperature-sensitive function than called for by initial optimal composition; for case 4 where $\gamma_A = \gamma_B$ but $\gamma_{A,d} < \gamma_{B,d}$ the preferable formulation includes more of the function which deactivates more rapidly. The optimal relation is a very complex one since there is a trajectory of optimal catalyst compositions with time of operation (or temperature level). Cases 2 and 4 are both systems with intrinsic selectivity for B–C, and in each case changes in catalyst composition which lead to increased catalyst life counter the intrinsic selectivity. Similar observations hold for the other cases of Table XIIIB; thus it appears that if one can experimentally identify the intrinsic selectivity of a reaction–deactivation system, changing catalyst composition counter to this will result in increased catalyst life. Whether the alteration in selectivity is desirable or not, of course, must rest on the merits of the individual situation.

The extension of the results discussed above to the nontrivial polyfunctional scheme, A $\rightleftarrows$ B $\rightarrow$ C (*80*) is relatively straightforward. Much depends on whether the temperature sensitivity of the reverse reaction, B $\rightarrow$ A, is large or small compared with the other reaction and deactivation steps. An example of this effect of the reversibility on catalyst life-composition relationships is shown in Figure 43, where the limit of $K = \infty$ represents the irreversible sequence. The top panel depicts behavior when the reverse reaction is not very temperature sensitive, and one can see that the shapes of the temperature contours and the trajectories of optimal catalyst composition are perturbed only slightly from those for the irreversible sequence, corresponding to case 1. Nonetheless, the effect of introducing reversibility into the system is to displace optimal catalyst formulations to those containing less of the U function in order to promote conversion (still a fixed requirement) to product C. As the temperature sensitivity of this reverse function is increased (bottom of the figure), the displacement becomes quite pronounced. The equilibrium constant in this instance is actually temperature independent, so the crucial relative temperature sensitivity involved in determining system behavior here is $(\gamma_{B,f}/\gamma_{B,r})$. Selectivity behavior is also in line with expectation based on perturbation of the irreversible sequence with reversibility of increasing temperature sensitivity. The ultimate selectivity in these systems, however, can be limited by the equilibrium step, at least in the rate at which an intrinsic selectivity is attained, so that the sharp limits of reactant/product or intermediate/product may not be observed over normal catalyst lifetime (*see* Ref. *80* for further information on the rather complex details of selectivity and catalyst life in these nontrivial polyfunctional reactions).

Fixed Bed Regeneration. The discussion of individual particle analysis has shown that many problems associated with catalyst regeneration, particularly those dealing with coke removal, are very important in their contributions to the over-all understanding of deactivation. This situation also is true for the macroscopic systems which we are now discussing. The regeneration of fixed bed reactors has been considered almost exclusively in terms of coke removal problems, in which it is desired not only to recover the lost activity of the catalyst but also to accomplish this without thermal damage (since the regeneration is highly exothermic) and with a minimum of process time. It is of particular interest to be able to estimate the thermal history of the reactor during regeneration since a knowledge of the behavior of the time-dependent thermal waves developed on operating parameters will allow one to avoid excessive heating of the bed and simultaneously minimize the time required for reactivation.

The history of those studies directed primarily toward the regeneration problem is one, essentially, of continuing refinement of the kinetics

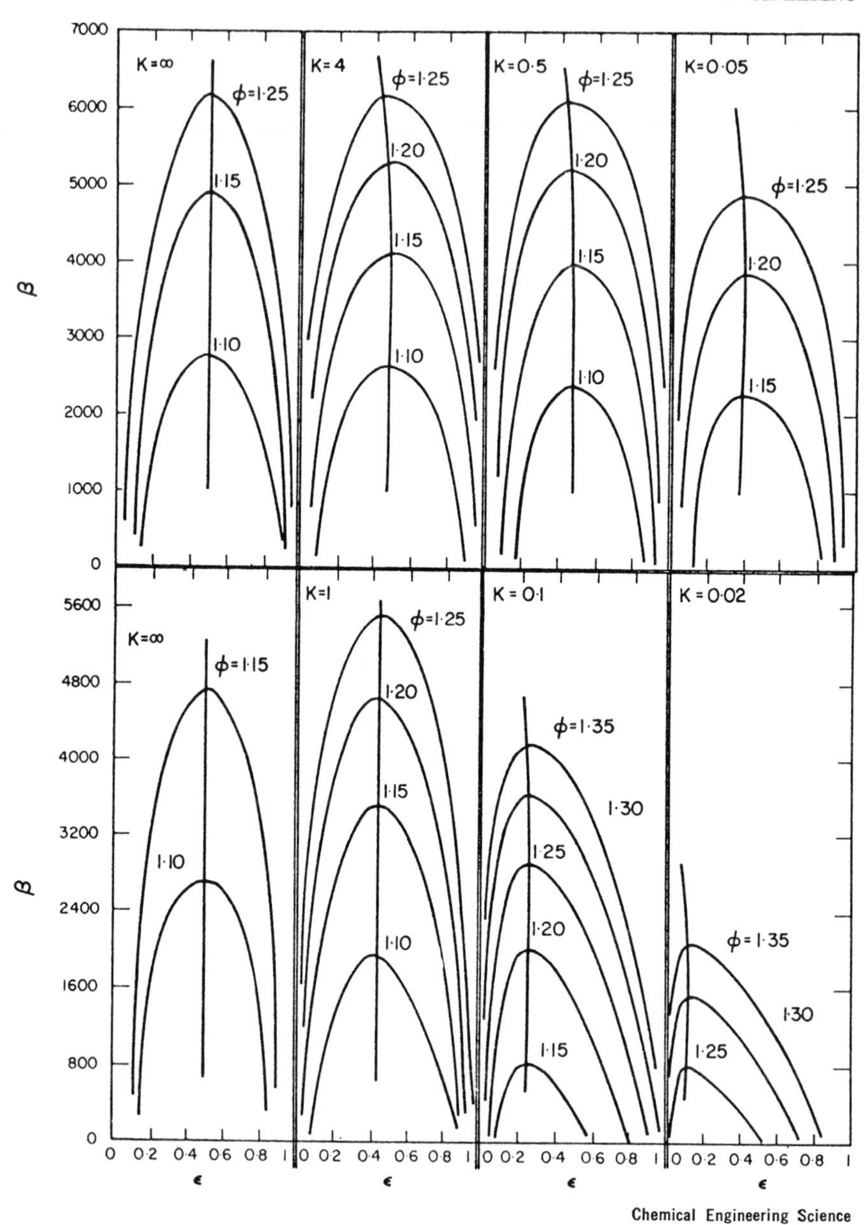

Figure 43. Reversibility effects on catalyst life: equivalent catalytic functions (79). Top: temperature insensitive reverse reaction: $\gamma_A = 20$, $\gamma_B = 20$, $\gamma_{B,r} = 5$, $\gamma_{A,d} = 5$, $\gamma_{B,d} = 5$. Bottom: temperature sensitive reverse reaction: $\gamma_A = 20$, $\gamma_B = 20$, $\gamma_{B,r} = 20$, $\gamma_{A,d} = 5$, $\gamma_{B,d} = 5$.

Parameters: $k^\circ_A S^\circ_1 \tau = k^\circ_B S^\circ_2 \tau = 2$
$k^\circ_{A,d} S^\circ_1 \tau = k^\circ_{B,d} S^\circ_2 \tau = 0.1$
$L^\circ / S^\circ_1 = M^\circ / S^\circ_2 = 0.003$
Conversion $= 0.6$

incorporated into a very simple reactor model. Since we are dealing with nonisothermal conditions, both heat and mass balances are required. However, the basic reactor model considered by all workers corresponds generally to that which we have just discussed: one-dimensional, plug flow with no density gradients. In addition, most workers have been concerned with adiabatic conditions in which interphase heat transfer rates between catalyst and reactant are large enough to preclude any significant temperature difference between the two phases. The original studies of van Deemter (*84, 85*) considered regeneration kinetics which were zero order and temperature independent. Johnson, Froment, and Watson (*86*) extended the treatment to a somewhat more realistic kinetic expression,

$$r_c = -kPy_{O_2}\frac{C_c}{C_c^o} \tag{121}$$

which they claim is an approximate representation of the rate when oxygen diffusion is the controlling mechanism but which we see is a form closely related to the intrinsic coke oxidation kinetics of Equation 43a. Johnson *et al.*, however, do not consider the temperature variation of r_c; hence the first-order relationship to oxygen concentration does confer a diffusion-like quality on their results. A subsequent study by Schulman (*87*) considers essentially the same model and kinetic form as Johnson *et al.* but considers rates to be controlled by intrinsic kinetics and assigns an Arrhenius temperature variation to k in Equation 121. The influence of nonadiabatic conditions on the thermal behavior of the reactor is also illustrated. More recently, Olson, Luss, and Amundson (*88*) presented results for the adiabatic regeneration in which the Weisz-Goodwin shell progressive parameters are used to express the kinetics of coke oxidation, and Ozawa (*89*) has given a semianalytical solution for low temperature regeneration with intrinsic kinetics included. A detailed consideration of numerical problems involved in the solution of more general model equations representing the fixed-bed regeneration problem has been given by Gonzalez and Spencer (*90*) which cannot be discussed in further detail here. Zhorov *et al.* (*91*) have also presented an approximate analytical solution to the problem.

Van Deemter's original analysis (1953) was relatively crude, assisted considerably by the assumption of constant reaction rate; nonetheless, a reasonable qualitative picture of the regeneration process is obtained. For the oxygen balance in the reactor:

$$\epsilon\frac{\partial C_{O_2}}{\partial\theta} + v\frac{\partial C_{O_2}}{\partial w} = -r_{O_2} = -U \tag{122}$$

and for coke:

$$r_c = -r_{O_2} = -U \text{ (per mole of } O_2) \tag{123}$$

where C_{O_2} is the oxygen concentration in moles/volume, v is the superficial gas velocity, w is bed length dimension, and ϵ is the bed porosity. For the heat balance:

$$[(1 - \epsilon)\rho_s c_s + \epsilon \rho_g c_g] \frac{\partial T}{\partial \theta} + \rho_g c_g v \frac{\partial T}{\partial w} = U \Delta H \tag{124}$$

with ΔH the heat of reaction per mole of oxygen, ρ_g and ρ_s the gas and solid densities, respectively, and c_g and c_s the gas and solid heat capacities, respectively. The boundary conditions are, for the initial stage of operation:

$$\begin{aligned} w = 0 \quad & C_{O_2} = C_{O_2} \\ & T = T_o \\ \theta = 0 \quad & (w > 0) C_{O_2} = 0 \\ & T = 0 \end{aligned} \tag{125}$$

The solution to Equation 124 is, by the method of characteristics:

$$T = T_o + U \Delta H w / \rho_g c_g v \tag{126}$$

for $0 \leqslant w \leqslant \alpha v \theta$

where

$$\alpha = \frac{\rho_g c_g}{(1 - \epsilon)\rho_s c_s + \epsilon \rho_g c_g}$$

This depicts the oxidation under these initial conditions as being confined to an initial region of the bed, $w < \alpha v \theta$, and the heat produced in this region is transported through the bed with the velocity αv. For $w > v\theta$, the temperature must correspond to the boundary condition, $T = 0$. Also for the initial operation, the solution to the oxygen balance, Equation 122, is:

$$C_{O_2} = C_{O_2}{}^o - Uw/v \tag{127}$$

for $0 \leqslant w \leqslant \alpha v \theta$

The coke distribution follows from the kinetics of Equation 123:

$$C_c = C_c{}^o - U(\theta - w/\alpha v) \tag{128}$$

Thus, all balance equations result in simple linear profiles. A zone $\alpha v \theta$ in length is that portion of the reactor in which regeneration is occurring.

The actual implications of these results in determining how the regenerating bed behaves depend entirely on the value of the inlet oxygen concentration, $C_{O_2}{}^o$. Equation 128 shows that the time for complete coke removal is:

$$\theta_1 = C_c{}^o/U \tag{129}$$

Further, oxygen is completely removed, according to Equation 127, when the reaction zone is:

$$w_o = C_{O_2}{}^o v/U \tag{130}$$

or at time:

$$\theta_o = w_o/\alpha v = C_{O_2}{}^o/\alpha U \tag{131}$$

Now, when $\theta_o > \theta_1$, the coke at the bed entrance will have reacted completely while oxygen remains; when $\theta_1 > \theta_o$ oxygen becomes the limiting reactant, and the zone of combustion becomes stabilized until the coke disappears. Obviously, bed behavior in the two cases will be completely different.

The simplest case to examine is the first, in which $\theta_o > \theta_1$. In such a case one can imagine typical concentration and temperature profiles within the reactor corresponding to those shown in Figure 44a. Here both the front and back of the oxidation zone move with a velocity, αv, equal to that of heat transport. Thus the oxidation zone is constant in size as shown, and the temperature profile is:

$$\begin{aligned} 0 \leqslant w \leqslant \alpha v(\theta - \theta_1),\ T = T_o \\ \alpha v(\theta - \theta_1) \leqslant w \leqslant \alpha v\theta,\ T = T_o + U\Delta H w/\rho_g c_g v \end{aligned} \tag{132}$$

The depth of the oxidizing zone is:

$$\alpha v \theta_1 = \alpha C_c{}^o v/U \tag{133}$$

and the time required for regeneration corresponds to that for the trailing edge of the oxidizing zone to pass through the bed:

$$\frac{w}{\alpha v} + \theta_1 = \frac{w}{\alpha v} + \frac{C_{O_2}{}^o}{U} \tag{134}$$

Similar behavior is found to some lower limit of inlet oxygen concentration, in which case the oxygen in the burning zone is consumed entirely before the coke is removed fully from the entrance to the bed—

i.e., $\theta_1 > \theta_o$. The general situation is illustrated in Figure 44b and c. At $\theta = \theta_o$, the depth of the burning zone, w_o, is given by Equation 130, and the zone will subsequently remain stationary in the bed until the coke at the bed entrance is depleted. At $\theta = \theta_1$, the coke at $w = 0$ has burned, and there is a discontinuity in coke concentration resulting at w_o (Figure 44c). We recall that U, the burning rate is non-zero only in the burning zone; the heat generated in this zone gives rise to the temperature profile of Equation 126, and although during the interval from θ_o to θ_1 the burning zone is stationary, the heat generated is removed from the zone at a velocity αv, and at a temperature level:

$$T = T_o + U \Delta H w_o / \rho_g c_g v \qquad (135)$$

as is shown in Figure 44c.

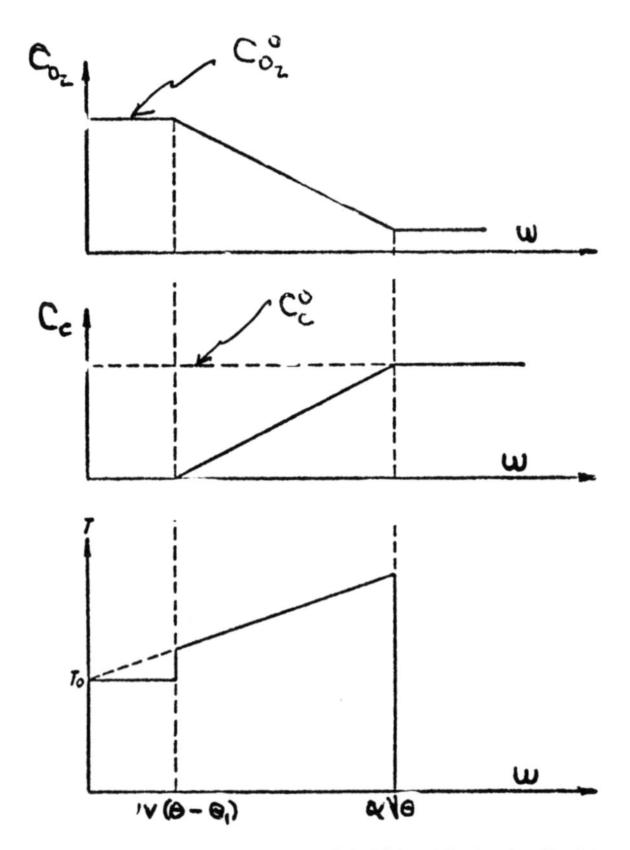

Industrial and Engineering Chemistry

Figure 44a. Oxygen concentration, coke concentration, and temperature. High oxygen concentration (84).

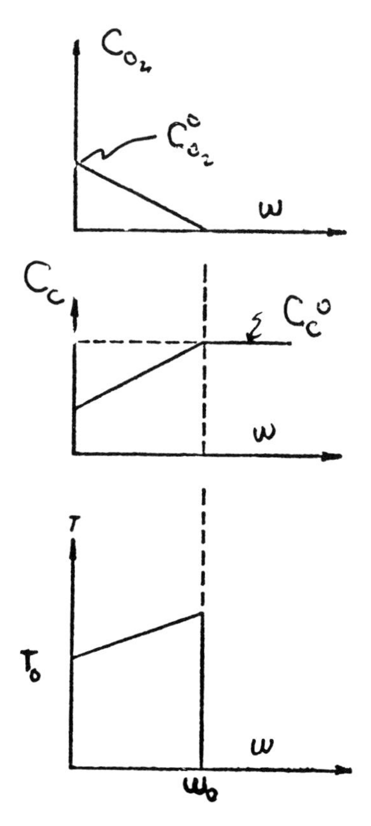

Industrial and Engineering Chemistry

Figure 44b. Oxygen concentration, coke concentration, and temperature; $\theta = \theta_o$ (84)

For $\theta > \theta_1$, the burning zone moves into the bed until the trailing edge coincides with the discontinuity of the coke concentration at w_o. In this interval the depth of the burning zone (determined only by the oxygen concentration) remains constant at w_o, and the thermal front passes through the bed such that:

$$0 \leqslant w \leqslant \alpha v(\theta - \theta_1), U = 0, T = T_o \quad (136)$$

$$\alpha v(\theta - \theta_1) \leqslant w \leqslant w_o + \alpha v(\theta - \theta_1), U \neq 0; T = T_o + U\Delta H w/\rho_g c_g v$$

$$w_o + \alpha v(\theta - \theta_1) \leqslant w \leqslant \alpha v\theta, U = 0; T = T_o + U\Delta H w_o/\rho_g c_g v$$

$$w > \alpha v\theta, U = 0, T = 0$$

The temperature always increases linearly in the burning zone while behind and ahead of it the temperature is constant (Figure 45a).

When the trailing edge of the zone reaches the coke discontinuity at w_o, it again becomes stationary until at $\theta = 2\theta_1$ all the coke has been removed (Figure 45b and c). Again heat is removed from the stationary zone, although this time at a temperature level $T = T_o + 2U\Delta H w_o/\rho_g c_g v$, and entering gas at T_o decreases the level of temperature in the region $w_o \leqslant w \leqslant \alpha v \theta_1$ of the zone (Figure 45c).

Such a stop and go process leads to very intricate temperature profiles within the bed; of course, this behavior is a product of the model assumptions, and in actual cases one expects the oxidation zone to move in a continuous fashion through the bed and the temperature profiles to assume the character of thermal waves passing through the bed (85). None-

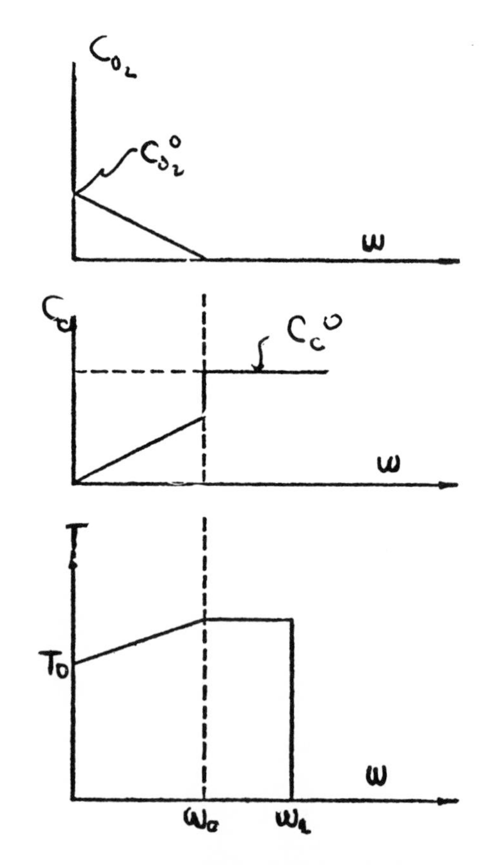

Industrial and Engineering Chemistry

Figure 44c. Oxygen concentration, coke concentration, and temperature; $\theta = \theta_1$ (84)

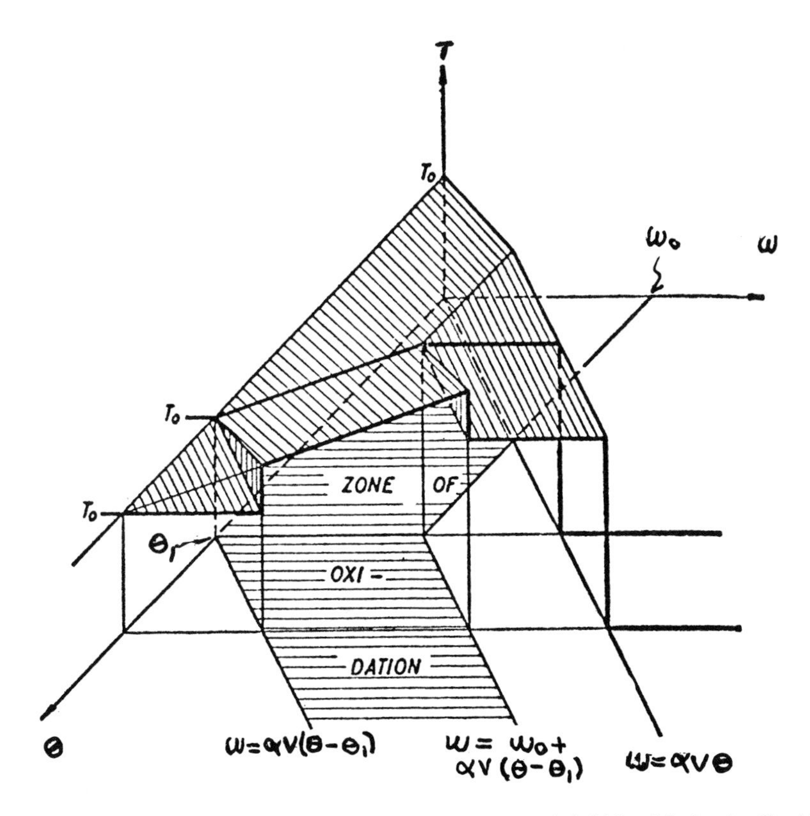

Industrial and Engineering Chemistry

Figure 45a. wTθ *diagram;* $\theta > \theta_1$ (84)

theless, some approximate operating limits can be obtained from this crude model. The mean regeneration time is:

$$\frac{w\theta_1}{\theta_o} + \tfrac{1}{2}(\theta_o + \theta_1) = \frac{C_c{}^o w}{C_{O_2}{}^o v} + \tfrac{1}{2}\left(\frac{C_c{}^o}{U} + \frac{C_{O_2}{}^o}{\alpha U}\right) \tag{137}$$

and the temperature profile is bounded by:

$$T \leqslant T_o + U\Delta H w/\rho_g c_g v \tag{138}$$

so the simple theory does yield results good for rough estimates of regenerator design parameters. A number of other workers have confirmed the form of Equation 138 for determining temperature rise inside the bed during regeneration. For the important case of low inlet oxygen concentration where the thermal transport velocity exceeds the combustion zone velocity, as discussed here, the asymptotic temperature limit is (85):

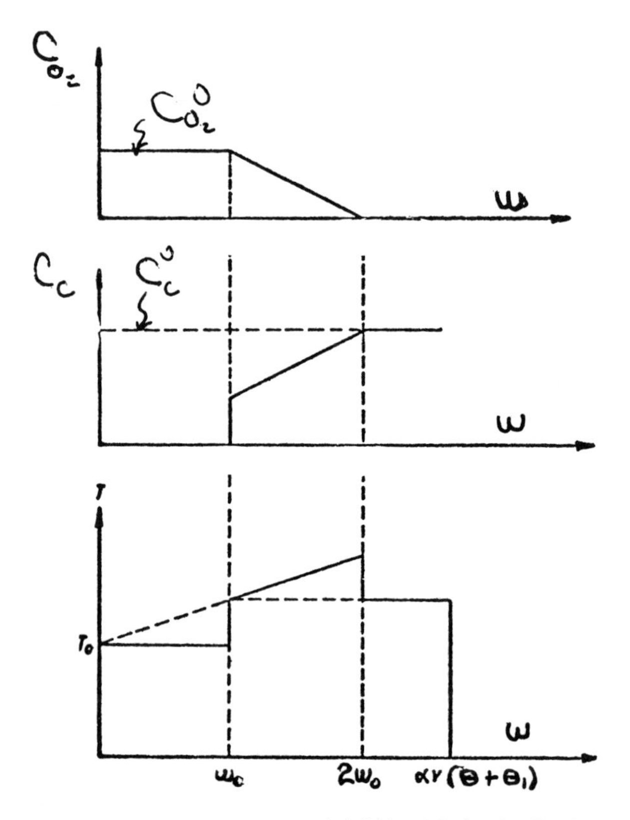

Industrial and Engineering Chemistry

Figure 45b. Oxygen concentration, coke concentration, and temperature; $\theta = \theta_0 + \theta_1$ (84)

$$\Delta T = T_{\max} - T_o = \frac{C_{O_2}{}^o \Delta H}{\rho_g c_g \left(1 - \dfrac{C_{O_2}{}^o}{\alpha C_c{}^o}\right)} \tag{138a}$$

As stated before, the analysis of Johnson *et al.* retains all the assumptions of the van Deemter study with the exception of the regeneration kinetics, which are given by Equation 121, where k is independent of temperature. The oxygen balance follows that of Equation 122; in terms of the notation of Johnson *et al.:*

$$\epsilon \frac{\partial C_{O_2}}{\partial \theta} + \epsilon v' \frac{\partial C_{O_2}}{\partial w} = -a \left(\frac{\rho_s}{M_c}\right) k P y_{O_2} \left(\frac{C_c}{C_c{}^o}\right) \tag{139}$$

in which

$$C_{O_2} = \frac{\rho_g}{M_g} y_{O_2}$$

$$v = G/\epsilon\rho_g$$

where ρ_g and ρ_s are density of the gas and the bed (bulk), respectively, in mass units, M_g and M_c molecular weights of gas and carbon, respectively, P is total pressure, a is a stoichiometric factor having a value of unity for combustion to CO_2, y_{O_2} is the mole fraction of oxygen, and G is the gas mass velocity. The carbon balance is:

$$r_c = \frac{\partial C_c}{\partial \theta} = -kPy_{O_2}\left(\frac{C_c}{C_c{}^o}\right) \tag{140}$$

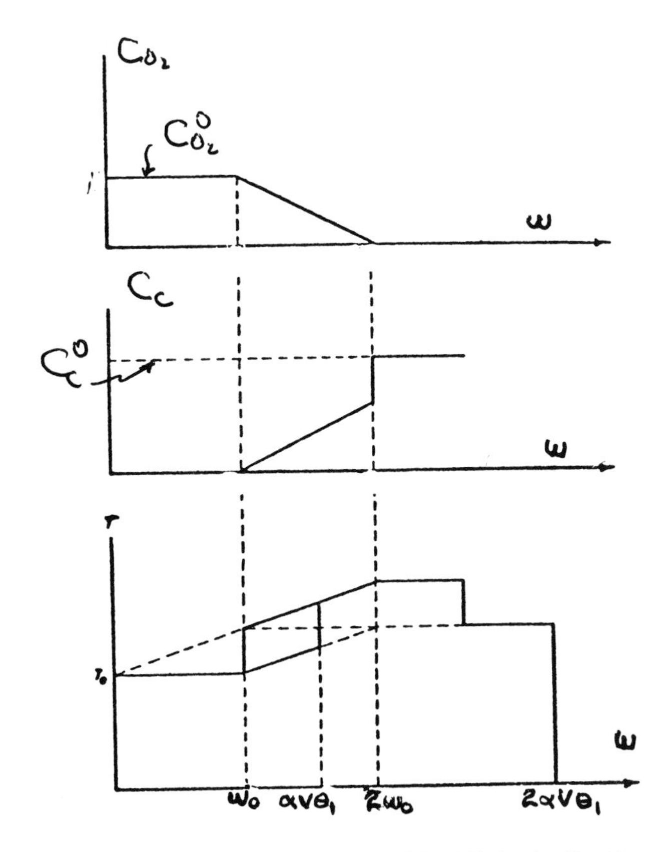

Industrial and Engineering Chemistry

Figure 45c. Oxygen concentration, coke concentration, and temperature; $\theta = 2\theta_1$ (84)

Boundary conditions for Equations 139 and 140 are:

$$
\begin{aligned}
&y = y_{O_2}{}^{o} && w = 0 \\
&C_c = C_c{}^{o} && \theta - w\left(\frac{\epsilon\rho_g}{G}\right) \leqslant 0
\end{aligned}
\tag{141}
$$

For the heat balance:

$$[\rho_s c_s + \epsilon\rho_g c_p]\frac{\partial T}{\partial\theta} + c_p G\frac{\partial T}{\partial w} = \rho_s \Delta H k P y_{O_2}\left(\frac{C_c}{C_c{}^{o}}\right) \tag{142}$$

The value of ΔH here is in mass units since the carbon concentrations used by Johnson *et al.* are in mass per mass of catalyst rather than the molal concentrations per unit volume used by van Deemter. The distinction in dimensions is really important only in proper interpretation of the magnitudes of parameters used in illustrative calculations, of course, since Equations 122 or 139 are valid in any dimensionally consistent set of units. Values of ΔH can be obtained for various stoichiometries of combustion from the results of Dart and Oblad (92). The nondimensional forms of Equations 139 and 142 are:

$$\frac{\partial y_{O_2}}{\partial z} = -a y_{O_2}\left(\frac{C_c}{C_c{}^{o}}\right) \tag{143}$$

$$\frac{y_{O_2}{}^{o}}{C_c{}^{o}} \cdot \frac{\partial C_c}{\partial \tau} = -y_{O_2}\left(\frac{C_c}{C_c{}^{o}}\right) \tag{144}$$

$$M\frac{\partial t}{\partial T} + H\frac{\partial t}{\partial z} = \frac{y_{O_2}C_c}{y_{O_2}{}^{o}C_c{}^{o}} \tag{145}$$

where:

$$y_{O_2} = y_{O_2}{}^{o},\ z = 0,\ \tau \geqslant 0$$

$$C_c = C_c{}^{o},\ \tau = 0,\ z \geqslant 0$$

$$t = 0,\ z = 0$$

$$H = \left(\frac{M_g c_p C_c{}^{o}}{M_c c_s y_{O_2}{}^{o}}\right),\ M = \left(1 - \frac{c_p}{c_s} \cdot \frac{\epsilon\rho_g}{\rho_s}\right)$$

$$t = T\frac{c_s}{\Delta H C_c{}^{o}},\ z = w\left(\frac{\rho_s}{G} \cdot \frac{M_g}{M_c} kP\right)$$

and

$$\tau = \left(\frac{y_{O_2}{}^{o}}{C_c{}^{o}} \cdot kP\right)\left(\theta - w\frac{\epsilon\rho_g}{G}\right)$$

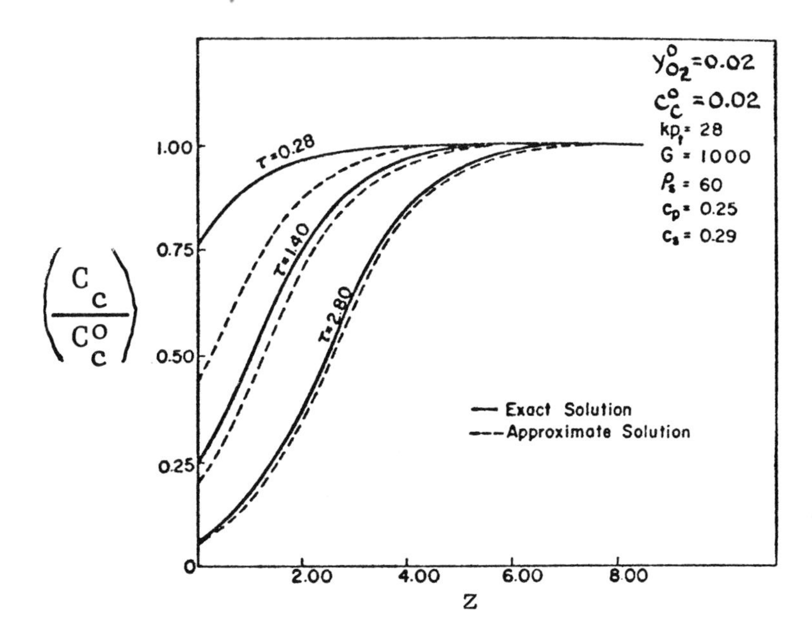

Figure 46a. Variation in carbon concentration down the bed (86)

This set yields a straightforward solution. Since the reaction rate is independent of temperature, the mass balance equations can be solved separately as:

$$\frac{C_c}{C_c^o} = \frac{1}{1 + e^{-az}(e^{\tau} - 1)} \tag{146}$$

$$\frac{y_{O_2}}{y_{O_2}{}^o} = \frac{1}{1 + e^{-\tau}(e^{az} - 1)} \tag{147}$$

These are continuous profiles through the bed, as shown in Figure 46a and b. In addition, the product of Equations 146 and 147 is equivalent to the heat generation term in the energy balance, Equation 145, giving the thermal generation profiles shown in Figure 46c. After a certain initial time period, the shape of each of the three profiles shown becomes invariant, forming a concentration or thermal wave which passes through the length of the bed. If one assumes that these fixed carbon and oxygen profiles are developed at $\tau = 0$, then:

$$\frac{C_c}{C_c^o} = \frac{1}{1 + e^{-az+\tau}} \tag{147a}$$

$$\frac{y_{O_2}}{y_{O_2}{}^o} = \frac{1}{1 + e^{az-\tau}} \tag{146a}$$

$$\frac{y_{O_2}C_c}{y_{O_2}C_c{}^o} = \frac{1}{2 + e^{az-\tau} + e^{-az+\tau}} \tag{148}$$

and these approximate solutions are also shown in Figure 46a-c. The heat balance equation may now be solved analytically also. Substituting Equation 148 into 145 and solving with indicated temperature boundary conditions one obtains:

$$t = \frac{1}{2(aH - M)}\left[\tanh\frac{az - \tau}{2} - \tanh\frac{z - (H/M)\tau}{2H/M}\right] \tag{149}$$

Temperature waves corresponding to this solution are illustrated in Figure 47a and b. The first term of Equation 149 measures the rate of translation of the burning zone while the coefficient of z in the second term measures the convective transport rate of a temperature wave in the absence of heat generation. For $(aH/M) = 1$, that is:

$$\left(1 - \frac{c_p}{c_s}\cdot\frac{\epsilon\rho_g}{\rho_s}\right)\left(\frac{y_{O_2}{}^o}{C_c{}^o}\right) = a\,\frac{c_p}{c_s}\cdot\frac{M_G}{M_s} \tag{150}$$

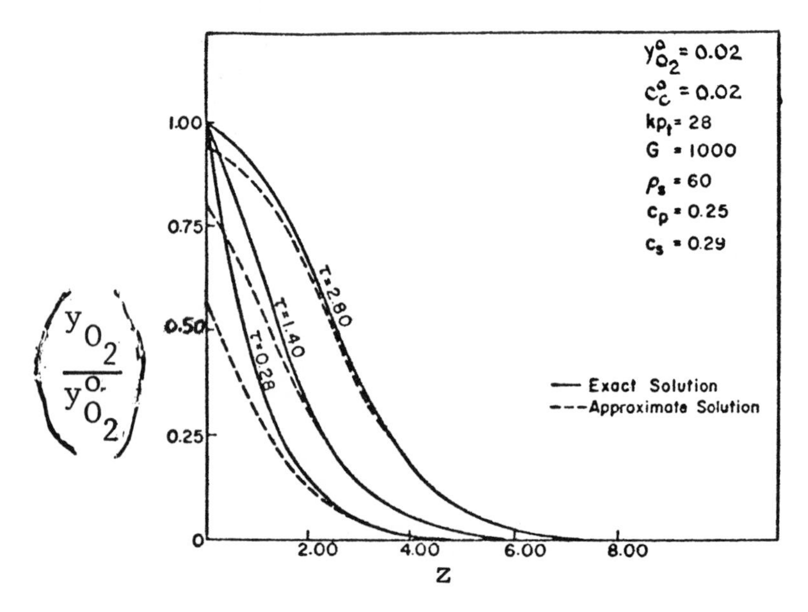

Chemical Engineering Science

Figure 46b. Variation in oxygen concentration down the bed (86)

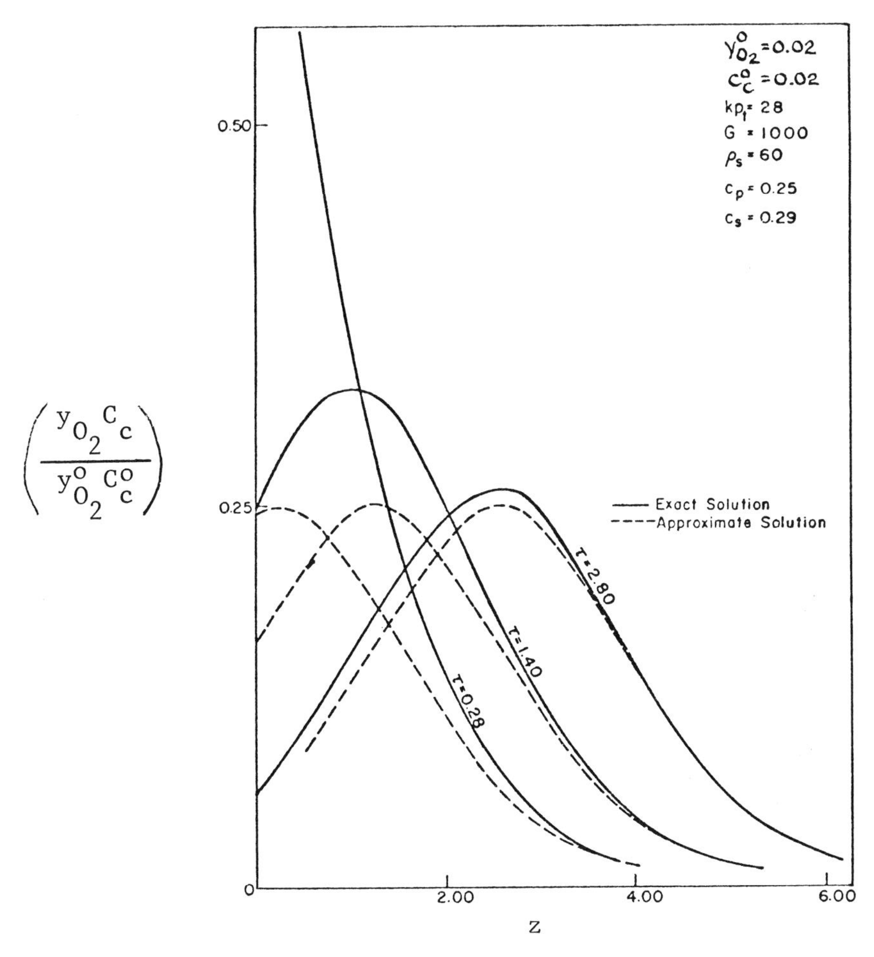

Chemical Engineering Science

Figure 46c. Variation in the product of oxygen and carbon concentration down the bed (86)

these two velocities are the same, and the limiting value of Equation 149 yields:

$$t = \frac{Mz}{H(2 + e^{az-\tau} + e^{\tau-az})} \tag{151}$$

This situation defines another type of limiting entrance oxygen concentration (differing somewhat in its physical interpretation from that described by van Deemter) which, it will be seen, has a pronounced effect on the maximum temperature within the bed.

The question of temperature maxima is an important one—indeed, one of the primary reasons for an analysis of this type. Two maxima are

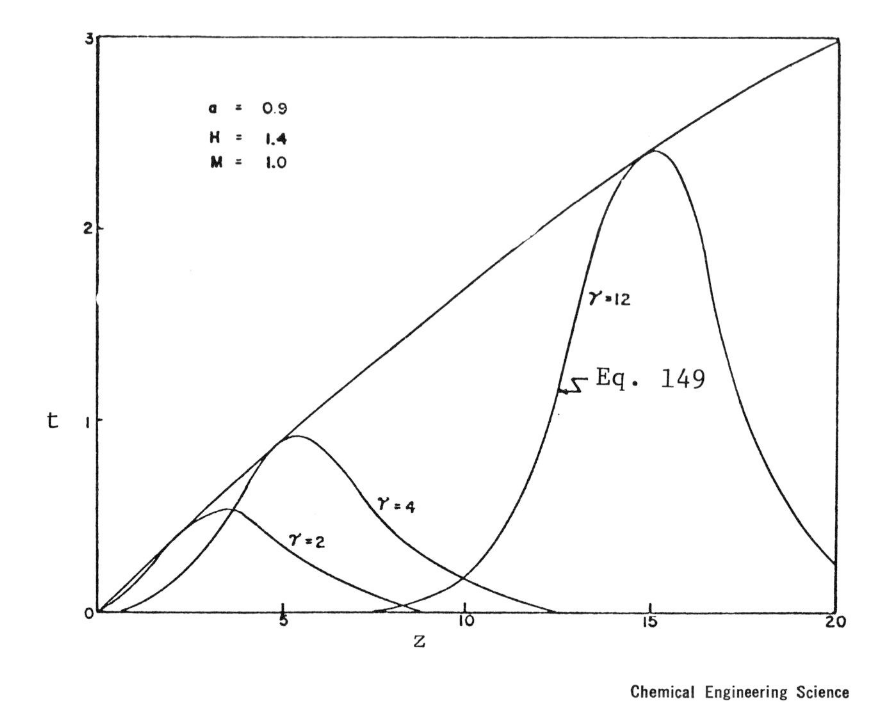

Chemical Engineering Science

Figure 47a. Reduced temperature as a function of position at various times (86)

identifiable: $t_{\max,z}$, which is the maximum temperature at a point z for any time τ, and $t_{\max,\tau}$, which is the instantaneous maximum temperature along the length of the reactor. By appropriate differentiation of Equation 149 we have:

$$t_{\max,z} = \frac{1}{(aH - M)}\left[\tanh \frac{zM}{4H}(aH - 1)\right] \tag{152}$$

$$t_{\max,\tau} = \frac{1}{2(aH - M)}\{\tanh[\Omega - \tfrac{1}{2}\ln X_1] - \tanh[-\Omega - \tfrac{1}{2}\ln X_2]\} \tag{153}$$

in which

$$\Omega = \frac{z}{4}\left(\frac{aH - M}{H}\right)$$

$$X_1 = \frac{\exp[(aH - M)/2Hz] - (aH - M)^{1/2}}{(aH/M)^{1/2}\exp[(aH - M)/2H] - 1}$$

$$X_2 = \frac{\exp[(aH - M)/2Hz] - (aH - M)^{1/2}}{(aH/M)^{1/2}\exp[(aH - M)/2H] - 1}$$

It is the value of $t_{\max,z}$ which is most important for design purposes since this establishes the limit on operation of the bed. Also, for the case of $(aH/M) = 1$ we have from Equation 151:

$$t_{\max,z} = \frac{Mz}{4H} \qquad (154)$$

These expressions for maximum temperature correspond to the simplified forms of Equations 146a and 147a and thus do not account for effects associated with the initial development of the profiles. However, Johnson *et al.* show that the errors associated with this assumption are relatively modest, and the approximation is a viable one for engineering purposes.

The regeneration time can be computed from the time required for the mid-point of the burning zone (*i.e.*, where $C_c/C_c^o = 0.5$) to reach the end of the bed, plus the time required to reduce (C_c/C_c^o) to some desired arbitrarily low level. The mid point time is az, and the additional time required to reduce coke to a final value $(C_c/C_c^o)_F$ from 0.5 is proportional to the log ratio $\ln[1 - (C_c/C_c^o)_F]/(C_c/C_c^o)_F$; thus:

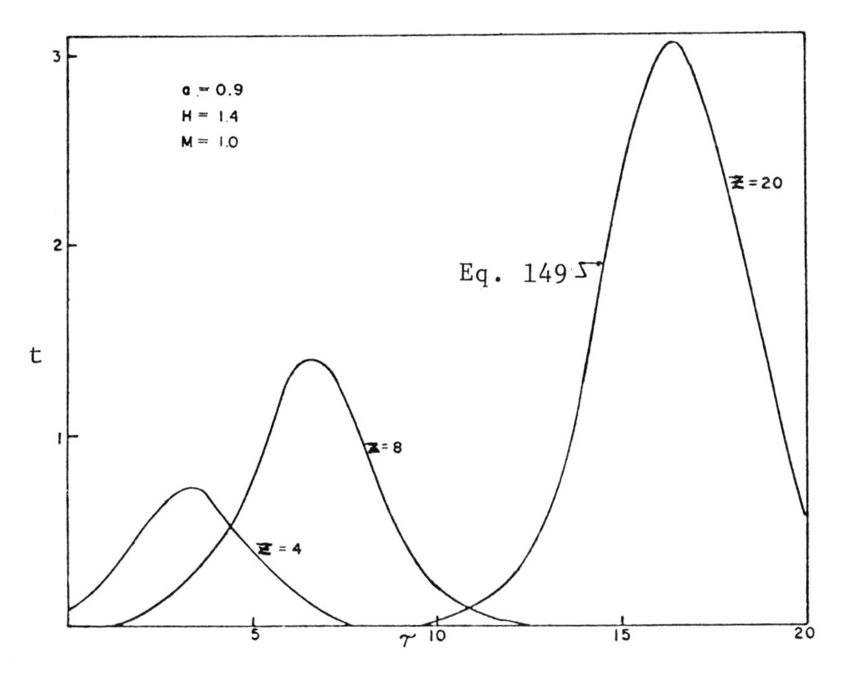

Figure 47b. Reduced temperature as a function of time at various positions (86)

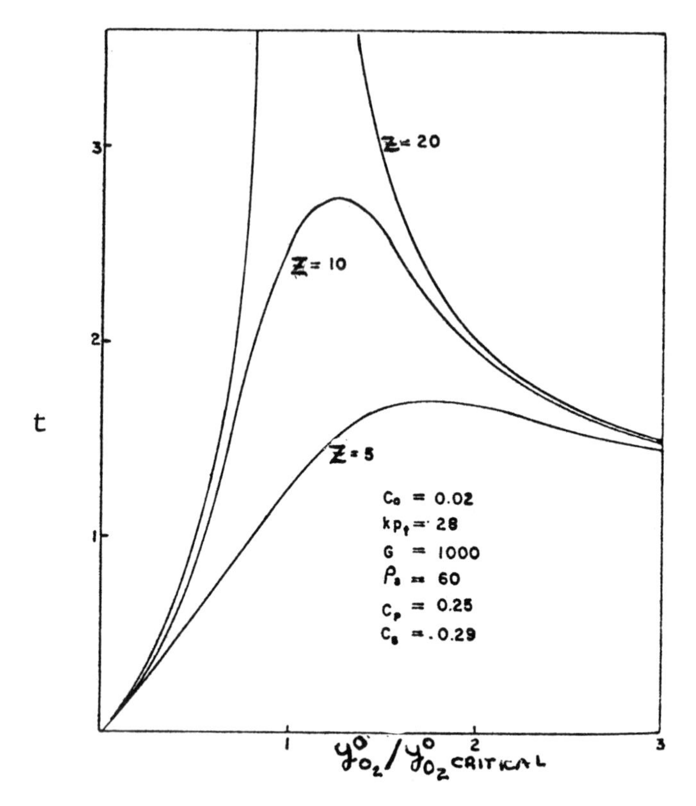

Figure 47c. Effect of the feed composition on maximum temperature in the bed (86)

$$\tau_{\text{regen}} = az + \ln \frac{1 - (C_c/C_c^o)_F}{(C_c/C_c^o)_F} \tag{155}$$

Inlet oxygen concentration has signal influence on the behavior of the bed. If $M \approx 1$, which is a reasonable approximation, it can be shown that for:

$$aH > 1, i.e. \left(\frac{y_{O_2}{}^o}{C_c{}^o}\right) < \frac{M_g}{M_c} \cdot \frac{c_p}{c_s} a$$

then the temperature wave $t_{\max,z}$ travels down the bed faster than the burning zone; for $aH = 1$ the two velocities are the same, and for $aH < 1$ the burning zone precedes the temperature maximum. Figure 47c shows the physical consequences of this. When the velocities of the burning zone and $t_{\max}$ are the same ($aH = 1$), a reinforcement occurs, causing very large temperature rises in the bed. This serves to define a "critical" oxygen–carbon ratio:

$$\left(\frac{y_{O_2}{}^o}{C_c{}^o}\right)_{\text{crit}} = a\,\frac{c_p}{c_s}\left(\frac{M_g}{M_c}\right) \tag{156}$$

such that operation at either greater or lower values than that specified by Equation 156 will avoid such hot spots.

Johnson *et al.* also present the appropriate theory for the reactor with isothermal cooling; the nonadiabatic conditions greatly complicate the analysis above. The interested reader is referred to the original paper for details of the solutions and associated appropriate simplifications.

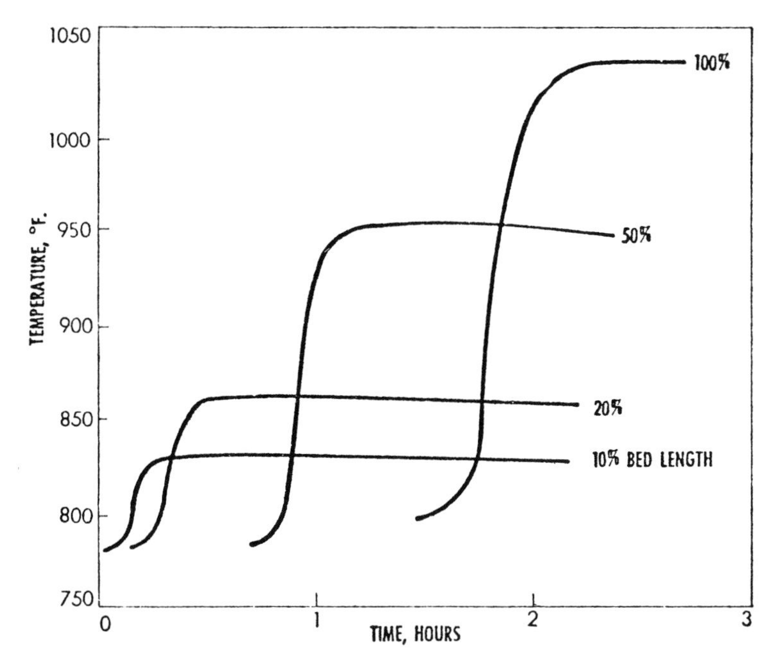

Industrial and Engineering Chemistry

Figure 48a. Temperature fronts in adiabatic regeneration (87)

Parameters for Figure 48a,b:

Inlet temps. = 780°F
(x_o) *Inlet O_2 mole fr.* = 0.01
(y_o) *Inlet wt. fr. C on Cat.* = 0.066
Bulk density of bed = 55.2 lb/ft³
Specific heat of cat. = 0.25 Btu/lb-°F
Specific heat of gas = 0.25
Bed length = 7.5 ft
Superficial gas vel. = 2700 ft/hr
Gas density = 0.212 lb/ft³
(P) *Total pressure* = 75 psia
Gas mol. wt. = 30
(k_o) *Rate constant at inlet* = 0.103

$$\frac{dy}{d\theta} = -kPxy$$

Schulman (87) also reported results for regeneration under non-adiabatic conditions, using a numerical technique to solve the conservation equations, considering the kinetics of the combustion reaction to depend on temperature level, and utilizing an effective wall heat transfer coefficient to surroundings to account for heat loss. Again, the reader is referred to the original paper for details of the computation; however, several specific results are of interest. Figure 48a shows for a typical example that in adiabatic regeneration, where the kinetics are considered temperature dependent, sharp temperature fronts can develop. Note the similarity of these temperature profiles to those predicted by van Deemter's analysis despite the completely different assumptions regarding regeneration kinetics. The maximum temperature of the front continually increases as the zone moves down the bed, reflecting the contribution of the heat of reaction to the gas temperature leaving the burning zone, as indicated by van Deemter. The waveform of temperature front, indicated in the results of Johnson *et al.*, is not apparent in Figure 48a. Actually, the temperature will drop slowly after the carbon burning diminishes; however, in the example the time required for complete regeneration is 30 hours, so the indicated 1.8 hours for temperature breakthrough corresponds only to about 6% regeneration as pointed out by Schulman. The effect of heat losses is shown in Figure 48b for the same parameters as the previous calculation. The actual temperature rise is both smaller in rate and magnitude, and the shape of the front is considerably altered; these are relatively sensitive effects since the total heat loss for the calculated example here is only about 15% of heat generated, yet we see about a 75°F difference in the temperature maximum and a displacement almost by a factor of 2 in the temporal behavior of the thermal wave.

Olson, Luss, and Amundson (88) included in their analysis more detail concerning the solid phase, writing separate heat balances for solid and gas phases and using the Weisz-Goodwin shell progressive model for burning kinetics with a parabolic intraparticle coke concentration profile. Two notable results are obtained from this study. First is the existence of large temperature transients (exceeding the maximum temperature rise of Equation 138a) which occur during the initial stages of bed regeneration. These transients are apparently the result of the initially high rate of reaction and heat generation occurring when the regeneration gas, relatively rich in oxygen at the bed inlet, contacts unregenerated catalyst in which the diffusion limitation characteristic of the shell progressive mechanism has not yet had opportunity to develop. A second result of interest is the close correspondence of the results obtained both for temperature profiles and for the length and concentration profiles of the burning zone to those of Johnson *et al.*, when a stationary state model was used in the calculations.

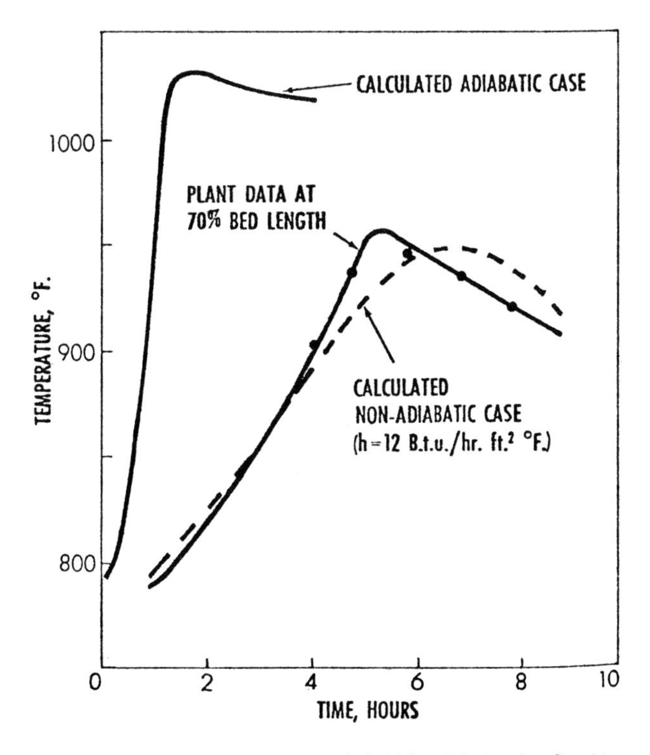

Figure 48b. Effect of heat loss on regeneration (87)

Some detail on these two points is shown in Figure 49. The initial temperature transients are associated with initial development of the combustion zone and are a function of the magnitude of the group $(PD_e/\sqrt{GR})$, where P is total pressure, D_e the effective diffusivity of oxygen within the pellet, R the pellet radius, and G the gas mass flow rate based on void volume. This group essentially expresses an internal to external mass transfer ratio since the external mass transfer coefficient is proportional to $\sqrt{G}$ and can be thought of as a mass Biot number. Some typical computational results are shown in Figure 49a for the parameter values enumerated on the figure. For the conditions of the calculation, each unit of dimensionless time τ is equivalent to about 6.5 sec, so the duration of the transient in the more severe cases is on the order of 10 minutes.

Detailed comparison with the results of Johnson *et al.* is given in Figures 49b and c. The major difference, aside from the fact that Olson *et al.* compute heat balances for both gas and solid phases, resides in the form of the kinetics employed. Thus we have for Johnson *et al.*:

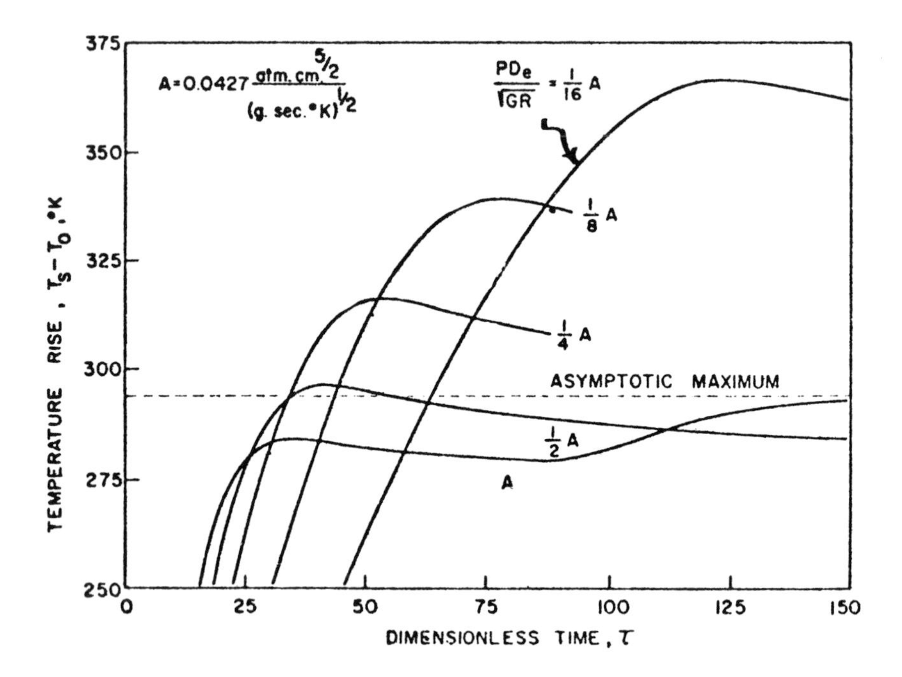

Industrial and Engineering Chemistry

Figure 49a. Initial maximum temperature transients. Diffusion-mass transfer ratio as parameter (88).

Parameters for Figures 49a,b,c:
Total pressure, $P = 3$ *atm.*
Inlet temperature, $T_o = 300°C$
Inlet mole fr. oxygen, $y°_{O_2} = 0.02$
Coke conc. distribution, $C_c = 0.06\left[0.067 + \left(\frac{r}{R}\right)^2\right]$
Bulk density, solid, $\rho_s = 1.10$ *gram/cc.*
Pellet radius, $R = 1/16$ *inch*
Gass. mass velocity, $G = 0.00957$ *gram/sq cm sec*
Oxygen eff. diffusivity, $D_e = 5.55 \cdot 10^{-4}\, T_s^{1/2}$ *sq cm/sec*
Solid heat capcity, $c_s = 0.2132 + 0.0851 \times 10^{-3}\, T_s$ *cal/g °K*
Gas heat capacity, $c_g = 0.2194 + 0.065\ 10^{-3}\, T_g$ *cal/g °K*
Gas mol. wt., $M = 30$ *g/g mole*
Bed porosity, $\epsilon = 0.38$
Stoichiometric coeff., $\alpha = 11.83$ *g coke/g mole* O_2
Heat of combustion, $-\Delta H = 93851 - 0.398\, T_s$ *cal/g mole* O_2
Interphase heat transfer c., $h = 2.023 \cdot 10^{-4} \sqrt{\frac{G}{\epsilon R}}\, T_g^{1/2}$ *cal/sq cm sec °K*
Interphase mass transfer c., $K = 6.855 \cdot 10^{-5} \sqrt{\frac{G}{\epsilon R}}\, T_g^{1/3}$ *g moles/sq cm sec*

$$\tau = \frac{3h(T_o)}{R\rho_s c_s(T_o)}\theta$$

$$r_c = \frac{dC_c}{d\theta} = -kPy_{O_2}\frac{C_c}{C_c^o} \tag{140}$$

and for the shell progressive mechanism used by Olson *et al.*:

$$r_c = \frac{dC_c}{d\theta} = -\frac{3\alpha D_e C_{O_2}}{R^2 \rho_g}\left[\left(\frac{C_c^o}{C_c}\right)^{1/3} - 1\right]^{-1} \tag{157}$$

where the coke concentrations in Equation 140 and those on the RHS of Equation 157 strictly refer to average values. In Equation 157, α is the stoichiometric coefficient, grams coke/gram-mole oxygen. The comparison of temperature profiles between the two models (model C being the stationary-state version of Olson *et al.*) is given on Figure 49b, and of concentration and temperature profiles within the burning zone on Figure 49c. The matching procedure involved aligning space coordinates by

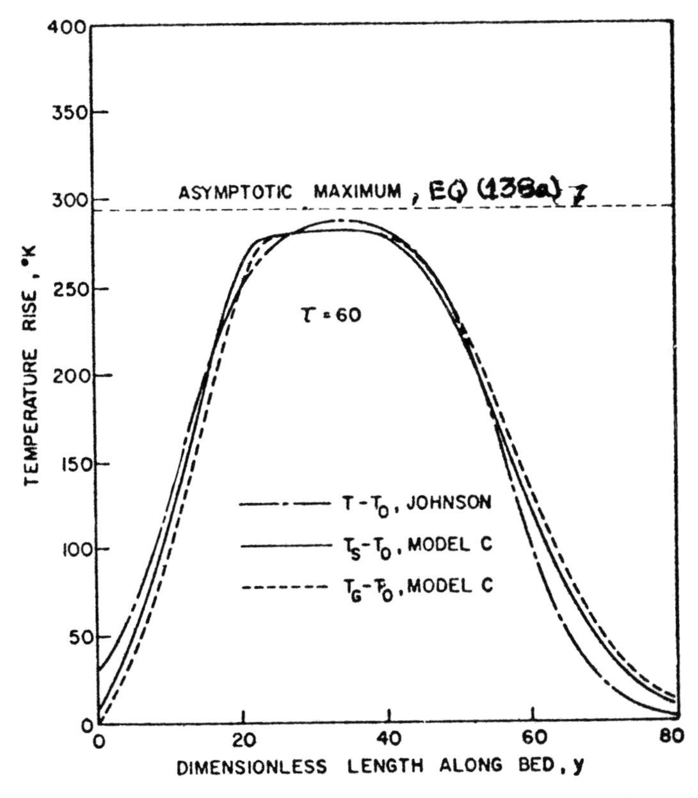

Industrial and Engineering Chemistry

Figure 49b. Temperature profiles at $\tau = 60$. Comparison of Johnson's model with model C (88).

forcing oxygen mole fraction curves to agree at a value of $(y_{O_2}/y_{O_2}{}^o) = 0.5$, and the value of k to be used with the kinetic expression of Johnson *et al.* determined by requiring the distance between $(y_{O_2}/y_{O_2}{}^o)$ of 0.25 and 0.75 be the same for both models.

The figure demonstrates that the results of the two computations are essentially indistinguishable. The major difference seems to be a more diffuse profile of both oxygen concentration and temperature near the leading edge of the burning zone in the Johnson *et al.* calculation—hardly of major importance. It should be noted that the Johnson model does not predict the initial temperature transients noted by Olson *et al.* and inclusion of separate heat balances for solid and gas phases makes little difference in the over-all computed results.

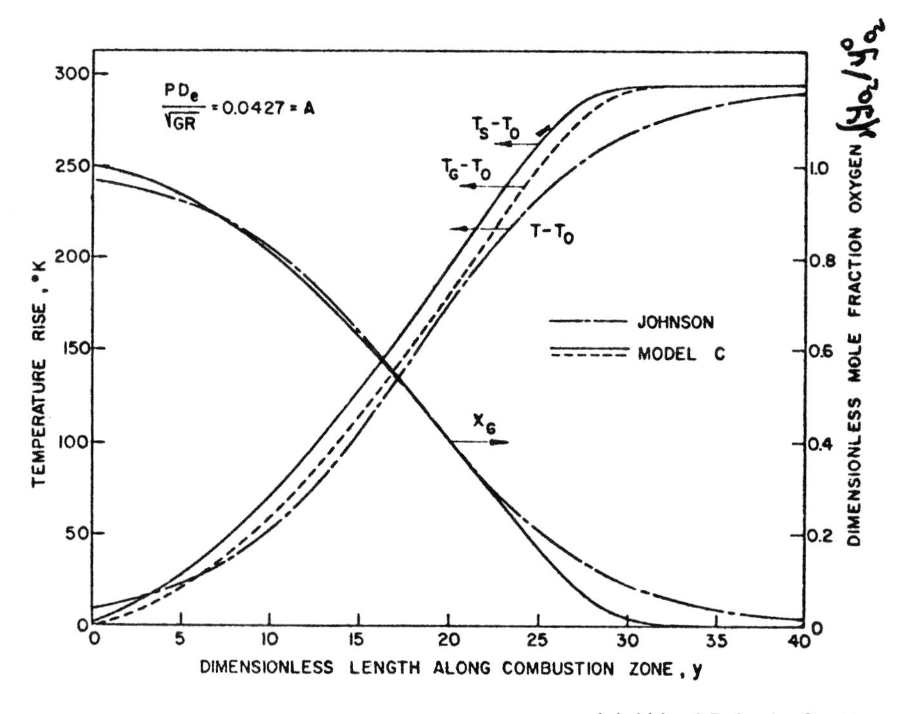

Figure 49c. Stationary state profiles. $\tau > 150$. *Comparison of Johnson's model with model C* (88)

The advent of zeolite catalysts, which are much more sensitive to thermal degradation than conventional silica–alumina, has led to recent interest in the analysis of low temperature regeneration (89). Here one is characteristically concerned with the intrinsic kinetics of coke burning, according to some expression such as Equation 43a, rather than the shell

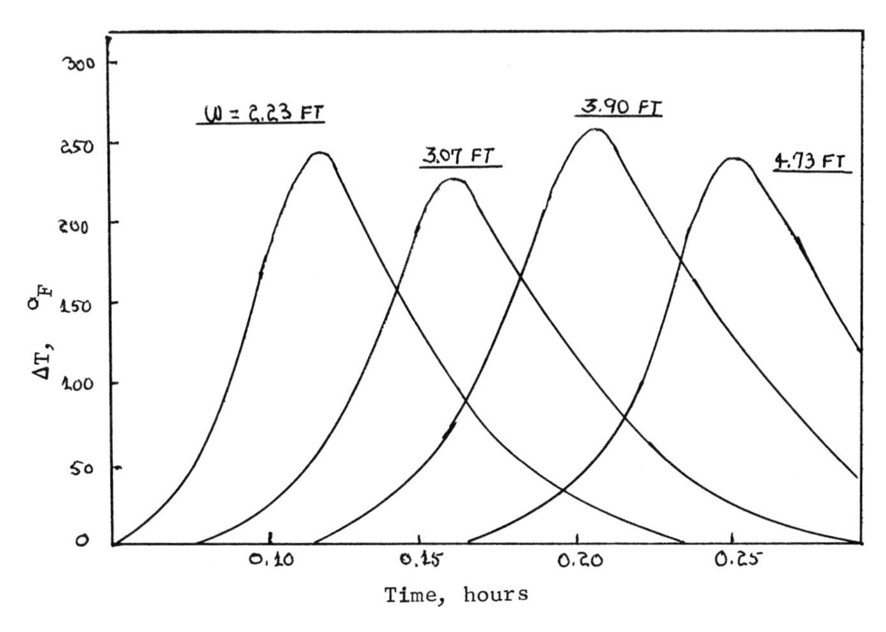

Chemical Engineering Science

Figure 50a. Thermal waves in coked catalyst regeneration (86)

progressive mechanism, and the major difference in the analysis arises from the fact that the reaction is now much more temperature sensitive. The activation energy for coke combustion as determined by Weisz and Goodwin (*48*) is on the order of 65,000 Btu/lb mole whereas diffusion rates are much less temperature sensitive (on the order of 5000 Btu/lb mole in terms of an Arrhenius function). Some considerable differences in reactor behavior are noted between regeneration in the intrinsic and shell progressive regimes. Notably, under conditions of initial operation, as the thermal wave passes through the bed, one can detect the development of a minimum in the coke concentration profile, arising from the opposing effects of temperature (increasing) and oxygen concentration (decreasing) along the bed length on combustion kinetics. This minimum eventually passes out the end of the bed; the computational example of Ozawa suggests, for reasonable values of the operating parameters, that the coke profile becomes monotone at approximately the same time the temperature wave breakthrough occurs. The Legendre transformation technique (*76*) which we have discussed previously, was used to obtain semianalytical solutions for the quasi-steady state period of operation.

A few experimental data are available dealing either directly with fixed bed regeneration or related topics. Comparison with actual plant operating data is given in the results of Schulman illustrated in Figure

48b. In addition, Johnson *et al.* carried out some nonadiabatic regeneration experiments on carbonized catalyst (one assumes silica–alumina, though no details concerning this are given) at a base temperature of 900°F in a 6-ft reactor, 1.05 in. id, with initial carbon levels on the order of 0.02 lb/lb catalyst and inlet oxygen mole fraction about 0.028. An example of the thermal waves measured are shown in Figure 50a; unfortunately not enough details are given concerning the experimentation to permit complete identification of all parameters. Nonetheless, one notes the characteristic development of the thermal wave and its approximate independence of position in the bed. Johnson *et al.* were able to fit these results with a nonadiabatic model, using the burning rate constant and reactor wall heat transfer coefficient as adjustable parameters.

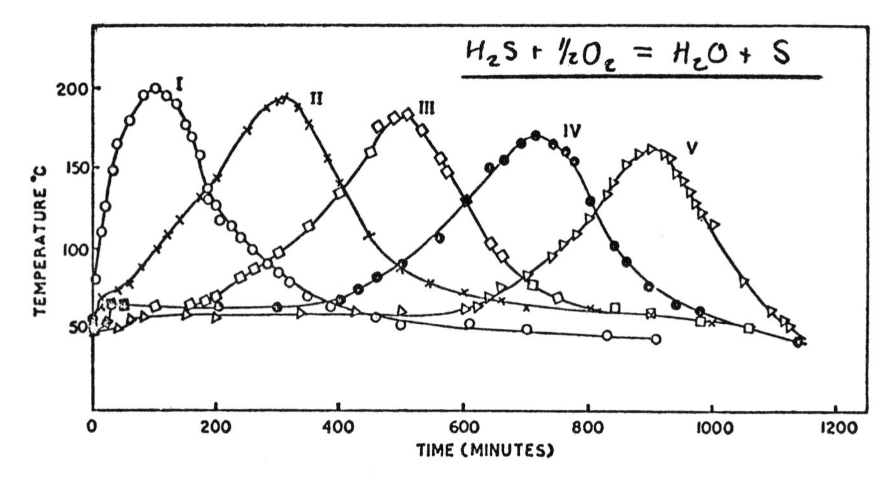

Figure 50b. Temperature profiles shown by the five thermometers for an initial reactor temperature of 50°C. I = 2.25 inches from inlet; 2.25 inch spacing (93).

A second experimental study demonstrating the propagation of thermal waves through the bed is that of Menon and Sreeramamurthy (93); they did not investigate catalyst regeneration but rather the poisoning of an active catalyst, charcoal in the H_2S oxidation by sulfur deposition. This reaction has a high heat of reaction ($\sim$53 kcal/mole), and consequently one can observe the active "front" passing through the bed as the catalyst is deactivated rapidly. The situation is more or less the reverse of the activation processes we have been considering; it corresponds formally to the passage of a reaction rate maxima through the bed, as indicated by the results of Froment and Bischoff shown in Figure 35d. The experimental results of Menon and Sreeramamurthy are shown in Figure 50b, for 13% H_2S and 9.8% O_2 (balance N_2) fed to a 1½-in. id glass reactor containing 2-4 mm particles of charcoal at 50°C. The temperature waves, representing passage of the zone of activity through the bed, are clearly shown in these results.

Optimization of Reactor Performance. As pointed out earlier, when one is concerned with the connections between reactor performance and catalyst deactivation mechanisms, one is generally not far from an optimization problem of some sort. While there is no extensive literature on the subject of optimization and control of reactors subject to catalyst decay, this is an area of considerable current research activity, much of it derived from problems originally proposed and discussed by Jackson (*94, 95*) with regard to optimal temperature control policies for deactivating reactor systems. Some difficulties arise in advancing a comprehensible discussion of this topic within reasonable length since some aspects of the elements of optimization and control theory involved may not be immediately familiar to the hopefully eclectic readers of a review such as this. It may, then, be desirable in some instances to supplement the present text and the original citations with a short review of optimization and/or control theory as available in current texts in the area.

Most optimization studies have considered the problem of maximization of the total amount of reaction in a tubular reactor by manipulating reactor temperature with time of operation (compensating for the decay in catalyst activity), either for a fixed time, fixed final catalyst activity, or free time with specified limits on ranges of conversion or reactor temperature or both. Variations in the exact specifications of the problem result in small variations in the type of analyses presented by various authors and in the specific results obtained. In general, however, the optimal policy for these reactors calls for an increase in temperature with time, initial levels of temperature being determined by the conditions of reactor operation or minimum temperature/conversion constraints, to a point at which a temperature limit is attained and operation continues to final conditions. The period of increasing temperature operation is generally at constant conversion. Such problems have been reported by Chou, Ray, and Aris (*96*), Ogunye and Ray (*97, 98, 99*), Crowe (*100*), and Szépe (*101*). In addition, isothermal batch reactor operation has been studied by Szépe and Levenspiel (*102*) for a fixed reaction time and specified final catalyst activity. Extensions of the basic problem have been given by Ogunye and Ray (*98*) and Paynter (*103*) for multibed reactors, by Ogunye and Ray (*98*) for adiabatic reactors, by Butt (*80*) for a generalized non-isothermal reactor example, and by Dalcorso and Bankoff (*104*) for a bifunctional catalyst.

Comparisons between various reactor types under conditions of catalyst deactivation have been given recently by Weekman (*27*) and, while these are not formal optimization problems, it is convenient to start our detailed discussion in this section with some of his major and fundamental results. The comparisons presented are specifically oriented toward reactor performance in cracking reactions and thus pertain particularly to second-order kinetics, which have been found to correlate cracking rates by a number of workers, and which also hold for data on zeolite catalyst as reported by Weekman. The rate of catalyst deactivation was correlated as first-order in activity—*i.e.*, a simple first-order decrease with time of utilization. Over-all kinetics are given then by:

$$r_A = k_A{}^o x^2 e^{-\lambda\theta} \quad (158)$$

in which τ is the time of reaction, x is the reactant mole fraction, and λ is a deactivation parameter which Weekman correlates as:

$$\lambda = \alpha\theta_c \quad (159)$$

where α is a decay velocity characteristic of the catalyst, and θ_c is the total time of decay. For the fixed bed reactor, the appropriate mass conservation equation to be solved is similar to Equation 73, where r_A is given by Equation 158. A separate equation for catalyst decay, as Equation 74 used by Froment and Bischoff, is not required since the time dependence of catalyst activity is given directly by the exponential correlation of Equations 158 and 159. Then:

$$\frac{dx}{dz} = -Ax^2 e^{-\lambda\theta} \quad (160)$$

where, in Weekman's nomenclature, A represents the ratio of specific reaction velocity to the vapor phase residence time (a measure of the extent of reaction), and $z = w/W$. In terms of measurable quantities:

$$A = \frac{\rho_o}{\rho_l}\left(\frac{k_A{}^o}{S}\right) \quad (161)$$

where ρ_o is initial charge density at reactor conditions, ρ_l the liquid charge density at room temperature, $k_A{}^o$ is defined by the kinetics of Equation 158, and S is the LHSV. With the boundary condition $x(0) = 1$ for all θ, the solution of Equation 160 is:

$$x = \frac{1}{1 + Aze^{-\lambda(\theta/\theta_c)}} \quad (162)$$

The observed conversion, however, will be the time-averaged value of x from Equation 162, evaluated at $z = 1$. Thus:

$$\bar{\epsilon} = 1 - \bar{x} = 1 - \int_0^1 x d(\theta/\theta_c)$$

$$\bar{\epsilon} = \frac{1}{\lambda} \ln \left[\frac{1 + A}{1 + Ae^{-\lambda}} \right] \quad (163)$$

[For an interesting account, somewhat tangential to our purposes at the present moment, of how such time averaging can falsify comparative kinetics and selectivity in reaction systems subject to catalyst decay, the reader is referred to Weekman (*26*).]

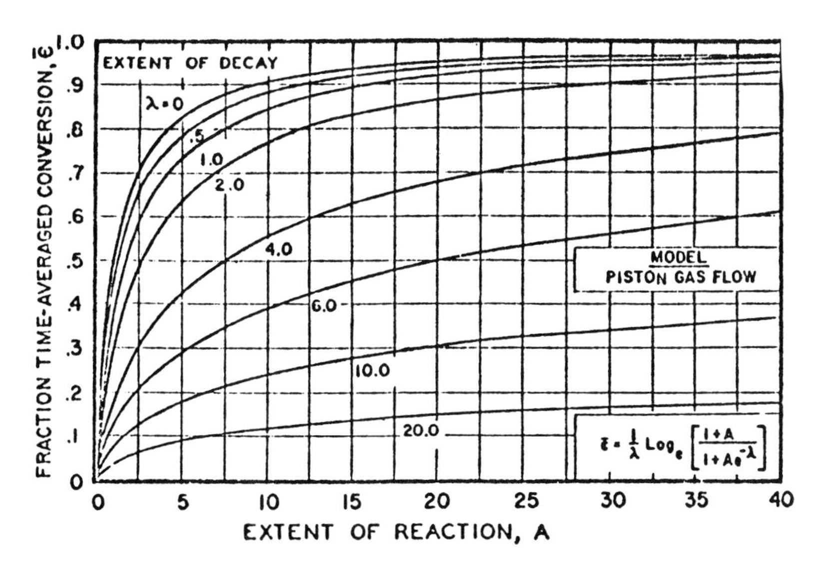

Figure 51a. Time-averaged conversion for fixed beds (27)

For the fixed bed reactor, the time of decay, θ_c, is the total time of catalyst utilization. Some results for the time averaged conversion computed from Equation 163 are shown in Figure 51a.

For moving beds with plug flow for both solid and fluid phases the time of catalyst decay is a function of position in the reactor and is given by the product of the catalyst residence time and the normalized position, z. In this case the reactor equation is:

$$\frac{dx}{dz} = -Ax^2e^{-\lambda z} \quad (164)$$

and the solution for bed-exit conversion is:

$$\epsilon = \frac{A(1 - e^{-\lambda})}{\lambda + A(1 - e^{-\lambda})} \tag{165}$$

Computed results from Equation 165 are given in Figure 51b.

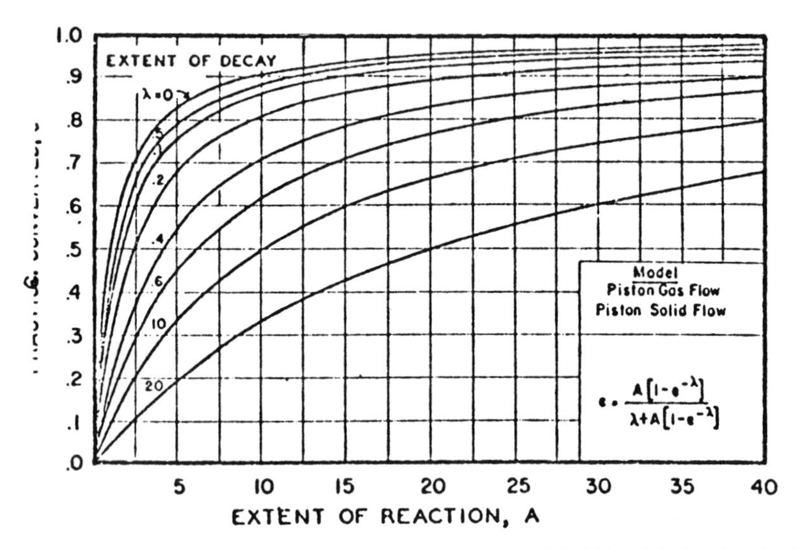

Industrial and Engineering Chemistry

Figure 51b. Conversion in moving beds (27)

Fluid beds present somewhat of a different problem than fixed or moving beds since one must deal with flow patterns for both fluid and solid phases which are much more complex. The particular model used by Weekman assumes plug flow of fluid, but perfect mixing of catalyst within the reactor, which represents a useful limiting case. In this case there is a distribution of catalyst ages within the reactor and hence some average activity level corresponding to the average age of catalyst. Denoting the internal age distribution function by $I(\theta)d\theta$, giving the fraction of total catalyst with ages between θ and $\theta + d\theta$, the average activity is defined by:

$$\bar{k}_A = k_A{}^o \int_0^\infty e^{-\lambda\theta} I(\theta)\, d\theta \tag{166}$$

For perfect mixing, $I(\theta)$ is given by $e^{-\theta}$, so:

$$\bar{k}_A = k_A{}^o/(1 + \lambda) \tag{167}$$

The reactor equation is now:

$$\frac{dx}{dz} = -\frac{A}{(1+\lambda)}x^2 \tag{168}$$

and

$$\epsilon = \frac{A}{1+\lambda+A} \tag{169}$$

The fluid bed conversions computed from Equation 169 are shown in Figure 51c. The relative performances of the three reactor systems are shown in Figures 51d-f. From Figure 51d it would appear that under any conditions of catalyst decay, moving beds are preferable to fixed beds in terms of conversion attained—a reasonable result in view of the different catalyst residence times in the two types of reactors (but one which assumes that the moving bed catalyst is always completely regenerated before reuse). Similarly, for moving beds and fluid beds, at least when the rate of decay is slow, the former gives larger conversions; however it can be seen in Figure 51e that there is little difference between the two when either the reaction rate parameter, A, or the deactivation parameter, λ, is large. The comparison between fixed and fluid beds, Figure 51f, is a bit more complicated; for low A and λ the fixed bed is preferable although for finite values of A this advantage holds only for a relatively narrow range of λ.

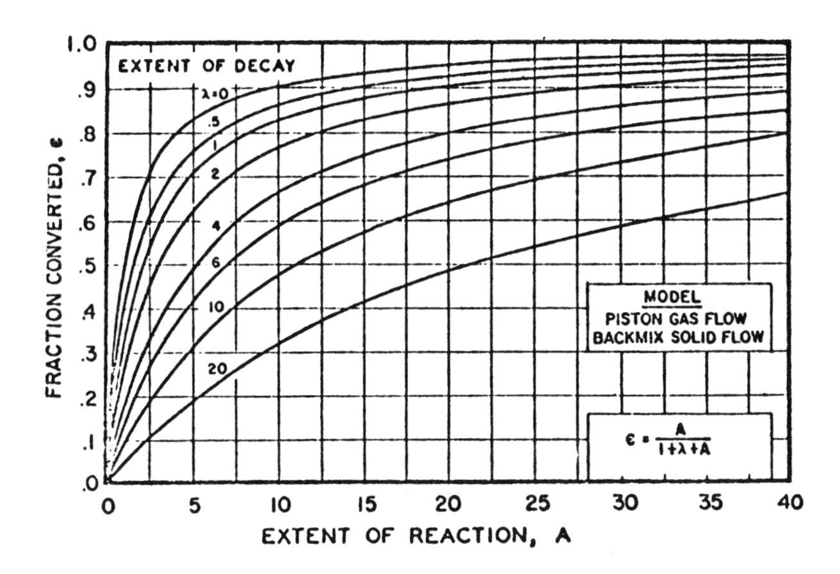

Industrial and Engineering Chemistry

Figure 51c. Conversion in fluid beds (27)

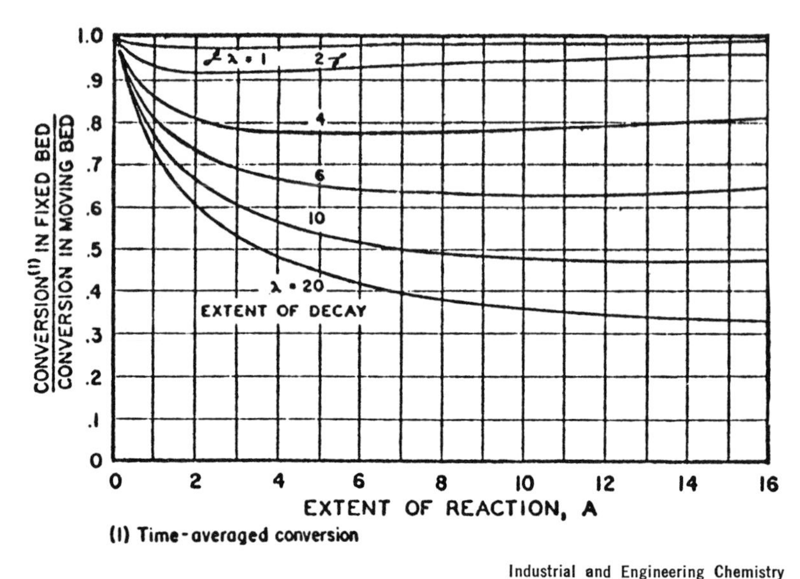

Industrial and Engineering Chemistry

Figure 51d. Ratio of fixed bed to moving bed conversion (27)

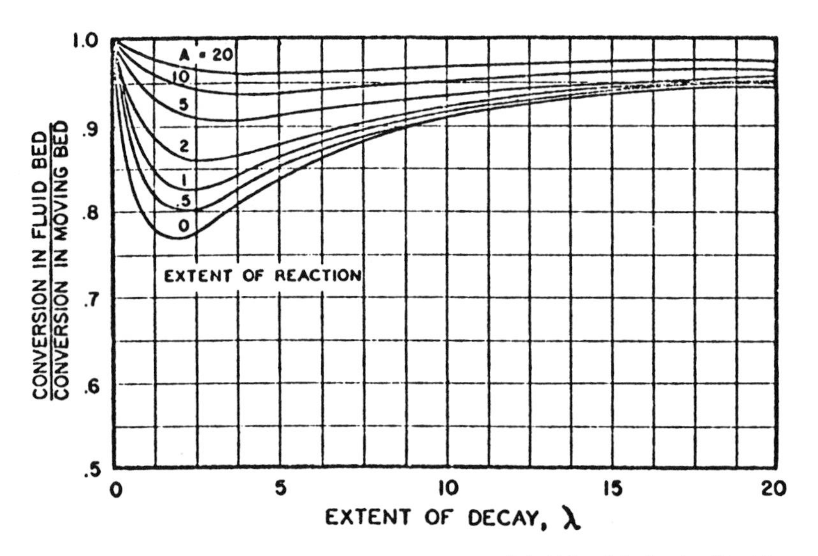

Industrial and Engineering Chemistry

Figure 51e. Ratio of fluid-bed to moving-bed conversion (27)

While in this and previous sections we are primarily concerned with fixed bed reactors, the comparison of types given by Weekman is valuable in telling us not that catalyst deactivation affects reactor performance (which we knew anyway) but in demonstrating that variations in reactor behavior with deactivation are not uniform. As a result, it is conceivable that considerations pertaining to reactor operation under the influence of poisoning could indicate the selection of reactor type and should, at the very least, always play a prominent part in preliminary process design and analysis. A further comparison of reactor types, dealing with cocurrent and countercurrent moving beds also been given by Szépe (*105*).

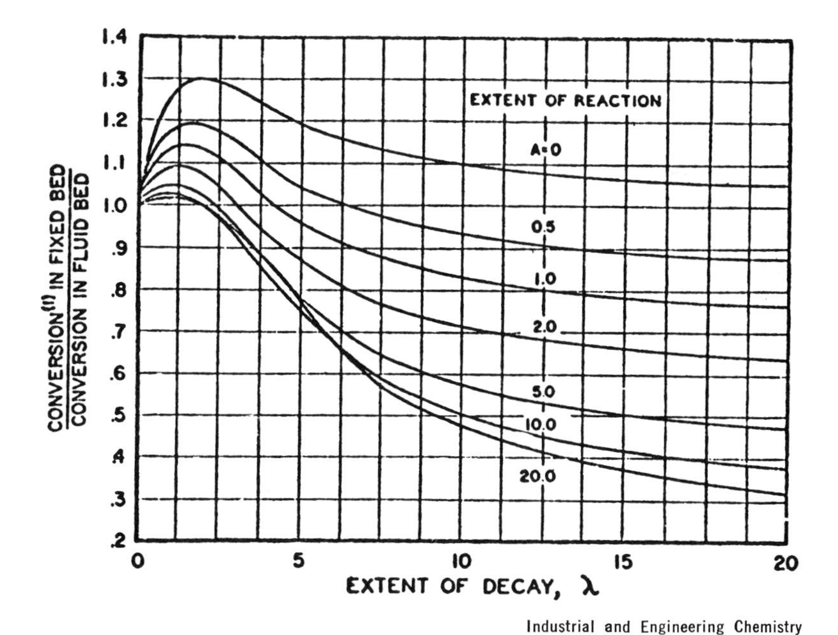

Industrial and Engineering Chemistry

Figure 51f. Ratio of fixed-bed to fluid-bed conversion (27)

More detailed optimization problems associated with reactors subject to catalyst deactivation seem to fall into two categories, (a) those dealing with design variables such as catalyst composition, distribution, reactor type and size, and (b) those dealing with control variables such as temperature *vs.* time of operation, conversion level, and selectivity limits. An over-all optimization would then involve aspects of both of these areas, and has been termed a design-control optimization by Paynter (*103*). We shall not attempt to discuss all the studies cited previously in the section since many of them are somewhat related. The studies of batch reactors by Szépe and Levenspiel (*102*) and of tubular reactors by Chou *et al.*

(*96*), both using the calculus of variations in analysis and both dealing with the control problem—stated as the temperature policy satisfying the stated objective function—are a good place to begin. In both cases the conversion-independent form of deactivation kinetics is used.

The objective function stated by Szépe and Levenspiel was to maximize the final conversion for a fixed time of reaction and specified final

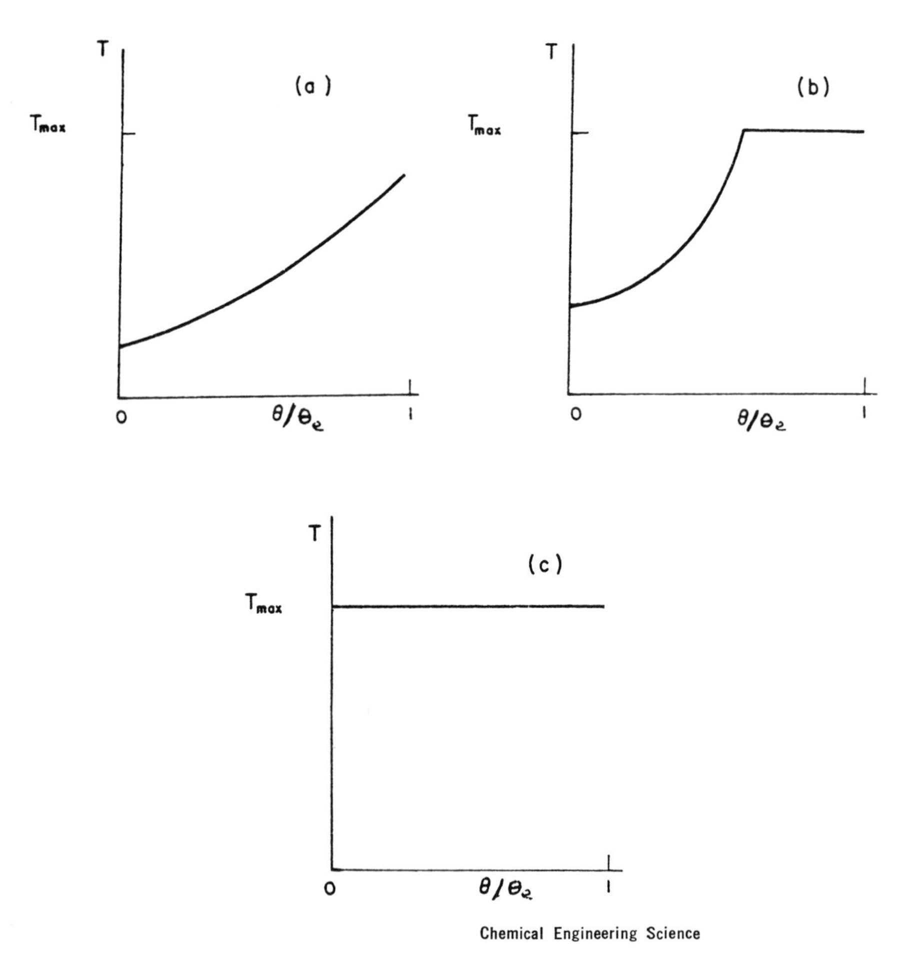

Figure 52. The structure of the optimal temperature policy in a batch reactor when an upper temperature limit is specified (102)

(a): For $E_{A,d} > E_A$ *(Equation 172b)*
(b): For $E_{A,d} > E_A$ *(Case I)*
(c): For $E_{A,d} > E_A$ *(Case II) or* $E_A > E_{A,d}$

catalyst activity by temperature variation with time of operation, subject to upper and lower bounds on allowable temperature. The optimal condition derived for a single irreversible reaction A → B conducted in an ideal batch reactor was:

$$\frac{d(k'_A)}{dt} - \frac{d(k'_{A_o}s)}{dt} = 0 \tag{170}$$

where the nomenclature corresponds to that of Equation 79, with s as the catalyst activity variable. This condition, that of constant activity, is one for which results were obtained by Butt and Rohan (77) for the case of a tubular reactor. The constant activity policy of Equation 170 is necessary but not sufficient, however; it was shown that an optimal result is obtained only if the activation energy of the deactivation reaction is greater than that of the main reaction. If the main reaction activation energy is greater than that of the deactivation reaction, that is that high temperature preferentially promotes the main reaction, it is obvious that continuous operation at the highest temperature permissible will give the maximum yield and the condition of Equation 170 is not optimal. For equal activation energies, all policies are the same.

An analytical solution for reactor temperature as a function of time can be obtained when $E_A < E_{A,d}$ from Equation 170 as:

$$\tau = \left(\frac{E_{A,d}}{R}\right) \ln \left(\frac{E_{A,d}}{E_A} \cdot k'_{A,d_o} \theta_e \left\{\left[1 - \left(\frac{s_e}{s_o}\right)^{\frac{E_{A,d}}{E_A}}\right]^{-1} - \theta\right\}\right) \tag{171}$$

in which $E_{A,d}$ is the activation energy of the deactivation reaction, k'_{A,d_o} is the preexponential factor for the rate constant of the deactivation reaction, θ_e and s_e are the fixed time of operation and final activity level, respectively. This policy calls for a monotonically increasing temperature with time of operation (Figure 52a). Since the argument of the logarithmic function must be positive, the following constraint applies to the parameters of the problem:

$$\frac{s_e}{s_o} \geqslant \left[\frac{1}{1 + (E_{A,d}/E_A) k'_{A,d_o} \theta_e}\right]^{\frac{E_A}{E_{A,d}}} \tag{172a}$$

or, if a temperature maximum is specified:

$$\frac{s_e}{s_o} \geqslant \left[\frac{1}{1 + (E_{A,d}/E_A) k'_{A,d_o} \theta_e \exp(-E_{A,d}/RT_{\max})}\right]^{\frac{E_A}{E_{A,d}}} \tag{172b}$$

If these constraints are not satisfied, then optimal policy will consist of either of the following two cases:

(a)

$$\exp[-k'_{A,d_o}\theta_e \exp(-E_{A,d}/RT_{\max})] \leqslant s_e/s_o \leqslant$$

$$[1 + (E_{A,d}/E_A)k'_{A,d_o}\theta_e \exp(-E_{A,d}/RT_{\max})]^{-E_A/E_{A,d}}$$

optimal policy is a rising temperature subarc followed by operation at $T_{\max}$, as shown on Figure 52b

(b)

$$\frac{s_e^*}{s_o} \leqslant \exp[-k'_{A,d_o}\theta_e \exp(-E_{A,d}/RT_{\max})]$$

optimal policy is to operate at maximum allowable temperature throughout process cycle. The term s_e^* differs from s_e since constant temperature operation will not permit the actual attainment of specified final cycle times and activity levels simultaneously.

A further refinement can involve the inclusion of a lower temperature limit as well; details of this are discussed by Szépe and Levenspiel. The type of optimal control policy is determined wholly by the relative magnitudes of the activation energies for main and deactivation reactions. Similar strong dependence on these parameters was noted by Butt and Rohan (*77*) in their studies of the constant conversion, temperature variable tubular reactor. It is reasonable to assert that when temperature is used as a control variable, the reaction and deactivation activation energies are by far the most important system parameters in determining control policy and reactor behavior. The dependence of the optimal policy on the form of kinetics employed is also significant; Lee and Crowe (*106*) have shown that the constant activity policy derived by Szépe and Levenspiel is optimal only when deactivation kinetics are independent of conversion.

The general formulation of the optimal control policy for tubular reactors was given by Chou *et al.* (*96*) as a variational calculus problem. For the normal plug flow reactor

$$\frac{\partial \xi}{\partial \theta} + \mathrm{v}\,\frac{\partial \xi}{\partial \mathrm{w}} = r_{\mathrm{A}} \quad (173)$$

in which ξ is the extent of reaction (conversion in the normal sense), v is the superficial linear velocity, and r_{A}, the rate of the main reaction is in general a function of ξ, the temperature, and catalyst activity s. For the activity variation:

$$\frac{\partial s}{\partial \theta} = r_{\mathrm{A},d}(\xi,s,T) \quad (174)$$

and for boundary and initial conditions:

$$\xi(0,\theta) = 0; \xi(w,0) = 0$$
$$s(w,0) = 1 \tag{175}$$

The objective is to maximize the total yield of the reactor:

$$Y = \int_0^{\theta_e} \xi(W,\theta)\,d\theta \tag{176}$$

for specified θ_e by temperature profile variation $(T(w,\theta))$. Again, one might expect constraints such as temperature bounds, $T_{min} \leqslant T \leqslant T_{max}$, and lower conversion limits, $\xi_W \geqslant \xi_{min}$.

The simplest problem studied by Chou *et al.*, and the one we shall describe here, considered the reactor to be isothermal, and the rate of catalyst decay to depend on temperature but not on ξ (*i.e.*, as suggested by the rate forms of Table X). Both temperature and conversion bounds are included, and it is assumed that the catalyst decay rate is slow so that $\left(\frac{\partial \xi}{\partial \theta}\right)$ in the conservation equation is once again removed from the problem. The maximization of Equation 176 is shown to imply a constant conversion, which in turn yields the condition:

$$\left(\frac{W}{v}\right) sk_A{}^o \exp(-E_A/RT) = \text{constant} = \sigma = \ln\left(\frac{1}{1-\xi_W}\right) \tag{177}$$

for the case of A → B with deactivation kinetics given as:

$$\frac{\partial s}{\partial \theta} = -k^o{}_{A,d} \exp(-E_{A,d}/RT)s \tag{178}$$

Equation 177 can be rewritten:

$$-\left(\frac{E_{A,d}}{E_A}\right)(s)^{E_{A,d}/E_A - 1}\left(\frac{ds}{d\theta}\right) = \frac{k^o{}_{A,d}E_{A,d}}{E_A}\left[\frac{\sigma}{k^o{}_A \frac{W}{v}}\right]^{\frac{E_{A,d}}{E_A}} \tag{179}$$

This can be solved for $s(\theta)$:

$$s(\theta) = [1 - \beta\theta]^{E_A/E_{A,d}} \tag{180}$$

where β is the RHS of Equation 179. Solving Equation 178 for $s(\theta)$ and writing in the temperature dependence explicitly on substitution in Equation 180 we can compute the optimal temperature history directly:

$$T(\theta) = \left(\frac{E_{A,d}}{R}\right) \ln \left[k^o{}_{A,d}(1 - \beta\theta)/(\beta E_{A,d}/E_A)\right] \tag{181}$$

The initial temperature of operation from Equation 181 at $\theta = 0$ is:

$$T(0) = \left(\frac{E_{A,d}}{R}\right) \ln \left[k^o{}_{A,d} \left(\frac{E_{A,d}}{E_A}\right)\right] \tag{181a}$$

Combining Equations 181 and 181a:

$$\frac{E_{A,d}}{RT(0)} - \frac{E_{A,d}}{RT(\theta)} = \ln \left(\frac{1}{1 - \beta\theta}\right) \tag{182}$$

and from Equation 182 we can compute the time of operation associated with optimal temperature control up to some allowable limit on reactor temperature:

$$\theta_c = \frac{1}{\beta}\left\{ 1 - \exp\left[- \frac{E}{R}\left(\frac{1}{T(0)} - \frac{1}{T_{max}}\right)\right]\right\} \tag{183}$$

The total conversion associated with this operation is:

$$Y = \xi_W \theta_c \tag{184}$$

If we define $\xi_W{}^*$ as the maximum extent of reaction obtainable with fresh catalyst at the maximum temperature of operation, T_{max}:

$$\xi_W{}^* = 1 - \exp\left(-k_A{}^* \cdot \frac{W}{v}\right) \tag{185}$$

the value of the optimal conversion level to choose—*i.e.*, $\xi_W(\text{opt})$ for maximizing yield, is given in Figure 53a, showing how this conversion depends on activation energy ratio and $\xi_W{}^*$ (*i.e.*, T_{max}). Figure 53b gives corresponding values of the yield-related function η:

$$\eta = k^*{}_{A,d} Y \tag{186}$$

A cross plot of Figures 53a and b may be used to compute estimates of optimal reactor size, which is a variable involved in the definition of $\xi_W{}^*$. In this case one must balance increased reactor size and cost against increased productivity; an example of this relationship between conversion level, temperature limit and productivity is shown in Figure 53c for specified $(E_A/E_{A,d})$.

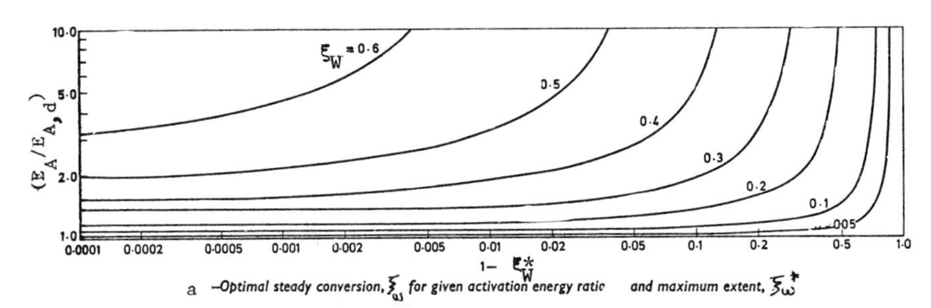

Figure 53a. Optimal steady conversion, ξ_W for given activ tion energy ratio and maximum extent, ξ_W (96)*

The analysis of Chou *et al.* is extended to isothermal reactors with arbitrary residence time distribution, in the limit of perfect mixing reducing to the problem considered by Szépe and Levenspiel, and to a non-isothermal reactor *via* the device of splitting the reactor into a number of segments, each isothermal but at differing temperature levels.

Further detailed studies of the optimization of reactors undergoing catalyst decay have primarily involved numerical analyses based on the maximum principle following, generally, the methods outlined by Volin and Ostrovskii (*107, 108, 109*). The work of Ogunye and Ray (*98*) will serve as our principal example of such investigations. The basic equations and assumptions used by Chou *et al.* are employed (latitude for

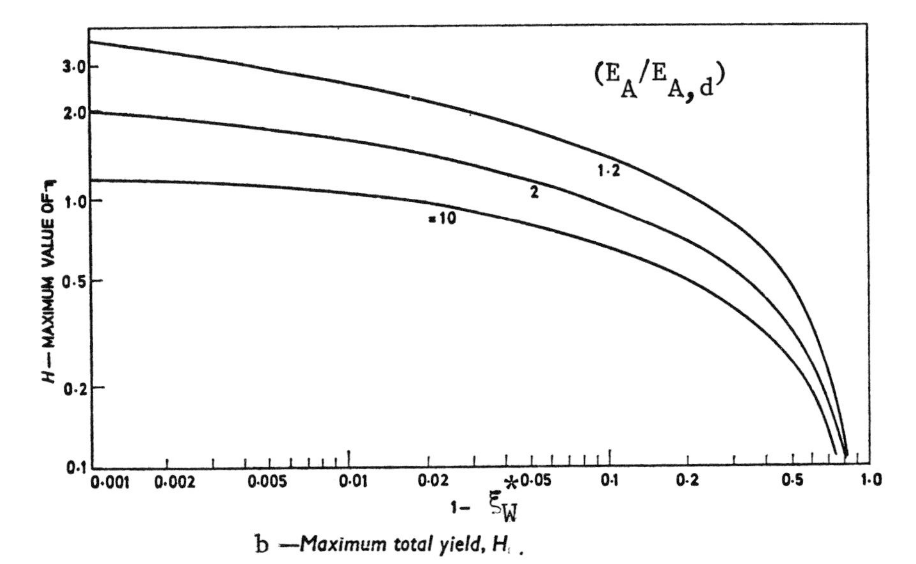

Figure 53b. Maximum total yield, H (96)

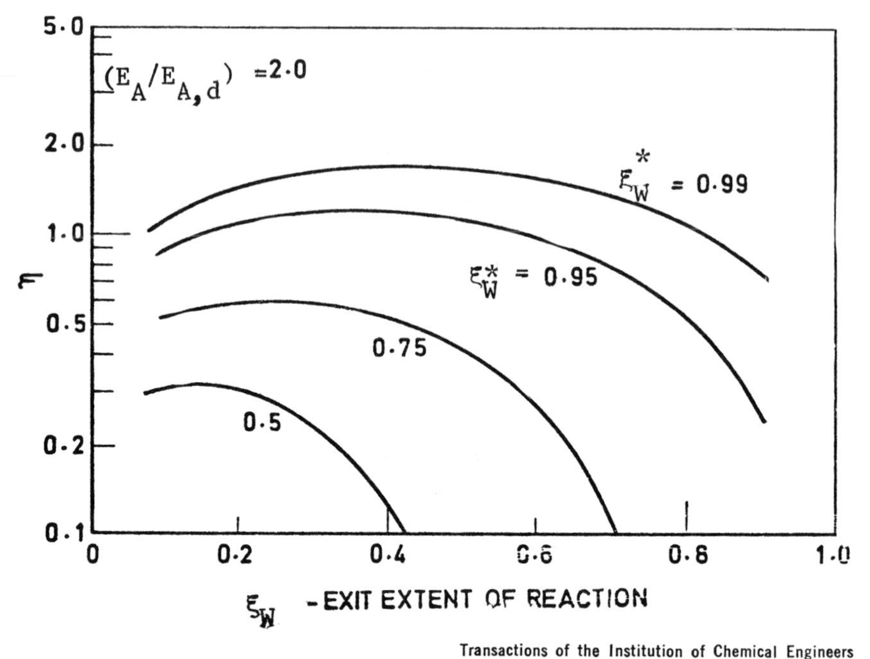

Figure 53c. Total yield as a function of ξ_W (96)

additional catalyst functions, reactions, and pressure variation, is provided in the original paper but are not derived in detail for the present discussion), as well as a separate heat balance allowing for wall heat transfer:

$$v\,\frac{\partial T}{\partial w} = r_T(\xi,s,T,\mathbf{u}) \tag{187}$$

with $T(0,\theta) = T_o$ and r_T a total rate function, $\mathbf{u}$ some control action affecting temperature level. The set of conservation equations may be generalized as:

$$\frac{\partial x_i}{\partial z} = f_i(\mathbf{x},\mathbf{y},\mathbf{u}) \qquad x_i(0,\tau) = v_i(\tau) \tag{188}$$

$$0 \leqslant z \leqslant 1, 0 \leqslant \tau \leqslant 1$$

where x_i is a state variable such as conversion or temperature, τ is a normalized time of operation (θ/θ_e), z the normalized position within the reactor and $\mathbf{x},\mathbf{y},\mathbf{u}$ state and control variable vectors associated with extent of reaction, catalyst activity and temperature control, respectively. The deactivation equation, kinetics as per the form of Table X, is generalized as:

$$\frac{\partial s_j}{\partial \tau} = \frac{\partial y_j}{\partial \tau} = g_j(\mathbf{x},\mathbf{y},\mathbf{u}) \qquad y_j(z,0) = w_j(z)$$
$$0 \leqslant z \leqslant 1, 0 \leqslant \tau \leqslant 1 \tag{189}$$

The initial condition on y allows for a distribution of catalyst activity initially along the reactor length, and optimization with regard to this condition would constitute the design portion of the design-control optimization mentioned earlier.

For a fixed operating time, θ_e, then the objective of the optimization can be written generally as:

$$I = \int_0^1 \int_0^1 G(\mathbf{x},\mathbf{y},\mathbf{u},\mathbf{v},\mathbf{w})\, dz d\tau \tag{190}$$

where G is a function to be specified according to the needs of the problem at hand. The assumption of fixed time, θ_e, does not overly restrict the problem considered since only a univariate search would be required to compute the optimal cycle time. The general optimization problem, then, is maximization of I, subject to the constraints of Equations 188 and 189. Constraints on the control variables may also be encountered—*i.e.*:

$$\mathbf{v}_{\min} \leqslant \mathbf{v} \leqslant \mathbf{v}_{\max};\ \mathbf{u}_{\min} \leqslant \mathbf{u} \leqslant \mathbf{u}_{\max};\ \mathbf{w}_{\min} \leqslant \mathbf{w} \leqslant \mathbf{w}_{\max} \tag{191}$$

The conditions required for optimality are as follows:

(I) For the controls $\mathbf{u}$:

(a) $u_k = u_k(z,\tau)$ and unconstrained

$$\frac{\partial H}{\partial u_k} = \frac{\partial G}{\partial u_k} + \lambda_i \frac{\partial f_i}{\partial u_k} + u_j \frac{\partial g_j}{\partial u_k} \tag{192}$$

should vanish on $0 \leqslant z \leqslant 1, 0 \leqslant \tau \leqslant 1$

(b) $u_k = u_k(z,\tau)$ and constrained

$$H = G + \lambda_i f_i + \mu_j g_j \tag{193}$$

should be a maximum with respect u_k

(c) $u_k = u_k(z)$ and unconstrained

$$\int_0^1 \frac{\partial H}{\partial u_k}\, d\tau = 0 \tag{194}$$

(d) $u_k = u_k(z)$ and constrained

$$\int_0^1 H dt = \text{maximum with respect to } u_k \tag{195}$$

(e) $u_k = u_k(\tau)$ and unconstrained

$$\int_0^1 \frac{\partial H}{\partial u_k}\, dz = 0 \tag{196}$$

(f) $u_k = u_k(\tau)$ and constrained

$$\int_0^1 H dz = \text{maximum with respect to } u_k \tag{197}$$

(II) For the controls or operating variables $\mathbf{v}$:

(a) $v_i = v_i(\tau)$ and unconstrained

$$\frac{\partial H_1}{\partial v_i} = \int_0^1 \frac{\partial G}{\partial v_i}\, dz + \lambda_i(0,\tau) = 0 \tag{198}$$

(b) $v_i = v_i(\tau)$ and constrained

$$H_1 = \int_0^1 G dz + \lambda_j(0,\tau) v_j = \text{maximum} \tag{199}$$

with respect to v_i

(c) v_i independent of τ and unconstrained

$$\int_0^1 \frac{\partial H_1}{\partial v_i}\, d\tau = 0 \tag{200}$$

(d) v_i independent of τ and constrained

$$\int_0^1 H_1 d\tau = \text{maximum with respect to } v_i \tag{201}$$

(III) For the design variables $\mathbf{w}$:

(a) $w_j = w_j(z)$ and unconstrained

$$\frac{\partial H_2}{\partial w_j} = \int_0^1 \frac{\partial G}{\partial w_j} d\tau + \mu_j(z,0) = 0 \quad (202)$$

(b) $w_j = w_j(z)$ and constrained

$$H_2 = \int_0^1 G d\tau + \mu_i(z,0) w_i(z) = \text{maximum with respect to } w_j \quad (203)$$

(c) w_j independent of z and unconstrained

$$\int_0^1 \frac{\partial H_2}{\partial w_j} dz = 0 \quad (204)$$

(d) w_j independent of z and constrained

$$\int_0^1 H_2 dz = \text{maximum with respect to } w_j \quad (205)$$

The adjoint variables μ_j and λ_i involved in Equations 192–205 are given by:

$$\frac{\partial \lambda_i}{\partial z} = -\left[\frac{\partial G}{\partial x_i} + \lambda_k \frac{\partial f_k}{\partial x_i} + \mu_j \frac{\partial g_j}{\partial x_i}\right] \quad (206)$$

$$\lambda_i(1,\tau) = 0, i = 1,2 \ldots r$$

$$\frac{\partial \mu_j}{\partial \tau} = -\left[\frac{\partial G}{\partial y_j} + \lambda_i \frac{\partial f_i}{\partial y_j} + \mu_k \frac{\partial g_k}{\partial y_j}\right] \quad (207)$$

$$\mu_j(z,1) = 0, j = 1,2 \ldots q$$

where $\mathbf{x}$ and $\mathbf{y}$ are not fixed at $z = \tau = 1$. The method of numerical solution recommended by Ogunye and Ray involves iteration on the control vector, using a gradient technique. The sequence of steps involved in the algorithm is as follows:

(a) Assume values for $\mathbf{u},\mathbf{v},\mathbf{w}$
(b) Solve conservation equations such as 188 and 189
(c) Solve adjoint equations (backwards in z and τ) 206 and 207.

The results of (b) and (c) give the current value of I.

(d) Compute a correction to the control **u,v,w** from values of H, H_1, and H_2 according to:

$$\delta u_k = \epsilon \frac{\partial H}{\partial u_k}$$

$$\delta v_i = \epsilon \frac{\partial H_1}{\partial v_i} \tag{208}$$

$$\delta w_j = \epsilon \frac{\partial H_2}{\partial w_j}$$

where ϵ is a scalar constant and $u_k = u_k(z,\tau)$, $v_i = v_i(\tau)$, $w_j = w_j(z)$.

(e) Solve conservation equations and adjoint equations to obtain new value of I.

(f) Depending on whether new I is improvement or not, either increase or decrease control increments of Equation 208 and repeat calculations.

The gradient technique used is said by Ogunye and Ray to perform very well with no significant convergence problems. Convergence, however, does seem to be a difficulty in some of these cases (*94, 103*), and alternatives to gradient methods may well be desirable in some instances.

For an example of the application of the techniques just described, consider the simplest illustration of Ogunye and Ray, the isothermal reactor with reaction A → B → C, single function catalyst, in which temperature *vs.* time of operation is used to maximize the production of B subject to a temperature maximum constraint on the reactor. The catalyst deactivation kinetics are:

$$\frac{\partial y_1}{\partial \tau} = \frac{\partial s}{\partial \tau} = \theta k^o{}_{A,d} \exp\left(-\frac{E_{A,d}}{RT}\right) \cdot s^2 = g_1 \tag{209}$$

and $s(z,0) = 1$. The equations for the extent of reaction are:

$$\frac{\partial x_1}{\partial z} = \frac{\partial \xi_A}{\partial z} = s\left(\frac{W}{v}\right) k^o{}_A \exp\left(-\frac{E_A}{RT}\right)(1 - \xi_A) = f_1 \tag{210}$$

$$\frac{\partial x_2}{\partial z} = \frac{\partial \xi_B}{\partial z} = s\left(\frac{W}{v}\right) k^o{}_B \exp\left(-\frac{E_B}{RT}\right)(R_o + \xi_A - \xi_B) = f_2 \tag{211}$$

where R_o = ratio of intermediate to reactant concentration in reactor feed. The control policy, temperature, is involved in Equations 209–211 in the exponential term. If we define:

$$u_1 = \frac{RT}{E_A} \tag{212}$$

and constrain the temperature such that $u_1 \leqslant u_1^*$, then the objective function for maximum production of B is:

$$I = \int_0^1 \int_0^1 (f_1 - f_2)\, dz\, d\tau \tag{213}$$

In this example, $u_1 = u_1(T)$, corresponding to the condition of Equation 197. Control modification proceeds in the direction:

$$\delta u_1 = \epsilon \int_0^1 \frac{\partial H}{\partial u_1}\, dz \tag{214}$$

at each step of the computation cycle. Some of the results obtained for this case are shown in Figure 54. The optimal temperature control involves increasing temperature with time until the constraint is encountered. The initial operation (before the temperature constraint is encountered) period proceeds at approximately constant total conversion, thus giving for the more complex sequential reaction scheme approximately the same optimal behavior as that for the simple irreversible A → B scheme: a period of constant conversion operation followed by operation on the temperature constraint. This is for a fixed time of operation, and an over-all optimization involves a search in θ_c. Obviously from the nature of the computations required the form of main reaction or deactivation kinetics constitutes no limitation on the technique, and because of the generality of the mathematical formulation a variety of problems can be solved by this method. Ogunye and Ray demonstrate extensions of the analysis to adiabatic reactors, various main reaction schemes in isothermal and adiabatic reactors, the design-control problem involving catalyst composition distributions ($s(z,0) = s(z)$) for Equation 209, multiple bed reactors involving catalyst and feed distribution optimization, and to a detailed model of the vinyl chloride monomer reactor.

Multibed reactors have also been the subject of a recent study by Paynter (*105*) who developed an optimization calculation capable of handling very general forms of objective functions. The calculation is demonstrated for a simple reversible reaction conducted in a sequence of four adiabatic reactors. The value sought is the temperature entering each catalyst bed which will maintain a desired conversion with a minimum amount of catalyst over specified catalyst life. Further, the optimal distribution of catalyst among the four beds is to be determined. This is a bit more complicated than the problems we have considered so far, particularly in the form of the objective function of the optimization, which consists of a penalty function to be minimized (reflecting the deviation of actual conversion from that specified) and a cost related function also to be minimized (reflecting the cost of the total amount of catalyst in the four beds). Details of the algorithm and the computational example are given in the original paper. It is claimed that the generality of the objective function allowed permits a variety of design-control problems to be solved (witness the computational example), and the method perhaps will handle a greater variety of problems than any other currently available. Generality, however, appears always to extract its own price from admirers; in the present instance the toll is that of computer time. The example solved by Paynter required approximately 45 minutes on an IBM 360/65.

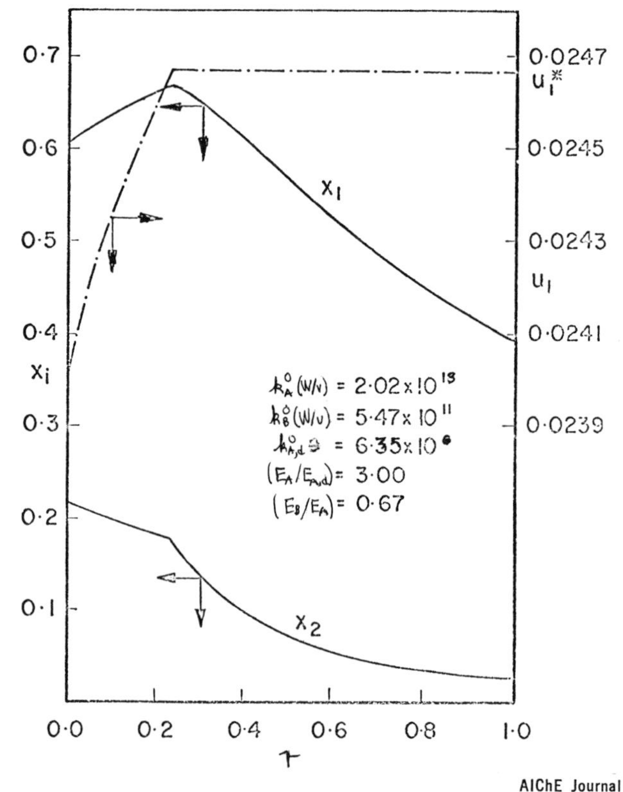

Figure 54. The isothermal reactor for A = B = C. Increase in yield is compared with entire operation at u_1* *is 2.4% (98).*

Reactor bed volume vs. iteration number

Iteration	Bed volume			
	Bed no. 1	Bed no.2	Bed no.3	Bed no.4
Initial	1 ft^3	1 ft^3	1 ft^3	1 ft^3
5	1·22 ft^3	1·23 ft^3	1·22 ft^3	1·22 ft^3
FINAL	1·31 ft^3	1·35 ft^3	1·35 ft^3	1·36 ft^3

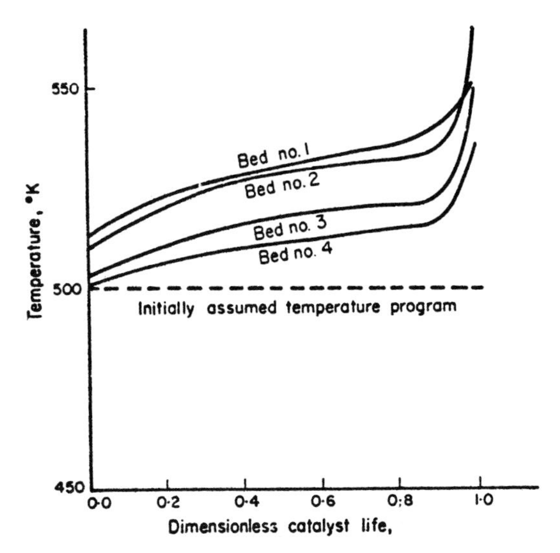

Chemical Engineering Science

Figure 55. Optimal temperature program for example problem (103). Parameters of the example problem: Reaction A = B; specified exit conversion of A = 0.75. Weighting factor (relative to unity) given to constant conversion portion of objective function; $k_A = 6900 \exp(-6000/T)$; $k_B = 3 \times 10^7 \exp(-10000/T)$; $(-\Delta H/\rho C_p)$ *for reaction mixture = 1000; base temperature = 1250°K.*

A final inquiry associated with these optimization problems is suggested in Figure 55, which sets forth the results obtained for the computational example of Paynter. The phenomena of major interest are the temperature histories of the individual beds and the temperature "profile" through the sequence of four beds at any given time of operation θ/θ_e. The latter are nonlinear, and their shape is time variant—very similar behavior (although decreasing rather than increasing with total reactor length) to that illustrated in Figure 41c required to implement the constant activity policy in the constant conversion, tubular reactor. The point is that such temperature policies, called for by the results of the optimization, would be in practice difficult to obtain since the nature of the temperature distribution and its absolute level vary with time of operation. It is thus proper to determine whether simpler temperature policies might give satisfactory suboptimal control and what the sensitivity of the overall operation is to variations in temperature profile. Some aspects of this have been investigated recently (*80*) with negative results. The specific problem was one similar to the fixed bed, constant conversion reactor investigated by Butt and Rohan; a maximum principle calculation was used to determine comparative rates of catalyst deactivation encountered for differing types of temperature profiles through the reactor. The objective function was maximum time of operation from initial conditions until the maximum temperature constraint was encountered by the system. The basic result of this study was that at least for simple reaction systems in which reversibility or selectivity is not involved, the aging of catalyst is remarkably insensitive to the nature of the temperature profile in the reactor. Calculations were carried out for a type I deactivation; the constant activity policy, which in general calls for an almost linear increase in temperature with bed length, does result in improved operation (with respect to isothermal conditions) when the activation energy of the main reaction is greater than that of the deactivation reaction ($E_A > E_{A,d}$). When $E_{A,d} > E_A$, however, isothermal reactor operation apparently results in the smallest rates of deactivation. In no case were cycle lengths increased by more than several percent by temperature profile manipulation. One is thus left with the impression that the cost of the sophisticated control required for temperature profile manipulation may not be commensurate with the benefits realized; determination of the optimum isothermal (but time variant) condition of operation may be sufficient for most purposes of practical optimization.

Conclusions

It is hoped that this trilevel view—microscopic, intermediate, and macroscopic—of catalyst deactivation has given the reader an appreciation not only of what has been done but what needs to be done. For example, we have detailed information on the mechanism of coke deposition and thus alteration of catalytic properties and structure as a result, strictly on the intermediate (intraparticle) scale. One can see immediately that an interesting and potentially important problem would be consideration of the catalyst not only as something in which chemical activity varies with time but in which also all the other properties and characteristics such as specific surface area, effective transport properties, etc., are time variant. The point here, of course, is not to suggest specific problems to investigate but to note the many problems associated with catalyst deactivation which have not been investigated or even formulated but which promise considerable return in product or process improvement. We have indeed progressed in reaction analysis and reactor design, and justification for the further pursuit of analytical studies of reaction system and process models can be found in the elegant statement of Aris (*110*):

> . . . they provide a firmer basis for advances in design. The mathematician in the engineer can say that if a given model is adequate and such and such conditions are met certain behavior will follow; it is the engineer in the mathematician who must judge that adequacy or appraise the desirability of the behavior.

The events we try to describe by modeling and analysis are fundamentally chemical. Accordingly, there must be a hard core of realism in representing reaction and deactivation schemes and in the process models which are built around them if the interests of the real world are to be served. Catalyst deactivation seems to be a highly individual activity—each reaction and catalyst behaves in its own particular way, and it is necessary to have specific information on these details before design or optimization can be undertaken with confidence. Fortunately, there is nothing particularly mysterious about the kinetics of catalyst deactivation (not the same may be said about mechanism), and in general the analyst may regard such events simply as adding additional complexity to the design by the introduction of new variables (*i.e.*, activity), additional chemical reactions, and more rate equations. As in many other problems of chemical reaction engineering, these factors are complications but not barriers. As we rapidly approach the maturity of computer applications to such problems, it is not detailed complications which should give pause but re-examination of basic reaction and process models so that we may successfully negotiate that narrow way defined on the one side by the mathematical neatness of Occam's razor and on the other by the complexity of nature.

Acknowledgment

The author is indebted to G. F. Froment, V. W. Weekman, Jr., and S. G. Bankoff for helpful discussion and to Chevron Research Co. and Universite Libre de Bruxelles for support of some of the work reported here. A considerable debt to the prior work of Szepe and Levenspiel should be evident.

Addendum

Fouling of Zeolites (p. 286).[a] It is of possible importance to note that the particular crystalline structure of various sorts of zeolites can lead to mechanisms of fouling which are related not to the formation of carbonaceous deposits but to the formation of large intermediate or product molecules which are immobilized within the fine pore structure. Such a phenomenon is a consequence of the shape selectivity of these catalysts as discussed elsewhere (*111*). These immobile molecules then block pores and active surfaces from participation in catalysis, and while the mechanism of deactivation is not that of coking in the sense we have discussed here, the symptoms are similar. Degradation of intermediates such as these is promoted by higher temperatures, however, so this type of deactivation would decrease in importance with temperature elevation, in contrast to coking behavior. Some aspects of this problem have been studied from both experimental and theoretical points recently by Tan and Fuller (*112*).

Thermal Sintering (p. 313). The problem is complicated by the fact that simple measurements of bulk catalyst temperatures or consideration of "equilibrium" conditions may not reflect the actual circumstances leading to thermal sintering, particularly for the case of supported metals. Luss (*113*) has shown recently, by order of magnitude calculations, that temperature differences between metal crystallites and the support surface for typical exothermic conditions and particle dimensions can reach hundreds of degrees (°C) for short periods of time following individual reaction events on the metal. The approximations in the model used for these calculations are admittedly crude, but the message is clear nonetheless. It is quite possible to supply the requisite energy for the activated mechanisms of sintering in a pointwise fashion even though over-all conditions would indicate strongly otherwise; further, it may be for this reason that some degree of crystallite growth is almost invariably noted in supported metal catalysts which have been used for any length of time in exothermic processes.

Sintering of Supported Metals (p. 313). Some additional data on the sintering of supported metals, amenable to kinetic interpretation, have been reported by Campbell (*114*) for Cu–ZnO supported on alumina, which is a catalyst for the low temperature shift reaction, and by Racz *et al.* (*115*) for Ni supported on low surface area mullite. Figure A1a

[a]These page numbers indicate the section of the main text to which these discussions correspond.

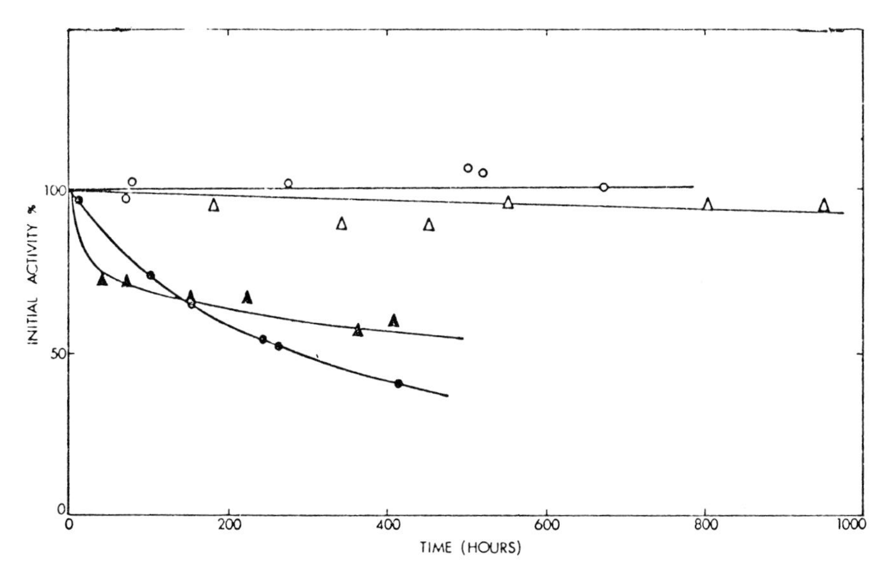

FigureA1a. Thermal stability of different formulations. ○,△: *standard 52-1 (Cu: ZnO:* Al_2O_3*);* ▲: *Cu: ZnO, 1:2;* ○: *poorly formulated Cu: ZnO:* Al_2O_3*, same composition as 52-1* (115).

gives some of Campbell's results for the rate of deactivation of the shift catalyst. Activity was shown to be directly proportional to Cu surface area, so these results indicate the rate of sintering of the metal. Surface area was determined on discreet samples using an oxygen chemisorption method which, unfortunately, is not described in detail. With the exception of formulation 1, the kinetics of sintering are again satisfactorily correlated with a second-order rate law, as shown on the figure for sample 2. Second-order correlations for the sintering of Ni–mullite, both metal and support, from the data of Racz *et al.* are shown on Figure A1c. In these experiments the sintering was conducted at 900°C with samples previously stabilized at 600°C. The Ni areas were determined by hydrogen chemisorption and total surface by BET method; there really are insufficient data here to establish the validity of the correlation, but it can be stated qualitatively that the second-order fit illustrated is superior to a first-order correlation for both metal and support.

Reversible Chemisorption of Poison (p.330). The analysis of Gioia (*116*) and Gioia and Greco (*117*) deals with a type I reversible Langmuir adsorption of poison. The adsorption rate of poison is taken to be diffusion limited; however the rate of the main reaction is not as limited. Over-all, for the calculation of the time varying rate of reaction we have:

$$D_L \frac{\partial^2[L]}{\partial y^2} = \frac{\partial[L]}{\partial \theta} + \frac{\partial[\bar{L}]}{\partial \theta} \tag{1}$$

$$\frac{[\bar{L}]}{[\bar{L}]_\infty} = \frac{K_L[L]}{1 + K_L[L]} = 1 - s \tag{2}$$

$$r_A = k_A(1 - [\bar{L}]/[\bar{L}]_\infty)\,[A]_o = k_A[s]\,[A]_o \tag{3}$$

in which D_L is the poison diffusivity, y is the distance into the particle, $[L]$ and $[\bar{L}]$ is the poison concentration in the gas phase and on the surface, respectively, $[\bar{L}]_\infty$ is the saturation poison concentration, K_L is the adsorption equilibrium constant for poison on the surface, and k_A is the rate constant of the main reaction. Since the main reaction is not diffusion limited in this analysis, one may solve Equations 1 and 3 for the intraparticle poison profile and use this result to determine the point values of reaction rate as a function of time according to Equation 3. In general, the solution of Equations 1–3 as given above, regardless of the form of boundary conditions imposed, requires numerical methods. Solutions for two limiting cases are given by Gioia corresponding to the conditions $K_L[L]_o >> 1$ and $K_L[L]_o << 1$, respectively. In each case the form of Equation 2 becomes linear, and an analytical solution is possible

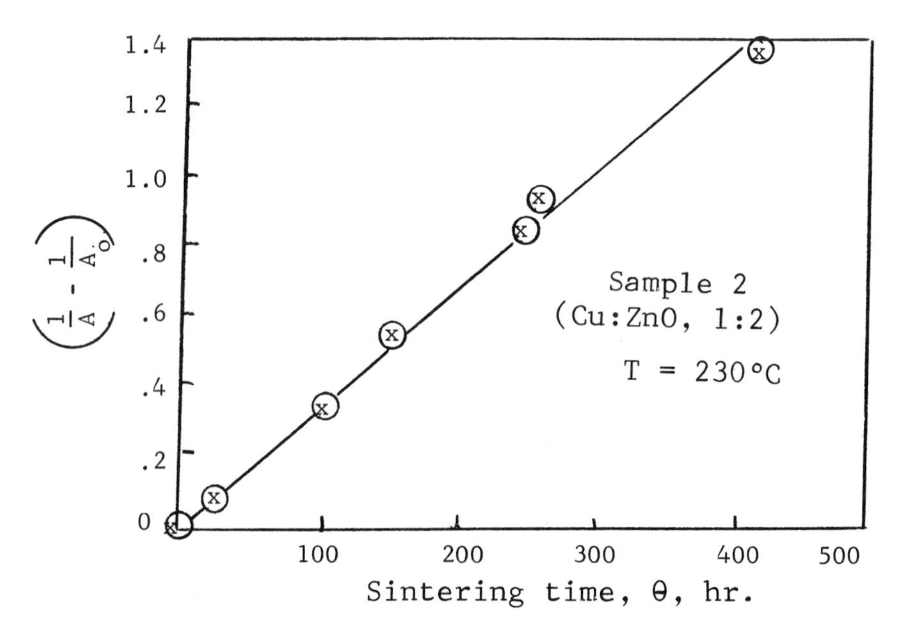

Figure A1b. Second-order correlation for Sample 2, Cu:ZnO, 1:2. T = 230°C.

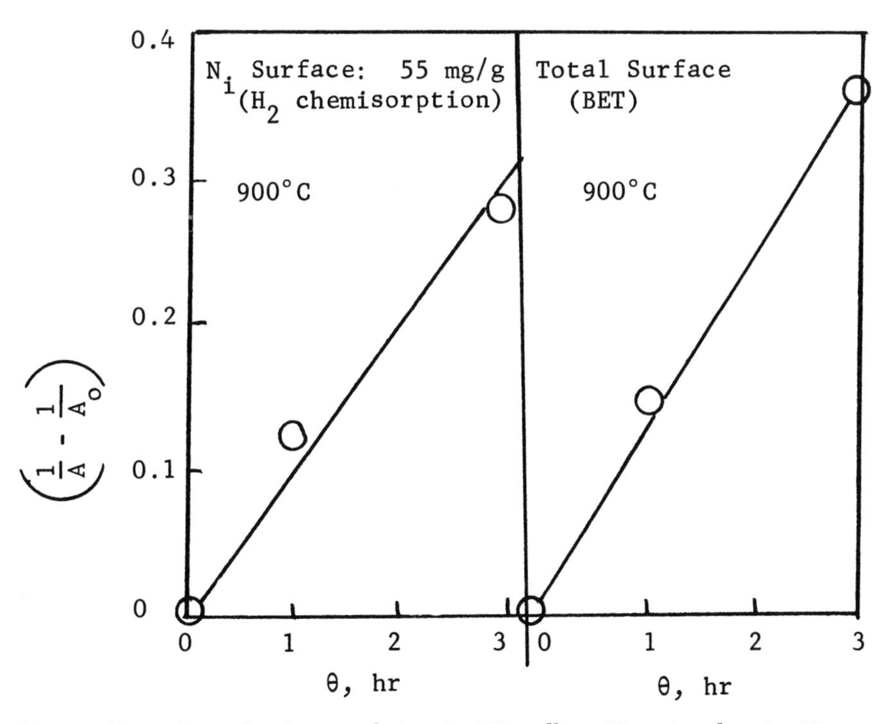

Figure A1c. Second-order correlation for Ni-mullite. Sintering data for Ni on mullite (55 mg/g) from Racz et al. (115) *as follows:*

Time	Ni Area, m^2/g	Total Area, m^2/g
0	*1.35*	*2.80*
1 hr–900°C	*0.51*	*2.00*
3 hr–900°C	*0.29*	*1.40*
1 hr–1100°C	*0.03*	*1.40*

provided the semi-infinite slab boundary conditions are used. These specify:

$$t = 0, \qquad [L] = 0$$

$$y = 0, \qquad [L] = [L]_o \tag{4}$$

$$y \rightarrow \infty, \qquad [L] = \text{finite}$$

The resulting solutions for poison profiles are:

(1) $K_L[L]_o \ll 1; 1 - s = K_L[L]$

$$(1 - s) = K_L[L]_o \operatorname{erfc}(y/2\sqrt{\sigma\theta}) \tag{5}$$

in which $\sigma \equiv D_L/([\bar{L}]_\infty K_L + 1)$.

$$(2)\ K_L[L]_o \gg 1;\ (1 - s) = 1 \text{ for } 0 \leq y \leq \lambda. \tag{6}$$

and $(1 - s) = K_L[L]_o \operatorname{erfc} [(y - \lambda)/2\sqrt{\sigma\theta}]$ for $\lambda \leq y \leq \infty$.

In this case a completely inactive zone penetrates inside the slab to a depth λ, which is given by Gioia and Greco (*117*) as:

$$\lambda = (2D_L[L]_o\theta/[\bar{L}]_\infty)^{1/2} \tag{7}$$

This type of progressive deactivation involving two zones of differing activity within the particle is the topic of an extensive discussion below.

The over-all results, in terms of apparent reaction rate as a function of time of utilization, can be obtained as follows. For $K_L[L]_o << 1$ and no diffusion limitation of the main reaction, the flux of poison entering the catalyst at any time is:

$$N_L = -D_L\left(\frac{\partial[L]}{\partial y}\right)_{y=0} \tag{8}$$

The total amount up to time θ which has entered the particle is given by integrating Equation 8 with respect to time, on obtaining the indicated derivative from Equation 5. The over-all rate is then:

$$r_A = k_A\left(1 - \frac{aK_L[L]_o}{\sqrt{\pi}}\sqrt{4\sigma\theta}\right)[A]_o \tag{9}$$

where a is the surface area (external) effective in poison transport. For $K_L[L]_o >> 1$ in the limit where the unpoisoned portion of the particle is taken to be completely active:

$$(1 - s) = (\lambda/W) \tag{10}$$

where W is the dimension of the particle. The rate of reaction is:

$$r_A = k_A\left(1 - \sqrt{\frac{2D_L[L]_o\theta}{[\bar{L}]_\infty W^2}}\right)[A]_o \tag{11}$$

This result for the rate is the limiting case, in which the interior profile of Equation 6 becomes essentially discontinuous from $s = 0$ to $s = 1$ at λ.

The specific experiments reported by Gioia are a bit difficult to interpret directly in terms of the model given above, and it is the writer's opinion that a number of questions are left unanswered in this work. Nonetheless, the study is valuable in demonstrating the complexities arising in deactivation—particularly that such reaction systems are adaptive both in parameter values and model equations; thus the interpretation offered by Gioia is not altered here. Experimental data were obtained for the water vapor-poisoned hydrogenation of ethylene on a copper–magnesia catalyst. The principal data of interest are measurements of conversion in a fixed bed reactor as a function of time of operation. Typical experimental conditions and conversion results for one run are shown in Figure A2a, plotted *vs.* time of catalyst utilization, θ. For very low conversions, one would expect a linear correlation of x with $\sqrt{\theta}$ since under these conditions the rate modification would be translated directly into a conversion modification. From Equation 9, for example, for the conversion of ethylene:

$$x = x_o\left(1 - \frac{aK_L[L]_o}{\sqrt{\pi}}\sqrt{4\sigma\theta}\right) \tag{12}$$

The data show that the response of conversion to time of operation consists of two regions. The second region is linear with respect to $\sqrt{\theta}$, as shown in Figure A2b; however, the correlation is not valid in general for the initial period of the response. The reason appears to lie in the fact that the initial catalyst activity was sufficiently large that heat generation rates were significant and there was an interphase gradient of temperature between particle and gas bulk phase. This gradient, of

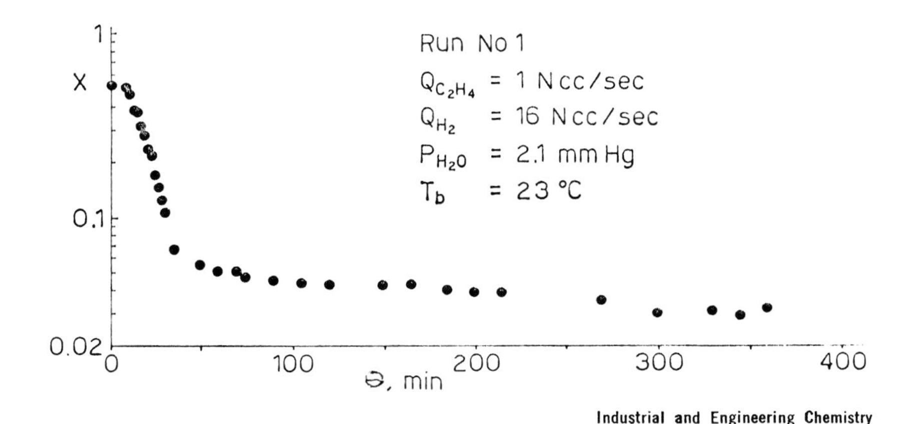

Figure A2a. Poisoning of particles (116)

Run conditions for Figures A2a,b: Hydrogen flow, 16 N cm³/sec; ethylene flow, 1 N cm³/sec; partial pressure of water, 2.1 mm Hg; reactor bulk temperature, 296°K; catalyst surface temperature initially, 354°K; catalyst density, 1.1 gm/cm³; initial conversion, 0.516; extrapolated conversion, x_o, *0.062; initial effectiveness factor, 0.20; extrapolated effectiveness factor, 1; catalyst half-thickness, 0.25 cm; reactor diameter, 2 cm; particle Reynolds number, 4.3 (based on mass flow over total cross section).*

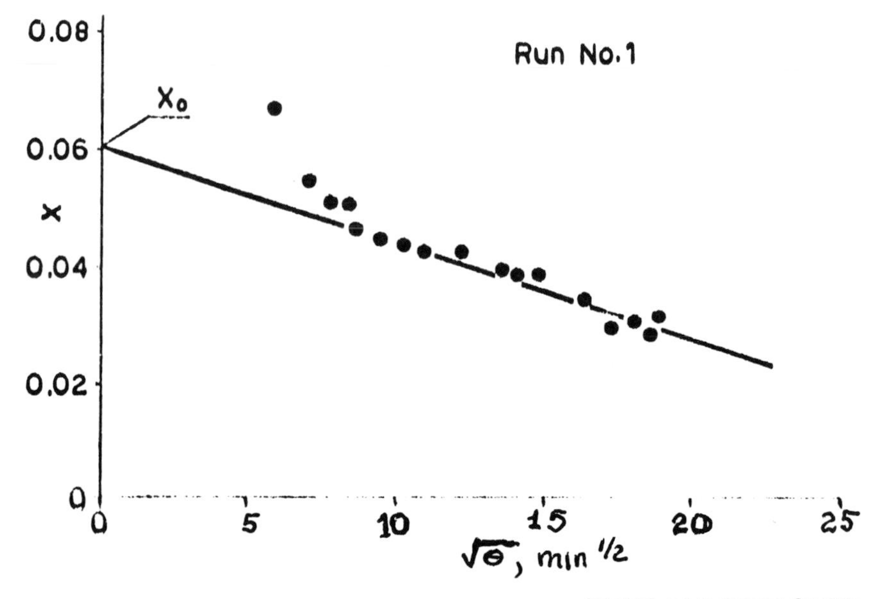

Figure A2b. Example of data correlation (116)

course, produces an exponential amplification of rate or, conversely, as the catalyst loses activity the rate of heat generation decreases exponentially as the catalyst particle temperature is lowered. The initial portion of the catalyst deactivation curve of Figure A2a is thus taken to be the response to a thermal transient which disappears as the catalyst deactivates. Gioia shows how surface temperatures can be calculated from the measured conversion data and thus is able to calculate heat transfer coefficients for the experiments. These are in reasonable agreement with values obtained from published correlations, so the general interpretation of the importance of initial nonisothermality appears to be valid. The $\sqrt{\theta}$ correlation of activity is a result of the diffusion limited penetration of poison, and this result is insensitive to the form of isotherm involved for the poison. The important lesson of this experiment is similar to that given by the work of Murakami *et al.*: reaction systems in which catalyst deactivation occurs are entities which change their parameters, and sometimes their very nature, as time proceeds. The two examples here show in particular this type of essential elusiveness when thermal factors are involved.

Poisoning Reaction—Diffusion Controlled (p. 382). A further extension of the analysis of Wheeler and Robell has been given by Haynes (*118*) who has reported a solution for the type I system in which the poisoning reaction, as well as the main reaction, is diffusion controlled. This is a situation which would seem more typical of most processing involving hydrocarbon reactions. In accord with the prior results of Masamune and Smith (*41*) the shell model is used as an approximation for poisoning with the type I mechanism. [It is worthwhile to repeat Haynes' caution that the "shell model" and the "shell progressive model" refer to related but distinct mechanisms. The former permits diffusion and reaction in the poison-free zone whereas in the latter all areas of the poison-free zone are equally accessible to the reactant]. A new parameter, h_L, the Thiele modulus for the poisoning reaction, is introduced, and the approximation is generally valid for $h_L \geqslant h_o$. In the following discussion the dimensionless variables and most of the notation used by Haynes are retained; the similarity to the terms used by Wheeler and Robell are pointed out. For the shell model of a spherical pellet the fraction of original activity, s, is related to the radius of the poison-free zone by:

$$s = 1 - \xi^3$$

where ξ is the ratio r_i/R of the poison-free zone to total radius. The effectiveness factor as a function of the degree of poisoning is obtained by mass balances on the poison and poison-free zones as:

$$\eta_L = \frac{3}{h_L^2 \xi^3}\left[\frac{h_L\xi \coth h_L\xi - 1}{h_L(1 - \xi)\coth h_L\xi + 1}\right] \tag{1}$$

The equation for poisoning kinetics and the conservation equation giving the poison distribution in the bed are:

$$\frac{\partial(C_{p,g}/C^o_{p,g})}{\partial z} = \frac{3N_t(C_{p,g}/C^o_{p,g})}{h_L^2}\left[\frac{h_L\xi \coth h_L\xi - 1}{h_L(1 - \xi)\coth h_L\xi + 1}\right] \tag{2}$$

$$\frac{\partial \xi}{\partial \tau} = \frac{(C_{p,g}/C^o_{p,g})}{\xi^2 h_L^2}\left[\frac{h_L\xi \coth h_L\xi - 1}{h_L(1 - \xi)\coth h_L\xi + 1}\right] \tag{3}$$

where N_t is a number of adsorption transfer units as defined by Wheeler and Robell, z is dimensionless reactor length, and τ is a dimensionless time defined as $\tau \equiv k_{Ads}C^o_{p,g}t/C_p^\infty$ in the nomenclature of Wheeler and Robell. As $h_L \rightarrow 0$, the above equations reduce to the diffusion-free forms; however, it can be seen that numerical solution will generally be required when such is not the case. Given the poison distribution in the bed as a function of time from solution of the above, the fraction of original activity in the bed as a function of time and position is:

$$(s)_{bed} = \left[\frac{h_o\xi \coth h_o\xi - 1}{h_o(1 - \xi)\coth h_o\xi + 1}\right]\left[\frac{1}{h_o \coth h_o - 1}\right] \tag{4}$$

If k is the first-order rate constant for the main reaction, the conversion relationship corresponding to Equation 70 may be obtained from:

$$\ln\left(\frac{C_W}{C_o}\right) = -k'\left(\frac{W}{v}\right)\int_o^1 (s)_{\text{bed}}\, dz \tag{5}$$

where $k' = k\left[\frac{3}{h_o^2}(h_o \coth h_o - 1)\right]$ is the effective rate constant of the unpoisoned catalyst.

The solution to these equations is given by Haynes for a range of parametric values in the form of generalized plots, permitting numerical or graphical evaluation of the activity or conversion variation with time according to Equation 5. The properties of the system which one needs to know are the true adsorption and rate constants, k and k_{Ads}, the effective diffusivities of reactant and poison in the catalyst pore structure (hence h_o and h_L) and the limiting poison capacity, C_p^{∞}.

Analytical Solutions for Simple Selectivity Problems (p. 389). A number of simpler selectivity problems, particularly for type II deactivations, can be solved using the method of Ozawa or in some cases, even direct integration. An example of particular interest is that of first-order reaction and first-order deactivation:

$$\left.\begin{array}{ll}(\text{U}) & A + S_1 \rightarrow B + S_1 \\ (\text{V}) & A + S_2 \rightarrow C + S_2\end{array}\right\} \text{(main reactions)}$$

$$\left.\begin{array}{ll}(\text{U}) & A + S_1 \rightarrow A{\cdot}S_1 \\ (\text{V}) & A + S_2 \rightarrow A{\cdot}S_2\end{array}\right\} \text{(fouling reactions)}$$

For an isothermal, fixed bed reactor with plug flow, the conservation equations for this reaction system assume the following non-dimensional forms:

$$\frac{\partial a}{\partial z} = -K_{\text{A},1}as_1 - K_{\text{A},2}as_2 \tag{1}$$

$$\frac{\partial s_1}{\partial \theta} = -K_{\text{d},1}as_1 \tag{2}$$

$$\frac{\partial s_2}{\partial \theta} = -K_{\text{d},2}as_2 \tag{3}$$

where s_1 and s_2 are the activity variables for the U and V functions, respectively, a is dimensionless reactant concentration $[\text{A}]/[\text{A}]_o$, $K_{\text{A},1}$

and $K_{A,2}$ are rate constants for the two main reactions, and $K_{d,1}$ and $K_{d,2}$ are rate constants for the deactivations. These rate constants are defined as follows:

$$K_{A,1} = k_{A,1}\tau(1 - \varepsilon) \tag{4}$$

$$K_{A,2} = k_{A,2}\tau(1 - \varepsilon)$$

$$K_{d,1} = k_{d,1}\tau\rho \tag{5}$$

$$K_{d,2} = k_{d,2}\tau\rho$$

in which τ is the total residence time in the reactor based on empty cross section, ϵ is bed void fraction, and ρ is the molal volume of the gas feed (assumed not to change with position). As usual, z is the normalized length, w/W, and θ is the time of operation in terms of the number of residence times. The form of Equation 1, with the time derivative omitted, invokes the assumption of slow deactivation. From Equations 2 and 3 we have:

$$\frac{ds_1}{ds_2} = \frac{K_{d,1}}{K_{d,2}}\left(\frac{s_1}{s_2}\right) = n\left(\frac{s_1}{s_2}\right)$$

so:

$$s_1 = s_2{}^n \tag{6}$$

The three original equations may be reduced to two using this result:

$$\frac{\partial a}{\partial z} = -K_{A,1}as_2{}^n - K_{A,2}as_2$$

$$\frac{\partial s_2}{\partial \theta} = -K_{d,2}as_2$$

If we define $r \equiv (K_{A,1}/K_{A,2})$, we may write the pair as:

$$\frac{\partial a}{\partial z} = -K_A a s_2 \tag{7}$$

$$\frac{\partial s_2}{\partial \theta} = -K_{d,2}as_2 \tag{8}$$

where $K_A = K_{A,2}\ (rs_2{}^{n-1} + 1)$. Analytical solution to the problem requires some form of linearization for K_A in Equation 7. It is easiest to assign a value for s_2 which represents the average over the time internal involved in the solution, thus, permitting the evaluation of K_A. As the time of operation increases, new averages for s_2 and new values of K_A

are determined on each time increment. With K_A taken as a constant, Equations 7 and 8 are exactly of the form required for the Legendre transformation solution. The boundary and initial conditions are:

$$a = 1, \quad z = 0, \quad \theta > 0$$

$$s_2 = 1, \quad \theta = 0, \quad 0 \leq z \leq 1 \tag{9}$$

and the final solutions are:

$$s_2 = -\frac{\exp(z)}{1 - \exp(z)[1 + \exp(K_{d,2}\theta/K_A - z)]} \tag{10}$$

$$s_1 = s_2{}^n \tag{11}$$

$$a = s_2 \exp(K_{d,2}\theta/K_A - z) \tag{12}$$

$$c = \frac{[C]}{[C]_o} = K_{A,2} + \frac{K_{A,2}\exp(K_{d,2}\theta/K_A)}{1 - \exp(K_{d,2}\theta/K_A) - \exp(z)} \tag{13}$$

$$\text{Selectivity} = c/a$$

The result for product c, Equation 13, is obtained from the solutions for a and s_2:

$$\frac{dc}{dz} = K_{A,2}as_2$$

$$c = K_{A,2}\exp(K_{d,2}\theta/K_A)\int_o^z \frac{\exp(z)\,dz}{\{[(1 - \exp(K_{d,2}\theta/K_A)] - \exp(z)\}^2} \tag{14}$$

Other solutions can be obtained for various kinds of reaction and deactivation kinetics in type II systems. We are particularly interested in behavior when the kinetics of deactivation are second order in activity and do not depend on conversion level—a state which appears to be roughly characteristic of sintering deactivation—and also in the behavior of systems in which the kinetics of the main reactions are zero order. In fact, analytical solutions to the activity and selectivity under isothermal conditions can be found easily for many of these cases, and no approximations are required. Table AI gives a summary of some solutions for activity and selectivity in type II systems under isothermal conditions.

A restricted class of nonisothermal problems can be treated by the methods outlined above. In general, reactor operation pertains to adiabatic conditions, and parametric values are limited to being the same for both functions or to the asymptotic case in which the parameter pertaining to one function (—*e.g.*, heat of reaction or activation energy) is much greater than that for the second function. The treatment of nonisothermal problems associated with single function deactivation or with catalyst regeneration by these methods, as discussed below, involves fewer restrictions and may be of more general applicability (*76*).

Table AI. Some Analytical Solutions for Activity and Selectivity in the Deactivation of Type II Systems in Fixed Beds

Main Reaction Zero Order, Deactivation Zero Order[a]

$$\frac{\partial a}{\partial z} = -K_{A,1}s_1 - K_{A,2}s_2$$

$$\frac{\partial s_1}{\partial \theta} = -K_{d,1}$$

$$\frac{\partial s_2}{\partial \theta} = -K_{d,2}$$

$$s_1 = 1 - K_{d,1}\theta$$

$$s_2 = 1 - K_{d,2}\theta$$

$$a = 1 - K_{A,1}z(1 - K_{d,1}\theta) - K_{A,2}z(1 - K_{d,2}\theta)$$

$$c = K_{A,2}z(1 - K_{d,2}\theta)$$

Main Reaction Zero Order, Deactivation First Order

$$\frac{\partial a}{\partial z} = -K_{A,1}s_1 - K_{A,2}s_2$$

$$\frac{\partial s_1}{\partial \theta} = -K_{d,1}s_1$$

$$\frac{\partial s_2}{\partial \theta} = -K_{d,2}s_2$$

$$s_1 = \exp(-K_{d,1}\theta)$$

$$s_2 = \exp(-K_{d,2}\theta)$$

$$a = 1 - K_{A,1}z \exp(-K_{d,1}\theta) - K_{A,2}z \exp(-K_{d,2}\theta)$$

$$c = K_{A,2}z \exp(-K_{d,2}\theta)$$

Main Reaction Zero Order, Deactivation Second Order (Sintering)

$$\frac{\partial a}{\partial z} = -K_{A,1}s_1 - K_{A,2}s_2$$

Table AI. Continued

Main Reaction Zero Order, Deactivation Second Order (Sintering) (Continued)

$$\frac{\partial s_1}{\partial \theta} = -K_{d,1}(s_1)^2$$

$$\frac{\partial s_2}{\partial \theta} = -K_{d,2}(s_2)^2$$

$$s_1 = 1/(1 + K_{d,1}\theta)$$

$$s_2 = 1/(1 + K_{d,2}\theta)$$

$$a = 1 - \frac{K_{A,1}z}{1 + K_{d,1}\theta} - \frac{K_{A,2}z}{1 + K_{d,2}\theta}$$

$$c = \frac{K_{A,2}z}{1 + K_{d,2}\theta}$$

Main Reaction First Order, Deactivation Zero Order

$$\frac{\partial a}{\partial z} = -K_{A,1}s_1 a - K_{A,2}s_2 a$$

$$\frac{\partial s_1}{\partial \theta} = -K_{d,1}$$

$$\frac{\partial s_2}{\partial \theta} = -K_{d,2}$$

$$s_1 = 1 - K_{d,1}\theta$$

$$s_2 = 1 - K_{d,2}\theta$$

$$a = \exp(m_1 z)$$

$$c = K_{A,2}(1 - K_{d,2}\theta)\frac{\exp(m_1 z)}{m_1}$$

$$\text{where } m_1 = -K_{A,1}(1 - K_{d,1}\theta) - K_{A,2}(1 - K_{d,2}\theta)$$

Main Reaction First Order, Deactivation First Order

$$\frac{\partial a}{\partial z} = -K_{A,1}s_1 a - K_{A,2}s_2 a$$

$$\frac{\partial s_1}{\partial \theta} = -K_{d,1}s_1$$

Table AI. Continued

Main Reaction First Order, Deactivation First Order (Continued)

$$\frac{\partial s_2}{\partial \theta} = -K_{d,2}s_2$$

$$s_1 = \exp(-K_{d,1}\theta)$$

$$s_2 = \exp(-K_{d,2}\theta)$$

$$a = \exp(m_2 z)$$

$$c = K_{A,2} \exp(-K_{d,2}\theta). \frac{\exp(m_2 z)}{m_2}$$

$$m_2 = -K_{A,1} \exp(-K_{d,1}\theta) - K_{A,2} \exp(-K_{d,2}\theta)$$

Main Reaction First Order, Deactivation First Order and Conversion Dependent

$$\frac{\partial a}{\partial z} = -K_{A,1}s_1 a - K_{A,2}s_2 a$$

$$\frac{\partial s_1}{\partial \theta} = -K_{d,1}s_1 a$$

$$\frac{\partial s_2}{\partial \theta} = -K_{d,2}s_2 a$$

$$s_1 = s_2{}^n$$

$$s_2 = -\frac{\exp(z)}{1 - \exp(z)[1 + \exp(K_{d,2}\theta/K_A - z)]}$$

$$a = s_2 \exp(K_{d,2}\theta/K_A - z)$$

$$c = K_{A,2} + \frac{K_{A,2} \exp(K_{d,2}\theta/K_A)}{1 - \exp(K_{d,2}\theta/K_A) - \exp(z)}$$

$$K_A = K_{A,2}\,[r(s_2)_{\mathrm{avg}}{}^{n-1} + 1]$$

$$r = K_{A,1}/K_{A,2};\ n = K_{d,1}/K_{d,2}$$

Main Reaction First Order, Deactivation Second Order (Sintering)

$$\frac{\partial a}{\partial z} = -K_{A,1}s_1 a - K_{A,2}s_2 a$$

$$\frac{\partial s_1}{\partial \theta} = -K_{d,1}(s_1)^2$$

Table AI. Continued

Main Reaction First Order, Deactivation Second Order (Sintering) (Continued)

$$\frac{\partial s_2}{\partial \theta} = -K_{d,2}(s_2)^2$$

$$s_1 = 1/(1 + K_{d,1}\theta)$$

$$s_2 = 1/(1 + K_{d,2}\theta)$$

$$a = \exp\left(- \frac{K_{A,1}z}{1 + K_{d,1}\theta} - \frac{K_{A,2}z}{1 + K_{d,2}\theta}\right)$$

$$c = \frac{-K_{A,2}}{1 + K_{d,2}\theta}\left(\frac{1}{m_3} - \frac{\exp\,(m_3 z)}{m_3}\right)$$

$$m_3 = -\left(\frac{K_{A,1}}{1 + K_{d,1}\theta} - \frac{K_{A,2}}{1 + K_{d,2}\theta}\right)$$

[a] For boundary and initial conditions *see* Equation 9.

Optimization of Catalyst Composition (p. 451). The question of optimizing catalyst composition has been suggested but not treated in detail in several studies cited in this discussion, and we conclude the remarks on optimization by considering this particular problem. Miertschin and Jackson (*119*) have reported specific results on the optimum catalyst replacement policy, length of operating cycle, and temperature program for simple reactor types operated with constant exit conversion. To allow temperature policy to be a free control variable if the conversion level is invariant, the feed rate to the reactor must be varied with time. Below we summarize the results for the simplest example of Miertschin and Jackson—an irreversible first-order reaction with conversion independent deactivation kinetics carried out in a CSTR. In this case the reactor performance equations are:

$$\frac{dy}{dt} = fx = s[r(x,T)] \tag{1}$$

$$\frac{ds}{dt} = -k_{A,d}s^n \tag{2}$$

$$r(x,T) = k_A{}^o(1 - x)\, e^{-E_A/RT} \tag{1a}$$

$$k_{A,d} = k^o{}_{A,d}\, e^{-E_{A,d}/RT}$$

where $f = F/V$, $y = Y/V$, x is conversion, V is reactor volume, F is molar rate of feed addition, Y moles of product, s is activity and E_A and $E_{A,d}$ activation energies for main and deactivation reactions, respectively. The objective function expresses profit in terms of selling price of product minus raw material cost and catalyst replacement cost. The first two terms are:

$$(\text{product} - \text{raw material}) = \alpha Y(\theta) - \beta Q(\theta) \qquad (3)$$

where $Y(\theta)$ and $Q(\theta)$ are moles of product and feed corresponding to a period of operation θ, and x and β are the respective values per mole.

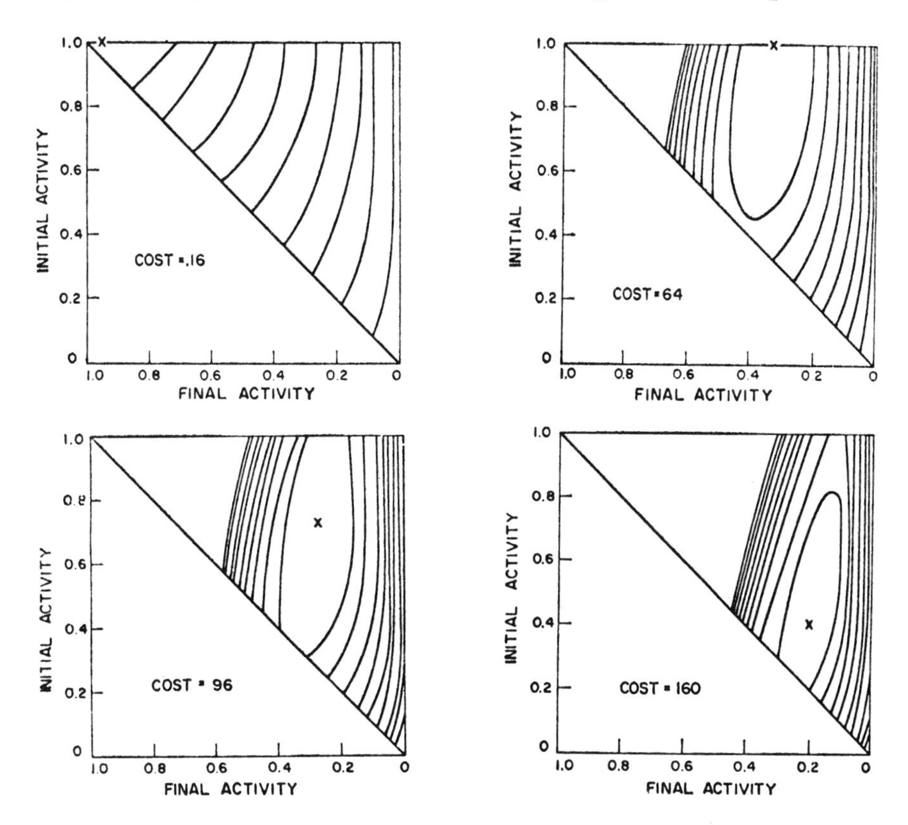

Canadian Journal of Chemical Engineering

Figure A3. Contour maps of objective function in the (ψ_f,ψ_o)–plane for different costs of complete catalyst renewal. Other constants in table. Contours at intervals of 10% of maximum value of objective function (119).

Cost of complete catalyst renewal:
(a) \$0.16 per liter of reactor volume
(b) \$64.00 per liter of reactor volume
(c) \$96.00 per liter of reactor volume
(d) \$160.00 per liter of reactor volume

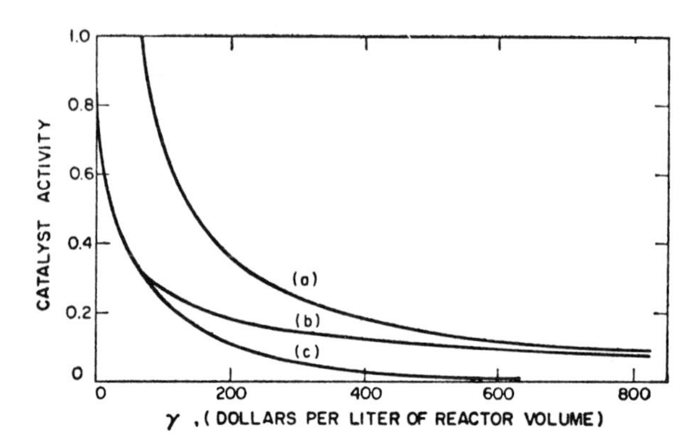

Canadian Journal of Chemical Engineering

Figure A3e. Curve (a), optimal ψ_o; curve (b), optimal ψ_f with optimal ψ_o; curve (c) optimal ψ_f with $\psi_o = 1$. Optimum temperature policy isothermal. (119).

The catalyst replacement cost is given by a function pertaining strictly to the replacement of a discarded fraction of spent catalyst with fresh catalyst:

$$\text{(catalyst replacement)} = \gamma V\left(\frac{s_o - s_f}{1 - s_f}\right) \quad (4)$$

where γ is the catalyst cost per reactor volume, s_o is initial (replacement) activity level, and s_f is the activity at θ. The over-all objective function, P', assuming replacement time is small compared with operating time, is then:

$$P' = \frac{V(\alpha - \beta/x)}{\theta}\left\{y(\theta) - \frac{\gamma}{\alpha - \beta/x}\left(\frac{s_o - s_f}{1 - s_f}\right)\right\} \quad (5)$$

A simple temperature policy is optimal in this case: temperature is time invariant and at the upper limit, T_{max}; additional cases are treated by Miertschin (*120*). Since the temperature is fixed, the complete problem can be solved in a relatively simple way because only optimization with respect to s_o and θ need be considered. For $n = 2$ in Equation 2, we have at constant temperature:

$$s(t) = \frac{s_o}{1 + k_{A,d}s_o t} \quad (6)$$

and this may be used with Equations 1 and 1a, recalling that $r(x,T_{max})$ is constant, to give:

$$y(t) = \frac{r}{k_{A,d}} \ln (1 + k_{A,d} s_o t) \tag{7}$$

Since s_f is a uniquely defined function of the time of operation θ from Equation 6 at $t = \theta$, we may write Equation 7 as:

$$y(\theta) = \frac{r}{k_{A,d}} \ln \left(\frac{s_o}{s_f}\right) \tag{8}$$

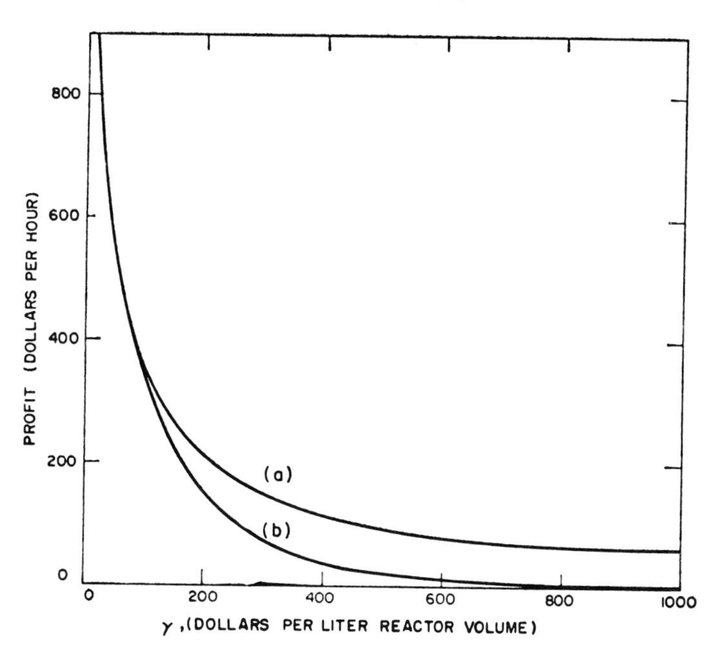

COSTS AND PHYSICAL CONSTANTS FOR EXAMPLE WITH ISOTHERMAL OPTIMAL TEMPERATURE POLICY

k_A	2.224×10^{12} mole/liter sec	V	1000 liter
$k_{A,d}$	5.0×10^6 hr^{-1}	n	2.0
E_A/R	3×10^4 °K	T_{MAX}	1000 °K
$E_{A,d}/R$	2×10^4 °K	α	\$0.05
x	0.9	β	\$0.027

Canadian Journal of Chemical Engineering

Figure A3f. Curve (a), maximum profit with optimal ψ_o; curve (b), maximum profit with $\psi_o = 1$. [Optimum temperature policy isothermal.] (119).

Substituting for $y(\theta)$ and θ from Equations 8 and 6, respectively, in the objective function, we have:

$$P' = V\left(\alpha - \frac{\beta}{x}\right)\left[\frac{k_{A,d}s_o s_f}{s_o - s_f}\left\{\frac{r}{k_{A,d}}\ln\left(\frac{s_o}{s_f}\right) - \frac{\gamma}{\alpha - \beta/x}\left(\frac{s_o - s_f}{1 - s_f}\right)\right\}\right] \quad (9)$$

where r refers to the rate at the maximum temperature allowed the system. The function P' is then maximized with respect to s_o and s_f (*i.e.*, θ).

The results of calculations on this system are shown in Figure A3 for typical parameter values as shown. Figure A3ad shows a series of contour maps of the function, contours drawn at 10% increments of the value of the maximum, reflecting only the effects caused by changes in replacement catalyst value. The locus of the maximum takes an extensive journey about the $s_o - s_f$ plane, and for higher catalyst replacement costs the optimal initial activity level is far removed from completely reactivated ($s_o = 1$) material. Some other aspects of the solution are illustrated in Figures A3e and f. The first compares the various catalytic activity relationships as a function of replacement cost, and the second compares the difference in maximum profit occasioned by the fact that $s_o < 1$ at high values of γ.

We have previously noted that conversion-independent deactivation rates, such as Equation 2, may lack some of the generality of kinetics such as those of Equation 16; however, the second-order law as used in this study is an adequate correlation for sintering kinetics.

Literature Cited

(1) Maxted, E. B., *Advan. Catalysis* (1951) **3**, 129.
(2) Voorhies, A., Jr., *Ind. Eng. Chem.* (1945) **37**, 318.
(3) Szépe, S., Levenspiel, O., *Proc. European Fed, 4th Chem. Reaction Eng., Brussels*, Pergamon Press (1970).
(4) Mills, G. A., Boedeker, E. R., Oblad, A. G., *J. Amer. Chem. Soc.* (1950) **72**, 1554.
(5) Pines, H., Haag, W. O., *J. Amer. Chem. Soc.* (1960) **82**, 2471.
(6) Parravano, G., *J. Amer. Chem. Soc.* (1953) **75**, 1448.
(7) Maxted, E. B., Marsden, A., *J. Chem. Soc.* **1945**, 766.
(8) Hayward, D. O., Trapnell, B. M. W., "Chemisorption," 2nd ed., Butterworth and Co., 1964.
(9) Bond, C. G., *Discussions Faraday Soc.* (1966) **41**, 200.
(10) Pease, R. N., Stewart, L., *J. Amer. Chem. Soc.* (1925) **47**, 1235.
(11) Clay, R. D., Petersen, E. E., *J. Catalysis* (1970) **16**, 32.
(12) Peri, J. B., *J. Phys. Chem.* (1965) **69**, 231.
(13) Dabrowski, J. E., Butt, J. B., Bliss, H., *J. Catalysis* (1970) **18**, 297.
(14) Callahan, J. L., Grasselli, R. K., *AIChE J.* (1963) **9**, 755.
(15) Blanding, F. H., *Ind. Eng. Chem.* (1953) **45**, 1186.
(16) Rudershausen, C. G., Watson, C. C., *Chem. Eng. Sci.* (1954) **3**, 110.
(17) Eberly, P. E., Jr., Kimberlin, C. N., Miller, W. H., Drushel, H. V., *Ind. Eng. Chem., Process Design Develop.* (1966) **5**, 193.
(18) Prater, C. D., Lago, R. M., *Advan. Catalysis* (1956) **8**, 293.

(19) Ozawa, Y., Bischoff, K. B., *Ind. Eng. Chem., Process Design Develop.* (1968) **7**, 67.
(20) Pozzi, A. L., Rase, H. F., *Ind. Eng. Chem.* (1958) **50**, 1075.
(21) Appleby, W. G., Gibson, J. W., Good, G. M., *Ind. Eng. Chem., Process Design Develop.* (1962) **1**, 102.
(22) Ramser, J. N., Hill, P. B., *Ind. Eng. Chem.* (1958) **50**, 117.
(23) Haldeman, R. G., Botty, M. C., *J. Phys. Chem.* (1959) **63**, 489.
(24) Levinter, M. E., Panchenkov, G. M., Tanatarov, M. A., *Intern. Chem. Eng.* (1967) **7**, 23.
(25) Eberly, P. E., Jr., Kimberlin, K. N., Jr., *Proc. Intern. Symp. Chem. Reaction Eng., 1st, Washington, D. C.* (June 1970).
(26) Weekman, V. W., Jr., *Ind. Eng. Chem., Process Design Develop.* (1969) **8**, 385.
(27) *Ibid.* (1968) **7**, 90.
(28) Weekman, V. W., Jr., Nace, D. M., *AIChE J.* (1970) **16**, 397.
(29) Kingery, W. D., Berg, M., *J. Appl. Phys.* (1955) **26**, 1205.
(30) MacIver, D. S., Tobin, H. H., Barth, R. T., *J. Catalysis* (1963) **2**, 485.
(31) Blakely, J. M., Mykura, H., *Acta Met* (1961) **9**, 23.
(32) Herrman, R. H., Adler, S. F., Goldstein, M. S., deBaun, R. M., *J. Phys. Chem.* (1961) **65**, 2189.
(33) Maat, H. J., Moscou, L., *Proc. Intern. Congr. Catalysis, 3rd,* p. 1277, North Holland, 1965.
(34) Spenadel, L., Boudart, M., *J. Phys. Chem.* (1960) **64**, 204.
(35) Gwyn, J. E., ADVAN. CHEM. SER. (1972) **109**, 513.
(36) Bondi, A., Miller, R. S., Schlaffer, W. G., *Ind. Eng. Chem., Process Design Develop.* (1962) **1**, 196.
(37) Wheeler, A., "Catalysis," P. H. Emmet, Ed., Vol. II, Reinhold, New York, 1955.
(38) Satterfield, C. D., "Mass Transfer in Heterogeneous Catalysis," MIT Press, Cambridge, 1970.
(39) Balder, J. R., Petersen, E. E., *Chem. Eng. Sci.* (1968) **23**, 1287.
(40) Dougharty, N., *Chem. Eng. Sci.* (1970) **25**, 489.
(41) Masamune, S., Smith, J. M., *AIChE J.* (1966) **12**, 384.
(42) Ozawa, Y., Bischoff, K. B., *Ind. Eng. Chem., Process Design Develop.* (1968) **7**, 72.
(43) Murakami, Y., Kobayshi, T., Hattori, T., Masuda, M., *Ind. Eng. Chem., Fundamentals* (1968) **7**, 599.
(44) Richardson, J. T., personal communication (June 1970).
(45) Gioia, F., Gibilaro, L. G., Greco, G., Jr., *Chem. Engr. J.* (1970) **1**, 9.
(46) Chu, C., *Ind. Eng. Chem., Fundamentals* (1968) **7**, 509.
(47) Ausman, J. M., Watson, C. C., *Chem. Eng. Sci.* (1962) **17**, 323.
(48) Weisz, P. B., Goodwin, R. B., *J. Catalysis* (1966) **6**, 227.
(49) *Ibid.* (1963) **2**, 397.
(50) Weisz, P. B., *Z. Physik Chem.* (1957) **11**, 1.
(51) Wicke, E., Kallenbach, R., *Kolloid Z.* (1941) **97**, 135.
(52) Weisz, P. B., Schwartz, A. B., *J. Catalysis* (1962) **1**, 399.
(53) Haggerhammer, W. A., Lee, R., *Trans. Am. Soc. Mech. Engrs.* (1947) 779.
(54) Carberry, J. J., Gorring, R. L., *J. Catalysis* (1966) **5**, 529.
(55) White, D., Carberry, J. J., *Can. J. Chem. Eng.* (1965) **43**, 334.
(56) Bischoff, K. B., *Chem. Eng. Sci.* (1963) **18**, 711.
(57) Wen, C. Y., *Ind. Eng. Chem.* (1968) **60** (9), 34.
(58) Wen, C. Y., Wang, S. C., *Ind. Eng. Chem.* (1970) **62** (9), 30.
(59) Sagara, M., Masamune, S., Smith, J. M., *AIChE J.* (1967) **13**, 1226.
(60) Weisz, P. B., Hicks, J. S., *Chem. Eng. Sci.* (1962) **17**, 265.

(61) Wilson, W. B., Good, G. M., Deahl, T. J., Brewer, C. P., Appleby, W. G., *Ind. Eng. Chem.* (1956) **48**, 1982.
(62) Massoth, F. E., *Ind. Eng. Chem., Process Design Develop.* (1967) **6**, 200.
(63) Massoth, F. E., Menon, P. G., *Ind. Eng. Chem., Process Design Develop.* (1969) **8**, 383.
(64) Gulbransen, E. A., Andrew, E. A., *Ind. Eng. Chem.* (1952) **44**, 1034, 1039.
(65) Walker, P. L., Jr., Rusinko, F., Jr., Austin, L. G., *Advan. Catalysis* (1959) **11**, 134.
(66) Petersen, E. E., *AIChE J.* (1960) **6**, 488.
(67) Gwyn, J. E., Colthart, J. D., *AIChE J.* (1969) **15**, 932.
(68) Anderson, R. B., Whitehouse, A. M., *Ind. Eng. Chem.* (1961) **53**, 1011.
(69) Wheeler, A., Robell, A. J., *J. Catalysis* (1969) **13**, 299.
(70) Bohart, G., Adams, E., *J. Amer. Chem. Soc.* (1920) **42**, 523.
(71) Froment, G. F., Bischoff, K. B., *Chem. Eng. Sci.* (1961) **16**, 189.
(72) *Ibid.* (1962) **17**, 105.
(73) Van Zoonen, D. D., *Proc. Intern. Congr. Catalysis, 3rd*, p. 1319, North Holland, 1965.
(74) Olson, J. H., *Ind. Eng. Chem., Fundamentals* (1968) **7**, 185.
(75) Bischoff, K. B., *Ind. Eng. Chem., Fundamentals* (1969) **8**, 665.
(76) Ozawa, Y., *Chem. Eng. Sci.* (1970) **25**, 529.
(77) Butt, J. B., Rohan, D. M., *Chem. Eng. Sci.* (1968) **23**, 489.
(78) Butt, J. B., *Proc. European Symp. Chemical Reaction Eng., 4th, Brussels*, Pergamon, New York, 1970.
(79) Butt, J. B., *Chem. Eng. Sci.* (1970) **25**, 801.
(80) Butt, J. B., *Chem. Eng. J.* (1971) **2**, 90.
(81) Levenspiel, O., "Chemical Reaction Engineering," Wiley, New York, 1962.
(82) Weisz, P. B., *Actes Congr. Intern. Catalyse, 2e*, p. 937, Editions Technip, Paris, 1961.
(83) Gunn, D. J., Thomas, W. J., *Chem. Eng. Sci.* (1965) **20**, 89.
(84) van Deemter, J. J., *Ind. Eng. Chem.* (1953) **45**, 1227.
(85) *Ibid.* (1954) **46**, 2300.
(86) Johnson, B. M., Froment, G. F., Watson, C. C., *Chem. Eng. Sci.* (1962) **17**, 835.
(87) Schulman, B. L., *Ind. Eng. Chem.* (1963) **55** (12), 44.
(88) Olson, K. E., Luss, D., Amundson, N. R., *Ind. Eng. Chem., Process Design Develop.* (1968) **7**, 96.
(89) Ozawa, Y., *Ind. Eng. Chem., Process Design Develop.* (1969) **8**, 378.
(90) Gonzalez, L. O., Spencer, F. H., *Chem. Eng. Sci.* (1963) **18**, 753.
(91) Zhorov, Yu. M., Panchekov, G. M., Laz'yan, Yu. I., *Zh. Fiz. Khim.* (1967) **41**, 1574.
(92) Dart, J. C., Oblad, A. G., *Chem. Eng. Progr.* (1949) **45**, 111.
(93) Menon, P., Sreeramamurthy, R., *J. Catalysis* (1967) **8**, 95.
(94) Jackson, R., Pirie, J. M., Eds., "Optimum Temperature Gradients in Tubular Reactors with Decaying Catalyst," p. 33, London Institute of Chemical Engineering, 1965.
(95) Jackson, R., Pirie, J. M., *Trans. Inst. Chem. Engrs.* (1967) **45**, 160.
(96) Chou, A., Ray, W. H., Aris, R., *Trans. Inst. Chem. Engrs.* (1967) **45**, 153.
(97) Ogunye, A. F., Ray, W. H., *Trans. Inst. Chem. Engrs.* (1968) **46**, 225.
(98) Ogunye, A. F., Ray, W. H., *AIChE J.* (1971) **17**, 43, 365.
(99) Ogunye, A. F., Ray, W. H., ADVAN. CHEM. SER. (1972) **109**, 497.
(100) Crowe, C. M., *AIChE J.*, in press.

(101) Szépe, S., Ph.D. dissertation, Illinois Institute of Technology, Chicago (1966).
(102) Szépe S., Levenspiel, O., *Chem. Eng. Sci.* (1968) **23**, 881.
(103) Paynter, J. D., *Chem. Eng. Sci.* (1969) **24**, 1277.
(104) Dalcorso, J. P., Bankoff, S. G., *Proc. Intern. Congr. Chem. Reaction Eng., 1st, Washington, D. C.* (June 1970).
(105) Szépe, S., ADVAN. CHEM. SER. (1972) **109**, 510.
(106) Lee, S.-I., Crowe, C. M., *Chem. Eng. Sci.* (1970) **25**, 743.
(107) Volin, Y. M., Ostrovskii, G. M., *Automatika Telemech.* (1964) **25**, 1414.
(108) *Ibid.* (1965) **26**, 1197.
(109) Volin, Y. M., Ostrovskii, G. M., *Prikl. Mat. Mekh.* (1965) **29**, 708.
(110) Aris, R., *Am. Scientist* (1970) **58**, 419.
(111) Chen, N. Y., Weisz, P. B., *Chem. Eng. Progr., Symp. Ser.* (1967) **73**, 86.
(112) Tan, C. H., Fuller, O. M., *Can. J. Chem. Engr.* (1970) **48**, 174.
(113) Luss, D., *Chem. Engr. J.* (1970) **1**, 311.
(114) Campbell, J. S., *Ind. Eng. Chem., Process Design Develop.* (1970) **9**, 588.
(115) Racz, G., Szekely, G., Huzar, K., Olah, K., *Period. Polytech. (Chemical Engineering) Bucharest* (1971) **15**, 111.
(116) Gioia, F., *Ind. Eng. Chem., Fundamentals* (1971) **10**, 204.
(117) Gioia, F., Greco, G., Jr., *Quad. Eng. Chim. Ital. (Milan)* (1970) **6**, 11.
(118) Haynes, H. W., Jr., *Chem. Eng. Sci.* (1970) **25**, 1615.
(119) Miertschin, G. N., Jackson, R., *Can. J. Chem. Engr.* (1970) **48**, 702.
(120) Miertschin, G. N., Chemical Engineering Dept. Report, Rice University, Houston, Tex. (1969).

Supplementary Bibliography

Deactivation of Fischer-Tropsch Synthesis Catalysts

Mittasch, A., *Advan. Catalysis* (1950) **2**, 90.
Shultz, J. F., Hofer, L. F. E., Karn, F. S., Anderson, R. B., *J. Phys. Chem.* (1962) **66**, 501.
Karn, F. S., Shultz, J. F., Kelly, R. E., Anderson, R. B., *Ind. Eng. Chem., Process Design Develop.* (1963) **2**, 43; (1964) **3**, 33.
Anderson, R. B., Karn, F. S., Shultz, J. F., *J. Catalysis* (1965) **4**, 56.

Added Detail on Various Aspects of Coking

Crawford, P. B., Cunningham, W. A., *Petrol. Refiner* (1956) **35** (1), 169.
Tyuryaev, M. D., *J. Appl. Chem. USSR* (1939) **12**, 1462.
Wilson, J. L., den Herder, M. J., *Ind. Eng. Chem.* (1958) **50**, 305.
Blue, R. W., Engel, C. J., *Ind. Eng. Chem.* (1951) **43**, 494.
Plank, C. J., Nace, D. M., *Ind. Eng. Chem.* (1955) **47**, 2374.
Germain, J. E., Maurel, R., *Compt. rend.* (1958) **247**, 1854.
Herington, E. F. K., Rideal, E. J., *Proc. Roy. Soc. (London)* (1945) **A184**, 434.
Dobychin, D. P., Klibanova, T. M., *Zh. Fiz. Khim.* (1959) **33**, 869.
Melik-Zada, M. M., Musaev, M. R., Buzova, N. G., Safaralieva, I. G., *Kinetica i Kataliz* (1961) **2** (5).

Poisoning of Acidic Function Catalysts

Oblad, A. G., Milliken, Jr., T. H., Mills, G. A., *Advan. Catalysis* (1951) **3**, 199.
Pines, H., Manassen, J., *Advan. Catalysis* (1966) **16**, 49.
Hirschler, A. E., *J. Catalysis* (1963) **2**, 428; (1966) **6**, 1.
Peri, J. B., *J. Phys. Chem.* (1965) **69**, 231.
MacIver, D. S., Tobin, H. H., Barth, R. T., *J. Catalysis* (1963) **2**, 485.

Wilmot, W. H., Barth, R. T., MacIver, D. S., *Proc. Intern. Congr. Catalysis, 3rd,* p. 1288, North Holland, 1965.
Hirschler, A. E., Hudson, J. O., *J. Catalysis* (1964) **3,** 239.
Medema, J., Houtman, J. R. W., *J. Catalysis* (1966) **6,** 322.
Brouwer, D. M., *J. Catalysis* (1962) **1,** 22.
MacIver, D. S., Wilmot, W. H., Bridges, J. M., *J. Catalysis* (1964) **3,** 502.
Rooney, J. J., Pink, R. C., *Trans. Faraday Soc.* (1962) **58,** 1632.
Parera, J. M., Figoli, N. S., *J. Catalysis* (1969) **14,** 303.
Parera, J. M., Hillar, S. A., Vincenzini, J. C., Figoli, N. S., *Symp. Catalysis, 3rd, Edmonton, Alberta* (Oct. 1969); *J. Res. Inst. Catalysis, Hokkaido Univ.* (1968) **16** (2) 525.
Forni, L., Zanderighi, L., Cavenaghi, C., Carra, S., *J. Catalysis* (1969) **15,** 153.

RECEIVED October 21, 1970.

Contributed Papers

Perturbation Analysis of Optimal Decaying Catalyst Systems: Fixed-Time, Regular Perturbations

J. P. DALCORSO and S. G. BANKOFF, Chemical Engineering Department, Northwestern University, Evanston, Ill. 60201

The problem of the optimal temperature policy for an isothermal tubular reactor packed with a catalyst whose activity depends on its past temperature history has been the subject of several investigations. These *a priori* calculations are clearly useful, however, only as a first estimate since in many cases competing side-reactions have been neglected or the effective bifunctionality of the catalyst has been ignored. Furthermore, the parameters of the governing equations are never known precisely, so that an optimal feedback control, based upon sequential estimation of the parameters from the current measurements, is almost always necessary. Other factors which call for a modification of policy would include variations in feed composition, raw materials, and/or product prices. These problems fall into the class of regular perturbation problems in that the dimension of the state equation set remains unchanged as a small parameter assumes the value zero. The procedures for the unconstrained problem are well-known. The corresponding algorithm for the constrained policy problem is developed here. The sensitivity equations for a decaying catalyst problem dealing with a single irreversible reaction, upon taking into account a small secondary reaction (either consecutive or simultaneous) are then written down as an application of this development. The sensitivity quantities can be used for optimal feedback control in the following way. One introduces a Riccati transformation linking x_ϵ and λ_ϵ, derives conventional Riccati equations for the matrices $P(t)$ and $Q(t)$, and precomputes these matrices. These can be used, again in the conventional manner, to determine the control correction, based upon a measured deviation in the state variables δx, from the *a priori* values. Alternatively one can determine these deviations from a new estimate of the system parameters, ϵ, where $\epsilon = 0$ corresponds to the values of the parameters on the *a priori* reference path since, to the first order, $\delta x = x_\epsilon \epsilon$, etc. As shown by numerical studies, the sensitivity coefficients, x_ϵ, u_ϵ, and λ_ϵ, for path variables, and particularly $t_{1\epsilon}$, for the constraint entry time, allow an estimate of the maximum permissible deviations of the system parameters within which a first-order perturbation scheme is permissible.

As an example of a regular perturbation of the reference solution, we consider the case when an unwanted product is formed by a small additional reaction, $A \xrightarrow{k_1} B \xrightarrow{k_2} C$. Let α_1 and α_2 be the extent of the main and of the parasitic reactions, both of which proceed under the influence of the same catalyst. The dimensionless concentration of B at the reactor exit is then

$$F = \frac{c_B}{c_0}\Big|_{s=1} = \frac{1}{1 - \varepsilon a}\left(e^{-\varepsilon x \bar{k}_2 \tau} - e^{-x \bar{k}_1 \tau}\right) \tag{1}$$

where $\epsilon = k_2(T_o)/k_1(T_o)$ is a small parameter, and x is the catalyst activity, assumed to follow an m-th order Arrhenius decay law:

$$\dot{x} \equiv \frac{dx}{dt} = -\gamma \bar{k}_0 x^m \equiv f(x,u);\ x(0) = 1 \tag{2}$$

Here $a = k_2/k_1$ does not become small as $\epsilon \to 0$.

The objective function to be maximized is

$$J = \int_0^\theta F(x,u,\varepsilon)\, dt \tag{3}$$

subject to a maximum temperature constraint,

$$C(u) \equiv u(t) - U \leq 0 \tag{4}$$

Let $\lambda(t)$ be introduced as an adjoint variable, and $h \equiv F_o + \lambda f$ as the Hamiltonian function. For this control-constrained problem, we define an augmented Hamiltonian function by

$$H = F_0 + \lambda f + \mu C = 1 - e^{-\bar{k}_1 x \tau} - \lambda \gamma \bar{k} x^m + \mu(u - U) \tag{5}$$

The necessary conditions for optimality are then

$$H_u = \frac{E_0}{RT_0^2 u^2}\left(p x \tau \bar{k}_1 \exp(-x \bar{k}_1 \tau) - \gamma \lambda \bar{k}_0 x^m\right) + \mu = 0 \tag{6}$$

$$\frac{d\lambda}{dt} = -H_x = -\bar{k}_1 \tau \exp(-x \bar{k}_1 \tau) + m \lambda \bar{k}_0 \gamma x^{m-1}; \tag{7}$$

where $\lambda(\theta) = 0$; $p = E_1/E_o$. The Riccati functions P and Q now satisfy the following equations:

$$I_2: t_1 \le t \le t_f: \quad \begin{array}{ll} \dot{P} = -2Ph_{x\lambda} - h_{xx} & P(t_f) = 0 \\ \dot{Q} = -Qh_{x\lambda} - h_{x\varepsilon} & Q(t_f) = 0 \end{array} \tag{8}$$

$I_1: 0 \le t \le t_1$

$$\begin{aligned} \dot{P} &= P^2 \frac{f_u h_{u\lambda}}{h_{uu}} - P(f_x + h_{x\lambda} - \frac{h_{ux}}{h_{uu}}(h_{u\lambda} + f_u)) - h_{xx} + \frac{h_{ux}^2}{h_{uu}} \\ \dot{Q} &= Q(h_{x\lambda} - \frac{h_{u\lambda}}{h_{uu}}(Pf_u + h_{xu})) + \frac{h_{u\varepsilon}}{h_{uu}}(Pf_u + h_{xu}) - h_{x\varepsilon} \end{aligned} \tag{9}$$

where P and Q are continuous at the constraint entry time, t_1. The feedback control algorithm is then:

$$t \varepsilon I_1: \delta u = -H_{uu}^{-1}[(H_{ux} + f_u^T P)\delta x + (H_{u\varepsilon} + f_u^T Q)\varepsilon] \tag{10}$$

$$t \varepsilon I_2: \delta u = -C_u^{-1} C_x \delta x \tag{11}$$

The estimated change in constraint-entry time, dt_1, is given by:

$$dt_1 = \frac{(\mu_x x\varepsilon + \mu_\lambda \lambda \varepsilon + \mu_u u\varepsilon + \mu \varepsilon)\Big|_{\substack{\varepsilon=0 \\ t=t_1}}}{\left.\frac{dh_u}{dt} C_u^{-1} + h_u \left(\frac{dC_u^{-1}}{dt}\right)\right|_{\substack{\varepsilon=0 \\ t=t_1^+}}} \tag{12}$$

Numerical calculations indicate that the use of the corrected control always gives higher yields than the uncorrected one, but the improvement is hardly worthwhile because of the characteristic flatness of the optimum with respect to control in problems of this type. On the other hand, one finds that the constraint entry time sensitivity function, $t_{1\epsilon}$, depends strongly upon the ratios of the activation energies of the decay, main, and parasitic reactions (E_o, E_1, E_2). Typically, the parameter error, ϵ^*, which produces a 5% change in constraint entry time decreases by a factor of 10^{-3} as $q \equiv E_2/E_o$ varies from 0.1 to 1.5. Thus, for a highly temperature-sensitive parasitic reaction ($E_2/E_o = 1.5$), $\epsilon^* \sim 10^{-3}$ for $p \equiv E_1/E_o = 0.5$, whereas $\epsilon^* = 0.1$ when $p = 1.5$. Especially in the former case but also in latter case the *a priori* calculation of the optimal policy can be expected to be in serious error since the activation energies are rarely known to within 10%. Successive re-linearizations as new estimates become available are thus required, and the estimation procedures should be fairly precise. On the other hand, if $E_2/E_o = 0.5$, $\epsilon^* \sim 0.1$ for $p = 0.5$, while $\epsilon^* = 1$ for $p = 1.5$. The estimation procedures can thus be fairly rough if the parasitic reaction is relatively insensitive to temperature changes.

The Optimal Control of a Vinyl Chloride Monomer Reactor Experiencing Catalyst Decay

A. F. OGUNYE and W. H. RAY, Department of Chemical Engineering, University of Waterloo, Canada

The method of producing vinyl chloride *via* the acetylene route is described in detail by Geiger (*1*). Acetylene is passed with excess hydrogen chloride gas over a stationary catalyst of mercuric chloride deposited on activated carbon. The reactors may be made of steel 8–10 ft in diameter and 10–15 ft in height with about 150–500 tubes of small diameter. The tubes are filled with catalysts of about ¼-inch diameter and containing 8–10% $HgCl_2$. The mixed reactant gases are usually passed through fresh catalyst at about 120°C. The heat of reaction is −26,000 cal/gram mole of vinyl chloride produced. As a result of this high heat of reaction, hot spots are developed inside the reactors, and the mercuric chloride sublimes, leaving the inactive carbon support. The optimization of this reacting system when there is catalyst decay is a distributed parameter problem and is a practical example showing the value of the theory and computational techniques developed previously (*2, 3*).

Reaction Rate Expression

The following mechanisms have been proposed by Gelbshtein *et al.* (*4, 5, 6*) after extensive studies on the complexes formed by the metallic chlorides of mercury, cadmium, zinc, and bismuth with acetylene and hydrogen chloride

$$HgCl_2 + HCl \overset{K_1}{\rightleftarrows} HgCl_2 \cdot HCl$$

$$HgCl_2 + C_2H_2 \overset{K_2}{\rightleftarrows} HgCl_2 \cdot C_2H_2$$

$$HgCl_2 \cdot HCl + HCl \overset{K_3}{\rightleftarrows} HgCl_2 \cdot 2HCl$$

$$HgCl_2 \cdot HCl + C_2H_2 \overset{K_4}{\rightleftarrows} HgCl_2 \cdot C_2H_2 \cdot HCl$$

$$HgCl_2 \cdot C_2H_2 \cdot HCl + HCl \overset{K_5}{\rightarrow} HgCl_2 \cdot HCl + C_2H_3Cl$$

where K_1, K_2, K_3, K_4 are equilibrium constants, and K_5 is a velocity constant. On the basis of these reaction mechanisms Gelbshtein *et al.* (*4*) proposed the rate of reaction to be given by

$$r = \frac{kP^2(1 - x)(\xi - x)}{(1 + \xi - x)\{(k' + P)(\xi - x) + k'\}} \tag{1}$$

where $k = K_5K_4K_3^{-1} = 1.05 \times 10^5 \exp\{-6000/RT\}$ moles C_2H_2/atm–hr–m^3

$k' = K_3^{-1} = \exp\{(14.6T - 7700)/RT\}$ atm

P = total pressure, atm, T = temperature of the system, °K

x = conversion

ξ = molar ratio of hydrogen chloride to acetylene in feed

Reactor Model

The catalyst decay as a result of sublimation caused by the reactor hotspots, may be regarded as an evaporation or a desorption process. By assuming a monolayer coverage of the carrier by mercuric chloride catalyst and that the catalyst activity is proportional to the number of active sites covered by the mercuric chloride, the model for the catalyst decay becomes

$$\frac{\partial y}{\partial t} = -\rho \exp\{-\Delta H_E/RT\}y$$
$$y(z,0) = W(z) \tag{2}$$
$$0 \leq t \leq 1,\ 0 \leq z \leq 1$$

where the spatial and temporal coordinates have been normalized to lie between 0 and 1, ΔH_E = heat of evaporation of mercuric chloride, y = catalyst activity, ρ = dimensionless preexponential factor for the decay reaction rate.

With the assumptions stated and justified in Ogunye and Ray (7), the following equations describe the extent of reactions, temperature, and pressure along the reactor length:

$$\frac{\partial x}{\partial z} = \frac{\beta \exp(-E_1/RT)yP^2(1 - x)(\xi - x)}{V_o(1 + \xi - x)\{(k' + P)(\xi - x) + k'\}} \tag{3}$$
$$0 \leq z \leq 1,\ 0 \leq t \leq 1$$

where the initial extent and C_2H_2 feed rate are given by

$$x(0,t) = 0$$
$$V_o = 1{,}000\chi(26 + 36.5\xi)^{-1} \tag{4}$$

$$\frac{\partial T}{\partial z} = \left\{\frac{(-\Delta H)\ \beta \exp\{-E_1/RT\}yP^2(1 - x)(\xi - x)}{(1 + \xi - x)[(k' + P)(\xi - x) + k']} - h_oA'L(T - U_c)\right\}\{V_1C_p\}^{-1};\ T(0,t) = V_2(t) \tag{5}$$

where the velocity of the flowing mixture is given by

$$V_1 = \frac{\chi RT}{P}\left\{\frac{1 + \xi - x}{26 + 36.5\xi}\right\}; \quad (6)$$

and C_p is given (8) by

$$C_p = [0.469 + 0.311\xi - 0.18x](1 + \xi - x) \quad (7)$$

$$\frac{\partial P}{\partial z} = \frac{-\chi(1 + \xi - x)RTL}{g_c dp(26 + 36.5\xi)} \frac{1 - \varepsilon_o}{P\varepsilon^3_o}\left\{\frac{(1 - \varepsilon_o}{dp} \nu + 1.75\chi\right\} \quad (8)$$

$$0 \leq z \leq 1, 0 \leq t \leq 1$$

where $P(0,t) = V_3(t)$

The Optimization Problem. If we assume as did Ogunye and Ray (2, 3) that the optimal operating time can be determined separately, we will choose our controls to maximize profits over the period of operation, $0 \leqslant t \leqslant 1$. The objective we choose has the form

$$I = \int_0^1 \chi\{x(1,t) - C_1[1 - x(1,t)] - C_2[\xi - x(1,t) - C_3x(1,t) \quad (9)$$

$$-C_4x(1,t)\}dt - a\int_0^1 w(z)dz$$

where the first term represents the value of the vinyl chloride product, the second and third terms the cost of separating reactants from the product, the fourth and fifth terms the cost of reactants, and the last term the cost of catalyst.

The controls which we assume we can manipulate are:

(a) The coolant temperature, $U_c(t)$
(b) The inlet temperature and pressure, $V_2(t)$, $V_3(t)$
(c) The molar ratio of HCl to C_2H_2, $\xi(t)$
(d) The feed rate (mass velocity) to the system, $\chi(t)$
(e) The initial catalyst distribution, $w(z)$

where it is assumed that there are fixed upper and lower bounds on these controls.

Computational Results

The maximum principle computational algorithms described in detail in Ogunye and Ray (2, 3) and Ogunye (9) were used to synthesize the controls. The method of characteristics with fourth-order Runge-Kutta method was used to integrate the state and adjoint equations. The set of constant parameters used for the numerical studies are given in Ogunye and Ray (7).

The policy when the inlet temperature and pressure as well as the mass velocity of the reacting system were kept constant throughout the cycle of operation at 150°C, 1.5 atm, and 1.5 kg/m² sec respectively calls for a gradual elevation of the cooling jacket temperature during the time of operation (*cf.* Ref. 7).

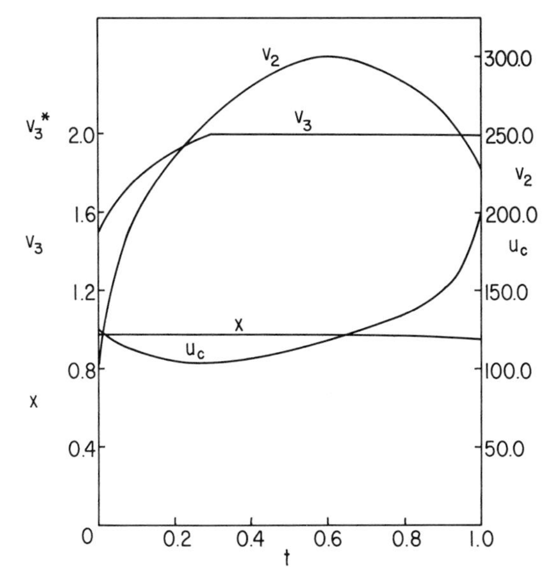

Figure 1. The optimal policy under U_c, V_2, V_3 *control*

Figure 1 shows the result of optimizing with respect to U_c, V_2, and V_3. An upper bound of 2 atm was picked for V_3 for this case. The optimal policy is again nearly constant conversion with the pressure, coolant, and jacket temperature being raised to compensate for falling catalyst activity.

Figures 2 and 3 show the results obtained when the catalyst distribution problem is considered simultaneously with the U_c, ξ, and χ policies. In this problem it is necessary initially to distribute the catalyst optimally throughout the bed. This can result in substantial savings by eliminating the hot spots in the reactor. If the average catalyst cost over the cycle of operation is a, the optimal policy will be pure catalyst loading only for $a \leqslant 0.076$. For example, with the cost of the catalyst, $a = 0.4$, the optimal policy (*cf.* Figure 3) calls for a high catalyst activity at the entrance of the reactor to attain a reasonable reaction temperature, followed by low catalyst activity to prevent hot spots. The increased activity

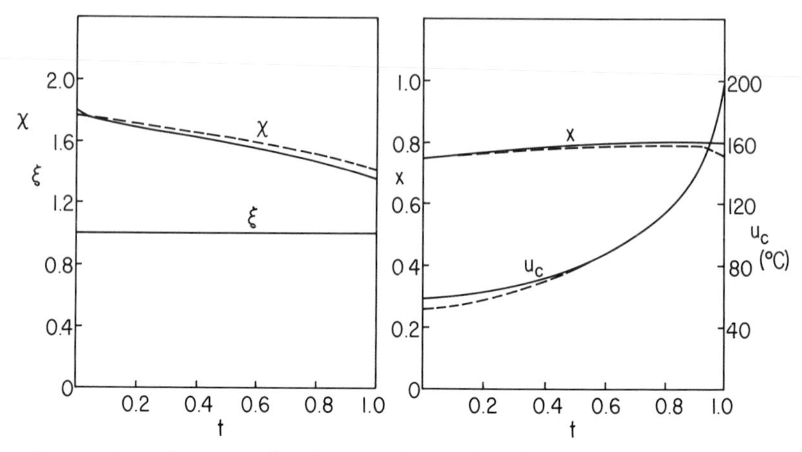

Figure 2. The optimal policy for the catalyst distribution problem

at the outlet of the reactor is required to aid the falling reaction rate. The final decrease in activity probably indicates that the reactor is too long and catalyst at the end is being wasted.

The hot spot temperature rise with pure catalyst loading is 350°C. When the optimal catalyst distribution is considered, the hot spot temperature rise is only 150°C. Even though the total production resulting from the reduced catalyst loading was found to be 9% lower than the case with pure catalyst loading, the savings in the cost of the catalyst gives an increase in the objective of 35%.

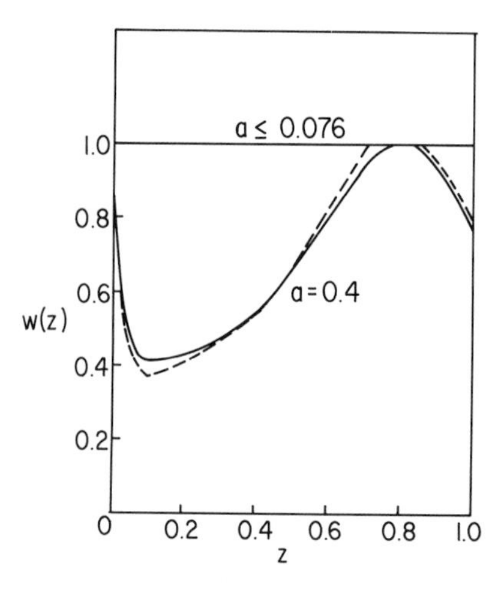

Figure 3. Optimal catalyst distribution

Conclusions

The models for both the main reaction and the decay rate are based on fundamental kinetic studies and embody few empirical assumptions. All the optimal policies were found from two starting points with only the differences noted in Figures 1, 2, and 3. Computational requirements for a complete optimization were modest, ranging from 5 minutes (on an IBM 360/75) for the results in Figure 1 to 15 minutes for the results in Figures 2 and 3.

One interesting result is that constant conversion appears to be a hallmark of all the optimal policies even though this system is much more complicated than the very simple ones in which it can be shown analytically that constant conversion is best (*10, 11*). A second result of equal interest is the sine wave-like shape of the optimal catalyst loading profile in Figure 3—an unexpected result which satisfies physical intuition. Finally, the control engineer should be amazed at the unbelievably rapid convergence of the gradient techniques for this problem, because the optimal control policy for most of the control variables was a singular one.

Acknowledgments

This research was supported by the National Research Council of Canada. We are also indebted to the University of Waterloo Computing Centre for providing computing time for this study.

(1) Geiger, M. G., Ph.D. Thesis, Purdue University (1954).
(2) Ogunye, A. F., Ray, W. H., *A.I.Ch.E. J.* (1971) **17,** 43.
(3) *Ibid.,* p. 365.
(4) Gelbshtein, A. I., Shcheglova, G. G., Khomenko, A. A., *Kinetika i Kataliz* (1963) **4,** 625.
(5) Gelbshtein, A. I., Siling, M. I., *Kinetika i Kataliz* (1963) **4,** 303.
(6) Gelbshtein, A. I., Siling, M. I., Sergeeva, G. A., Shcheglova, G. G., *Kinetika i Kataliz* (1963) **4,** 149.
(7) Ogunye, A. F., Ray, W. H., *Ind. Eng. Chem., Prod. Design Develop.* (1970) **9,** 619.
(8) Yablonskii, G. S., Kamenko, B. L., Gelbshtein, A. I., Slinko, M. G., *Intern. Chem. Eng.* (1968) **8** (1), 6.
(9) Ogunye, A. F., Ph.D. Thesis, University of Waterloo (1969).
(10) Chou, A., Ray, W. H., Aris, R., *Trans. Inst. Chem. Eng.* (1967) **45,** T153.
(11) Szepe, S., Ph.D. Thesis, Illinois Institute of Technology (1966).

Deactivation of Mordenite, Aluminum-Deficient Mordenite, and Faujasite Catalysts in the Cracking of Cumene

PAUL E. EBERLY, JR. and C. N. KIMBERLIN, JR., Esso Research Laboratories, Humble Oil & Refining Co., Baton Rouge, La. 70821

Mordenite type zeolites have become of increasing interest as catalysts and adsorbents in the petroleum industry (*1*). Basically, the pore structure of mordenite consists of parallel adsorption tubes each having an approximately elliptical opening with a major and minor diameter of 6.95 and 5.81 A., respectively. This structure prohibits motion of molecules from one main tube to the other. Consequently, mordenite is susceptible to loss of much of its adsorptive capacity by the presence of small amounts of impurities (*2*). By varying synthesis conditions it is possible to prepare mordenites which have quite different adsorptive properties because of some, as yet undefined, change in structure. Mordenites have been classified as "large-port" or "small-port," depending on whether or not they adsorb large molecules such as benzene and cyclohexane (*3*). Further varieties of mordenite can be produced by removing aluminum from the structure by strong acid treatment with HCl. This study compares the cracking and sorption properties of H-mordenite with a conventional SiO_2–Al_2O_3 ratio to those of a highly aluminum-deficient mordenite.

Results

Properties of the mordenites used in this study are given below:

Catalyst	H–M(12)	H–M(64)
Composition, wt %		
Na_2O	0.1	0.0
Al_2O_3	12.5	2.6
SiO_2	87.4	97.4
Relative x-ray cryst., %	100	90
Surface area, m^2/gram	540	602

Hydrogen-mordenite with a conventional SiO_2–Al_2O_3 ratio of 12, H-M(12), was prepared from the Na form by exchange with NH_4NO_3. To prepare the aluminum-deficient mordenite, H-M(64), the hydrogen-form was extracted for several hours with 5*N* HCl. The removal of 80% of the aluminum results in surprisingly little change in the x-ray diffraction pattern. The relative crystallinity of H-M(64) is 90% that of H-M(12). Its surface area and pore volume are larger.

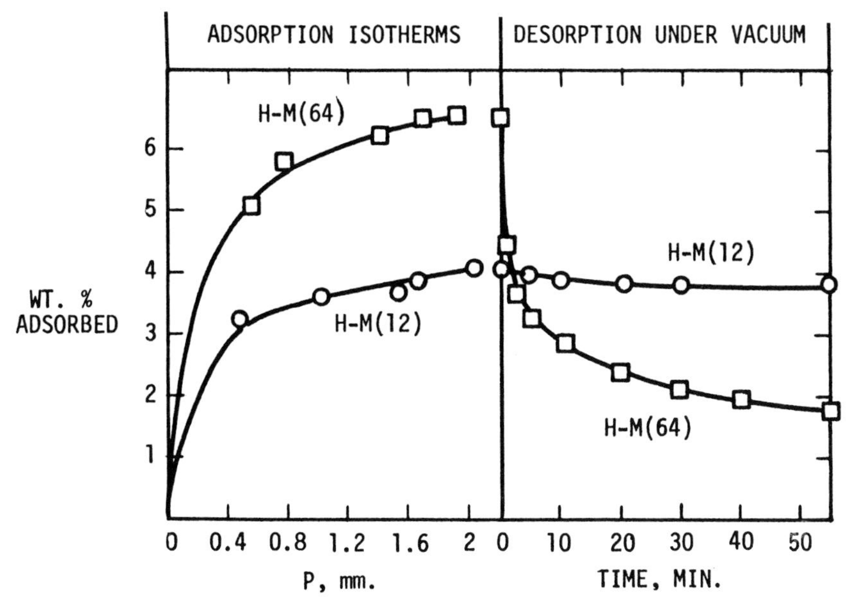

Figure 1. Cumene adsorption at 200°F

The adsorption of cumene was studied at 200°F. Results are shown in Figure 1. The isotherms on the left show that both materials are effective adsorbents even at pressures below 2 mm. This behavior is characteristic of zeolites having pores of molecular dimensions. H-M(64) has a higher capacity for cumene. Adsorption equilibria are established more rapidly on this material than on H-M(12). The desorption curves show that cumene is very difficult to remove from conventional mordenite.

In cumene cracking experiments, the activity (A) of the catalyst is defined as

$$A = \frac{1}{W} \log_{10} \frac{100}{100-C} \tag{1}$$

where W is the weight of the catalyst and C is mole % benzene in the aromatic fraction. The activity varied with time (t) according to Equation 2:

$$A = at^n \tag{2}$$

This relationship has previously been reported for amorphous silica-alumina catalysts (*4, 5*). Results on the two mordenites are shown in Figure 2. H-M(64) has nearly twice the activity for cumene cracking. The difference in the slopes (n) is not considered significant.

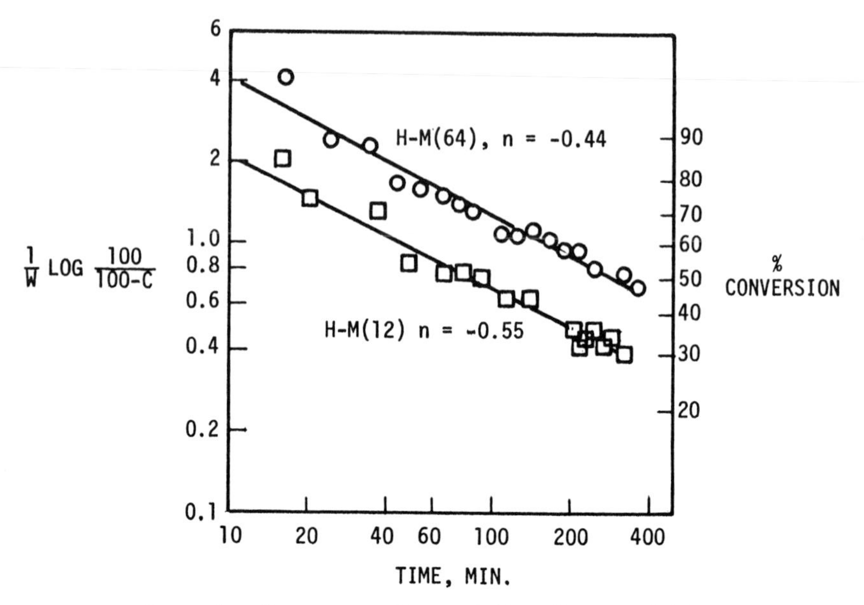

Figure 2. Cumene cracking at 0.33 w/hr/w and 450°F

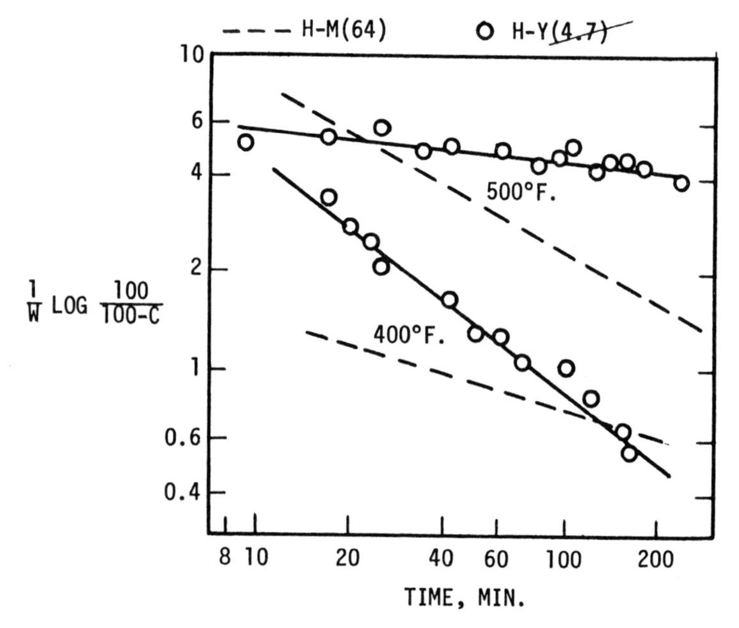

Figure 3. Cumene cracking at 0.33 w/hr/w

Runs were also made on hydrogen-faujasite, H-Y, which is a key component in many of the newer, commercial cracking catalysts. The results are shown in Figure 3. At 400°F, the activity undergoes a severe decline with time on stream. H-Y has a considerably higher adsorption capacity, and at this low temperature we suspect that the propylene formed is being adsorbed rapidly and covering the active sites. A previous report demonstrated that propylene on H-Y undergoes polymerization, dehydrogenation, and cyclization to form aromatic ring structures at 400°F (*6*). Apparently, this does not occur to the same extent on mordenite catalysts. At 525°F; the activity is quite high, and deactivation is much less severe.

Discussion

The fact that most of the aluminum can be removed from mordenite with only minor changes in the x-ray diffraction pattern is worthy of comment. It suggests that perhaps the lattice has reformed so that the original alumina tetrahedra are now replaced by silica tetrahedra. This process would give a crystal having the same diffraction pattern with the peaks shifted slightly to larger angles corresponding to a lower unit cell dimension (the Si–O bond is shorter than the Al–O bond). This removal of aluminum from the lattice eliminates the need for cations or protons for charge neutralization and would therefore be expected to lower acidity. At the same time, however, acid leaching removes extraneous materials from the mordenite channels resulting in larger adsorptive capacity and a greatly decreased resistance to adsorptive diffusion. This decreased resistance is also aided by lowering the acidity of the surface.

With this in mind, the higher activity of H-M(64) can be attributed largely to a lowering in diffusion resistance. At any given time in the cracking cycle, the activity follows an Arrhenius type relationship with temperature, and the apparent activation energy amounts to 11 kcal/mole for both H-M(12) and H-M(64). This indicates that the reaction is strongly diffusion limited since values for other more open catalysts are in the range of 20–35 kcal/mole. Diffusion resistance in H-M(64) is considerably less than that in H-M(12) as seen by the adsorption measurements, but this does not increase the apparent activation energy above that for H-M(12). Katzer (*7*) observed that counterdiffusion of benzene and cumene did not occur on a H-mordenite with conventional SiO_2–Al_2O_3 ratio. This is in line with our results on the difficulty of desorbing material from H-M(12) and suggests a low effectiveness factor for this catalyst.

(1) Burbidge, B. W., Keen, I. M., Eyles, M. K., ADVAN. CHEM. SER. (1971) **102**, 400.
(2) Beecher, R., Voorhies, A., Jr., Eberly, P. E., Jr., *Ind. Eng. Chem. Prod. Res. Develop.* (1968) **7**, 203.
(3) Sand, L. B., *Conf. Mole. Sieves, Soc. Chem. Ind.* (1968) 71.
(4) Voorhies, A., *Ind. Eng. Chem.* (1945) **37**, 318.
(5) Eberly, P. E., Kimberlin, C. N., Jr., Miller, W. H., Drushel, H. V., *Ind. Eng. Chem. Process Design Develop.* (1966) **5**, 193.
(6) Eberly, P. E., Jr., *J. Phys. Chem.* (1967) **71**, 1717.
(7) Katzer, J. R., Thesis Digest, Massachusetts Institute of Technology, Sept. 1969.

Cocurrent and Countercurrent Moving Bed Reactors

STEPHEN SZÉPE, Department of Energy Engineering, University of Illinois at Chicago Circle, Chicago, Ill. 60680

Reactions subject to catalyst deactivation may be conducted in stationary, in fluidized, or in moving beds. This paper analyzes the operation of moving bed reactors. The process involved is a single reaction over a deactivating solid catalyst, with reactants and products being in a single fluid phase of constant density. The rate equations considered for the reaction and deactivation kinetics are separable (*1*), having the following forms at any given temperature

$$r = kaf(x) \tag{1}$$

$$\rho = \kappa\psi(a)\phi(x) \tag{2}$$

A specific case of such kinetics [$f(x) = (1 - x)^2$; $\psi(a) = a$, and $\phi(x) =$ constant] has been analyzed for cocurrent reactors by Weekman (*2*). In the present study, arbitrary functional forms are allowed for all three terms, with the only restriction that

$$\begin{aligned} f(x) &> 0 \text{ in } x_o \leq x \leq x_f \\ \psi(a) &> 0 \text{ in } a_o \geq a \geq a_f \\ \phi(x) &> 0 \text{ in } x_o \leq x \leq x_f \end{aligned} \tag{3}$$

Compared with the other two types of reactions, moving beds have an additional degree of freedom—namely, they may be operated either cocurrently or countercurrently. A major objective of this analysis is to develop some criterion or criteria for exploiting this potential advantage —*i.e.*, to determine the influence of the functional forms of $f(x)$, $\psi(a)$, and $\phi(x)$ on the selection of the preferred contacting scheme.

The reaction system described by Equations 1–3 is analyzed in an ideal moving bed reactor in which, by definition, both the fluid and the catalyst streams are in plug flow, having space times of τ_f and θ_f, respectively.

It is shown that the steady state isothermal operation of ideal moving bed reactors may be studied conveniently for either contacting scheme and for all functional forms of $f(x)$, $\psi(a)$, and $\phi(x)$ by following the trajectory in the conversion–activity plane. Such a trajectory will show the corresponding values of these two variables at various points along the length of the reactor. Under restriction No. 3 any trajectory will be strictly monotonic. Moreover, in a plane of transformed coordinates defined by $X(x) = \int_{x_{ref}}^{x} [\phi(x)/f(x)]\, dx$ and $A(a) = \int_{a_{ref}}^{a} [a/\psi(a)] da$, the trajectory will be a straight line, having a slope of $-\alpha = -k\tau_f/\kappa\theta_f$ in cocurrent operation, and a slope of $\alpha' = k'\tau_f'/\kappa'\ {}_f'$ in countercurrent operation (countercurrent operations are denoted by prime signs). Cocurrent and countercurrent operations are compared by the solution of two general problems.

Problem 1. Which type of contacting, cocurrent or countercurrent, will utilize the catalyst more effectively? If $\alpha' = \alpha$, $x_o' = x_o$, $a_o' = a_o$, and if the two contacting schemes are to yield identical used catalyst activities ($a_f' = a_f$), which scheme will result in a higher conversion?

One may expect that the solution to this problem will depend strongly on the functional forms of $f(x)$, $\psi(a)$, and $\phi(x)$ and that useful criteria can be developed only for some limited class of functions such as power equations. It can be proved, however, that a much simpler and more general solution holds—namely, for all functions satisfying restriction No. 3:

> Theorem 1. Cocurrent and countercurrent processes operated under conditions of $\alpha' = \alpha$, $x_o' = x_o$, $a_o' = a_o$ and $a_f' = a_f$ will result in identical conversions; $x_f' = x_f$.

In addition to providing the answer to the catalyst utilization problem, this result has another important application. For countercurrent operations, the inherent mixed boundary conditions will generally necessitate trial and error computation. Nevertheless, for countercurrent moving bed reactors characterized by Equations 1–3, the trial and error method may be circumvented by using the above result combined with the proper choice of the independent variable.

Theorem 1 by no means implies that the space times $\tau f'$ and τ_f belonging to these identical conversions are also equal. In fact, the rational selection of the preferred type of contacting should be made by evaluating and comparing space time requirements. This problem is formulated in the following way.

Problem 2. Which type of contacting, cocurrent or countercurrent, will utilize the reactor more effectively? If $\alpha' = \alpha$, $x_o' = x_o$, $a_o' = a_o$, and if the two contacting schemes are to yield identical conversions $(x_f' = x_f)$, which scheme will require less space time?

This reactor utilization problem is more difficult to solve than the first one, and it is advisable to attack it in two steps. It is known (*3, 4*) that some important optimization problems have simple analytical solutions for the special case of conversion-independent deactivation [$\phi(x) =$ constant]. Consequently, at first this special case is considered.

A simple and general solution is obtained again—namely, for all $f(x)$ and $\psi(a)$ satisfying requirement No. 3:

Theorem 2. If $\phi(x) =$ constant, then cocurrent and countercurrent processes operated under conditions of $\alpha' = \alpha$, $x_o' = x_o$, $a_o' = a_o$, and $x_f' = x_f$ will require identical space times; $\tau_f' = \tau_f$.

Finally, for the more general case of $\phi(x) \neq$ constant, the following result is obtained.

Theorem 3. For cocurrent and countercurrent processes operated under conditions of $\alpha' = \alpha$, $x_o' = x_o$, $a_o' = a_o$, and $x_f' = x_f$, $\tau_f' \leqslant \tau_f$ *if* $\frac{d\phi(x)}{dx} \leqslant o$ in $x_o \leqslant x \leqslant x_f$, and $\tau_f' \geqslant \tau f$ if $\frac{d\phi(x)}{dx} \geqslant o$ in $x_o \leqslant x \leqslant x_f$.

Thus, if $\phi(x)$ is a monotonically decreasing function, the preferred operation is countercurrent. In turn, cocurrent operation is preferred in case of a monotonically increasing $\phi(x)$.

Finally, if the function $\phi(x)$ is not monotonic, a single criterion depending only on the functional form of $\phi(x)$ cannot be developed; in this case, the proper choice will also depend on the initial and final conversions.

The last result, Theorem 3, underlines the importance of the conversion-dependent term of the deactivation rate equation. A reasonably good knowledge of this term is also essential in all other deactivation problems, such as in the design and optimization of stationary and fluidized bed reactors. Unfortunately, this term is often masked, and it is by far the most difficult to determine by conventional methods. An experimental method devised to eliminate, or at least to reduce this masking has already been proposed (*5*). The above result suggests that another and probably more sensitive way of determining $\phi(x)$ may be through kinetic studies made in laboratory moving bed reactors.

Nomenclature

a = catalyst activity
f = conversion-dependent term of the reaction rate equation
k = reaction rate constant
r = reaction rate
x = conversion
α = a parameter defined as $k\tau_f/\kappa\theta_f$
θ = catalyst space time
ρ = deactivation rate
τ = fluid space time
ϕ = conversion-dependent term of the deactivation rate equation
ψ = activity-dependent term of the deactivation rate equation

Subscripts

f = outlet condition
o = inlet condition
ref = reference condition

Superscripts

$'$ = countercurrent operation

(1) Szépe, S., Levenspiel, O., *Proc. European Symp. Chem. Reaction Eng., 4th, Bruxelles, Sept. 1968.*
(2) Weekman, V. W., Jr., *Ind. Eng. Chem., Process Design Develop.* (1968) **7**, 90.
(3) Szépe, S., Levenspiel, O., *Chem. Eng. Sci.* (1968) **23**, 881.
(4) Szépe, S., Ph.D. Thesis, Illinois Institute of Technology (1966).
(5) Szépe, S., *Proc. Ann. Meetg. AIChE, 62nd, Washington, D. C., Nov. 1969.*

Effect of Operating Variables on the Permanent Deactivation of Cracking Catalyst

J. E. GWYN, Shell Oil Co., Houston Research Laboratory, P. O. Box 100, Deer Park, Tex. 77536

A mathematical model for the loss of surface area of cracking catalysts in commercial units has been derived by considering the rate of coke laydown and the temperature rise from coke burning. This model is based on and is compatible with the extensive laboratory deactivation data. A generalized deactivation rate equation was developed to fit the laboratory data at various temperatures, steam partial pressures, and surface areas. This equation was used in connection with coke-burning calculations to obtain effective deactivation rates and equilibrium surface areas in a commercial unit. In normal operations, the most significant parameters are the difference between coke on spent catalyst and coke on regenerated catalyst and a temperature rise coefficient.

Observation of Deactivations in Commercial Cracking Units

Deactivation tests in commercial units indicated that individual unit conditions were much more important than the particular catalyst used. The surface areas, at 30 days in various units, of test catalysts (synthetic and clay extended) were nearly equal to the steady-state surface area of respective inventory catalyst.

Procedure for Model Development

It was apparent from deactivation rate data that the temperatures of the reactors and regenerators are not high enough in themselves to cause the degree of deactivation of cracking catalysts observed. A combination of high coke on catalyst and high oxygen concentration during coke burnoff appears necessary to attain individual particle temperatures at which deactivation becomes significant. Therefore, a general rate equation for the loss of surface area was developed from laboratory data. The particle temperature during regeneration used in this equation is expressed as a function of the coke added during a pass through the reactor (the incremental coke). Since the incremental coke on any catalyst particle depends on its residence time in the reactor, an expression for a weighted average regeneration temperature is calculated by integration over the residence time distribution. This expression is inserted into the deactivation rate equation, which is then integrated over the distribution of total residence times of catalysts in the unit to obtain the average surface area of the inventory.

Generalized Deactivation Rate Equation

Except for persistence effects from initial moisture content and for some minor effects of catalyst pore volumes, the deactivation rates can be expressed in terms of temperature, surface area, and steam partial pressure by

$$-\frac{dS}{dt} = F_b P_v^2 e^{a(T-T_0)} \quad (1)$$

where S = surface area in m²/grams

t = total catalyst residence time in the unit in hours

T = temperature of deactivation, °F

P_v = steam partial pressure in atmospheres

For a high alumina catalyst, designated G, and T_0 in °F

$$a = -0.00720 + 0.3879 \cdot S^{-1/2} \quad (2)$$

$$T_0 = 1681.17 - 2.043 \cdot S \quad (3)$$

F_b is the fraction of time that a particle is at the elevated temperature. The elevated temperature that a particle encounters upon mixing spent catalyst with air—*e.g.*, in the transfer line—is a function of coke make and may be represented approximately as:

$$T = b\Delta C + T_m \quad (4)$$

where T_m is the effective catalyst–air mix temperature.

Miller *et al.* (*1*) demonstrated that coke burning within a catalyst particle is oxygen diffusion limited. The coefficient was based on their numerical solutions at the level of particle temperature rise of interest; $b = 100°F/\%C$ at 0.2 atm partial pressure of O_2 and for a coke composition of 8% hydrogen (basis carbon).

The coke laydown on a catalyst particle is expressed as (*2*):

$$\Delta C = K_c S \left(\frac{\theta}{r}\right)^{1/2} \quad (5)$$

where $K_c = \dfrac{2\Delta C_{avg}}{\sqrt{\pi}\, S_{inv}}$

r = the mean residence time in the reactor

θ = the individual particle residence time in the reactor

ΔC_{avg} = the average coke laydown in the reactor

S_{inv} = the average inventory surface area

π = 3.1416

The integration over the residence time distribution in the reactor gave the following effective deactivation rate for catalysts of surface area, S:

$$-\frac{dS}{dt} = F_b P_v{}^2 \pi b K_c a S e^{\left[a(T_m - T_0) + \left(\frac{bK_c aS}{2}\right)^2\right]} \quad (6)$$

or

$$-\frac{dS}{dt} = K_d \gamma a S e^{\left[a(T_m - T_0) + \left(\frac{\gamma a S}{2}\right)^2\right]} \quad (7)$$

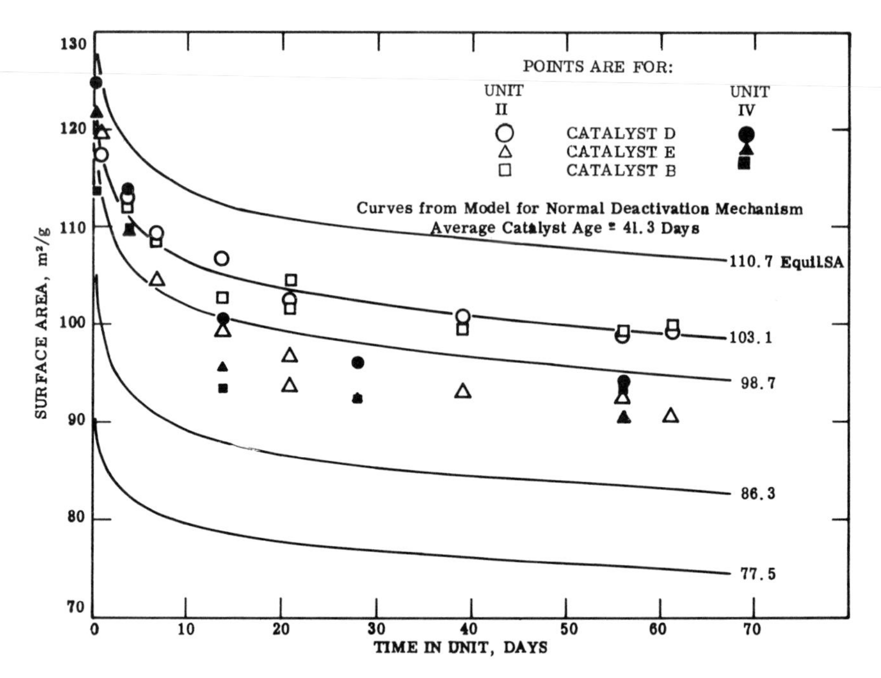

Figure 1. Comparison of catalyst deactivation by coke burning (normal) model with deactivation in catalytic cracking units

where $K_d = F_b P_v^2 \sqrt{\pi}$

$\gamma = bK_c$

Thus, there are three constants, K_d, γ, and T_m, which are functions of operating conditions.

The comparison of decline of given batches of catalysts in different inventories is given in Figure 1. The drop in surface area from about 500 m²/gram of the fresh catalyst to about 120 m²/gram in the first day is predicted well by the model. Long term deactivations were also generally consistent with experimental data.

Equilibrium Surface Area

To obtain the steady-state surface area, the age in the unit of the individual catalyst portion, ϕ, was introduced. The surface area of a catalyst portion at a given age is obtained by:

$$S_\phi = S_0 - \int_0^\phi \left(-\frac{dS}{dt}\right)_s dt \qquad (8)$$

The equilibrium (steady-state) surface area of the inventory is obtained by integration over the catalyst age distribution.

$$S_{eq} = \int_0^\infty S\phi \frac{1}{R} e^{-\phi/R} d\phi \tag{9}$$

where R = average catalyst age in the unit, and a well-mixed vessel residence time distribution is assumed.

The equilibrium surface areas for various unit parameters and R = 1000 hours are shown in Figure 2.

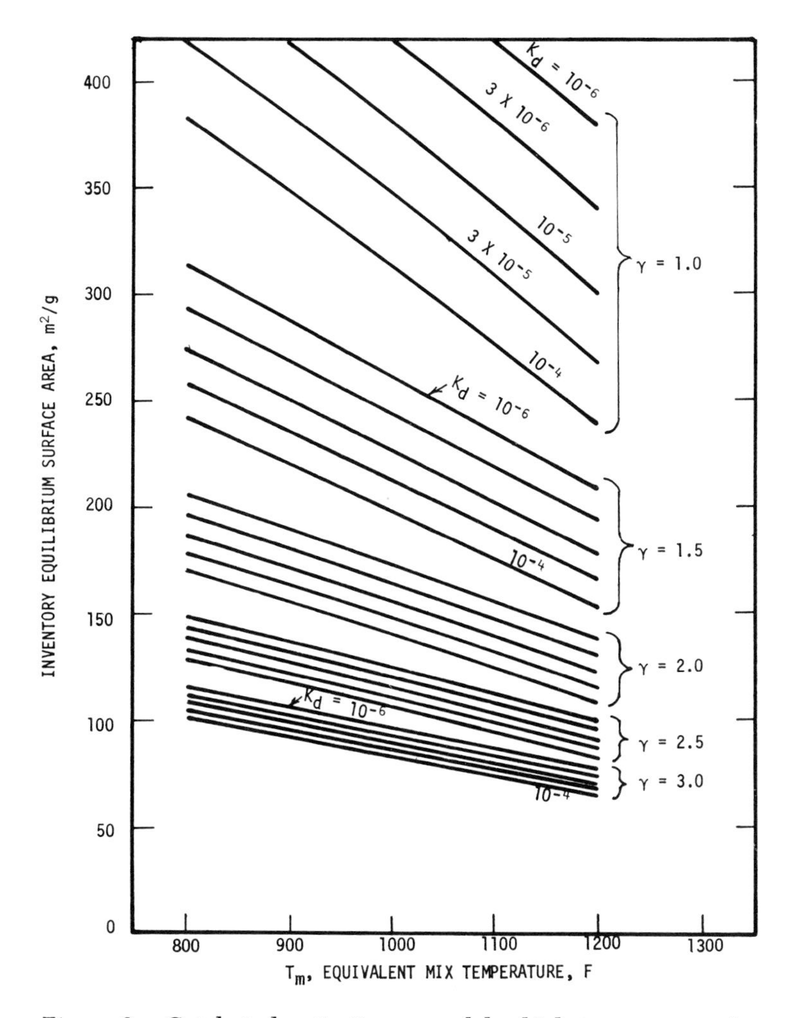

Figure 2. Catalyst deactivation caused by high temperature rise during regeneration burning

Evaluation of Parameters

The parameters for a specific catalytic cracking unit were evaluated by assuming:

(1) A particle burning time of 10 msec per pass (*1*)

(2) Partial pressure of steam equal to 0.8 of stripper pressure

(3) $\Delta C_{avg} = 1\%$

(4) Entrained hydrocarbon and prior carbon (on regenerated catalyst) add to the particle temperature and are included as $T_m = T_{cat/air\,mix} + b(C_r + C_e)$, where C_r is carbon-on-regenerated catalyst and C_e is unstripped hydrocarbon (as %C)

For this cracking unit:

$$K_d = 5.3 \times 10^{-5}$$

$$bK_c = 2.50$$

$$T_m = 1045°F$$

From Figure 2, S_{eq} = 105 m^2/gram (compared with 110 m^2/gram in inventory). Thus, the model closely approaches the CCU conditions during normal deactivation.

Significance of Parameters

Of the three parameters, the term $b \times K_c$ has the predominant effect on equilibrium surface area. The term b increases with oxygen partial pressure. The value of K_c reflects the coking tendencies at a given surface area. The parameter T_m reflects the effective mix temperature of the catalyst in the oxygen-rich area of regeneration. Countercurrent flow results in a high mix temperature but permits removal of entrained hydrocarbon before the main regeneration burning region is reached. The parameter K_d represents the frequency of regeneration and the steam partial pressure. The effect of the parameter is small and the assumption of persistence of stripper steam effects does not greatly change the results. Evaluating existing or proposed fluid catalytic cracking units in terms of these parameters should suggest means of minimizing deactivation of catalysts.

(1) Miller, R. S., Bondi, A. S., Schlaffer, W. G., *Ind. Eng. Chem., Process Develop.* (1962) **1** (3), 196–203.
(2) Blanding, F. H., *Ind. Eng. Chem.* (1953) **45,** 1186.

8

Industrial Process Kinetics and Parameter Estimation

HANNS HOFMANN

University of Erlangen, Nuremberg, West Germany

A review is given on principal objectives, some experimental techniques and up to date knowledge in planning of experiments for model screening and parameter estimation in chemical reaction kinetics. These lead to a general strategy for analyzing industrial process kinetics. Selected examples show the trend in kinetic description of homogeneous and heterogeneous industrial reactions.

This review focuses on (1) general statements on industrial process kinetics and parameter estimation, (2) some experimental techniques in chemical kinetics, (3) up-to-date knowledge in planning experiments for model discrimination and parameter estimation, and (4) a few selected papers concerning kinetics of important industrial processes. For a complete review of recent publications on industrial reaction kinetics reference is made to the excellent regular annual reviews on chemical reaction engineering (*1, 2, 3*). The kinetics of transport phenomena, which sometimes are important in industrial processes, are not discussed here.

General Statements

Kinetic studies provide the designer with reliable rate data to predict reactor performance and to select optimum operating conditions. In the chemical industry—big single train plants, short depreciation times, and worldwide competition—there is no doubt about the importance of this subject. No wonder many national and international symposia on chemical reaction engineering have special sessions on kinetics for industrial processes and techniques for parameter estimation (*4, 5, 6, 7, 8, 9*).

In the early days of chemical reaction engineering, investigations of process kinetics were often limited in usefulness because idealized operating conditions were chosen to find the "true" reaction mechanism or because the influence of impurities, catalyst deactivation, undesired side reactions, etc. were neglected in describing the process.

With a better understanding of chemical reaction modeling the situation changed, and an increasing number of papers now deal with the kinetics and parameter values of commercial processes. Nevertheless, some reserve remains in industrial companies, claiming process kinetics as the real knowhow; there are more publications concerning process kinetics from universities than from industry; therefore, even today, it is sometimes hard to judge the usefulness of published results, especially since some are contradictory. At the least more industrial feedback is needed.

Developing the kinetic model of a chemical reaction requires the solution to the following problems. Starting with experimental results from differential or integral laboratory reactors, pilot plants, or even commercial reactors the chemical reaction engineer must (Figure 1):

(a) Find a function f for the dependence of a value y (here the reaction velocity or the degree of conversion as the time integral of the reaction velocity) from the vector $\mathbf{x}$ of the independent variables such as mole fractions of the reactants, temperature, pressure, catalyst concentration etc.

(b) Estimate the vector $\boldsymbol{\kappa}$ of the parameters such as rate constants, absorption constants, or equilibrium constants, in the chosen function, to describe the experimental findings quantitatively.

(c) Test the significance of the postulated variables, sometimes in connection with the first two steps, to find the minimum number of variables necessary to describe the process accurately.

These steps can be done simultaneously or sequentially as indicated in Figure 1.

All these steps are well known from classical chemical kinetics. Standard textbooks recommend that one find a linear representation of the data (Figure 2) to derive the parameter values from the intercept and the slope of the least-squares line fitting the data. Today, however, applied statistics and nonlinear regression give us methods for designing and analyzing sequential experiments to gain maximum information from all three steps from a given number of experiments or vice-versa to obtain the desired information with minimum expense. Moreover, since all experimental results are affected by errors, we are able to draw detailed conclusions about the uncertainty of the results.

Since deficiencies in the design of experiments never can be compensated by an elaborate analysis of the results (whereas even a simple analysis of well planned experiments give significant results), the advantage of statistically designed experiments can no longer be overlooked. Activities in this special field are facilitated by computer techniques.

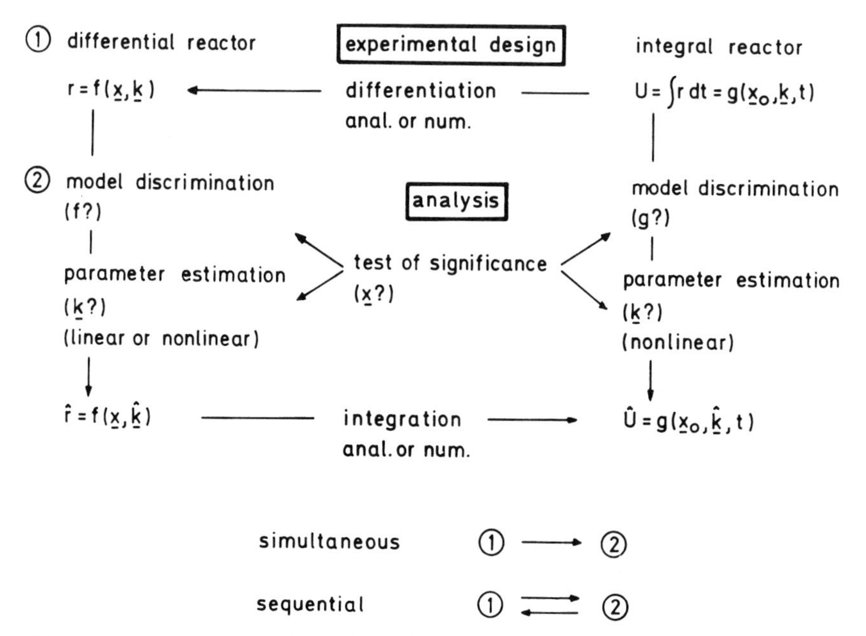

Figure 1. Experimental and mathematical steps in process kinetics and parameter estimation

$$r = k_n \cdot c^n \qquad r = \frac{k \cdot p_A}{1 + K \cdot p_A} \qquad \frac{dc}{dt} = k_n \cdot c^n \quad (n \neq 1)$$

$$\log r = \log k_n + n \cdot \log c \qquad p_A/r = 1/k + (K/k)p_A \qquad \frac{1}{c^{n-1}} - \frac{1}{c_o^{n-1}} = k_n \cdot t$$

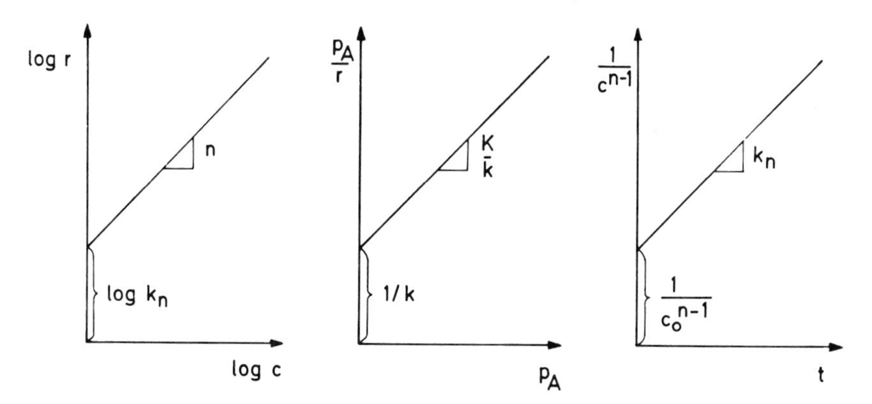

Figure 2. Types of linear plots in chemical kinetics

With growing experience in data logging and digital control, kinetic modeling and parameter estimation will be possible in the not too distant future through on-line experimentation aided by process computers.

At this stage of development the question arises as to the adequacy in generating, analyzing, and applying kinetic data for design purposes. Since digital computers yield accurate numerical calculations for analysis and design, the limiting step in the above-mentioned chain seems to be the data generation by experiments. There are two possibilities for improving this step:

(1) More precise measuring techniques to minimize experimental error (this point will not be stressed here).

(2) Guided experimentation to obtain the most precise information from the experiments.

Experimental Generation of Kinetic Data

In principle, industrial process kinetics can be studied on laboratory, pilot, or commercial scale. On the laboratory scale experimental conditions can be chosen to separate mass transfer and chemical phenomena as well as to guarantee isothermal conditions—*e.g.*, in a differential reactor. At the same time dose problems and difficulties in the analytical determination of the reactants may arise because of the small quantities of products or by-product. Additionally, the advantages of a differential reactor may be overcompensated by the small degree of conversion, which renders model screening more difficult because of the unavoidable experimental error. Most of these disadvantages can be eliminated by an integral reactor, provided that isothermal conditions can be fulfilled (*10*). Without too much expense for equipment, this seems to be possible only for reactions with an enthalpy of less than 20 kcal/mole.

Considering this situation, generation of kinetic data in an adiabatic integral laboratory reactor seems to be progressing. Not much has been published about this type of reactor, but the results so far are encouraging (*11, 12*), at least for reactions not too kinetically complex or too exothermic.

On the pilot or commercial scale we find the reverse situation because analytical difficulties normally do not arise. Unfortunately, there it is more or less impossible to separate mass transfer and chemical phenomena, and heterogenous models must be used with all their complications. Again, adiabatic operating conditions are favored. Data analysis is facilitated because of the linked energy and material balances and because of the absence of radial temperature and concentration gradients. However, the flexibility of a commercial reactor needed to cover the region of interest is poor. Nevertheless, valuable information on process kinetics can be drawn from commercial scale reactors, and this possibility is not used enough.

true rate (unknown)

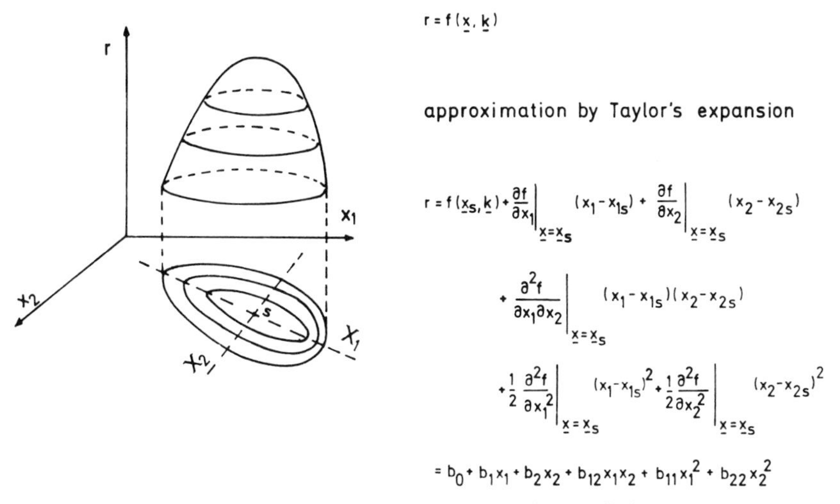

Figure 3. Rate surface approximation by Taylor's expansion

Model Building and Selection

To find the most suitable function f to describe the experimental results or more especially to correlate the reaction rate and the independent variables is the main problem in industrial process kinetics. The rate equation must be detailed enough to describe minor influences, but it should not be too complex to be useful in reactor design. Empirically developed models, despite their simplicity, are of limited value in extrapolation. Therefore, theoretically founded models are desirable.

For industrial processes the deduction of a theoretically based rate expression from the true reaction mechanism, if possible at all, normally leads to a much too complex expression for reaction engineering purposes, even with simplifications like the Bodenstein theorem of quasi-stationarity. Moreover, in the case of newly developed processes, the physicochemical fundamentals often are not developed enough to understand the true reaction mechanism. Therefore, simplified reaction schemes must be used, taking advantage of the principle of the rate-determining step (*13*) to develop a so-called mechanistic model of the reaction.

In this situation, as a first step, response surface methods (*14*) can be used to explore the dependence of the reaction rate on the independent variables. This exploration yields:

(a) An empirical rate equation in form of a power series sometimes useful for preliminary designs,

(b) An indication of possible mechanistic rate expressions.

Power function model:

$$r = k_{\infty}(\exp -E/RT)\, x_1^{\alpha} \cdot x_2^{\beta} \cdots\cdots$$

Hyperbolic model:

$$r = \frac{k\,(x_1^{\alpha} x_2^{\beta} - \frac{1}{K} x_3^{\gamma} x_4^{\delta})}{(K_0 + \sum K_{ij}\, x_i^{m} x_j^{p})^{n}}$$

Figure 4. Typical model equations in chemical kinetics

The basic idea is to approximate the unknown rate expression, forming a hypersurface in the n-dimensional space of the independent variables (Figure 3), by a Taylor expansion about a stationary point, using only terms with first- and second-order derivatives. To explore the shape of this hypersurface, the experimental data are fitted to this (empirical) model equation by linear regression, determining the p parameter values b_i. To get maximum information from the data, the desired experiments in this stage should be planned—*e.g.*, as factorial, regular simplex, or central composite design (*15*). Even if errors in the factor levels are unavoidable (*16*), these designs are optimal. More insight into the shape of the response surface and the importance of several variables is provided by a transformation of the second degree polynomial into its canonical form, eliminating linear and mixed terms (*14*). Confidence intervals or tests of significance are helpful in selecting the desired minimum number of independent variables or variable combinations (*17*). However, one should avoid certain pitfalls in statistical analysis and explanation of data (*18, 19*). The most refined version of linear regression is the stepwise multiple regression technique, guided by the value of the multiple correlation coefficient and executed automatically on a digital computer (*20, 21*).

Having some idea about the significant variables, their combinations, and the shape of the response surface, mechanistic models can be established with a better background. Nevertheless, in nearly all cases more than one reaction model can fit the experimental data at the same confidence level; hence, the problem of model screening arises (*12*).

Typical model equations, suitable for chemical reactions, are nonlinear in the parameters and are of the power function or hyperbolic type (Figure 4). The power function model normally is preferred for homogeneous reactions and the hyperbolic model for heterogeneous surface reactions. There are some indications, however, that power function models can also be used successfully with heterogeneous surface reactions (*22, 23, 24*). Alternatively, based on a "steady-state" hypothesis, the hyperbolic model has some justification in homogeneous reactions.

Rival models can be screened with several discrimination criteria (Figure 5), developed from statistics and information theory (*25, 26, 27*). Starting from a given set of preliminary data—*e.g.*, from the exploration of the response surface—a sequence of experiments can be planned according to these criteria at settings of the independent variables for which on can expect the response of rival models to show their maximum divergence and for which the response is relatively well known, taking account of the limits of error, at the same time not giving too much weight to suspect models. Most of these design points are located at the boundary of the operability region (*30*).

Single response, two rival models

$$D_{12} = \underset{\underline{x}}{\mathrm{Max}} \left(\frac{1}{2} \Pi_1 \Pi_2 \left\{ \frac{(\sigma_2^2 - \sigma_1^2)^2}{(\sigma^2 + \sigma_1^2)(\sigma^2 + \sigma_2^2)} + (r_1 - r_2)^2 \left[\frac{1}{\sigma^2 + \sigma_1^2} + \frac{1}{\sigma^2 + \sigma_2^2} \right] \right\} \right)$$

Single response, m rival models

$$D_{ij} = \underset{\underline{x}}{\mathrm{Max}} \left(\frac{1}{2} \sum_{j=1}^{m} \sum_{i=j+1}^{m} \Pi_{i,n-1} \cdot \Pi_{j,n-1} \left\{ \frac{(\sigma_j^2 - \sigma_i^2)^2}{(\sigma^2 + \sigma_j^2)(\sigma^2 + \sigma_i^2)} + (r_j - r_i)^2 \left[\frac{1}{\sigma^2 + \sigma_j^2} + \frac{1}{\sigma^2 + \sigma_i^2} \right] \right\} \right)$$

Multiple response, m rival models

$$D_{ij} = \underset{\underline{x}}{\mathrm{Max}} \left(\frac{1}{2} \sum_{j=1}^{m} \sum_{i=j+1}^{m} \Pi_{i,n-1} \cdot \Pi_{j,n-1} \left\{ \mathrm{trace} \left[\underline{A}_{in} \underline{A}_{jn}^{-1} + \underline{A}_{jn} \underline{A}_{in}^{-1} - 2 \underline{I}_r \right] + (\underline{r}_i - \underline{r}_j)^T (\underline{A}_{jn}^{-1} + \underline{A}_{in}^{-1}) (\underline{r}_i + \underline{r}_j) \right\} \right)$$

Figure 5. Criteria for model discrimination

The maximization of these design criteria normally will be done by some optimization technique, like search methods or other combined hill-climbing methods (*28, 29*). Of course, this can be done rationally only by a digital computer having a large memory, especially for multiresponse situations.

It is surprising to see how powerful these screening criteria are. If the data are accurate enough, regardless of the preliminary data, two or three additional experiments designed with these criteria are usually sufficient to raise the normalized *a posteriori* probability for one of the rival models up to 1, whereas the others can be excluded as improbable

Single response, two rival models

$$D = \Pi_{1,n-1} \cdot \Pi_{2,n-1} \left[\left(\ln A_1 - \ln A_2 + \frac{(r_1 - r_2)^2}{2\sigma^2} \right) \{A_1 - A_2\} \right]$$

$$\text{where} \quad A_i = \exp(S_{i,n-1}/2\sigma^2) \cdot 2^{-\nu_i/2} \sqrt{\frac{|\underline{X}^T_{i,n-1} \cdot \underline{X}_{i,n-1}|}{|\underline{X}^T_{i,n} \cdot \underline{X}_{i,n}|}}$$

Single response, m rival models

$$D = \sum_{i=1}^{m} \sum_{j=i+1}^{m} \Pi_{i,n-1} \cdot \Pi_{j,n-1} \left[\left(\ln A_i - \ln A_j + \frac{(r_i - r_j)^2}{2\sigma^2} \right) \{A_i - A_j\} \right]$$

Figure 6. Revised criteria for model discrimination

(*30*). Moreover, screening becomes more effective with multiresponse data because of the higher information value of the response. However, in analyzing multiresponse data one should always be assured by an eigen-value eigen-vector analysis that linear relationships among the responses are absent to avoid meaningless results (*32*). Recent publications give some revision of the statistical background of the criteria (Figure 6) (*33*) and adaptation to special error distributions as well as to blocked designs (*34*). Despite the fact that the above-mentioned model screening criteria have been developed, in the last five years little has been published about their use for modeling industrial reactions. One reason could be the fact that at least medium sized digital computers are needed to design screening experiments, and routine programs for model screening are not generally available.

Even for those who don't have a digital computer, new methods have been developed which are superior to the classical least-squares linear plot techniques. First, in these plots the data should be weighted suitably or transformed suitably to produce a constant error variance (*35, 36*). Secondly, to discriminate between two or more rival models (*37*), one can use diagnostic parameters, either inherently present in the model (*38*) (intrinsic parameters) or especially introduced (nonintrinsic parameters).

The basic idea with these nonintrinsic parameters is to define a simple linear function of the observed rates and its corresponding values calculated from the different model equations and to determine (for example by a linear least-squares analysis) whether the value of a discrimination parameter λ is significantly positive or negative and thus to judge between rival models (*39*).

With the intrinsic parameters, which are of a simpler functional form than the entire model equation, one can prepare linear plots also for high conversion data, similar to the well-known linearized initial rate plots for low degrees of conversion (*40*). All these together facilitate the selection of an adequate model. Analysis of variance and/or an inspection of a plot of residuals gives a final check of model adequacy. Whereas the analysis of variance which yields the ratio of the lack-of-fit mean square to the pure error mean square is only an over-all test for the goodness of fit of a model, more subtle model inadequacies can be detected by analyzing the residuals. Plots of residuals *vs.* predicted values or independent variables, showing non-randomness of the data, give clear indications about the nature of model inadequacies (*41*).

Moreover, in nonlinear model-building activities, parametric residuals built with the above-mentioned intrinsic parameters can be used in a similar way as the multiple correlation coefficient in stepwise linear multiple regression for adaptive model building to find a model with a minimum number of parameters which represents the data adequately. Residual analysis of intrinsic parameters will indicate how the model might be modified to yield a more satisfactory one (*42*).

Planning Experiments for Precise Parameter Estimation

Once model building and discrimination have been accomplished, the next stage is to obtain precise estimates of the parameters in this model. Again the general procedure is to generate an initial set of data, analyze these data by nonlinear regression techniques to determine the best estimates of the parameters of the selected model, then extremize some criteria to design additional, simultaneously, or sequentially analyzed data until the desired accuracy is reached. This strategy, developed by Box and co-workers (*43*), also uses the computer extensively. There is a continuous exchange of information between the data source and the computer, indicating the advantage of on-line experimentation.

The fundamental idea for precise parameter estimation is to find experimental settings which will reduce as much as possible the size of a joint confidence region for all p parameters of the model. This hypervolume in the parameter space is preferable to individual confidence intervals because the latter provides no information about correlations among the parameters.

For nonlinear models used in reaction kinetics, only approximate confidence regions can be obtained by first-order Taylor series expansion of the model equation in terms of the parameters. The minimization of this joint confidence volume is essentially equivalent to the maximization of a determinant Δ, formed by the $n \times p$ derivative matrix and its transpose (Figure 7), where n represents the number of experiments and p the number of parameters in the model. Actually this matrix is the moment matrix of residuals, assuming that the experimental errors are approximately normally distributed and the response relationship is approximately linear near the initial least-squares estimates of the parameters. Since these derivatives are functions of the parameters itself, initial estimates of these parameters must be supplied, and the efficiency of the design depends on these estimates.

Box and Lucas (1959), all parameters, single response

$$\Delta = \max_{\underline{x}} \left| \underline{X}^T \underline{X} \right| \quad \text{resp.} \quad \max_{\underline{x}} \left| \underline{X} \right| \quad \text{if } n = p$$

Draper and Hunter (1966), all parameters, multiple response

$$\Delta_r = \max_{\underline{x}} \left| \sum_{i=1}^{r} \sum_{j=1}^{r} \sigma^{ij} \underline{X}^{(i)} \underline{X}^{(j)} \right|$$

Hunter, Hill and Henson (1969), subset of parameters, single response

$$\Delta_s = \max_{\underline{x}} \left| Y_{11} - Y_{12} Y_{22}^{-1} Y_{21} \right| \quad \text{where} \quad \left[X^T X \right] = \begin{bmatrix} Y_{11} & Y_{12} \\ Y_{21} & Y_{22} \end{bmatrix}$$

Figure 7. Criteria for precise parameter estimation

There are many modifications of the original Box-Lucas criterion (*43*)—*e.g.*, for multiresponse situations (*44*), for the case where prior information is available (*45*), or for the case where more than p trials are planned (*46*). Expansions are made for the case where the variance matrix is not constant but a known function of the independent variables and for the case of replicated experimental designs (*34*), for cost optimal designs, and for continuous measurements. In particular the criterion for multiresponse data (*45*) provides an effective method to estimate even in very complicated cases (*30*). It demonstrates a parametric sensitivity which allows a precision of estimation not possible with other techniques.

For parameter estimation in industrial processes the most useful modification seems to be the ability to design simultaneous or sequential experiments, focused on the precise estimation of a subset of the more crucial parameters of a model (*47*) because the economics or the design of a process often depend principally on only one or two of the model parameters. For parameter estimation, as in model discrimination, the sequential technique is preferrable because every additional piece of information gained from step to step can be used to improve the design in the following step. Published results (*30*) show how the number of experiments can be reduced drastically with sequentially designed experiments. This point is most important for expensive experimentation and can reduce experimental costs greatly.

The maximization of the determinant Δ will be done again by some optimization method as in model screening, including eventually inequality constraints which reflect physical constraints that the parameter values must satisfy. Since the efficiency of such an optimization depends on the starting value, the initial estimate of operating conditions is important. Sometimes a grid search in the total space of operability or at least in the most interesting region can provide helpful information, indicating the sensitivity of the design to changeable variations.

At several steps in model screening and parameter estimation nonlinear least-squares estimates of the parameters are involved. Most of the algorithms for this estimation are based on a gradient method, like the Gauss, the Marquardt, or the Davidon-Fletcher-Powell method (*49*). Bard (*50*) has found in one study that the Gauss method is the most efficient. For all these methods standard routine programs are available (*51, 52, 53*), providing the user with much information about the precision and correlation of the estimates.

A final remark should be made about the assumptions implicit in the different criteria because this violation may lead to wrong conclusions (*54*). One of the most important assumptions is the constant error variance whereas the common case is a constant percent error. In this case the logarithm of the data should be fitted to the model instead of the data.

Complementary to this estimation of parameters in nonlinear algebraic models is the estimation of states and parameters for nonlinear dynamic systems, taking into account the local or time dependence of the system—*e.g.*, the conversion time function (*49*). There are a few examples where this "filter" theory is used for chemical reactions, but none of them concerns industrially important processes. Nevertheless, these methods will probably become more important in the future as a powerful tool for parameter estimation in reacting systems.

General Strategy for Industrial Process Kinetics and Parameter Estimation

Taking together the above-mentioned methods, an ideal, powerful strategy for model screening and parameter estimation in industrial processes can be developed, assuming the data can be generated free of transport processes or that their influence can be kept constant during any scale-up procedure:

(a) Starting with factorial, simplex, or central composite designed experiments the response surface of the reaction rate is explored by linear regression of the data and transformation of the regression equation into its canonical form, using a single response for this first step. At this stage significant variables can be selected.

(b) The information thus obtained and any additional information about the type of reaction should be used to establish some mechanistic reaction schemes, leading to several possible model equations. A first, selection between these rival models is possible, using the classical linear plot technique, supplemented by diagnostic methods. The final decision should be made according to the results of a few additional experiments, designed with the above-mentioned discrimination criteria.

(c) For the most suitable model equation the parameter values should be determined by nonlinear regression. The precision of the estimated parameters can be improved by sequentially designed additional experiments using the parameter precision criteria. If possible, multiresponse data should be used to accelerate the improvement.

Model discrimination and parameter estimation can be joined in a single criterion to tackle both problems simultaneously (*55*). The new criterion consists of a weighted addition of the separate criteria, where the weights change as progress is made on the two objectives. Starting with a high weight for model discrimination this will be lowered as the probability, associated with one of the models grows, in favor of the weight for the parameter estimation for this model. This combined criterion can be extended to the situation where the error variance is not constant and where it is expedient to design the experiments in blocks (*34*).

This strategy seems to be safe and rational in cases where no preliminary information about the reacting system can be gained. It is linked strongly to the use of a digital computer and demands routine programs. It cannot, however, replace the ingenuity and experience of a well trained chemical reaction engineer, who has additional fundamental insight into reacting systems and is able to abbreviate some of the steps or to find alternative solutions (*56*) either in the sequence of experiments or in the use of mathematical techniques. The contributions of R. Tanner, Sheanlin Liu, H. Zeininger and U. Onken as well as that of J. M. H. Fortuin (pp. 535–552) are examples.

Selected Examples for Industrial Process Kinetics

Only a few examples have been published for applying the above-mentioned sequential procedure in reaction modeling and parameter estimation to industrial processes. Isomerization of *n*-pentane on a commercial Pt–Al_2O_3 reforming catalyst (*30*) seems to be one of the most complete examples for testing these procedures on real data. It was shown that a substantial savings in experimental expense is possible. However, care must be taken to start model screening with different combinations of preliminary runs.

Apart from polymerization processes, one of the most complex theoretical models published concerns the thermal cracking of hydrocarbon mixtures (*57*). It consists of six reaction types: initiation or chain generation, hydrogen extraction, radical addition to unsaturated hydrocarbons, decomposition or internal rearrangement of radicals, termination or radical recombination, and molecular reactions of dimerization or co-dimerization with unsaturated hydrocarbons. The total reaction scheme composed of more than 2000 over-all reactions considers the presence of 114 hydrocarbons in the main reactions and 12 radicals participating in the chain propagation and termination steps. With this model, the product distribution in light naphtha pyrolysis (*58*) can be predicted with an accuracy of about 1 wt %. The kinetic scheme of alkylation of isobutane with propylene by J. R. Langley and R. W. Pike (pp. 571–576) seems to have a similar degree of complexity.

The fractional conversion in the steam cracking of naphtha can be calculated with satisfactory accuracy using a reaction model consisting of "only" 13 pseudo-first-order elementary reactions (*59*). Moreover, pseudo-first-order kinetics with respect to the fraction remaining unconverted seems adequate for hydrocracking gas and coal oils (*60*) as well as for desulfurization and dehydrogenation reactions.

The general rule that the noncatalytic pyrolysis of low molcular weight hydrocarbons follows a pseudo-first-order kinetic law is confirmed by K. K. Robinson and E. Weger (pp. 557–561). In contradiction to this statement thermal cracking of high molecular weight paraffins demands other than first-order kinetics (*61*).

Catalytic petrochemical processes like catalytic cracking, involving the formation of carbonium ions or oxidation of methanol on molybdenum catalysts (*62*) can only be described by more complex kinetic models. Fortunately, the type of the rate equation seems to be the same for all hydrocarbons in a homologous series as was shown for example for the competitive–noncompetitive model in catalytic hydrogenation of olefins (*63*). Only the values of the rate constants and adsorption constants change with a change in the number of C atoms in the molecule. Furthermore, regardless of the underlying hypotheses—Langmuir-Hinshelwood or redox mechanism—in all these reactions a hyperbolic type of rate equation seems to be indicated.

On the other hand, for the well known water–gas shift reaction, it has not been possible to make a clear decision whether a rate law of the potential type (*23*, *24*), or a hyperbolic equation (*64*) is preferable. However, there are kinetic models which include the effects of sulfur content in the reaction mixture as well as the influence of pressure (*65*). Summarizing the numerous publications on petrochemical process kinetics, it seems that over-all kinetic models exist for a variety of processes so that the design engineer has only to fit the model parameters to his own data. Sometimes only a single factor like the specific surface [cm^2/gram] must be adapted to the catalyst in use (*66*).

Ammonia synthesis (*67*), SO_3 synthesis (*68*), and methanol synthesis (*11*) have been studied intensively to find an optimal rate equation. Despite the merits of theoretically founded models like the Tempkin-Pyzhev equation for ammonia synthesis or the Eklund equation for SO_2 oxidation, hyperbolic model equations seem to be more flexible in fitting the data, as shown recently for SO_2 oxidation (*69*) or for methanol synthesis (*70*).

Even for the chlorination reactions in pyrite cinder purification a similar, but more complex type of hyperbolic model presented by A. Cappelli and A. Collina (pp. 562–564) seems to be very successful. Purely from the standpoint of model building, the hyperbolic type of rate equation always has the advantage of possessing more parameters for adaptation.

Conclusion

A generally applicable strategy for kinetic modeling and parameter estimation exists, and only two fundamentally different types of rate equations are needed to describe the elementary steps of industrially important reactions. Both strategies and models are open for refinement, but no need is seen for a fundamental change. An exception seems to be the case of rapid flame reactions, like the one presented by H. A. Herbertz, H. Bockhorn, and F. Fetting (pp. 553–557). In this case our knowledge of kinetic models for design purposes is poor. One reason can be the complex radical reaction scheme of these reactions and the extremely high reaction velocity, offering many experimental difficulties in testing possible models. Also in ionic reactions we are just beginning to understand their kinetics, having now a generalized relaxation method (*71*) at hand as a powerful tool for this study. Fortunately, the need for better knowledge in kinetics of this type of reactions is not always too important since the reactor design in this case often depends to a greater degree on nonreaction parameters like heat transfer coefficients, etc. Nothing is said here about gas–solid, liquid–solid, or solid–solid reactions, but it seems that up to now there has not been enough understanding of the nature of these reactions to make a useful generalization.

Literature Cited

(1) Weekman, V. W. Jr., "Annual Review on Chemical Reaction Engineering," *Ind. Eng. Chem.* (Feb. 1969) **61,** 53.
(2) Mukherjee, S. P., Doraiswamy, L. K., "Reaction Kinetics and Reactor Design," *British Chem. Eng.* (Jan. 1967) **12,** 70.
(3) Siemes, W., "Fortschritte der Verfahrenstechnik 1966/67," Vol. 8, pp. 212–279, Verlag Chemie, Weinheim, 1969.
(4) *Proc. European Symp. Chem. Reaction Eng., 6th, Brussels, Sept. 1968.*
(5) *Proc. All-Union Conf. Modeling Ind. Reactors, 3rd, Kiev, May 1968.*
(6) *Proc. Chisa Congr., 3rd, Mariánské Lázně, Sept. 1969.*
(7) *Proc. Symp. Mathematical Models Heterogeneous Catalysis, Königstein, Sept. 1969;* published in *Ber. Bunsenges.* (1970) **74,** 81.
(8) *Proc. Symp. Working Party European Fed. Chem. Eng., on Routine Computer Programs, Use Computers Chem. Eng., 3rd, Florence, April 1970.*
(9) *Chem. Eng. Conf., Melbourne/Sidney, Aug. 1970.*
(10) Barnard, J. A., Mitchell, D. S., *J. Catalysis* (1968) **12,** 376–385, 386–397.
(11) Bakemeier, H., Laurer, P. R., Schroder, W., *Chem. Eng. Progr., Symp. Ser. No. 98* (1970) **66,** 1.
(12) Lumpkin, R. E., Smith, W. D., Douglas, J. M., *Ind. Eng. Chem., Fundamentals* (1969) **8,** 407.
(13) Boudart, M., "Kinetics of Chemical Processes," pp. 59 ff, 81 ff, Prentice Hall, Englewood Cliffs, 1968.
(14) Kittrell, J. R., Erjavec, J., *Ind. Eng. Chem., Process Design Develop.* (July 1968) **7,** 321.
(15) Box, G. E. P., Hunter, J. S., *Ann. Mathem. Statistics* (1957) **28,** 195.
(16) Draper, N. R., Beggs, W. J., Dept. of Statistics, University of Wisconsin, *Tech. Rept. No. 209,* June 1969.
(17) Andersen, L. B., *Chem. Eng.* (June 10, 1963) 223.
(18) Mayer, R. P., Stowe, R. A., *Ind. Eng. Chem.* (1969) **61** (5), 42.
(19) Stowe, R. A., Mayer, R. P., *Ind. Eng. Chem.* (1969) **61** (6), 12.
(20) Efroymson, M. A., "Mathematical Methods for Digital Computers," pp. 191–203, Wiley, New York, 1965.
(21) Breaux, H. J., *Comm. Assoc. Comp. Mash.* (1968) **11** (8), 556.
(22) Bakemeier, H., Detzer, H., Krabetz, R., *Chem. Ing. Tech.* (1965) **37,** 422, 429.
(23) Moe, J. M., *Chem. Eng. Progr.* (1962) **58,** 33.
(24) Auer, E., Bakemeier, H., Detzer, H., Krabetz, R., *Chem. Ing. Tech.* (1964) **36,** 774.
(25) Hunter, W. G., Reiner, A. M., *Technometrics* (1965) **7,** 307.
(26) Box, G. E. P., Hill, W. J., *Technometrics* (1967) **9,** 517.
(27) Hunter, W. G., Hill, W. J., Dept. of Statistics, University of Wisconsin, *Tech. Rept. No. 65,* June 1966.
(28) Box, M. J., Davies, D., Swann, W. H., "Nonlinear Optimization Techniques," ICI Monograph No. 5, Oliver & Boyd, Edingburgh, 1969.
(29) Fletcher, R., "Optimization," Academic, London and New York, 1969.
(30) Froment, G. F., Mezaki, R., *Chem. Eng. Sci.* (1970) **25,** 293.
(31) Hunter, W. G., Mezaki, R., *Can. J. Chem. Eng.* (1967) **45,** 247.
(32) Box, G. E. P., Hunter, W. G., Erjavec, J., McGregor, J. F., Dept. of Statistics, University of Wisconsin, *Tech. Rept. No. 212,* July 1969.
(33) Box, G. E. P., Henson, T. L., Dept. of Statistics, University of Wisconsin, *Tech. Rept. No. 211,* July 1969.
(34) Box, M. J., "The Future of Statistics," p. 241, Academic, New York, 1968.
(35) Johnson, R. A., Standal, N. A., Mezaki, R., *Ind. Eng. Chem. Fundamentals* (1968) **7,** 181.
(36) Johnson, R. A., Mezaki, R., *Can. J. Chem. Eng.* (1969) **47,** 517.
(37) Kittrell, J. R., Hunter, W. G., Mezaki, R., *A.I.Ch.E. J.* (1966) **12,** 1014.

(38) Kittrell, J. R., Mezaki, R., *A.I.Ch.E. J.* (1967) **13**, 389.
(39) Mezaki, R., Kittrell, J. R., *Can. J. Chem. Eng.* (1966) **44**, 285.
(40) Yang, K. H., Hougen, O. A., *Chem. Eng. Progr.* (1950) **46** (3), 146.
(41) Draper, N. R., Smith, H., "Applied Regression Analysis," Wiley, New York, 1966.
(42) Kittrell, J. R., Hunter, W. G., Mezaki, R., *A.I.Ch.E. J.* (1966) **12**, 1014.
(43) Box, G. E. P., Lucas, H. L., *Biometrica* (1959) **46**, 77.
(44) Draper, N. R., Hunter, W. G., *Biometrica* (1966) **53**, 525.
(45) Box, G. E. P., Draper, N. R., *Biometrica* (1965) **52**, 355.
(46) Atkinson, A. C., Hunter, W. G., *Technometrics* (1968) **10**, 271.
(47) Hunter, W. G., Hill, W. J., Henson, Th. L., *Can. J. Chem. Eng.* (1969) **47**, 76.
(48) Mezaki, R., Butt, J. B., *Ind. Eng. Chem., Fundamentals* (1968) **7**, 120.
(49) Seinfeld, J. H., *Ind. Eng. Chem.* (1970) **62**, 32.
(50) Bard, Y., IBM New York Scientific Center *Rept. No.* **320-2955** (Sept. 1968).
(51) Marquardt, D. L., *Soc. Ind. Appl. Math.* (1963) **2**, 431.
(52) Booth, G. W., Peterson, T. I., IBM Share Program No. 687 (1968).
(53) Bard, Y., Program 360 D 13.6.003, IBM Hawthorne, New York (1967).
(54) Hunter, W. G., *Ind. Eng. Chem., Fundamentals* (1967) **6**, 461.
(55) Hill, W. J., Hunter, W. G., Wichern, D. W., *Technometrics* (1968) **10** (1), 145.
(56) Himmelblau, D. M., Jones, C. R., Bischoff, K. B., *Ind. Eng. Chem., Fundamentals* (1967) **6**, 539.
(57) Dente, M., Ranzi, E., Antolini, G., Losco, F., *Symp. Working Party European Fed. Chem. Eng. Routine Computer Programs, 3rd, Florence, April 1970.*
(58) Knaus, J. A., Patton, J. L., *Chem. Eng. Progr.* (Aug. 1961) **57**, 57.
(59) Schwarz, P., DiMassa, I., Ragain, V., *Symp. Working Party European Fed. Chem. Eng. Routine Computer Programs, 3rd, Florence, April 1970.*
(60) Quader, S. A., Hill, G. R., *Ind. Eng. Chem. Process Design Develop.* (1969) **8**, 462.
(61) Woinsky, S. G., *Ind. Eng. Chem. Prod. Res. Develop.* (1968) **7**, 529.
(62) Dente, M., Poppi, R., Pasquon, I., *Chim. Ind.* (1964) **46**, 1326.
(63) Sawyer, D. N., Mezaki, R., *A.I.Ch.E. J.* (1967) **13**, 1221.
(64) Ruthven, D. M., *J. Chem. Eng.* (1969) **47**, 327.
(65) Ting, A. P., Wan, S. W., *Chem. Eng.* (May 19, 1969) **76**, 185.
(66) Dente, M., Cappelli, A., Collina, A., *Quad. Ing. Chim. Ital.* (1967) **3** (7/8), 137.
(67) Dyson, D. C., Simon, J. M., *Ind. Eng. Chem., Fundamentals* (1968) **7**, 605.
(68) Weychert, S., Urbanek, A., *Intern. Chem. Eng.* (1969) **9**, 39.
(69) Collina, A., Corbetta, A., Cappelli, A., *Symp. Working Party European Fed. Chem. Eng. Routine Computer Programs, 3rd, Florence, April 1970.*
(70) Cappelli, A., Collina, A., Dente, M., ADVAN. CHEM. SER. (1972) **109**, 562.
(71) Abresman, R. W., Kim, Y. G., *Ind. Eng. Chem., Fundamentals* (1969) **8**, 216.

RECEIVED July 30, 1970.

Contributed Papers

Determination of Kinetic Parameters in Ordinary and Partial Differential Equations

SHEAN-LIN LIU, Mobil Research and Development Corp., Research Department, Central Research Division, P. O. Box 1025, Princeton, N. J. 08540

A problem commonly occurring in chemistry and chemical engineering is the determination of the parameters (reaction rate constants, effective diffusivity, etc.) of a kinetic model from experimental data. These parameters can be estimated by linear and nonlinear least-squares procedures (*1–7*). Frequently, however, some reaction paths are insensitive to the values of one or more parameters. When this occurs, the values for the insensitive parameters obtained by least squares may be consistent with the data but be quite different from the true values.

The purpose of this paper is to investigate the effects of changes in the initial conditions of reaction systems upon the determination of parameters. A modified Gauss-Newton nonlinear least-squares technique, which uses the equations of variation (*2*, *8*) and can handle constraints on parameters, is described. Sometimes a single reaction path is appropriate for determining parameters; some single reaction paths may be insensitive to one or more parameters. Usually one does not *a priori* know which particular reaction path would give a satisfactory set of parameters. We have observed that reliable parameters can be obtained by simultaneous fitting to several sets of experimental data which have been obtained under differing experimental conditions. Computations for simultaneous fitting to several sets of data are less sensitive to experimental errors than computations for fitting to single reaction path. Three examples, two with actual experimental data, are given for illustration. They are: (1) butene isomerization, (2) a hypothetical four-component system, and (3) zeolite A crystallization.

The computation method can be applied to distributed parameter systems (involving partial differential equations). Using Begley and Smith's experimental data (*14*), the effective radial diffusivity and the over-all heat transfer coefficient for a fixed bed reactor are calculated.

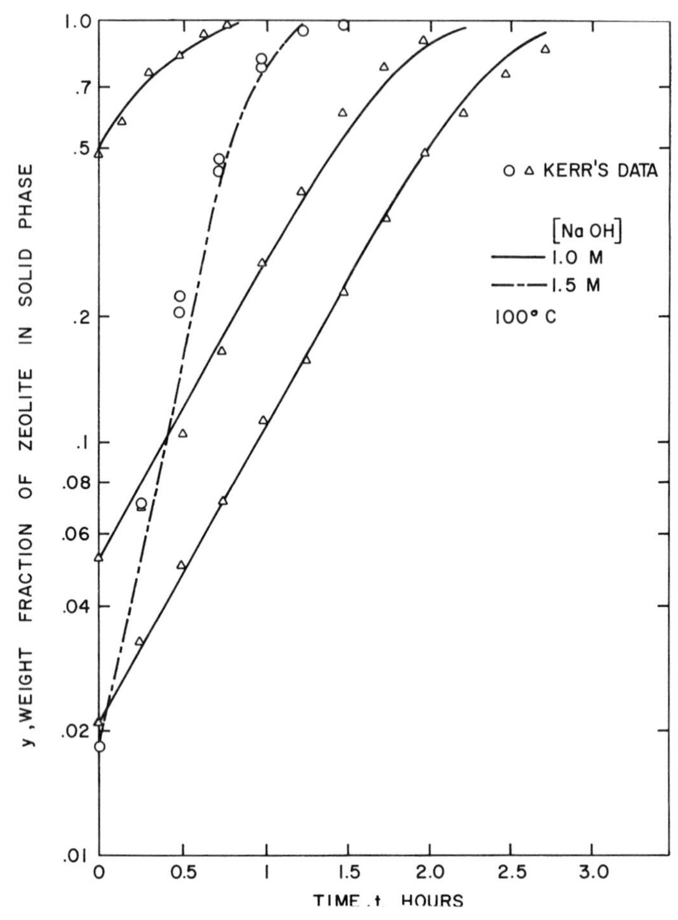

Figure 1. Batch growth of zeolite A

Examples

Zeolite A Crystallization. Using Kerr's batch data, Liu (*12*) has developed the following mathematical model:

$$\frac{dy}{dt} = [\mathrm{OH}]^{k_1}\left[\frac{k_2(1-y)}{k_3+(1-y)}\right]y \tag{1}$$

with

$$y = \frac{z}{z+s}$$

where s is the concentration of suspended amorphous solid particles on a weight basis, z is the concentration of zeolite A crystals, y is the weight fraction of zeolite in the solid phase, [OH] is the concentration of alkali in the aqueous phase; k_1, k_2, and k_3 are the three kinetic parameters to be determined. Using the present curve-fitting procedure, the following values of kinetic constants are obtained by fitting the reaction paths obtained by Kerr (*11*).

$$\begin{aligned} k_1 &= 2.36 \\ k_2 &= 2.25 \text{ hr}^{-1} \\ k_3 &= 0.36 \end{aligned} \tag{2}$$

Figure 1 shows the computed reaction paths using the above values of k_i and Kerr's experimental data. The experimental conditions, [NaOH], and temperature are also shown in Figure 1. The agreement between the nonlinear least-squares fit and the experimental results are quite good, which indicates that a reasonable mathematical model has been obtained.

Distributed Parameter System. The present computation method can be applied to distributed parameter systems. Begley and Smith (*14*) considered the effective radial thermal conductivity in a fixed-bed reactor and measured the temperature distributions at various radial positions and depths. The mathematical model is:

$$\frac{\partial y_1}{\partial x} = K_1 \left(\frac{1}{r} \frac{\partial y_1}{\partial r} + \frac{\partial^2 y_1}{\partial r^2} \right) \tag{3a}$$

$$\begin{aligned} x &= 0;\ y_1 = y_{1e}(r) \\ r &= 0;\ \frac{\partial y_1}{\partial r} = 0 \\ r &= 1;\ K_2 (y_1 - y_w) = -K_1 \left(\frac{\partial y_1}{\partial r} \right) \end{aligned} \tag{3b}$$

where y_1 is the temperature, r is the radial variable, x is the axial variable, K_1 is the effective radial thermal conductivity, and K_2 is the over-all wall heat-transfer coefficient. Using the actual experimental data obtained by Begley and Smith (*14*), the values of K_1 and K_2 are estimated.

Following our computation method, the equations of variation can be obtained by differentiating Equations 3a and 3b with respect to K_1 and K_2.

Let

$$y_2 = \frac{\partial y_1}{\partial K_1},\ y_3 = \frac{\partial y_1}{\partial K_2} \tag{3c}$$

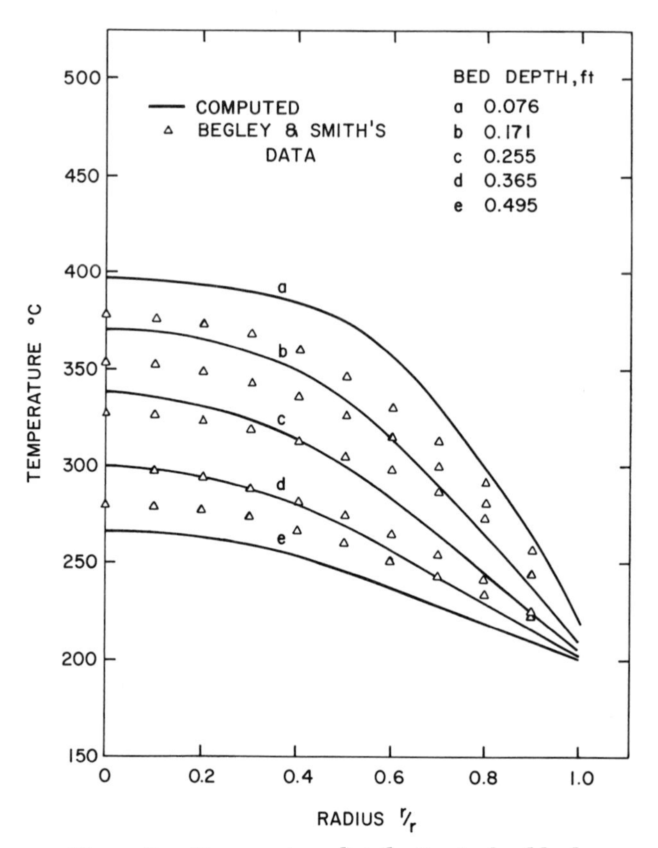

Figure 2. Temperature distribution in fixed bed

Then from Equations 3a and 3b, one obtains

$$\frac{\partial y_2}{\partial x} = \left(\frac{1}{r}\frac{\partial y_1}{\partial r} + \frac{\partial^2 y_1}{\partial r^2}\right) + K_1\left(\frac{1}{r}\frac{\partial y_2}{\partial r} + \frac{\partial^2 y_2}{\partial r^2}\right) \tag{4a}$$

$$\frac{\partial y_3}{\partial x} = K_1\left(\frac{1}{r}\frac{\partial y_3}{\partial r} + \frac{\partial^2 y_3}{\partial r^2}\right) \tag{4b}$$

$$x = 0;\ y_2 = 0,\ y_3 = 0$$

$$r = 0;\ \frac{\partial y_2}{\partial r} = 0,\ \frac{\partial y_3}{\partial r} = 0 \tag{4c}$$

$$r = 1;\ K_2 y_2 = -\left(\frac{\partial y_1}{\partial r}\right) - K_1\left(\frac{\partial y_2}{\partial r}\right)$$

$$(y_1 - y_w) + K_2 y_2 = -K_1\left(\frac{\partial y_3}{\partial r}\right)$$

Given initial estimates of K_1 and K_2, iterative computations are performed according to the modified Gauss-Newton method. During each iteration, the values of y_1, y_2, and y_3 are obtained by solving the parabolic partial differential Equations 3a through 4c by a stable explicit finite-difference method developed by the author (*13*). The values of K_1 and K_2, which yield the minimum value of the sum of the squares of the differences between the observed and computed temperatures at various radial positions and depths, are obtained as follows:

$$K_1 = 0.7143$$

$$K_2 = 12.32$$

Figure 2 shows the computed temperature distributions and the experimental data. Note that very close experimental data cannot be expected because the effective radial thermal conductivity is a complicated function of reactor properties (*15*), but a constant value is assumed for K_1 in the mathematical model, Equation 3a. Parameters for other distributed systems (—*e.g.*, axial diffusion coefficients in tubular reactors) can be estimated by the present computation method.

(1) Hartley, H. O., *Technometrics* (1961) **3**, 269.
(2) Howland, J. L., Vaillancourt, R. J., *J. SIAM* (1961) **9**, 165.
(3) Bard, Y., *IBM New York Scientific Center Rept.* **320-2902** (1967).
(4) Lapidus, L., Peterson, T. I., *AIChE J.* (1965) **11**, 891.
(5) Marquardt, D. W., *J. SIAM* (1963) **11**, 431.
(6) Kittrell, J. R., Hunter, W. G., Watson, C. C., *AIChE J.* (1965) **11**, 1051.
(7) Himmelblau, D. M., Jones, C. R., Bischoff, K. B., *Ind. Eng. Chem. Fundamentals* (1967) **6**, 539.
(8) Coddington and Levinson, "Theory of Ordinary Differential Equations," McGraw-Hill, New York, 1955.
(9) Hildebrand, F. B., "Introduction to Numerical Analysis," McGraw-Hill, New York, 1956.
(10) Wei, J., Prater, C. D., *Advan. Catalysis* (1962) **13**, 204–390.
(11) Kerr, G. T., *J. Phys. Chem.* (1966) **70**, 1047.
(12) Liu, S. L., *Chem. Eng. Sci.* (1969) **24**, 57.
(13) Liu, S. L., *AIChE J.* (1969) **15**, 334.
(14) Begley, J. W., Smith, J. M., M.S. thesis, Purdue University, 1951.
(15) Smith, J. M., "Chemical Engineering Kinetics," McGraw-Hill, New York, 1956.

The Kinetics of the Oxethylation of Nonylphenol under Industrial Conditions

H. ZEININGER and U. ONKEN, Farbwerke Hoechst AG, 6230 Frankfurt/Main 80, Germany

Ethylene oxide combines with organic hydroxyl compounds, its three-membered ring being split in the process, according to Reaction 1.

$$R{-}OH + \underset{\diagdown O \diagup}{CH_2{-}CH_2} \rightarrow R{-}O{-}CH_2{-}CH_2{-}OH \quad (1)$$

Since the reaction product contains a hydroxyl group, further ethylene oxide molecules are added, whereby poly(ethylene glycol) ether derivatives are formed:

$$R{-}O{-}CH_2{-}CH_2{-}OH + (n-1)\ \underset{\diagdown O \diagup}{CH_2{-}CH_2} \rightarrow R{-}(OCH_2{-}CH_2)_n{-}OH \quad (2)$$

This polyaddition, usually designated as oxyethylation, is used to produce nonionogenic surface-active agents from alkylphenols or alcohols and ethylene oxide (*12*). With these products the number n of added ethylene oxide molecules (degree of oxethylation) usually lies between 5 and 20. Alkaline compounds are used mostly as catalysts; the best are the alkali salts of the reactants, used in concentrations of a few parts per thousand. The reaction temperatures lie above 120°C. Products are mixtures of substituted polyglycol ethers containing various numbers of oxethylene units. The molecular weight distribution of these products is determined by the kinetics of the reaction. In connection with the theoretical investigation of a continuous process for polyoxethylation calculations of the molecular weight distributions were necessary. For this work (to be published in *Chem.-Ing.-Techn.*), a knowledge of the kinetics of the reaction was required.

Several investigations (*1–10*) have been done on the kinetics of the oxethylation of phenol, substituted phenols, and alcohols; however, most of these studies are concerned with the simple addition of ethylene oxide to a phenol or alcohol. Moreover, most of these investigations were carried out under conditions which did not correspond to those of industrial oxethylation (low reaction temperatures, excess ethylene oxide, use of solvents). Only the experiments of Satkowski and Hsu (*6*) were conducted under industrial conditions. However, these authors did not investigate the kinetics of the liquid-phase reaction but measured the rate of consumption of ethylene oxide. They concluded that during the polyaddition of ethylene oxide onto alcohol the speed of reaction after the addition of the first ethylene oxide molecule to the alcohol is practically independent of the number of added ethylene oxide molecules.

Experimental

Since the results of the studies conducted so far did not allow any certain conclusions as to the rate of reaction during the polycondensation of ethylene oxide onto alkylphenols under industrial conditions, we conducted our own kinetic investigations. Nonylphenol was chosen as the primary component. We applied a static method, whereby a small amount of ethylene oxide was injected into catalyst-containing nonylphenol which had been prooxethylated to various degrees. The drop in ethylene oxide concentration as a function of time was observed. Samples were taken in which the ethylene oxide concentration was analyzed volumetrically by reaction with hydrochloric acid in pyridine according to Reaction 3.

$$\underset{\diagdown_{O}\diagup}{CH_2{-}CH_2} + HCl \rightarrow \underset{OH\quad Cl}{CH_2{-}CH_2}$$

Measurements were carried out between 120° and 175°C; the degree of oxethylation was between 0.85 and 7.2. Sodium nonylphenolate was used as the catalyst. The initial concentrations of ethylene oxide were 0.12–0.18 mole/liter, and of the catalyst 0.1–1.0 mole % (related to nonylphenol), which corresponds to 0.03–0.017 mole/liter.

Results

To represent the reaction rate as a function of concentration, the concentration unit of mole/liter was used. Industrial oxethylation is a reaction in a concentrated solution of one of the reaction components (*viz.*, the hydroxyl component) whose mole volume increases greatly during the reaction. For example, oxethylated nonylphenol with a degree of oxethylation of 5 has a molar volume of nearly twice that of the starting component, nonylphenol. These large differences in molar volumes must be accounted for when giving concentrations in mole/unit volume.

The reaction rate showed a first-order dependence on both ethylene oxide and catalyst concentrations:

$$-\frac{d[EO]}{dt} = k\,[EO][Kat] \tag{4}$$

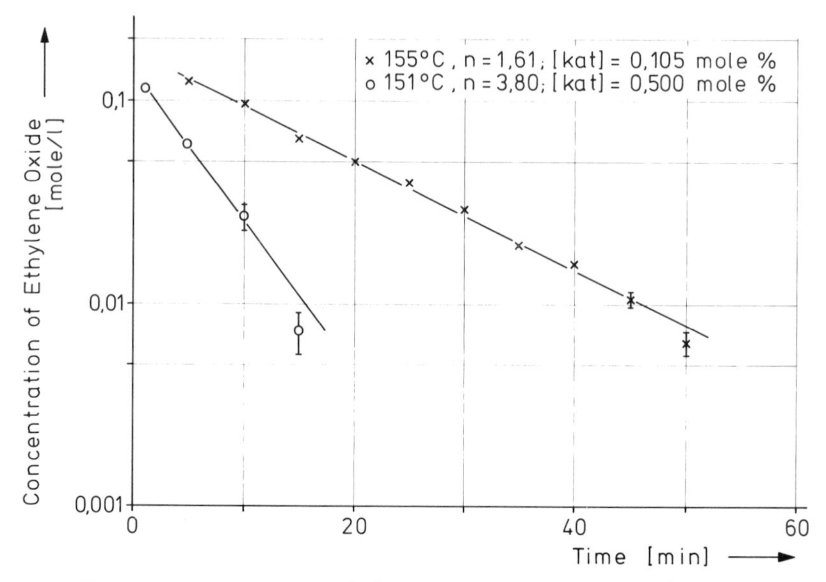

Figure 1. Decrease in ethylene oxide concentration with time

where

t = time in minutes

k = rate constant [min^{-1}, mole^{-1}, liter]

[EO][Kat] = concentration of ethylene oxide or catalyst [mole/liter]

The first order with respect to ethylene oxide resulted directly from the linear dependence of the logarithm of concentration on time. In Figure 1 this is shown for two experiments. The k values obtained from Equation 4 for the measurements at various temperatures, degrees of oxethylation ($n > 1$), and catalyst concentrations could be fitted with sufficient accuracy by the following Arrhenius equation:

$$k = C \exp\left[-\frac{A}{RT}\right] [\text{min}^{-1}, \text{mole}^{-1}, \text{liter}] \qquad (5)$$

where

$C = 10^{10.9}$

A = activation energy = 19.0×10^3 cal/mole

R = gas constant = 1.9865 cal/mole °K

T = absolute temperature [°K]

This is demonstrated in Figure 2. On the one hand, this proves that the reaction is first order with respect to catalyst concentration. On the other hand, it means that the reaction rate within the accuracy of measurement is independent of the degree of oxethylation of the hydroxyl component, except for the first stage of the ethylene oxide addition to form glycol ether ($n = 1$), which is more complicated (*2, 3, 7, 8, 10*).

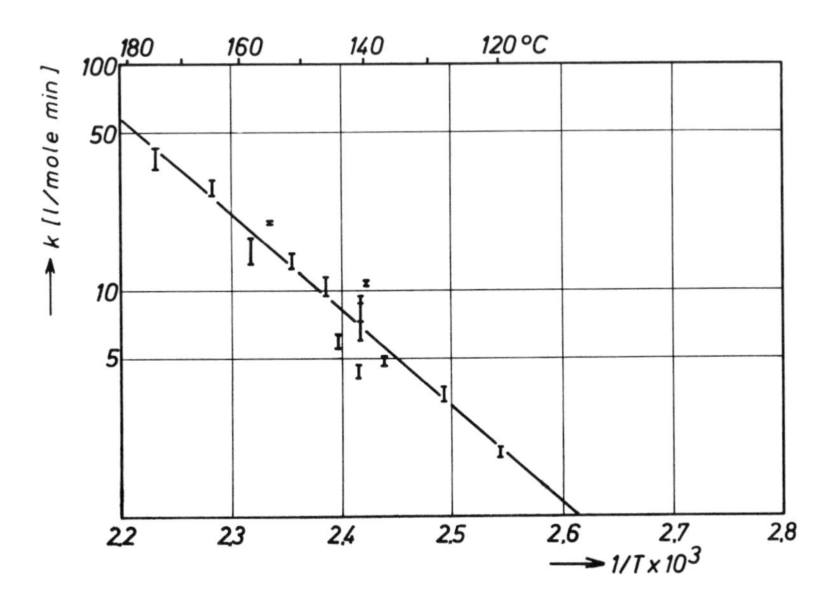

Figure 2. Rate constant of oxethylation

Comparison with Measurements on a Flow Apparatus

A comparison with earlier measurements of the conversion during the oxethylation of nonylphenol in a pilot flow apparatus with a relatively wide residence time spectrum was of interest. The apparatus (Figure 3) consisted of an unstirred tank (a) and a tubular heat exchanger (c) connected in series with (a) through which the reaction mixture flowed between the tubes. The reaction volume was divided about equally between the tank (a) and the exchanger (c). To keep backmixing in the vessel to a minimum, a baffle plate (b) was placed at the inlet, which prevented the incoming jet of reaction mixture from impinging upon the liquid already present in the vessel.

Measurements of the conversion of ethylene oxide at 156° and 167.5°C and degrees of oxethylation of *ca.* 3 and 8.5, gave (assuming plug flow without backmixing) rate constants which were on an average less than 10% and at most 25% lower than the values obtained from kinetic measurements. This is experimental support for theoretical in-

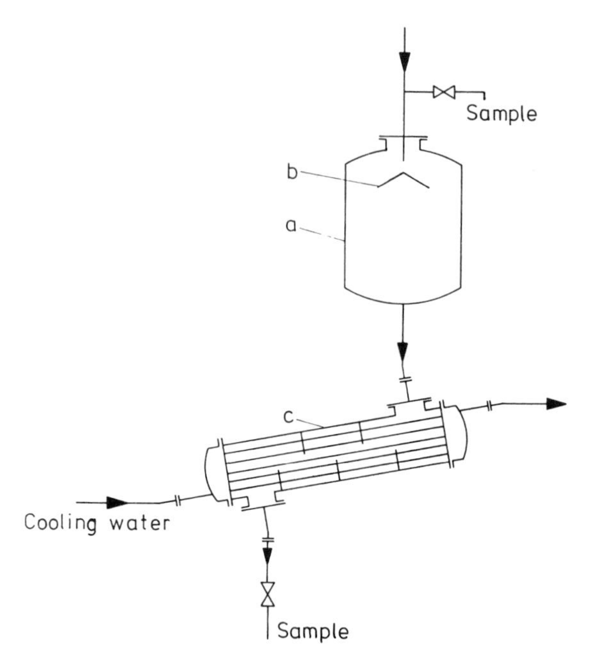

Figure 3. Flow system

vestigations in the literature (*11*), according to which reactors even with a relatively wide residence time spectrum can be considered an ideal tube as a first approximation for reactions of first or pseudo-first-order.

The latter case applies to the experiments in the flow apparatus since only the concentration of ethylene oxide changed when the reaction mixture passed through the apparatus, whereas the concentrations of the hydroxyl component and of the catalyst were practically constant.

Conclusion

The kinetics of polyoxethylation may be expressed by simple equations. Therefore, the reaction is suitable for studying residence time behavior of more complicated reactor systems, especially with the use of gas chromatographic analysis for the reaction products.

Acknowledgment

We thank Dr. Helmes for measuring the densities of oxethylated nonylphenols. The comparison experiments were supported by Dr. Raizner, head of the Hoechst TH plant at that time.

(1) Boyd, D. R., Marle, E. R., *J. Chem. Soc. (London)* (1914) **105**, 2117.
(2) Patat, F., Cremer, E., Bobleter, O., *Monatsh. Chem.* (1952) **83**, 322.
(3) Patat, F., Cremer, E., Bobleter, O., *J. Polymer Sci.* (1954) **12**, 489.
(4) Gee, G., Higginson, W. C. E., Levesley, P., Taylor, K. J., *J. Chem. Soc. (London)* **1959** 1338.
(5) Gee, G., Higginson, W. C. E., Merrall, G. T., *J. Chem. Soc. (London)* **1959**, 1345.
(6) Satkowski, W. B., Hsu, C. G., *Ind. Eng. Chem.* (1957) **49**, 1875.
(7) Patat, F., Wojtech, B., *Makromol. Chem.* (1960) **37**, 1.
(8) Lebedev, N. N., Shvets, V. F., *Kinetics Catalysis* (1965) **6**, 706.
(9) Lebedev, N. N., Baranov, Y. I., *Kinetics Catalysis* (1966) **7**, 545.
(10) Lebedev, N. N., Shvets, V. F., *Kinetics Catalysis* (1968) **9**, 418.
(11) Kramers, H., Westerterp, K. R., "Elements of Chemical Reactor Design and Operation," p. 80, University Press, Amsterdam, 1963.
(12) Schönfeldt, N., "Óberflächenaktive Anlagerungsprodukte des Athylenoxids," Wiss. Verlag, Stuttgart, 1959.

Kinetics of and Parameter Estimation for the Treatment Steps in the Waste Heat Boiler and the Absorption Tower of the Nitric Acid Process

J. M. H. FORTUIN, Dutch State Mines, Central Laboratory, P. O. Box 18, Geleen, The Netherlands

A treatment step of a chemical process relates to some treatment of a system in an installation. A process engineering model (PEM) of a treatment step consists of a set of equations and is the mathematical formulation of the representative physical and chemical changes of the system passing through the installation in the steady state—*i.e.*, a continuous process step expressed in formulae. The set of equations comprise variables and parameters relating to the system, the installation, the environment, the treatment conditions, the inlet and outlet requirements, and the constraints.

In developing a PEM a process engineer uses the laws of conservation, the theory of transport phenomena, thermodynamics, reaction kinetics, and additional information. This information must be completed with reasonable assumptions—*e.g.*, with respect to the flow regime. For a PEM each parameter must have a physical meaning, and the parameter values determined in large-scale equipment must have physically acceptable values.

The Nitric Acid Process

In the nitric acid process ammonia in air is oxidized on Pt-gauze at about 850°C. The resulting gas mixture is cooled, first to about 200°C in a waste-heat boiler, and then to about 50°C in a condenser-cooler. Finally, the nitrous gases are oxidized, absorbed, and cooled in one or more packed towers or tray columns. In each treatment step of the process physical and chemical changes take place.

Waste Heat Boiler

In the waste-heat boiler a gas mixture of about 850°C is cooled to about 200°C. The NO in this gas is oxidized partially to NO_2. In designing nitric acid plants to operate at higher pressures (*e.g.*, $>$ 5 atm), one has to bear in mind that the degree of oxidation of the gas in the outlet may be about 50% and higher. This means that a considerable amount of extra heat will have to be removed over and above the quantity corresponding to the sensible heat of the gas mixture alone. For a correct dimensioning of such a heat exchanger it is necessary to know both where and how much heat will be released in the oxidation. The model for the waste-heat boiler is based upon a one-dimensional enthalpy balance, component mole balances, and reaction kinetics, assuming that the oxidation of nitric oxide is the only reaction that occurs. The information about the gas flow, the apparatus, and the cooling water is supplemented with reasonable assumptions (*e.g.*, piston flow). In the boiler and economizer the heat transfer coefficient is treated as an adjustable parameter whose value is determined by fitting temperature data measured in an industrial waste-heat boiler. Generally, these adjusted values agree satisfactorily with those obtained by extrapolating the results measured in small scale experiments. These experiments were carried out at room temperature with a cross flow of air and water as a cooling medium in small tube bundles of similar geometry.

With regard to the model we may state that the assumptions are physically reasonable, and the adjusted parameter has a physically acceptable value (in our case about 90 $Wm^{-2}K^{-1}$). Therefore, the requirements made on a PEM are satisfied. Consequently, the model is considered a suitable basis for the design of this type of waste-heat boilers and for optimizing this kind of design.

Absorption Column

The absorption of nitrous gases in a tray column for producing nitric acid comprises repeating absorption and oxidation steps. For either type of step a separate model can be developed. The model for the absorption step relates to the gas–liquid dispersion on the tray, the model for the oxidation step to the gas phase above the dispersion. On each tray the nitrous gases are absorbed by the liquid, with simultaneous formation of liquid nitric acid and gaseous nitrogen monoxide. This reaction is called the heterogeneous nitric acid reaction. In the gas compartment over the tray, the nitrogen monoxide is oxidized subsequently to nitrogen dioxide. This is referred to as the oxidation reaction. Further, at lower temperatures ($< 100°C$) NO_2 can form N_2O_4 according to the dimerization reaction.

The heat of oxidation and acid formation is removed *via* cooling coils immersed in the gas–liquid dispersion on the trays.

Assumptions. To construct the model we assumed that:

(a) The nitric acid reaction is an equilibrium reaction.

(b) Equilibrium is normally not reached, owing to transport limitations in the absorption step.

(c) The oxidation reaction is relatively slow and, hence, does not go to completion in the oxidation step.

(d) The equilibrium of the dimerization reaction is reached with infinite rapidity.

(e) The column is isothermal. (In the complete model for calculating an optimum design use has also been made of heat balance, pressure drop, transport equations, and tray hydrodynamics.)

Absorption Step. With regard to the absorption step it is assumed that the liquid phase is perfectly mixed and that the oxidation reaction does not occur. The plate efficiency of the nitrous gas absorption is defined by

$$\eta = \frac{n_o - n_a}{n_o - n_e}$$

where n = effective molar flow rate of nitrous gas ($n = n_{NO} + n_{NO_2} + 2\, n_{N_2O_4}$). The subscripts refer to the feed of the absorption step (o), the gas leaving the absorption step (a), and to equilibrium (e).

Oxidation Step. With regard to the oxidation step we assume that the gas phase is perfectly mixed and that the oxidation reaction proceeds according to the Bodenstein equation:

$$r_{NO} = -\, k_1 p^2{}_{NO} p_{O_2}$$

Consequences for the Absorption Step. For the nitrous gas escaping absorption, the effective molar flow rate equals:

$$n_a = n_o \left(1 - \eta\, \frac{\xi_o - \xi_e}{1.5 - \xi_e}\right)$$

This is calculated from the equation for n, ξ, and η and the stoichiometry of the reactions—*i.e.*:

$$dn_{NO} = -(1/3)d(n_{NO_2} + 2n_{N_2O_4})$$

The degree of oxidation equals:

$$\xi_a = 1.5 - \frac{1.5 - \xi_o}{n_a/n_o}$$

where the degree of oxidation, ξ is:

$$(n_{NO2} + 2n_{N2O4})/n$$

Consequences for the Oxidation Step. The effective molar flow rate is constant:

$$n_b = n_a$$

The degree of oxidation is equal to:

$$\xi_b = \xi_b\,(\xi_a;\, k_1;\, \tau;\, p_{O_2}$$

where b refers to the gas leaving the oxidation step, τ is the residence time, and p_{O_2} is the partial pressure of oxygen.

The Over-all Efficiency. E_o for ν plates can be obtained from:

$$E_o = \frac{(n_o)_{j=1} - (n_b)_{j=\nu}}{(n_o)_{j=1}}$$

where j = plate number.

Conclusions. The foregoing shows that the over-all efficiency can be determined if the ξ_e and the η of each tray are known. ξ_e can be determined from the chemical and physical equilibria, the pressure balance, and the stoichiometry of the nitric acid reaction. The plate efficiency η must be determined empirically as a function of tray geometry and process conditions. In calculating η one must also use equations describing the transport of nitrous components from the gas to the liquid phase (*1*). In doing so we used the transfer area per unit volume of dispersion as an adjustable parameter. By fitting pilot plant data, reasonable physical values are obtained (in our case about 100 m^{-1}). Consequently, the transport model can be considered an adequate tool for designing absorption columns in nitric acid plants and for optimizing this kind of design—*e.g.*, optimization of the tray spacing (*2*).

(1) Kwanten, F. J. G., Wijnands, J. J. H., unpublished work.
(2) Hoftyzer, P. J., Kwanten, F. J. G., "Gas Purification Processes of Gas Absorption Towers," Part B, Butterworth, London.

Estimating Kinetic Rate Constants Using Orthogonal Polynomials and Picard's Iteration Method

ROBERT D. TANNER, Merck Sharp & Dohme Research Laboratories, Rahway, N. J. 07065

A new technique is proposed for estimating rate constants in systems of nonlinear differential equations. The method does not require integration of differential equations or repetitive iterations. The computations, moreover, are simple and require only modest amounts of computer capacity and time.

The method is based on the well-developed mathematical tools of Picard iteration and least-squares orthogonal polynomials. Over small time domains, infinite Picard polynomials are uniformly convergent to the differential equations describing the kinetic model. Similarly, large degree least-squares, orthogonal polynomials, such as Legendre polynomials, converge uniformly to data which approach a continuum in a small time domain. Equating the coefficients of the Picard and the orthogonal polynomials gives, therefore, a sound method for mapping the differential equations with unknown parameters into a set of algebraic relationships containing these parameters. Since low degree orthogonal polynomials, widely spaced data points, and a finite number of significant figures in the data are used in practice, the algebraic equations must reflect a transmission loss from this mapping process. This loss is minimized, however, by an extrapolation scheme which predicts the orthogonal polynomial coefficients for a zero width time domain.

The rate constants are extracted from the entire kinetic time profile in a piecewise fashion over successive time domains. Concentration data are not required for all of the variables, but to retain a meaningful amount of kinetic information, more data points for each of the remaining variables must take up the slack. Another tradeoff occurs in determining the width of the fitting domain. From one standpoint, this domain must be narrow enough for the Picard and orthogonal polynomials to retain their attractive convergence properties. On the other hand, the domain needs to be wide enough to separate the concentrations at the borders to avoid ill-conditioned or singular algebraic equations.

No assumptions are made concerning the linearity of the rate constants. Therefore, it appears that this technique will apply to differential equations describing many other physical models as well. Conceptually, at least, there appear to be no size or computational limitations in applying the method to large systems of nonlinear differential equations other than the usual inadequacies in the data or the model.

A biochemical example is given to develop and explain the method. Simulated data provide the orientation required to choose reasonable domain widths as well as the degrees of the fitting polynomials. The usual stationary state assumption, often used in kinetics, need not be invoked. In addition, all of the rate constants are obtained, not just ratios of these parameters.

We choose for the enzyme–substrate model:

$$\mathrm{E} + \mathrm{S} \underset{k_2}{\overset{k_1}{\rightleftarrows}} \mathrm{ES} \overset{k_3}{\rightarrow} \mathrm{E} + \mathrm{P}$$

where E = enzyme

S = substrate

ES = enzyme–substrate complex, and

P = product

t = time

The estimation process is illustrated for this model by the substrate concentration function over the first time domain. This polynomial, for the second Picard iteration, is:

$$(\mathrm{S})_2 = \mathrm{S}^* - k_1\mathrm{S}^*\mathrm{E}^*t + k_1\mathrm{S}^*\mathrm{E}^*[k_1\mathrm{S}^* + k_1\mathrm{E}^* + k_2]\frac{t^2}{2} - [k_1{}^3(\mathrm{S}^*)^2(\mathrm{E}^*)^2]\frac{t^3}{3}$$

for $0 \leqslant t \leqslant T$, equating the linear coefficients gives the following rela-concentrations, and T is the fitting time domain. Since the orthogonal polynomial for the same conditions can be expressed as:

$$\mathrm{S}(t) = \mathrm{S}^* + at + bt^2 + ct^3 + \ldots.$$

for $0 \leqslant t \leqslant T$ equating the linear coefficients gives the following relationship:

$$a(m, T) \simeq -k_1\mathrm{S}^*\mathrm{E}^*$$

where m is the degree of $\mathrm{S}(t)$. Extrapolating a to zero gives:

$$a(m) = \lim_{T \to 0} a(m, T) \simeq -k_1\mathrm{S}^*\mathrm{E}^*$$

To test this scheme for estimating the best value for a and subsequently the k's and E^*, we pick data simulated from the defining parameters, $\mathrm{S}^* = 1$, $\mathrm{E}^* = 0.1$, and $k_1 = k_2 = k_3 = 1$. Fitting 11 of these data points, spaced at one time unit intervals, with S(t) gives the typical family of curves for $a(m)$, as shown in Figure 1. This graph shows that from the set of extrapolated a's at integer m's, $a(4) = -.095$ comes closest to the precise vale of a (*i.e.*, $a = -0.1$).

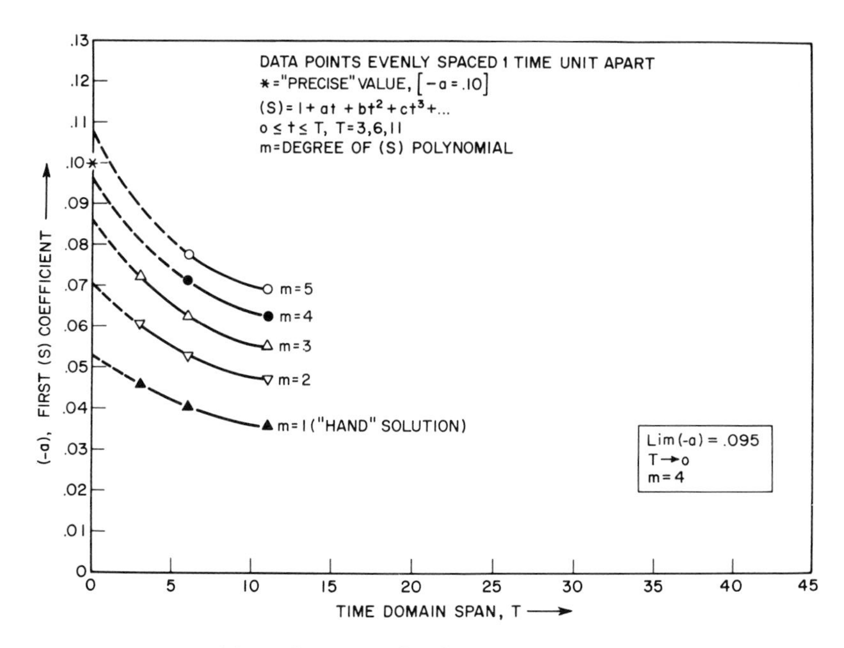

Figure 1. First (S) coefficient vs. *fitted time domain span with the degree of the orthogonal least-squares polynomial as a parameter—first time domain,* $0 \leqslant t \leqslant 11$

Continuing in this way, we fit both S and P data segmentally over the entire time domain. In this way, we obtain a sufficient number of algebraic equations to estimate all of the unknown parameters. For this three-place data example, the following parameter estimates were obtained:

$$k_1 = 0.945$$

$$k_2 = 0.874$$

$$k_3 = 1.06$$

$$E^* = E_o = 0.101$$

Simulating the differential equations describing the model with these estimated parameters gives the curves shown in Figure 2. These functions give a faithful reproduction of the data, which indicates that the model is insensitive to parameter perturbations within a 10% range.

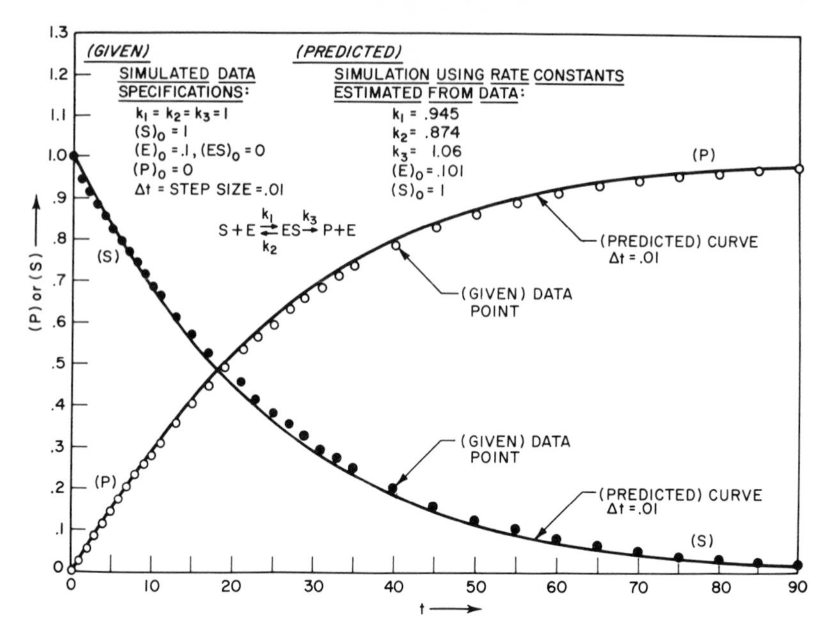

Figure 2. Computer simulation of the classical enzymatic reaction model

Since only moderate computing efforts are required in this method, the accuracy of the results can be improved by other algorithms. If the parameters estimated using this technique are modified using an analog computer, the initial searching directions on the analog machine may be taken from the extrapolation scheme. Alternatively, gradient search schemes can complement this method since they work only when the starting vector of parameters is sufficiently close to a stationary point. When this separation distance from the solution is small, however, gradient methods converge rapidly and with great accuracy to the more precise parameter vector. This method, therefore, appears to be a general technique for initializing parameter estimation procedures which employ search routines.

Synthesis of Hydrogen Cyanide by Light Hydrocarbons and Ammonia in a Flame Reaction

H.-A. HERBERTZ, H. BOCKHORN, and F. FETTING, Institut für Chemische Technologie der Technischen Hochschule, Darmstadt, Germany

Analogous to several processes which use a flame reaction to perform endothermic conversions at high temperatures (—*e.g.*, the production of acetylene and/or ethylene) the reaction of ammonia with propane under oxygen addition was studied in a specially constructed burner. Different burner arrangements were used to study the influence of mixing the feed components, of temperature, of the C/N ratio, and of the residence time on the HCN yield under turbulent flow conditions. Guibet and van Tiggelen (*1*) obtained a yield of 50–60% (CNH_3 converted to HCN) in small laminar premixed flat flames, which are unsuitable for large scale equipment.

The burner was a cylindrical tube, at the bottom of which was a plate with 19 holes; around these were annuli. In this apparatus diffusion flames could be burned from the burner plate. Downstream 10 cm from this burner plate 12 holes in the wall of the burner tube allowed 12 jets to be introduced transversely into the flow from the bottom. On the top of the burner quartz tubes were installed to provide space for the reaction to develop. Samples of the product were sucked into a water-cooled probe (inner cross section 1 mm^2), quenched, and analyzed (for details *see* Ref. *2*).

Using this device for a two-stage process stoichiometric hydrogen/oxygen diffusion flames were established at the bottom of the burner plate. Jets of a stoichiometric mixture of ammonia and propane were introduced transversely into the stream of the flue gases of these diffusion flames (at temperatures up to 2000°C). Cold ammonia and propane mixes in extremely fast (*3, 4*).

In this case the yield was limited to 20% and was nearly independent of residence time and depended only weakly on temperature. By increasing the C/N ratio from 1 to 8, the yield could be increased to 35% (Table I).

CH radicals are involved in HCN production (*5, 6*). It was thought that not enough CH and C_2 radicals were present under these conditions which correspond to pyrolysis at high temperature. Other investigations show that these radicals are present in rich hydrocarbon/oxygen flames (*7, 8*). Consequently, when oxygen was added to the propane/ammonia mixture, the yield of HCN increased. Interestingly the highest yield (75%) was obtained if no oxygen was introduced at the bottom. This

Table I. Influence of C/N Ratio on HCN Yield

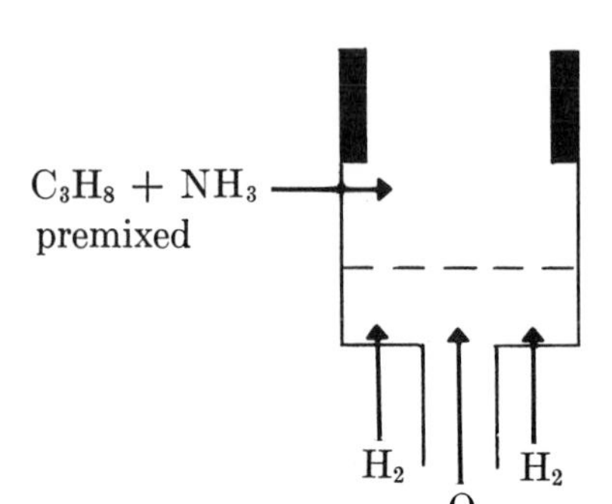

Total throughput, (Nm^3/hr)	15			
Ratio of volumetric flow rate vertically/ radially	3	5	7	7
C/N ratio	1	1	1	7
Conversion of NH_3%	45	70	88	57
Yield of HCN (based on N),%	18	22	15	35
Yield of carbon-containing products (based on C), %				
HCN	17	23	14	8
CO	16	45	68	15
CO_2	20	10	12	42
CH_4	22	15	3	12
C_2H_2	23	17	4	30

means that propane/ammonia/oxygen jet flames burned perpendicularly in a stream of cold hydrogen. This may be considered a one-stage process.

If there were no flow of cold hydrogen the yield decreased by 20%. Instead of hydrogen, cold carbon monoxide could be used for the gas flow from the bottom with almost the same result. In studying the influence of C/O ratio on the yield of HCN a slight maximum around C/O $\sim$ 0.79 was found. An increase in the C/N ratio is followed by an increase in yield (Table II).

Similar results could be obtained when cold ammonia was used for the cold stream from the bottom and no ammonia was added to the jet flames in the upper part of burner. Rich premixed propane/oxygen flames burned in an atmosphere of cold ammonia giving yields of up to 75%. The highest yields could be reached with a high C/O ratio and a high C/N ratio. In Table III the yield of hydrogen cyanide, the conversion of NH_3, and the yield of carbon-containing products for jet flames burning in a cross-flow of ammonia are compared with those for jet flames burning in a concurrent flow of ammonia and for premixed propane/oxygen/ammonia flames. To synthesize hydrogen cyanide in jet flames burning in a concurrent flow of ammonia and in premixed flames only the lower part of the burner was used. The premixed propane/oxygen/ammonia flames burned at the nozzles of the burner plate or premixed propane/oxygen flames burned at the nozzles in a concurrent flow of ammonia from the annuli. Table III shows that the yield of hydrogen cyanide for jet flames burning perpendicularly in a stream of cold ammonia is about 10% higher compared with premixed propane/ammonia/oxygen flames using the same C/N and C/O ratio.

While some of the results are easily understood, some are not. One key to such understanding is the behavior of sooting flames. Soot can be observed in rich propane/oxygen flames if a critical C/O ratio of ~ 0.47 is exceeded. In all experiments the C/O ratio was between 0.7 and 0.9; nevertheless, no soot was observed if ammonia was present. If ammonia were omitted, soot appeared. This suppression of soot formation by ammonia is independent of the type of hydrocarbon used (9). In flames with aliphatic hydrocarbons soot is formed by the polymerization of CH radicals and other fragments like C_2 and C_3 radicals and lower poly-

Table II. Influence of C/O Ratio on HCN Yield

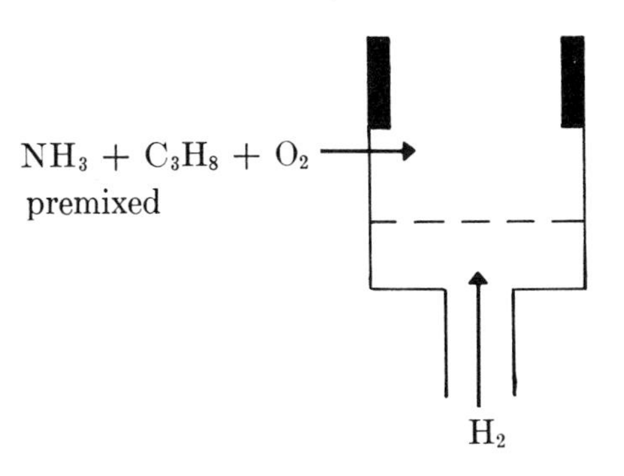

Total throughput, (Nm/hr)	~ 8					
Ratio of volumetric flow rate vertically/radially	~ 0.5					
C/N ratio	5	5	5	3	4	5
C/O ratio	0.72	0.75	0.8	0.75	0.75	0.75
Conversion of NH_3, %	88	85	80	81	83	85
Yield of HCN (based on N), %	69	74	75	65	70	74
Yield of carbon-containing products (based on C), %						
HCN	15	18	18	20	19	18
CO	68	65	61	53	62	65
CO_2	10	5	5	15	10	5
CH_4	3	5	6	5	5	5
C_2H_2	8	10	11	5	8	10

Table III. Results using Cross Flow and Concurrent Flow of Ammonia

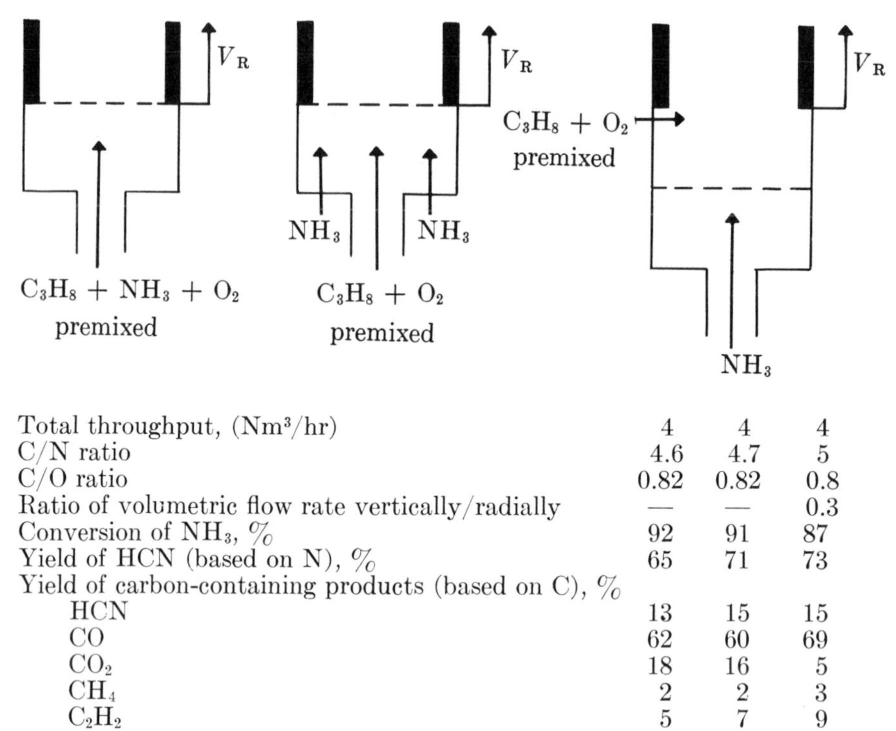

Total throughput, (Nm^3/hr)	4	4	4
C/N ratio	4.6	4.7	5
C/O ratio	0.82	0.82	0.8
Ratio of volumetric flow rate vertically/radially	—	—	0.3
Conversion of NH_3, %	92	91	87
Yield of HCN (based on N), %	65	71	73
Yield of carbon-containing products (based on C), %			
HCN	13	15	15
CO	62	60	69
CO_2	18	16	5
CH_4	2	2	3
C_2H_2	5	7	9

acetylenes (*10*). It seems reasonable to assume that soot formation can be blocked if NH_3 or fragments of NH_3 react with the lower radical-type first steps of polyacetylenes or presoot particles and form HCN. Concentration profiles of rich hydrocarbon/oxygen flames at lower pressure showed (*10*) that hydrocarbon radicals which are essential for HCN formation have their highest concentrations in regions where oxygen is no longer present. Thus, ammonia should be added to this part of the flamefront. This took place in the arrangement where propane/oxygen flames burned perpendicularly in a stream of cold ammonia, and in this case the HCN yield is higher than in premixed propane/oxygen/ammonia flames where ammonia must be transported through the hot oxidation zone to the spot where HCN is formed. For the case where premixed propane/oxygen flames burn in concurrent flow of ammonia the HCN yield is slightly lower than for premixed propane/oxygen flames burning perpendicularly in a flow of ammonia because the intermixing of ammonia into the jet flames is not as fast and ammonia is mixed with hot recirculating flame gases before it is sucked up by the jet flames (for detailed discussion *see* Ref. *11*).

(1) Guibet, J. G., van Tiggelen, A., *Rev. Inst. Franc. Petrole Ann. Combust. Liquides* (1963) **18**, 1284.
(2) Herbertz, H.-A., Ph.D. Thesis, TH Darmstadt, 1969.
(3) Patrick, M. A., *Trans. Inst. Chem. Engr.* (1967) **45**, T16.
(4) El-Zein, R., M.S. Thesis, Darmstadt, 1968.
(5) Janin, J., Mathais, M., *Rev. Inst. Franc. Petrole Ann. Combust. Liquides* (1964) **19**, 523.
(6) Janin, J., Mathais, M., *Comm. Colloq. Spectrosc. Intern. IX, Lyon, 1961,* S. 274.
(7) Bleekrode, R., Nieuwpoort, W. C., *J. Chem. Phys.* (1965) **43**, 3680.
(8) Jessen, D. F., Gaydon, A. G., *Symp. (Intern.) Combust., 12th, Poitiers (Frankreich) 1968,* S. 481.
(9) Bockhorn, H., M.S. Thesis, Darmstadt, 1969.
(10) Homann, K. H., Wagner, H. Gg., *Ber. Bunsenges. Phys. Chem.* (1965) **69**, 200.
(11) "Combustion Science and Technology," to be published.

Reaction Kinetics of High Temperature Copyrolysis

KEN K. ROBINSON and ERIC WEGER, Monsanto Co., 800 N. Lindbergh Blvd., St. Louis, Mo. 63166

Hydrocarbons, when subjected to conventional pyrolysis conditions, yield primarily lower molecular weight compounds. Scott (*1*) found, however, that by using extremely high temperatures and short residence times, one could also form significant amounts of higher molecular weight compounds. These particular reaction conditions promote high concentrations of free radicals which subsequently lead to increases in the rate of radical combination. The highest yields of radical combination products are achieved by copyrolyzing a methyl free radical generator, such as propane, with olefins which decompose to relatively stable radicals.

Free radicals are very reactive (reaction rate constant is known to be quite large) but are normally present in extremely low concentrations. This causes the rate of radical combination to be small under conventional pyrolysis conditions. However, Scott (*1, 2*) found in his recent pyrolysis studies that significant amounts of radical combination products were formed. For example the copyrolysis of ethylene–acetone, propylene–propane, and toluene–propane mixtures produced propylene, 1-butene, and ethylbenzene, respectively with the methyl radical playing an important role in these thermal reforming systems. The specific reaction conditions (characterized by temperatures around 1000°C and reaction times of the order of milliseconds) are believed to promote the reaction by causing high concentrations of free radicals. The authors believe that the radical formation zone must be coupled with a rapid cooling zone to attain a favorable equilibrium yield of the higher molecular weight compounds.

In this investigation, propylene–propane mixtures were pyrolyzed at temperatures around 1100°C for 1 msec in an annular quartz flow reactor to promote the formation of 1-butene. This is believed to occur by a process in which the methyl radical from propane combines with the allyl radical from propylene.

Experimental

The reactor consisted of a quartz annulus which was jacketed by a glowing carbon jacket (inductively heated by a high frequency generator) and a quench leg where the hot gases leaving the annular pyrolysis zone were cooled with nitrogen or methane. The temperature in the pyrolysis zone was measured with a movable platinum–platinum–10% rhodium thermocouple situated inside an axial thermowell. Corrections to the thermocouple readings were made for induction and radiation effects. The temperature at the quenching junction was measured with a stainless steel sheathed chromel-alumel thermocouple.

The reactor was jacketed by two larger concentric tubes, 6½ inches long, which had cooling water circulating through the annulus between them. The innner tube was constructed from quartz and was 2½ inches in diameter, while the outer tube was borosilicate glass and 3½ inches in diameter. Stainless steel plates fitted with silicone rubber gaskets held the concentric tubes in place. A 3¾-inch diameter induction coil was constructed from ¼-inch copper tubing and heated the carbon jacket with power supplied by a Lepel model T25–3, 25-kilowatt generator with a frequency range between 180 and 450 kc.

The reaction kinetics were determined by a technique developed by Towell and Martin (3) in which the reactor is operated differentially with the nonisothermal temperature profile held constant. Power law rate expressions were substituted into a differential rate equation and integrated (subject to differential reactor operation) to yield the following mass balance equation:

$$F_{A_0} z_A = \bar{x}_A{}^n \bar{x}_B{}^m \left[aA(\pi/R)^{n+m} \int_0^l \frac{\exp(-\Delta E/RT)}{T^{n+m}} \, dl \right] \tag{1}$$

By holding temperature constant from one set of experiments to the next, the terms in the brackets remain constant, and the equation can be expressed as:

$$F_{A_0} z_A = K \bar{x}_A{}^n \bar{x}_B{}^m \tag{2}$$

This equation is transformed easily to a linear form by taking the logarithm.

The reaction orders for the propane, propylene, and 1-butene rate equations were determined by plotting conversion or formation rate as a function of the average mole fraction on logarithmic coordinates and then calculating the slope.

Results and Discussion

Propane and propylene were pyrolyzed separately to establish general trends in the product distribution. No 1-butene was formed when propane was pyrolyzed alone and only very small amounts when propylene was pyrolyzed separately. For the latter case, the 1-butene yield (moles 1-butene formed per mole propylene fed) was less than 0.5% for 10% propylene conversion. However, when propane and propylene were copyrolyzed, 1-butene formation was increased significantly.

Of course, in addition to 1-butene, substantial quantities of methane, ethylene, and hydrogen were formed. 1-Butene was formed selectively at low propylene conversions and then decreased exponentially with increasing conversion. For example, propylene selectivity went from 55% at 3% propylene conversion to around 14% at 30% conversion. The highest 1-butene yield achieved was 5 moles of 1-butene formed per 100 moles of propylene fed at 45% propylene conversion.

The decomposition kinetics of propylene and propane were determined to be first order over a large range of propylene and propane concentrations. The decomposition kinetics were studied under copyrolysis reaction conditions in which both hydrocarbons were fed simultaneously to the reactor. The first-order dependence of propylene breaks down at low propane concentrations where propane appears to promote propylene decomposition. When propylene is pyrolyzed separately, the propylene molecule must generate its own chain carrier species before it will decompose at an appreciable rate. On the other hand, when propane is added to the pyrolysis reaction system, one has an additional source of free radials which accelerate propylene decomposition by interaction in a typical chain propagation step. Table I illustrates that the addition of small amounts of propane leads to much higher propylene conversions.

Table I. Propane Promotion of Propylene Decomposition

Run	*Maximum Temperature,* °*C*	*Feed Composition,* *%* C_3H_6	*%* C_3H_8	C_3H_6 *Conversion,* *%*
1	1109	29.5	0.0	1.6
2	1109	29.0	0.5	9.7
3	1147	32.4	0.0	10.4
4	1147	28.7	28.7	23.2

The rate of 1-butene formation is a function of the hydrocarbon concentration in the pyrolysis zone of the reactor (which determines the steady-state concentration of free radicals) and the quench-to-feed ratio (which affects both the temperature and concentration of free radicals present in the quench zone). To take into account the dilution effect, the 1-butene formation rate was correlated as a function of hydrocarbon concentration in the quench zone.

The rate of 1-butene formation was found to be directly proportional to propylene concentration in the quench zone and independent of propane over a reasonably wide range of reaction conditions. The molar formation rate of 1-butene was plotted as a function of propylene mole fraction in the quench zone for three sets of data and was correlated by three straight lines having a slope of approximately 1. Therefore, the kinetic rate equation for 1-butene may be expressed as:

$$\frac{d[C_4H_8]}{dt} = k\,[C_3H_6] \tag{3}$$

Since propane improves the selectivity of propylene to 1-butene, one would expect some dependence of 1-butene rate on propane. However, if the allyl radical is destroyed primarily by combining with the methyl radical to form 1-butene, its combination rate will be equal to the rate at which propylene decomposes to generate allyl radicals. For extremely low propane concentrations, 1-butene formation will exhibit a weak dependence on propane. This is expected, naturally, since propane serves as a methyl free radical source. A set of experiments was conducted in which propylene was copyrolyzed with very small amounts of propane, and the results are summarized in Table II.

Table II. 1-Butene Formation at Low Propane Concentrations

	Base Kinetic Conditions at 1109°C			
Run No.	5	6	7	8
Mole % in feed				
Propylene	29.2	27.4	26.8	23.3
Propane	0.0	0.7	1.2	2.5
Butene-yield, %	0.3	0.4	0.7	0.8

It appears that propane plays a dual role in propylene–propane copyrolysis. In addition to its promotional effect on propylene decomposition, it also influences 1-butene rate when present in low concentrations. A rate equation for 1-butene formation applicable over a wide range of reaction conditions (*i.e.*, inclusive of both low and intermediate propane concentrations) would no doubt be much more complicated than Equation 3. It is, in fact, possible that the power law form of the rate equation would not adequately represent the reaction kinetics. Determination of kinetics in a nonisothermal reactor, however, imposes the restriction that a power law rate equation be used since it is transformed easily to a linear form by taking the logarithm. Hence, it was felt that the development of a more definitive rate equation was beyond the scope of the study. A simplified 12-step reaction mechanism has been proposed, which is consistent with the decomposition kinetics and explains the particular form for the 1-butene rate expression. Briefly it consists of the following steps:

(1) Initiation. Propane decomposing to a methyl and ethyl radical and propylene decomposing to the allyl radical and hydrogen.

(2) Propagation. Transfer type reactions of propane and propylene with the methyl radical and atomic hydrogen. Allyl and propyl radicals decomposing respectively to allene, ethylene, and propylene plus the associated free radicals. Ethyl radical abstracting hydrogen from propane.

(3) Termination. Disproportionation of methyl and propyl radical to methane and propylene. Combination of methyl and allyl radical to form 1-butene.

Based on this reaction mechanism the free radical concentrations were calculated (numerically on a digital computer) as a function of time for an isothermal reactor (1100°C). After 10^{-5} sec the free radicals had reached a steady state concentration indicating that the reaction system is a long chain process. Further details may be found in Robinson's latest pyrolysis studies. (*4, 5*).

(1) Scott, E. J. Y., "Free Radical Combination Reactions Involving the Methyl Radical at 1000° to 1200°C," *Ind. Eng. Chem.* (1967) **6,** 67.
(2) Scott, E. J. Y., "Reaction of Alkanes with Toluene at 800° to 1200°C," *Ind. Eng. Chem.* (1967) **6,** 72.
(3) Towell, G. O., Martin, J. J., "Kinetic Data from Non-isothermal Experiments: Thermal Decomposition of Ethane, Ethylene, and Acetylene," *AIChE J.* (1961) **7,** 693.
(4) Robinson, K. K., "Thermal Reforming by Means of Propylene-Propane Co-Pyrolysis," Ph.D. Thesis, Washington University, St. Louis, 1970.
(5) Robinson, K. K., Weger, E., *Ind. Eng. Chem., Fundamentals* (1971) **10,** 205.

Kinetic Models of the Chlorination Reactions for Pyrite Cinder Purification

A. CAPPELLI, A. COLLINA, G. SIRONI,[1] and B. VIVIANI,[1] Montecatini Edison, Direzione Centrale delle Ricerche, Milano, Italy

During the development of an industrial process for pyrite cinder purification, a mathematical model was prepared for designing and simulating a fluid bed reactor in which nonferrous metals are eliminated as volatile chlorides. Pyrite cinder purification takes place in a fluid bed reactor to which magnetic cinders are fed at a suitable degree of reduction, together with a gaseous mixture of air and chlorine.

The most important of the many chemical reactions occurring are reoxidation of the cinders, desulfuration, dearsenification, and chlorination and volatilization of nonferrous metals. The reactions which, from the viewpoint of plant design and operation, appear to be most critical are the chlorination reactions of nonferrous metals and, in particular, of the metals present in the largest quantities in the cinders fed in—*i.e.*, copper, zinc, and lead. The work necessary for developing the model is outlined in the following three stages.

(1) Study of the kinetics relative to chlorination reactions of nonferrous metals by systematic experimentation in a fixed bed laboratory-scale reactor. The following volatilization reactions have been considered:

$$Cu + \tfrac{1}{2} Cl_2 \rightarrow CuCl$$

$$Zn + Cl_2 \rightarrow ZnCl_2$$

$$Pb + Cl_2 \rightarrow PbCl_2$$

The kinetics were studied under the following operating conditions: reaction temperature 850°–950°C; concentration of chlorine in air 0.1–5%; time 1–30 minutes. Experimentation has been performed so as to operate in differential conditions as regards the chlorine; in this way it has been possible to process the experimental data independently for each of the three reactions. The kinetic equations proposed are of the type:

$$V_{Me} = \frac{K_1 \exp(-E/RT) X^n}{1 - K_2 X_{Me}} \; \frac{Y_{Cl2}}{1 + K_3 Y_{Cl2}}$$

The values of the parameters n, K_1, K_2, K_3, and E were estimated using nonlinear regression method (*1, 2*).

[1] Montecatini Edison, Direzione Centrale delle Ricerche, Istituto, Donegani, Novara, Italy.

(2) Building of a mathematical model of a fluid bed reactor and determination of unknown parameters utilizing experimental data obtained on the prepilot scale in a fluid bed reactor (100 mm diameter). To characterize the behavior of the fluid bed reactor a two-phase type model has been chosen, to be exact (*3, 4*): (a) a dense phase, which contains all the solid of the fluid bed and where the chemical reactions take place and (b) a lean phase, consisting exclusively of gas. A gas interchange by cross-flow takes place between the phases (*4*).

The following hypotheses were made for the dense phase. Gas flow rate fed directly to the dense phase is negligible compared with the total gas flow rate fed into the reactor. This flow rate, which is generally considered equal to that corresponding to the minimum fluidizing velocity, is in fact, less than 10% of the total flow rate normally utilized. Flow of the solid through the phase is completely segregated. Each particle retains its identity from the reactor inlet to outlet and follows the residence time distribution characteristic of the fluid bed. It is assumed that the gas, which comes in contact with the particles of solid within the dense phase, is characterized by an average chlorine content equal for all particles. Carryover of fines has been neglected assuming that all the cinders fed in are discharged from the lower portion of the reactor.

The following hypotheses were made for the lean phase. Its behavior is described by a dispersed plug flow model, taking into account the cross-flow rate. The variation of the total flow rate between the reactor inlet and outlet has been neglected. The segregation of gas within the lean phase is assumed to be nil. Finally, the bed is considered isothermal.

The average Cu, Zn, and Pb content in the purified cinders is calculated, keeping in mind the residence time distribution of the solid determined experimentally. The three parameters of the model (Péclet number of the lean phase and two parameters relative to the cross flow) were evaluated using experimental data obtained on the prepilot scale, with a nonlinear regression method (*1, 2*). The model was developed using a Univac 1108, and the computer time required was about 8 hours.

(3) Check of the mathematical model by experimentation in a pilot scale fluid bed reactor and determination of the scale factors (reactor diameter 1000 mm). To perform a first check, some *a priori* considerations were made on the behavior of the pilot reactor which permits preliminary estimates of the model parameters. From the viewpoint of the residence time distribution of the solids, it has been assumed that the behavior of the pilot reactor is similar to that of an ideally mixed tank reactor. As for the Péclet number relative to the lean phase, a value characteristic of good mixing has been taken. With regard to the values of the parameters relative to the cross flow, the same values obtained for the prepilot reactor have been utilized.

Table I. Experimental and Calculated Data for Pilot Plant Fluid Bed Chlorination

Test No.	X_{Cu} *% exper.*	X_{Cu} *% calc.*	X_{Zn} *% exper.*	X_{Zn} *% calc.*	X_{Pb} *% exper.*	X_{Pb} *% calc.*
P 1	0.040	0.047	0.080	0.070	0.015	0.022
P 2	0.037	0.038	0.062	0.032	0.010	0.015
P 3	0.043	0.057	0.157	0.107	0.015	0.028
P 4	0.040	0.044	0.070	0.059	0.010	0.019

Table I compares the experimental data and the calculated results. On considering the hypotheses made, the agreement is very satisfactory. It could be improved once the residence time distribution of the solid in the pilot reactor has been measured and on the basis of more numerous experimental data. For this purpose a series of systematic tests is now being performed on the pilot plant, in operation at Scarlino-Follonica (for Russian pyrite cinders and Rio-Tinto type Spanish pyrite cinders) which will permit us to draw final conclusions for sizing the industrial plant reactor. This research work is being carried out in cooperation with Dorr-Oliver Inc. (U.S.A.). The systematic use of the model is expected to contribute to the correct solution of some essential scaling-up problems in the design and operation of industrial reactors for pyrite cinder purification.

(1) Ferraris, G. Buzzi, *Ing. Chim. It.* (1968) **14,** 171.
(2) *Ibid.,* p. 180.
(3) Kunii, D., Levenspiel, O., "Fluidization Engineering," Wiley, New York, 1969.
(4) May, W. G., *Chem. Eng. Progr.* (1959) **55,** 49.

Surface-Catalyzed Reactions in Turbulent Pipe Flows

S. S. RANDHAVA, Union Carbide Corp., Tarrytown Technical Center, Tarrytown, N. Y. 10591

D. T. WASAN, Illinois Institute of Technology, Chicago, Ill. 60616

Flow reactor systems lend themselves to the study of rates of rapid surface reactions. Operation in turbulent flows is often desirable to avoid masking the true surface kinetics by diffusional effects since the increased supply of reactant from the bulk stream to the solid surface makes it difficult for the catalytic surface to deplete the labile species significantly (*1–4*).

This study analyzes numerically an arbitrary order heterogeneous reaction which occurs on the surface of a turbulent flow tubular reactor. The treatment is directed primarily toward obtaining solutions for the axial wall concentrations and the radial concentration distributions for pertinent cases. The transverse velocities generated by molecular diffusion toward the catalytic wall have also been incorporated.

The case used is the following. Consider a fluid in fully developed turbulent flow entering a section with unchanging velocity and concentration profiles. The section is a catalytic tube of arbitrary length; it is assumed that the intrinsic chemical kinetics at the surface are well described by a power function of the local reactant concentration. Within the turbulent flow field reactant transport occurs both by convection and a combination of molecular and eddy diffusion.

At steady state, time-averaging the species conservation equation results in:

$$U \frac{\partial W_A}{\partial x} + v \frac{\partial W_A}{\partial r} = \frac{1}{r} \frac{\partial}{\partial r} \left[r(D + E) \frac{\partial W_A}{\partial r} \right] \tag{1}$$

In the present case where the labile species diffuse through an inert gas, the transfer mechanism can be approximated as that of diffusion through a gas which exhibits no net transfer in the direction of the labile species transfer—*i.e.*, the flux of the inert species B can be assumed to be zero. Strictly speaking, the net flux of the inert component B is zero only at the wall and the center line. In this analysis we assume that the condition $N_B = 0$ is true at every radial position in the pipe.

From this condition, using Ficks' law, the relationship between the crossflow diffusion velocity and the concentration gradient is:

$$v = \frac{(D + E)}{1 - W_A} \frac{\partial W_A}{\partial r} \tag{2}$$

It is assumed that this crossflow diffusion velocity does not affect the momentum profiles.

Since the diffusion coefficient D can be considered constant and the eddy diffusion E is a function of the radial position in the pipe, Equations 1 and 2 can be combined to give:

$$U \frac{\partial W_A}{\partial x} + \frac{(D + E(r))}{1 - W_A} \left[\frac{\partial W_A}{\partial r} \right]^2 = \frac{1}{r} \frac{\partial}{\partial r} \left[r(D + E(r)) \frac{\partial W_A}{\partial r} \right] \tag{3}$$

Equation 3 is a nonlinear, second-order equation subject to the following boundary conditions:

(1) The inlet composition of the labile species is uniform.

$$W_A(0, r) = W_{A_0} \tag{4}$$

(2) The concentration profile is symmetric around the radial axis.

$$\frac{\partial W_A}{\partial r}(x, 0) = 0 \tag{5}$$

(3) Since the crossflow velocity vanishes along the catalytic wall and the eddy diffusivity is negligible in this region, the use of Fick's law of molecular diffusion yields the following balance for an n-th order reaction:

$$\rho_t D \frac{\partial W_A}{\partial r} = -\frac{k_w \rho_t^{\,n}}{M^{(n-1)}} W_A^{\,n} \tag{6}$$

To specify the problem completely it is necessary to determine the variation of the time-averaged velocity and eddy diffusivity as a function of the radial position. For the present analysis, the eddy diffusivity and velocity distribution functions recently developed and used by Wasan *et al.* (*5, 6*) are used.

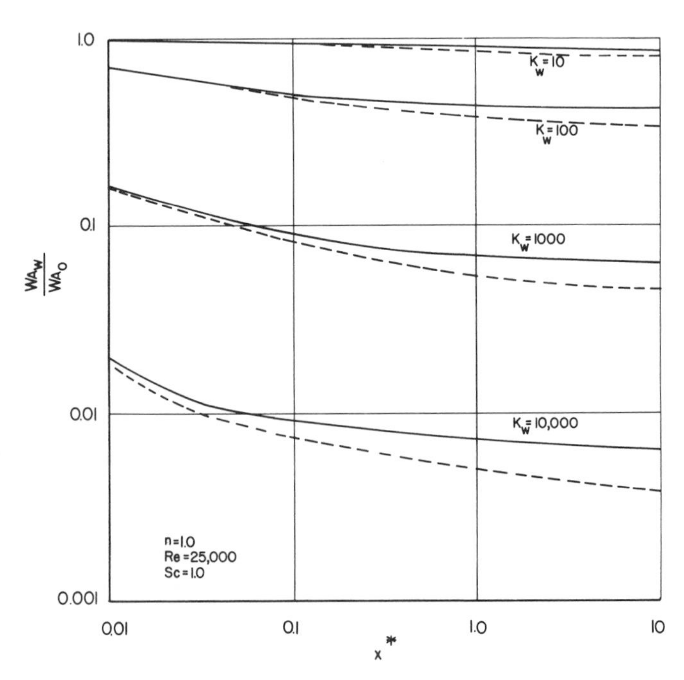

Figure 1. Effect of dimensionless rate constants on axial wall concentration distributions

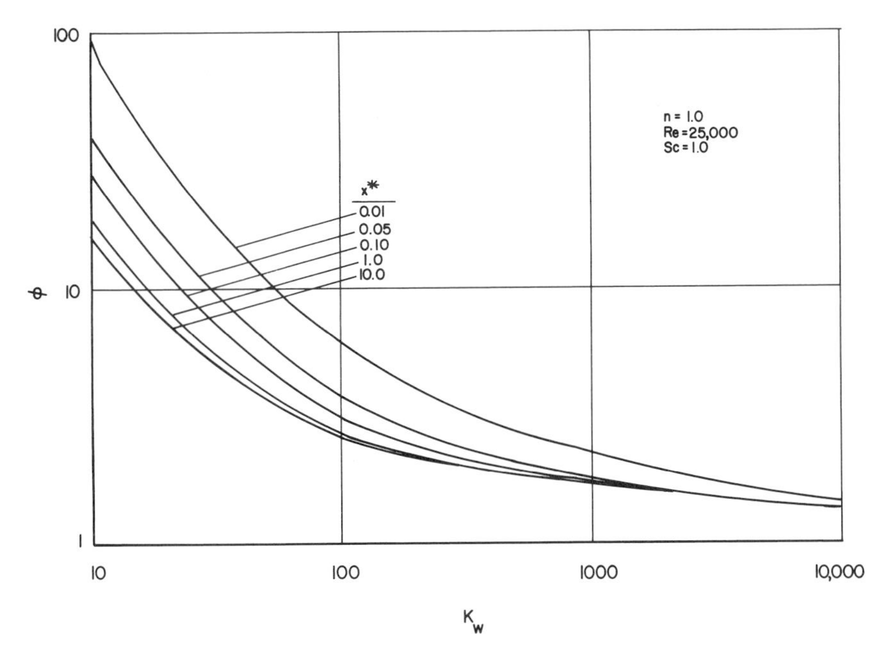

Figure 2. Change in ϕ as a function of K_w *for various values of the aspect ratio*

Equation 3 and the boundary conditions were non-dimensionalized, and an implicit finite difference technique was used to solve these equations.

Our present results apply to a constant density system (reactant is present indilute amounts) with zero pressure gradient reactors of finite length. Concentrations and local Stanton numbers were computed over a range of reaction orders, dimensionless rate constants, and Reynolds and Schmidt numbers. Since most phenomena of surface-catalyzed reactions occur within a short distance, particular attention was devoted to determining the extent of changes in the entrance region. Comparison with the existing solution of Wissler and Schecter (*2*) for a first-order reaction has shown the validity of the technique.

Changes in the wall and radial concentrations as a function of the dimensionless rate constants and reaction orders were shown. The effect including the diffusional radial velocity contribution is substantial, especially for higher values of the rate constant (*see* Figure 1).

Increasing Reynolds numbers caused an increase in the wall concentrations, W_{A_w}, the effect becoming more pronounced downstream. Increasing Schmidt numbers drastically decreased the W_{A_w} proportionally.

Stanton numbers were computed using the standard definition. They decrease rapidly with an increase in K_w and eventually converge toward values predicted by assuming pure diffusion-controlled phenomena (*see* Figure 2). A substantial effect of downstream aspect ratios has also been noted. There is a noticeable effect caused by changes in reaction orders, but in this case the effect of the downstream aspect ratios is negligible.

Nomenclature

D	molecular diffusivity
E	eddy diffusivity
k_w	heterogeneous rate constant
K_w	dimensionless heterogeneous rate constant, $\frac{k_w R}{D} \frac{\rho_t^{n-1}}{M^{n-1}}$
M	molecular weight of diffusing species A
N_B	mass flux of species B
n	heterogeneous reaction order
r	radial coordinate
R	pipe radius
U	axial velocity
v	radial velocity
W_A	mass fraction of species A
W_{A_w}	mass fraction of species A at catalytic wall
W_{A_0}	inlet mass fraction of species A
x	axial coordinate
ρ_t	total density of mixture
Re	Reynolds number
Sc	Schmidt number
ϕ	Stanton number for mass transfer with reaction/Stanton number for pure mass transfer

(1) Satterfield, C. N., Resnick, H., Wentworth, R. L., *Chem. Eng. Progr.* (1954) **50,** 460.
(2) Wissler, E. H., Schecter, R. S., *Chem. Eng. Sci.* (1962) **17,** 937.
(3) Rosner, D. E., *Chem. Eng. Sci.* (1964) **19,** 1.
(4) Solbrig, C. W., Gidaspow, D., *Intern. J. Heat Mass Trans.* (1968) **11,** 155.
(5) Wasan, D. T., Tien, C. L., Wilke, C. R., *A.I.Ch.E. J.* (1963) **9,** 567.
(6) Wasan, D. T., Wilke, C .R., *A.I.Ch.E. J.* (1968) **14,** 577.

Isomerization of *n*-Pentane over a Commercial Pt on Al_2O_3 Catalyst

JORGEN HENNINGSEN, Technical University of Denmark, 2800 Lyngby, Denmark

The kinetics of the *n*-pentane isomerization over a commercial Pt on Cl-activated Al_2O_3 catalyst has been studied under the following experimental conditions: temperature = 420°–480°C; hydrogen pressure = 12–36 kp/cm²; *n*-pentane pressure = 0.04–4 kp/cm². The experiments were performed at low conversion, thus allowing direct initial rate calculations. In all experiments the concentration of pentenes in the reactor effluent was measured, and it was observed that equilibrium concentration between *n*-pentane/*n*-pentenes and isopentane/isopentenes was established, thus supporting the theory that the olefin isomerization in the generally accepted reaction sequence (*1*):

$$n\text{-pentane} \underset{-H_2}{\overset{Pt}{\rightleftarrows}} n\text{-pentene} \overset{Al_2O_3}{\rightleftarrows} \text{isopentene} \underset{+H_2}{\overset{Pt}{\rightleftarrows}} \text{isopentane}$$

is the rate-determining step.

A total of 54 experiments were performed corresponding to a complete combination of three temperatures, three hydrogen pressures, and six pentane pressures. Fitting all data to a simple power-function model gave the following rate equation (for the forward reaction):

$$r_f = 10^{9.94} \times e^{-24600/RT} \times p_{nC_5}{}^{0.69} \times p_{H_2}{}^{-0.86}$$

with a residual standard error between observed and predicted rate values of 27%.

Re-examining the data in groups corresponding to low, intermediate, and high surface coverages showed high pentane reaction order (0.9–1) and low activation energy at high temperature and low pentane pressure, and low pentane reaction order (*ca.* 0.5) and high activation energy at low temperature and high pentane pressures. Since these observations agree with what could be expected for the influence of adsorbed paraffin molecules on the alumina, the data were fitted to the corresponding rate equation by nonlinear regression

$$r_f = \frac{k_o \times e^{-E/RT} \times p_{nC_5} \times p_{H_2}{}^{\gamma}}{1 + K_{ads,O} \times e^{-\Delta H_{ads}/RT} \times p_{C_5}}$$

giving the following estimates (with 95% confidence limits):

$$
\begin{aligned}
\log_{10} k_0 &= 8.50 \pm 1.04 \\
E &= 19.0 \pm 3.4 \text{ kcal} \\
\gamma &= -0.85 \pm 0.11 \\
K_{ads,\,0} &= (0.037 \pm 0.209) \cdot 10^{-4} \ (\text{kp/cm}^2)^{-1} \\
\Delta H_{ads} &= -17.8 \pm 4.0 \text{ kcal}
\end{aligned}
$$

and reducing the residual standard error to 20%.

Plotting the differences between observed and predicted rates *vs.* experiment number (time) showed an apparent increase in catalyst activity throughout the investigation period (3 weeks). Including a term in the rate equation to account for this increase in activity reduced the residual standard error to 16%, leaving all the parameter estimates practically unchanged but reducing the confidence limits by approximately 20%, ($\Delta H_{ads} = -17.9 \pm 3.1$ kcal).

The agreement between the theoretical considerations and the estimated parameters can be summarized as follows.

(1) The over-all temperature dependence (or activation energy) is surprisingly low, remembering that an increased temperature will increase both the pentene equilibrium concentration and the pentene isomerization rate constant. A supplementary determination of only the activation energy has indicated values of 50–60 kcal *vs.* 20–30 kcal found in this investigation.

(2) The deviation of the hydrogen reaction order from -1 may be explained in at least three ways, all in agreement with the theoretical consideration outlined:

(a) By assuming a certain degree of surface heterogeneity, which will, in general, lower the reaction order in the numerator in Langmuir's rate expressions (*cf.*, the Freundlich equation).

(b) By assuming that the third (hydrogenation) step in the reaction sequence is partly rate-determining. The pentene concentrations in some of the low hydrogen pressure experiments make this explanation possible, although the analytic accuracy was not sufficient for a significant statement.

(c) By a certain influence of the pore diffusion on the over-all reaction rate. Carrying out the reaction in the strongly diffusion-limited range will cause the apparent hydrogen reaction order to drop to half the value obtained under nondiffusional-limited conditions, owing to reasons similar to those halving the activation energy. Effectiveness factor calculations based upon pore size distribution measurements showed, however, that effectiveness factors close to unity could be expected in all experiments. Furthermore, the observed rates in the experiments with the highest apparent intrinsic rate constants were not lower than expected from the model determined on basis of all 54 experiments. Consequently, this explanation was rejected as reason for the deviation of the hydrogen reaction order from -1.

(3) The determination of ΔH_{ads} is typical for what one would expect for chemisorption of paraffins on Al_2O_3. However, the pre-exponential factor in the adsorption term is not significantly different from zero. This

is because this parameter is correlated to the other parameters. Attempts to lower this correlation by reparametrization of the model were not successful.

In conclusion, we showed that the rate equation agreed with the assumption of a dual function mechanism, the pentene isomerization being rate determining, but that it was not possible to correlate the data only by means of the (calculated equilibrium) pentene partial pressures as suggested by other investigators (*2*, *3*). The most obvious explanation appears to be an influence on the isomerization rate by adsorbed pentane molecules.

(1) Sinfelt, J. H., *Advan. Chem. Eng.* (1964) **5,** 37.
(2) Sinfelt, J. H., Hurwitz, H., Rohrer, J. C., *J. Phys. Chem.* (1960) **64,** 892.
(3) Froment, G. F., Mezaki, Reiji, *Chem. Eng. Sci.* (1970) **25,** 293–301.

The Kinetics of Alkylation of Isobutane with Propylene

J. RANDOLPH LANGLEY[1] and RALPH W. PIKE, Reacting Fluids Laboratory, Department of Chemical Engineering, Louisiana State University, Baton Rouge, La.

Commercial catalytic alkylation of isobutane with C_3 to C_5 olefins is an important source of high octane motor fuel. Both sulfuric and hydrofluoric acid are used as catalysts; however, the sulfuric acid process is the more widely used. Because liquid–liquid complex reactions are involved, theoretical knowledge of the process has lagged behind commercial development.

The most significant theoretical contribution has been the Schmerling carbonium ion mechanism (*1*). Other researchers have sought to modify certain portion of this theory; however, it still remains the most comprehensive description of the alkylation reactions available. In addition, many research contributions have been allied more closely to the effect of operating variables on the product quality. The work in this area has been so voluminous that several written surveys have been devoted to it (*2*, *3*, *4*).

[1] Present address: U. S. Army Engineer Reactors Group, Ft. Belvoir, Va.

Theory for the Kinetic Analysis

Since the mechanism of Schmerling (*1*) is the most comprehensive one available, it was the basis for the mathematical model used in this study. This mechanism describes possibles routes of formation of C_3 through C_8, C_{10}, and C_{11} saturated hydrocarbon species. The most plausible of these reactions was chosen for each species while considering modifications proposed by authors of later works such as Hofmann and Schriesheim (*5*), Kennedy (*6*), and others (*2, 7, 8*). Specifically, the reactions to form the nonane fraction was a modification. Also, the route of decane formation was tailored to account for the yields of this species over the range of temperatures investigated. In addition, the question of dimethylhexane formation *via* allylic ions as proposed by Hofmann and Schriesheim (*5*) was a considered modification. However, as experimental information was lacking to discern between this theory and Schmerling's proposed self-alkylation of isobutane, only the latter was retained to account for all C_8 formation. The 17 reactions used are shown in Table I.

Since the reactions are considered to occur in the acid (catalyst) phase, the model was derived considering all the reactants (carbonium ions and unsaturates) and products to be in solution in this phase. The concentrations of the various species were measurable in the hydrocarbon phase only, however. Thus, mass transfer relationships obtained from a physically similar but nonreacting system (*10*) along with solubility data were used to estimate the acid-phase concentrations. The ionic and unsaturate concentrations within this phase were derived assuming proportional relationships among them and their parent species.

Rate equations were written for the modified Schmerling mechanism. The steady-state assumption of zero production of reaction intermediates was applied, and a mathematical model was derived by solving the steady-state rate equations for the rate constants. This resulted in a total of 17 rate constants and 17 rate equations of the form:

$$k_i = \frac{f(r_{C_j}, r_{C_k} \ldots)}{g((C_k), (C_1) \ldots)}$$

where k_i is the reaction rate constant, r_j is the rate of formation of species C_j per unit volume of catalyst, and C_k is the concentration of species C_k in moles per unit volume of catalyst. The species concentrations are those of the saturated products, reactants, olefinic intermediates, or ionic intermediates. The rate constant expressions are shown in Table II.

Table I. Results of the Least-Square Fits of the Rate Constants over the Range 81°–135°F and with a 95% H_2SO_4 Catalyst

Reaction Rate	*Rate Law*	*Frequency Factor*[a]	*Activation Energy*[b]
Initiation			
$C_3^{2-} + HX \xrightarrow{k_1} C_3^+X^-$	kAB	1.01×10^7	2.35
$C_3^+X^- + iC_4 \xrightarrow{k_2} C_3 + iC_4^+X^-$	kAB	2.07×10^{11}	0.0
Primary			
$iC_4^+X^- + C_3^{2-} \xrightarrow{k_{11}} iC_7^+X^-$	kAB	1.99×10^{17}	2.36
$iC_7^+X^- + iC_4 \xrightarrow{k_5} iC_7 + iC_4^+X^-$	kAB	4.20×10^{10}	0.0
Self-Alkylation			
$iC_4^+X^- \xrightarrow{k_9} iC_4^{2-} + HX$	kA	3.92×10^4	0.40
$iC_4^+X^- + iC_4^{2-} \xrightarrow{k_{10}} iC_8^+X^-$	kAB	5.63×10^{19}	4.10
$iC_8^+X^- + iC_4 \xrightarrow{k_6} iC_8 + iC_4^+X^-$	kAB	5.35×10^{10}	0.0
Destructive Aklylation			
$iC_7^+X^- \xrightarrow{k_{12}} iC_7^{2-} + HX$	kA	7.49×10^5	1.08
$iC_7^{2-} + iC_4^+X^- \xrightarrow{k_{13}} iC_5^{2-} + C_6^+X^-$	kAB	3.64×10^{21}	5.73
$iC_5^{2-} + HX \xrightarrow{k_{14}} iC_5^+X^-$	kAB	1.62×10^{11}	3.69
$iC_5^+X^- + iC_4 \xrightarrow{k_3} iC_5 + iC_4^+X^-$	kAB	3.29×10^{10}	0.0
$C_6^+X^- + iC_4 \xrightarrow{k_4} iC_6 + iC_4^+X^-$	kAB	4.04×10^{10}	0.0
$iC_7^+X^- + C_3^{2-} \xrightarrow{k_{15}} C_{10}^+X^-$	kAB	3.72×10^{17}	2.59
$C_{10}^+X^- + iC_4 \xrightarrow{k_8} C_{10} + iC_4^+X^-$	kAB	6.68×10^{10}	0.0
$iC_5^{2-} + iC_4^+X^- \xrightarrow{k_{16}} C_9^+X^-$	kAB	4.26×10^{19}	2.65
$C_9^+X^- + iC_4 \xrightarrow{k_7} C_9 + iC_4^+X^-$	kAB	6.02×10^{10}	0.0
$C_{10}^+X^- \xrightarrow{k_{17}} iC_5^{2-} + iC_5^+X^-$	kA	4.45×10^{11}	8.40

[a] cm^3/gm mole sec or $sec.^{-1}$
[b] kcal/gram mole.

Experimental Results

To obtain the necessary data isobutane was alkylated with propylene using a sulfuric acid catalyst. A continuous-flow, stirred-tank reactor, which was shown to be ideally mixed *via* tracer tests, was used.

Measurements were made using a nominal 95% H_2SO_4 catalyst concentration to test the model. In this series of experiments the range of other process variables was: temperature, 65°–135°F; propylene feed concentration, 12.5–22.4 wt %; and olefin space velocities, 0.104–0.184 vol olefin/vol catalyst-hr. The percent acid in emulsion and the flat blade turbine speed were held constant at 60% and 1700 rpm respectively.

Another set of measurements were made using a 90% H_2SO_4 catalyst to determine the qualitative effects of acid concentration. The only other varible to be perturbed here was temperature, varying from 65° to 105°F. The percent acid in emulsion and turbine speed were the same as in the runs at the higher catalyst concentration. The olefin feed concentration was 12.5 wt %, and the olefin space velocity was 0.105 vol olefin/vol catalyst-hr. In all the experiments a hydrocarbon residence time of 52 minutes was used, consistent with commercial practice.

The reactor effluence was analyzed on a programmed temperature gas chromatograph with squalane-coated, capillary column and flame ionization detector. The results were the composition of the hydrocarbon species present in the weight percent and are shown in Table III.

Table II. Rate Constant Expressions

$$k_1 = r_{C_3}/(C_3^{2-})(HX) \qquad (1)$$

$$k_5 = r_{iC_7}/(iC_7^+X^-)(iC_4) \qquad (2–8)$$

$$k_9 = r_{iC_8}/(iC_4^+X^-) \qquad (9)$$

$$k_{10} = r_{C_8}/(iC_4^+X^-)(iC_4^{2-}) \qquad (10)$$

$$k_{11} = [r_{iC_7} + r_{C_{10}} + (r_{iC_5} + r_{C_6} + r_{C_9})/2]/(iC_4^+X^-)(C_3^{2-}) \qquad (11)$$

$$k_{12} = r_{C_6}/(iC_7^+X^-) \qquad (12)$$

$$k_{13} = r_{C_6}/(iC_7^{2-})(iC_4^+X^-) \qquad (13)$$

$$k_{14} = (r_{iC_5} + r_{C_6} - r_{C_9})/2(iC_5^{2-})(HX) \qquad (14)$$

$$k_{15} = (r_{iC_5} - r_{C_6} + r_{C_9} + 2r_{C_{10}})/2(iC_7^+X^-)(C_3^{2-}) \qquad (15)$$

$$k_{16} = r_{C_9}/(iC_5^{2-})(iC_4^+X^-) \qquad (16)$$

$$k_{17} = (r_{iC_5} - r_{C_6} + r_{C_9})/2(C_{10}^+X^-) \qquad (17)$$

Table III. Weight Percent Product Distribution Based on Gas Chromatograph Analyses

Catalyst Concentration, % H_2SO_4	*95*	*95*	*90*	*90*	*95*	*95*	*90*	*95*	*95*
Temperature, °F	*65*	*65*	*65*	*81*	*81*	*105*	*105*	*120*	*135*
Olefin Space Velocity, hr.$^{-1}$	0.104	0.118	0.101	0.104	0.185	0.131	0.105	0.117	0.104
Run Number	*2*	*9*	*5*	*1*	*6*	*7*	*4*	*8*	*3*
C_3	1.69	1.59	2.40	1.98	3.88	2.83	2.28	2.87	3.20
iC_4	67.75	69.83	79.10	74.85	49.85	67.66	72.22	68.51	76.41
iC_5	0.70	0.97	0.76	0.65	2.66	2.07	1.26	2.20	1.93
23 DMB	0.72	0.63	0.46	0.48	1.61	1.01	0.75	1.02	0.69
2 MP	0.52	0.42	0.26	0.20	0.84	0.60	0.38	0.65	0.51
BMP	0.16	0.19	0.12	0.09	0.37	0.27	0.16	0.29	0.23
Total C_6	1.00	1.24	0.84	0.77	2.82	1.88	1.29	1.96	1.43
24 DMP	5.16	5.24	3.54	4.68	13.95	8.57	7.39	8.20	6.01
23 DMP	12.13	10.68	7.19	9.27	13.22	8.13	8.48	7.88	5.16
3 MC_6	0.30	0.37	0.22	0.21	0.67	0.64	0.50	0.87	0.90
Total C_7	17.59	16.29	10.95	14.16	27.84	17.34	16.37	16.95	12.07
224 TMP	1.47	1.38	0.85	0.87	2.85	2.35	1.58	2.00	1.31
25 DMH	0.56	0.54	0.34	0.42	0.74	0.59	0.45	0.71	0.59
223 TMP	0.48	0.54	0.30	0.34	0.78	0.63	0.38	0.74	0.57
234 TMP	0.61	0.57	0.37	0.43	0.78	0.68	0.66	0.55	0.37
233 TMP	0.59	0.58	0.34	0.39	1.00	0.89	0.67	0.75	0.52
23 DMH	0.23	0.23	0.15	0.21	0.39	0.29	0.23	0.31	0.27
24 DMH	0.18	0.13	0.17	0.08	0.12	0.07	0.04	0.07	0.14
Other C_8	0.39	0.31	0.16	0.24	0.56	0.31	0.21	0.32	0.41
Total C_8	4.51	4.28	2.58	2.98	7.22	5.81	4.22	5.45	4.18
225 TMC_6	0.30	0.46	0.23	0.31	0.64	0.38	0.27	0.33	0.25
224 TMC_6	Trace	Trace	Trace	Trace	0.01	Trace	Trace	Trace	Trace
334 TMC_6	Trace	Trace	Trace	Trace	0.01	0.01	Trace	Trace	0.01
235 TMC_6	0.04	0.05	0.03	0.04	0.09	0.05	0.04	0.05	0.04
334 TMC_6	0.03	0.06	0.03	0.05	0.05	0.02	0.02	0.03	0.01
Other C_9	1.10	0.80	0.38	0.53	0.05	0.18	0.17	0.12	0.13
Total C_9	1.47	1.37	0.67	0.92	1.30	0.64	0.50	0.53	0.44
2235 TMC_6	0.04	0.13	0.07	0.10	0.13	0.04	0.05	0.03	0.02
224 TMC_7	0.47	0.36	0.23	0.39	0.43	0.14	0.20	0.10	0.07
225 TMC_7	0.24	0.19	0.12	0.25	0.23	0.08	0.19	0.06	0.05
Other C_{10}	4.54	3.75	2.29	3.10	3.77	1.49	1.42	1.31	1.04
Total C_{10}	5.29	4.43	2.71	3.84	4.56	1.75	1.86	1.50	1.18

Results and Conclusions

The mathematical model was valid in the range 81°–135°F using a 95% H_2SO_4 catalyst since the rate constants obeyed the Arrhenius theory. The resulting first-order rate constants ranged from 2×10^4 to 1×10^6 sec^{-1}; the second-order rate constants ranged from 2×10^5 to 1×10^8 cc/gm mole-sec.

Owing to the assumptions of constant mass transfer coefficient and constant isobutane concentration (*9*), the activation energies corresponding to the rate constants k_2 through k_8 were zero. This is shown with the results of the other activation energies and frequency factors in Table I.

The predictions obtained from the model included: (1) yield (lb products/lb C_3^{2-} fed) increased with both temperature and olefin feed concentration increase; (2) the dimethylpentane product concentration remained constant with increase in temperature but increased with olefin feed concentration; (3) the octanes (chiefly trimethylpentanes) decreased with increase of both temperature and olefin feed concentration. In general, owing to the combined effects of reduced concentrations of octanes and increased yields of heavier products, lower temperatures, and olefin feed concentrations favored a more commercially desirable product.

The results obtained at 65°F and 95% H_2SO_4 indicated a significant departure from the predictions of the mathematical model. It is believed that this was caused by a change in reaction mechanism, and high rates of formation of C_9 and C_{10} were found experimentally. There was an apparent change in selectivity of the catalyst when the concentration was lowered to 90% H_2SO_4. This resulted in increased rates of formation of C_9 and C_{10} and decreased rates of formation of C_5, C_6, and C_8 from that predicted by the model.

These results and predictions generally agree with those of Shlegeris and Albright (*8*). Areas of disagreement such as the yield of the heavy products and conversion can be explained by the differences in the degree of mixing used and the consistency of catalyst character with respect to H_2O content and type of diluent (*9*).

(1) Schmerling, L., "The Chemistry of Petroleum Hydrocarbons," Vol. 3, Chap. 54, Reinhold, New York, 1955.
(2) Cupit, C. R., Gwyn, J. E., Jernigan, E. C., *Petro/Chem. Eng.* (Dec. 1961) **47**; (Jan. 1962) **49.**
(3) Jones, E. K., *Advan. Catalysis* (1958) X, 165.
(4) Putney, D. H., *Advan. Petrol. Chem. Refining* (1959) **2,** 315.
(5) Hofmann, J. E., Schriesheim, A., *J. Am. Chem. Soc.* (1962) **84,** 953, 957.
(6) Kennedy, R. M., *Catalysis* (1958) **VI.**
(7) Mosby, J. F., Ph.D. Thesis, Purdue University (1964).

(8) Schlegeris, R. J., Albright, L. F., *Ind. Eng. Chem., Process Design Develop.* (1969) 8 (1).
(9) Langley, J. R., Ph.D. Thesis, Louisiana State University (Baton Rouge) (May 1969).
(10) Malloy, J. B., Taylor, W. C., *Ann. Meetg. Am. Inst. Chem. Engrs., 57th, Boston* (1964), Preprint 44c.

9

Some Problems in the Analysis of Transient Behavior and Stability of Chemical Reactors

RUTHERFORD ARIS

Department of Chemical Engineering, University of Minnesota,
Minneapolis, Minn. 55455

Analysis of the transient behavior of chemical reactors must consider the relationship between its possible "steady states." These often take the form of points in the state space of the reactor corresponding to conditions being steady everywhere, but there can also be invariant sets or limit cycles which correspond to a regular oscillation in the conditions at each point. Corresponding to each invariant set there is region of the state space which is attracted into the steady state or limit cycle, and regions of attraction to stable invariant sets are separated by the lower dimensional regions corresponding to the unstable ones. The stirred tank reactor demands this analysis, but simplifications are required. Among such are Hlavacek's lumping technique, the method of simplified modelling, and the proximate steady-state hypothesis.

Intensive research on the transient behavior and stability of chemical reactors began with the publication of van Heerden's paper on "Autothermic Processes," (*1*). Indeed, if there are ignition years, as there are ignition temperatures, before which little is done in a given research area and after which a finite jump is made to a much higher level of activity, then there must be an ignition year among the first of the 1950's. Van Heerden was about to draw attention to the fundamental importance of the heat balance and the insight to be gained by finding the steady state from the intersection of the heat generation curve and heat removal line. Amundson (*2*) was already working with Bilous (*3*) in using the stability analysis of nonlinear mechanics and in explaining the phenomena of parametric sensitivity. If some industrious bibliographer would count the number of relevant papers published per year and plot it against the date, he would surely find a curve very like that of the rate of heat generation in the stirred tank as a function of temperature. That the curve has yet passed its inflection point, let alone approached an equilibrium value, is at least open to question and discussion by those concerned with the information explosion.

It would be impossible to review the present state of activity in the area of interest without either becoming intolerably dull in listing a comprehensive catalog of papers or vacuously superficial in trying to describe the diverse questions of interest that have been studied. Between the Scylla of superficiality and the Charybdis of crashing boredom a track may perhaps be found by considering some of the problems to be faced by any attempt to give a theoretical account of reactor behavior. The magnitude of these difficulties more than justifies the various simplifications that chemical engineers have devised, and if they may not always have been aware of the vastness of the problems that they were avoiding, this ingenuousness does not diminish the admiration due to their ingenuity nor to the good engineering sense that solutions exhibit.

Luss has recently discovered a remarkable paper that predates some of van Heerden's results by almost 35 years. In *Chemical and Metallurgical Engineering* (Sept. 15, 1918) appears a paper by Frans G. Liljenroth on the "Starting and Stability Phenomena of Ammonia-Oxidation and Similar Reactions" (*4*). It contains seven pages and seven figures innocent of any equation beyond a few arithmetical statements. It is addressed from Washington, D. C., and the final paragraph disclaiming any reference to specific data on the part of the curves and figures given suggests that the author was employed in industry. He was a member of the American Chemical Society from 1919 to 1932, when he resigned his membership. His address at first was Wilmington, Del., and he seems to have returned to Stockholm in 1927. From the enquiries by T. R. Keane it appears that Liljenroth was a duPont employee from 1916 to 1920, but the influence of his paper has not yet come to light. Certainly van Heerden was completely unaware of it, for in his 1953 paper he understandably comments that "it is remarkable that the elementary considerations given in this paper . . . have not been presented before." Yet almost all those elementary considerations are given by Liljenroth in the context of the ammonia oxidation burner. Multiple steady states are shown to exist, the instability of the middle steady state is demonstrated, and stability for the high temperature steady state is claimed (using the familiar, if inadequate, argument from the crossing of the heat generation and heat removal lines), the start-up of the reactor is discussed, as are the stability and sensitivity of the optimal steady state and the influence of feed conditions and poisons. All in all it is a remarkable paper, and it seems to have been completely ahead of its time. The only comparable situation in the literature of chemical reaction engineering is the derivation of the correct boundary conditions for the tubular reactor, with which we quite appropriately associate the name of Danckwerts (5), but which in fact goes back to a paper by Langmuir in 1908 (6).

The Stirred Tank Reactor

The stirred tank reactor was one of the earliest systems to be considered, and the long series of papers by Amundson and co-workers (7–21) shows how fecund it is, even in its simplicity, of important developments.

Consider a single reaction $\Sigma \alpha_j A_j = 0$ taking place in a stirred tank of volume V. The feed consists of some or all of the species A_j in molar concentrations c_{jf}, temperature T_f, and total volume flow rate q. Let $\mathbf{c} = (c_1, \ldots c_S)$ denote the set of concentrations c_j within the well-mixed reactor and T the temperature. The rate of reaction $r(\mathbf{c}, T)$ per unit volume may be defined so that under these constant volume conditions the rate of formation of A_j by the single reaction is $\alpha_j r$. A balance for species A_j gives

$$V\frac{dc_j}{dt} = q(c_{jf} - c_j) + \alpha_j Vr(\mathbf{c}, T) \; ; 1 \leq j \leq S \tag{1}$$

If the reaction is exothermic, heat is removed at a rate of Q^*, C_p denotes the heat capacity per unit volume of the reaction mixture (assumed constant), and ΔH is the heat of reaction, a heat balance gives

$$VC_p\frac{dT}{dt} = qC_p(T_f - T) + (-\Delta H)Vr(\mathbf{c}, T) - Q^* \tag{2}$$

In writing this heat balance the heat capacity of the vessel has been ignored. This however has a stabilizing effect so that by ignoring it the analysis is not overlooking a dangerous effect. To analyze the problem fully would be to consider the transient flow of heat in the vessel walls, a much more complicated problem involving the partial differential equations of a distributed system.

The S Equations 1 and 2 can be solved when c_{jo} and T_o, the initial concentration and temperature are specified. They form a system of ordinary nonlinear differential equations and present few difficulties to the modern computer. As always, however, it is important not to start computing before a thorough, qualitative understanding of the solution has been obtained, and to this end certain simplifications may be made.

If the c_{jf} are constant in time and c_{jo} is a compatible composition—*i.e.*, $(c_{jo} - c_{jf}) = \alpha_j \xi_o$, then the first S equations can be replaced by a single equation by defining the extent of reaction as

$$\xi = (c_j - c_{jf})/\alpha_j \tag{3}$$

Setting $\theta = V/q$ and suppressing reference to the constants α_j and c_{jf} in the reaction rate expression $r(\xi, T)$ gives

$$\theta \frac{d\xi}{dt} = -\xi + \theta\, r\,(\xi, T) \tag{4}$$

The naturally occurring scale of time is θ, so that a dimensionless variable $\tau = t/\theta$ may be chosen, and from a naturally occurring scale of concentration (such as $c_f = \Sigma c_{jf}$) there arises the dimensionless extent of reaction $x = \xi/c_f$. It then appears that $\rho = \theta r/c_f$ will be the dimensionless reaction rate, so that

$$\frac{dx}{d\tau} = -x + \rho\,(x, y) \tag{5}$$

where y is the dimensionless temperature which must now be defined. The temperature scale intrinsic to the reaction system is the adiabatic temperature rise

$$\Delta T_{ad} = (-\Delta H)c_f/C_p \tag{6}$$

Thus letting

$$y = T/\Delta T_{ad},\ y_f = T_f/\Delta T_{ad} \tag{7}$$

gives

$$\frac{dy}{d\tau} = y_f - y + \rho(x, y) - Q(y, y_c) \tag{8}$$

where $Q = Q^*/qc_f(-\Delta H)$ is a function of temperature within the reactor, y, some coolant temperature, y_c, and other parameters of the cooling configuration. These last terms can often be combined into one proportionally constant, κ, and the equation for temperature may be written as

$$\frac{dy}{d\tau} = y_f + \kappa y_c - (1 + \eta)y + \rho(x, y) \tag{9}$$

Thus, Equations 5 and 9 subject to

$$x = x_o,\ y = y_o,\ \text{when}\ \tau = 0 \tag{10}$$

describe the system in the most economical way possible. The only remaining parameters are κ and those implicit in $\rho(x,y)$; for example, for an irreversible first-order reaction ρ might have the form $\alpha(1 - x) \exp (-\delta/y)$, involving two kinetic parameters.

These are ordinary differential equations, and by the standard existence and uniqueness theorems we know that if $\rho(x,y)$ is continuous and satisfies a Lipschitz condition in a bounded region of x,y,τ space, there exists a unique solution continuously dependent on its initial conditions. If ρ is equally well behaved with respect to its parameters, the solution depends continuously on them also.

The steady-state analysis of these equations proceeds as follows. If the reactor described by these equations is operating in a steady state, x and y (say, x_s and y_s) do not change with time, and they may be found as the solutions to the simultaneous equations

$$0 = - x_s + \rho(x_s, y_s) \tag{11}$$

$$0 = y_f + \kappa y_c - (1 + \kappa)y_s + \rho(x_s, y_s) \tag{12}$$

The first of these may often be solved readily in the form

$$x_s = X(y_s) \tag{13}$$

and the second then becomes

$$(1 + \kappa)(y_s - y^*) = X(y_s) \tag{14}$$

where

$$y^* = (y_f + \kappa y_c)/(1 + \kappa) \tag{15}$$

In physical terms the left side of Equation 14 represents the total heat removal, partly obtained by heating up the feed and partly by deliberate cooling; the right side is proportional to the rate of heat generation by reaction. Clearly the former is a linear function of y_s, whereas the latter often takes the form shown in Figure 1 where $z = X(y)$ is plotted for a typical exothermic reversible reaction. The left side of the equation is a straight line of slope $(1 + \kappa)$ through the point $y = y_c$, $z = y_c - y_f$, and it is clear that for any pair (y_c, y_f) there may be a range of κ, $\kappa_* \leqslant \kappa \leqslant \kappa^*$, for which there is not one but three steady states. The lower part of Figure 1 illustrates the dependence of the steady state on the parameters. Since y_f and y_c combine with κ to give y^* according to Equation 15, we need consider only the two-dimensional parameter space of κ and y^*. Loci of constant y_s are hyperbolic curves

$$(1 + \kappa)(y_s - y^*) = X(y_s)$$

with asymptotes $y^* = y_s$ and $\kappa = 1$. Several of these are shown in the lower part of the figure with labels corresponding to the points *A* through *G* in the upper part. The family has an envelope *LMN* touched by its members between *C* and *E*, and for (κ, y^*) within the region *LMN* there are three possible steady states. The familiar argument of van Heerden shows that states on the part of the $X(y)$ curve between *C* and *E* will be unstable if $(1 + \kappa)$ is less than $X'(y)$ at $y = y_s$. Even if we exclude unstable steady states there are still two whose possible stability makes them candidates for physically realizable states. Thus we see that the steady state is not a well-defined function of the parameters despite the fact that the solution of the transient equations, which presumably tends to a steady state, is unique and depends continuously on the parameters. This apparent paradox is resolved when we recall that the theorem on continuous dependence only guarantees that a small enough change in the parameters will produce only a small change in the trajectory over a finite interval of time, and that an infinite time is required to reach the steady state.

The dependence on the initial conditions is shown best by using the phase plane. In the x,y plane the solution may be represented by a curve along which τ is a parameter. Thus, for a unique steady state we might have a phase plane such as that in the upper part of Figure 2. Here each trajectory leads to the unique steady state (x_s,y_s) whatever its starting point (x_o,y_o). In the lower part, however, we have a typical phase plane for three steady states, labelled *B*, *D*, and *F* to correspond with the possible combination in Figure 1. *D* is an unstable saddle point, and the two trajectories, *HD* and *ID*, (themselves unstable) that approach it divide the (x,y) plane in two. If (x_o,y_o) lies to the left of *HI*, the trajectory goes to *B*, and if it lies to the right, it goes to *F*. A trajectory starting sufficiently close to *I* will remain around *ID* for an arbitrarily long finite interval of time, but it will ultimately diverge to either *B* or *F*. Thus, though the steady state is not uniquely defined as a function of the parameters it is uniquely defined as a function of the initial point (x_o,y_o).

The character of the steady state will be determined by three functions of y_s and κ:

$$\left.\begin{aligned} \phi &\equiv (1 - \rho_x) + (1 + \kappa) - \rho_y \\ \chi &\equiv (1 - \rho_x)\ (1 + \kappa) - \rho_y) \\ \psi &\equiv (\rho_x + \rho_y - \kappa)^2 - 4\kappa\rho_y) \end{aligned}\right]$$

where $\rho_x = \rho_x(x_s,y_s)$, $\rho_y = \rho_y(x_s,y_s)$. The value of χ is positive when the curve of heat generation has a smaller slope than the heat removal line, the latent roots are real when ψ is positive and ϕ distinguishes between roots with positive or negative real parts. The character of the steady state is related to these three criteria in Table I.

Table I. Characteristics of the Steady State

	$\psi > 0$		$\psi < 0$	
	$\phi > 0$	$\phi < 0$	$\phi > 0$	$\phi < 0$
$\chi > 0$	stable node	unstable node	stable focus	unstable focus
$\chi < 0$	saddle	point	—	—

Thus for each point on the curve in the upper part of Figure 1 the values ρ_x and ρ_y and hence the character of the steady state are determined for each point in the lower part of the figure. We will not pursue the representation here. Hlavacek has given a remarkably full and detailed discussion of the behavior of the stirred tank (*22*).

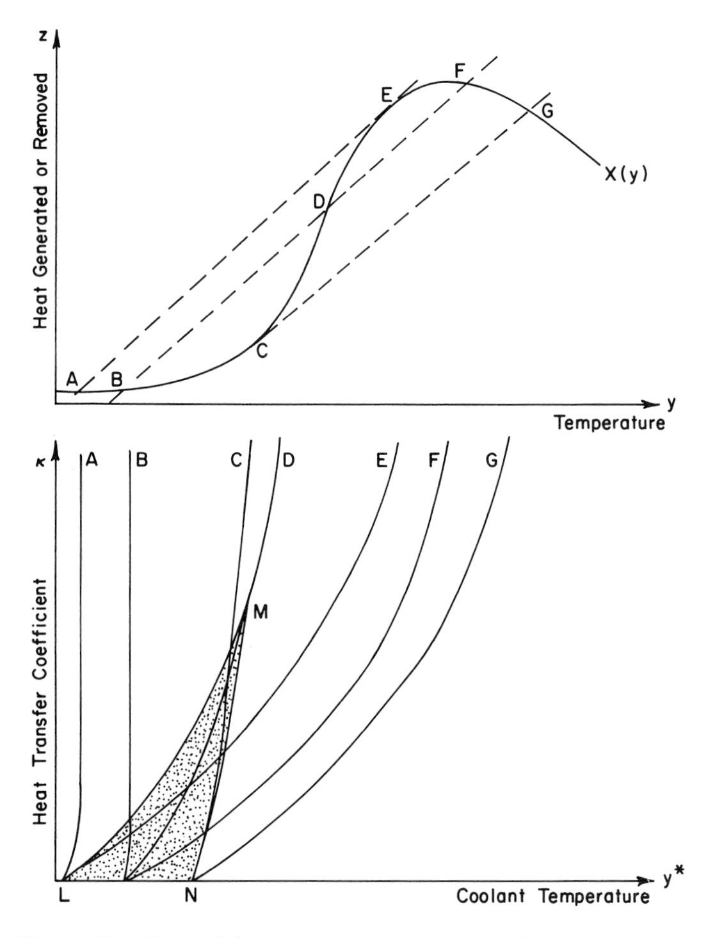

Figure 1. Typical heat generation curve and loci of given steady state in the plane of κ and y*

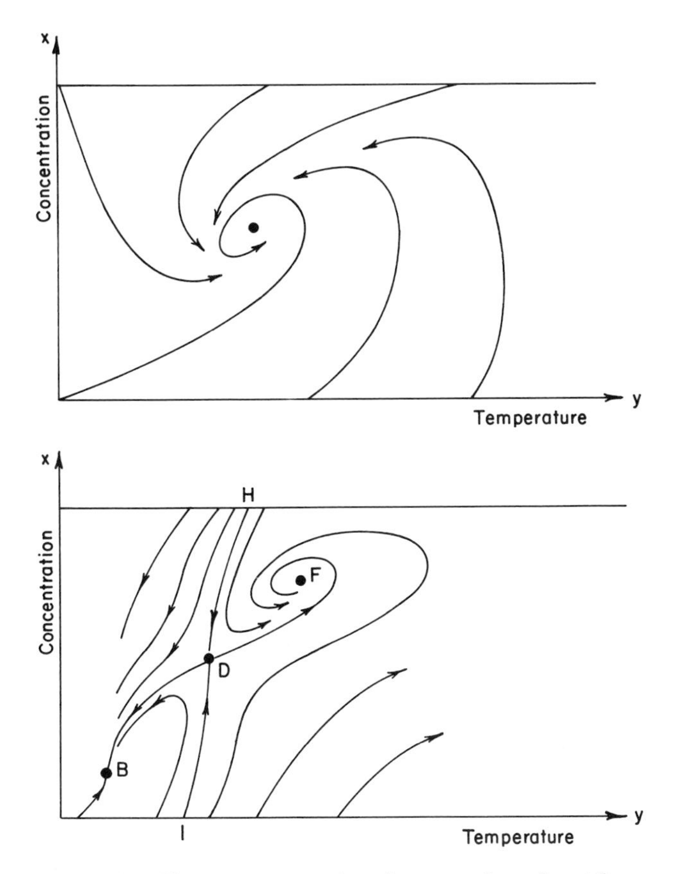

Figure 2. Phase portraits for the stirred tank with one and three steady states

Let κ be constant and y_c and y_f be the same, but let their common value $y_c = y_f = y^*$ take on any value in a certain range. Thus y^* could be represented as a third dimension, and the x,y planes could be stacked together in the order of y^* to give a pseudo-phase space (pseudo because y^* is a parameter and not a dependent variable). Figure 3 shows such a block in x, y, y^* space. The locus of steady states lies on the cylinder $x = X(y)$, and the locus of saddle points (CE) is characterized by the fact that along this part of the locus the value of y^* decreases as y increases. The phase plane section corresponding to the situation with three steady states is also shown.

The steady state reached depends on the history of the start-up of the reactor, and it is worth asking what can be said about this. Suppose, for simplicity, κ is constant and y^* varies as a function of time; then

$$\frac{dx}{d\tau} = -x + \rho(x, y)$$

$$\frac{dy}{d\tau} = (1 + \kappa)y^*(\tau) - (1 + \kappa)y + \rho(x, y)$$

These are no longer autonomous, and the solution depends not only on x_o and y_o but also on the whole history of the function $y^*(\tau)$. If the x,y plane is denoted by E and the space of bounded continuous functions $y^*(\tau)(0 \leqslant \tau < \sigma)$ by F_σ, then the state $x(\sigma),y(\sigma)$ is defined on the

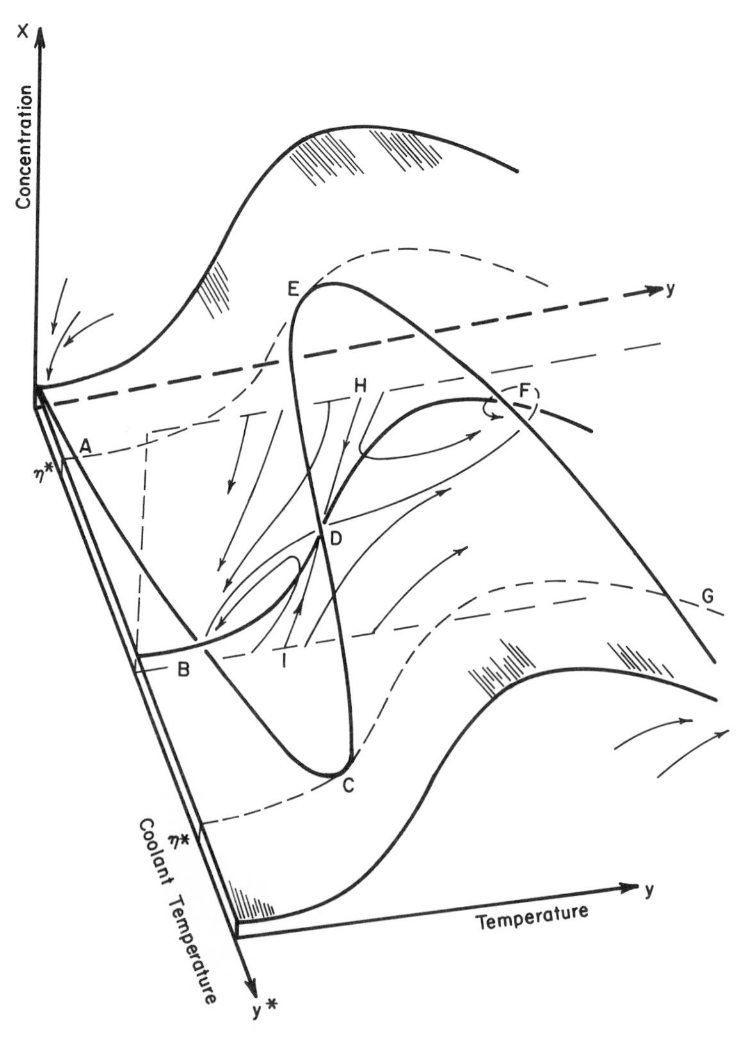

*Figure 3. A block of phase portraits for fixed κ and varying y**

Cartesian product space $E \times F_\sigma$. Suppose we let σ tend to infinity but restrict attention to functions $y^*(\tau)$ which are asymptotically constant, calling this space F; it is clear that the steady state that is finally achieved is only uniquely defined on $E \times F$. If I is the subspace of F containing all the constant functions, Figure 3 represents the situation in $E \times I$. This figure is made up by stacking the phase planes for constant y^* into a block, the plane for a given y^* being perpendicular to the y^* axis at the coordinate y^*. The locus $x = X(y)$ becomes the curved cylindrical surface, and the possible steady states lie on the curve $ABCDEFG$. Obviously $x_s = x_s(x_o,y_o,y^*)$, $_*y_s = y_s(x_o,y_o,y^*)$. We observe that if $y^* < \eta$ or $y^* > \eta^*$ where η_* and η^* are the two values of

$$\eta = y - [X(y)/(1 + \kappa)] \tag{17}$$

for the two roots of the equation

$$X'(y) = (1 + \kappa) \tag{18}$$

then the steady state is unique, and x_s,y_s are really only functions of y^*. Thus

$$x_s = x_s(y^*),\ y_s = y_s(y^*),\ (x_o,\ y_o)\varepsilon E$$

if

$$y^* < \eta_* \text{ or } y^* > \eta^*$$

If $\eta_* < y^* < \eta^*$, there are three steady states, and E may be partitioned into three subspaces $E_1(y^*)$, $E_2(y^*)$, $E_3(y^*)$ such that if $(x_o,y_o)\epsilon E_i(y^*)$, $x_s = x_{si}(y^*)$. For example, in the lower half of Figure 2, x_{s1}, x_{s2}, and x_{s3} are respectively B, D, and F and E_1, E_2, and E_3 are the open region of the plane to the left of HI, the line HI itself, and the open region to its right, respectively.

This much is straightforward and describes completely the behavior of (x_s,y_s) in $E \times I$. The real challenge is to describe the situation in $E \times F$. Certain things are intuitively clear. First, if $y(\tau) \to y_\infty$ as $\tau \to \infty$ and $y_\infty < \eta_*$ or $y_\infty > \eta^*$, then since the steady state is unique in these regions, we should have $x_s = x_s(y_\infty)$, $y_s = y_s(y_\infty)$. Moreover if $(x_o,y_o)\varepsilon$ $E_i(y^*(o))$ and $E_i(y(\tau)) \subseteq E_i(y(\tau'))$ for all $\tau' > \tau \geqslant 0$, then we should expect that $x_s = x_{si}(y_\infty)$, $y_s = y_{si}(y_\infty)$. Beyond this, however, it is difficult to see clearly, and the problem is to partition the function space F according to the steady state that is achieved. If κ is allowed to vary as well as y^* and the space of asymptotically constant functions $\kappa(\tau)$ is denoted G, the steady state is uniquely defined only in the product space $E \times F \times G$.

Two Distributed Reactor Systems

The simple case of the stirred tank has been examined in detail because it illustrates the type of problem that arises in analyzing transient behavior. The situation could be elaborated by considering the possibility of variable feed composition which would force the consideration of all the $(S + 1)$ equations for concentrations and temperature. The initial space would then be the physically proper region of $(S + 1)$ dimensional Euclidean space and $c_{jf}(t)$, $T_f(t)$ would be taken from suitable function spaces. When we turn to even the simplest distributed systems we are faced with function spaces and their Cartesian products from the outset. Thus, the equations for the concentrations $c_j(\mathbf{r},t)$ and temperature $T(\mathbf{r},t)$ in a single catalyst particle within the region $\mathbf{r}\epsilon B$ bounded by the piece-wise smooth surface ∂B (as shown in Figure 4) are

$$\frac{\partial c_j}{dt} = \nabla(D_{je}\nabla c_j) + \alpha_j r_e(\mathbf{c}, T) \tag{19}$$

$$C_p \frac{\partial T}{\partial t} = \nabla(K_e \nabla T_j) + (-\Delta H)\, r_e(\mathbf{c}, T) \tag{20}$$

where D_{je} is the effective diffusion coefficient of A_j, K_e the effective thermal conductivity, C_p the heat capacity per unit volume. The symbol r_e denotes the effective reaction rate per unit volume and is the product of the reaction rate per unit area and the catalytic area per unit volume. The initial conditions are

$$c_j(\mathbf{r}, o) = c_{jo}(\mathbf{r}),\ T(\mathbf{r}, o) = T_o(\mathbf{r}) \tag{21}$$

and the boundary conditions

$$\left.\begin{aligned} D_{je}\frac{\partial c_j}{\partial n} &= k_j(c_{jf} - c_j) \\ K_e \frac{\partial T}{\partial n} &= h(T_f - T) \end{aligned}\right] \quad \text{on } \partial B \tag{22}$$

where k_j and h are mass and heat transfer coefficients and $\partial/\partial n$ is differentiation along an outward normal to ∂B. The conditions under which these can be simplified to two equations are much more restrictive, and it would not be to the point to go into the reduction here. Again it is known that there may be multiple steady states; in this case three or five might be expected under circumstances, and this number can often be extended by suitable parameter choices. The steady states satisfy

$$\left.\begin{aligned} 0 &= \nabla(D_{je}\nabla c_j) + \alpha_j r_e(\mathbf{c}, T) \\ 0 &= \nabla(K_e \nabla T) + (-\Delta H) r_e(\mathbf{c}, T) \end{aligned}\right] \tag{23}$$

together with the boundary conditions given by Equation 22. If c_{jf} and T_f are constant, the steady state is a point in the space of vector functions $(c_1(\mathbf{r}), \ldots c_S(\mathbf{r}), T(\mathbf{r}))$ which satisfy Equations 22 and 23. This space is a subspace of the space of physically admissible vector functions $(c_1(\mathbf{r}), \ldots c_S(\mathbf{r}), T(\mathbf{r}))$, itself a subspace of $C_B^+ \times C_B^+ \ldots \times C_B^+$ ($(S + 1)$ factors), where C_B^+ denotes the space of non-negative continuous functions defined on the domain B. The steady state that is reached will, under general conditions, depend on three sets of quantities:

(a) The constant parameters—*e.g.*, α_j, K_e etc.;

(b) The boundary conditions—*i.e.*, $c_{jf}(t)$, $T_f(t)$;

(c) The initial state—*i.e.*, $c_{jo}(\mathbf{r})$, $T_o(\mathbf{r})$.

(We will not consider that c_{jf} and T_f can also be functions of position on ∂B).

A similar situation obtains for the flow reactor for which there is convective flow through the region B. The general equations are

$$\frac{\partial c_j}{\partial t} + \nabla\cdot(\mathbf{v}(\mathbf{r})c_j) = \nabla(D_j \nabla c_j) + \alpha_j \mathbf{r}(\mathbf{c}, T) \tag{24}$$

$$C_p \frac{\partial T}{\partial t} + \nabla\cdot(C_p \mathbf{v}(\mathbf{r}) T) = \nabla(K_e \nabla T) + (-\Delta H) r(\mathbf{c}, T) \tag{25}$$

The boundary ∂B is divisible into three distinguishable parts (*see* the lower part of Figure 4); the inlet ports, ∂B_i, the exit ports, ∂B_e and the wall, ∂B_w. The wall is usually stationary and impermeable to matter so that the boundary conditions are

$$\frac{\partial c_j}{\partial n} = 0, \ K_e \frac{\partial T}{\partial n} = h(T_w - T) \text{ on } \partial B_w \tag{26}$$

whereas

$$v(c_{jf} - c_j) = D_{je} \frac{\partial c_j}{\partial n}, \ vC_p(T_f - T) = K_e \frac{\partial T}{\partial n} \text{ on } \partial B_i \tag{27}$$

and

$$\frac{\partial c_j}{\partial n} = 0, \ \frac{\partial T}{\partial n} = 0 \text{ on } \partial B_e \tag{28}$$

The same three classes of quantities as were given above may be discerned here with the added complication that the wall temperature T_w may be a function of time t and come within the second class. These two situations are basically comparable in complexity provided that an effec-

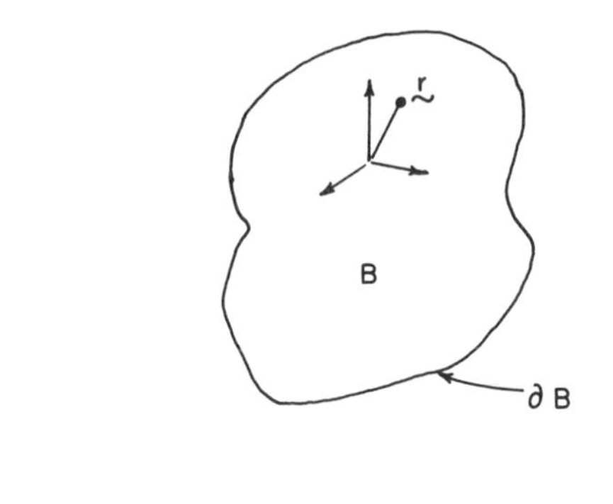

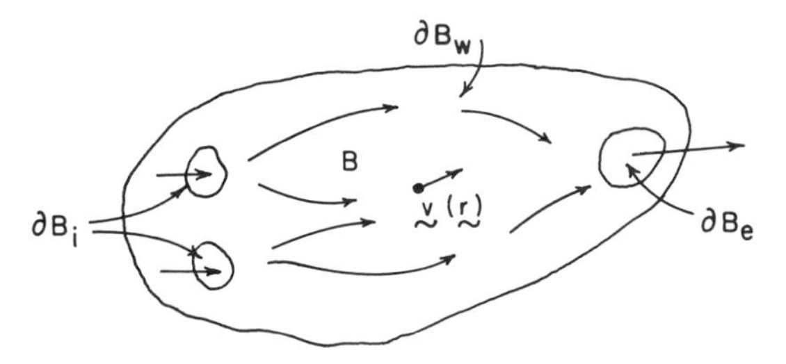

Figure 4. Distributed reactor configurations

tive reaction rate per unit volume may be defined as a function of the local concentrations and temperature.

The steady state is found by solving the equations with all time derivatives set equal to zero. In this context we might hope to devise effective reaction rate, r_e, for the packed bed by using the concept of the effectiveness factor. This factor summarizes the calculation of the steady state for the catalyst particle by determining the ratio of the actual rate of reaction to the rate that would prevail if the concentrations and temperature throughout the particle were constant and equal to c_{jf} and T_f, respectively. It is a function of (a) the parameters and (b) the boundary conditions which are constant for any one particle but vary from particle to particle throughout the reactor. However, the steady state and the effectiveness factor (a function of the steady state) are not always uniquely defined by these two sets of quantities so that it may not be possible to formulate an effective reaction rate in this pseudo-homogeneous model of the packed bed. The packed bed then presents itself as a problem of the next higher order of magnitude since to find its steady state, the transient equations must be solved, and this involves taking into account not only (a) the parameters and (b) the boundary conditions, but also (c) the initial state of each particle. Vastly simplifying principles will therefore have to be found before a tolerably complete resolution of this problem is attained.

The Abstract Problem of Describing Transients and Steady States

In general the state of the reactor at time t may be specified by a point $\mathbf{x}(t)$ in some space R. This may be a subspace of V_n the space of n dimensional vectors with real components (as was the case for the stirred tank reactor, $n = 2$), or a space of vectors of real valued functions defined over a certain region of physical space B (as for the catalyst particle), or the Cartesian product of such spaces as for the packed bed. We shall generally suppress reference to the spatial independent variables and concentrate on the time dependence. The state will satisfy a system of equations

$$S : \dot{\mathbf{x}} = M[\mathbf{x}] \tag{29}$$

where M is generally a nonlinear differential or perhaps an integrodifferential operator in R. It is subject to certain boundary conditions that may involve functions of time.

$$C_t : 0 = N_t[\mathbf{x}] \quad . \tag{30}$$

The space of initial states $\mathbf{x}_o$ is a subspace of R, and we denote it by

$$\mathbf{x}_o = \mathbf{x}(0) \in I \subseteq R \quad . \tag{31}$$

If the equations model the reactor adequately, we may expect that the triple (S, C_t, I) will lead to a unique solution $\mathbf{x}(t) \in R$.

Within R we distinguish certain invariant subspaces $W_i \equiv W_i(S; C_t)$. These are non-intersecting subsets having the property that if $\mathbf{x}_o \in W_i$, then $\mathbf{x}(t) \in W_i$. The number of these subspaces (*i.e.*, the range of the index i) itself depends on S and C_t. The commonest case of an invariant subspace is the point $\mathbf{x}_s \in R$ representing a steady state, but a limit cycle also qualifies as an invariant subspace. Under certain circumstances we can think of a moving limit cycle that generates a space of higher dimensions. If we limit the class of boundary conditions to those that are asymptotically constant (and accordingly drop the suffix on C) the set of invariant subspaces W_i may be denoted by $\mathbf{W} = \mathbf{W}(S; C)$. The index i may be an integer, but there are circumstances in which it may be regarded as a continuous variable. Such a case was given by Amundson and Liu (*23*) in their model of the packed bed. (A simpler mechanical example is the family of solutions of $x + \omega^2 x = 0$ the equation for a simple harmonic oscillator.)

A solution $\mathbf{x}(t)$ of $(S, C, \mathbf{x}_o)$ has the property that either it lies entirely within an invariant subspace or else it approaches one of them as $t \to \infty$. The first possibility will be the case if $\mathbf{x}_o$ lies in a particular W_i by virtue of the definition of an invariant subspace. The invariant subspace W_i is called asymptotically stable if $\mathbf{x}(t) \rightarrow W_i$ whenever $\mathbf{x}_o$ lies in some neighborhood, $N(W_i)$, of W_i. It is asymptotically unstable if there are points $\mathbf{x}_o$ in any $N(W_i)$ for which $\mathbf{x}(t)$, the solution of $(S,C,\mathbf{x}_o)$ does not approach W_i. Let the invariant subspaces of these two classes be enumerated by V_j and U_k, respectively. Then it is evident that the initial space is partitioned into subspaces I_i such that $\mathbf{x}(t) \rightarrow W_i$ if $\mathbf{x}_o \epsilon I_i$. The uniqueness theorem ensures that these are disjoint subsets of I and therefore of R. The fact that $W_i \subseteq I_i$ shows that each I_i is connected since W_i is connected and could not belong to two parts of a disconnected subspace. If the initial subspaces belonging to the V_j and U_k are denoted by J_j and K_k, respectively, each J_j must be of the same dimension as R; otherwise it could not include a neighborhood of its V_j. On the other hand, each K_k must be of lower dimension. It may be shown that the K_k are closed sets separating adjacent J_j. To refer again to the upper part of Figure 2, $R \equiv I$ is the strip $0 \leqslant x \leqslant 1$, $y \geqslant 0$. The only invariant subset is the point (x_s,y_s), and since this is globally stable, $J_1 \equiv I \equiv R$. In the lower part of Figure 2 the set W consists of the points B, D, and F with $V_1 = B$, $V_2 = F$, and $U_1 = D$. J_1 is the part of R strictly to the left of HI, J_2 is the part of R strictly to its right, and K_1 is the curve HI itself. Two other possible configurations may be seen in Figure 5. In the upper part two limit cycles Δ and Γ, the latter unstable, surround the stable steady state A. Here $V_1 = A$, $V_2 = \Delta$, $U_1 = \Gamma$; $J_1 \equiv$ interior of Γ, $J_2 \equiv$ exterior Γ, $K_1 = \Gamma$. In the lower part there are two steady states D and F, the former lying on a closed loop Γ. Then $V_1 = B$, $U_1 = D$, $U_2 = \Gamma$; $J_1 =$ exterior of Γ except the curve ID, $K_1 =$ the curve ID, $K_2 = \Gamma$ and its interior. We observe that the steady-state system

$$\Sigma : 0 = M[\mathbf{x}]$$

subject to

$$C : 0 = N[\mathbf{x}]$$

can have multiple solutions which are invariant subspaces of $(S;C;I)$. However $(\Sigma;C)$ only generates those members of $\mathbf{W}$ which are points in R. Finally, we observe that like $\mathbf{W}$ the subspaces I_i depend on C. If C_t is only asymptotically constant, we have the same set $\mathbf{W}(S;C)$, but now its members partition the Cartesian product space $I \times C_t$.

This is not the place to go into great detail about the existence and uniqueness of theorems that are available. For the parabolic systems of equation that frequently occur Wake (*24*) has proved a very general theorem. He considers systems

$$F_i(\mathbf{x}, t, u, \nabla u_i, \nabla\nabla u_i, \partial u_i/\partial t) = 0$$

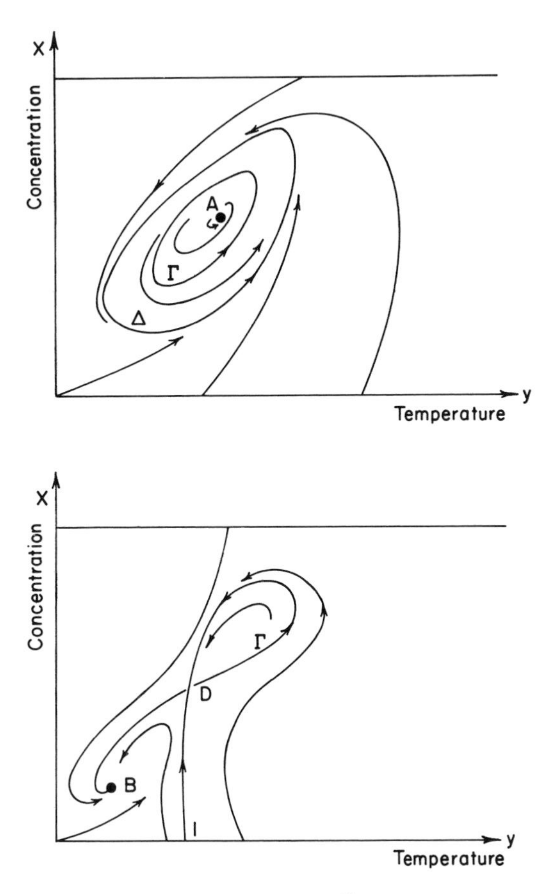

Figure 5. Phase portraits illustrating more complex relations between invariant subspaces

$i = 1, \ldots m$, for a set of m functions $u_i(\mathbf{x},t)$. Here Δu_i and $\Delta\Delta u_i$ denote the sets of first and second derivatives of u_i with respect to the space variables $x_1, \ldots x_n$ so that the equations are coupled only through the u's and not through their derivatives. The system is parabolic if $\partial F_i/\partial(\partial u_i/\partial t) < 0$ and the matrices Φ_i whose j,k^{th} elements are $\partial F_i/\partial(\partial^2 u_i/\partial x_j \partial x_k)$ are all positive definite. If each F_i is analytic in its arguments (other than $\mathbf{x},t$), two solutions having the same initial and boundary conditions are identical if they are continuous and have bounded derivatives. As always existence is harder to prove than uniqueness, but Pao (25) has some useful results and references.

The Equivalent Lumped System. The complexity this analysis reveals obviously cries out for some simplifying principles or methods.

One of the most ingenious of these is the notion of the equivalent stirred tank advanced by Hlavacek. This is described with reference to the equations for the catalyst particle (*19–22*). [In the long and interesting series of papers on the modeling of chemical reactors that Hlavacek has written with Marek, Kubicek, and others (*26–42*) there are many clever methods and fruitful ideas. (As this series of papers presents some difficulties to the bibliographer I have ventured to list them all in the order which Hlavacek regards as canonical.)]

The equation

$$\frac{\partial c_j}{\partial t} = \nabla(D_{je}\nabla c_j) + \alpha_j r_e(\mathbf{c}, T)$$

is difficult to solve because of the nonlinear reaction term. Disregarding this term for the moment and setting

$$c_j = c_{jf} + u_j \tag{32}$$

the linear equations

$$\frac{\partial u_j}{\partial t} = \nabla(D_{je}\nabla u_j) \text{ in } B \tag{33}$$

$$D_{je}\frac{\partial u_j}{\partial n} + k_j u_j = 0 \text{ on } \partial B \tag{34}$$

$$u_j(r, o) = c_{jo}(r) - c_{jf} \tag{35}$$

can be solved by the straightforward, albeit intricate, methods of linear analysis. If D_{je} is constant, and the domain B has a characteristic dimension R, the following dimensionless variables may be chosen

$$\tau_j = D_{je}t/R^2,\ \hat{\rho} = \mathbf{r}/R,\ \mu_j = k_jR/D_{je} \tag{36}$$

so that

$$\frac{\partial u_j}{\partial \tau_j} = \nabla^2 u_j \text{ in } B,\ \frac{\partial u_j}{\partial \mu} + \mu_j u_j = 0 \text{ on } \partial B$$

The solution of this equation is

$$u_j(\hat{\rho}, \tau_j) = \sum_{n=1}^{\infty} a_{jn}\phi_{jn}(\rho) \exp - \lambda_{jn}\tau_j \tag{37}$$

where the (positive) eigenvalues λ_{jn} and the corresponding eigenfunctions ϕ_{jn} depend on the shape of B and the parameter μ_j, and the constants a_{jn} depend on the initial solute distribution. If only the first term in Equation 37 were needed in the solution, we would have

$$\frac{\partial u_j}{\partial \tau_j} = \nabla^2 u_j = -\lambda_{j1} u_j$$

or

$$\frac{\partial c_j}{\partial t} = D_{je} \nabla^2 c_j = \lambda_{j1} D_{je} (c_j - c_{jf})/R^2 \tag{38}$$

In general the later terms will not vanish since this would require the special initial condition in which $c_{jo} - c_{jf}$ would be proportional to ϕ_{j1}, but because λ_{j1} is the smallest eigenvalue, the first term is quickly dominant. Suppose that the last term in Equation 38 were used as an approximation to the Laplacian and were inserted in Equation 19 to give

$$\frac{dc_j}{dt} = \frac{\partial c_j}{\partial t} = (\lambda_{j1} D_{je}/R^2)(c_{jf} - c_j) + \alpha_j r_e(\mathbf{c}, T) \tag{39}$$

This is just the equation of a stirred tank with holding time $(R^2/D_{je}\lambda_{j1})$, and the partial differential equation has been replaced by an ordinary differential equation whose behavior is much more easily described. A similar reduction of Equation 20 gives

$$\frac{dT}{dt} = (\lambda_1 K_e/R^2 C_p)(T_f - T) + [(-\Delta H)/C_p] r_e(\mathbf{c}, T) \tag{40}$$

where λ_1 is the first eigenvalue of the problem

$$\frac{\partial u}{\partial \tau} = \nabla^2 u, \frac{\partial u}{\partial \nu} + \mu u = 0, \mu = hR/K_e \tag{41}$$

This remarkable reduction of the partial differential equations to the familiar equations of the stirred tank will be complete if it can be assumed that

$$\lambda_{j1} D_{je} = \lambda_1 K_e / C_p, j = 1, \ldots S \tag{42}$$

for then the factor corresponding to the holding time will be the same in each. We may let

$$\tau = \lambda_{j1} D_{je} t / R^2, x = (c_j - c_{jf})/\alpha_j c_f, y = T/\Delta T_{ad} \tag{43}$$

and

$$\rho(x, y) = R^2 r_e / \lambda_{j1} D_{je} c_f ,$$

and obtain Equations 5 and 8

$$\begin{aligned} \frac{dx}{d\tau} &= -x + \rho(x, y) \\ \frac{dy}{d\tau} &= y_f - y + \rho(x, y) \end{aligned} \tag{44}$$

There is no term in Q, the heat removal, since a catalyst particle is internally adiabatic and the transfer at the surface corresponds to the feed and effluent from the stirred tank. This final reduction to complete conformity with the stirred tank equations is not strictly necessary since the stability analysis of the $(S + 1)$ Equations 39 and 40 is not out of the question. It is easier to justify a partial reduction by assuming that all the $\lambda_{j1} D_{je}$ are equal but different from $\lambda_1 K_e / C_p$. Then with

$$Le = \lambda_1 K_e / C_p \lambda_{j1} D_{je}, \ \Delta T_{ad} = (-\Delta H) c_f \lambda_{j1} D_{je} / \lambda_1 K_e$$

the equation for y becomes

$$\frac{1}{Le} \frac{dy}{d\tau} = y_f - y + \rho(x, y) \tag{45}$$

The multiplicity of steady states can be examined by looking at Equations 44 and 45 with the left sides set equal to zero,

$$y - y_f = x = \rho(x, y)$$

Thus uniqueness for all y_f is ensured if $\rho_x + \rho_y < 1$ on the entire curve. This condition is also necessary for if $\rho_x + \rho_y \geqslant 1$ between y_* and y^*, then for y_f between $y^* - X(y^*)$ and $y_* - X(y_*)$ there will indeed be three steady states. The stability of the steady states can also be investigated as was done before. Some of the estimates of the conditions for uniqueness and stability obtained in this way are extraordinarily good estimates of the conditions obtained by exact numerical integration of the partial differential equations.

However ingenious this method of approximation may be and however good the results it has attained, it leaves a lot to be desired in any theoretical justification. A justification and examination of its limitations is possible on the lines of Eckhaus' stability theory, but this is not pursued here (*43*). The work of Finlayson (p. 50) also bears closely on these ideas.

Model Systems. A second approach to the complexities of the distributed reactor is made by gaining experience of the behavior of simplified systems for which analytic or easily computable results are available. Liu and Amundson analyzed the packed bed (*23*) and showed that an individual particle could have one of two stable states at certain positions and times. From four nonlinear partial differential equations it is difficult to abstract immediately the possible selection rules, and it may be well to look at a simpler, if somewhat artificial, system. The S-shaped curve representing the heat generation in the particle as a function of temperature will be simplified to a step function, which could represent heat generation by a zeroth order reaction triggered at a certain ignition temperature. Since only heat transfer and generation are to be considered, the ignition temperature may be taken as the zero of temperature and the step function proportional to the Heavyside function,

$$H(S) = \begin{bmatrix} 0, S < 0 \\ 1, S > 0 \end{bmatrix}$$

Letting S be the temperature in the stationary component where the heat is generated (corresponding to the catalyst particles) and taking T to be the temperature in the flowing stream (corresponding to the reaction mixture) the dependent variables $S(z,t)$ both functions of position and time, must satisfy

$$\frac{\partial T}{\partial t} + V\frac{\partial T}{\partial z} = h(S - T)$$

$$\frac{\partial S}{\partial t} = h(T - S) + \delta H(S)$$

where V is the velocity, h a heat transfer coefficient, and δ is the temperature rise from heat generation. However, by altering the units of all the variables, these three constants may be made to have unit values, and the simplest form of the equation is

$$T_t + T_z = S - T \qquad (46)$$

$$S_t = T - S + H(S) \qquad (47)$$

where the suffixes conveniently denote partial derivatives. [This system of equations evolved from a brief but lively discussion with V. Mizel, M. Gurtin, and others of the Mathematics department at Carnegie-Mellon University, where I spent a stimulating week in 1967 at the invitation of H. L. Toor, head of the Chemical Engineering Department.] This system of equations has inlet conditions

$$T(o, t) = T_i(t) \qquad (48)$$

and initial conditions

$$T(z, o) = T_o(z), S(z, o) = S_o(z) \qquad (49)$$

The same kind of continuum of steady-state solutions may be observed here as was found by Liu and Amundson. Let T_i be constant and all time derivatives vanish; then the second equation can be written

$$S - T = H(S) \qquad (50)$$

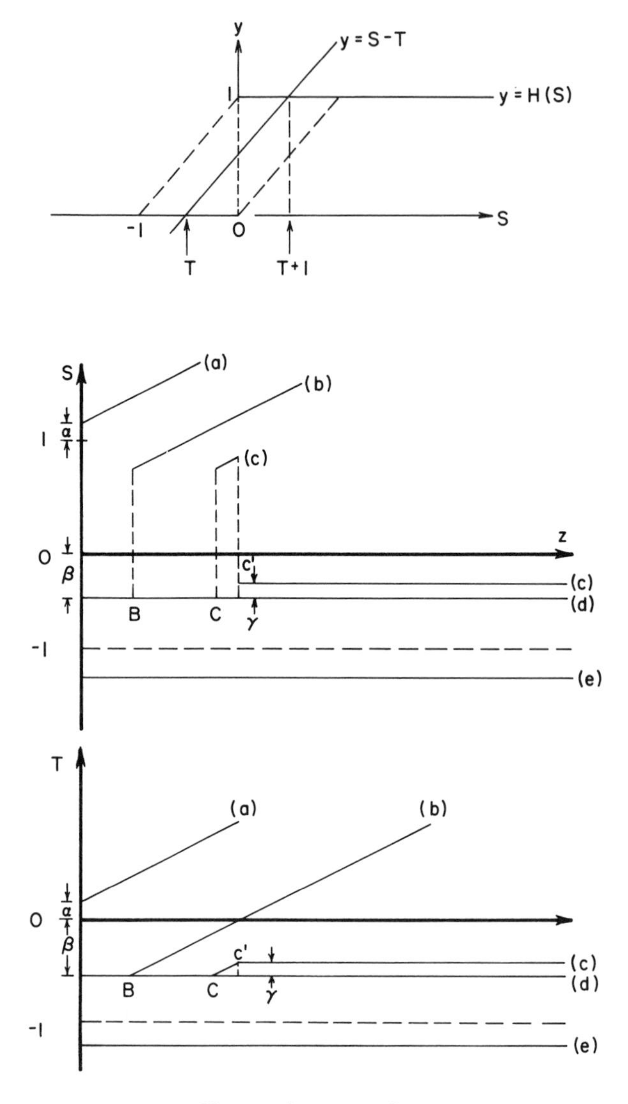

Figure 6. Possible steady-state solutions of Equations 46 and 47

which is of the form of a heat balance, shown in the upper part of Figure 6. Clearly if $T < -1$, there is a unique solution $S = T$, and if $T > 0$, there is a unique solution $S = T + 1$; however, if $-1 \leqslant T \leqslant 0$, either solution is possible, even when we reject the third solution $S = 0$ as unstable. Equation 46 thus has the form

$$T_z = H(S) = \begin{bmatrix} 0, \; T < -1 \\ 0 \text{ or } 1, \; -1 \leq T \leq 0 \\ 1, \; 0 < T \end{bmatrix} \qquad (51)$$

In the intermediate range the choice is arbitrary as far as the steady state is concerned, and the steady state that will actually be reached depends on the initial conditions and history of the feed. Some possibilities are shown in the lower part of Figure 6. Thus, (a) and (e) correspond to the unique cases of $T_i > 0$ or $T_i < -1$. If T_i is in the range $(-1{,}0)$ then a solution such as (d) is possible with $S = T$ and no heat generation. On the other hand, if at any point such as B we elect to take the root $S = T + 1$ and remain with it until $T > 0$, it must persist, and we have a solution like the curve, or broken line, (b). The switch B may be made at any point in the bed. If the switch to the root $S = T + 1$ is made at a point such as C, but before T becomes positive the root $S = T$ is resumed, then we have a solution such as (c). The heat generated in the segment CC' is sufficient to raise the temperature of the stream from the level (d) to the level (c) but not sufficient to ignite the process permanently.

The full transient equations must be analyzed to know what form of steady state is reached from given initial conditions. This has been done by Schruben (*44*) in some detail, and the full results will be published elsewhere. This study shows that a steady state like Curve c of Figure 6 can be obtained from a spiked perturbation of a uniform steady state like (d), as shown in Figure 7. Thus, the continuum of steady states has a quasi-stable character.

The original state may be regained after a minor perturbation, but larger perturbations may cause the reactor to return to another steady state perhaps arbitrarily close to the first. In Figure 7 the initial state $S = T = -0.5$ is perturbed by a spike in both S and T. The spike in T moves with the stream and decays. The spike in S turns into a sloping segment starting at the point where S became positive. It has a temporary influence on T in the form of a small spike on the ramp of T. In Figure 8 a perturbation is made in T only, and here the ramp is displaced somewhat to the right corresponding to the fact that it takes a short time for the temperature S to reach a positive value and ignite the reaction.

Quasi- and Proximate-Steady State Hypotheses. The full solution of the transient equations for the packed bed is impossible to attain. It would require at the very least the solution of the transient equations of transport and transfer in the fluid phase coupled with the solution of the transient equations for each catalyst particle. Some investigators have therefore lumped the resistances to transfer at the surface of the particle. Others have claimed, with some justice, that the particle is isothermal

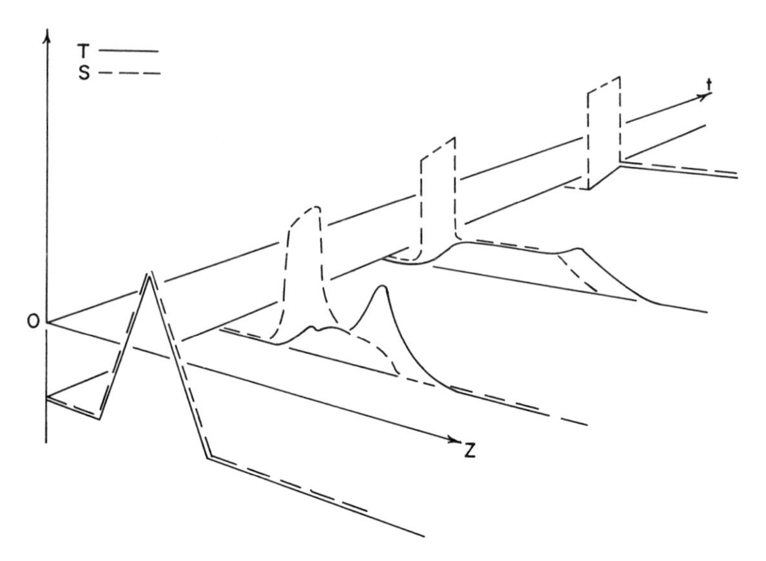

Figure 7. Development of a partially ignited steady-state solution from initial conditions with S = T

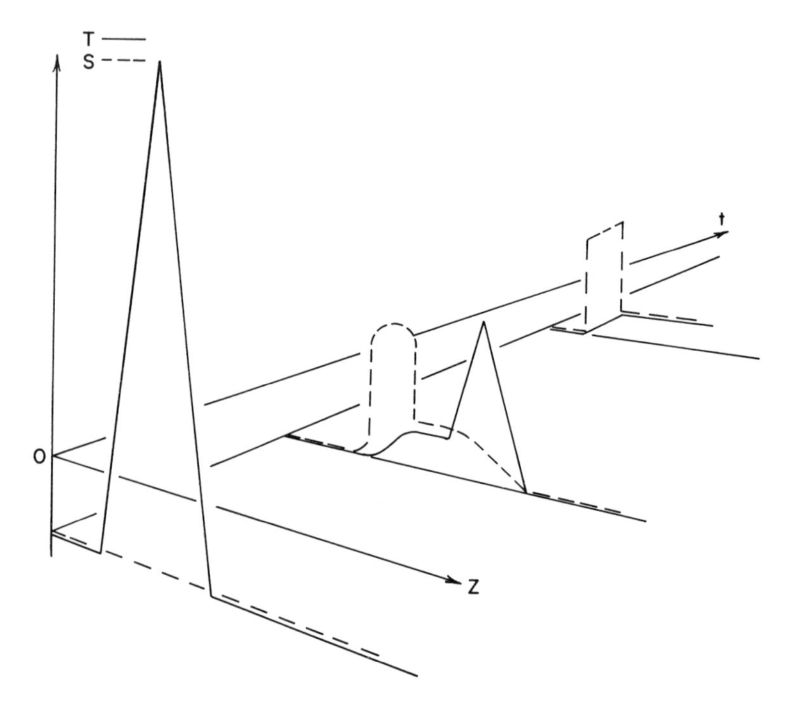

Figure 8. Development of a partially ignited steady-state solution from a perturbation in T *only*

(*45*). The latter assumption simplifies the situation considerably since then each particle often has a unique steady state for given surface conditions. If it is relaxed, there can be five steady states for certain fluid conditions, of which three will be stable (*46*). Even retaining nonisothermality within the particle it can be argued that the individual particles achieve their steady state much more rapidly than does the bed as a whole (*47*). Thus a quasi-steady state hypothesis may be made, and the effectiveness factor for the steady state of the particle is introduced into the transient equations for the bed. As long as there is uniqueness this causes no trouble, but when two or three stable steady states are possible, some selection rule is necessary. If we exclude discontinuous changes, it should be possible to make a proximate steady-state hypothesis and assert that in general the steady state of each particle takes to another stable branch of the solution only when a bifurcation point is reached, and that it then goes to the "nearest" stable steady state. This hypothesis will need some refinement in the light of the experience being gained in current research programs.

To implement such a program it would be necessary to have a rapid way to calculate the effectiveness factor of a catalyst particle since to do this at each point of the bed would require the solution of a two-point boundary value problem for each particle. Liu (*48*) has given a method for codifying unique effectiveness factors, and more recently Jouven (*47*) has been able to do the same for non-isothermal effectiveness factors. Thus by Jouven's subroutine the effectiveness factor for a sphere in which a first-order irreversible reaction is taking place can be called up 300 times faster than it can be calculated. The accuracy is quite satisfactory over the full range of Weiss and Hicks' (*49*) original calculations. With respect to the surface conditions there can be at most three values of the effectiveness factor in this range; however the program can be used to find effectiveness factors for any given fluid conditions, and here there may be as many as five, three of which are stable.

Conclusion

Many important aspects of the transient behavior of reactors have not been mentioned here. Thus, though the phase plane gives a full description of the start-up of the reactor, there are interesting problems concerned with the optimization of the start-up control. Limit cycles have appeared in passing, and these have been examined systematically in the context of the cyclic operation of reactors by Douglas, Horn, Bankoff, and others. They have in fact shown that the time average of the cyclic output may be more desirable than the output obtainable from the steady state about which the system is oscillating. We have also mentioned stability but have not reviewed the means available for its analysis.

It is hoped that what has been said about the general features of stirred tank behavior, its abstraction to cover more general types of reactor, and the grounds for strengthening the analogy are sufficient to illustrate one aspect of the place of mathematical analysis in the study of chemical systems.

Acknowledgment

This study has grown from a long term interest in the transient behavior of chemical reactors. Its early development was encouraged by a grant from the Research Corp. and its later progress greatly assisted by the continuing support of the National Science Foundation. The author is also indebted to one of the reviewers for some helpful comments.

Nomenclature

A_j	jth chemical species
C_p, c_p	heat capacities per unit volume, mass
$\mathbf{c}$	vector of concentrations
c_j	concentration of A_j
D_{je}	effective diffusivity of A_j
$H(S)$	Heavyside step function
h	heat transfer coefficient
I	initial space
I_i, J_j, K_k	initial spaces associated with W_i, V_j, and U_k
K_e	effective thermal conductivity
k_j	mass transfer coefficient for A_j
Le	Lewis number
M, N	operators in general system
Q	dimensionless rate of heat removal
Q^*	rate of heat removal from stirred tank
q	flow rate
R	state space
$\mathbf{r}$	vector of position in distributed system
r	reaction rate
r_e	effective reaction rate
S	number of reacting species
$S(z,t)$	wall temperature in model system
T	temperature
t	time
U_k, V_j, W_i	invariant subspaces
V	volume
$\mathbf{v}(\mathbf{r})$	velocity vector in distributed system
$X(y)$	steady-state extent for given y
x	dimensionless extent in stirred tank

$\mathbf{x}$	state of system
y	dimensionless temperature
y^*	$(y_f + \kappa y_c)/(l + \kappa)$
α_j	stoichiometric coefficient of A_j
ΔH	heat of reaction
ΔT_{ad}	adiabatic temperature rise
θ	holding time of stirred tank
κ	$Q/(y - y_c)$
λ,μ	eigenvalues
μ_j	dimensionless external transfer coefficient
ν	denotes normal derivative in dimensionless coordinates
ξ	extent of reaction
ρ	dimensionless reaction rate
$\hat{\rho}$	vector of position in dimensionless coordinates
τ	dimensionless time
ϕ,χ,ψ	stability criteria for stirred tank

Subscripts:

f	feed conditions
s	steady state
o	initial value
x,y	partial derivatives with respect to x,y
∞	final value

Literature Cited

(1) van Heerden, C., *Ind. Eng. Chem.* (1953) **45**, 1242.
(2) Amundson, N. R., *De Ingenieur* 1955, **37,** *Chem. Tech.* 8.
(3) Amundson, N. R., Bilous, O., *A.I.Ch.E. J.* (1955) **1**, 513; (1956) **2**, 117.
(4) Liljenroth, F. G., *Chem. Met. Eng.* (1918) **19**, 287.
(5) Danckwerts, P. V., *Chem. Eng. Sci.* (1953) **2**, 1.
(6) Langmuir, I. J., *J. Am. Chem. Soc.* (1908) **30**, 1742.
(7) Amundson, N. R., Aris, R., *Chem. Eng. Sci.* (1958) **7**, 121.
(8) *Ibid.*, p. 132.
(9) *Ibid.*, p. 148.
(10) Amundson, N. R., Aris, R., Nemanic, D. J., Tierney, J. W., *Chem. Eng. Sci.* (1959) **11**, 199.
(11) Amundson, N. R., Schmitz, R. A., *Chem. Eng. Sci.* (1963) **18**, 265.
(12) *Ibid.*, p. 391.
(13) *Ibid.*, p. 415.
(14) *Ibid.*, p. 447.
(15) Amundson, N. R., Warden, R. B., Aris, R., *Chem. Eng. Sci.* (1964) **19**, 149.
(16) Amundson, N. R., Warden, R. B., Aris, R., *Chem. Eng. Sci.* (1964) **19**, 173.
(17) Amundson, N. R., Goldstein, R. P., *Chem. Eng. Sci.* (1965) **20**, 195.
(18) *Ibid.*, p. 449.
(19) *Ibid.*, p. 477.
(20) *Ibid.*, p. 501.
(21) Amundson, N. R., Luss, D., *Chem. Eng. Sci.* (1967) **22**, 267.
(22) Hlavacek, V., Kubicek, M., Jelinek, J., "Stability and Oscillatory Behaviour of the CSTR."
(23) Amundson, N. R., Liu, S-L, *Ind. Eng. Chem., Fundamentals* (1962) **1**, 200.

(24) Wake, G. C. J., *Diff. Eqns.* (1969) **6**, 36.
(25) Pao, C. V., *Arch. Rat. Mech. Anal.* (1969) **35**, 16.
(26) Hlavacek, V., Marek, M., *Collect. Czech. Chem. Commun.* (1967) **32**, 3291.
(27) *Ibid.*, p. 3309.
(28) Hlavacek, V., Marek, M., *Chem. Eng. Sci.* (1966) **21**, 493.
(29) *Ibid.*, p. 501.
(30) Hlavacek, V., Marek, M., *Collect. Czech. Chem. Commun.* (1967) **32**, 4004.
(31) *Ibid.*, (1968), **33**, 506.
(32) *Ibid.*, p. 718.
(33) Hlavacek, V., Marek, M., *Chem. Ing. Tech.* (1968) **40**, 1086.
(34) Hlavacek, V., Marek, M., *Chem. Eng. Sci.* (1968) **23**, 865.
(35) Hlavacek, V., Marek, M., Kubicek, M., *Chem. Eng. Sci.* (1968) **23**, 1083.
(36) Hlavacek, V., Marek, M., John, T. M., *Collect. Czech. Chem. Commun.* (1969) **34**, 3664.
(37) *Ibid.*, p. 3868.
(38) Hlavacek, V., Marek, M., *Collect. Czech. Chem. Commun.*, in press.
(39) Hlavacek, V., Marek, M., Kubicek, M., *J. Catalysis* (1969) **15**, 17.
(40) *Ibid.*, p. 31.
(41) Hlavacek, V., Hofman, H., *Chem. Eng. Sci.* (1970) **25**, 173.
(42) *Ibid.*, p. 189.
(43) Eckhaus, W., "Studies in Nonlinear Stability Theory," Springer Tracts in Natural Philosophy Vol. 6, New York, 1965.
(44) Schruben, D., M.S. Thesis, Department of Chemical Engineering, University of Minnesota, 1970; *J. Chem. Eng.*, in press.
(45) McGreavy, C., Cresswell, D., *Can. J. Chem. Eng.* (1969) **47**, 583.
(46) Hatfield, B., Aris, R., *Chem. Eng. Sci.* (1969) **24**, 1213.
(47) Jouven, J. G., Ph.D. Thesis, Department of Chemical Engineering, University of Minnesota, 1970.
(48) Liu, S-L, "The Influence of Intraparticle Diffusion in Fixed Bed Catalytic Reactors," *A.I.Ch.E. J.* (1970) **16**, 742.
(49) Weiss, P., Hicks, J., *Chem. Eng. Sci.* (1962) **17**, 265.

RECEIVED July 21, 1970.

Contributed Papers

Dynamic Operation of a Fixed Bed Reactor

ROBERT L. KABEL and GILLES H. DENIS, The Pennsylvania State University, University Park, Pa. 16802

In recent years understanding and characterization of the unsteady state has advanced almost explosively. Now exploitation of transient behavior can be contemplated. In this paper the vapor-phase dehydration of ethyl alcohol to diethyl ether in a fixed bed reactor is examined. The catalyst is an ion exchange resin in the acid form.

Steady-state reaction rate data have been correlated well by a Langmuir-Hinshelwood model in which the surface reaction is the rate-controlling step, the adsorption processes being about an order of magnitude faster. The use of Langmuir adsorption theory in this case is supported by the agreement between adsorption equilibrium constants obtained from the Langmuir-Hinshelwood correlation and also directly from adsorption measurements. Water and ethanol are adsorbed strongly on the hydrogen ion catalytic site whereas ether is essentially not adsorbed. The suitability of Langmuir-Hinshelwood kinetics to this reaction system is attributed to the rather ideal nature of the catalyst. This allows the extension of the theory into the transient regime with some confidence in its validity.

An elaborate mathematical model, based on the above characteristics was formulated to predict the response of the reactor outlet concentrations to step changes in flow rate, temperature, feed concentration, and total pressure. Experiments were performed to check the predictions for flow rate and temperature variations. The model was in qualitative agreement with all experimental observations and in quantitative agreement with many of them. These results make it clear that the presence of significant adsorption effects, differing for the various reaction components, is the key to the unusual transient behavior.

Adsorbed-phase concentrations play a major role in determining the rate of a solid-catalyzed vapor-phase reaction. Large amounts of ethanol enhance the rate, whereas water suppresses it. In this reaction system at steady state the adsorbed-phase concentrations achieve nearly equilibrium values, and the reaction rate is fixed at any point in the reactor. In transient operation the adsorbed-phase concentrations vary according to the imposed perturbation and thus affect the rate of reaction. Consider a sudden increase in flow rate. At the low flow steady state more water and less ethanol are adsorbed than at the final steady state. Upon the imposition of the higher flow rate, the reaction rate will be low, corresponding to the catalyst conditions of the initial steady state. This effect coupled with the new low residence time produces a lower transient production rate than will exist at the final steady state. Since unsteady-state operation alters the reaction rate, one wonders if a cleverly selected regime of periodic variation of one or more process parameters might not allow reactor performance to be improved beyond that obtained in normal steady state operation.

In interpreting unsteady state behavior it is useful to think first of a process in which the new steady state is attained instantly following a perturbation. This is equivalent to the concept of slow (infrequent) switching between two set points, wherein the transient period becomes negligible compared with the steady-state period at a given set point. The results from an instant steady-state model of a reactor cycled between two set points would correspond to mixing the products of two steady state reactors operating separately at the given set points. With more frequent switching the transient effects can be made dominant. The performance in such a case can differ from either true steady state or cyclic instant steady state operation.

Table I gives production rates for steady-state and cycled operation. Computed data are shown; however, experimental work confirms the results. Steady state results were calculated for a flow rate of 0.265 cc of liquid ethanol/min and 109.4°C. The flow rate cycle comprised 18 minutes at 0.56 cc/min and 54 minutes at 0.167 cc/min, giving a time average flow rate equal to the steady state value. In the temperature cycle, equal 36-minute periods at 104.4° and 114.4°C were used.

Table I. Production Rates for Steady-State and Cycled Operation

Basis of Calculation	*Production Rate × 10^4 Moles of Ether or Water/min* — *Flow Cycling*	*Temperature Cycling*
True steady state	4.53	4.53
Instant steady-state model	4.306	4.76
Transient model	4.40	4.95

For flow cycling, the instant steady-state model (slow switching case) gives a worse performance than true steady-state operation. The more rapid switching case (transient model) improves on the slow switching performance by taking advantage of the transient adsorption effects but does not overcome the original deficit with respect to steady state. When reaction rate decreases with reaction progress (as in this case), operation with a variety of residence times gives a lower conversion than with a constant residence time at the same bulk average flow rate. The decrement in performance illustrated in the present study will be the more common occurrence. However cycling of flow rate could be valuable in reaction systems having rates which increase with increasing extent of reaction. Examples of such behavior can be found in autocatalytic reactions and in intermediate reactions occurring in a sequence of reactions in a series.

For temperature cycling the exponential effect of temperature on reaction rate improves the value calculated with the instant steady-state model over the true steady-state value. This much improvement could have been achieved by operating two steady-state reactors, one at the high temperature and one at the low temperature. However, the transient model predicts a still greater improvement which cannot be achieved by any steady-state operation at the same time average temperature. This additional benefit is the result of the effect of temperature on the adsorption processes which are active in this transient process.

The potential of the unsteady state in bringing about an effective shift in the rate of a single catalytic reaction is clear. Similarly the relative predominance of two or more competing reactions should be alterable also. Thus, depending upon the specified objective such dynamic operations are also expected to be effective in improving product distribution or catalyst selectivity.

Stability Studies of Single Catalyst Particles

C. McGREAVY and J. M. THORNTON, University of Leeds, England

In heterogeneous catalytic systems, the identification of potential instability is usually confined to establishing criteria for the existence of multiple solutions for the steady-state equations describing the behavior of individual catalyst pellets. Even where these bounds are given accurately and not by the various equations based on gross simplifications, the estimates are usually conservative. In particular, unstable behavior may not develop when the pellet passes transiently through a set of conditions having multiple solutions in the steady state.

The possibility of passing transiently into the steady-state multiple solution region and returning to a stable steady state has been investigated for a particular system based on the oxidation of benzene to maleic anhydride. In the presence of excess air this reaction may be represented schematically as:

$$\begin{array}{l} A \rightarrow \mathrm{B} \rightarrow \mathrm{C} \\ \;\;\; \rightarrow \mathrm{C} \end{array}$$

where C represents overoxidation to CO, CO_2, and water. In the limiting case of temperature and reaction runaway, this overoxidation occurs almost completely. When this happens, the reaction may be regarded as a single irreversible first-order reaction. Experimental data on the kinetic parameters over the relevant range of conditions have enabled a preliminary analysis to be made.

Previous investigations have shown that for the normal range of operating conditions, the physical parameters for real systems enable convenient simplifications of the model to be made. The most important is that there are no appreciable temperature gradients inside the catalyst pellet, although the temperature difference across the external fluid film may be significant. This approximation leads to:

$$K_T \frac{dt}{d\tau} = T(\tau) - t + f(t, \tau)$$

where K_T is a constant, τ is time, T is the dimensionless fluid temperature which may vary with time, t is the dimensionless pellet temperature, and $g(t, \tau)$ is a nonlinear function which involves the analytic solution of the equation describing diffusion inside the pellet and is therefore a function of the fluid phase concentration which may vary with time. (An equation of this form also results if the full scheme of consecutive and parallel reactions is considered to be occurring within the catalyst pellet.)

The critical factors which determine whether an excursion into the multiple steady-state region will result in instability depend on the exact trajectory followed by the transient response. Step changes give some insight as to the way that runaway develops but do not help to overcome these difficulties. A convenient way of examining the characteristics is to look at the frequency response, especially in the phase–plane of fluid–solid temperatures. Such a plot enables the influence of the magnitude and frequency of the perturbation to be assessed.

The information obtained from such investigations is useful in deciding the effect of interaction between pellets in a packed bed reactor. Whether an instability will grow or not depends on whether the damping effects of the interaction can prevent the perturbation from propagating.

No detailed analytical solution of the transient equation can be given at this time since the behavior of the pellet is governed by the nonlinearities, and any attempts to linearize the equation results in loss of the important features of the response. However, a graphical approach gives considerable insight into the behavior around the multiple solution region, particularly when the pellet is changing from one state to another, and can be useful for analyzing particular systems.

When the catalyst pellet is subjected to sinusoidal perturbations of various frequencies, the highest frequency produced the response with the smallest amplitude, the least distortion, and the largest relative time lag. As the frequency decreases, a longer time per cycle is spent in the region of potential runaway, and the pellet becomes increasingly less stable. The distortion of the response also becomes more pronounced owing to the highly nonlinear effect of temperature on the reaction rate; this causes high peaks in the pellet temperature as the fluid temperature rises. At very low frequencies the fluid conditions remain in the unstable region long enough for runaway to occur, and film mass transfer control results.

A general conclusion which may be drawn from the results is that the fluid temperature cannot be manipulated normally to return the pellet to its initial state after temperature runaway has occurred, but that this may always be done by adjusting the concentration, indicating that if a reactor is to be operated near the non-unique region, multivariable control techniques may be essential.

The Measurement and Prediction of Instability in Certain Exothermic Vapor-Phase Reactors

S. F. BUSH, Imperial Chemical Industries, Ltd., Central Instrument Research Laboratory, Bozedown House, Whitchurch Hill, Reading, England

The experiments and mathematical analysis abstracted here from Ref. *1* were prompted by the observation, on the industrial scale, of a sudden instability in certain vapor-phase chlorine–hydrocarbon reactions. Reactions of this sort are often carried out in reactors of the form shown in Figure 1 where the velocities have been deduced from calculation and measurement on the laboratory scale. Under normal conditions the reaction of chlorine to completion occurs in the region below the injection point, but under some conditions the reaction ceases abruptly, and unreacted chlorine leaves this region to react eventually elsewhere in a dangerous and uncontrolled manner.

The steady reaction states observed cannot be regarded as unstable to infinitesimal disturbances since reaction is typically maintained for periods which are long compared with the time for such disturbances to pass through the system. The sudden cessation of reaction can be explained, however, as the result of a sufficiently large disturbance being applied to a system which is locally stable.

The stability of the systems studied is characterized as the time in seconds for which a given percentage reduction in the flow of the leaner feed component (usually chlorine) will just permit the system to recover. Experiments of this sort have been carried out in laboratory reactors of the type in Figure 2 with methyl chloride and methane as the hydrocarbon feedstocks. The over-all stoichiometry of the main reactions from these feedstocks is summarized by

$$CH_{(4-i)}Cl_i + Cl_2 = CH_{(3-i)}Cl_{(i+1)} + HCl,\ i = 0 \text{ to } 3 \qquad (1)$$

Also shown in Figure 2 is a measured steady-state temperature distribution. Three main sets of experiments have been carried out. The first measured steady-state distributions of temperature and composition. The second measured the minimum wall temperature necessary (under given conditions) to start the reaction in the base of the reactors. In the third set the steady states were disturbed by reducing the chlorine supply for known periods of time. A fourth set of experiments has examined the start-up of the reactions under automatic and manual control.

Typical disturbance traces are reproduced in Figure 3. In Curve A the reaction is extinguished, but an oscillatory recovery is obtained. Curve B shows no recovery. Curve C shows the reaction starting under automatic control, with little or no overshoot; manual starts can give overshoots of 50°–80°C.

A theory capable of predicting the dynamic behavior of reaction systems of the general type described, at different scales of operation, must be based on the relevant fluid dynamic and chemical kinetic processes as demonstrated for a simpler physical system in an earlier paper (*2*). The structure of turbulent flow within the reactors is determined principally by the ratios l/d, D/d (Figure 1). Four main regions are identified; the entraining jet, the impingement boundary layer, the wall flow, and the irregular vortex region between the sides and the jet. Expressions for the fluid velocities and eddy diffusivities in these regions

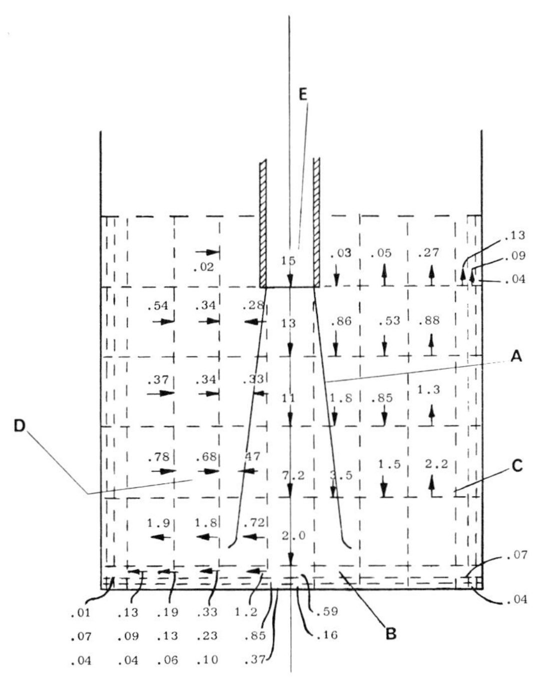

Figure 1. Cell division of laboratory reactor and typical average velocities (ft/sec) across cell boundaries in axisymmetric flow

A: measured jet boundary
B: boundary layer
C: wall flow
D: vortex region of irregular flow
E: inlet point

have been derived and verified experimentally in hydraulic models and in the case of the boundary layers by heat transfer experiments. Using these expressions, velocity diagrams of the form shown in Figure 1 are generated, enabling different configurations to be studied, including those of intense mixing (large l/d, D/d) and of direct flow ($l >> D$).

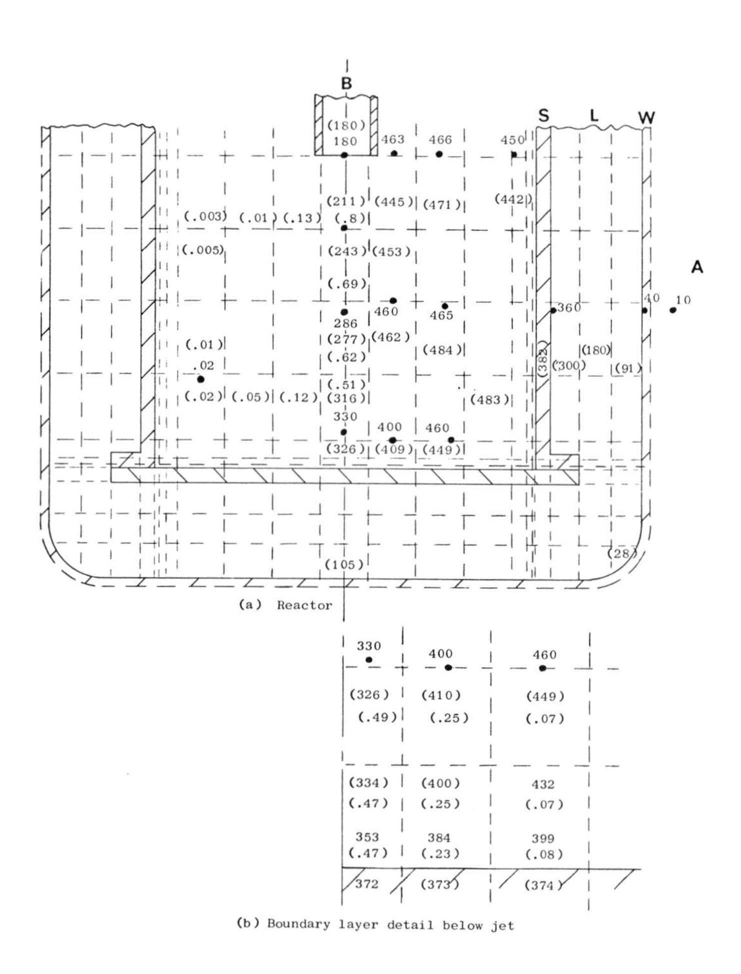

Figure 2. Temperature and composition distributions. Measured temperatures in °C (three-digit values). Fractions of initial Cl_2 remaining (two-digit values, e.g., .02). Calculated values in parentheses at cell centers.

A: atmosphere
B: inlet
S: reactor shell and lining
L: lagging
W: outer boundary layer

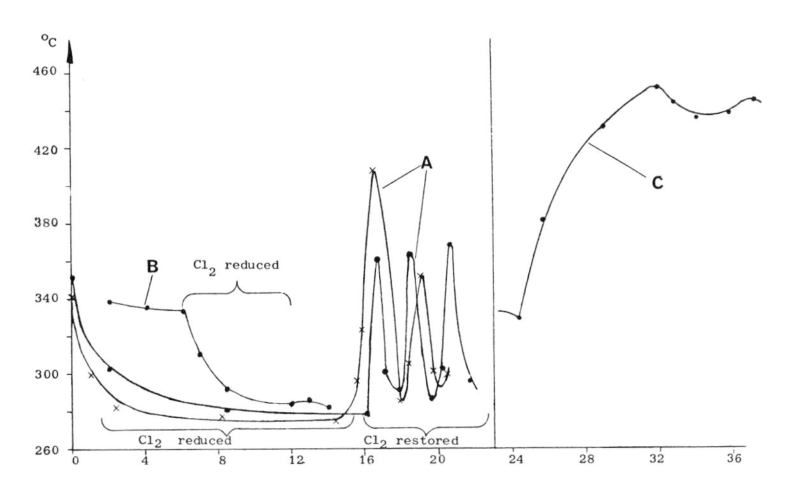

Figure 3. Stability test on laboratory scale showing experimental points · and computed points, x.

A: CH_3Cl feedstock, C_2 m.f. 0.23, flow rate 0.018 gram moles/sec
B: CH_4 feedstock, Cl_2 m.f. 0.20, flow rate 0.03 gram moles/sec
C: CH_3Cl start-up under automatic control at 24 sec

The postulated mechanism of Reaction 1 is of the standard free radical type, with initiation and termination shown by experiment to be predominantly at the walls. All told 72 elementary processes of the type

$$Cl_2^{(a)} + M \rightarrow Cl^{(a)} + Cl\cdot + M$$

$$CH_4 + Cl\cdot \rightarrow CH_3\cdot + HCl$$

$$Cl_2^{(a)} + CH_2Cl\cdot \rightarrow CH_2Cl_2 + Cl^{(a)}$$

are postulated, where (a) denotes wall-adsorbed species. Since propagation rates greatly exceed initiation rates, the observed rates of Reaction 1 from left to right can be found in the form

$$R_i = \frac{k_i[M][CH_{(4-i)}Cl_i][Cl_2]}{\sum_{j=0}^{3} \beta j[CH_{(4-j)}Cl_j]} \quad i = 0 \text{ to } 3 \tag{2}$$

β_j and k_i were determined by experiment in continuous stirred reactors (*e.g.*, Ref. *1*). Typically, $\beta_o = 2.0$; $k_o = 1.13 \times 10^{15} \exp(-33{,}800/RT)$ cc/mole sec. The forms (Equation 2) which are typical of a wall initiated reaction can only apply if Cl˙ atoms diffuse sufficiently rapidly from the walls. A necessary condition is $(gA\tau/V)$ greatly to exceed unity, where g is a boundary layer mass transfer coefficient and τ is the over-all residence time, which gradually becomes more difficult to satisfy as the scale increases.

The mathematical analysis is carried out by first discretizing the system into small cells (Figure 1); then using the fluid dynamics and chemical kinetics one derives conservation equations for each cell. The equation for chlorine mole fraction (y) in the (m, n) th cell is $(u_{m-1,n}, v_{m,n-1} \geqslant 0)$:

$$\frac{dy_{mn}}{dt} = u_{m-1,n}\frac{A'_n}{V_{mn}}(y_{m-1,n} - y_{mn})\frac{\rho_{m-1,n}}{\rho_{mn}} + u_{m,n-1}\frac{A_{n-1}}{V_{mn}}(y_{m,n-1} - y_{mn})\frac{\rho_{m,n-1}}{\rho_{mn}} - \sum_{\text{all faces}}\frac{D_{mn}A_n}{\Delta r V_{mn}}(y_{mn} - y_{m,n+1}) - \sum_{i=0}^{3}\frac{R_i}{\rho_{mn}} \quad (3)$$

and similarly for T_{mn}. Although transient density changes do not greatly affect the results (*see also* Ref. 2), provision is made in the theory for coupling the fluid flow equations directly into the conservation Equations 3. The cell formulation was the only method of discretizing the analysis which gave a numerically stable result and also allowed for the rapid space rates of change of velocity. Typically 100–200 cells are employed, each generating several equations of the type of Equation 3. Full solutions are obtained by implicit methods, but approximate solutions are obtained from simplified models.

Some results obtained from the theory are compared with experiment in Table I and in Figures 2 and 3. In Figure 2 note the heat flow from the wall into the boundary layer below the jet.

Table I. Theoretical and Experimental Results

Flow Rate, gram moles/sec	T_1	y_i	*Pressure, atm*	*Experimental Wall Start-up Temps., °C*	*Computed Wall Start-up Temps., °C*
.0091	210	.23	1.8	360	365
.0122	211	.31	2.5	350	350
.0200	177	.23	2.5	383	385

In Ref. *1* the experimental results and analysis indicated here are developed into a theory which appears to predict well the steady state and dynamic behavior of vapor-phase chlorinations in reactors where partially reacted gas is entrained in the inlet jet. The theory applies in a variety of geometries and to different scales of operation.

(1) Bush, S. F., "Measurement and Prediction of Instability in Certain Exothermic Vapor-Phase Reactors," to be published.
(2) Bush, S. F., *Proc. Roy. Soc.* (1969) **A 309,** 1–26.

Unstable Behavior of Chemical Reactions at Single Catalyst Particles

H. BEUSCH, P. FIEGUTH, and E. WICKE, Institut für Physikalische Chemie, Universität Münster, Germany

Stepwise instabilities, corresponding to "ignition" and "extinction," as well as oscillatory behavior of chemical reactions have been observed in isothermal and nonisothermal heterogeneous catalytic systems. H_2 oxidation was chosen as the nonisothermal example while CO oxidation represented isothermal unstable behavior.

Nonisothermal Case

A vertical glass tube in an electric furnace with temperature control contained an active catalyst pellet, freely suspended in the gas flow or imbedded in a layer of inert particles. The reaction mixture, H_2 in air, was purified over molecular sieves; after preheating it entered the reactor from below. Thermocouples from 0.2 mm Ni–NiCr wires measured the temperature of the gas flow, T_g, and of the center of the catalyst pellet, T_c (the pellet surface temperature is only slightly different, Figure 2).

The catalyst particles were 8 mm spheres from silica-alumina containing 0.4% Pt; mean pore radius was about 10^{-4} cm (macropores only), permeability was 0.1, thermal conductivity was 0.5×10^{-3} cal/cm deg sec. The reaction proceeded with an activation energy (in the kinetic range) of 19 kcal/mole hydrogen.

The stepwise instabilities were studied first—*i.e.*, the transition from lower to upper stable states (ignition) and the reverse (extinction).

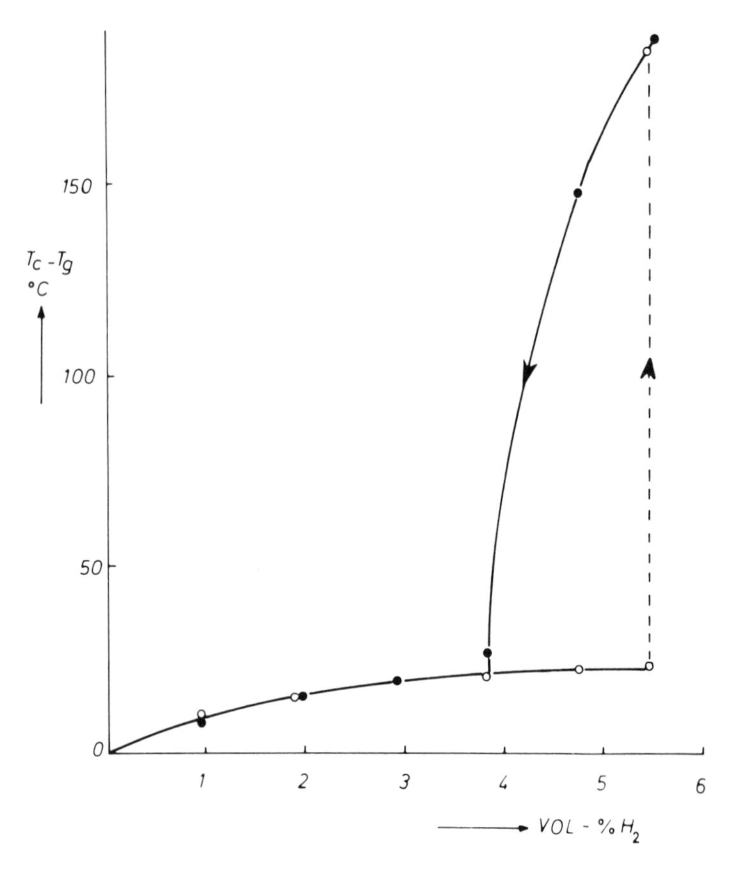

Figure 1. Stepwise transitions between lower and upper stable states of H_2 oxidation on single catalyst particle in a packed bed of inactive pellets. Linear gas flow rate (standard conditions, empty tube): $u_o = 1.7$ *cm/sec;* H_2 *percentage increasing: ○, decreasing: ●. Temperature of furnace and gas flow is 139°C.*

Overheating of the catalyst particle by chemical reaction was measured as function of H_2 concentration in the inlet mixture at constant gas temperature and gas flow rate. The results of measurements with the active particle imbedded in a 6-cm high column of inactive pellets at $T_g =$ 139°C are shown in Figure 1. With increasing H_2 concentration the steady-state excess temperature of the pellet increases gradually to a level of $T_c - T_g = \Delta T = 20°C$, and after 5.5% H_2 has been reached, it jumps suddenly to $\Delta T = 185°C$. If the H_2 concentration is lowered, the pellet temperature decreases again, running along the upper branch of a hysteresis loop, thereby marking a range of multiple steady states. Finally the temperature drops to the ascending branch; below this "extinction" point (3.8% H_2 in Figure 1) the pellet temperature can be adjusted up and down in the lower stable state.

At a higher gas flow temperature ignition occurs at lower H_2 concentration, and the area of the hysteresis loop—*i.e.*, the range of multiple steady states—is then smaller. At even higher T_g the hysteresis loop vanishes; the pellet excess temperature then increases linearly with the H_2 concentration since the external mass transfer is rate determining in this region. These results can be understood on the basis of a stability analysis of the catalyst pellet by means of the S-shaped heat generation function and the straight line characteristic of heat removal. On the other hand, the stability diagram can be constructed by evaluating the results of these measurements.

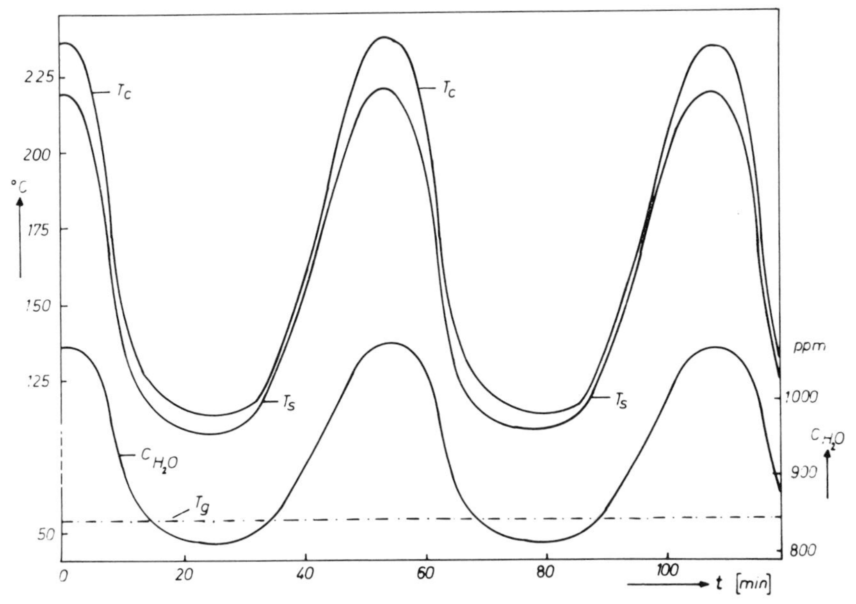

Figure 2. Oscillations of temperatures of a single catalyst particle (T_c *center,* T_s *surface) and of reaction rate* (C_{H_2O} *in the effluent).* T_g = *70°C; 3.14 vol %* H_2*;* u_o = *36 cm/sec.*

When the temperature of the gas flow is fixed at low values (*ca.* 30°–80°C), oscillations of the particle temperature and of the reaction rate (composition of the effluent) are observed. Figure 2 gives an example with the catalyst particle suspended freely in the gas stream, whose temperature was fixed at 53°C. (Oscillations could also be observed with the active particle imbedded in a layer of inactive pellets.) The mean value of the particle temperature is 170°C, and the oscillations extend over ±60°. In this case the temperature of the particle surface,

T_s, also was measured; it differs at its maximum by 15° from the center temperature. Both temperatures oscillate in phase with the reaction rate. The low frequency is surprising, the period amounting to nearly one hour. The following characteristics could be verified:

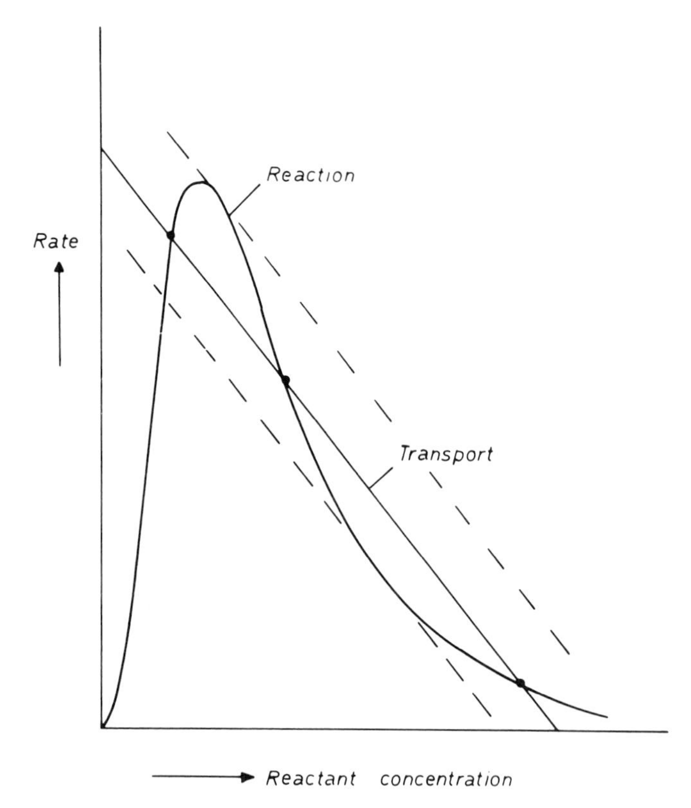

Figure 3. Range of multiple steady states for a reaction with self-poisoning superimposed by a transport process (schematic)

(1) Oscillations begin voluntarily. Once established, they run uniformly for days. They change slightly when the gas flow rate is varied.

(2) Superpositions of oscillations (double peaks) have been observed in some experiments.

(3) With increasing gas temperature the amplitudes decrease, and the oscillatory state passes to a stable steady state.

The reason for these oscillations and their mechanism is not clear. The long period and the absence of phase shifts seem to indicate, however, that the phenomenon does not arise from specific properties of the mass and heat balances of the catalyst pellet. More likely it originates from particular reaction mechanisms which produce long time periodic changes in the nature of the active catalyst surface.

Isothermal Case

If the rate of a chemical reaction goes through a maximum with increasing concentration of reactants or products (self-poisoning or autocatalysis), the reacting system may behave unstably even under isothermal conditions. An example is the catalytic oxidation of CO on Pd and on Pt, where the strong chemisorption of CO inhibits the reaction rate at higher CO partial pressures. This is shown schematically in Figure 3. The combination with a transport process, characterized by the straight balance line in Figure 3, gives rise to multiple steady states between the two broken lines.

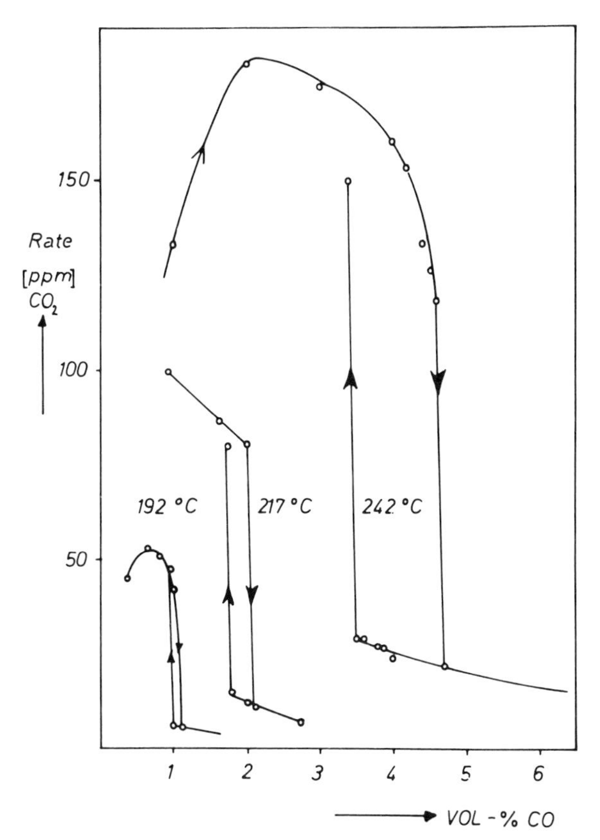

Figure 4. Stepwise transitions between upper and lower stable states of CO oxidation on single catalyst particle suspended freely in the gas stream at different temperatures; $u_0 = 6.1$ *cm/sec*

The experimental setup has been arranged similar to that described above. The catalyst pellet, a 3 × 3 mm cylinder from γ-Al_2O_3 with 0.3% Pt (BET surface 130 m^2/gram), was suspended freely in a mixture of air and CO, which streamed upward through the vertical reactor tube. The temperatures of the gas flow (fixed) and of the pellet were measured by thermocouples. Since the pellet temperature remained nearly constant during reaction rate changes, the course of the reaction rate was pursued by infrared analysis for CO_2 in the effluent.

At fixed values of gas temperature and gas flow rate, transitions between upper and lower stable states could be observed by changing the CO concentration in suitable ranges. Figure 4 gives some results obtained at 192°, 217°, and 242°C. With increasing CO concentration the reaction rate first goes through a maximum and then drops suddenly when a critical "quenching" CO concentration has been reached. The reverse transition occurs at a smaller ("ignition") concentration of CO; thus, a hysteresis loop is established between the upper and lower stable state, similar to the stepwise transitions in Figure 1. In those cases the unstable behavior originated from the influence of external heat and mass transfer; in the present case these transport processes have little effect.

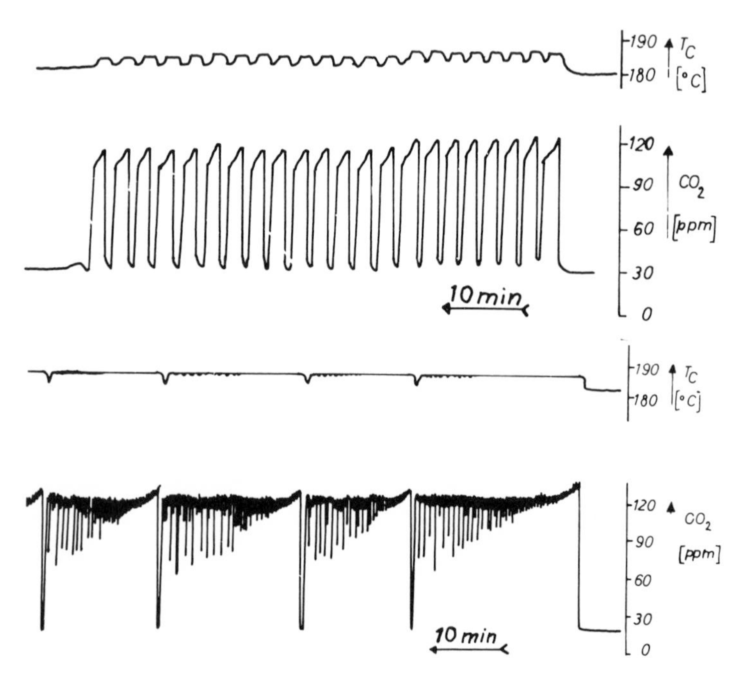

Figure 5. Oscillatory behavior of CO oxidation at a single catalyst particle (time runs from right to left). $T_g = 180°C$; $u_o = 3$ *cm/sec; 1 vol % CO.*

The stepwise transitions shown in Figure 4 are nearly independent of gas flow rate. They occur remotely from the mass transfer region, mainly in the kinetic range of the catalytic reaction. Hence, the transport process which, in combination with the reaction rate, is necessary for multiple steady states (Figure 3) cannot be external mass transfer. Presumably the process in question is CO chemisorption or some related process on the catalyst surface by which CO occupies the reaction sites. Similar stepwise transitions have been observed with a platinum wire screen, coated by platinum black, instead of the porous catalyst particle.

When the system is in the upper stable state and the gas flow temperature is lowered, the reaction rate may begin to oscillate. These oscillations, as Figure 5 shows, have periods of the order of minutes and are asymmetrical in shape (the reaction rate is represented as CO_2 content [ppm] in the effluent). In Figure 5 the line above the reaction rate in each cycle at first decreases slightly—flickering with high unresolved frequencies; then it suddenly breaks down and just as suddenly returns to its initial value, where it begins to descend again. In Figure 5 below the minimum values in the reaction rates the lines decrease gradually until a main breakdown occurs, after which the initial rate is restored quickly. If higher CO concentrations are applied, the minima become deeper; finally, when the quenching concentration has been reached, the reaction rate remains at small values of the lower stable state without further oscillation. This behavior seems to indicate that breaks in the reaction rate are brought about by the blocking of reaction sites with CO. In the restorative process the active sites may be recovered by burning away the blocking CO molecules with oxygen. Presumably several states of chemisorbed CO molecules participate in these phenomena; a detailed mechanism cannot yet be presented. The breaks in reaction rate give rise to small minima of pellet temperature, at a maximum about 2° as shown in Figure 5. At higher temperatures the frequencies of oscillation increase and the amplitudes decrease; above about 250°C no further oscillations could be observed.

Nonlinear Functional Analysis and Reactor Stability

C. E. GALL, University of Waterloo, Ontario Canada

The fixed point theorems of Tychonoff and Schauder are useful in deriving regions of asymptotic stability for chemical reactors. A method suggested by A. Stokes derived from Tychonoff's fixed point theorem is used to derive a region of asymptotic stability for a CSTR (continuous stirred tank reactor) example. The RAS obtained is larger than that obtained previously using Krasovskii's theorem and the Liapunov second method.

Schauder's inclusion theorem, an extension of Schauder's fixed point theorem to partially ordered spaces, is applicable to higher order systems such as a chain of CSTR's in order to derive a region of asymptotic stability with considerably less computation than previously known methods. The concept can be extended readily to the determination of a region of asymptotic stability for tubular reactors.

Stokes' Theorem (2)

This is a form of Tychonoff's fixed point theorem. The CSTR example is that considered by Berger and Perlmutter (*1*). The following are equations (*2*) in that reference with substitutions made from the data given.

$$\frac{d\hat{\eta}}{dt} = 10^8(\hat{y} + .61) \exp\left[-185/(\hat{\eta} + 10.18)\right] - .778 - 2.1\hat{\eta}$$

$$\frac{d\hat{\eta}}{dt} = -10^8(\hat{y} + .61) \exp\left[-185/(\hat{\eta} + 10.18)\right] + .778 - 2\hat{y}$$

The nonlinear term above may be written in terms of a MacLaurin's series with remainder as:

$$[1.272\ (1 + 1.785\hat{\eta}) + \alpha\hat{\eta}^2]\ (\hat{y} + .61)$$

where

$$2.2 < \alpha < 8.3$$

A somewhat more convenient form is obtained by letting $x_1 = \hat{\eta} + \hat{y}$, $x_2 = \hat{\eta}$ and adding equations and by letting $\beta = [(.61 + y)\alpha - 2.27]$.

This gives:

$$\dot{x}_1 = -2x_1 - .1x_2$$
$$\dot{x}_2 = 1.27x_1 - 3x_2 + (2.27x_1 + \beta x_2)x_2$$

or using the fundamental matrix solution:

$$x = e^{At}x_0 + \int_0^t e^{A(t-\tau)}\,(2.27x_1 + \beta x_2)x_2 \begin{vmatrix} 0 \\ 1 \end{vmatrix} d\tau = T_b x(t)$$

Then we can show

$$||T_b x(t)|| \leq 1.1e^{-2t}\{||x_0|| + \int_0^t e^{2\tau}|2.27x_1 + \beta x_2||x_2|d\tau\}$$

$$\leq 1.1e^{-2t}\{||x_0|| + (2.27 + |\beta|) \int_0^t e^{2\tau}||x||^2 d\tau\}$$

Thus $||x|| \leqslant g(t)$ where

$$\dot{g} \geq -2g + 1.1\,(2.27 + |\beta|)g^2$$
$$g(0) \geq 1.1\,||x_0||$$

This is asymptotically stable if

$$g(0) \leq \frac{2}{1.1(2.27 + |\beta|)}$$

Hence the CSTR is asymptotically stable if

$$||x_0|| \leq \frac{2}{(1.1)^2(2.27 + |\beta|)}$$

Now

$$|\beta| < .61 \times 8.3 - 2.27 = 5.06 - 2.27$$

Therefore, we are assured of asymptotic stability if

$$||x_0|| \leq .327$$

i.e.
$$\max\{|\hat{\eta}_0 + \hat{y}_0|, |\hat{\eta}_0|\} \leq .327$$

Therefore, we may start up with initial conditions anywhere in the RAS in the c *vs.* T plane bounded by the four lines:

$$(c - .077) = \pm \frac{.088}{17.6} (T - 550)$$

$$(c - .253) = \pm \frac{.088}{17.6} (T - 550)$$

Schauder Inclusion Theorem

This is a variation of Schauder's fixed point theorem (*3*). The continuous stirred tank vector example considered by Bilous and Ammundson (*4*) may be rearranged to:

$$\frac{dz}{dt} = 3.408 - .5678z + 7.86 \times 10^{14} y \exp(-227.1/z)$$

$$\triangleq a - bz + k(z)y$$

$$\frac{dy}{dt} = 0.5(1 - y) - 7.86 \times 10^{14} y \exp(-227.1/z)$$

$$\triangleq c - dy - k(z)y$$

The desire is to find a region in the z *vs.* y plane from which all trajectories terminate at the steady state. The solution to the system may be written as:

$$z(t) = z(0)e^{-bt} + \frac{a}{b}(1 - e^{-bt}) + e^{-bt} \int_0^t e^{b\tau} k(z(\tau))y(\tau)d\tau$$

$$y(t) = y(0) \exp\{-\int_0^t \alpha(\tau)d\tau\} + c \int_0^t \exp\{-\int_0^\tau \alpha(u)du\}d\tau$$

where $\alpha(\tau) = d + k(z(\tau))$.
In operator notation:

$$z = P_1(z,y) + r_1$$

$$y = N_2(z,y) + r_2$$

where

$$P_1[z,y](t) = e^{-bt} \int_0^t e^{b\tau} k(z(\tau))y(\tau)d\tau$$

$$N_2[z,y](t) = y(t)$$

$$r_1(t) = \frac{a}{b}(1 - e^{-bt}) + c_1 e^{-bt},\ c_1 = z(o)$$

$$r_2(t) = 0$$

Now $(z_1,y_1) \geqslant (z_2,y_2)$ will mean that

$$z_1(t) \geq z_2(t)$$
$$y_1(t) \geq y_2(t)$$

and from these it follows that

$$k(z_1(t))y_1(t)) \geq k(z_2(t))y_2(t))$$

so that

$$P_1(z_1,y_1) \geq P_1(z_2,y_2) \text{ (i.e., isotone)}$$

and

$$N_2(z_1,\ y_1) \leq N_2(z_2,\ y_2) \text{ (i.e., antitone)}$$

The problem now is to determine an interval, $I = [{}_oz,{}^oz] \times [{}_oy,{}^oy]$ which includes the iterates:

$$^1z = P_1({}^oz,\ {}^oy) + r_1$$
$$^1y = N_2({}_oz,\ {}_oy) + r_2$$
$${}_1z = P_1({}_oz,\ {}_oy) + r_1$$
$${}_1y = N_2({}^oz,\ {}^oy) + r_2$$

and so that

$${}_oz \leq {}_1z \leq {}^1z \leq {}^oz$$
$${}_oy \leq {}_1y \leq {}^1y \leq {}^oy$$

and which includes the single steady state of interest and no others.

Assume the interval with ${}^oz$, ${}_oz$, ${}^oy$, ${}_oy$ all constant functions of t to be determined. For this example we can show that

$$({}_oz,\ {}_oy) \leq ({}_1z,\ {}_1y) \leq ({}^1z,\ {}^1y) \leq ({}^oz,\ {}^oy)$$

if

$${}_oz \leq z(o) \leq {}^oz$$
$${}_oy \leq y(o) \leq {}^oy$$

and if

$${}^1z_s \equiv [a + k({}^oz)\,{}^oy]/b \leq {}^oz$$
$${}^1y_s \equiv c/[d + k({}_oz)] \leq {}^oy$$
$${}_1z_s \equiv [a + k({}_oz)\,{}_oy]/b \geq {}_oz$$
$${}_1y_s \equiv c/[d + k({}^oz)] \geq {}_oy$$

and

$$({}_1z_s,\ {}_1y_s) \leq ({}^1z_s,\ {}^1y_s)$$

Thus, if constants $({}^oz, {}^oy)$, $({}_oz, {}_oy)$ can be found such that

$${}_oz \leq {}_1z_s \leq {}^1z_s \leq {}^oz$$
$${}_oy \leq {}_1y_s \leq {}^1y_s \leq {}^oy$$

and such that

$$z_s \in [{}_oz,\ {}^oz]$$
$$y_s \in [{}_oy,\ {}^oy]$$

we are assured that trajectories initiating in the rectangle $[{}_oz, {}^oz] \times [{}_oy, {}^oy]$ will converge on the steady state (z_s, y_s). For the example described above a simple computer computation shows that:

$${}_oz = 0.001 \qquad {}^oz = 6.288119$$
$${}_oy = 0.000 \qquad {}^oy = 1.000$$

gives:

$${}_1z_s = 6.001 \qquad {}^1z_s = 6.288113$$
$${}_1y_s = 0.332 \qquad {}^1y_s = 0.998$$

For a two-stage model of continuous reactors with material feedback a region of stability in four-dimensional phase space may be determined in the same way. The equations may be written in the form:

$$\dot{z}_1 = a_1 z_2 - b_1 z_1 + k_1(z_1) y_1 + e_1$$
$$\dot{y}_1 = c_1 y_2 - d_1 y_1 - k_1(z_1) y_1 + f_1$$
$$\dot{z}_2 = a_2 z_1 - b_2 z_2 + k_2(z_2) y_2 + e_2$$
$$\dot{y}_2 = c_2 y_1 - d_2 y_2 - k_2(z_2) y_2 + f_2$$

and we need to determine constant functions

$$({}_o z_i,\ y_i),\ ({}^o z_i,\ {}^o y_i),\ i = 1,2$$

so that:

$$^1 z_1 = [a_1 {}^o z_2 + e_1 + k_1({}^o z_1)\, {}^o y_1]/b_1$$
$$^1 y_1 = (f_1 + c_1 {}^o y_2)/[d_1 + k_1({}_o z_1)]$$

$$^1 z_2 = [a_2 {}^o z_1 + e_2 + k_2({}^o z_2)\, {}^o y_2]/b_2$$
$$^1 y_2 = (f_2 + c_2 {}^o y_1)/[d_2 + k_2({}_o z_2)]$$
$$_1 z_1 = [a_1 {}_o z_2 + k_1({}_o z_1)\, {}_o y_1]/b_1$$
$$_1 y_1 = c_1 {}_o y_2/[d_1 + k_1({}^o z_1)]$$
$$_1 z_2 = [a_2 {}_o z_1 + k_2({}_o z_2)\, {}_o y_2]/b_2$$
$$_1 y_2 = c_2 {}_o y_1/[d_2 + k_2({}^o z_2)]$$

are in the hyper-rectangle

$$[{}_o z_1,\ {}^o z_1] \times [{}_o y_1,\ {}^o y_1] \times [{}_o z_2,\ {}^o z_2] \times [{}_o y_2,\ {}^o y_2]$$

which includes the single steady state of interest and no other and

$$({}_o z_i,\ {}_o y_i) \leq ({}_1 z_i,\ {}_1 y_i) \leq ({}^1 z_i,\ {}^1 y_i) \leq ({}^o z_i,\ {}^o y_i) \text{ for } i = 1,2.$$

With two stages each of size equal to that of the single reactor of the previous example, the same reaction rate expression and external heat transfer from the second stage the parameters for the iteration equations are:

$$a_1 = (1 - \lambda) \qquad a_2 = 1/\lambda$$
$$b_1 = 0.5678 \qquad b_2 = 2b_1$$
$$c_1 = a_1 \qquad c_2 = 0.5a_2$$
$$d_1 = 0.5 \qquad d_2 = 0.5$$
$$e_1 = 3.408\lambda \qquad e_2 = 0$$
$$f_1 = \lambda/2 \qquad f_2 = 0$$

and

$$k(z) = 7.86 \times 10^{14} \exp(-227.1/z)$$

For $\lambda = 0.989$ the iteration equations were programmed, and a region of asymptotic stability was determined to be the hyper-rectangle above with:

$$
\begin{aligned}
{}_oz_1 &= 6.124088 & {}^oz_1 &= 6.194783 \\
{}_oy_1 &= .8435543 & {}^oy_1 &= .9109129 \\
{}_oz_2 &= 5.453202 & {}^oz_2 &= 5.516600 \\
{}_oy_2 &= .8510970 & {}^oy_2 &= .9201147
\end{aligned}
$$

After 30 iterations it is found that:

$$
\begin{aligned}
6.160017 &\leq z_{1s} \leq 6.161305 \\
.8734471 &\leq y_{1s} \leq .8740321 \\
5.485973 &\leq z_{2s} \leq 5.486590 \\
.8816984 &\leq y_{2s} \leq .8822964
\end{aligned}
$$

Example. A simple tubular reactor for a first-order reaction can be described by two equations of the form

$$
\begin{aligned}
\lambda \dot{T} &= T'' - aT' + W(T_c - T) + Hk(T)c \\
\alpha \dot{c} &= c'' - bc' - k(T)c \\
\dot{T}(o, x) &= T_o(x) \\
c(o, x) &= c_o(x)
\end{aligned}
$$

with appropriate boundary conditions.

These may also be written in the form

$$
\begin{aligned}
T(x, t) &= P_1[T, c](x, t) + r_1(x, t) \\
c(x, t) &= N_2[T, c](x, t) + r_2(x, t)
\end{aligned}
$$

where P_1 is isotone and N_2 is antitone because the Green's function for the negative Laplacian operator is positive (5).

Then we may determine a region of asymptotic stability:

$${}_oT \leq T_o \leq {}^oT \qquad {}_oc \leq c_o \leq {}^oc$$

if we can find ${}_oc, {}^oc, {}_oT, {}^oT$, functions of spacial position only such that:

$$ {}_oT \leq {}_1T_s \leq T_s \leq {}^1T_s \leq {}^oT $$

$$ {}_oc \leq {}_1c_s \leq c_s \leq {}^1c_s \leq {}^oc $$

where (T_s, c_s) is the desired stable steady state and

${}^1T_s$ is the solution of

$$ T'' - aT' - WT' = -WT_c - Hk({}^oT)\,{}^oc $$

${}_1T_s$ is the solution of

$$ T'' - aT' - WT = -WT_c - Hk({}_oT)\,{}_oc $$

with initial condition: ${}_oT \leqslant T_o(x) \leqslant {}^oT$ and the given boundary conditions and

$$ {}^1c_s \text{ is the solution of } c'' - k({}_oT(x))c = 0 $$

$$ {}_1c_s \text{ is the solution of } c'' - k({}^oT(x))c = 0 $$

with initial condition: ${}_oc \leqslant c_o(x) \leqslant {}^oc$ and the given boundary conditions.

(1) Berger, Perlmutter, D. D., *A.I.Ch.E. J.* (1964) **10,** 233.
(2) Stokes, A., "Contributions to the Theory of Nonlinear Oscillations," J. P. LaSalle, Ed., Vol. V, Princeton University Press, Princeton, 1958.
(3) Collatz, L., "Functional Analysis and Numerical Mathematics," p. 358, Academic Press, New York, 1966.
(4) Bilous, Amundson, N. R., *A.I.Ch.E. J.* (1955) **1,** 513.
(5) Stakgold, I., "Boundary Value Problems of Mathematical Physics," Vol. II, Macmillan, New York, 1968.

10

The Kinetics of Biosystems: A Review

ARTHUR E. HUMPHREY

School of Chemical Engineering, University of Pennsylvania,
Philadelphia, Pa. 19104

The literature dealing with kinetics of biosystems is quite extensive. It covers such topics as activities of cellular organisms including the kinetics of growth, product formation, and death. Most of these kinetics are derived from consideration of basic enzyme behavior models. Among the literature on enzyme kinetics can be found an emergence of research dealing with immobilized enzyme systems. Some of the kinetic considerations involve highly complex and interactive systems. This is particularly true with sequential enzyme reactions and mixed population systems. Often these complex systems exhibit oscillatory behavior in the steady state. Of necessity, the several disciplines of biology, biochemistry, and chemical engineering will need to interact to produce useful models of biosystems behavior.

The literature dealing with the kinetics of biosystems is extremely extensive (*1, 2 3*). Since no single review could possibly do justice to the whole field, this review is necessarily limited in its consideration. Only those systems deemed of greatest immediate concern to chemists and engineers—*i.e.*, enzymes and single cellular activities—are considered. Special consideration is given to the oscillatory systems because they represent an area of intense kinetic interest at present.

Cellular Activities

This section on cellular activities is limited to kinetics of growth, product formation, and death. For unicellular cells the growth rate can be expressed in terms of the cell concentration, X, concentration of a growth limiting substrate, S (note: there are occasions when more than a single substrate can be limiting growth), and concentration of an inhibitor or a predator, *I*—*i.e.*,

$$\dot{X} = f(X,S,I) \tag{1}$$

Several expressions for Equation 1 can be found in the literature. For many situations the variables *X*, *S*, and *I* are highly coupled. Hence, expressions for cell growth rate are usually highly nonlinear.

For single cellular organisms undergoing binary fission, the growth can be expressed in terms of a growth rate constant, μ, and the cell concentration.

$$\dot{X} = \mu X \tag{2}$$

This is the so-called exponential growth. Other growth models have been proposed (*4, 5, 6, 7*). These involve

linear model

square model

$\frac{2}{3}$ *rds* model

cubic model

Each model applies to a specific physical situation. For example, linear growth (X = constant) occurs in some hydrocarbon growth cultures where limitation is caused by the rate of diffusion of substrate from essentially constant surface area oil droplets. In filamentous organisms where growth occurs from the tip, but nutrients diffuse throughout the filamentous cell mass, the 2/3 *rds* model fits some data.

Expressions for μ. One commonly used expression for the specific growth rate is the so called Monod equation:

$$\mu = \frac{\mu_{max} S}{K_s + S} \tag{3}$$

In reading the original paper by Monod (*8*) one finds that he attached much less significance to his model than subsequent workers. He stated that the character of the data for cell growth on a single limiting substrate suggested that it might be fitted by an equation of a form similar to that for simple enzyme kinetics. We now know that such an expression is a gross oversimplification of the true kinetics (*9, 10*). There are problems of inhibition which include substrate inhibition at high concentration, product inhibition, and even enzyme poisoning. Then there are reaction lags arising from the operation of repression and induction control mechanisms. Finally, there are problems of multisubstrate control, competition from other species for the substrate, commensalism, amensalism, and predation.

Figure 1 is a typical representation of the growth–substrate relationship. There are three regions where different models have been used to fit the data:

Region I. This is the region where $S << K_s$ and where essentially a linear relationship exists between the growth rate and substrate concentration. In many biological waste treatment processes, such as activated sludge, this behavior is approximated (*11*).

Region II. Here the single substrate limitation of the Monod-type applies best.

Region III. This is a region of high substrate concentration where the maximum growth rate should be achieved but where inhibition arising from either metabolic products of growth or substrate limits the rate.

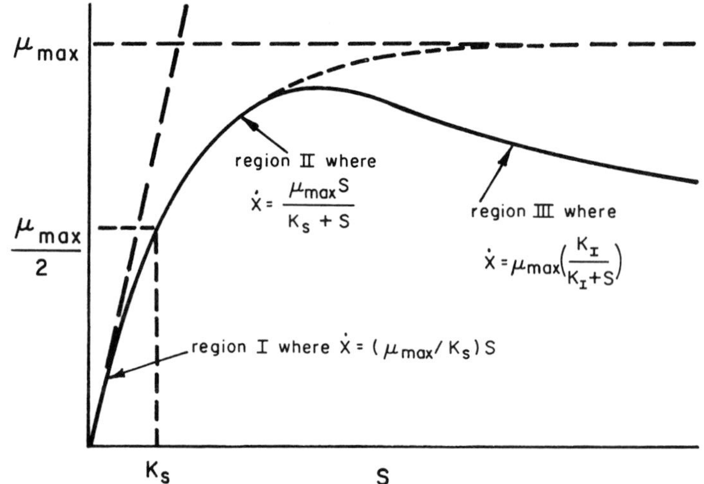

Figure 1. Effect of substrate concentration on growth

When no single substrate is limiting, the kinetics are much more complicated. Figure 2 illustrates the difference between single and multiple substrate limitation in a batch culture system. Several workers (*11*, *12*) have proposed a multiple substrate limitation model analogous to the nonobligatory independent binding of the two substrates enzyme model —*i.e.*,

$$\mu = \frac{\mu_{max_1} S_1}{K_{s1} + S_1} \cdot \frac{\mu_{max_2} S_2}{K_{s2} + S_2} \tag{4}$$

$$= \frac{\mu_{max_{12}}}{\left(1 + \frac{K_{s1}}{S_1}\right)\left(1 + \frac{K_{s2}}{S_2}\right)}$$

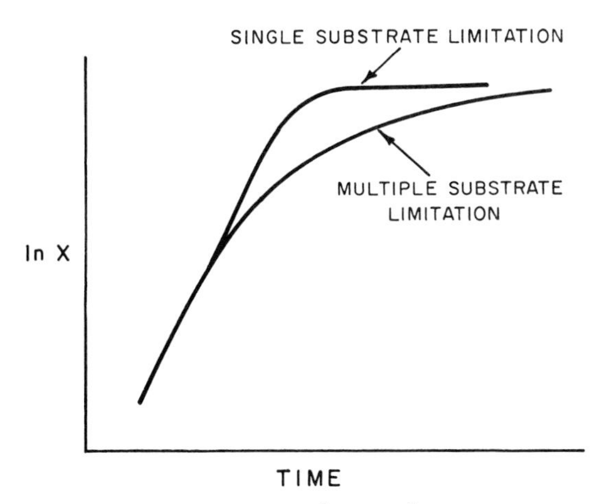

Figure 2. Batch growth curves

Unfortunately, most situations are not as simple as that depicted in Equation 4. Often one substrate is a repressor of the enzyme system for the other substrate. When this occurs, it is possible to use fermentation to separate or purify a system of mixed substrates.

Coupling between X and S (Yield). Monod has suggested (*13*) that at high growth rates and moderate substrate concentrations the coupling between cells and substrate can be expressed in terms of a yield, Y—*i.e.,*

$$\dot{X} = - Y\dot{S} \tag{5}$$

However, certain substrates, primarily those utilized for energy, are involved in maintenance. A cell can use glucose for endogenous respiration without growth. At low concentration, therefore, the yield expression must include a factor to account for utilization of the limiting substrate for maintenance.

$$\dot{S} = - \frac{\dot{X}}{Y} - rX \tag{6}$$

In this relationship Y is a yield constant in terms of the grams of substrate utilized per gram of cells produced, assuming no maintenance requirement, and r is the grams of substrate utilized for maintenance per gram of viable cells.

Structured Growth Models. A number of workers have suggested that growth models should be structured to include age distribution data (*14, 15, 16, 17*). The concept of age is most difficult to visualize with cells undergoing binary fission except for the idea of age in terms of the time since division. On the other hand yeast, which bud and scar when the bud breaks away from the mother cell, can be identified in terms of age by counting the number of bud scars. Table I compares some scar distribution data for yeast grown in continuous culture at various dilution rates with the ideal scar distribution. Ideal distribution would occur when the rate of budding is constant and independent of the number of scars.

The data tend to indicate that near ideal distributions are achieved at all growth rate conditions. Deviations from ideal distribution may be caused by a physical selection mechanism that operates on the exit stream. There is very little encouragement from these results to include an age distribution parameter in cell culture models. However, Herbert and others (*19, 20, 21*) (*see* Figure 3) have shown that there are distinct compositional changes in cells depending upon the environment and the rate at which they are grown. This has led a number of workers (*7, 11, 14, 15, 16*) to evolve models for cellular activities that are "structured." Various cell components have been used to represent different cellular capabilities. For example, RNA concentration is taken as indicative of the cell's capability for enzyme synthesis. Protein concentration is taken to indicate the cellular enzyme concentration. Such models undoubtedly will prove useful in correlating cellular activities with productivities.

Mixed Culture. Mixed culture systems are receiving considerable attention these days because of problems in biological waste control where naturally occurring mixed culture systems are utilized (*11*). Also, in the behavior of estuary microflora, prey–predation systems are at work.

The prey–predator system can be described by the following set of equations

$$\dot{S} = \frac{\mu_1 X_1}{Y_1} + S_o D_o - SD \quad (7)$$

$$\dot{X}_1 = \mu_1 X_1 - X_1 D - \frac{\mu_2 X_2}{Y_2} \quad (8)$$

$$\dot{X}_2 = \mu_2 X_2 - D X_2 \quad (9)$$

$$\mu_1 = f(S) \quad (10)$$

$$\mu_2 = f(X_1) \quad (11)$$

where S is the concentration of growth limiting substrate for the prey population, X_1 is the prey concentration, μ_1 is the specific growth rate of the prey, Y_1 is the yield of prey per unit of substrate consumed, X_2 is the predator concentration, μ_2 is the specific growth rate of the predator, Y_2 is the yield of predator per unit of prey consumed, D_o is the entering flow rate per unit system volume, and D is the leaving flow rate per unit system volume.

Table I. Scar Distribution Data for a Growing System (*18*)

		Dilution Rate, hr^{-1}		
Number of Scars	*Ideal*	*0.11*	*0.2*	*0.3*
0	50	57.9	59.1	56.5
1	25	17.3	18.2	18.8
2	12.5	10.1	9.1	11.3
3	6.25	6.1	5.5	5.0
4	3.125	3.6	3.1	3.1
>4	3.125	4.7	5.0	4.1

When the prey growth rate is directly proportional to the substrate concentration and the predator growth rate is directly proportional to the prey concentration—*i.e.*,

$$\mu_1 = a(S) \tag{10a}$$

$$\mu_2 = b(X_1) \tag{11a}$$

and a and b are constants, then Equations 7 through 9 yield the classical prey–predator model of Lotka-Volterra. This model can, under certain conditions, exhibit undamped oscillating behavior (*14, 22, 23*). Such behavior has been demonstrated experimentally by several investigators (*12, 23, 24, 25*). One such system involves *Colpoda stenii* feeding on *E. coli* (*23*) which is growing on a growth-limiting carbon substrate.

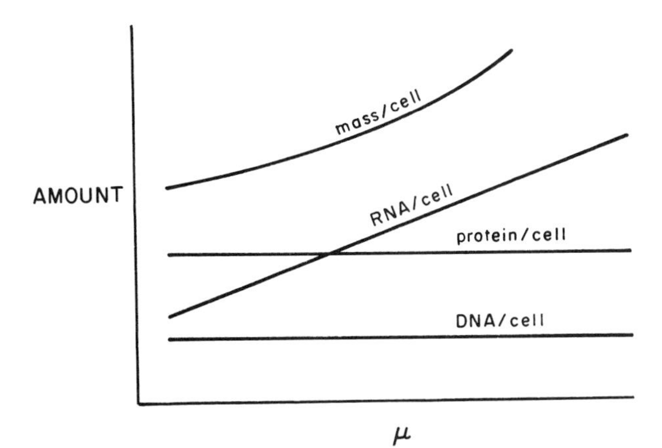

Figure 3. Compositional changes in cells continuously cultured on an energy-limiting substrate

The general set of equations for mixed culture growth involve:

$$\dot{S} = \frac{\mu_1 X_1}{Y_1} + S_o D_o - SD \tag{7}$$

$$X_1 = \mu_1 X_1 - X_1 D \tag{8a}$$

$$X_2 = \mu_2 X_2 - X_2 D \tag{9a}$$

$$\vdots \qquad \vdots \qquad \vdots$$

$$X_n = \mu_n X_n - X_n D$$

and

$$\mu_n = f(S_1, X_1 ---- X_{n-1}, I) \tag{12}$$

The growth rate μ_n, of any species, *n*, could be enhanced, inhibited, or not affected by any of the other species. Depending upon the particular model used to express the growth rate—*i.e.*, the form of Equation 12—it is possible to have competition, mutualism, commensalism, or amensalism.

Product Formation. Many models have been proposed for the production of various useful metabolites utilizing cellular activities. Three of these models have been reasonably useful in expressing the behavior of simple systems. These are the (1) growth associated model, (2) non-growth associated model, and (3) combined model.

When a substrate is converted stoichiometrically to a single product, *P*, one may relate the rate of product formation to the rate of growth by a simple constant, α, analogous to a yield constant—*i.e.*,

$$\dot{P} = \alpha \dot{X} \tag{13}$$

This is the so called growth associated model.

For cases where the rate of product formation depends only on cell concentration—*i.e.*, the cell has a constitutive enzyme system that controls the product formation rate—then

$$\dot{P} = \beta X \tag{14}$$

β is a proportionality constant, similar to the activity expression for an enzyme. It can be thought of as representing so many units of product-forming activity per mass of cells.

In 1959 Luedeking (*26*), proposed that Equations 13 and 14 could be combined to express the productivity of organic acid fermentations (*26*).

$$\dot{P} = \alpha \dot{X} + \beta X \tag{15}$$

or

$$\mu_p \equiv \dot{P}/X = \alpha \mu_x + \beta \tag{15a}$$

where μ_p is the specific product formation rate, and μ_x is the specific cell growth rate. Luedeking and others have successfully applied this model to both batch and continuous organic acid fermentations (*10, 26, 27, 28, 29, 30*). Figure 4 illustrates the behavior of the three models.

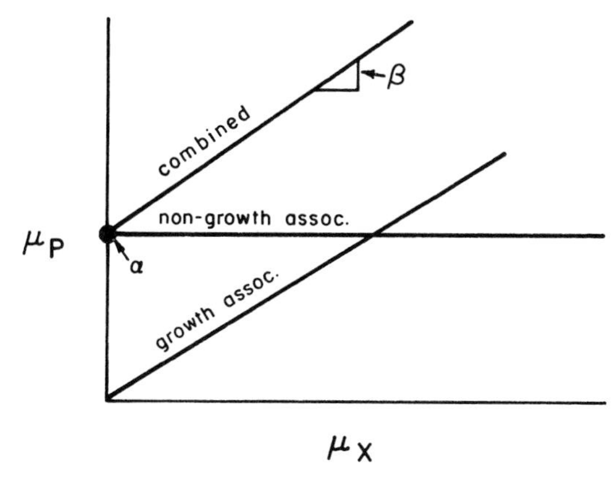

Figure 4. Comparison of product formation models

A more complicated use of the combined model has been made by Kono and Asai (*29, 30*), who treated the constants α and β as variables, having discrete and particular values for each of four phases of batch growth. These were lag, exponential, constant, and declining growth phases. Terui (*10*) has developed a series of kinetic models for enzyme production by microorganisms. These models consider such factors as preferential synthesis, salvage synthesis, and repression-derepression relationships. Perhaps the most elegant models, from a mathematical point of view, are those from the work of Ramkrishna, Fredrickson, and Tsuchiya (*16*). They evolved a number of models that consider such factors as cell age, cell structure, and function of the various cellular components. However, it is difficult to evaluate the constants in the models. Industrial experience with several antibiotic fermentation systems indicates that the most useful models will be those that involve environmental factors such as temperature, pH, dissolved oxygen, concentrations of inhibitors, repressors, and inducers, etc. because these can be used in computer controlled feedback loops on the process.

Kinetics of Death. Both the deterministic and probabilistic approach to modeling the death of microorganism can be found in the literature (*7, 31, 32*). Prokop and Humphrey (*32*) have written an extensive review on the various models. While the probabilistic model is certainly more exact, the uncertainties in the kinetic parameters of death for most microbial cultures, particularly those for mixed populations, do not warrant the extra complexity of the probabilistic models. For most sterilization practice design, the deterministic model is a good approximation.

For vegetative cell systems, the death kinetics can be expressed as

$$V \xrightarrow{k_1} D \tag{16}$$

or

$$-\dot{V} = +\dot{D} = k_1 V \tag{17}$$

where V is the viable cell concentration, D is the dead cell concentration, and k_1 is the specific death rate constant. This is the so-called exponential death relationship. It applies to most vegatative cell death regardless of whether the death is caused by moist heat, dry heat, electromagnetic radiation, or chemical disinfectants.

Organisms that form spores present a different death behavior pattern. Spores often die in a non-exponential fashion. Different investigators (*32, 33*) have suggested that both a resistant state, R, and an intermediate sensitive state, S, are involved. Several of these models are listed below.

Humphrey Model (*32*):

$$R \xrightarrow{\lambda_r} S \xrightarrow{k_1} D,\ \lambda_r > k_1 \tag{18}$$

both $R + S$ are viable, $S_o = 0$

$$\frac{R+S}{R_o} = \frac{\lambda_r}{\lambda_r - k_1}\left(\exp\left(-k_1 t\right) - \frac{k_1}{\lambda_r}\exp\left(-\lambda_r t\right)\right) \tag{19}$$

Tervi Model (*33*);

$$R \xrightarrow{\lambda_r} S,\quad R \xrightarrow{k_2} D,\quad S \xrightarrow{k_1} D,\quad k_1 > k_2 \tag{20}$$

both states $R + S$ are viable

$$\frac{S_o}{R_o + S_o} \equiv 1 - \gamma_r \equiv \gamma_s$$

$$K_2 \equiv \frac{\gamma_r(k_1 - k_2)}{k_2 + \lambda_r - k_1}$$

$$\frac{R+S}{R_o + S_o} = K_2 \exp\left[-\left(\lambda_r - k_2\right)t\right] + \left(1 - K_2\right)\exp\left(-k_1 t\right) \tag{21}$$

Brannen model (34):

$$N \underset{k_{-1}}{\overset{k_1}{\rightleftarrows}} V \overset{k_2}{\longrightarrow} D,\ k_1 > k_{-1} \tag{22}$$

V = viable + reproducing cells

N = viable but non-reproducing cells

$$\frac{V}{V_o} = \frac{1}{k_1 + k_{-1}} \exp\,(-k_2 t)\,\{k_1 \exp\,[(-k_1 + k_{-1})t] - k_{-1}\} \tag{23}$$

Conditions have been reported where each of these models appears to fit. Perhaps spores are so constituted that various environmental conditions can produce situations represented by each model.

Enzyme Systems

Reacting enzyme systems can involve enzymes in several forms. These are

(1) Coupled and sequential reactions such as one finds in living system.

(2) Purified enzymes in a homogeneous state. An example would be an enzyme such as cellulose held within a staged ultrafiltration reactor and catalyzing the hydrolysis of starch.

(3) Enzymes adsorbed on inert materials and used in repeated batch or column operation.

(4) Entrapped enzymes which are held within a polymer matrix such as an acrylamide gel.

(5) Insolubilized enzymes which are chemically bound to an inert porous matrix such as glass.

The kinetic behavior of the enzyme systems are similar to many of the non-biologically catalyzed chemical systems. Both homogeneous and heterogeneous systems occur (*2, 22, 35–43*).

Simple Enzyme Kinetics. A simple kinetic model for enzyme–substrate interaction that has been particular useful is that proposed by Michaelis-Menten (*1*). In this model the enzyme (E) reversibly combines with substrate (S) forming an enzyme-substrate complex (ES) which irreversibly decomposes to form product (P) and free enzyme—*i.e.*,

$$E + S \underset{k_{-s}}{\overset{k_{+s}}{\rightleftarrows}} ES \overset{k_{+p}}{\longrightarrow} E + P \tag{24}$$

If the substrate concentration is much greater than the enzyme concentration—*i.e.*, $S_o >> E_o$—and if the decomposition of the enzyme-substrate complex is the rate-limiting step—*i.e.*, $k_{+s} >> K_{+p}$—then

$$\dot{P} \doteq \dot{S} = k_{+p}\,(\mathrm{ES}) \tag{25}$$

or

$$\dot{P} = \frac{k_{+p}E_o(S)}{\dfrac{k_{-s} + k_{+p}}{k_{+s}} + (S)} = \frac{k_{+p}E_o(S)}{K_M + (S)} \tag{26}$$

where $\dot{P}$ is the velocity of the reaction, and $k_{+p}\ E_o$ is the maximum possible velocity of reaction. The reaction velocity is related linearly to the enzyme concentration.

Usually $k_{+p} << k_{-s}$, then

$$K_M \equiv \frac{k_{+p} + k_{-s}}{k_{+s}} \doteq \frac{1}{K_{eq}} = \frac{k_{-s}}{k_{+s}} \tag{27}$$

This relation between substrate and reaction velocity is shown in Figure 5 and is mathematically expressed as

$$v = \frac{v_{max}(S)}{K_M + (S)} \tag{28}$$

where v represents the reaction velocity and v_{max} represents the maximum reaction velocity.

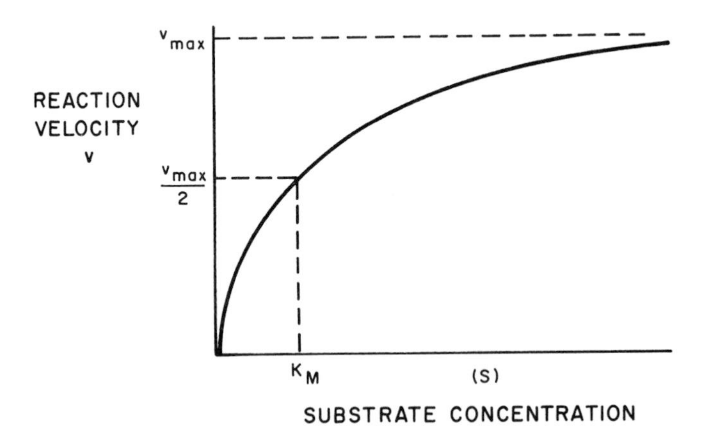

Figure 5. Simple Michaelis-Menten kinetics

Magnitude of K_M for Enzymes. In general K_M for the respiratory enzymes, those associate with sugar metabolism, is lower than K_M for the hydrolytic enzymes, those associate with primary substrate attack. The table below gives some typical data for a variety of enzymes.

Table II. K_M for Various Enzymes

Enzyme	*Substrate*	*K_M, Molarity*
Maltase	maltose	2.1×10^{-1}
Sucrase	sucrose	2.8×10^{-2}
Phosphatase	glycerophosphate	3.0×10^{-3}
Lactic dehydrogenase	pyruvate	3.5×10^{-5}

Complex Systems. Often, even an apparent one-step enzyme system does not follow Michaelis-Menten behavior. In certain systems such as the enzymatic oxidation of glucose to δ-gluconolactone, a regeneration of the enzyme step is involved (*45*).

$$E_0 + S \underset{k_{-1}}{\overset{k_1}{\rightleftarrows}} E_oS \overset{k_2}{\longrightarrow} E_r + \delta\text{-lactone} \tag{29}$$

$$E_r + O_2 \overset{k_4}{\longrightarrow} E_o + H_2O_2 \tag{30}$$

where E_o is the oxidized form of enzyme and E_r is the reduced form. Stop-flow experiments have shown that at pH = 5.6 and 25°C the observed apparent rate constant, k_{app}, for the appearance of lactone, could be represented by the following model (*45*).

$$\frac{1}{k_{app}} = \frac{1}{k_2} + \frac{1}{k_4[O_2]} + \frac{1}{\dfrac{k_1 k_2[G]}{k_{-1} + k_2}} \tag{31}$$

where $[O_2]$ is the dissolved O_2 concentration in the reacting system and [G] is the glucose concentration.

Enzyme Inhibition. Two common kinds of inhibition that occur in enzyme systems are (1) competitive or substrate analog inhibition and (2) reversible noncompetitive inhibition.

The competitive inhibition can be depicted by

$$E \underset{\pm S}{\overset{\pm I}{\rightleftarrows}} \begin{matrix} EI \text{ (inactive)} \\ ES \longrightarrow E + P \end{matrix} \tag{32}$$

and

$$EI \underset{k_{-I}}{\overset{k_{+I}}{\rightleftarrows}} E + I \tag{33}$$

where

$$K_I = \frac{k_{+I}}{k_{-I}} \tag{34}$$

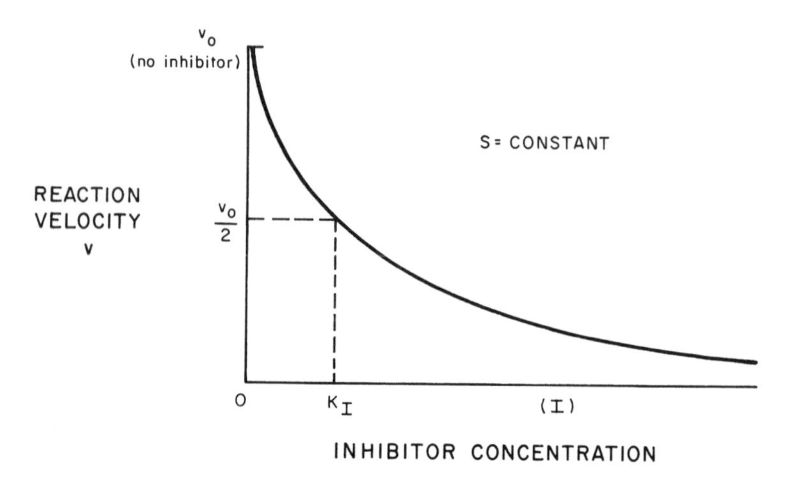

Figure 6. Effect of inhibitor on enzyme kinetics

This model leads to a reaction rate expression of the form

$$v = \frac{v_{\max}(S)}{K_M + (S) + \frac{K_M}{K_I}(\mathrm{I})} \tag{35}$$

This type of inhibition is frequently found in reactions where the product inhibits the enzyme. An example is glucose which is a competitive inhibitor of the action of invertase on sucrose.

The second kind of inhibition commonly encountered in enzyme systems is the reversible noncompetitive type. This can be depicted by

$$\begin{array}{ccccc} \pm I & & EI & & \pm S \\ & \nearrow\swarrow & & \searrow\nwarrow & \\ E & & & & EIS \longrightarrow EI + S \\ & \searrow\nwarrow & & \nearrow\swarrow & \\ \pm S & & ES & & \pm I \\ & & \downarrow & & \\ & & E+P & & \end{array} \tag{36}$$

This model leads to a reaction rate expression of the form

$$v = \frac{v_{\max}(S)}{K_M + (S)}\left[\frac{K_I}{K_I + (\mathrm{I})}\right] \tag{37}$$

The effect of inhibitor on the enzyme kinetics is shown in Figure 6. This kind of inhibition is typical of the product inhibition effect of organic acids such as acetate, propionate, and lactate on the hydrolytic enzymes.

Effect of Temperature and pH. Enzyme systems exhibit both temperature and pH optimal. The reason for the temperature optimum is that above a certain temperature the enzyme begins to exhibit appreciable rate of inactivation. One advantage to insolubilized systems is that this temperature optimal can be shifted to higher temperatures owing to a more stable character of the enzyme (*42, 46*). Since enzymes are ampholytes—*i.e.*, they have both acidic and alkaline groups—it is not surprising that pH has a pronounced effect on enzyme activity (*47*). However, what is frequently surprising is the narrow pH range over which the enzyme is active and how widely the pH optimal varies between enzymes. Table III compares the pH optimum for several representative enzymes.

One of the many difficulties with modeling and optimizing enzyme systems is the interaction between the environmental variables such as temperature and pH. Temperature changes will, for example, shift the pH optimum of an enzyme reaction. This is illustrated by the pH activity curve of wheat β-amylase at various temperatures (*see* Figure 7).

Behavior of Crude Enzyme Preparation. When enzymes are used in a crude non-purified form or when they are used as dried whole cell preparations, peculiar kinetics are exhibited. An activity history typical for a crude glucose oxidase preparation is illustrated in Figure 8.

Table III. pH of Maximal Enzyme Activity (*44*)

Enzyme	*pH Optimum*
Pepsin	2
Glutamic acid decarboxylase	6
Salivary amylase	7
Pancreatic carboxypeptidase	8
Arginase	10

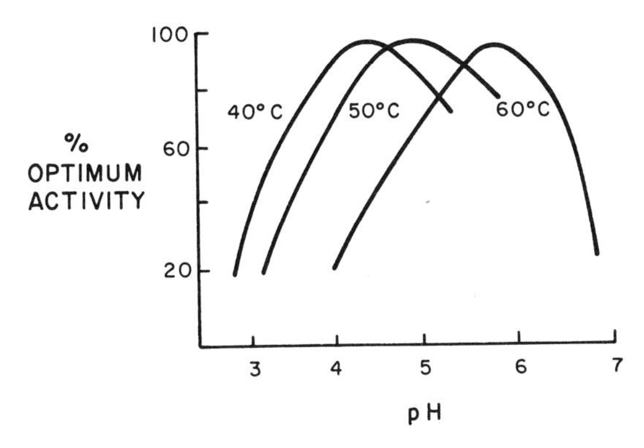

Figure 7. pH–activity curve for wheat β-amylase

This is characterized by three distinct periods: (1) an initial period in which induction of the activity is necessary, (2) a period of relatively constant activity, and (3) a final period of exponential decay. Why crude enzyme preparations have these peculiar activity histories is not always clear. In some instances metal poisons may be reacting with non-active site portions of the enzyme before they react with the active site. In other cases degradation of non-essential portions of the enzyme molecule may occur before essential portions.

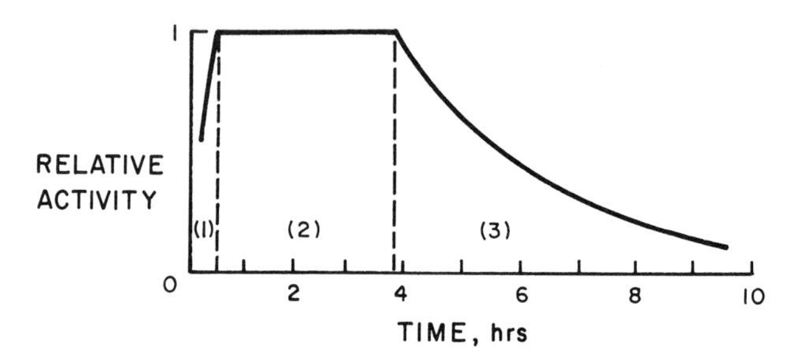

Figure 8. Activity history of a crude glucose oxidase preparation

Behavior of Adsorbed Enzyme Systems. In the case of enzymes adsorbed to an insoluble matrix, equilibrium is usually established between the adsorbed or attached enzyme and the free enzyme. When enzyme in this form is used in repeated batch processes where the solution is drawn off in each run, there is a constant fractional loss of enzyme per batch run. This loss can be expressed by

$$\Delta E \text{ per batch} = -k_B E \tag{38}$$

where E is the total enzyme concentration or activity in the batch reactor and k_B is the fractional enzyme loss per batch. This latter may be expressed by

$$k_B = \left[\frac{DV}{M} \Big/ \left(\frac{DV}{M} + 1\right)\right] w \tag{39}$$

where D is the equilibrium coefficient for the distribution of the enzyme between the inert matrix and solution phases, V is the amount of solution phase, M is the amount of inert matrix phase in the reactor, and w is the fractional amount of solution added or withdrawn in each repeated operation. This relation is depicted by Figure 9.

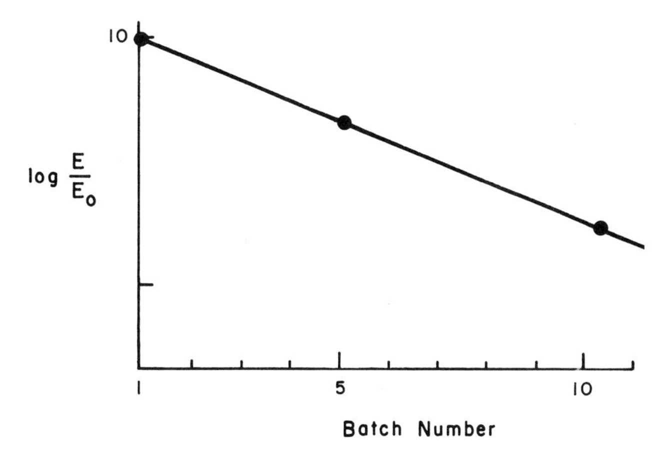

Figure 9. Repeated batch catalysis with an adsorbed enzyme system

The interesting feature of these adsorbed enzymes systems is their mixed function catalytic action. The free enzyme in solution is essentially homogeneous, while the catalytic action of enzyme adsorbed on the matrix is heterogeneous—*i.e.*, it can be pore diffusion limited, film diffusion limited, and/or reaction rate limited.

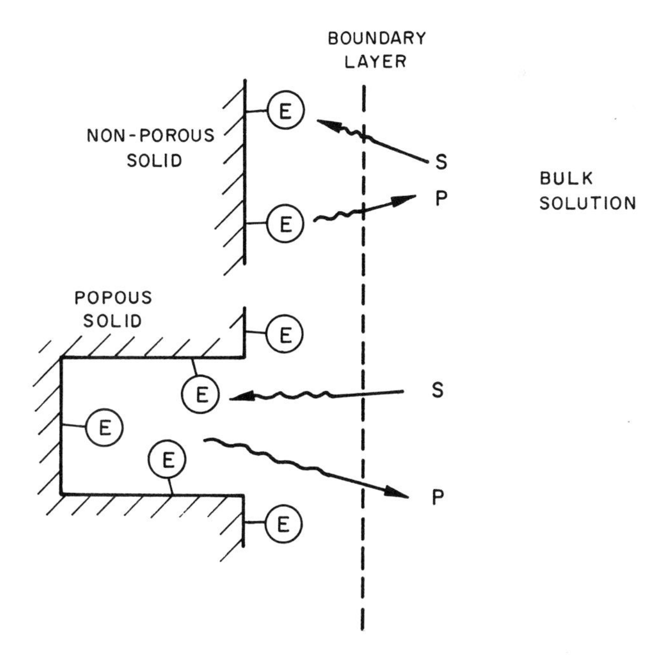

Figure 10. Insoluble enzyme systems

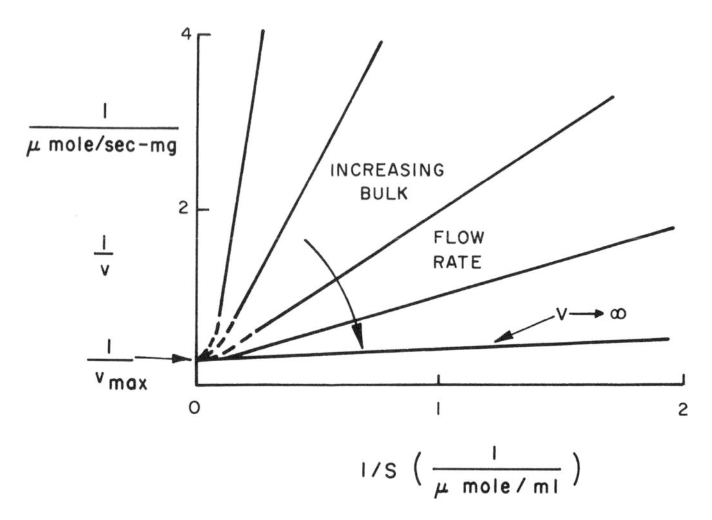

Figure 11. Non-porous solid bound enzyme kinetics (Lineweaver-Burk plot)

Behavior of Insoluble Enzymes. From a kinetic point of view insoluble enzyme systems have two important forms: (1) enzyme bound to the surface of a non-porous solid, and (2) enzyme bound within a porous solid. These forms are illustrated in Figure 10.

The case of the non-porous solid bound enzyme systems is approximated by systems in which the enzyme is bound on a membrane surface, in a very open and thin membrane, or on a smooth glass wall surface. The porous, solid bound enzyme systems are representative of those in which the enzyme is bound to porous (pore size = 500 to 1000 A) aminoalkylsilane glass beads through a diazotization reaction.

In the non-porous system the reaction rate can be controlled by one of three steps: (1) diffusion of the substrate to the enzymatically active surface from the bulk of solution, (2) enzymic reaction at the surface, or (3) diffusion of the reaction products back into the bulk of solution. Sharp and Lilly (42) have suggested that the approximate behavior of such system can be represented by

$$N_S = \frac{v_{\max} E_s S_B}{S_B + K_M + \frac{v_{\max} E_s}{D_S} \Delta X} \tag{40}$$

where N_S is the molar flux of substrate at the boundary layer, $v_{\max}$ = maximum specific activity of enzyme, E_s is the surface unit concentration of enzyme, S_B is the bulk concentration of substrate, ΔX is the diffusional boundary layer thickness and D_S is the substrate diffusivity. This equation leads to results of a type shown in Figure 11, a Lineweaver-Burk plot.

The Lineweaver-Burk plot is used by all enzyme workers to obtain the kinetic constants for enzyme reactions. However, Sharp and Lilly concluded that linear extrapolation of a Lineweaver-Burk plot cannot be used with insolubilized enzyme systems to obtain directly the kinetic constants v_{max} and K_M since the plot depends on flow conditions.

For enzymes bound to a porous solid, a more complicated situation results. The reaction rate can be controlled not only by the reaction and diffusion in the boundary layer but also by the diffusion rate within the pore.

Explicit solutions for this problem cannot be obtained. Rather, as in typical heterogeneous catalysis, an effectiveness factor can be defined for the reaction in terms of the ratio of the actual reaction rate to the rate at which the reaction would proceed if there were no diffusional limitations. Difficulty arises from the fact that while the kinetic constants can be obtained for the free enzyme in solution, it is not clear just how these values change once the enzyme is bound. Preliminary experiments indicate that the binding step—*i.e.*, the k_s/k_{-s} ratio in Equation 24—is greatly affected. What is needed is experimentation with enzymes bound at non-porous surfaces and reacting in high velocity flows where the boundary layer is essentially zero. This raises an important question, however: will high shear fields damage or otherwise inhibit enzyme reactions and provide still another kinetic consideration in these systems?

Enzymatic Action on Large Polymeric Surfaces. An interesting category of enzyme reactions is that represented by attack of enzyme on large polymeric surfaces. An example is the attack of egg lysozyme on the surface of a gram positive microorganism causing the cell to lyse. In this case the enzyme is causing the hydrolysis of the bond between the 1-carbon and bridge-oxygen of such cell wall polyglucosides as poly-(*n*-acetyl glucosamine) and oligomers of *N*-acetylmuramic acid.

The reaction is inhibited by the products—*i.e.*, the small polymers and the monomer. The kinetics can be described by

$$v = \frac{v_{max}(S)}{K_M + (S)} \cdot \prod_i \left[\frac{K_{I_i}}{K_{I_i} + I_i}\right] \tag{51}$$

where (S) denotes the cell wall substrate, and I_i represents the various low molecular weight product inhibitors. Table IV illustrates how the reaction rate is affected by the various molecular weight products.

This appears to be an interesting kinetic problem. Optimization of this reaction will be important to the economic recovery of protoplasmic protein and intracellular enzymes from single cells.

Table IV. Specific Activity of Lysozyme

Substrate	*Relative Rate*	K_M, *Molarity*
N–acetylglucosamine (NAG)	0	5.0×10^{-2}
di–NAG	0.003	1.8×10^{-4}
tri–NAG	1	6.6×10^{-6}
tetra–NAG	8	9.5×10^{-6}
penta–NAG	4000	9.4×10^{-6}
hexa–NAG	30,000	6.2×10^{-6}

Oscillatory (Dynamic) Systems

Oscillatory behavior in biological systems is quite common. It can be caused in cellular populations by combinations of competition and commensalism or amensalism or by prey-predator situations (*12, 23, 24, 25*). In single cellular systems it can also be caused by feed back or feed forward control of key enzyme systems (*9, 48*). Higgins has presented a very detailed theoretical discussion of the subject (*22*).

One such oscillatory behavior is the genetic-metabolic control model by Jacob and Monod (*20*)

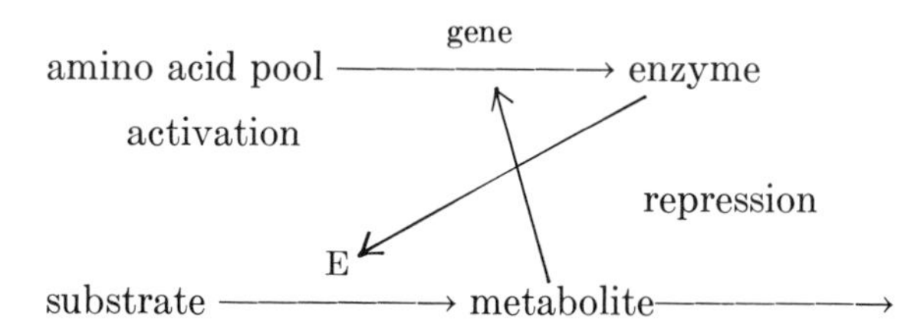

However, this control can act directly without gene regulation by back activation (Equation 43) or forward inhibition (Equation 44) of an enzyme in the reaction train.

$$S \longrightarrow A_1 \xrightarrow{\text{activation}} A_2 \longrightarrow \qquad (43)$$

$$S \longrightarrow A_1 \xrightarrow{\text{inhibition}} A_2 \longrightarrow \qquad (44)$$

The Lotka-Voltera prey-predator model could be depicted as a back activation of hosts (H) on substrate (S) and of predator (P) on host.

$$S \xrightarrow{\text{activation}} H \xrightarrow{\text{activation}} P \longrightarrow \text{dies} \qquad (45)$$

Single cell systems can also exhibit oscillatory behavior. Pye (*9, 48*) has shown by the use of a fluorescence trace of $DPNH_2$ in a buffered yeast cell suspension, that upon feeding 100 mmoles of glucose and producing an anaerobic state, damped oscillation in the $DPNH_2$ content arise. He has suggested that glycolytic flux pulses probably arise from alternate activation and inactivation of phosphofructokinase.

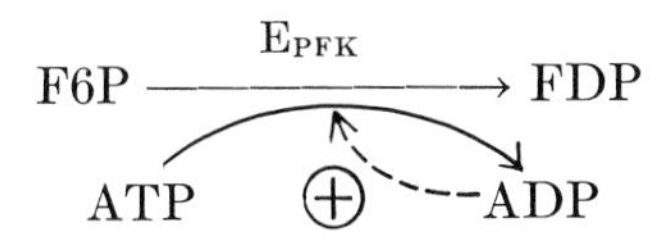

Detailed studies of these oscillatory biological systems certainly present a challenge if one is to understand control systems that operate within cellular systems and if population dynamics of waste reactors and estuaries are to be predicted.

Summary

The kinetics of biological systems offer an array of systems and mechanisms at least as vast in their diversity and probably more complex than strictly chemical systems. As an area of study it presents a real challenge to the chemist and chemical engineer.

Literature Cited

(1) Bernhard, S., "The Structure and Function of Enzymes," Benjamin, New York, 1968.
(2) Gutfreund, H., "An Introduction to the Study of Enzymes," Wiley, New York, 1966.
(3) Solomons, G. L., *Appl. Chem.* (1967) **52,** 505.
(4) Erickson, L. E., Humphrey, A. E., Prokop, A., *Biotechnol. Bioeng.* (1969) **11,** 449.
(5) Erickson, L. E., Humphrey, A. E., *Biotechnol. Bioeng.* (1969) **11,** 467.
(6) Erickson, L. E., Humphrey, A. E., *Biotechnol. Bioeng.* (1969) **11,** 489.
(7) Fredrickson, A. G., *Biotechnol. Bioeng.* (1966) **8,** 167.
(8) Monod, J., *Ann. Rev. Microbiol.* (1949) **3,** 371.
(9) Pye, K., *Stud. Biophys. (Berlin)* (1966) **2,** 75.
(10) Terui, G., Okazaki, M., Kinoshita, S., *J. Ferment. Technol.* (1967) **45,** 497.
(11) Anderson, J. S., Washington, D. R., *Chem. Eng. Progr. Symp. Ser.* (1966) **62,** 60.
(12) Bungay, H. R., *Chem. Eng. Progr. Symp. Ser.* (1968) **64,** 10.
(13) Monod, J., *Ann. Inst. Pasteur* (1966) **79,** 390.
(14) Eakman, J. M., Fredrickson, A. G., Tsuchiya, H. M., *Chem. Eng. Progr. Symp. Ser.* (1966) **62,** 37.
(15) Fredrickson, A. G., Tsuchiya, H. M., *AIChE J.* (1968) **9,** 459.
(16) Ramkrishna, D., Fredrickson, A. G., Tsuchiya, H. M., *Biotechnol. Bioeng.* (1967) **9,** 129.
(17) Shu, P., *J. Biochem. Microbial. Technol. Eng.* (1961) **3,** 95.

(18) Beran, K., Malek, I., Streiblova, E., Lieblova, J., in "Microbial Physiology and Continuous Culture," C. G. T. Evans, R. E. Strange, D. W. Tempest, Eds., p. 57, Her Majesty's Stationary Office, London, 1967.
(19) Herbert, D., in "Microbial Reaction to Environment," G. G. Meynell, H. Gooder, Eds., University Press, Cambridge, 1961.
(20) Jacob, F., Monod, J., *J. Mol. Biol.* (1961) **3**, 318.
(21) Powell, E. O., *J. Gen. Microbiol.* (1956) **15**, 492.
(22) Higgins, J., *Ind. Eng. Chem.* (1969) **59**, (5) 19.
(23) Proper, A., Garver, J., *Biotechnol. Bioeng.* (1966) **8**, 287.
(24) Bungay, H. R., Bungay, S. A., *Advan. Appl. Microbiol.* (1968) **10**, 269.
(25) Bungay, H. R., Krieg, N. R., *Chem. Eng. Progr. Symp. Ser.* (1966) **62**, 68.
(26) Leudeking, R., Piret, E. L., *J. Biochem. Microbiol. Technol. Eng.* (1959) **1**, 393.
(27) Aiyar, A. S., Leudeking, R., *Chem. Eng. Progr. Symp. Ser.* (1966) **62**, 55.
(28) Koga, S., Burg, C., Humphrey, A. E., *Appl. Microbiol.* (1967) **15**, 493.
(29) Kono, T., Asai, T., *Biotechnol. Bioeng.* (1969) **11**, 19.
(30) Kono, T., Asai, T., *Biotechnol. Bioeng.* (1969) **11**, 293.
(31) Aiba, S., Toda, K., *J. Ferment. Technol.* (1966) **44**, 301.
(32) Prokop, A., Humphrey, A. E., in "Kinetics of Disinfection," M. A. Benarde, Ed., Chap. 3, Academic, 1970.
(33) Komenushi, S., Terui, G., *J. Ferment. Technol.* (1967) **45**, 764.
(34) Brannen, J. P., *Math. Biosci.* (1968) **2**, 165.
(35) Butterworth, T. A., Wang, D. I. C., Sinskey, A. J., "Abstracts of Papers," 158th National Meeting, ACS, Sept. 1969, MICR 005.
(36) Ghose, T. R., Kostick, J. A., "Abstracts of Papers," 158th National Meeting, ACS, Sept. 1969, MICR 004.
(37) Goldstein, L., Levin, Y., Katchalski, E., *Biochemistry* (1964) **3**, 1913.
(38) Kay, G., *Process Biochem.* (1968) **3**, (8) 36.
(39) Neilands, J. B., Stumpf, P. K., "Outlines of Enzyme Chemistry," Wiley, New York, 1966.
(40) Self, D. A., Kay, G., Lilly, M. D., *Biotechnol. Bioeng.* (1969) **11**, 337.
(41) Sharp, A. K., Kay, G., Lilly, M. D., *Biotechnol. Bioeng.* (1969) **11**, 363.
(42) Sharp, A. K., Lilly, M. D., *Chem. Eng. (London)* (1968) **12**, Jan./Feb.
(43) Wilson, R. J. H., Lilly, M. D., *Biotechnol. Bioeng.* (1969) **11**, 349.
(44) Reed, G., "Enzymes in Food Processing," Academic, New York, 1966.
(45) Bright, H. J., Appleby, M., *J. Biol. Chem.* (1969) **244**, (13) 3625.
(46) Silman, I. H., Katchalski, E., *Ann. Rev. Biochem.* (1966) **35**, 873.
(47) Phillips, D. C., "The Three-dimensional Structure of an Enzyme Model," *Sci. Am.* (Nov. 1966) 78.
(48) Pye, K., *Can. J. Bot.* (1969) **47**, 271.

RECEIVED June 12, 1970.

Contributed Papers

The Generation of Continuous Oscillations in Biochemical Reaction Sequences

E. KENDALL PYE, Biochemistry Department, University of Pennsylvania Medical School, Philadelphia, Pa. 19104

In the living cell primary metabolites are converted to final end products by a sequence of relatively simple reactions in which the product of one reaction is the substrate for the next reaction in the sequence. The behavior of such a reaction sequence and the mechanism by which it is controlled is, of course, determined by the characteristics of the individual reactions involved. Without exception all important reactions in metabolism are mediated by enzymes, and for this reason their kinetic characteristics are very different from simple non-catalyzed reactions. Most enzymatic reactions display simple Michaelis-Menten type kinetics in which the velocity of the reaction, at high substrate concentrations, is asymptotic to a certain maximal velocity. Other enzymatic reactions—those mediated by K-type allosteric enzymes—have a sigmoidal rather than a hyperbolic relationship between velocity and substrate concentration, indicating cooperativity in the binding of substrate molecules to the enzyme. The affinity of the enzyme for the substrate molecule, and therefore the velocity of the reaction, can be enhanced or decreased by the binding to the enzyme of small molecule "effectors," which are usually metabolites. Effectors can also control the activity of V-type allosteric enzymes, but in this case cooperativity of substrate binding is not observed, and the presence of effectors either increases or decreases the maximal velocity of the reaction.

Apart from the remarkable control properties of allosteric enzymes the relative reversibility of enzymatic reactions plays an important role in the regulation of flux through reaction sequences. For readily reversible reactions the build-up of product will lower the net forward velocity of the reaction, but, on the other hand, the physiologically "irreversible" reactions are essentially unaffected by product accumulation. This feedback inhibition by mass action effect is an important feature of control within enzymatic reaction sequences and is the major effect determining that in the steady state the velocity through each individual reaction is the same as through the sequence as a whole.

Most reaction sequences in metabolism consist of many different types of individual enzymatic reactions ranging from the readily reversible to the practically irreversible and from the simple Michaelis-Menten type to the allosteric type. We have studied the glycolytic pathway in the yeast S. *carlsbergensis* to observe how a typical metabolic reaction sequence behaves when perturbed from a steady state and to determine its particular control characteristics. This reaction sequence, shown in part in Figure 1, converts glucose to pyruvate by a series of 10 enzymatic reactions which involve both allosteric and Michaelis-Menten type enzymes as well as highly reversible and "irreversible" reactions. Pyruvate can be converted further to ethanol *via* the intermediate acetaldehyde. Among the features of this pathway which makes it a convenient model system is that it occurs entirely within an apparently homogeneous phase within the cell (the cytoplasm), and the concentration of a cofactor in the sequence (DPNH) can be monitored continuously and non-destructively within the cell and in cell-free extracts by optical methods (both absorption and fluorescence).

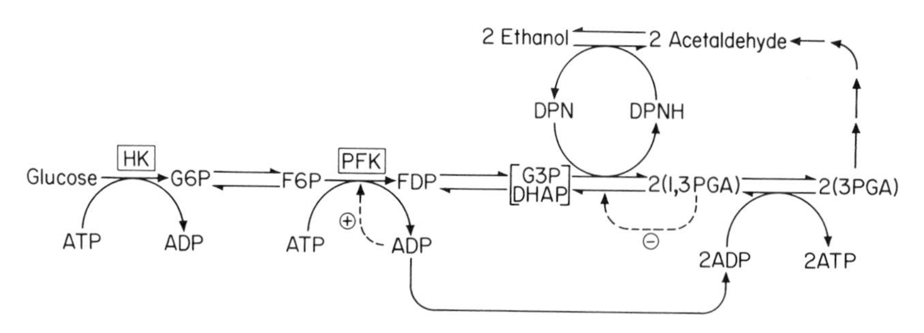

Figure 1. Major portion of the glycolytic reaction sequence in yeast. Hexokinase (HK), phosphofructokinase (PFK), and pyruvate kinase (not shown) catalyze essentially irreversible reactions. Most of the other reactions are readily reversible. PFK is allosterically activated by its product ADP.

When intracellular DPNH is monitored by fluorescence in a suspension of yeast cells in aerobic buffer (Figure 2), the DPNH concentration is low. (Fluorescence in this case measures the relative changes in DPNH concentration rather than absolute values). When glucose (100 m*M*) is added to the cells, there is a rapid increase in the DPNH concentration, then a slight fall, followed by an approximate steady state.

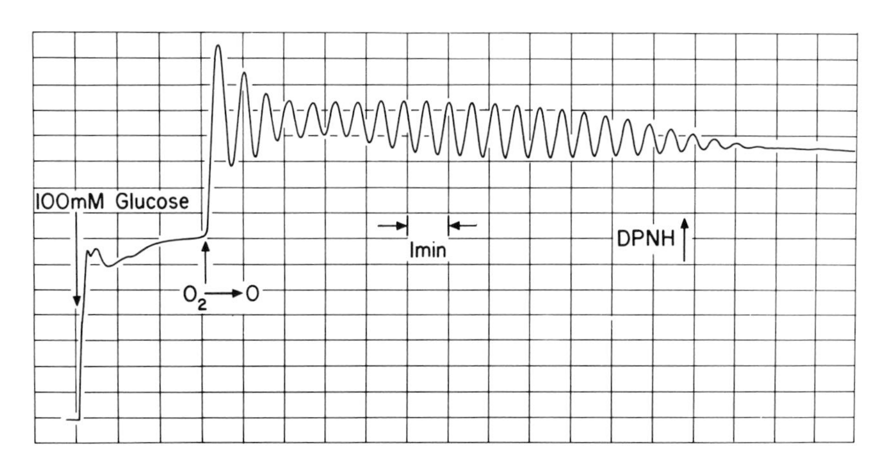

Figure 2. Record of the DPNH concentration, monitored by fluorescence, in a suspension of yeast cells, in buffer, at 25°C. Time proceeds from left to right. The trace only shows relative changes in DPNH concentration (see *text).*

The sharp increase observed on adding glucose is caused by an increased flux through the glycolytic sequence giving rise to an increased rate of DPNH production. The slight fall possibly represents the stimulation of the alcohol dehydrogenase reaction, which is one of the reactions utilizing DPNH, by the slightly delayed appearance of the second substrate of this reaction, acetaldehyde. The steady state in the DPNH concentration indicates a balance between the reactions generating DPNH and those reactions utilizing it.

This steady state of DPNH concentration is relatively short lived because the cells, which are respiring, eventually exhaust the oxygen from the surrounding medium and become anaerobic. The onset of anaerobiosis causes a marked increase in the flux through glycolysis (the Pasteur effect), and DPNH oxidation by the mitochondria ceases. Figure 2 shows that the onset of anaerobic conditions in the yeast suspension is accompanied by a sharp increase in the DPNH concentration which then proceeds to oscillate with a relatively stable period of approximately 35 sec. The train of cycles is sustained (24 cycles in all) and does not show continuous damping except over the last nine cycles. Two essentially independent pools of DPNH occur in yeast, cytoplasmic and mitochondrial, but considerable evidence indicates that it is the glycolytically associated cytoplasmic pool which is oscillating. Analysis has shown that all the intermediates in the glycolytic reaction sequence oscillate with the same frequency as the DPNH but not necessarily in phase with it, nor with the same relative amplitude. Assays of the major end products of anaerobic glucose metabolism (CO_2 and ethanol) demonstrate that the oscillations are accompanied by a pulsing of the glycolytic flux with the period of rapid product formation occurring during the top half of the DPNH cycle and the slow rate of product formation (approximately 20% of the fast rate) occurring during the bottom half of the cycle.

Theoretical studies of the behavior of enzymatic reaction sequences have shown that oscillations can occur when negative feedback loops exist in the sequence, but it has also been shown that sustained oscillations can only be generated in such a system when there is a significant number of intermediate reactions between the source and the site of the feedback and furthermore when there is a significant degree of cooperativity for the binding of the feedback substance. Examination of the phase relationships of the various oscillating intermediates indicates that this type of feedback system is not the mechanism for the glycolytic oscillations in yeast because the substrate fructose-6-phosphate (F6P), and the product, fructose-1,6-diphosphate (FDP) of the phosphofructokinase (PFK) mediated reaction appear to be approximately 180° out of phase—*i.e.*, F6P is maximal when FDP is close to minimal and vice-versa. This evidence indicates that the PFK reaction is the source of the oscillations which are generated by an alternate speeding up and slowing down of this reaction. Considerable evidence now indicates that the activity of the enzyme PFK is very much controlled by the level of the allosteric activator adenosine diphosphate (ADP) which is both an important component of energy metabolism and a product of the PFK reaction.

The mechanism which has been established experimentally for these glycolytic oscillations (*1*) and which was also predicted from analog computer studies (*2*) involves three essential features. These are (a) a product-activated reaction, (b) a relatively constant flux into this reaction which is lower than the velocity of the activated enzyme reaction but higher than the velocity of the deactivated enzyme reaction, and (c) an enzymatic reaction to remove the activator. For the glycolytic oscillations the product-activated reaction is the PFK reaction activated by ADP; the feed-in reaction is the formation of F6P, *via* G6P, from glucose which apparently enters the cell at the correct velocity; and the removal of the activator, ADP, is accomplished by the enzymes pyruvate kinase and 3-phosphoglyceric acid kinase for which ADP is a substrate.

Some interesting observations have arisen during this study. For example many different waveforms are possible, especially in cell-free extracts which oscillate at a lower frequency than the intact cells. These different waveforms which range from almost square-wave type to pulsed type and even to double-periodic waveforms (*3*) probably arise because of variations in the velocity relationships between the three essential parameters just mentioned.

Further, there appears to be good evidence for a strong metabolic synchronization between individual yeast cells in the suspension. This was demonstrated by mixing two equal populations of yeast cells which were oscillating 180° out of phase (*1*). If no metabolic synchronization occurred between the cells it was reasoned that the resultant oscillation from the combined population would be small or nonexistent. In reality after two cycles of lower amplitude the oscillations returned with full amplitude, indicating that all the cells were oscillating in synchrony.

A general conclusion from this study is that the control characteristics of a metabolic reaction sequence can be highly complex, especially when allosteric enzymes and feedback loops exist within that sequence. This complexity presumably provides living organisms with considerable flexibility in their metabolic response to various changes in environmental and physiological conditions.

(1) Pye, E. K., "Biochemical Mechanisms Underlying the Metabolic Oscillations in Yeast," *Can. J. Botany* (1969) **47,** 271.
(2) Higgins, J., "The Theory of Oscillating Reactions," *Ind. Eng. Chem.* (1967) **59,** 18.
(3) Chance, B., Pye, K., Higgins, J., "Waveform Generation by Enzymatic Oscillators," *IEEE Spectrum* (1967) **4,** 79.

Use of Glucose Oxidase in a Polyacrylamide Gel as a Fuel Cell Catalyst

N. L. NAGDA, C. C. LIU, L. B. WINGARD, JR., Department of Chemical Engineering, University of Pittsburgh, Pittsburgh, Pa. 15213

The development of suitable catalysts is a major problem in designing fuel cells. Oxidative enzymes are prime candidates, but the detailed methods of incorporating them into the electrode structure are critical design factors that remain to be evaluated. Since enzymes are water soluble and function essentially only in aqueous solution, an enzyme must be immobilized to keep it within the electrode assembly. Some of the resulting problems are electrode polarization caused by concentration gradients, possible changes in activity and kinetics of the immobilized *vs.* free enzyme, method for supplying enzyme cofactors, and the efficiency of electron transfer from enzyme to external circuit.

Table I. Measured Voltage at Constant Currents for Enzyme Electrode[a]

Constant dc Current, ma	*Glucose/Lactone Half-Cell Voltage, Reference Saturated Calomel, volt*
6	−0.74
4	−0.72
3	−0.70
2	−0.685
1	−0.66
0 (extrapolated)	−0.65

[a] Phosphate buffer-1*M*; pH-6.48; initial concentrations: glucose, 0.01*M*; lactone, 0.01*M*; temperature-37°C.

Glucose oxidase catalyzes the oxidation of β-D-glucose to D-glucono-δ-lactone and the concurrent reduction of oxygen to hydrogen peroxide. The enzyme requires flavin adenine dinucleotide as cofactor. Drake and co-workers (*1*) evaluated a series of catalysts, including glucose oxidase, for the oxidation of glucose in a fuel cell reactor. They concluded that metal chelate or alloy catalysts showed more near term promise than did the enzyme. However, the inherent advantages of enzyme specificity and mild reaction conditions warrant continued work on this approach.

The current obtainable in a glucose oxidase fuel cell is proportional to the rate of oxidation of glucose. Using a slightly simplified form of the kinetic sequence of Bright and Gibson (*2*), Wingard and Liu (*3*) calculated a current of 27 ma with concentrations of $5.5 \times 10^{-4}M$ glucose, $8.6 \times 10^{-4}M$ oxygen, and $6 \times 10^{-9}M$ enzyme in an arbitrary working electrode volume of 10 ml. At 100 times all of the above concentrations the calculated current was 540 ma.

The work reported is part of a larger study to evaluate the effects of enzyme immobilization and electrode structure on the efficiency of an enzyme-catalyzed fuel cell. This portion of the study was designed to provide a basis of comparison for future evaluations of electrode variables. Glucose oxidase was trapped in a polyacrylamide gel–platinum gauze matrix, following the method of Hicks and Updike (*4*). Pulverized samples of the enzyme–gel mixture showed only about 25% the activity of the free enzyme, as determined in a Warburg unit. This discrepancy most likely was caused by incomplete trapping of the enzyme although the rate of oxygen transfer in the Warburg may have influenced the results.

Constant current voltametry was used to evaluate the performance of the enzyme–gel electrode. This involved measuring the half-cell potential of the glucose–gluconolactone reaction at constant current. Using a two-compartment H-cell, a fixed dc current was passed between the enzyme electrode and an auxiliary platinum electrode. The half-cell potential was read between the enzyme electrode and a reference saturated calomel electrode. The cell electrolyte consisted of a 0.01*M* solution each of glucose and of lactone, buffered at the desired pH. This procedure was repeated at a series of constant currents and over a pH range of 3.2 to 10.3.

Table I lists the measured potentials for the runs at pH 6.48 and 37°C. The change in voltage with increased current denotes additional concentration polarization. At a current density of 1 ma/cm^2 (obtained at 4 ma) the measured voltage of −0.72 agreed fairly well with the value of −0.62 obtained by Drake (*1*) using a platinum foil electrode and the same reactants. Extrapolation of the data in Table I to zero current gave an open-circuit voltage of −0.65 volt. This compared favorably with a value of −0.68 volt calculated from the Nernst equation.

$$E_{\text{at } X} = E_{\text{at } 7} - 0.0133 \ln \frac{[\text{glucose}]}{[\text{lactone}]} - 0.061\ (\text{pH} - 7)$$

where $E_{\text{at}X}$ = half-cell potential at desired pH, referenced to a saturated calomel electrode

$E_{\text{at}7}$ = −0.605 volt at pH 7

[glucose] = 0.01*M*

[lactone] = 2.7×10^{-6} molar at pH 6.48 and initially 0.01*M*

The initial lactone concentration was corrected for hydrolysis to gluconic acid and subsequent dissociation of the acid.

Table II. Zero-Current Voltages

Electrolyte	*Extrapolated from Experimental Data,*	*Calculated from Nernst Equation,*
pH	*volt*[a]	*volt*[a]
3.2	−0.50	−0.40
4.45	−0.57	−0.50
5.45	−0.61	−0.59
6.48	−0.65	−0.68
7.2	−0.73	−0.74
7.38	−0.71	−0.76
7.61	−0.74	−0.79
10.3	−0.89	−0.96

[a] = Reference to saturated calomel electrode.

Table II lists the measured (extrapolated) and calculated zero-current potentials over the pH range 3.2 to 10.3. The experimental data gave essentially a linear plot with a slope of 0.06 compared with a slope of 0.08 for the calculated values. The experimental and calculated values agreed within 12% over the pH range of about 4.0 to 10.0. Since the free enzyme has its maximum activity at pH 5.6 and retains at least 50%

of this activity over a pH range of 3.0 to 7.5, the agreement between experimental and calculated voltages is good over the pH range of expected enzyme activity. As expected, the experimental (extrapolated) zero-current values were independent of the amount of enzyme in the electrode.

The results indicate that considerable concentration polarization would be expected with a fuel cell using the enzyme–gel electrode described herein. However, the objectives of demonstrating the suitability of constant current voltametry and of establishing a reference level for comparison of subseqeunt changes in enzyme–electrode variables have been met.

(1) Drake, R. F. *et al.*, "First Annual Summary Report Implantable Fuel Cell for an Artificial Heart," *U.S. Govt. Rept.* **PB-177-695** (1968) 1–127.
(2) Bright, H. J., Gibson, Q. H., "The Oxidation of 1-Deuterated Glucose by Glucose Oxidase," *J. Biol. Chem.* (1967) **242,** 994–1003.
(3) Wingard, Jr., L. B., Liu, C. C., "Development of a Glucose Oxidase Fuel Cell," *Proc. Intern. Conf. Med. Biol. Eng., 8th,* Chicago, July 1969.
(4) Hicks, G. P., Updike, S. J., "The Preparation and Characterization of Lyophilized Polyacrylamide Enzyme Gel for Chemical Analysis," *Anal. Chem.* (1966) **38,** 726–730.

Bioenergetic Analysis of Yeast Cell Growth

SHUICHI AIBA and MAKOTO SHODA, Institute of Applied Microbiology, The University of Tokyo, Tokyo, Japan

Biochemical reaction kinetics have been discussed mainly from two contrasting approaches. One is the engineering approach, which pays particular attention to the macroscopic rates of cell mass synthesis and metabolite production. The other is the biochemical approach, which deals principally with enzymatic pathways *in vivo* and/or *in vitro*. Since the gap between the two aproaches should not be left untouched, further experimentation and theoretical consideration are needed to elucidate the kinetic nature of biochemical reactions.

Using a respiration-deficient mutant of baker's yeast, we obtained data for Equations 1 and 2 which showed non-competitive inhibition of the fermentation by ethanol (*1*):

$$\frac{dX}{dt} = \frac{\mu_o}{1 + \frac{p}{K_p}} \cdot \frac{S}{K_s + S} \cdot X \quad (1)$$

$$\frac{dp}{dt} = \frac{\nu_o}{1 + \frac{p}{K'_p}} \cdot \frac{S}{K'_s + S} \cdot X \quad (2)$$

To apply these rate equations to the aerobic cultivation of a wild strain of baker's yeast, another equation on the rate of sugar consumption is required. Amount of ATP, ΔATP which will be produced as energy when sugar is consumed, can be subdivided into two terms relevant to respiration and fermentation.

$$\Delta ATP = (\Delta ATP)_R + (\Delta ATP)_F \quad (3)$$

Assuming that the value of Y_{ATP} is constant, Equation 3 can be arranged as follows:

$$\Delta X = \Delta X_R + \Delta X_F \quad (4)$$

Equation 4 suggests that the cell material synthesized as a result of consumption of a specific amount of sugar (energy source) is also subdivided into the cell formed by the energy originating from respiration and the material ascribable to fermentation. Equation 4 is rearranged to:

$$\frac{dX}{dt} = \frac{dX_R}{dt} + \frac{dX_F}{dt} \quad (5)$$

Regarding the substrate consumption rate, the following equation is derived from the concept mentioned previously.

$$\frac{dS}{dt} = \frac{dS_R}{dt} + \frac{dS_F}{dt} \quad (6)$$

To demonstrate the relationship between the corresponding terms in the right-hand side of Equations 5 and 6, the conversion efficiencies, $(Y_{x/s})_R$, $(Y_{x/s})_F$, $(Y_{P/x})_F$, and $(Y_{P/s})_F$ are defined below.

$$\frac{dX_R}{dt} \equiv (Y_{x/s})_R \left(-\frac{dS_R}{dt}\right)$$

$$= (Y_{x/s})_R \left[-\frac{dS}{dt} - \left(-\frac{dS_F}{dt}\right)\right]$$

$$= (Y_{x/s})_R \left[-\frac{dS}{dt} - \left\{\frac{1}{(Y_{p/s})_F} \cdot \frac{dp}{dt}\right\}\right] \quad (7)$$

$$\frac{dX_F}{dt} \equiv \frac{1}{(Y_{p/x})_F} \cdot \frac{dp}{dt}$$

$$= (Y_{x/s})_F \cdot \frac{1}{(Y_{p/s})_F} \cdot \frac{dp}{dt} \tag{8}$$

Substituting Equations 7 and 8 into the right side of Equation 5 and rearranging,

$$-\frac{dS}{dt} = \frac{1}{(Y_{x/s})_R} \cdot \left[\frac{dX}{dt} + \left\{\frac{(Y_{x/s})_R - (Y_{x/s})_F}{(Y_{p/s})_F}\right\}\frac{dp}{dt}\right] \tag{9}$$

If the application of the previous kinetic equations (*i.e.*, Equations 1 and 2) to the aerobic cultivation of the yeast cells in batch is acceptable, Equations 1, 2, and 9 should represent the aerobic pattern.

The values of the various empirical constants in Equations 1, 2, and 9 were determined separately by referring to experimentation.

(a)

$$\mu_o = 0.497 \text{ hr}^{-1}$$

$$K_s = 14.2 \text{ grams/liter}$$

(b)

$$\nu_o = 0.216 \text{ hr}^{-1}$$

$$K'_s = 28.4 \text{ grams/liter}$$

Batch cultivation of the wild strain of baker's yeast was preceded by a continuous culture (chemostat) using molasses (fermentable sugar = limiting substrate) as the carbon source to determine μ_o, K_s, ν_o, and K'_s. The continuous preculture was suspended to transfer the cells to the catch culture, whose initial medium composition was exactly the same as that of the fresh medium in the chemostat.

(c)

$$K_p = 55 \text{ grams/liter}$$

$$K'_p = 12 \text{ grams/liter}$$

These values were taken from previous work in which glucose instead of molasses was used as the carbon source (*1*).

(d)

$$(Y_{x/s})_R = 1.62$$

$$(Y_{x/s})_F = 0.10$$

$$(Y_{p/s})_F = 0.35$$

The value of $(Y_{x/s})_R$ was determined from the aerobic and chemostat preculture (carbon source = molasses) whereas values of $(Y_{x/s})_F$ and $(Y_{p/s})_F$ were from the previous anaerobic cultivation of the respiration-deficient mutant of baker's yeast (*1*).

Proper calculation of the kinetic pattern with Equations 1, 2, and 9 by using the values from Equations a–d and using a digital computer was in favorable agreement with the batch and aerobic culture data of a wild strain.

The kinetic equations secured from a specific experimentation of the respiration-deficient mutant of baker's yeast was confirmed to be applicable to predicting the kinetic pattern of aerobic culture of the wild strain.

Nomenclature

K_s, K'_s = empirical constants, grams/liter
K_p, K'_p = empirical constants, grams/liter
p = product (ethanol) concentration, grams/liter
S = substrate (carbonaceous source) concentration, grams/liter
t = time, hours
X = cell mass concentration, grams/liter
Y_{ATP} = cell material synthesized per mole of ATP produced ($\div$ 10.5 grams dry cell mass per mole ATP) (2)
$(Y_{p/x})_F$ = yield factor defined by $\Delta p/\Delta X_F$ pertaining to fermentation
$(Y_{p/s})_F$ = yield factor defined by $\Delta p/-\Delta S_F$ pertaining to fermentation
$(Y_{x/s})_F$ = yield factor defined by $\Delta X/-\Delta S_F$—*i.e.*, cell mass synthesized per unit amount of energy source devoted to fermentation
$(Y_{x/s})_R$ = yield factor defined by $\Delta X_R/-\Delta S_R$—*i.e.*, cell mass synthesized per unit amount of energy source devoted to respiration

Subscripts
F = fermentation
R = respiration

Greek letters
μ_0 = specific growth rate at $p = 0$, hr^{-1}
ν_0 = specific rate of ethanol production at $p = 0$, hr^{-1}

(1) Aiba, S., Shoda, M., *J. Ferm. Technol.* (1969) **47**, 790.
(2) Kormancikova, V., Kovac, L., Vigova, M., *Biochem. Biophys. Acta* (1969) **180**, 9.

Growth of Yeast and Lactic Acid Bacteria in a Common Environment

R. D. MEGEE, III, H. M. TSUCHIYA, and A. G. FREDRICKSON, University of Minnesota, Minneapolis, Minn. 55455

Pure culture populations of microorganisms rarely exist in nature. Generally, several biological species coexist in a given environment at a specified time. The individual fluctuations in one population are often a function of the other species as well as the environment. The associations and interactions are referred to as symbiosis—*i.e.*, an association of two or more different species living together in a close spatial and physiological relationship. Symbiosis does not mean mutualism as the latter relationship is sometimes called; mutualism is a class of interactions in symbiosis. The nature of the interactions between the member populations and the environment in symbiotic associations are determined by the characteristics of each of the pure populations and that of their common environment.

This report is concerned with our work on symbiotic relationships that develop under certain environmental conditions in the mixed population of *Saccharomyces cerevisiae* NRRL Y-567 and *Lactobacillus casei* NRRL B-1445. We have used mathematical models to direct the experimental program. With this approach, hypotheses concerning the interactions between the populations and their common environment are made, consequences of the model are drawn, and predictions of the model are tested experimentally. The validity of our models is tested experimentally in growth situations other than the source of the numerical constants. Modeling is focused on interactions that result from the activities of both populations. Given models for the pure populations, a general model is constructed to represent the dynamics of mixed population by incorporating symbiotic variables or interactions.

Experimental work was carried out in batch and continuous growth cultures. The latter were conducted in chemostats with 100-ml growth chambers modified from those of Novick and Szilard (*1*). The growth medium was a standard riboflavin bioassay modified with respect to initial pH and $CaCl_2$ content (*2*). This medium adequately supported the anaerobic growth of both organisms at pH 5.5 and 33°C with the addition of glucose, riboflavin, and gaseous carbon dioxide. Anaerobic conditions were maintained in continuous growth with a gas stream containing nitrogen and 1% carbon dioxide. All growth requirements of both populations were supplied in excess except for glucose and riboflavin which are present in varying proportions.

Our original assumptions concerning the system were as follows:

(1) *S. cereviasiae* grows well in the basal medium on adding glucose but in the absence of riboflavin.

(2) The yeast synthesizes riboflavin in excess of its own metabolic requirements; the excess riboflavin eventually passes to the environment.

(3) *L. casei* grows only if riboflavin and glucose are present. The extent of growth and the vitamin concentration are related quantitatively. In the mixed population a commensal relationship should develop in the vitamin free medium, similar to the study of *S. cerevisiae* and *P. vulgaris* (*3*). By commensalism we mean that the rate of growth of the bacteria is faster than that in a pure population. We assume that this riboflavin interaction is the characteristic feature of the commensalism relationship in this mixed population.

(4) Competition for the common limiting substrate (glucose) should develop if sufficient riboflavin is present, similar to the yeast systems studied by Gause (*4*).

Our interest concerns the major interactions between the mixed populations and the environment. Hence, we restrict our models to non-segregated and unstructured (*5*), making it possible to coordinate directly modeling and experimental work. Thus, we assume that each population is characterized sufficiently by one dependent variable—a mass concentration. Therefore, we cannot apply our models to a highly transient growth situation. Growth is modeled as a series of homogeneous quasi-chemical reactions similar to those proposed by Monod (*6*), Herbert *et al.* (*7*), others, and ourselves (*5*). These reactions obey a definite stoichiometry and rate expression. Further, we make the usual assumptions of no wall growth, and "perfectly" mixed growth vessels. The models are discussed below.

Yeast

Our model of the population dynamics of *S. cerevisiae* in the present growth medium is based on the following observations: (1) Growth is limited entirely by glucose, and a definite stoichiometry exists between the extent of growth and the initial glucose concentration. The glucose concentration is such that alcohol inhibition is minimized. (2) The yeast consumes a portion of the available glucose to "maintain" the viability of the population. We call this mechanism exogenous maintenance since no weight loss per individual cell is observed. This idea formed the basis of models by Marr *et al.* (*8*) and Ramkrishna *et al.* (*9*). (3) Riboflavin is produced as a growth-associated product.

These assumptions lead to the following mass reactions between the yeast and the environmental concentrations of glucose and riboflavin:

$$Y + a_s S \rightarrow 2Y + \beta V + \ldots \tag{1}$$

$$Y \xrightarrow{k_1} Y' \xrightarrow{k_2} N_y \tag{2}$$

$$Y' + b_s S \rightarrow Y \tag{3}$$

where a_s, b_s, and β are stoichiometric coefficients expressed in grams of substrate per gram of biomaterial. Reaction 1 represents the stoichiometric consumption of glucose, S, leading to the formation of a unit mass of viable biomass, Y. A stoichiometric amount of riboflavin, V, is produced during growth. An approach similar to the activated complex theory of chemical kinetics is used to model maintenance. Here the complex takes the form of a physiological state, Y′, which is assumed to exist between Y and non-viable biomaterial, N_y. Y′ is formed by a first-order rate process and is characterized by substrate consumption that does not lead directly to growth. If the exogenous substrate is present, Y′ consumes a stoichiometric amount, presumably for maintenance energy, and returns to Y. If absent, Y′ passes irreversibly to N_y. These events are modeled by Equations 2 and 3.

Bacteria

The growth of *L. casei* in the test medium may be limited by one, or both, of two substrates, glucose or riboflavin—*i.e.*, the final extent of growth can be related directly to the exhaustion of one, or both, of the two substrates. However, if the "non-limiting" substrate is not in large excess, its concentration may influence the rate of growth. The key to a model of growth in this system is that both glucose and riboflavin are consumed by *L. casei* simultaneously. This is different from the diauxie effect studied by Monod (*10*).

Our experimental work has demonstrated that a model of bacterial growth need not include effects of maintenance, loss of viability, and inhibition in this growth medium. This does not mean these events do not occur, but if they do, their effect is minimal. The absence of these mechanisms is not, of course, true in general and may hold only in our experimental environment. Thus, a sufficient model for *L. casei* need only consist of growth effects.

Growth of *L. casei* is an interaction between the bacterial mass, B; glucose, S; and riboflavin, V, and leads to the formation of fresh biomaterial. To model growth, we assume that both substrates are consumed stoichiometrically:

$$B + c_s S + \alpha V \rightarrow 2B + \gamma L + \ldots \tag{4}$$

where c_8, α, and γ are stoichiometric coefficients. Here we show the stoichiometric production of lactic acid, L, from glucose. Although we shall not consider lactic acid in the model, it must be included in models of other growth situations. Specifically, in lightly buffered medium lactic acid formation acts to decrease the pH of the environment, and for a range of initial pH values the growth of S. *cerevisiae* is a strong function of $[H^+]$. Thus, under these conditions mutualism should develop in a riboflavin free medium and has been observed.

Mathematical Model

Reactions 1–4 form the basis of our model of symbiosis in the system *L. casei–S. cerevisiae.* The mathematical model is constructed by making the usual hypotheses about the kinetics and stoichiometry of the reactions. A set of four coupled, nonlinear, ordinary differential equations is thereby obtained, and constitutes the model.

Determination of Model Constants. Since there are no direct interactions, such as predation or parasitism, in the symbiotic association of interest here, constants may be determined from batch growth experiments on the separate populations. This was done for both populations, and the models gave a satisfactory "fit" of the data taken.

Testing the Model: Symbiosis. The model, with constants determined as stated, was applied to steady-state chemostat data on symbiotic growth of the two populations. Three of the constants had to be adjusted somewhat to obtain a satisfactory "fit." These three constants deal with the riboflavin interaction, and the necessity for adjustment reflects inaccuracies in the measurement of low concentrations of riboflavin. Batch symbiotic growth experiments were also done, and the model described these experiments also.

(1) Novick, A., Szilard, L., *Science* (1950) **112,** 715.
(2) Roberts, E. C., Snell, E. E., *J. Biol. Chem.* (1946) **163,** 499.
(3) Shindala, A., Bungay III, H. R., Kreig, N. R., Culbert, K., *J. Bacteriol.* (1965) **89,** 693.
(4) Gause, G. F., "The Struggle for Existence," Hatner, New York, 1934.
(5) Tsuchiya, H. M., Fredrickson, A. G., Aris, R., *Advan. Chem. Eng.* (1966) **6,** 125.
(6) Monod, J., "Recherches sur la Croissance des Cultures Bactériennes," Herman et Cie, Paris, 1942.
(7) Herbert, D., Ellsworth, R., Telling, R. C., *J. Gen. Microbiol.* (1956) **14,** 601.
(8) Marr, A. G., Nelson, E. H., Clark, D. J., *Ann. N. Y. Acad. Sci.* (1963) **102,** 536.
(9) Ramkrishna, D., Fredrickson, A. G., Tsuchiya, H. M., *J. Gen. Appl. Microbiol.* (1966) **72,** 311.
(10) Monod, J., *Ann. Rev. of Microbiol.* (1949) **3,** 371.

Extended Culture: The Growth of *Candida utilis* at Controlled Acetate Concentrations

V. H. EDWARDS,[1] M. J. GOTTSCHALK, A. Y. NOOJIN, III,[2] A. L. TANNAHILL, and L. B. TUTHILL, School of Chemical Engineering, Cornell University, Ithaca, N. Y. 14850

A variety of techniques and equipment were developed for the submerged culture of microorganisms. Batch and certain forms of semicontinuous culture techniques have been an integral part of the fermentation arts since antiquity, and batch culture is still the most widely practiced laboratory technique. Continuous-flow microbiological culture techniques were first used in certain industrial fermentations such as the "vinegar generator" (early 19th century, Schutzenbach), and continuous yeast propagations, as well as biological waste treatment processes such as the trickling filter (1899, C. J. Whittaker), and the activated sludge process (1913, Fowler and Mumford). Monod and Novick and Szilard are usually credited with initiating the widespread use of the continuous culture technique in the microbiological laboratory about 1950, although a number of prior laboratory applications have been cited. During the past 20 years, continuous culture has been researched extensively. The now well-known potential advantages offered by steady-state continuous culture of a constant environment and a constant phenotype (assuming a constant genotype) has led to its widespread adoption in laboratory studies of microorganisms. The technique has yet to revolutionize industrial fermentations; instead, batch and semicontinuous culture techniques are still used in most industrial fermentations.

The quest for a better understanding of microorganisms and for their more efficient use continues to lead to new or revitalized culture techniques such as synchronous culture, dialysis culture, dense culture, continuous phased culture, and multistage tower culture, to cite a few examples.

The semicontinuous culture technique summarized here and described in detail elsewhere (*1*) was developed to study growth at controlled inhibitory substrate concentrations and to measure the effects of the cell age distribution on the physiological properties of a culture. The technique bears a close operational resemblance to the current industrial fermentation processes that use serial feeding and should thus be additionally useful in scale-up.

[1] Present address: National Science Foundation, Division of Engineering, Room 336, 1800 G St., N.W., Washington, D. C. 20550.
[2] Present address: Shell Chemical Co., Deer Park, Tex.

Qualitative Features of Extended Culture

In the batch culture technique, sterile growth medium is inoculated with a culture of microorganisms. Once adapted to their new environment the microorganisms grow and proliferate until limited by nutrient depletion, product accumulation, and other unfavorable changes in their surroundings. These batch processes are dynamic by nature, and the concentrations of all substrates vary constantly with time (with the possible exception of oxygen). Conversely, in the single-stage homogeneous continuous culture technique, the process may be maintained at various steady states by continuously supplying nutrient media to a vessel and removing culture at an equal volumetric rate. In the absence of substrate inhibition or changes in the organism, these steady states are stable and unique. Each steady state corresponds to a balanced degree of conversion of consumables to cells and other products. Also, the cell population in the culture vessel must be multiplying geometrically although the growth and division of individual cells within the population may be other than geometric. Other things being equal, the cell age distribution in continuous culture will be different from that in an exponential batch culture. At equal specific growth rates, the continuous culture population will tend to be skewed (approximately exponentially) to contain a higher proportion of cells which have been formed recently by division, because "older" cells will be exposed to the continuous removal of a fraction of the culture for a longer time.

The semicontinuous culture technique proposed here can provide a constant environment like that found in steady-state continuous culture, while maintaining a cell age distribution within the culture different from that found in a continuous culture. The environmental composition in the microbial culture is measured continuously by one or more analytical instruments, and the values of various substrate concentrations are transmitted to a composition controller and a composition programmer. The controller adds concentrated solutions of one or more substrates from an appropriate number of reservoirs to maintain the concentration of each substrate equal to the value specified by the programmer. In the simplest case, one or more substrate concentrations are simply maintained at constant values, thereby attaining a constant environment in the absence of substantial by-product formation by the culture. Obviously, the duration of the steady-state environment is ultimately limited by the ever-increasing mass of cells and their products, but the duration can often be prolonged much beyond that attainable with ordinary batch culture techniques. The name "extended culture" is suggested to distinguish semicontinuous cultures programmed for a constant environment from various other semicontinuous culture techniques.

The concentration of one or a few substrates is often maintained approximately constant in industrial fermentations by periodic addition of the appropriate material. This is practiced most frequently to obtain higher final product concentrations when lower rates and a reduction in yields result from high initial substrate and/or precursor concentrations.

Martin and Felsenfeld (*2*) developed a technique similar to extended culture which they named the exponential gradient generator. Their device supplies a limiting substrate to a culture vessel at a rate that increases exponentially with time. If the culture consumes substrate exponentially (*e.g.*, by exponential growth), this has the net effect of maintaining a constant substrate concentration. A mathematical framework for analyzing programmed semicontinuous cultures has been set forth elsewsere (*1*).

Results

In this initial study of extended culture, the feedback control of a single limiting substrate (acetate) was achieved by the well-known indirect method of controlling pH with a carbon source. Mixtures of acetic acid and sodium acetate were used to control pH. By choice of the proper ratio of sodium acetate to acetic acid, the concentration of non-carbonate dissolved carbon could be maintained constant while simultaneously controlling pH.

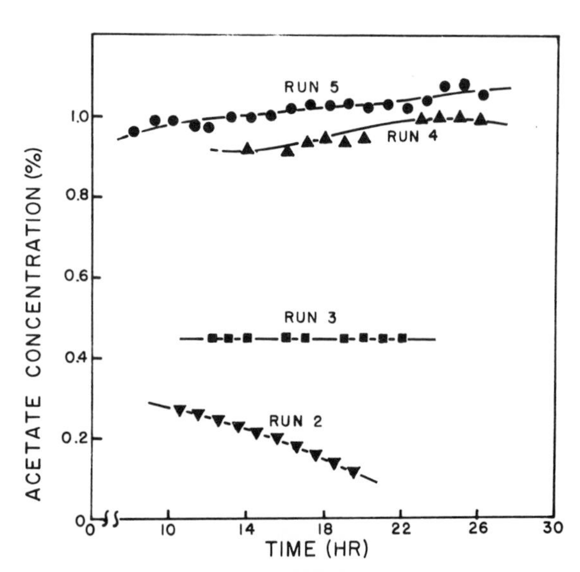

Figure 1. Acetate concentration histories from four extended culture experiments. Acetate concentration is expressed as %(w/v) acetic acid.

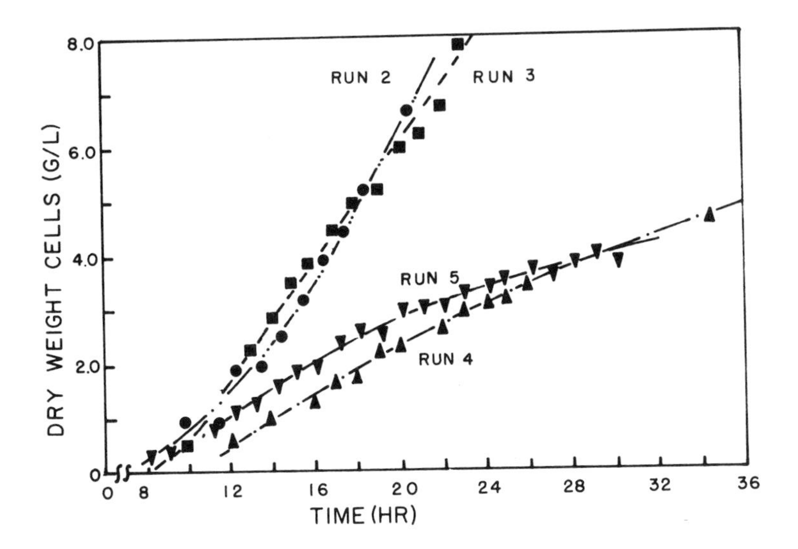

Figure 2. Cell dry weight concentration histories from four extended culture experiments

Four experimental runs conducted in the manner described elsewhere (*1*) are summarized here. Figures 1 and 2 show substrate concentration and cell dry weight concentration histories for the runs. Figure 1 shows that the first experimental objective of maintaining a constant acetate concentration was met successfully in Run 3 and nearly so in Runs 4 and 5. In Run 2 the acetate concentration declined steadily. This arises from two causes: carbonate produced by the cells and ammonium consumed by the cells affect the pH significantly. Thus in Run 2, where the pH was controlled by adding 40% acetic acid, the acetate replacement could not keep up with the acetate consumption. In the other runs, the pH was controlled with 40% acetic acid solutions which also contained the following amounts of sodium acetate trihydrate (Run 3, 85.7 grams/liter; Run 4, 200 grams/liter; Run 5, 150 grams/liter). The amounts of acetate salt were calculated *a priori* from estimated rates of acid and base exchange and from dissociation constants of the relevant species.

Figure 2 shows the reduced growth rates and final cell concentrations obtained at the higher sodium acetate concentrations. This effect was expected from earlier work which showed that *Candida utilis* is partially inhibited by sodium acetate concentrations above about 0.5% (w/v). Figure 2 also suggests that growth was much more nearly linear than exponential. In fact, the only cell concentration history that is strongly concave upward is from Run 2, which had a low and continuously declining acetate concentration. By combining the data in Figures 1 and 2 with periodic measurements of the acid reservoir level during each run, linear growth rates, specific growth rates, cumulative acetate consumption, cell yield, and specific rates of acetate consumption were calculated. Polynomials were fitted to the various sets of data for each run using an IBM 360/65 computer and a least-squares technique. The fitted curves are compared with the data in Figures 1 and 2. With X (cell dry weight concentration), S (acetate concentration), F (flow rate of acetic acid/sodium acetate feed), and sample removal rate known (fitted polynomial) functions of time, the theory (*1*) was used to calculate instantaneous values of the rate parameters listed above.

Preliminary continuous culture data and Warburg respiration rates were collected for comparison with the results obtained in the extended culture experiments (*1*).

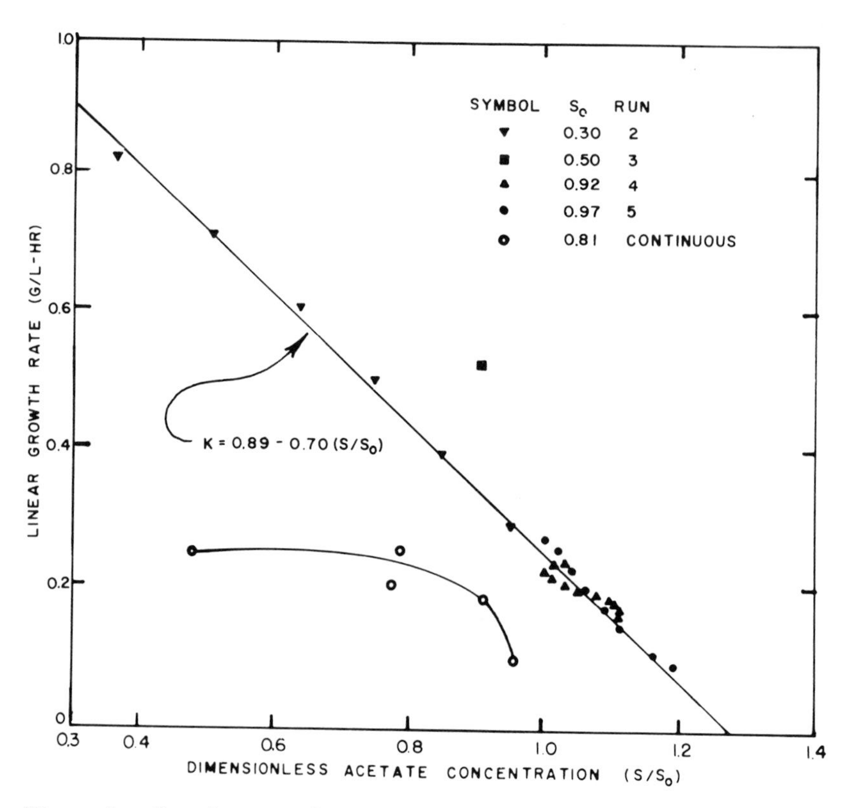

Figure 3. Correlation of linear growth rates with dimensionless acetate concentration

Conclusions

(1) The feasibility of semicontinuous culture of *Candida utilis* (ATCC 9226) at controlled acetate concentrations has been demonstrated under conditions which are more nearly representative of many industrial fermentations than either batch or continuous processes. It is proposed that semicontinuous culture in which at least one substrate concentration is maintained constant be called extended culture. The control or programming of multiple environmental factors can be achieved by adding suitable hardware and may offer substantial advantages both in process optimization and in fundamental studies of microbial cultures.

(2) A theoretical framework has been presented and applied that permits evaluation of kinetic parameters in semicontinuous culture (*1*).

(3) In the growth of *C. utilis* at inhibitory concentrations of sodium acetate, longer periods of growth and higher cell concentrations were achieved by controlling the acetate concentration through serial feeding. The cell yield data appear to correlate by standard methods (*3*), although acetate did not restrict growth under the conditions of most experiments reported here and some cell yields from acetic acid were abnormally high (*1*).

(4) Growth was more nearly linear than exponential during each extended culture experiment, despite approximately constant acetate concentrations. For given acetate concentrations, the decline in linear growth rate in extended culture was correlated linearly with the ratio of the instantaneous acetate concentration to the initial acetate concentration (Figure 3). Limitation by ferrous iron is thought to be responsible, with reduced oxygen and nitrogen levels perhaps playing secondary roles, although further experiments are necessary to test these tentative conclusions. Diffusion-limited transport of acetic acid may also be involved.

(5) Cell age distributions are different in semicontinuous and continuous culture experiments although environments can in principle be kept identical. Comparison of semicontinuous and continuous culture data at similar acetate and cell concentrations showed large differences in linear growth rate which may be a result of the differences in the cell age distribution in the two cultures (Figure 3).

(1) Edwards, V. H. *et al.*, *Biotechnol. Bioeng.*, in press.
(2) Martin, R. G., Felsenfeld, G., *Anal. Biochem.* (1964) **8,** 43–53.
(3) Marr, A. G. *et al.*, *Ann. N. Y. Acad. Sci.* (1963) **144,** 536–548.

INDEX

INDEX

D

N

O

T

U

V

W

X

Y

Z